AF686125

Recent Advances in Fluid Mechanics: Feature Papers, 2024

Recent Advances in Fluid Mechanics: Feature Papers, 2024

Guest Editors

D. Andrew S. Rees
Giuliano De Stefano

MDPI

Basel • Beijing • Wuhan • Barcelona • Belgrade • Novi Sad • Cluj • Manchester

Guest Editors

D. Andrew S. Rees
Department of Mechanical Engineering
University of Bath
Bath
UK

Giuliano De Stefano
Department of Engineering
University of Campania Luigi Vanvitelli
Aversa
Italy

Editorial Office
MDPI AG
Grosspeteranlage 5
4052 Basel, Switzerland

This is a reprint of the Special Issue, published open access by the journal *Fluids* (ISSN 2311-5521), freely accessible at: https://www.mdpi.com/journal/fluids/special_issues/F0U62YR361.

For citation purposes, cite each article independently as indicated on the article page online and as indicated below:

Lastname, A.A.; Lastname, B.B. Article Title. *Journal Name* **Year**, *Volume Number*, Page Range.

ISBN 978-3-7258-4455-5 (Hbk)
ISBN 978-3-7258-4456-2 (PDF)
https://doi.org/10.3390/books978-3-7258-4456-2

© 2025 by the authors. Articles in this book are Open Access and distributed under the Creative Commons Attribution (CC BY) license. The book as a whole is distributed by MDPI under the terms and conditions of the Creative Commons Attribution-NonCommercial-NoDerivs (CC BY-NC-ND) license (https://creativecommons.org/licenses/by-nc-nd/4.0/).

Contents

About the Editors

D. Andrew S. Rees

Dr Andrew Rees has a 1st class honours degree in Mathematics from Imperial College, London, and a PhD in Applied Mathematics from the University of Bristol. He has been the author of 195 research papers and 78 conference papers and, in addition to being the Editor-in-Chief of Fluids, is an Associate Editor for a variety of other journals. His research centres on convective instabilities in fluid mechanics. He has been at the Department of Mechanical Engineering at the University of Bath since 1990 and has recently been responsible for teaching Mathematics to first and second year Mechanical, Aeronautical, and Automotive Engineering students. He won the Mary Tasker award for teaching excellence in 2006.

Giuliano De Stefano

Giuliano De Stefano is a full professor of Fluid Dynamics at the Engineering Department of the University of Campania Luigi Vanvitelli (Italy). He earned his laureate degree cum laude in Aeronautical Engineering from the University of Naples Federico II in 1994, followed by a PhD in Aerospace Sciences and Technologies from the Second University of Naples in 1998. His research mainly focuses on computational fluid dynamics, with a particular emphasis on the numerical modelling and simulation of turbulent flows. He has actively participated in several international research projects and collaborated with international academic institutions such as the University of Colorado Boulder (USA), Skolkovo Institute of Science and Technology (Moscow), and Queen's University (Canada). He is the author/coauthor of over 130 scientific publications, including numerous articles in leading journals in fluid dynamics. He has been a member of the Editorial Board of the journal Fluids since 2020 and has served as Guest Editor for Special Issues on wavelets and fluid mechanics. He is also an active participant in major scientific conferences and has contributed to technology transfer initiatives with prominent aerospace and mechanical engineering companies. He mentored several PhD students and is currently serving as the Coordinator of the PhD Programme in Industrial and Information Engineering (41st cycle) at the University of Campania.

fluids

MDPI

Editorial

Recent Advances in Fluid Mechanics: Feature Papers, 2024

Giuliano De Stefano [1,*] and D. Andrew S. Rees [2]

[1] Engineering Department, University of Campania Luigi Vanvitelli, Via Roma 29, 81031 Aversa, Italy
[2] Department of Mechanical Engineering, University of Bath, Bath BA2 7AY, UK; ensdasr@bath.ac.uk
* Correspondence: giuliano.destefano@unicampania.it

1. Premise

The present Special Issue consists of a collection of feature articles by distinct investigators and research groups discussing new findings or cutting-edge developments concerning different aspects of fluid mechanics. This Special Issue comprises ten studies covering some of the latest advances in the field, relevant to both theoretical and applied research contexts. The different papers present in-depth analyses, commentaries, and studies of a diverse set of topics within the fluids domain. The main themes include the following:

Reynolds number analysis;
Computational fluid dynamics;
Turbulent flows;
Machine learning and fluid mechanics;
Fluids engineering;
Multidisciplinary optimization in aerodynamics.

Specifically, nine high-quality review papers have been published, along with one original research article. All these works provide detailed insights, opinions, or perspectives on current issues or important trends. In the following section, before presenting the methodologies, the key findings, and the conclusions of each contributing paper, a very brief overview of recent developments and future perspectives in their relevant fields is offered.

2. Feature Articles

2.1. Reynolds Number Analysis

The Reynolds number analysis remains a fundamental tool in fluid mechanics, with ongoing research expanding its applications and enhancing its theoretical foundations. Classically, the Reynolds number serves as a critical parameter in determining flow regimes, particularly in distinguishing between laminar and turbulent flows. Research in this area has advanced significantly, particularly through high-fidelity numerical simulations, e.g., refs. [1,2]. These efforts have also led to the development of improved intermittency and transition models that enhance the accuracy of predicting flow transitions, especially within Reynolds-averaged Navier–Stokes (RANS) frameworks. Additionally, studies on aerodynamic applications have integrated Reynolds number considerations to better predict aircraft performance [3], while emerging standardization efforts aim to unify its application in fields like biomechanics and nanotechnology. Despite recent advances, the influence of this parameter in complex geometry flows (for instance, in the presence of strong curvature or multi-element configurations) is not yet fully understood. Moreover, predicting transition Reynolds numbers under realistic turbulence intensities and environmental conditions still poses challenges, limiting the reliability of current models. There is also a need for better integration of Reynolds number effects across different physical scales and disciplines to foster cross-field consistency [4]. The following two review papers are dedicated to the analysis of this crucial parameter in fluid mechanics.

The work by Saldana et al. [5] explores the historical development and contemporary relevance of the Reynolds number. The paper aims to trace the origins of this parameter, its definition, and its evolution over time, highlighting its significance in various applications across different fields of fluids science and engineering. The study emphasizes the broad applicability of this parameter beyond traditional fluid dynamics, including its relevance in areas such as aerodynamics, microfluidics, and even astrophysics. Recent advancements and reinterpretations of this important number are discussed, including its adaptations for non-Newtonian fluids and granular flows. The authors highlight the importance of the Reynolds number in practical engineering applications, such as the design of pipelines, aircraft, and various fluid mechanics systems, demonstrating its role in optimizing performance and ensuring safety. Finally, they advocate for a deeper understanding of this parameter and its implications in modern fluid dynamics studies.

Similarly, but differently, the work by Tamburrino and Niño [6] examines the universal presence of the Reynolds number and its fundamental role in fluid dynamics, together with its implications across various scientific and engineering fields. This review paper aims to highlight the significance of this parameter as a dimensionless quantity that characterizes flow behavior and its universal applicability in very different contexts. The work explores the application of the Reynolds number for studying a vast range of fluid phenomena, from microscopic scales such as cellular motion to macroscopic scales, such as turbulent aerodynamic and hydrodynamic flows, and even intergalactic dynamics. The recent advancements in the study of this flow parameter are discussed, including its adaptations for non-Newtonian fluids and the exploration of its effects in complex flow scenarios. In conclusion, the Reynolds number represents a fundamental concept in fluid dynamics with widespread implications, with the authors emphasizing the need for continued research to explore its effects in new emerging fields, while reinforcing its relevance in both theoretical and practical contexts.

2.2. Computational Fluid Dynamics

The research findings in computational fluid dynamics (CFD) methodologies have direct implications for various scientific and technological fields. For instance, in aerospace engineering, accurate simulations of turbulent compressible flows are critical for design and analysis. Recent advances in shock-capturing methodologies for high-speed flows apply high-order schemes supplied with improved shock sensors, such as weighted essentially non-oscillatory (WENO) schemes, implicit shock tracking techniques, and adaptive artificial diffusivity, to accurately capture shocks and their interaction with fluid turbulence, e.g., refs. [7,8]. Despite these advances, challenges remain in accurately capturing shock–turbulence interactions at high Reynolds and Mach numbers, especially in three-dimensional, unsteady, and hypersonic regimes. Moreover, computational complexity and numerical dissipation still limit the fidelity of CFD simulations. Future research should focus on developing more efficient and robust high-order schemes capable of handling very complex flows, creating dynamic shock sensors that further reduce numerical dissipation, improving fluid–solid interface modeling under shock impact, and enabling high-fidelity simulations of hypersonic flows. The following two papers review high-order accurate schemes for compressible turbulent flows with applications in geosciences and aerospace engineering.

The work by Assis et al. [9] focuses on the development and evaluation of high-order shock-capturing schemes for simulating high-speed flows. The authors investigate WENO schemes for simulating the Euler equations under canonical coordinates and the Navier–Stokes equations under generalized coordinates, providing a detailed analysis of their performance, and highlighting their robustness in handling shock waves and discontinuities [10]. Tested for three canonical supersonic flows (upstream of a cylinder, over a blunt wedge, and over a compression ramp), these schemes demonstrated the capability to achieve machine precision, indicating their effectiveness in capturing the complex gas dynamics. Basically, the present results show that these schemes can significantly improve the accuracy and efficiency of simulations compared to traditional low-order methods. One can conclude that high-order accurate shock-capturing schemes are a valuable tool for achieving convergence towards steady states in high-speed flows. However, the importance of continued research to further refine these methods and explore their potential applications in more complex problems should be emphasized.

The work by Yee et al. [11] presents a comprehensive overview of research aimed at improving the predictability and reliability of numerical simulations for compressible flows involving shock waves. The paper focuses on integrating nonlinear dynamical approaches with newly developed high-order numerical methods to quantify numerical uncertainty in simulations of compressible turbulence. In particular, the construction of high-order entropy-conserving, momentum-conserving, and kinetic energy-preserving numerical methods is discussed, while emphasizing the importance of accurate discretization in capturing the complexities of nonlinear systems [12]. As a key finding, the solution space of discrete genuinely nonlinear systems is significantly larger than that of corresponding continuous systems, which can lead to discrepancies in numerical solutions. In fact, traditional uncertainty quantification methods, which often rely on linearized analyses, may not adequately capture the true behavior of fluid flows. The paper highlights the necessity of adaptive numerical methods and the blending of multiple schemes to manage numerical dissipation and enhance stability, particularly in the presence of shock waves [13]. The authors conclude that a nonlinear approach to numerical uncertainty quantification is essential for improving the reliability of CFD calculations. They advocate for continued development of high-order methods and adaptive techniques to better address the challenges posed by complex compressible flows.

The color gradient method (CGM) is a variant of the lattice Boltzmann method (LBM) used for simulating two-phase flow of immiscible fluids through porous media, e.g., ref. [14]. Recent developments have focused on improving the accuracy and stability of the method, where key advances include enhancing isotropy in the color gradient calculation to better handle very high-density ratios, which is critical for realistic two-phase flow modeling [15]. Additionally, numerical improvements in recoloring operators and adaptive grid schemes have been explored to reduce spurious velocities and improve interface capturing. However, despite some progress, gaps remain in fully understanding and optimizing CGM for real-world multiphase flows, especially under extreme conditions such as highly turbulent regimes. Challenges include accurately modeling fluid acceleration effects, viscosity contrasts, and interface dynamics in porous media. Future developments should lead to more robust formulations that maintain accuracy at very high density and viscosity ratios, using improved numerical schemes for interface tracking to minimize spurious currents (and ensure mass conservation), extending the method to complex geometries and porous media with realistic boundary conditions [16], and coupling with advanced turbulence models and machine learning techniques for enhanced simulation fidelity.

In this context, the work by Zahid and Cunningham [17] aims to critically review and evaluate various algorithms and methodologies associated with the CGM approach. The authors compare different methods for modeling fluids with varying viscosities, emphasizing the importance of selecting appropriate algorithms for accurate simulations. The paper discusses methods for calculating the color gradient, which is essential for modeling the interface between different fluid phases. External forces or accelerations acting on the fluids are also analyzed, with comparisons made between different forcing schemes. The authors highlight the necessity of proper algorithm selection for conducting benchmark simulations, specifically mentioning bubble tests and layered Poiseuille flow as common scenarios for validation. The review indicates that future works will be exploring more complex phenomena such as fluids with widely differing viscosities, wettability effects, and extending the CGM to three-dimensional simulations.

2.3. Turbulent Flows

Accurately predicting the evolution of turbulent flows of practical interest remains a big challenge in CFD, with scientists and engineers constantly developing methods and models to simulate the complex behavior of turbulent flows. In this framework, the integration of classical methodologies such as large eddy simulation (LES) with reduced-order modeling (ROM) appears promising, e.g., refs. [18,19]. Particularly, some recent advances in LES-ROM coupling focus on hybrid models combining physics-based ROM techniques with machine learning (ML) to better capture turbulent flow dynamics. Though LES-inspired closures improve ROM accuracy by modeling unresolved scales, challenges remain in computational cost, generalizability, and the representation of higher modes in proper orthogonal decomposition (POD). Future research should develop efficient nonlinear closures, enhance ML model robustness across flow regimes, incorporate physics constraints to reduce data needs, and extend methods to complex turbulent flows.

The work by Quaini et al. [20] reviews and discusses approaches for effectively bridging these two methodologies to enhance the computational efficiency and accuracy of simulations, particularly for convection-dominated flows. The authors explore various spatial filtering techniques that can be applied to LES to obtain ROM methods, discussing the mathematical foundations of these approaches and their implications for modeling complex fluid behaviors. As a key finding, the integration of LES and ROM can significantly reduce computational costs while maintaining essential flow features, making it suitable for real-time applications and complex geometries. The authors highlight the importance of appropriate filtering techniques in ensuring the stability and accuracy of ROM derived from LES data. They present case studies demonstrating the effectiveness of the proposed methodologies in capturing key flow characteristics in various applications, including biomedical and engineering contexts. The paper also discusses possible future directions for research, emphasizing the need for further development of hybrid models that can seamlessly integrate LES and ROM, also calling for more extensive validation studies to establish the reliability of these approaches in practical scenarios. In conclusion, bridging the two methods through spatial filtering techniques offers a promising pathway for advancing the simulation of complex flows in science and engineering.

For turbulent flows over rough surfaces, recent developments have leveraged advanced simulation techniques such as wall-modeled large-eddy simulations (WM-LES) and direct numerical simulations (DNSs) to investigate complex roughness effects. Some interesting studies have been conducted for flow over heterogeneous roughness, revealing slow turbulence recovery downstream of rough-to-smooth transitions [21,22], while other recent works have highlighted the significant role of porous substrate depth in turbulence penetration and drag characteristics, e.g., ref. [23]. Despite these advances, gaps remain in understanding the detailed mechanisms of turbulence recovery after roughness transitions for heterogeneous roughness patterns, while the interplay between microscale roughness geometry and macroscale permeability in porous substrates is not fully resolved. As these facts limit predictive capabilities for engineering and environmental applications, predictably, future research will focus on multi-scale modeling that combines different approaches to capture the turbulence behavior in complex flow scenarios.

The review work by Hantsis and Piomelli [24] analyzes how scalar quantities are transported in turbulent flows over rough surfaces. Numerical simulations are utilized to study the velocity and pressure fields in conjunction with scalar transport, exploring different roughness geometries and their effects on the scalar field. The study investigates the existence of a condition analogous to the fully rough regime for scalar transport, similar to what is observed in momentum transfer [25]. The authors highlight the importance of the roughness sublayer, which extends a few roughness heights above the roughness crest, in determining scalar transport characteristics. The paper discusses how Townsend's similarity hypothesis, which applies to momentum transport, can also extend to scalar transport, suggesting similitudes in the mechanisms governing these processes, and implying the differences due to roughness are limited to the roughness sublayer. Novel findings regarding the interactions between turbulent structures and scalar fields in the presence of roughness are presented, contributing to a deeper understanding of transport phenomena in practical engineering applications. The review concludes with a discussion of relevant open questions, emphasizing that, beyond extending the range of simulated Reynolds and Prandtl numbers, the systematic coverage of different roughness types and topologies would be required, as scalar transport appears to remain sensitive to the geometrical details.

2.4. *Machine Learning and Fluid Mechanics*

Recently, there has been a tremendous increase in the application of machine learning as a new and efficient mathematical tool in fluid mechanics research. In particular, deep reinforcement learning (DRL) has become popular for active flow control, optimization, and automation, leveraging its ability to handle complex, nonlinear, and high-dimensional problems without explicit physical models [26,27]. DRL frameworks use neural networks to learn control policies through interactions with fluid systems, enabling adaptive and efficient flow manipulation. For example, important studies demonstrate the success of this technique in stabilizing vortex shedding, reducing drag, and suppressing vortex-induced vibrations, e.g., ref. [28]. Despite these advances, however, challenges remain in sensor placement and data efficiency, the interpretability of learned policies, and robustness across varying flow regimes. Moreover, training DRL agents is computationally demanding, and transferring strategies from simulations to real-world turbulent flows is still limited. As embedding physical constraints and ROM techniques into DRL frameworks is underexplored but promising for improving generalization and reducing data needs, future research should focus on this aspect to enhance learning efficiency and interpretability.

The work by Kim et al. [29] presents a comprehensive review of the application of DRL in fluid dynamics. The authors categorize the diverse applications of DRL, discussing the methodologies used and the challenges faced in implementing these techniques for fluids. The paper explores recent advancements in applying DRL to fundamental fluids engineering problems, for instance, flow control, including vortex shedding and drag reduction in turbulent flows, demonstrating its capability to adaptively manage complex fluid behaviors. The potential of DRL in shape optimization, where it can be used to enhance aerodynamic performance by optimizing the geometry of airfoils and other structures, is highlighted. Additionally, the automation of CFD processes using DRL is emerging as a promising area, allowing for more efficient simulations and mesh generation, thus reducing computational complexity. In addition, emerging research trends are discussed, such as the use of multi-agent reinforcement and transfer learning to improve the generalizability and efficiency of DRL applications in fluid dynamics. In conclusion, DRL offers significant potential for advancing fluids research by providing innovative solutions to very complex problems. However, the need for further research addressing challenges related to the DRL implementation and expanding its applicability across various flow scenarios should be emphasized.

2.5. Fluids Engineering

The fluid dynamics of disease transmission is an emergent research area, where rich multi-scale flow physics, from interfacial to multiphase complex flows, play crucial roles [30]. The complex scenario includes the generation and aerosolization of virus-laden respiratory droplets from a host, airborne dispersion and deposition on surfaces, and subsequent inhalation of these bioaerosols by unsuspecting recipients [31]. Fluid mechanics is also key to preventative measures such as the use of face masks, ventilation of indoor environments, and even social distancing [32]. Recent advances with investigative techniques, including new developments in computational and experimental methods, as well as the physical interpretation of results, enhance the knowledge and understanding of fluid flow phenomena in engineered systems and processes in this particular area.

The work by Sankurantripati and Duchaine [33] focuses on fluids engineering technology for improving indoor air quality to mitigate the spread of airborne diseases. Their contribution reviews various ventilation strategies and the role of portable air purifiers using ultraviolet (UV) light in controlling airborne pathogens. Both natural and mechanical systems, and their effectiveness in reducing pathogen concentrations in indoor environments, are analyzed, specifically examining the technologies used in portable air purifiers that incorporate UV-C radiation. As a key finding, effective ventilation systems are demonstrated to significantly lower pathogen concentrations by improving airflow in enclosed spaces, where the integration of portable purifiers with existing ventilation systems is recommended for enhanced control. The study highlights challenges in the practical implementation of UV air purifiers, including the need for further research to both validate their efficacy in real-world applications and optimize their design for different environments. In particular, the importance of understanding the interaction between UV dosage, airflow rates, and the effectiveness of air purifiers is emphasized. The authors conclude that combining effective ventilation strategies with portable purifiers serves as a viable solution for improving indoor air quality and reducing the risk of airborne disease transmission. However, they advocate for continued research to better understand the dynamics of indoor air quality and the role of advanced purification technologies.

2.6. Multidisciplinary Optimization in Aerodynamics

Modern aircraft design technologies increasingly face various complex operating conditions, and one must consider the requirements of different disciplines to achieve good overall performance. Recent advancements in multidisciplinary design optimization (MDO) for aircraft combine high-fidelity CFD analysis with structural, propulsion, and operational models to address complex design challenges, e.g., refs. [34,35]. In particular, the shape design of stealth aircraft needs to give consideration to both aerodynamic and stealth performance, while balancing these two different disciplines [36].

The original research work by Thoulon et al. [37] focuses on the development of a methodology for simultaneously optimizing the aerodynamic performance and stealth characteristics of a fighter aircraft. The paper presents a gradient-based bidisciplinary optimization approach that balances aerodynamic efficiency and low observability for simplified aircraft design. The authors utilize a computational framework that incorporates shape parameterization, through computer-aided design (CAD) modeling, and optimization algorithm, with the sequential least squares quadratic programming method being employed to find optimal designs based on multiple criteria. The optimization process involves both aerodynamic performance, which is evaluated through drag calculations using RANS calculations, and stealth characteristics, where the radar cross-section is analyzed to minimize the visibility of the aircraft to radar systems. The study identifies a Pareto front illustrating the trade-offs between the two different goals, allowing designers to select optimal configurations based on specific mission requirements. The proposed optimization methodology effectively balances diverse requirements, providing valuable insights for the design of next-generation aircraft.

3. Conclusions

This editorial concisely captures the essence of this Special Issue, entitled *Recent Advances in Fluid Mechanics: Feature Papers, 2024*, while offering a brief overview of the relevant research areas of the different contributions. The guest editors consider the present collection to be a precious forum to review and disseminate important research findings and share innovative ideas in the broad field of fluid mechanics. In the same vein, with the next Special Issue of the series, entitled *Feature Reviews for Fluids 2025–2026*, the editors will continue to present both review articles of archival significance and new theoretical, numerical, and experimental studies of lasting scientific value for the fluid dynamics research community.

Conflicts of Interest: The authors declare no conflicts of interest.

References

1. Nabae, Y.; Inagaki, K.; Kobayashi, H.; Gotoda, H.; Fukagata, K. Large-eddy simulation of high-Reynolds-number turbulent channel flow controlled using streamwise travelling wave-like wall deformation for drag reduction. *J. Fluid Mech.* **2025**, *1003*, A2. [CrossRef]
2. Araya, G.; Lagares-Nieves, C.J.; Jansen, K. Reynolds number dependency in compressible crossflow-jet problems. In Proceedings of the AIAA Science and Technology Forum and Exposition 2025, Orlando, FL, USA, 6–10 January 2025.
3. Smith, L.; Taira, K. The effect of Reynolds number on the separated flow over a low-aspect-ratio wing. *J. Fluid Mech.* **2024**, *992*, R2. [CrossRef]
4. Brynjell-Rahkola, M.; Duguet, Y.; Boeck, T. Route to turbulence in magnetohydrodynamic square duct flow. *Phys. Rev. Fluids* **2025**, *10*, 023903. [CrossRef]
5. Saldana, M.; Gallegos, S.; Gálvez, E.; Castillo, J.; Salinas-Rodríguez, E.; Cerecedo-Sáenz, E.; Hernández-Ávila, J.; Navarra, A.; Toro, N. The Reynolds number: A journey from its origin to modern applications. *Fluids* **2024**, *9*, 299. [CrossRef]
6. Tamburrino, A.; Niño, Y. The universal presence of the Reynolds number. *Fluids* **2025**, *10*, 117. [CrossRef]
7. Li, Q.; Yokoi, K.; Xie, Z.; Omar, S.; Xue, J. A fifth-order high-resolution shock-capturing scheme based on modified weighted essentially non-oscillatory method and boundary variation diminishing framework for compressible flows and compressible two-phase flows. *Phys. Fluids* **2021**, *33*, 056104. [CrossRef]
8. Liang, T.; Fu, L. A new high-order shock-capturing TENO scheme combined with skew-symmetric-splitting method for compressible gas dynamics and turbulence simulation. *Comput. Phys. Commun.* **2024**, *302*, 109236. [CrossRef]
9. Assis, J.C.; Santos, R.D.; Schuabb, M.S.; Falcão, C.E.G.; Freitas, R.B.; Alves, L.S.B. Convergence towards high-speed steady states using high-order accurate shock-capturing schemes. *Fluids* **2024**, *9*, 133. [CrossRef]
10. Wang, Y.; Chen, J.; Wang, Y.; Zeng, Y.; Ke, S. A WENO-based upwind rotated lattice Boltzmann flux solver with lower numerical dissipation for simulating compressible flows with contact discontinuities and strong shock waves. *Appl. Sci.* **2024**, *14*, 11450. [CrossRef]
11. Yee, H.C.; Sweby, P.K.; Sjögreen, B.; Kotov, D.V. A nonlinear approach in the quantification of numerical uncertainty by high-order methods for compressible turbulence with shocks. *Fluids* **2024**, *9*, 250. [CrossRef]
12. Kuja, Y.; Kawai, S. High-order accurate kinetic-energy and entropy preserving (KEEP) schemes on curvilinear grids. *J. Comput. Phys.* **2021**, *442*, 110482.
13. Ge, X.; Vasilyev, O.V.; De Stefano, G.; Hussaini, M.Y. Wavelet-based adaptive unsteady Reynolds-averaged Navier–Stokes simulations of wall-bounded compressible turbulent flows. *AIAA J.* **2020**, *58*, 1529–1549. [CrossRef]
14. Leclaire, S.; Reggio, M.; Trépanier, J.-Y. Isotropic color gradient for simulating very high-density ratios with a two-phase flow lattice Boltzmann model. *Comput. Fluids* **2011**, *48*, 98–112. [CrossRef]

15. Ba, Y.; Liu, H.; Li, Q.; Kang, Q.; Sun, J. Multiple-relaxation-time color-gradient lattice Boltzmann model for simulating two-phase flows with high density ratio. *Phys. Rev. E* **2016**, *94*, 023310.
16. Leclaire, S.; Parmigiani, A.; Malaspinas, O.; Chopard, B.; Latt, J. Generalized three-dimensional lattice Boltzmann color-gradient method for immiscible two-phase pore-scale imbibition and drainage in porous media. *Phys. Rev. E* **2017**, *95*, 033306. [CrossRef] [PubMed]
17. Zahid, F.; Cunningham, J.A. Review of the color gradient lattice Boltzmann method for simulating multi-phase flow in porous media: Viscosity, gradient calculation, and fluid acceleration. *Fluids* **2025**, *10*, 128. [CrossRef]
18. Xie, X.; Mohebujjaman, M.; Rebholz, L.G.; Iliescu, T. Data-driven filtered reduced order modeling of fluid flows. *SIAM J. Sci. Comput.* **2018**, *40*, B834–B857. [CrossRef]
19. Hijazi, S.; Stabile, G.; Mola, A.; Rozza, G. Data-driven POD-Galerkin reduced order model for turbulent flows. *J. Comput. Phys.* **2020**, *416*, 109513. [CrossRef]
20. Quaini, A.; San, O.; Veneziani, A.; Iliescu, T. Bridging large eddy simulation and reduced-order modeling of convection-dominated flows through spatial filtering: Review and perspectives. *Fluids* **2024**, *9*, 178. [CrossRef]
21. Salomone, T.; Piomelli, U.; De Stefano, G. Wall-modeled and hybrid large-eddy simulations of the flow over roughness strips. *Fluids* **2023**, *8*, 10. [CrossRef]
22. Jensen, R.M.; Forooghi, P. A study of turbulent flow over patchy roughness. *J. Fluid Mech.* **2025**, *1006*, A28. [CrossRef]
23. Hao, Z.; García-Mayoral, R. Turbulent flows over porous and rough substrates. *J. Fluid Mech.* **2025**, *1008*, A1. [CrossRef]
24. Hantsis, Z.; Piomelli, U. Numerical simulations of scalar transport on rough surfaces. *Fluids* **2024**, *9*, 159. [CrossRef]
25. Li, Q.; Bou-Zeid, E. Contrasts between momentum and scalar transport over very rough surfaces. *J. Fluid Mech.* **2019**, *880*, 32–58. [CrossRef]
26. Garnier, P.; Viquerat, J.; Rabault, J.; Larcher, A.; Kuhnle, A.; Hachem, E. A review on deep reinforcement learning for fluid mechanics. *Comput. Fluids* **2021**, *225*, 104973. [CrossRef]
27. Viquerat, J.; Meliga, P.; Larcher, A.; Hachem, E. A review on deep reinforcement learning for fluid mechanics: An update. *Phys. Fluids* **2022**, *34*, 113101. [CrossRef]
28. Rabault, J.; Kuchta, M.; Jensen, A.; Réglade, U.; Cerardi, N. Artificial neural networks trained through deep reinforcement learning discover control strategies for active flow control. *J. Fluid Mech.* **2019**, *865*, 281–302. [CrossRef]
29. Kim, I.; Jeon, Y.; Chae, J.; You, D. Deep reinforcement learning for fluid mechanics: Control, optimization, and automation. *Fluids* **2024**, *9*, 216. [CrossRef]
30. Bourouiba, L. The fluid dynamics of disease transmission. *Annu. Rev. Fluid Mech.* **2021**, *53*, 473–508. [CrossRef]
31. Mittal, R.; Ni, R.; Seo, J.-H. The flow physics of COVID-19. *J. Fluid Mech.* **2020**, *894*, F2. [CrossRef]
32. Krishnaprasad, K.A.; Salinas, J.S.; Zgheib, N.; Balachandar, S. Fluid mechanics of air recycling and filtration for indoor airborne transmission. *Phys. Fluids* **2023**, *35*, 013344. [CrossRef]
33. Sankurantripati, S.; Duchaine, F. Indoor air quality control for airborne diseases: A review on portable UV air purifiers. *Fluids* **2024**, *9*, 281. [CrossRef]
34. Wahler, N.F.; Maruyama, D.; Elham, A. Credibility-based multidisciplinary design optimization of electric aircraft. In Proceedings of the AIAA Science and Technology Forum and Exposition 2023, National Harbor, MD, USA, 23–27 January 2023.
35. Ruh, M.L.; Warner, M.; Hwang, J.T. Toward large-scale multidisciplinary design optimization of aircraft with CFD and mixed-fidelity system models. In Proceedings of the AIAA Science and Technology Forum and Exposition 2025, Orlando, FL, USA, 6–10 January 2025.
36. Li, M.; Chen, J.; Zhou, H.; Chen, Y.; Bai, J. Adjoint-based aero/stealth optimization design for UAV. *Acta Aeronaut. Astronaut. Sin.* **2024**, *45*, 530010.
37. Thoulon, C.; Roge, G.; Pironneau, O. Gradient-based aero-stealth optimization of a simplified aircraft. *Fluids* **2024**, *9*, 174. [CrossRef]

Disclaimer/Publisher's Note: The statements, opinions and data contained in all publications are solely those of the individual author(s) and contributor(s) and not of MDPI and/or the editor(s). MDPI and/or the editor(s) disclaim responsibility for any injury to people or property resulting from any ideas, methods, instructions or products referred to in the content.

fluids

MDPI

Review

The Reynolds Number: A Journey from Its Origin to Modern Applications

Manuel Saldana [1,2,*], Sandra Gallegos [1], Edelmira Gálvez [3], Jonathan Castillo [4], Eleazar Salinas-Rodríguez [5], Eduardo Cerecedo-Sáenz [5], Juan Hernández-Ávila [5], Alessandro Navarra [6] and Norman Toro [1]

1 Faculty of Engineering and Architecture, Universidad Arturo Prat, Iquique 1110939, Chile; chichined@gmail.com (S.G.); notoro@unap.cl (N.T.)
2 Departamento de Ingeniería Química y Procesos de Minerales, Universidad de Antofagasta, Antofagasta 1270300, Chile
3 Departamento de Ingeniería Metalúrgica y Minas, Universidad Católica del Norte, Antofagasta 1270709, Chile; egalvez@ucn.cl
4 Departamento de Ingeniería en Metalurgia, Universidad de Atacama, Copiapo 1531772, Chile; jonathan.castillo@uda.cl
5 Academic Area of Earth Sciences and Materials, Institute of Basic Sciences and Engineering, Autonomous University of the State of Hidalgo, Pachuca 42184, Mexico; salinasr@uaeh.edu.mx (E.S.-R.); eduardoc@uaeh.edu.mx (E.C.-S.); herjuan@uaeh.edu.mx (J.H.-Á.)
6 Department of Mining and Materials Engineering, McGill University, 3610 University Street, Montreal, QC H3A 0C5, Canada; alessandro.navarra@mcgill.ca
* Correspondence: masaldana@unap.cl; Tel.: +56-9-5383-4174

Abstract: The Reynolds number (Re), introduced in the late 19th century, has become a fundamental parameter in a lot of scientific fields—the main one being fluid mechanics—as it allows for the determination of flow characteristics by distinguishing between laminar and turbulent regimes, or some intermediate stage. Reynolds' 1895 paper, which decomposed velocity into average and fluctuating components, laid the foundation for modern turbulence modeling. Since then, the concept has been applied to various fields, including external flows—the science that studies friction—as well as wear, lubrication, and heat transfer. Literature research in recent times has explored new interpretations of Re, and despite its apparent simplicity, the precise prediction of Reynolds numbers remains a computational challenge, especially under conditions such as the study of multiphase flows, non-Newtonian fluids, highly turbulent flow conditions, flows on very small scales or nanofluids, flows with complex geometries, transient or non-stationary flows, and flows of fluids with variable properties. Reynolds' work, which encompasses both scientific and engineering contributions, continues to influence research and applications in fluid dynamics.

Keywords: fluid mechanics; laminar and turbulent flows; modeling; multiphase flows; fluid dynamics; nanofluids; flows in complex geometries

1. Introduction

The Reynolds number (Re) is a dimensionless number that establishes the relation between inertial and viscous forces [1,2], and thus is essential for flow characterization (laminar, transitional or turbulent) [3]. Although Re has no unique definition and its physical meaning varies depending on the context [1], it has a strong influence in fields such as aerodynamics and aeronautics, particularly in transitional flows, where its effects can be non-monatomic and highly sensitive to several parameters [4]. The scaling behavior of turbulent boundary layers is closely related to the dependencies of Re [5]. Generally, the flow in pipes is laminar when Re < 2000 and turbulent when Re > 2000–2300, with an intermediate transitional range, but varies depending on the source consulted [6]. Interestingly, Re can be interpreted as a curvature ratio, which potentially offers insights into the design of high-performance exchangers [7] or can be used to help understand the dynamics of the early universe [8].

The concept has a rich history [9] and is essential in various applications of fluid dynamics, including boundary layer theory [10], microfluidics, or fluid physics at the nanometric scale, where other dimensionless numbers such as the Péclet, capillary, and Knudsen numbers also become important [11], as well as mixing processes, chaotic advection, and turbulence [12]. Understanding Re and related concepts is fundamental for analyzing complex phenomena on a wide range of scales, from nanoliters to geophysical flows [12,13].

This review attempts to synthesize the state of art on Re, providing an overview of most relevant findings and highlighting both theoretical innovations and practical applications. The review will also compare the theoretical and experimental approaches that have emerged over time, evaluating their impact and accuracy in different contexts. Another goal is to demonstrate the applicability of Re in various fields, emphasizing its importance in solving complex problems. Finally, this review aims to offer a solid knowledge base that will be useful for both teaching and scientific dissemination, facilitating the understanding and use of the concept in future research and professional settings.

The organization of the work follows the following sequence: in the historical background, an analysis of the origin of Re is provided, mentioning the first studies and how the need to define this dimensionless number arose. Then, the definition and theoretical foundations are presented, where the formula for the Reynolds number and its basic applications are exposed. In the classical and modern applications section, studies in laminar and turbulent flows and applications in fields such as mechanical engineering, chemistry, biomedicine, and aeronautics are reviewed, along with recent studies on CFD simulations. In the experimental studies and empirical studies section, experimental studies where Re is measured or calculated are analyzed and compared with theoretical studies, identifying areas of agreement and discrepancies, in addition to studying how Re behaves in more complex flows, such as non-Newtonian fluids or multiphase flows. In recent advances and developments, a review of recent studies that expand the use of the Reynolds number in emerging areas, such as nanotechnology, bioengineering, or advanced numerical simulations, is presented along with mentions of innovations in Reynolds number theory and how it has been reinterpreted or refined. Discussions present known limitations and ways to address them, along with controversies and debates regarding them. Finally, the main findings and comments about future research are summarized in the conclusions.

2. Historical Background

2.1. Origins of Reynolds Number

The Reynolds number, a crucial dimensionless parameter in fluid mechanics, originated from the work of Osborne Reynolds on pipe flow in the late 19th century [9,14]. The concept has been applied to various fluid flow scenarios, from microscopic to macroscopic scales [15]. Reynolds' original paper, which introduced the averaging of flow equations, became fundamental in turbulence modeling [16], although both the definition and physical interpretation may vary depending on the application or source used [2,7]. The need to define the dimensionless Reynolds number arose from a desire to understand and predict the behavior of fluids in different flow situations. Until the end of the 19th century, scientists and engineers realized that the behavior of fluids contained many repeating patterns; however, there was no clear method to provide a relationship between such patterns and their physical characteristics—velocity of fluid, viscosity, object, or conduit dimensions. Transitions between smooth, regular flow—subsequently labeled laminar flow—and chaotic, disordered flow—subsequently called turbulent flow—already existed; however, without a proper mechanism, predicting when these changes would occur was very hard. In 1883, British engineer and physicist Osborne Reynolds performed experiments in which he measured how a fluid (water) moved through a cylindrical tube. Reynolds observed that flow behavior not only depends on fluid velocity, but also on additional factors, such as density, viscosity, and pipe diameter. This led him to define a combination of these variables to characterize the flow [2].

Reynolds's discovery was essential because it provided a way to characterize and predict the type of flow under different conditions. This allowed researchers to classify flows as laminar or turbulent without having to perform experiments for each specific situation, i.e., this mathematical relationship allowed researchers to describe the behavior of fluids in a wide range of applications.

2.2. Concept Evolution

Re has evolved significantly since its inception until today. Originally introduced to explain the transition from laminar to turbulent flow in pipes [17], it has become fundamental to characterize flow behavior [1]. Reynolds' work on velocity decomposition into mean and fluctuating components laid the foundation for turbulence modeling [16], and although its generic definition is given as the relationship between inertial and viscous forces [14], its interpretation varies depending on the flow regime [18]. Recent research has explored novel perspectives, such as a relationship of curvatures in spacetime [7] and their role in the transition to turbulence [15]. Despite its long history dating back to Osborne Reynolds and Theodore von Kármán [9], the concept continues to be refined and applied in various fields of fluid mechanics.

Below is a timeline of Re evolution, highlighting key milestones in its development and significance [3,19,20]:

Before the 19th century—early concepts of fluid mechanics: The first attempts to understand fluid motion were made by Greek, Roman, and Arab scholars. Archimedes, for example, studied buoyancy, but the concept of fluid flow remained largely qualitative. Leonardo da Vinci, on the other hand, created the first sketches and observations of fluid flow patterns, including concepts that nowadays are recognized as turbulence.

1700—development of fluid mechanics: Bernoulli developed fluid mechanics by relating fluid flow with pressure and velocity, while Leonhard Euler developed the basic equations of motion for inviscid fluids, now known as Euler's equations. The limitations of both equations were that they did not take viscosity into account.

1800—introduction of viscosity and laminar flow: Navier introduced the concept of viscosity to fluid flow, improving understanding of how fluid layers resist motion, while Stokes expanded on Navier's equations and derived what are known as the Navier–Stokes equations, which describe how viscous fluids behave under different types of forces. These equations take viscosity into account and form the basis of most modern fluid dynamics studies.

1880s—Osborne Reynolds and the birth of the Reynolds number: Reynolds performed his experiments on fluid flow through pipes, where he observed a transition between laminar (smooth) and turbulent (chaotic) flow, establishing Re, a dimensionless quantity that predicts the onset of turbulence based on the ratio of inertial forces to viscous forces. This formula allowed the universal classification of flow regimes. Reynolds also showed that, at low values (typically below 2000), the flow is laminar, and at high values (above 4000), the flow becomes turbulent, declaring a transition region between these values.

Early 1900s—expansion and application of the Reynolds number: Prandtl introduced the concept of the boundary layer (influenced by Re), which was crucial to understanding how fluids interact with solid surfaces and led to significant advances in aerodynamics. Theodore von Kármán then applied the Reynolds number in the study of vortex shedding behind cylindrical objects, contributing to the body of knowledge in fluid dynamics, and finally, engineers began using Re in the design of aircraft wings, propellers, and ships, and applied it to the study of lift, drag, and fluid resistance.

Mid-20th century—computational fluid dynamics (CFD) and expanded use: After WWII, Re became a fundamental tool in the fields of aerodynamics, hydrodynamics, and industrial fluid systems. Engineers used it in wind tunnel tests and large-scale engineering projects (dams, pipelines, cooling systems, etc.), and with the rise of CFD, Re became an integral part of the computer simulation and modeling of complex fluid flows, allowing phenomena to be explored in regions where physical experiments were impractical or impossible.

21st century—advanced applications and research: Re still remains a critical factor in advanced research and applications that range from biomedical engineering, where it is applied to model the blood flow in arteries (the variation in Re helping in the prediction of diseases like atherosclerosis), to high-Re experiments in the aerospace industry that enable the design of hypersonic vehicles and improvements in the efficiency of commercial air travel, as well as micro- and nanofluidics, where an extremely low Re characterizes laminar flows in applications related to drug delivery and lab-on-a-chip technologies.

Key developments throughout the timeline: The concept of dimensionless numbers expanded beyond the Reynolds number, giving rise to others such as the Froude, Mach, and Prandtl numbers, each of which serves a unique function in fluid dynamics and related fields. The evolution of Re reflects the growing understanding of fluid mechanics, moving from basic observational studies to a quantitative and predictive tool that supports many of the most advanced scientific and engineering projects in the modern world. Re is key in a variety of engineering and science disciplines, and its relevance varies depending on the field and type of flow being studied.

2.3. Bibliometric Analysis

The study of the Reynolds number has experienced significant growth in the last century, reflected in the increase in publications in indexed journals (based on SCOPUS bibliographic references), particularly in the last three decades. Bibliometric analysis indicates that 73.6% are original articles and 24.3% are conference papers (the remaining 2.1% correspond to reviews, reports, books, etc.) (see Figure 1a), focusing mainly on areas such as engineering, physics and astronomy, chemical engineering, mathematics, energy, and materials science (see Figure 1b). This expansion in research has led to a deeper understanding and broader applications of the Reynolds number, from its initial conception by Osborne Reynolds in 1883 to its current use in diverse fields (from the study of fluid flow mechanics in pipes to the study of spacetime curvature). Technological advances in experimental methods and computer simulations have allowed researchers to study more complex flows and transition regimes with greater precision. This is evidenced by the exponential increase in scientific work on the subject (see Figure 1c), extending the applicability of the Reynolds number from classical hydrodynamics to modern aerodynamics, microfluidics, and even the study of astrophysical flows.

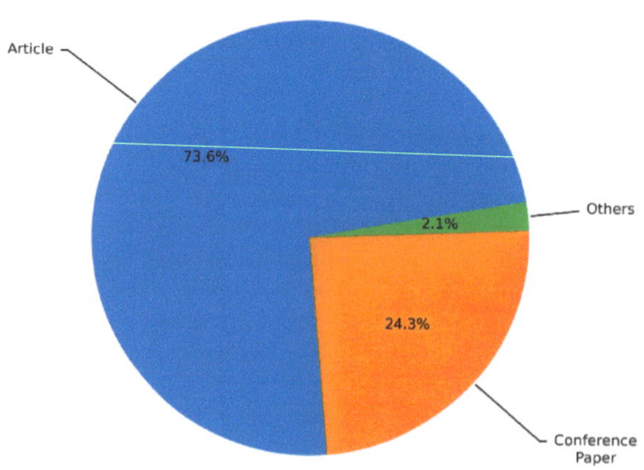

(a)

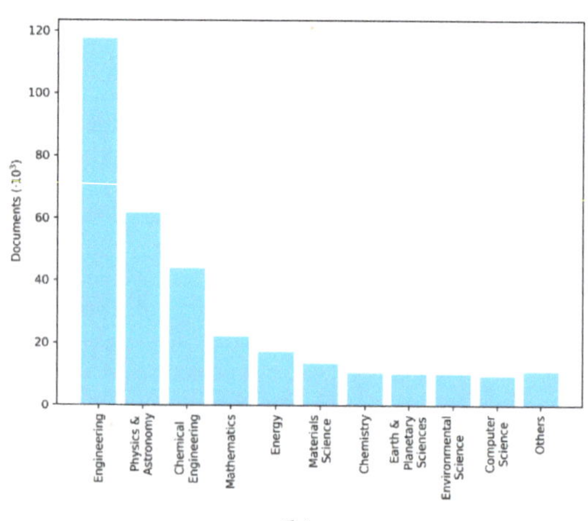

(b)

Figure 1. *Cont.*

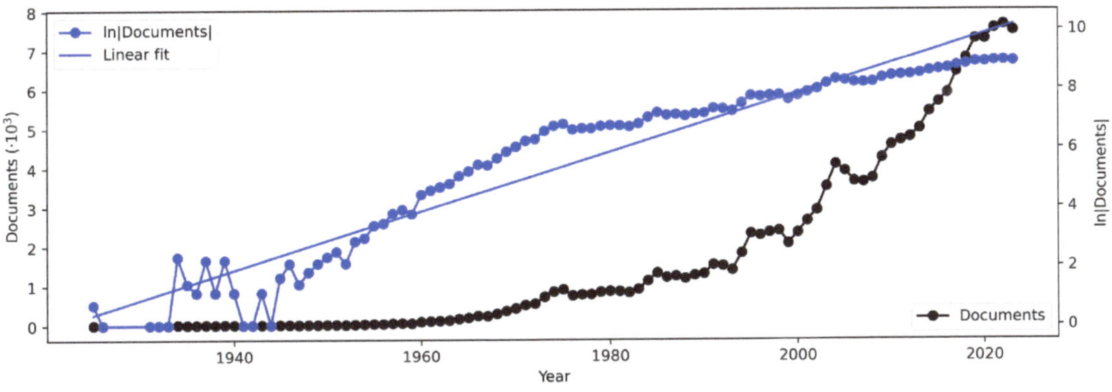

(**c**)

Figure 1. Distribution of documents by type (**a**), field of study (**b**), and temporal variation (**c**) with respect to the Reynolds number in the SCOPUS bibliographic reference database.

3. Definition and Theoretical Foundations

3.1. Mathematical Definition

Re derivation is based on an analysis of the Navier–Stokes equations and the scaling of the characteristic quantities of the flow. The Navier–Stokes equation for an incompressible flow is shown in Equation (1).

$$\rho\left(\frac{\partial u}{\partial t} + u \cdot \nabla u\right) = -\nabla p + \mu \nabla^2 u \quad (1)$$

where u is the velocity vector, ρ is the fluid density, p is pressure, μ is the dynamic viscosity, and $\nabla^2 u$ is a viscous diffusion term. Key terms in Equation (1) are inertial forces $\left[\rho\left(\frac{\partial u}{\partial t} + u \cdot \nabla u\right)\right]$ and viscous forces $\left[\mu \nabla^2 u\right]$. Then, to analyze these equations in terms of a dimensionless parameter, characteristic flow magnitudes are considered: characteristic length (L); characteristic velocity (U); characteristic time (L/U); and characteristic velocity gradient (U/L).

Substituting these scales into the Navier–Stokes terms, the scaling of the inertial forces and the viscous forces are presented in Equations (2) and (3), respectively.

$$\rho u \cdot \nabla u \sim \rho \frac{U^2}{L} \quad (2)$$

$$\mu \nabla^2 u \sim \mu \frac{U}{L^2} \quad (3)$$

Since Re is defined as the ratio between the inertial and viscous forces, replacing the characteristic scales, its mathematical definition is presented in Equation (4).

$$Re = \frac{\text{inertial forces}}{\text{viscous forces}} = \frac{\rho \frac{U^2}{L}}{\mu \frac{U}{L^2}} = \frac{\rho UL}{\mu} = \frac{\rho \cdot v \cdot L}{\mu} = \frac{v \cdot L}{\nu} \quad (4)$$

where ρ is fluid density $\left(\frac{kg}{m^3}\right)$, $v = U$ is the characteristic velocity of the flow $\left(\frac{m}{s}\right)$, L is a characteristic length, such as the pipe diameter (m), μ is the fluid dynamic viscosity $(Pa \cdot s)$, and $\nu = \frac{\mu}{\rho}$ is the fluid kinematic viscosity $\left(\frac{m^2}{s}\right)$.

3.2. Physical Interpretation

Reynolds number physical interpretations aim to determine the regime or behavior in which a fluid is found, which can be categorized as: laminar, turbulent, or in transition.

- Re < 2000: indicates laminar flow, where the fluid moves in ordered layers and viscous forces dominate (see Figure 2a).

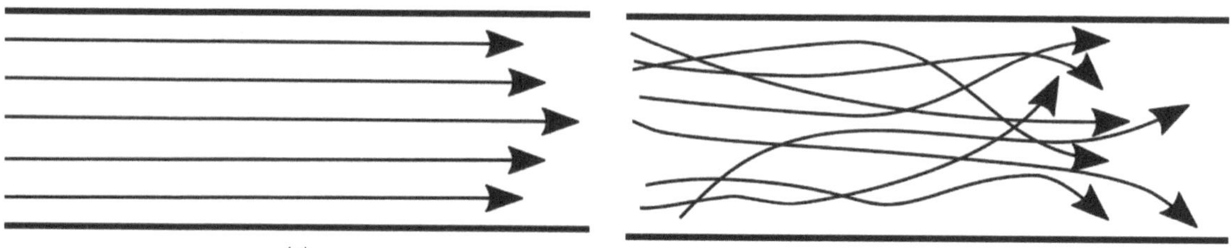

Figure 2. Diagrams of laminar (**a**) and turbulent (**b**) flow regimes.

- Re > 4000: indicates turbulent flow, where inertial forces predominate and the flow is chaotic (see Figure 2b).
- 2000 < Re < 4000: correspond to a transition regime between laminar and turbulent.

The Reynolds number allows for the characterization of flow behavior without having to completely solve the Navier–Stokes equations, acting as a key criterion in the design of experiments and applications in engineering.

3.3. Basic Applications

Re has a fundamental role in a variety of applications in hydro- and aeromechanics [21]. Its importance spans aquatic locomotion, biological systems, boundary layer transitions, impact dynamics, engineering systems, and turbulent flow behaviors. Some of the basic applications of the Reynolds number are the study of flow dynamics in pipes, aircraft design, automotive, heat exchangers, industrial mixers, air conditioning flows, hydrodynamics of ships and submarines, and mineral processing. The main basic applications of the Reynolds number are listed below.

3.3.1. Fluid Flow in Pipes

As stated above, the Reynolds number characterizes the flow regime—laminar or turbulent. Therefore, understanding its implications is essential to optimize fluid transport systems in pipes. Laminar versus turbulent flow is understood when Re < 2000, where the flow is typically laminar, characterized by a smooth and orderly fluid motion, while as Re increases beyond this threshold, the flow transitions to turbulence, marked by a chaotic and irregular fluid motion [22]. The literature indicates that the transition point can vary depending on factors such as fluid properties and pipe geometry, but Re remains a fundamental indicator of flow behavior [23]. Additionally, it is noted that, at higher Re, the azimuthal and streamwise structures of the flow exhibit distinct scale behaviors, influencing the turbulence characteristics and energy distribution within the pipe [23].

The relationship between Re and velocity fluctuations is significant, and higher Re leads to greater variation in streamwise and spanwise velocities, affecting overall flow stability and efficiency [24]. Although more basic applications consider the study of the correlation between pressure drops in pipe flows [9], other studies have explored other phenomena, including Re effects in direct numerical simulations of turbulent flow in pipes and comparison with channels and minimal layers [25], flows through diaphragm orifices in pipes where the orifice functions as a flow meter [26], and flows in pipes with high [27], extreme [28], or variable [29] Reynolds numbers. Additional studies have addressed low-Re flows in helical pipes [30] and the evaluation of laminar flows with high Re, considering internal flows, external flows, and flows with special situations such as flow over an obstacle in a wall layer, marginal separation, and unsteady breakup separation [31]. Other studies have examined turbulence bounded by high-Re walls, zero pressure gradient boundary layers, and flows in fully developed pipes and channels [32], along with the scaling properties of turbulent flow in low-Re pipes, where the influence of Re on turbulence statistics such as Re stresses, skewness, and kurtosis is simulated and studied, and where discrepancies in turbulence characteristics at low Re compared to high-Re predictions are also highlighted [33].

Although the transition to turbulence in pipe flow historically occurs at a Reynolds number of around 2000, this limit can even be delayed by up to more than 100,000 under specific conditions where care is taken to inject the liquid into the interior in a uniform manner. In the work developed by Dabiri et al. [34], it is shown that a laboratory jet flow maintains laminar flow beyond a Reynolds number of 116,000, which is confirmed by high-speed videography and flow velocimetry, where the laminar nature of the flow is maintained despite environmental perturbations. The quantified spatial development of the velocity distribution within the jet, which tends to assume a "top-hat" configuration as the distance downstream enlarges, seemingly facilitates the maintenance of laminar flow. These findings imply the presence of a flow regime that is empirically attainable, wherein turbulence may be evaded at an elevated Re.

Finally, instabilities in flow studies arise when a disturbance, however small, grows over time instead of dissipating. These instabilities are usually associated with the physical properties of the flow, the geometry of the system, the boundary conditions, and the problem parameters, such as the Reynolds number. The physical origins of the instabilities can have their genesis in the competition between forces, the interaction with the geometry (where vorticity concentrations or inflection points in the velocity profile are generated, facilitating the amplification of disturbances), and the transition from laminar to turbulent flow.

The stability of a flow can be studied by linear perturbation theory. Assuming that the steady base flow $U(x,t)$ is perturbed by a small perturbation $u'(x,t)$, the total flow is given by Equation (5):

$$u(x,t) = U(x,t) + u'(x,t) \tag{5}$$

where $u'(x,t) \ll U(x,t)$. Then, substituting into the Navier–Stokes equations and linearizing (discarding quadratic terms in u'), Equation (6) is obtained, where $\mathcal{L}(U)$ is a linear operator that depends on base flow U.

$$\frac{\partial u'}{\partial t} = \mathcal{L}(U)u' \tag{6}$$

Assuming modal type solutions as presented in Equation (7), where k is the wave vector and ω is the complex frequency, the stability criterion is given by the following: if $\mathrm{Im}(\omega) > 0$, the disturbance grows exponentially $\left(e^{\mathrm{Im}(\omega)t}\right)$, and the flow is unstable; and if $\mathrm{Im}(\omega) \le 0$, the flow is stable.

$$u'(x,t) = \hat{u}(x)e^{i(kx-\omega t)} \tag{7}$$

3.3.2. Aircraft Design

In aeronautics, Re significantly influences the explanation of flow behavior, particularly in transitional flows, affecting vehicle aerodynamics [4]. It significantly affects aircraft design and aerodynamic performance, because as the number increases, the maximum lift coefficient increases, while the minimum drag coefficient decreases [35]. The Reynolds number also affects the pressure distribution, shock wave position, and boundary layer development in supercritical airfoils used in transport aircraft [36,37], and in the case of large aircraft, the effects of Re must be taken into account during design and optimization [37]. Airfoils with a low Reynolds number are crucial for unmanned aerial vehicles and high-altitude aircraft [38], with airfoil designs varying across Reynolds number ranges, from insect flight to supersonic aircraft. In turbine blade design, on the other hand, the Reynolds number affects flow physics phenomena including boundary layer evolution and wake shedding [39].

Finally, useful tools to investigate the effects of Re are numerical simulations and wind tunnel tests, with computational methods allowing simulations of higher Reynolds numbers comparable to real flight conditions, stressing the models in order to analyze their limits [40,41].

3.3.3. Automotive

Re plays a crucial role in automotive aerodynamics and vehicle performance as it significantly influences drag coefficients, whereby higher Re generally leads to lower drag and higher efficiency [42,43]. The effects of Re on vehicle dynamics extend to several aspects including the directionality of motion in self-propelled systems [44] or leveraging the effects of Re to substantially reduce drag, as demonstrated in studies on short blunt bodies [45] and computational models [46]. Re facilitates the study of turbulence control mechanisms, such as: the study of the asymmetric shear layer force, where the use of a small square cylinder to disturb the flow has been tested, determining that it is possible to reduce resistance by up to 8% by improving flow re-insertion and reducing wake asymmetry [47]; the reduction of turbulent resistance by a constant force near the wall [48]; and the use of thermal bands in the direction of the flow to reduce resistance by surface friction, improving the formation of streaks at low speed and suppressing cross-flow fluctuations [49]. These findings have potential applications for automotive design, since they can lead to a more effective management of turbulence and to significant reductions in aerodynamic resistance, improving fuel efficiency and equipment performance.

On the other hand, the effect of turbulence on local drag forces in automotive computational models is significant, as various control strategies can lead to substantial drag reduction. Integrating turbulence management techniques can optimize aerodynamic performance, particularly in complex flow scenarios, so understanding and manipulating Re effects through various design strategies and flow control techniques can lead to significant improvements in automotive performance and efficiency across different transportation domains.

3.3.4. Heat Exchangers

The Reynolds number is a key factor influencing heat exchangers performance, affecting both heat transfer and fluid flow characteristics. The reviewed literature indicates that, as the Reynolds number increases, the heat transfer coefficient generally improves while the friction factor decreases [50]. Furthermore, studies on various types of heat exchangers, including shell-and-tube and multi-grid fin designs, have shown that performance can be improved by manipulating geometric parameters and flow conditions [51,52]. Along the same lines, a case study indicates that higher Reynolds numbers improve fluid dynamics, reduce concentration polarization, and enhance membrane performance in electrodialysis desalination [53]. Therefore, in transitional flow regimes, optimal Reynolds numbers can be identified to balance heat transfer and pressure drops [54]. Nanofluids can significantly increase thermal efficiency in heat exchangers [55], and particle size and volume fraction strongly influence their thermophysical properties [56]. The improvement in the thermal efficiency of nanofluids is given by the following conditions: improved thermal conductivity, because nanofluids—fluids containing nanoparticles—have a higher thermal conductivity compared to conventional fluids, allowing for more efficient heat transfer in heat exchangers; the solid volume fraction of nanoparticles significantly affects thermal properties—for example, increasing the graphite concentration in nanofluids improves thermal conductivity, leading to better heat transfer efficiency; the use of nanofluids, especially those based on carbon nanotubes (CNT), has been shown to improve the heat transfer efficiency of plate and plate-fin heat exchangers; and finally, nanofluids also possess unique thermophysical properties that contribute to their ability to improve the thermohydraulic performance of heat exchangers [55]. Another more recent study corroborates that Re significantly influences the heat transfer coefficient and the Nusselt number, improving heat transfer and thermal performance in nanofluid applications [22]. Additionally, the stability of nanofluids is crucial, since their instability can reduce system performance considerably [57].

On the other hand, optimal thermodynamic performance of horizontal ground-based heat exchangers in ground-source heat pump systems has been analyzed using various configurations [58], and geothermal heat pump systems, which use the ground as a heat source or sink, have been found to be highly efficient and are becoming popular in heating, ventilation and air conditioning (HVAC). Horizontal circuits are more cost-effective than vertical wells if sufficient land is available, but thermodynamic approaches still need to be improved to evaluate their performance.

3.3.5. Industrial Mixers

The Reynolds number plays a crucial role in the performance and efficiency of industrial mixers, influencing flow patterns, mixture quality, and operational parameters and/or responses, so understanding its effects on mixing dynamics can improve the design and application of various mixing technologies. From the literature review, it is evident that experiments show that, as Re increases, the islands of poor mixing decrease, indicating the directly proportional relationship between mixture quality and mixer speed [59], while simulations of large eddies in stirred tank reactors show that flow structures and vortex formations vary significantly with Re, affecting the mixing dynamics [60].

Continuing the analysis regarding the flow characteristics, studies on partitioned pipe mixers indicate that the three-dimensionality of the velocity field decreases with increasing Re, leading to a more chaotic mixing [61], while in the case of studies of more bounded flows, such as jets, Re is a factor that influences decay rates and turbulence structures in mixing operations [62]. Additionally, high values of Re can also improve thermal mixing characteristics [63].

3.3.6. Air Conditioning Flow

Re plays a crucial role in the dynamics, performance, and efficiency of air conditioning flow and heat transfer as it influences flow characteristics, heat transfer, and sterilization processes, so studying and understanding its effects can lead to optimized designs and improved operating results. In compact heat exchangers, new models have been developed to predict friction factors and heat transfer coefficients over a wide range of Reynolds numbers [64]. For finned tube bundles, correlations between friction factors and Reynolds numbers have been derived to improve the estimation of cooling capacity [65]. In inclined semicircular ducts, a decrease in Re improves heat transfer rates, particularly in specific orientations, indicating that Re is a key factor in duct design for HVAC applications [66]. And finally, recent research on microtubes indicates that as Re increases, both the friction factor and heat transfer efficiency decrease, suggesting that optimal Re values are critical to maximizing heat transfer in HVAC systems [67].

3.3.7. Hydrodynamics of Ships and Submarines

Hydrodynamics of ships and submarines are significantly influenced by Re [68], affecting drag, wake, maneuvering characteristics, performance, and stability. As aquatic vehicles have improved their speed capabilities, recent research in ship and submarine hydrodynamics has focused on improving numerical simulation methods and understanding complex flow phenomena at high Reynolds numbers, finding applications in the field of technological development including the development of flexible coatings that improve hydrodynamic performance [69]. Therefore, understanding the effects of Re is essential to optimize the design and operational efficiency of this type of vehicle.

The variation of hydrodynamic coefficients with Re is significant, particularly at low values. Studies show that using fitted functions instead of constant values yields higher accuracy in estimating these coefficients for underwater vehicles [70], while numerical simulations of submarines, such as the SUBOFF model, reveal that hydrodynamic forces and moments are sensitive to changes in Re, affecting maneuverability and stability [71]. Research on subsea pipelines, on the other hand, indicates that gas leaks alter wake structures and hydrodynamic forces, and that the interaction of gas bubbles and vortices is strongly dependent on Re [72]. The behavior of turbulent junction flows, common in subsea appendages, shows that while mean flow characteristics may not vary significantly with Re, turbulence intensity and heat transfer efficiency do increase, affecting overall performance [73].

Studies have explored the use of dense gases for model testing [74], as the use of highly compressed air and other dense gases allows for full-scale Re testing of submarines and aircraft, and the implementation of boundary layer transition models for naval applications [75]. Another application of boundary layer transitions studies how the Reynolds number affects the transition of boundary layers over underwater bodies, which impacts design considerations in marine engineering [76].

Advances in computational fluid dynamics (CFD), to be studied in more detail in the next section, have allowed more accurate simulations of turbulent flows around underwater vehicles at high Re [77], with hybrid URANS/LES methods showing potential for certain applications [78].

3.3.8. Mineral Processing

The Reynolds number plays a critical role in mineral processing, influencing fluid behavior and particle transport in a variety of applications, and its impact is seen in processes such as crystallization, mixing, and permeability, which are essential to optimize mineral recovery and product quality. In microfluidic systems, low Reynolds numbers indicate laminar flow, allowing for the precise control of fluids and particles [11]. In solids and gas flow, the Reynolds number affects particle motion and drag in viscous regimes [79]. In mineral suspensions, rheological properties (including Re) are influenced by particle concentration and surface chemistry, affecting flow characteristics [80].

Re affects the kinetics of nucleation and crystal growth in membrane crystallization, where higher Re enhances mass and heat transfer, increasing permeate flux and modifying interfacial supersaturation, which in turn regulates bulk nucleation rates [81]. This relationship allows for refined control over scale-up and crystallization processes. In planetary mixers, on the other hand, increasing Re correlates with improved mixing quality and experiments show that islands of poor mixing decrease as Re increases, indicating that higher flow rates enhance material interface interactions [59], which is vital for achieving uniformity in mineral processing. Additionally, Re also influences apparent permeability in spatially periodic arrays, where higher Re leads to orientation-dependent permeability changes. This affects fluid flow characteristics essential for efficient mineral extraction [82].

While the Reynolds number is essential to understanding fluid dynamics, its limitations in highly turbulent flows or complex geometries can complicate predictions, requiring advanced modeling techniques to capture the nuances of complex environments, such as near-collisionless plasmas like solar wind [83]. In addition to generating prediction errors, excessive turbulence can also lead to operational challenges, such as increased equipment wear and non-ideal flow patterns, which can affect negatively processing results.

4. Classic and Modern Applications

4.1. Classical Applications in Hydrodynamics and Aerodynamics

The study of the Reynolds number in hydrodynamics has evolved significantly. In the design of pipelines, a field in which it is required to know a priori whether the flow will be laminar or turbulent, Re is the criterion par excellence for characterizing flow regimes, as developed by Ryan and Johnson [84], who formulated a criterion to characterize non-

Newtonian fluid regimes. Additionally, Bankston [85] studied the transition of hydrogen and helium gas flow in heated pipelines from turbulent to laminar, concluding that the transition occurs as the heat flow increases and that the transition significantly affected the heat transfer coefficients compared to completely turbulent flow. Spriggs [86] and Barr [87] also analyzed the transition from laminar to turbulent flow, exploring the conditions and implications of this change and providing information on its effects on flow behavior and practical implications for engineering processes.

One of the foundational works on low-Reynolds-number hydrodynamics was presented by Happel and Brenner [88], who explored its applications in several fields, including sedimentation, fluidization, particle size classification, dust and mist collection, filtration, centrifugation, polymer and suspension rheology, flow through porous media, colloidal science, aerosol and hydrosalt technology, lubrication theory, blood flow, Brownian motion, geophysics, and meteorology, among others. Complementing the above, Hinch provides an elementary introduction to the conservation principles governing incompressible fluids, laying the foundation for understanding flow behavior at low Re [89].

Khayat and Eu [90] examined the influence of the Reynolds number on the steady flow properties in cylindrical Couette flow, showing that fluids can behave nearly inviscidly beyond a critical Re. Szwalek and Larsen [91] explored the impact of Re (in turbulent flows) on the hydrodynamic coefficients in vortex-induced forced vibrations, revealing that while certain coefficients showed minimal dependence on Re (the force in phase with velocity, the force in phase with acceleration, and the mean drag coefficient, with Re up to 36,000), trends were observed in specific instability regions, where the hydrodynamic coefficients did not appear to be stable and can be considered to be highly Re dependent. Dewsbury et al. [92], on the other hand, examined the drag coefficient of ascending solid spheres in non-Newtonian fluids over a wide range of Re ($135 < Re < 55{,}000$), noting a significant decrease in drag as Reynolds numbers increased, particularly beyond 20,000.

In Meister et al. [93], sensitivity of Re is investigated in the context of the weakly compressible smoothed particle hydrodynamics method, discussing aspects such as physical viscosity and the origin of Couette flow instability, exploring their effects on the convergence properties of the method. Furthermore, a new instability is identified in the flow of pipes with expanding diameters, and both instabilities are observed to occur even at low Re, suggesting that they are caused by high-frequency particle oscillations, unrelated to turbulence. Other applications of Re to hydrodynamics consider measurements of flow dynamics in pipe bends for different regimes [94].

Re plays a crucial role in marine engineering applications, in particular in understanding ship drag, and wake and boundary layer transitions. Studies have shown that Re affects ship drag and wake characteristics, with implications for meshing requirements in computational fluid dynamics (CFD) simulations [95,96], and applications in marine engineering problems [97]. Reynolds-averaged Navier–Stokes (RANS) equations are commonly used to simulate flows at the full scale of Re, which can reach 10^9 in naval applications [98]. Several turbulence models have been developed to address the Reynolds number stresses generated by the RANS equations [99]. These models have been applied to analyze marine thrusters [100], submarine models [101], and marine rudders [102]. Recent research has focused on the implementation of boundary layer transition models in CFD codes for naval applications, addressing the challenges of predicting turbulent flows with transition and separation [75]. Hence, understanding hydrodynamic resistances is crucial to optimize ship design and performance [103]. Additionally, other interesting studies include CFD large eddy simulations (LES) focusing on ship hydrodynamics at different Re. These show detailed simulations of turbulent boundary layers and wake characteristics without external turbulence disturbances [104], or, on the other hand, the application of CFD to ship hydrodynamics and naval engineering design [105].

Re also plays a determinant role in aerodynamics and early aviation. It influences laminar flow separation, transitional flow behavior, and the aerodynamic characteristics of airfoils [4], critical for a wide variety of flying vehicles, from birds and bats to human-powered aircraft and remotely piloted vehicles [106]. At low Re, typical of small air vehicles and micro air vehicles, laminar flow prevails, affecting airfoil performance and design considerations [107,108]. In airfoil scaling applications, dynamic scaling ensures that flight dynamics are proportionally tuned, and aerodynamic scaling ensures that the scaled aircraft—particularly its airfoil—maintains the same characteristics as the full-sized aircraft, despite differences in Re caused by changes in length, velocity, and air properties [109]. In contrast, for large air vehicles, the aerodynamics exhibit large Reynolds numbers. Aerodynamic hysteresis, where the lift and drag coefficients become history-dependent near stall angles, is common for round-tip airfoils with low Re [110]. The Reynolds number affects the maximum lift coefficient, with higher values generally increasing lift and decreasing drag [35,111]. Understanding these effects is critical to designing efficient airfoils for a variety of applications—from insect-sized vehicles to full-scale aircraft—to achieve low drag while maintaining lift performance by optimizing boundary layer development and transition location [112]. Wind tunnel testing must account for Reynolds number effects to accurately predict full-scale aircraft performance [113].

4.2. Classical Applications in Pipe and Canal Engineering

Re plays a crucial role in understanding flow dynamics and sediment transport in sewer systems. It helps determine the flow regimes, from laminar to turbulent, that affect airflow in sewer headspaces [114] and wastewater entrainment [115]. Re is essential for analyzing turbulent coherent structures around sediment reduction plates [116] and formulating mathematical expressions for roughness functions and sediment transport initiation curves [117]. In sedimentation tanks, Reynolds and Froude numbers are used to optimize design and improve efficiency [118]. Hydraulic models can examine stormwater overflow performance at different Re [119]. Understanding these applications is vital for the effective design and management of sewer systems [120].

Regarding water supply networks, Re influences flow regimes and pipe friction. It is used to distinguish between laminar, transitional, and turbulent flows [121]. Although for PVC pipes, different equations apply depending on the Reynolds number ranges and pipe diameters [122], the Hazen–Williams equation is commonly used in hydraulic engineering, with limitations based on Re and pipe size [123]. Re is essential for modeling instantaneous residential water demands and predicting contaminant movement in distribution systems [124], while it is also considered in identifying pipe roughness for both steady-state and dynamic models [125]. Understanding Re is important to optimize water supply system design [126] and analyze the efficiency, reliability, and vulnerability of a water supply network system [127].

Re also plays a key role in predicting pressure losses in industrial systems. For high Re piping systems, traditional correlations may be too conservative, requiring new methodologies for more accurate predictions [128]. In duct components, both size and Re influence pressure loss factors, contrary to previous assumptions [129]. The effects of temperature and pressure on fluid properties must be considered for accurate pressure loss calculations in hydraulic systems [130]. In ventilation systems, the effects of a low Reynolds number are significant, requiring suitable turbulence models for accurate simulations [131], while for non-Newtonian fluids, new relationships between pressure drop and flow rate, as well as Reynolds number expressions, have been derived [132,133].

4.3. Modern Applications in Aeronautical and Automotive Engineering

Re is determinant in the design of wings and fuselages of aircraft and UAVs. Studies have investigated its effects on aerodynamic characteristics over various Reynolds number ranges, from low (200,000–500,000) for UAVs [134] to high (75 million) for large aircraft [135]. Research has shown that increasing Re (generally) improves aerodynamic performance, with higher maximum lift coefficients and lower minimum drag coefficients [35]. In flying wing designs, Re affects drag under cruise conditions and influences self-alignment characteristics [136]. Wind tunnel testing and CFD simulations have been employed to study the effects of Re on several aircraft configurations, including wing-fuselage junctions [137], blended wing bodies [135], and helicopter fuselages [138]. These studies provide valuable data for improving aerodynamic design codes and understanding complex flow phenomena in aircraft design [139].

Regarding aerodynamic optimization of racing cars, Re affects drag, downforce, and overall performance. Studies have shown that variations in Re can significantly affect single-element wings, with up to a 320% difference in downforce observed over a small range of Re [140]. However, multi-element configurations show less sensitivity to Reynolds number changes [140,141]. Optimization techniques have been employed to improve vehicle performance by modifying nose shapes and airfoils [142,143], while wing configuration, including shape and angle, also influences aerodynamic efficiency [144]. Additionally, for Formula One cars, front wing optimization is critical as it contributes approximately 25% of the total downforce. These studies collectively demonstrate the importance of considering Re effects in race car design, from ultra-low Reynolds numbers for micro-air vehicles to higher ranges for full-scale racing cars [42].

Low-Reynolds-number aerodynamics are important for the design and performance of micro aerial vehicles (MAVs), drones, and autonomous underwater vehicles (AUVs) [68,108,145]. These vehicles typically operate in Re ranges below 200,000, where laminar flow prevails and unique aerodynamic challenges arise [145,146]. Factors such as rotor spacing, angle of attack, and turbulence significantly affect the aerodynamic characteristics of these vehicles [68,147]. CFD simulations and experimental studies have revealed the importance of considering transitional flow models in addition to fully turbulent models for accurate performance predictions [148], while biomimetic approaches, inspired by insect flight, offer promising solutions to address the challenges of low-Reynolds-number flight [149]. Understanding these aerodynamic phenomena is essential to improve the efficiency, endurance, and maneuverability of MAVs, drones, and AUVs [146,150].

Reynolds numbers affect significantly the aerodynamic characteristics of supersonic transport vehicles in various flight regimes [151]. In fact, under transonic and subsonic conditions, increasing Reynolds numbers lead to changes in pitching moment, stability, and control effectiveness [152–154], delaying the leading edge separation and pitching moment exit, while improving stabilizer and rudder effectiveness [155]. In supersonic flows, Reynolds number sensitivity affects the formation of shock waves in turbulent shear layers, influencing mixing and combustion processes [156]. The effects are particularly pronounced in transitional flows, where Re variations can cause striking changes in aerodynamic behavior. These phenomena are attributed to changes in boundary layer development, shock wave–boundary layer interactions, and flow separation characteristics [36]. Understanding these effects is critical to accurately predicting flight vehicle performance and optimally designing supersonic transports.

4.4. Modern Applications in Nanotechnology and Microfluidics

Investigations on the Reynolds number in fluid flows through microchannels reveal complex behavior influenced by channel geometry and flow conditions. Studies show that conventional fluid mechanics theory can often predict flow behavior in microchannels with hydraulic diameters larger than 100 μm [157,158]; however, significant scale effects are observed for channels with heights ≤ 0.3 mm [159]. The transition to turbulence in microchannels occurs at lower Reynolds numbers than in macroscale flows, with critical values ranging from 200 to 700 [160] or even as low as 40 [161]. Factors affecting flow through microchannels include channel geometry, surface roughness, viscous dissipation and mainly, and interfaces such as channel walls (that is the reason the critical Re are smaller for small channels) [162], while new approaches to characterize microchannel flows include the introduction of new dimensionless numbers that combine the Reynolds number with geometric parameters [163]. On the other hand, CFD simulations are valuable tools to investigate microchannel flow behavior under various conditions [164].

Microfluidic devices operating at low Re face challenges in achieving efficient mixing and flow control. Recent studies have shown that turbulence can occur in microfluidics at Re ∼1 under electrokinetic forcing, exhibiting features such as the Obukhov–Corrsin spectrum [165,166]. To improve mixing efficiency, researchers have explored several designs, including T-shaped micromixers with obstacles and slots [167], magnetic bead-based mixing [168], and flexible thermoplastic film mixers [169]. Flow rectification at low Re has been achieved using microfluidic bead-based diodes with circular microchannels [170] and single-layer "domino" diodes [171]. The interplay between fluid inertia and elasticity in microfluidic inlet flows has been studied using dilute polymer solutions.

With respect to biomedical engineering, on the other hand, the Reynolds number has important implications, particularly in fluid dynamics and mixing processes. In microfluidic devices, higher Re results in better mixing and smaller channel widths [172]. For blood flow in cerebral vessels, increasing Re leads to higher pressure drops and wall shear stress [173]. Re is essential to reconcile discrepancies between Doppler and catheter pressure measurements in aortic stenosis [174,175]. In small centrifugal blood pumps, lower Re flows deviate from conventional pump affinity laws, necessitating consideration of Re for accurate scaling [176]. Re also influences cell adhesion in biofouling, with a negative linear correlation observed between adhesion density and the product of the designed roughness index and Re for both bacterial cells and algal spores [177]. In stenosed artery models, the effect of lesion eccentricity on flow behavior decreases as Re increases [178]. In the case of intracranial aneurysms, both the Re and Womersley number affect vortex ring propagation, with the vortex ring location being proportional to Re/Wo^2 [179]. Understanding the effects of Re is crucial for designing biomedical devices and analyzing biological fluid dynamics [180,181]. The above-cited studies demonstrate the broad applications of Re in biomedicine, from microscale fluid dynamics to macroscale physiological processes.

Other applications in biology consider the study of hydrodynamic synchronization at low Reynolds numbers, referring to the phenomenon in which fluid interactions synchronize the movements of microscopic objects such as cilia or flagella, which affect bacterial pumping and swimming efficiency [182], and the study of hydrodynamics and hydrodynamic interactions in micron motility, including the propulsion mechanism itself, the synchronized movement of flagella in flagellar bundles, the movement of cilia in cilia arrays, and collective behaviors [183]. Another investigation studies how weakly non-linear self-sustained oscillators synchronize in a fluid environment at low Re, helping to understand how small-scale physical and biological systems synchronize in fluid environments, which offers insights into natural phenomena such as cell motility or the coordination of microorganisms [184]. In blood flow dynamics, the Reynolds number is vital to model the flow through arteries, where it can be used to predict flow patterns and resistance, which is critical for medical diagnoses and treatments [185].

Research on nanoscale fluid manipulation at low Re reveals unique phenomena and challenges. Molecular dynamics simulations show deviations from classical Navier–Stokes predictions, particularly near channel walls [186]. Nanoscale surface roughness significantly affects flow patterns, leading to perturbations extending to the channel midplane [187]. Alternating current electric fields can effectively manipulate suspended nanowires despite extremely low Re [188]. Three-dimensional nanostructures in microfluidic devices improve mixing efficiency at low Re [189]. The rotational motion of elongated nanoscale objects under external torque can be described analytically and exhibits multiple steady states [190]. Furthermore, computational methods such as molecular dynamics simulations provide valuable insights into fluid behavior at the subcontinuum level [191]. These findings have implications for various applications, including nanorobotics in medical settings [192] and the design of microfluidic devices [11].

4.5. Other Classic and Modern Applications

Additional applications of the Reynolds number in engineering and science can be found in fields such as mechanical engineering, chemical engineering, bridge and dam design, astrophysics, nanotechnology, renewable energy engineering (wind and offshore turbines), aquatic locomotion, impact dynamics, and aerospace exploration, among others.

In mechanical engineering, applications are found in turbomachines, where Re affects the overall efficiency and performance [193,194]. In centrifugal compressors, Re affects the efficiency, head, and flow, requiring correction formulas for accurate performance prediction [195]. In microaxial flow fans, the effects of Re are not prohibitive, allowing for dimensional analysis and performance scaling [196]. Re also influences wax deposition in oil production systems, affecting flow assurance [197]. RANS modeling is widely used in the industry for CFD simulations, although challenges still exist regarding the reliability of the turbulence model [198]. Considering the above, understanding the effects of Re is essential to optimize design and performance in various mechanical engineering applications.

From the literature review, a collection of articles can be found where various applications of Re in chemical engineering are explored. Micromixing technologies are classified based on chemical, biological, and analytical applications [199]. Re effects are studied in turbulent reactive flows, affecting mixing processes and chemical reactions [200,201]. Turbulence fundamentals are applied to reactor modeling and scaling, addressing complex geometries and chemistries [202]. Magnetic resonance is used to study hydrodynamics in chemical engineering systems, including rheology and reactor flows [203]. And finally, Re correlations are investigated for dense slurry systems with coaxial mixers [204].

Regarding the construction of bridges and dams, or hydraulic structures as a whole, the effects of Re on the aerodynamics of these structures have been widely studied due to its importance in wind engineering and design. Research shows that Re can significantly influence aerodynamic force coefficients, Strouhal numbers, and pressure distributions on bridge decks, even those with well-defined edges [205,206]. These effects are particularly pronounced in streamlined and double-box girder bridges [207–209]. High-pressure wind tunnels and CFD have been employed to investigate the effects of Re over a wide range, from 10^4 to 10^6 [210,211]. Studies on specific bridges, such as the Great Belt East Bridge, Sutong Bridge, and Stonecutters Bridge, have demonstrated the importance of considering the effects of Re in the design [205,212]. Furthermore, Re can affect vortex shedding and the effectiveness of mitigation devices [213]. The above findings underline the need to carefully consider the effects of Re in bridge design and wind tunnel testing.

Reynolds numbers also play a crucial role in astrophysical turbulence and magnetic reconnection. High Re in astrophysical environments give rise to turbulent flows [214,215], which affect diverse phenomena from stars to accretion disks. Turbulent reconnection, in which the magnetic field topology changes in turbulent plasmas, is particularly significant in astrophysical environments [216]. This process can be self-driven and has implications for solar flares, gamma-ray bursts, and particle acceleration, while numerical simulations and observational data support turbulent reconnection theories [216,217]. The concept of magnetic flux freezing, fundamental to many plasma theories, is challenged by turbulent reconnection [216]. Scaling laws and similarity criteria have been developed to study astrophysical turbulence in laboratory experiments and simulations, analyzing the relevance of the Reynolds number in astrophysical systems, and studying its role in understanding turbulence and dissipation in high-energy environments [218,219], emphasizing the importance of capturing the full spectral range of high-Reynolds-number flows. It is worth noting that a problem similar to astrophysical magnetic instabilities is that of tokamaks, and that there have been efforts in the magnetic control of this type of plasma through the application of artificial intelligence algorithms [220].

Re influences several aspects of nanoparticle and nanofluid behavior. It affects the manipulation of nanowire lengths in suspension [221] and allows for size control of hybrid lipid-polymer nanoparticles by controlled microvortices [222]. Reynolds number considerations are essential in nanorobotics simulations [223], and it also influences nanodrug delivery in microfluidic channels [224].

Regarding Re applications in the field of renewable energy engineering, Re significantly influences the performance and wake characteristics of wind turbines. Studies show that increasing Re leads to higher power coefficients, torque coefficients, and optimal tip speed ratios in horizontal-axis wind turbines [225,226]. For vertical-axis turbines, performance becomes independent of Re at $Re_D \approx 10^6$ [227]. Low Re conditions, common in small-scale turbines, can result in reduced efficiency [228]; however, intentional roughness on blade surfaces can improve performance at low Re [229]. Novel blade designs, such as parabolic-bladed Savonius rotors, show improved aerodynamic performance at low Re ranges [230]. In the case of semi-passive fin energy harvesters, higher Re leads to higher energy extraction [231].

Re also plays a determinant role in aquatic locomotion at various scales, helping to analyze the forces acting on aquatic animals such as fish by aiding in understanding swimming efficiency and modes of movement [232]. At low values of Re (<100), viscous forces predominate and organisms use flagellar or ciliary propulsion [233]. At intermediate values of Re (100–102), both viscous and inertial forces contribute to thrust and drag [234]. At high values of Re (>102), inertial forces predominate and organisms employ wave or pulsed jet propulsion [235,236]. Swimming efficiency and mechanisms vary with Re, affecting optimal body shapes and locomotion strategies [237]. Scaling laws such as $Re{\sim}Sw^{\alpha} = (\omega AL/\nu)^{\alpha}$ unify locomotion across taxa [235], and biomimetic studies have explored finning blades and pulsed jet propulsion at different Re, revealing insights into propulsion efficiency [238,239]. Understanding these principles contributes to the development of biologically inspired aquatic vehicles and to the evolution of swimming adaptations [240].

Regarding the study of impact dynamics, it is noted that events such as droplet impact on solid surfaces, on the other hand, are influenced by several factors, including Reynolds number, surface wettability, and viscosity [241,242]. At high Re, inertia dominates, generating self-similar pressure fields and propagation patterns, and the impact force initially follows a square root pattern, which matches a known theory on pressure fields, while as the velocity decreases, viscous forces become more important and affect the way droplets deform upon impact [243]. Then, when viscous forces become more significant, this affects the impact force and collision processes [242]. On the other hand, ultra-high velocity impacts introduce compressibility effects, altering flow fields and pressure distributions [244]. The number of fingers in the splash patterns scales with impact Reynolds numbers [245], while for highly viscous and rheologically complex liquids, splashing and rebounding are inhibited at high Oh numbers [246]. The Reynolds number variation can also be used for the characterization of the in situ modal response of pipe structures [247]. The impact of droplets on spherical targets exhibits distinct temporal phases and is affected by Re and the target-to-droplet size ratio [248].

Re also plays a critical role in aerospace exploration, influencing several aspects of fluid dynamics and vehicle performance. Low-Reynolds-number airfoils are important for applications such as micro air vehicles and Mars exploration [249]. Re effects are also critical in the design of turbomachinery for space power units [250] and the ground wind load testing of launch vehicles [251]. The ability to accurately simulate flight Reynolds numbers in wind tunnels is crucial for predicting aerodynamic performance, especially in hypersonic regimes [252]. Understanding Re effects is essential for designing airfoils and airfoils in a variety of environments from Earth to other planets [253,254].

5. Advances and Recent Development

Re has diverse applications in different scientific fields. In biology, it helps to analyze complex biological phenomena [232,255]. In aeronautics, it significantly influences flow behavior, particularly in transitional flows [4]. Reynolds number effects are crucial in turbulent wall-bounded flows, as they affect the amounts of mean and statistical turbulence [256]. In superfluid physics, liquid helium wind tunnels can reach exceptionally high Reynolds numbers [257]. In the case of lipid vesicles, increasing Reynolds numbers alter their dynamics in shear flow [258]. New applications include the relationship between the Reynolds number and spacetime curvature [7] and the use of deep reinforcement learning for flow control in different Reynolds number regimes [259]. Understanding the effects of the Reynolds number is essential for accurate flow prediction and control, especially in external flows with low Reynolds numbers [260].

More recent studies explore new applications of Re in several fields. In aerodynamics, studies investigate its effects on bionic flaps [261] and surface pattern fabrics for noise and drag reduction [262]. In fluid mechanics, Re is applied to characterize drilling fluid flow regimes [263] and define the local Reynolds number for CFD simulations [264]. The concept extends to finance, where it is used to identify explosive behavior in stock indices [265] and as a volatility indicator in stochastic time series [266]. In automotive engineering, Re effects are optimized to reduce aerodynamic drag and improve performance [42]. The study of micropolar nanofluids shows that Re affects velocity profiles and thermal properties, which is crucial for applications such as drug delivery and blood flow [185]. Variations in Re affect shock wave interactions in turbulent boundary layers, thereby affecting flow separation and pressure fluctuations, which are critical for aerospace applications [267]. These diverse applications demonstrate the versatility and importance of Re for understanding complex systems across multiple disciplines.

Re has been widely studied since its original formulation, but in recent decades there have been significant theoretical advances and reinterpretations that have refined its understanding and applications in various areas of science and engineering. Some of the most notable advances are presented below:

1. Interpretations in Turbulent Flows: New research has refined how turbulence scales are scaled under high turbulence conditions [268], numerical simulations and experiments have been improved to capture the fine structures of turbulence [269,270], and the use of techniques such as large-scale turbulence and Navier–Stokes direct simulations has allowed for a better understanding of transitions [271,272]. Additionally, subgrid models [273,274] and large eddy simulations (LES) [275] have allowed for a more accurate description of meso- and micro-level flows, improving prediction in aerospace, marine, and meteorological applications.
2. Non-Newtonian Flows and Complex Fluids: The Reynolds number has been formulated and generalized for non-Newtonian fluids, such as polymers and gels, because they exhibit non-linear viscosity behaviors [276]. This has allowed for better predictions in fields such as bioengineering [277]. Additionally, some rheological phenomena that occur at a low Reynolds number in a viscoelastic fluid have not been described entirely; for example, the elusive vortices produced by a rotating sphere in a viscoelastic fluid at a low Reynolds number described firstly by Giesekus [278], and later studied numerically by Garduño et al. [279,280].
3. Microfluidics and Nanofluidics: Microfluidics and nanofluidics developments have required reinterpretations of Re for flows at these scales, where surface effects, surface tension, and intermolecular forces dominate over inertial and viscous forces. A new theory that incorporates corrections for flows in geometric confinements at nanometer and micrometer scales has been developed, where Re tends to be extremely low [281].
4. Multiscale Applications: With the rise of multiscale simulations, which attempt to integrate fluid behaviors on scales ranging from molecular to macroscopic, the Reynolds number has been refined to account for effects at multiple levels [282–284].
5. New Interpretations in Biological and Natural Systems: Advances have been made in the interpretation of Re in biological and natural systems. In biological systems, such as insect flight or the movement of microorganisms, Re is low, meaning that viscous forces predominate [255]. In geophysical flows, such as oceanic or atmospheric flows, the Reynolds number is extremely high, leading to turbulent behavior on large scales [285].
6. Reynolds Number Corrections for Extreme Conditions: In the study of flows under extreme conditions, such as high pressure or temperature, or in environments with fluids with exotic properties, work has been done on adjustments to the Reynolds number. To study the fluid dynamics in planetary atmospheres [286], such as those of

Mars or Venus [287], corrections to Re have been proposed that take into account the low density or viscous properties of the gases present. In fluids under extremely low temperatures, such as superfluids [288–290], the Reynolds number has been modified to reflect quantum effects in fluid dynamics [291].

Recent additional theoretical advances on Re have explored its importance at various scales and contexts. At high Reynolds numbers, challenges remain in understanding wall-bounded turbulent flows, including the extent of the logarithmic overlay layer and the universality of model parameters [292]. Theoretical estimates of critical Re have been derived using stochastic equations [293], and new models have been developed to estimate the effects of surface geometry on laminar boundary layers [294]. Researchers have also proposed redefining Reynolds number based on the interplay between macroscopic and microscopic phenomena, potentially offering new insights into turbulence [15].

Research on extreme Re flows has focused on understanding the scaling laws, coherent structures, and modeling techniques that characterize these turbulent systems. These studies have revealed new insights into the behavior of turbulent flows at high Re, particularly in terms of mean velocity profiles, friction factors, and the role of coherent structures.

At extreme Re, the mean velocity profile (MVP) of turbulent pipe and channel flows transitions to new scaling laws. This transition is characterized by complete similarity asymptotics in bulk coordinates and incomplete similarity asymptotics in inner coordinates. A multiscale polynomial model has been proposed to approximate the MVP, providing excellent agreement with experimental data in both bulk and inner coordinates [295]. A novel power-law scaling for the friction factor in extreme Re pipe flows has been identified by Anbarlooei et al. [296]. This scaling is based on the behavior of coherent structures that dominate momentum exchange in meso-layer regions. This new scaling law aligns well with experimental data from facilities like the Princeton Superpipe, suggesting a shift in the mechanism of turbulent momentum transfer at these high Re [296]. Studies on extreme turbulent circulation events have also highlighted the role of multifractality breaking in turbulent intermittency. This concept helps describe the asymptotic form of circulation probability distribution functions (cPDFs) [297]. Exact coherent structures (ECS) have been shown to play a fundamental role in transitional and turbulent wall flows, with new insights into their interactions and sustainment mechanisms at high Reynolds numbers [298]. Finally, advances in mathematical methods and computational capacity are paving the way for high-fidelity models of wall turbulence at large Re. These models aim to predict flow properties accurately while capturing the underlying physics based on the Navier–Stokes equations [299].

Research by Chen and Sreenivasan [24] indicates that at high Re, a plateau emerges in the velocity fluctuation variations, suggesting a decoupling of Re effects from wall normal profiles. This finding improves the understanding of turbulence dynamics in channel and pipe flows. In shock wave interactions, Laguarda et al. [300] demonstrate that while the mean flow characteristics remain stable across varying Re, the dynamics of velocity statistics change significantly, indicating a complex relationship between Re and flow behavior. Ashoor et al. [301] explore Re in microvasculature, revealing its role in diagnosing vascular conditions.

Another work of interest was developed by Martínez et al. [302], whose study illustrates how microscopic systems can operate similarly to macroscopic engines while also presenting unique challenges and opportunities due to random (or stochastic) thermal fluctuations. In the work of Martinez et al., the empirical realization of a Carnot engine utilizing a single optically confined Brownian particle as its working medium is delineated, offering an extensive investigation into the energetics of the engine and scrutinizing the fluctuations associated with finite-time efficiency, thereby demonstrating that the Carnot limit may be surpassed with a limited number of non-equilibrium cycles. Mirroring its macroscopic equivalent, the energetics of the proposed Carnot apparatus reveals fundamental characteristics that one would anticipate to manifest in any microscopic energy transducer operating with thermal reservoirs at disparate temperatures.

Future perspectives on Re encompass its implications in various fluid dynamics contexts, particularly in biomedical and turbulence applications, highlighting its pivotal role in fluid dynamics at various scales. Recent studies highlight the nuanced behavior of Re in wall turbulence fluctuations and its diagnostic potential in microvasculature. At low Re, airfoil designs are being developed for planetary exploration [303], while high Re flows present challenges in mixing and turbulent simulation [304]. The RANS approach remains relevant for industrial applications, and potential hybrid RANS/LES strategies are emerging [305]. The criticality of the Reynolds number in dynamic testing requires analytical extrapolation tools to make accurate full-scale predictions [306]. Studies on turbulent boundary layers reveal complex scale behaviors and internal hierarchies of motions [5,299]. Advanced mathematical methods and computational capabilities are enabling the development of high-fidelity models based on Navier–Stokes equations [299]. Interestingly, Re has been conceptualized as a curvature ratio, which could inform heat exchanger design and link hydrodynamics to gravity theory [7].

Additionally, some researchers argue that the traditional focus on Re may overlook other critical parameters that influence flow behavior, suggesting the need for a broader framework in fluid dynamics studies [14,15,88,307,308].

6. Discussions

The Reynolds number plays a crucial role in fluid dynamics and aeronautics, influencing flow behavior at various scales. Although it is very useful in several scientific fields, it has some limitations that must be taken into account when using it: model simplification—assuming laminarity or turbulence depending on a critical value, instead of a gradual multifactorial behavior; application in non-homogeneous media, such as in fluids with suspended particles or in porous media, where Re may not be suitable since it does not consider interactions with particles or effects of the medium; evaluation in complex or unconventional geometric conditions, where Re may not provide an accurate representation of the flow behavior dynamics; validity ranges limited to ideal flows, which implies an impossibility to describe flows with forces dominated by electromagnetics effects or non-viscous effects; limitations associated with non-Newtonian fluids, whose viscosity varies with the deformation rate; and limitations related to scales, where the effects of interactions at the molecular level can exceed predictions based on Re, making it inaccurate in these contexts.

Limitations such as low-Reynolds-number phenomena require specialized turbulence models to accurately predict near-wall structures [309]. Limitations in velocity distribution calculations arise at high Re, particularly above 100,000 [310], while scale effects in physical modeling of free surface flows can be minimized by respecting Reynolds and Weber number constraints [311]. High-Reynolds-number flows present challenges for CFD, especially in predicting flow over rough surfaces [312]. Airfoil design, on the other hand, is significantly affected by scale effects, characterized by the chord Reynolds number [106]. The concepts of self-similarity and Reynolds number invariance can help to exclude significant scale effects in Froude modeling [313], and the transition behavior varies widely over Reynolds number ranges, complicating aerodynamic predictions for complex configurations [4].

There are several topics of discussion regarding the interpretation and use of Re to determine the critical value for the transition between laminar and turbulent flows, with several studies that propose slightly different values [314]. There is also debate about the applicability and usefulness of Re in more complex flows than the ideal cases of pipes or flat plates [291]. The interpretation and use of Re in non-Newtonian fluids has also been the subject of debate, since the viscosity is not constant (conventional Re assumption) [315]. There are discussions about the limitations of using the Reynolds number to predict complete similarity between flows at different scales [316]. In the emerging field of microfluidics or nanofluidics, there are debates about how to define and apply the Reynolds number at very small scales [11]. The application of Re in multiphase flows is an active research area with different approaches proposed [317]. There are also debates on how to interpret and use the Reynolds number in high-velocity compressible flows [318].

Regarding future developments, some of the main research approaches and directions consider applications in complex and multiphysical flows, where the extension or applicability of Re to more complex systems involving multiple physical phenomena is studied [319], or applications to microfluidics and nanofluidics, where the behavior of fluids at very small scales is studied [320]. For applications to non-Newtonian fluids and complex flows, new approaches are being developed to apply the Reynolds number concept [321], while for applications to turbulence and transition, a detailed understanding of the transition to turbulence and the behavior at very high Reynolds numbers is sought [322]. For applications to advanced numerical simulations, advances in computational capabilities are applied to develop more detailed simulations of high-Reynolds-number flows [323]. New models and approaches are being developed to apply Re in multiphase flows [324]. Re also has applications in the biomedical industry, such as the in study of blood flow [325]. It can also be applied to active fluids and biological systems, such as bacterial suspensions [326], and in advanced experimental methods, where the behavior of flows at extreme Re or in difficult-to-access conditions is studied [327].

Additional applications of Re consider its application to the study of renewable energies, as in the design of wind turbines and marine power generators, in the development of hypersonic vehicles and space exploration, or in the research of new materials and their interactions with fluids. In wind energy, Re is crucial in the design of wind turbines, since it affects the aerodynamics of the blades and the overall efficiency of the system [328,329]. Regarding hydroelectric energy, Re is fundamental to understand and optimize the flow in water conduction systems and in the design of hydraulic turbines [330]. In the field of tidal energy, it is important in the design of tidal energy devices to maximize the energy extraction efficiency [331]. In the field of solar thermal energy or geothermal systems, in solar concentrating systems, the number is relevant for the design of heat exchangers and heat transfer fluid systems [331,332]. In biomass processing and biofuel production, Re is relevant for the design of reactors and mixing systems [333]. In the design of devices to harness wave energy, Re is important to optimize the interaction between the device and the fluid [334]. Finally, in microhydraulics, or small-scale generation systems, Reynolds number is crucial for the design of microturbines and drive systems [335].

Regarding the development of hypersonic vehicles and space exploration, Re influences aerodynamic design, heat transfer, and other critical aspects. In hypersonic aerodynamics, Re is fundamental to understand and predict the flow behavior around hypersonic vehicles [336] and is addressed in studies on heat transfer in the atmospheric reentry of space vehicles [337], the design of rockets and launch vehicles, where Re is crucial to optimize aerodynamics and efficiency [338], and on the aerodynamics of planetary exploration vehicles, where atmospheric conditions are different [339]. In the development of propulsion systems for space travel, Re is relevant for the design of nozzles and combustion chambers [340]. In hypersonic wind tunnels, Re is crucial in the design and operation [341]. It is also relevant to the study of the following areas: the fluid–structure interaction of vehicles under hypersonic conditions [342]; the dynamics of rarefied gases, such as the upper layers of the atmosphere and space [343]; the modeling of atmospheric flows on other planets for exploration missions [344]; the design of heat shields to protect vehicles during atmospheric reentry [345]; and finally, magnetohydrodynamics in space propulsion, where the magnetic Re is relevant in the study of magnetohydrodynamic propulsion for space travel [346].

The Reynolds number has been utilized in research on new materials and their interactions with fluids, which is crucial in the development of advanced materials with specific properties for applications in various areas. In superhydrophobic materials, Re is crucial to understanding how these materials interact with fluids, especially in relation to friction reduction and self-cleaning [347]. Regarding to the behavior of nanofluids, Re is especially determinant in applications in heat and energy transfer [348]. To understand the dynamics of flows through porous media, it is crucial in filtration and energy storage applications [349]. Re is also used for studying the effectiveness of antifouling coatings, used to prevent the accumulation of organisms on submerged surfaces [350], and for studying smart materials that change their properties depending on flow conditions, characterized by a certain Re [351]. Re is used in research on materials that reduce drag in fluids at high Re for applications in energy efficiency [352], the study of nanocomposites designed to improve heat transfer in fluid systems [353], the development of materials for biomedical applications [354], the study of the ability of self-cleaning materials to repel contaminants under different flow conditions [355], research on materials that can actively modify flow characteristics at different Re [356], and the design of materials for efficient CO_2 capture in gas flows [357].

These applications demonstrate the importance of the Reynolds number in various aspects of renewable energy development, hypersonic vehicle development, and space exploration, and its use in research into new materials. Research in areas such as hypersonic or space vehicle development continues to develop, seeking to improve understanding and the ability to design more efficient and safer vehicles and systems for space exploration and hypersonic flight.

7. Conclusions

Recent research on the effects of Re on fluid dynamics has revealed significant trends and discoveries over the last decades. Studies have shown that the influences of the Reynolds number are particularly strong in transitional flows, with a transition occurring over a wide range of values, which is a function of the fluid dynamics under study. Turbulent flows at high Re exhibit a universal scaling behavior for velocity fluctuations, similar to the scaling of the mean velocity distribution, while the approach for asymptotically high Re is slow, and the logarithmic law remains fundamental in the description of mean flow. Very large scale motions become increasingly important at higher Re, interacting with smaller scales near the walls.

The convergence between classical and modern approaches in the study of Re represents an area of interest in contemporary fluid mechanics, because it combines established theoretical foundations with new technologies and analysis methods. Some key points of this convergence are given by the following conditions: the integration of classical theory and advanced numerical simulations, which enable the validation and extension of classical Re theories to previously inaccessible regimes; the refinement of the transition to turbulence, which improves understanding of the transition; the extension and adaptation of the classical concept to the application to complex non-Newtonian fluids; the reinterpretation of classical Re theory to micro- and nanofluidics applications; the development of advanced experimental methods allowing for the validation and refinement of the classical theories; the interdisciplinary applications of the classical Re concept to modern fields such as bio-fluidics and active matter physics (such as blood flow and locomotion of microorganisms); the refinement of the classical theory and its adaptation to extremely high Re flows; the application of modern complex systems or system dynamics approaches to gain a better understanding of classical phenomena related to Re; the application of classical concepts to modern technologies related to renewable energy development and the environment; the integration of classical theory with modern fluid–structure interaction models; the combination of classical concepts with modern flow control techniques; and the application of modern data analysis and machine learning techniques to extract new insights from classical data. Current trends also include advancements in understanding Re effects on interface phenomena such as droplet and bubble formation processes, as well as the application of the concepts in large-scale geophysical flows such as ocean and atmospheric currents.

The convergence between classical and modern approaches is leading to a richer and more nuanced understanding of Re and its role in fluid dynamics, while the integration of classical theories with new technologies or analysis methods is opening the door to new research and applications, allowing us to address more complex and challenging problems in fluid science and mechanics. The importance of Re in future research areas in fluid mechanics and related fields lies in its ability to characterize the dynamics of flow behavior and its applicability in a wide range of scales and phenomena. Some areas where Re will be crucial in future research are listed as follows:

1. Understanding turbulence at very high Reynolds numbers.
2. Advanced microfluidics and nanofluidics: fluid behavior at very small scales, where surface effects dominate.
3. Complex fluids and soft materials: application to non-Newtonian fluids, complex suspensions, and soft materials.
4. Active fluids and active matter: study of systems where particles can self-propel, such as in bacterial suspensions, swarms of robots, or collaborative robotics in fluid environments.
5. Fluid–structure interactions: understanding how flexible structures interact with flows at different Re, designing more efficient and resilient structures.
6. Multiphase and multiscale flows: application in systems involving multiple phases or scales.
7. Fluids in extreme conditions: fluid behavior under extreme conditions of pressure or temperature, relevant to geophysical and astrophysical applications.
8. Shape optimization and flow control: using Re in aerodynamic and hydrodynamic shape design and optimization.
9. Flows in biological systems: application in biological systems, from blood flow to locomotion of organisms.
10. Transition to turbulence: a detailed understanding of how and why flows transition from laminar to turbulent.
11. Advanced numerical simulations: development of more advanced numerical methods will allow simulating flows at extreme Re in more complex geometries.
12. Applications in energy and environment: development of more efficient energy technologies and understanding of environmental phenomena.

Despite the extensive research on Re developed during the last decades, there are still various gaps in its knowledge and understanding, areas that give important opportunities for future research. The main identified gaps are that the mechanism of when a laminar flow transitions to turbulence at some critical Re is not fully known, or there is a gap in knowledge regarding how flows behave at extreme Re, as mentioned for astrophysical and geophysical applications. The applicability of Re in complex non-Newtonian fluids, such as polymers or concentrated solutions, is still a challenging issue; the interaction between the effects of Re and compressibility in high-velocity flows is not fully known, and the definition and application of Re in complex multiphase flows, such as liquid-gas mixtures or particle suspensions, also remains an open question. The applicability of Re on very small scales, where surface effects dominate, is also not fully known. The application in self-propelled particle systems or active fluids represents an emerging area with a lot of open questions. The understanding of its effect on fluid–structure interactions under unsteady conditions is incomplete, the exact influence of surface roughness at critical Re and when reaching the transition to turbulence is not fully characterized, the application in complex geometries and how it relates to flow structures is not fully known, and the influence of external fields, such as magnetic or electric fields, on the flow behavior for various Re is also not known.

Author Contributions: Conceptualization, M.S. and N.T.; methodology, M.S., A.N. and N.T.; validation, S.G., E.G., J.C., E.S.-R., E.C.-S. and J.H.-Á.; formal analysis, M.S., S.G., E.G., J.C. and E.S.-R.; investigation, M.S., J.C., E.S.-R., E.C.-S., J.H.-Á., A.N. and N.T.; writing—original draft preparation, M.S.; writing—review and editing, M.S., E.C.-S., J.H.-Á., A.N. and N.T. All authors have read and agreed to the published version of the manuscript.

Funding: This research received no external funding.

Data Availability Statement: No new data were created or analyzed in this study. Data sharing is not applicable to this article.

Acknowledgments: Manuel Saldana acknowledges the infrastructure and support from Doctorado en Ingeniería de Procesos de Minerales at the Universidad de Antofagasta.

Conflicts of Interest: The authors declare no conflicts of interest.

References

1. Uruba, V. Reynolds Number in Laminar Flows and in Turbulence. *AIP Conf. Proc.* **2019**, *2118*, 020003. [CrossRef]
2. Smith, L.H. Some Comments on Reynolds Number. *J. Eng. Power* **1964**, *86*, 225–226. [CrossRef]
3. Batchelor, G.K. *An Introduction to Fluid Dynamics*; Cambridge University Press: Cambridge, UK, 2000; ISBN 9780521663960.
4. Bushnell, D.M.; Yip, L.P.; Yao, C.-S.; Lin, J.C.; Lawing, P.L.; Batina, J.T.; Hardin, J.C.; Horvath, T.J.; Fenbert, J.W.; Domack, C.S. *Reynolds Number Influences in Aeronautics*; NTRS—NASA Technical Reports Server: Hampton, VA, USA, 1993.
5. Klewicki, J.C. Reynolds Number Dependence, Scaling, and Dynamics of Turbulent Boundary Layers. *J. Fluids Eng. Trans. ASME* **2010**, *132*, 094001. [CrossRef]
6. Hillebrandt, W.; Kupta, F. An Introduction to Turbulence. In *Lecture Notes in Physics*; Springer: Berlin/Heidelberg, Germany, 2008; Volume 756, pp. 1–20. [CrossRef]
7. Delplace, F. Reynolds Number and Spacetime Curvature. *Fluid Mech. Open Access* **2016**, *3*, 1000125. [CrossRef]
8. Giovannini, M. Reynolds Numbers in the Early Universe. *Phys. Lett. B* **2012**, *711*, 327–331. [CrossRef]
9. Krause, E. Application of Numerical Techniques in Fluid Mechanics. *Aeronaut. J.* **1974**, *78*, 337–354. [CrossRef]
10. Gersten, K. Introduction to Boundary-Layer Theory. In *Recent Advances in Boundary Layer Theory*; CISM International Centre for Mechanical Sciences; Springer: Vienna, Austria, 1998; Volume 390, pp. 1–8. ISBN 978-3-7091-2518-2.
11. Squires, T.M.; Quake, S.R. Microfluidics: Fluid Physics at the Nanoliter Scale. *Rev. Mod. Phys.* **2005**, *77*, 977–1026. [CrossRef]
12. Ottino, J.M. Mixing, Chaotic Advection, and Turbulence. *Annu. Rev. Fluid Mech.* **1990**, *22*, 207–254. [CrossRef]
13. Faber, T.E.; Lumley, J.L. Fluid Dynamics for Physicists. *Am. J. Phys.* **1996**, *64*, 1214–1215. [CrossRef]
14. Patel, Y.; Sanyal, A.; Sharma, D.; Rai, S.; Basu, D.; Patel, S.; Jyoti Sen, D. Dimensionless Reynolds Number as a Dimension for Fluid Mechanics in Rheology. *J. Drug Discov. Ther.* **2017**, *5*, 25–30.
15. Bodmann, B.E.J.; Vilhena, M.T.; Zabadal, J.R.; Beck, D. On a New Definition of the Reynolds Number from the Interplay of Macroscopic and Microscopic Phenomenology. In *Integral Methods in Science and Engineering*; Birkhäuser Boston: Boston, MA, USA, 2011; pp. 7–14. [CrossRef]
16. Launder, B.E. First Steps in Modelling Turbulence and Its Origins: A Commentary on Reynolds (1895) 'On the Dynamical Theory of Incompressible Viscous Fluids and the Determination of the Criterion'. *Philos. Trans. R. Soc. A Math. Phys. Eng. Sci.* **2015**, *373*, 20140231. [CrossRef]
17. Kilgore, R.A. Evolution and Development of Cryogenic Wind Tunnels. In Proceedings of the 43rd AIAA Aerospace Sciences Meeting and Exhibit—Meeting Papers, Reno, NV, USA, 10–13 January 2005; pp. 7133–7144. [CrossRef]
18. Uruba, V. On Reynolds Number Physical Interpretation. *AIP Conf. Proc.* **2018**, *2000*, 020019. [CrossRef]
19. Rouse, H. The Origins of Fluid Mechanics. *J. Eng. Mech.* **1987**, *113*, 66–71. [CrossRef]
20. Tokaty, G.A. *A History and Philosophy of Fluid Mechanics*, 1st ed.; Dover: New York, NY, USA, 1994; ISBN 0486681033.
21. Prandtl, L.; Tietjens, O.K.G. *Applied Hydro- and Aeromechanics: Based on Lectures of L. Prandtl*; Dover: New York, NY, USA, 1957; ISBN 9780486603759.
22. Basavaraj, K.; Elangovan, K.; Shankar, S.; Kulkarni, A. Satyanarayan Effects of Reynolds Number on the Volume Concentration of Al_2O_3 in Ethylene Glycol. *Int. J. Veh. Struct. Syst.* **2023**, *15*, 351–355. [CrossRef]
23. Zanoun, E.S.; Dewidar, Y.; Egbers, C. Reynolds Number Dependence of Azimuthal and Streamwise Pipe Flow Structures. *J. Fluid. Mech.* **2023**, *973*, A1. [CrossRef]
24. Chen, X.; Sreenivasan, K.R. Reynolds Number Asymptotics of Wall-Turbulence Fluctuations. *J. Fluid Mech.* **2023**, *976*, A21. [CrossRef]
25. Chin, C.; Monty, J.P.; Ooi, A. Reynolds Number Effects in DNS of Pipe Flow and Comparison with Channels and Boundary Layers. *Int. J. Heat Fluid Flow* **2014**, *45*, 33–40. [CrossRef]
26. Mech, A.M.I.E. Flow through Pipe Orifices at Low Reynolds Numbers. *Proc. R. Soc. Lond. Ser. A Contain. Pap. A Math. Phys. Character* **1930**, *126*, 231–245. [CrossRef]
27. Zagarola, M.V.; Smits, A.J.; Orszag, S.A.; Yakhot, V. Experiments in High Reynolds Number Turbulent Pipe Flow. In Proceedings of the 34th Aerospace Sciences Meeting and Exhibit, Reno, NV, USA, 15–18 January 1996. [CrossRef]
28. Hultmark, M.; Vallikivi, M.; Bailey, S.C.C.; Smits, A.J. Turbulent Pipe Flow at Extreme Reynolds Numbers. *Phys. Rev. Lett.* **2012**, *108*, 094501. [CrossRef]
29. Ng, H.C.H.; Monty, J.P.; Hutchins, N.; Chong, M.S.; Marusic, I. Comparison of Turbulent Channel and Pipe Flows with Varying Reynolds Number. *Exp. Fluids* **2011**, *51*, 1261–1281. [CrossRef]
30. Wang, C.Y. On the Low-Reynolds-Number Flow in a Helical Pipe. *J. Fluid Mech.* **1981**, *108*, 185–194. [CrossRef]
31. Smith, F.T. On the High Reynolds Number Theory of Laminar Flows. *IMA J. Appl. Math.* **1982**, *28*, 207–281. [CrossRef]

32. Smits, A.J.; McKeon, B.J.; Marusic, I. High-Reynolds Number Wall Turbulence. *Annu. Rev. Fluid Mech.* **2011**, *43*, 353–375. [CrossRef]
33. Manna, M.; Vacca, A. Scaling Properties of Turbulent Pipe Flow at Low Reynolds Number. *Comput. Fluids* **2001**, *30*, 393–415. [CrossRef]
34. Dabiri, J.O.; Mohebbi, N.; Fu, M.K. Persistent Laminar Flow at Reynolds Numbers Exceeding 100,000. *arXiv* **2022**, arXiv:2212.02488.
35. Shirbhate, V.S.; Siva Kumar, K.; Madhu Babu, K. Effect of Reynolds Number on Typical Civil Transport Aircraft. *Lecture Notes Mech. Eng.* **2021**, *53*, 471–480. [CrossRef]
36. Wang, Y.; Liu, D.; Xu, X.; Li, G. Investigation of Reynolds Number Effects on Aerodynamic Characteristics of a Transport Aircraft. *Aerospace* **2021**, *8*, 177. [CrossRef]
37. Liu, D.; Wang, Y.; Chen, D.; Peng, X.; Xu, X. Numerical Investigation on the Reynolds Number Effects of Supercritical Airfoil. *Procedia Eng.* **2012**, *31*, 103–109. [CrossRef]
38. Novokhatska, A. Design of an Airfoil by Mathematical Modelling Using Database. *J. Airl. Oper. Aviat. Manag.* **2022**, *1*, 19–26. [CrossRef]
39. Osusky, M.; Ledezma, G.; Sluyter, G.; Shankaran, S. Reynolds Number Impact on High Pressure Turbine Flows. *Proc. ASME Turbo Expo.* **2023**, *13C*, V13CT32A030. [CrossRef]
40. Drela, M. Low-Reynolds-Number Airfoil Design for the M.I.T. Daedalus Prototype- A Case Study. *J. Aircr.* **1988**, *25*, 724–732. [CrossRef]
41. Liu, D.; Xu, X.; Wei, Z.; Dehua, C. Investigation on the Reynolds Number Simulation of Supercritical Airfoil. In Proceedings of the 2013 4th International Conference on Digital Manufacturing and Automation, ICDMA 2013, Shinan, China, 29–30 June 2013; pp. 740–744. [CrossRef]
42. Gu, Y.; Liu, Y.; Shen, J.; Wang, D. Optimization of Reynolds Number's Effects on Aerodynamic Drag and Automotive Performance. *Theor. Nat. Sci.* **2023**, *13*, 144–149. [CrossRef]
43. Solmaz, H.; İçingür, Y. Drag Coefficient Determination Of A Bus Model Using Reynolds Number Independence. *Int. J. Automot. Eng. Technol.* **2015**, *4*, 146–151. [CrossRef]
44. Zhao, G.; Pumera, M. Reynolds Numbers Exhibit Dramatic Influence on Directionality of Movement of Self-Propelled Systems. *Phys. Chem. Chem. Phys.* **2012**, *14*, 6456–6458. [CrossRef] [PubMed]
45. Amromin, E.L. Vehicles Drag Reduction with Control of Critical Reynolds Number. *J. Fluids Eng.* **2013**, *135*, 101105. [CrossRef]
46. Qin, P.; Ricci, A.; Blocken, B. CFD Simulation of Aerodynamic Forces on the DrivAer Car Model: Impact of Computational Parameters. *J. Wind Eng. Ind. Aerodyn.* **2024**, *248*, 105711. [CrossRef]
47. Haffner, Y.; Li, R.; Meldi, M.; Borée, J. Drag Reduction of a Square-Back Bluff Body under Constant Cross-Wind Conditions Using Asymmetric Shear Layer Forcing. *Int. J. Heat Fluid Flow* **2022**, *96*, 109003. [CrossRef]
48. Xu, J.; Dong, S.; Maxey, M.R.; Karniadakis, G.E. Turbulent Drag Reduction by Constant Near-Wall Forcing. *J. Fluid Mech.* **2007**, *582*, 79–101. [CrossRef]
49. Fuaad, P.A.; Baig, M.F.; Khan, B.A. Turbulent Drag Reduction Using Active Control of Buoyancy Forces. *Int. J. Heat Fluid Flow* **2016**, *61*, 585–598. [CrossRef]
50. Gugulothu, R.; Sanke, N.; Gupta, A.V.S.S.K.S. Numerical Study of Heat Transfer Characteristics in Shell-and-Tube Heat Exchanger. In *Lecture Notes in Mechanical Engineering*; Springer: Singapore, 2019; pp. 375–383. [CrossRef]
51. Saleem, A.; Kim, M.H. CFD Analysis on the Air-Side Thermal-Hydraulic Performance of Multi-Louvered Fin Heat Exchangers at Low Reynolds Numbers. *Energies* **2017**, *10*, 823. [CrossRef]
52. Shinde, P.; Lin, C.X. A Heat Transfer and Friction Factor Correlation for Low Air-Side Reynolds Number Applications of Compact Heat Exchangers (1535-RP). *Sci. Technol. Built Environ.* **2017**, *23*, 192–210. [CrossRef]
53. Al-Amshawee, S.K.A.; Yunus, M.Y.B.M. Electrodialysis Desalination: The Impact of Solution Flowrate (or Reynolds Number) on Fluid Dynamics throughout Membrane Spacers. *Environ. Res.* **2023**, *219*, 115115. [CrossRef] [PubMed]
54. Amalia, R.; Safitra, A.G.; Ubudiyah, H. Numerical Study on Heat Transfer and Friction Factor Characteristics of Transition Flow in Shell and Tube Heat Exchanger. In Proceedings of the IES 2019—International Electronics Symposium: The Role of Techno-Intelligence in Creating an Open Energy System Towards Energy Democracy, Surabaya, Indonesia, 27–28 September 2019; pp. 283–286. [CrossRef]
55. Almurtaji, S.; Ali, N.; Teixeira, J.A.; Addali, A. On the Role of Nanofluids in Thermal-Hydraulic Performance of Heat Exchangers—A Review. *Nanomaterials* **2020**, *10*, 734. [CrossRef] [PubMed]
56. Mehta, B.; Subhedar, D.; Panchal, H.; Said, Z. Synthesis, Stability, Thermophysical Properties and Heat Transfer Applications of Nanofluid–A Review. *J. Mol. Liq.* **2022**, *364*, 120034. [CrossRef]
57. Wang, J.; Yang, X.; Klemeš, J.J.; Tian, K.; Ma, T.; Sunden, B. A Review on Nanofluid Stability: Preparation and Application. *Renew. Sustain. Energy Rev.* **2023**, *188*, 113854. [CrossRef]
58. Hou, G.; Taherian, H.; Song, Y.; Jiang, W.; Chen, D. A Systematic Review on Optimal Analysis of Horizontal Heat Exchangers in Ground Source Heat Pump Systems. *Renew. Sustain. Energy Rev.* **2022**, *154*, 111830. [CrossRef]
59. Clifford, M.J.; Cox, S.M.; Finn, M.D. Reynolds Number Effects in a Simple Planetary Mixer. *Chem. Eng. Sci.* **2004**, *59*, 3371–3379. [CrossRef]

60. Yoon, H.-S.; Chun, H.-H.; Ha, M.-Y. The Study of Flow Structure in a Mixing Tank for Different Reynolds Numbers Using LES. *Trans. Korean Soc. Mech. Eng. B* **2003**, *27*, 1290–1298. [CrossRef]
61. Mizuno, Y.; Funakoshi, M. Reynolds Number Dependences of Velocity Field and Fluid Mixing in Partitioned-Pipe Mixer. *J. Phys. Soc. Jpn.* **2005**, *74*, 1479–1489. [CrossRef]
62. Suresh, P.R.; Das, S.K.; Sundararajan, T. Influence of Reynolds Number on the Evolution of a Plane Air Jet Issuing From a Slit. *J. Fluids Eng.* **2007**, *129*, 1288–1296. [CrossRef]
63. Evrim, C.; Laurien, E. Effect of the Reynolds and Richardson Numbers on Thermal Mixing Characteristics. *Int. J. Heat Mass Transf.* **2021**, *169*, 120917. [CrossRef]
64. Awad, M.M.; Muzychka, Y.S. Models for Pressure Drop and Heat Transfer in Air Cooled Compact Wavy Fin Heat Exchangers. *J. Enhanc. Heat Transf.* **2011**, *18*, 191–207. [CrossRef]
65. Sotsu, M. Friction Factor and Reynolds Number Correlation for Finned Tube Bundle of the Air Cooler of Monju Reactor. *Nucl. Eng. Des.* **2022**, *396*, 111893. [CrossRef]
66. El-Abeedy, E.A.; Busedra, A.A. Effect of Reynolds Number on Inclined Heated Semicircular Ducts at Different Rotations. *Fluid Dyn. Mater. Process.* **2013**, *9*, 153–167. [CrossRef]
67. Qi, R.; Chen, F.; Xu, L.; Yu, J.; Chen, X. Effects of Reynolds Number on the Overall Characteristics of Flow and Heat Transfer in the Long Micro-Tube with Dimples. *Processes* **2022**, *10*, 2696. [CrossRef]
68. Mostafapour, K.; Nouri, N.M.; Zeinali, M. The Effects of the Reynolds Number on the Hydrodynamics Characteristics of an AUV. *J. Appl. Fluid Mech.* **2018**, *11*, 343–352. [CrossRef]
69. Gad-El-Hak, M. Compliant Coatings: A Decade of Progress. *Appl. Mech. Rev.* **1996**, *49*, S147–S157. [CrossRef]
70. Amini Foroushani, J.; Sabzpooshani, M. An Approach for the Estimation of Hydrodynamic Coefficients of an Underwater Vehicle in Off-Design Velocities. *J. Mar. Sci. Technol.* **2021**, *26*, 368–381. [CrossRef]
71. Pan, Y.C.; Zhang, H.X.; Zhou, Q.D. Numerical Prediction of Submarine Hydrodynamic Coefficients Using CFD Simulation. *J. Hydrodyn.* **2012**, *24*, 840–847. [CrossRef]
72. Zhu, H.; Hu, J.; Alam, M.M.; Ji, C.; Zhou, T. Hydrodynamic Characteristics and Wake Evolution of a Submarine Pipe with the Presence of Gas Leakage at a Low Reynolds Number of 160. *Phys. Fluids* **2022**, *34*, 083609. [CrossRef]
73. Elahi, S.S.; Lange, E.A.; Lynch, S.P. Effect of Reynolds Number on Turbulent Junction Flow Fluid Dynamics and Heat Transfer. *Int. J. Heat Mass Transf.* **2019**, *142*, 118328. [CrossRef]
74. Smits, A.J.; Zagarola, M.V. Applications of Dense Gases to Model Testing for Aeronautical and Hydrodynamic Applications. *Meas. Sci. Technol.* **2005**, *16*, 1710. [CrossRef]
75. Kim, D.; Kim, Y.; Li, J.; Wilson, R.V.; Ezequiel Martin, J.; Carrica, P.M. Boundary Layer Transition Models for Naval Applications: Capabilities and Limitations. *J. Ship Res.* **2019**, *63*, 294–307. [CrossRef]
76. Liu, J.; Liu, J.; Zhang, Y. Influence of Reynolds Number on the Natural Transition of Boundary Layers over Underwater Axisymmetric Bodies. *Phys. Fluids* **2023**, *35*, 044107. [CrossRef]
77. He, K.; Pan, Z.; Zhao, W.; Wang, J.; Wan, D. Overview of Research Progress on Numerical Simulation Methods for Turbulent Flows Around Underwater Vehicles. *J. Mar. Sci. Appl.* **2024**, *23*, 1–22. [CrossRef]
78. Kornev, N.; Shevchuk, I.; Abbas, N.; Anschau, P.; Samarbakhsh, S. Potential and Limitations of Scale Resolved Simulations for Ship Hydrodynamics Applications. *Ship Technol. Res.* **2019**, *66*, 83–96. [CrossRef]
79. Torobin, L.B.; Gauvin, W.H. Fundamental Aspects of Solids-Gas Flow: Part I: Introductory Concepts and Idealised Sphere Motion in Viscous Regime. *Can. J. Chem. Eng.* **1959**, *37*, 129–141. [CrossRef]
80. Boger, D.V. Rheology and the Minerals Industry. *Miner. Process. Extr. Metall. Rev.* **2000**, *20*, 1–25. [CrossRef]
81. Jikazana, A.; Campo, P.; McAdam, E.J. Hydrodynamics (Reynolds Number) Determine Scaling, Nucleation and Crystal Growth Kinetics in Membrane Distillation Crystallisation. *J. Memb. Sci.* **2023**, *685*, 121909. [CrossRef]
82. Edwards, D.A.; Shapiro, M.; Bar-Yoseph, P.; Shapira, M. The Influence of Reynolds Number upon the Apparent Permeability of Spatially Periodic Arrays of Cylinders. *Phys. Fluids A Fluid. Dyn.* **1990**, *2*, 45–55. [CrossRef]
83. Wrench, D.; Parashar, T.N.; Oughton, S.; de Lange, K.; Frean, M. What Is the Reynolds Number of the Solar Wind? *Astrophys. J.* **2024**, *961*, 182. [CrossRef]
84. Ryan, N.W.; Johnson, M.M. Transistion from Laminar to Turbulent Flow in Pipes. *AIChE J.* **1959**, *5*, 433–435. [CrossRef]
85. Bankston, C.A. The Transition From Turbulent to Laminar Gas Flow in a Heated Pipe. *J. Heat Transf.* **1970**, *92*, 569–579. [CrossRef]
86. Spriggs, H.D. Comments on Transition from Laminar to Turbulent Flow. *Ind. Eng. Chem. Fundam.* **1973**, *12*, 286–290. [CrossRef]
87. Barr, D. The Transition from Laminar to Turbulent Flow. *Proc. Inst. Civ. Eng.* **1980**, *69*, 555–562. [CrossRef]
88. Happel, J.; Brenner, H. *Low Reynolds Number Hydrodynamics*; Mechanics of Fluids and Transport Processes; Springer: Dordrecht, The Netherlands, 1983; Volume 1, ISBN 978-90-247-2877-0.
89. Hinch, E.J. Hydrodynamics at Low Reynolds Numbers: A Brief and Elementary Introduction. In *Disorder and Mixing*; NATO ASI Series; Springer: Dordrecht, The Netherlands, 1988; Volume 152, pp. 43–56. [CrossRef]
90. Khayat, R.E.; Eu, B.C. Generalized Hydrodynamics and Reynolds-Number Dependence of Steady-Flow Properties in the Cylindrical Couette Flow of Lennard-Jones Fluids. *Phys. Rev. A* **1989**, *40*, 946. [CrossRef]
91. Szwalek, J.L.; Larsen, C.M. Reynolds Number Effects on Hydrodynamic Coefficients for Pure In-Line and Pure Cross-Flow Forced Vortex Induced Vibrations. In Proceedings of the International Conference on Offshore Mechanics and Arctic Engineering—OMAE, Honolulu, HI, USA, 31 May–5 June 2009; Volume 5, pp. 439–452. [CrossRef]

92. Dewsbury, K.H.; Karamanev, D.G.; Margaritis, A. Rising Solid Sphere Hydrodynamics at High Reynolds Numbers in Non-Newtonian Fluids. *Chem. Eng. J.* **2002**, *87*, 129–133. [CrossRef]
93. Meister, M.; Burger, G.; Rauch, W. On the Reynolds Number Sensitivity of Smoothed Particle Hydrodynamics. *J. Hydraul. Res.* **2014**, *52*, 824–835. [CrossRef]
94. Enayet, M.M.; Gibson, M.M.; Taylor, A.M.K.P.; Yianneskis, M. Laser-Doppler Measurements of Laminar and Turbulent Flow in a Pipe Bend. *Int. J. Heat Fluid Flow* **1982**, *3*, 213–219. [CrossRef]
95. Xiao, Z.; Guo, L.; Feng, D.; Lai, S.; Hu, Z. Numerical Study on the Scale Effects of the Ship Resistance and Form Factor. In Proceedings of the 2018 IEEE 8th International Conference on Underwater System Technology: Theory and Application (USYS 2018), Wuhan, China, 1–3 December 2018. [CrossRef]
96. Matsuda, S.; Katsui, T. Hydrodynamic Forces and Wake Distribution of Various Ship Shapes Calculated Using a Reynolds Stress Model. *J. Mar. Sci. Eng.* **2022**, *10*, 777. [CrossRef]
97. Parolini, N. Computational Fluid Dynamics for Naval Engineering Problems. Ph.D. Thesis, École Polytechnique Fédérale de Lausanne, Lausanne, Switzerland, 2004.
98. Eça, L.; Kerkvliet, M.; Toxopeus, S.L. Using Validation Metrics to Assess RANS Turbulence Models Performance at Full Scale Reynolds Numbers. In Proceedings of the ASME International Mechanical Engineering Congress and Exposition, Proceedings (IMECE), New Orleans, LA, USA, 29 October–2 November 2023; Volume 9, p. 13. [CrossRef]
99. Speziale, C. A Review of Reynolds Stress Models for Turbulent Shear Flows. In *Final Report Institute for Computer Applications in Science and Engineering*; NTRS—NASA Technical Reports Server: Hampton, Virginia, USA, 1995.
100. Valentine, D. *Reynolds-Averaged Navier-Stokes Codes and Marine Propulsor Analysis*; Naval Surface Warfare Center: Bethesda, MD, USA, 1993.
101. Gross, A.; Kremheller, A.; Fasel, H. Simulation of Flow over Suboff Bare Hull Model. In Proceedings of the 49th AIAA Aerospace Sciences Meeting including the New Horizons Forum and Aerospace Exposition, Orlando, FL, USA, 4–7 January 2011; American Institute of Aeronautics and Astronautics: Reston, VA, USA, 2011.
102. Sener, M.Z.; Aksu, E. The Numerical Investigation of the Rotation Speed and Reynolds Number Variations of a NACA 0012 Airfoil. *Ocean Eng.* **2022**, *249*, 110899. [CrossRef]
103. Ferreira, J.W.; Vanessa, L.; Da, M.; Judice, R.; Lucas Heldt, L. Resistências Hidrodinâmicas Observados Em Projetos Navais. *Rev. Ciênc. Exatas E Tecnol.* **2021**, *16*, 2–6. [CrossRef]
104. Tatsuo Nishikawa, by; Yoshinobu Yamade, M.; Masaru Sakuma, N.; Chisachi Kato, N. Application of Fully-Resolved Large Eddy Simulation to KVLCC2-Bare Hull Double Model at Model Ship Reynolds Number. *J. Jpn. Soc. Nav. Archit. Ocean. Eng.* **2012**, *16*, 1–9. [CrossRef]
105. Zhang, Z.; Liu, H.; Zhu, S.; Zhao, F. Application of CFD in Ship Engineering Design Practice and Ship Hydrodynamics. *J. Hydrodyn. B* **2006**, *18*, 315–322. [CrossRef]
106. Lissaman, P.B.S. Low-Reynolds-Number Airfoils. *Annu. Rev. Fluid Mech.* **1983**, *15*, 223–239. [CrossRef]
107. Mueller, T.J.; DeLaurier, J.D. Aerodynamics of Small Vehicles. *Annu. Rev. Fluid Mech.* **2003**, *35*, 89–111. [CrossRef]
108. Hemanth, V.; Kushal, K.S.; Varun, S.S. Aerodynamic Study on Low Reynolds Number Aerofoil. *ACS J. Sci. Eng.* **2024**, *4*, 39–57. [CrossRef]
109. Heine, B.; Mack, S.; Kurz, A.; Gross, A.; Fasel, H. Aerodynamic Scaling of General Aviation Airfoil for Low Reynolds Number Application. In Proceedings of the 38th Fluid Dynamics Conference and Exhibit, Seattle, WA, USA, 23–26 June 2008; American Institute of Aeronautics and Astronautics: Reston, VA, USA, 2008; pp. 2008–4410.
110. Hu, H.; Yang, Z.; Igarashi, H. Aerodynamic Hysteresis of a Low-Reynolds-Number Airfoil. *J. Aircr.* **2007**, *44*, 2083–2086. [CrossRef]
111. Spreitr, J.; Steffen, P. *Effect of Mach and Reynolds Numbers on Maximum Lift Coefficient*; NTRS—NASA Technical Reports Server: Hampton, VA, USA, 1946.
112. Selig, M.S.; Maughmer, M.D.; Somers, D.M. Natural-Laminar-Flow Airfoil for General-Aviation Applications. *J. Aircr.* **1995**, *32*, 710–715. [CrossRef]
113. Dryden, H.L. Turbulence, Companion of Reynolds Number. *J. Aeronaut. Sci.* **1934**, *1*, 67–75. [CrossRef]
114. Edwini-Bonsu, S.; Steffler, P.M. Dynamics of Air Flow in Sewer Conduit Headspace. *J. Hydraul. Eng.* **2006**, *132*, 791–799. [CrossRef]
115. Edwini-Bonsu, S.; Steffler, P.M. Air Flow in Sanitary Sewer Conduits Due to Wastewater Drag: A Computational Fluid Dynamics Approach. *J. Environ. Eng. Sci.* **2004**, *3*, 331–342. [CrossRef]
116. Li, Z.; Wang, B.; Wang, F.; Sun, B.; Li, L. Flow Dynamics and Turbulent Coherent Structures around Sediment Reduction Plates of a Sewer System. *J. Environ. Manag.* **2024**, *366*, 121594. [CrossRef] [PubMed]
117. da Silva, A.M.A.; Bolisetti, T. A Method for the Formulation of Reynolds Number Functions. *Can. J. Civ. Eng.* **2000**, *27*, 829–833. [CrossRef]
118. Cahyanto, A.D.; Ilman, I.N.; Putra, A. Penentuan Bilangan Reynold Dan Froude, Redesign Zona Sedimentasi Rectangular Dengan Plate Settler Dan Tube Settler, Water Treatment Plant (Wtp), Kawasan Industri Dan Perumahan "X", Kabupaten Bekasi. *J. Poli-Teknol.* **2023**, *22*, 71–83. [CrossRef]
119. Halliwell, A.R.; Barr, D.I.H.; Saul, A.J.; Burrows, R.; Oakley, H.R.; Ali, K.H.M. Discussion. The Use of Hydraulic Models to Examine the Performance of Storm-Sewage Overflows. *Proc. Inst. Civ. Eng.* **1981**, *71*, 277–282. [CrossRef]
120. Soares-Frazão, S. Wastewater Hydraulics. *J. Hydraul. Res.* **2011**, *49*, 842–843. [CrossRef]

121. Altowayti, W.A.H.; Othman, N.; Tajarudin, H.A.; Al-Dhaqm, A.; Asharuddin, S.M.; Al-Gheethi, A.; Alshalif, A.F.; Salem, A.A.; Md Din, M.F.; Fitriani, N.; et al. Evaluating the Pressure and Loss Behavior in Water Pipes Using Smart Mathematical Modelling. *Water* **2021**, *13*, 3500. [CrossRef]
122. Carvajal, J.; Zambrano, W.; Gómez, N.; Saldarriaga, J. Turbulent Flow in PVC Pipes in Water Distribution Systems. *Urban Water J.* **2020**, *17*, 503–511. [CrossRef]
123. Liou, C.P. Limitations and Proper Use of the Hazen-Williams Equation. *J. Hydraul. Eng.* **1998**, *124*, 951–954. [CrossRef]
124. Buchberger, S.G.; Wu, L. Model for Instantaneous Residential Water Demands. *J. Hydraul. Eng.* **1995**, *121*, 232–246. [CrossRef]
125. Kaltenbacher, S.; Steinberger, M.; Horn, M. Applied Pipe Roughness Identification of Water Networks: Consideration of All Flow Regimes. *IEEE Trans. Control Syst. Technol.* **2023**, *31*, 676–691. [CrossRef]
126. Montalvo, I.; Izquierdo, J.; Pérez, R.; Tung, M.M. Particle Swarm Optimization Applied to the Design of Water Supply Systems. *Comput. Math. Appl.* **2008**, *56*, 769–776. [CrossRef]
127. Yazdani, A.; Jeffrey, P. Complex Network Analysis of Water Distribution Systems. *Chaos* **2011**, *21*, 016111. [CrossRef] [PubMed]
128. Sigg, K.C.; Coffield, R.D. Qualification of a Method to Calculate the Irrecoverable Pressure Loss in High Reynolds Number Piping Systems. In Proceedings of the ASME/JSME Joint Fluids Engineering Conference, Honolulu, HI, USA, 6–10 July 2003; Volume 36975, pp. 143–151. [CrossRef]
129. Koch, P. The Influence of Reynolds Number and Size Effects on Pressure Loss Factors of Ductwork Components. *Build. Serv. Eng. Res. Technol.* **2006**, *27*, 261–283. [CrossRef]
130. Savić, V.; Knežević, D.; Lovrec, D.; Jocanović, M.; Karanović, V. Determination of Pressure Losses in Hydraulic Pipeline Systems by Considering Temperature and Pressure. *J. Mech. Eng.* **2009**, *55*, 237–443.
131. Tawackolian, K.; Kriegel, M. Turbulence Model Performance for Ventilation Components Pressure Losses. *Build. Simul.* **2022**, *15*, 389–399. [CrossRef]
132. Dosunmu, I.T.; Shah, S.N. Pressure Drop Predictions for Laminar Pipe Flow of Carreau and Modified Power Law Fluids. *Can. J. Chem. Eng.* **2015**, *93*, 929–934. [CrossRef]
133. Dutta, P.; Nandi, N. Study on Pressure Drop Characteristics of Single Phase Turbulent Flow in Pipe Bend for High Reynolds Number. *ARPN J. Eng. Appl. Sci.* **2015**, *10*, 2221–2226.
134. El-Adawy, M.; Farid, A.H.; Hassan, M.A.; Abady, M.; Medhat, D.; Abdullatif, H.; Hassan, O.; Thabet, S.M. Low Reynolds Number Wing Design for Unmanned Aerial Vehicle: A Case Study. *Am. J. Eng. Appl. Sci.* **2022**, *15*, 264–273. [CrossRef]
135. Campbell, R.L.; Carter, M.B.; Pendergraft, O.C.; Friedman, D.M.; Serrano, L. Design and Testing of a Blended Wing Body with Boundary Layer Ingestion Nacelles at High Reynolds Numbers (Invited). In Proceedings of the 43rd AIAA Aerospace Sciences Meeting and Exhibit—Meeting Papers, Reno, NV, USA, 10–13 January 2005; pp. 7145–7171. [CrossRef]
136. Lin, P.; Wu, J.; Lu, L.; Xiong, N.; Liu, D.; Su, J.; Liu, G.; Tao, Y.; Wu, J.; Liu, X. Investigation on the Reynolds Number Effect of a Flying Wing Model with Large Sweep Angle and Small Aspect Ratio. *Aerospace* **2022**, *9*, 523. [CrossRef]
137. Kegerise, M.A.; Neuhart, D.H.; Hannon, J.A.; Rumsey, C.L. An Experimental Investigation of a Wing-Fuselage Junction Model in the Nasa Langley 14-by 22-Foot Subsonic Wind Tunnel. In Proceedings of the AIAA Scitech 2019 Forum, San Diego, CA, USA, 7–11 January 2019. [CrossRef]
138. Gatard, J.; Costes, M.; Kroll, N.; Renzoni, P.; Kokkalis, A.; Rocchetto, A.; Serr, C.; Larrey, E.; Filippone, A.; Wehr, D. High Reynolds Number Helicopter Fuselage Test in the ONERA F1 Pressurised Wind Tunnel. In *Proceedings of the Rotorcraft Forum*; Institute of Aerodynamics and Fluid Dynamics: Dresden, Germany, 1997.
139. Polharnus, E.C. *A Review of Some Reynolds Number Effects Related to Bodies at High Angles of Attack*; Newport News; NASA: Hampton, VA, USA, 1984.
140. Roberts, L.S.; Finnis, M.V.; Knowles, K. Characteristics of Boundary-Layer Transition and Reynolds-Number Sensitivity of Three-Dimensional Wings of Varying Complexity Operating in Ground Effect. *J. Fluids Eng.* **2016**, *138*, 091106. [CrossRef]
141. Mokhtar, W.A. A Numerical Study of High-Lift Single Element Airfoils with Ground Effect for Racing Cars. *SAE Int.* **2005**, *114*, 682–688. [CrossRef]
142. Hoque, M.A.; Rahman, M.S.; Rimi, K.N.; Alif, A.R.; Haque, M.R. Enhancing Formula Student Car Performance: Nose Shape Optimization via Adjoint Method. *Results Eng.* **2023**, *20*, 101636. [CrossRef]
143. Lei, J.; He, J. Adjoint-Based Aerodynamic Shape Optimization for Low Reynolds Number Airfoils. *J. Fluids Eng. Trans. ASME* **2016**, *138*, 021401. [CrossRef]
144. Eftekhari, H.; Al-Obaidi, A.S.M.; Eftekhari, S. The Effect of Spoiler Shape and Setting Angle on Racing Cars Aerodynamic Performance. *Indones. J. Sci. Technol.* **2020**, *5*, 11–20. [CrossRef]
145. Ravi, S. The Influence of Turbulence on a Flat Plate Aerofoil at Reynolds Numbers Relevant to MAVs. Ph.D. Thesis, RMIT University, Melbourne, VIC, Australia, 2011.
146. Genç, M.S.; Karasu, İ.; Açıkel, H.H.; Akpolat, M.T.; Genç, M.S.; Karasu, İ.; Açıkel, H.H.; Akpolat, M.T. Low Reynolds Number Flows and Transition. In *Low Reynolds Number Aerodynamics and Transition*; IntechOpen: London, UK, 2012. [CrossRef]
147. Shukla, D.; Komerath, N. Multirotor Drone Aerodynamic Interaction Investigation. *Drones* **2018**, *2*, 43. [CrossRef]
148. Pérez Gordillo, A.M.; Escobar, J.A.; Lopez Mejia, O.D. Influence of the Reynolds Number on the Aerodynamic Performance of a Small Rotor. *Aerospace* **2023**, *10*, 130. [CrossRef]
149. Liu, H.; Ravi, S.; Kolomenskiy, D.; Tanaka, H. Biomechanics and Biomimetics in Insect-Inspired Flight Systems. *Philos. Trans. R. Soc. B Biol. Sci.* **2016**, *371*, 20150390. [CrossRef]

150. Li, F.; Bai, P.; Shi, W.; Li, J.W. Low Reynolds Number Aerodynamics of Micro Air Vehicles. *Adv. Mech.* **2007**, *37*, 257–268. [CrossRef]
151. le Balleur, J.C.; Peyret, R.; Viviand, H. Numerical Studies in High Reynolds Number Aerodynamics. *Comput. Fluids* **1980**, *8*, 1–30. [CrossRef]
152. Owens, L.R.; Wahls, R.A.; Elzey, M.B.; Hamner, M.P. Reynolds Number Effects on Off-Design Stability and Control Characteristics of Supersonic Transports. *J. Aircr.* **2007**, *44*, 134–143. [CrossRef]
153. Wahls, R.A.; Owens, L.R.; Rivers, S.M.B. Reynolds Number Effects on a Supersonic Transport at Transonic Conditions. In Proceedings of the 39th AIAA Aerospace Sciences Meeting & Exhibit, Reno, NV, USA, 8–11 January 2001. [CrossRef]
154. Owens, L.; Wahls, R. Reynolds Number Effects on a Supersonic Transport at Subsonic High-Lift Conditions. In Proceedings of the 39th AIAA Aerospace Sciences Meeting & Exhibit, Reno, NV, USA, 8–11 January 2001. [CrossRef]
155. Owens, L.R.; Wahls, R.A.; Elzey, M.B.; Hamner, M.P. Reynolds Number Effects on the Stability & Control Characteristics of a Supersonic Transport. In Proceedings of the 40th AIAA Aerospace Sciences Meeting and Exhibit, Reno, NV, USA, 14–17 January 2002. [CrossRef]
156. Johnson, J.A.; Zhang, Y.; Johnson, L.E. Evidence of Reynolds Number Sensitivity in Supersonic Turbulent Shocklets. *AIAA J.* **1988**, *26*, 502–504. [CrossRef]
157. Kawaji, M.; Kawahara, A.; Chung, P.M.Y. Instantaneous Velocity Profiles and Characteristics of Pressure-Driven Flow in Microchannels. In Proceedings of the ASME 2001 International Mechanical Engineering Congress and Exposition, New York, NY, USA, 11–16 November 2001; pp. 131–141. [CrossRef]
158. Liu, D.; Garimella, S.V. Investigation of Liquid Flow in Microchannels. *J. Thermophys. Heat Trans.* **2004**, *18*, 65–72. [CrossRef]
159. Yao, J.; Yao, Y.F.; Patel, M.K.; Mason, P.J. On Reynolds Number and Scaling Effects in Microchannel Flows. *Eur. Phys. J. Appl. Phys.* **2007**, *37*, 229–235. [CrossRef]
160. Peng, X.F.; Peterson, G.P.; Wang, B.X. Frictional Flow Characteristics of Water Flowing through Rectangular Microchannels. *Exp. Heat Transf.* **1994**, *7*, 249–264. [CrossRef]
161. Pfund, D.; Rector, D.; Shekarriz, A.; Popescu, A.; Welty, J. Pressure Drop Measurements in a Microchannel. *AIChE J.* **2000**, *46*, 1496–1507. [CrossRef]
162. Koo, J.; Kleinstreuer, C. Liquid Flow in Microchannels: Experimental Observations and Computational Analyses of Microfluidics Effects. *J. Micromech. Microeng.* **2003**, *13*, 568. [CrossRef]
163. Tao, R.; Jin, Y.; Gao, X.; Li, Z. Flow Characterization in Converging-Diverging Microchannels. *Phys. Fluids* **2018**, *30*, 112004. [CrossRef]
164. Chaudhari, P.; Kapoor, A.; Awasthi, Y.; Thakur, A.K.; Kumar, R. Numerical Simulation of Fluid Flow in Microchannels with Induced Irregularities. *Int. J. Chem. React. Eng.* **2023**, *21*, 1443–1452. [CrossRef]
165. Wang, G.R.; Yang, F.; Zhao, W. There Can Be Turbulence in Microfluidics at Low Reynolds Number. *Lab Chip* **2014**, *14*, 1452–1458. [CrossRef] [PubMed]
166. Wang, G.; Yang, F.; Zhao, W.; Chen, C.P. On Micro-Electrokinetic Scalar Turbulence in Microfluidics at a Low Reynolds Number. *Lab Chip* **2016**, *16*, 1030–1038. [CrossRef] [PubMed]
167. Rasouli, M.R.; Abouei Mehrizi, A.; Lashkaripour, A. Numerical Study on Low Reynolds Mixing OfT-Shaped Micro-Mixers with Obstacles. *Chall. Nano Micro Scale Sci. Technol.* **2015**, *3*, 68–76. [CrossRef]
168. Shanko, E.S.; van de Burgt, Y.; Anderson, P.D.; den Toonder, J.M.J. Microfluidic Magnetic Mixing at Low Reynolds Numbers and in Stagnant Fluids. *Micromachines* **2019**, *10*, 731. [CrossRef] [PubMed]
169. Paya, N.; Dankovic, T.; Feinerman, A. A Microfluidic Mixer Fabricated From Compliant Thermoplastic Films. *J. Undergrad. Res. Univ. Ill. Chic.* **2008**, *2*. [CrossRef]
170. Sochol, R.D.; Lu, A.; Lei, J.; Iwai, K.; Lee, L.P.; Lin, L. Microfluidic Bead-Based Diodes with Targeted Circular Microchannels for Low Reynolds Number Applications. *Lab. Chip* **2014**, *14*, 1585–1594. [CrossRef]
171. Sochol, R.D.; Glick, C.C.; Lee, K.Y.; Brubaker, T.; Lu, A.; Wah, M.; Gao, S.; Hicks, E.; Wolf, K.T.; Iwai, K.; et al. Single-Layer "Domino" Diodes via Optofluidic Lithography for Ultra-Low Reynolds Number Applications. In Proceedings of the IEEE International Conference on Micro Electro Mechanical Systems (MEMS), Taipei, Taiwan, 20–24 January 2013; pp. 153–156. [CrossRef]
172. Abdul Hamid, I.S.L. Reynolds Number Effects in Designing a Micromixer for Biomems Application. Master's Thesis, Universiti Tun Hussein Onn, Johor, Malaysia, 2008.
173. Yan, S.R.; Sedeh, S.N.; Toghraie, D.; Afrand, M.; Foong, L.K. Analysis and Manegement of Laminar Blood Flow inside a Cerebral Blood Vessel Using a Finite Volume Software Program for Biomedical Engineering. *Comput. Methods Programs Biomed.* **2020**, *190*, 105384. [CrossRef] [PubMed]
174. Cape, E.G.; Jones, M.; Yamada, I.; VanAuker, M.D.; Valdes-Cruz, L.M. Turbulent/Viscous Interactions Control Doppler/Catheter Pressure Discrepancies in Aortic Stenosis: The Role of the Reynolds Number. *Circulation* **1996**, *94*, 2975–2981. [CrossRef]
175. Adams, J.C.; Jiamsripong, P.; Belohlavek, M.; McMahon, E.M.; Marupakula, V.; Heys, J.; Chaliki, H.P. Potential Role of Reynolds Number in Resolving Doppler- and Catheter-Based Transvalvular Gradient Discrepancies in Aortic Stenosis. *J. Heart Valve Dis.* **2011**, *20*, 159–164. [PubMed]

176. Day, S.W.; Lemire, P.P.; Flack, R.D.; McDaniel, J.C. Effect of Reynolds Number on Performance of a Small Centrifugal Pump. In Proceedings of the ASME/JSME Joint Fluids Engineering Conference, Honolulu, HI, USA, 6–10 July 2003; pp. 1893–1899. [CrossRef]
177. Magin, C.M.; Long, C.J.; Cooper, S.P.; Ista, L.K.; López, G.P.; Brennan, A.B. Engineered Antifouling Microtopographies: The Role of Reynolds Number in a Model That Predicts Attachment of Zoospores of Ulva and Cells of Cobetia Marina. *Biofouling* **2010**, *26*, 719–727. [CrossRef]
178. Javadzadegan, A.; Shimizu, Y.; Behnia, M.; Ohta, M. Correlation between Reynolds Number and Eccentricity Effect in Stenosed Artery Models. *Technol. Health Care* **2013**, *21*, 357–367. [CrossRef]
179. Asgharzadeh, H.; Borazjani, I. Effects of Reynolds and Womersley Numbers on the Hemodynamics of Intracranial Aneurysms. *Comput. Math. Methods Med.* **2016**, *2016*, 7412926. [CrossRef] [PubMed]
180. Shinbrot, T. *Biomedical Fluid Dynamics*; Oxford University Press: Oxford, UK, 2019; ISBN 9780198812586.
181. Nachtigall, W. Some Aspects of Reynolds Number Effects in Animals. *Math. Methods Appl. Sci.* **2001**, *24*, 1401–1408. [CrossRef]
182. Golestanian, R.; Yeomans, J.M.; Uchida, N. Hydrodynamic Synchronization at Low Reynolds Number. *Soft Matter* **2011**, *7*, 3074–3082. [CrossRef]
183. Winkler, R.G. Low Reynolds Number Hydrodynamics and Mesoscale Simulations. *Eur. Phys. J. Spec. Top.* **2016**, *225*, 2079–2097. [CrossRef]
184. Leoni, M.; Liverpool, T.B. Hydrodynamic Synchronization of Nonlinear Oscillators at Low Reynolds Number. *Phys. Rev. E Stat. Nonlin Soft Matter Phys.* **2012**, *85*, 040901. [CrossRef]
185. Acharya, S.; Nayak, B.; Mishra, S.R. Illustration of the Reynolds Number on Micropolar Nanofluid Flow through a Permeable Medium Due to the Interaction of Thermal Radiation. *Waves Random Complex. Media* **2022**, 1–18. [CrossRef]
186. Mi, X.B.; Chwang, A.T. Molecular Dynamics Simulations of Nanochannel Flows at Low Reynolds Numbers. *Molecules* **2003**, *8*, 193–206. [CrossRef]
187. Jaeger, R.; Ren, J.; Xie, Y.; Sundararajan, S.; Olsen, M.G.; Ganapathysubramanian, B. Nanoscale Surface Roughness Affects Low Reynolds Number Flow: Experiments and Modeling. *Appl. Phys. Lett.* **2012**, *101*, 184102. [CrossRef]
188. Fan, D.L.; Zhu, F.Q.; Cammarata, R.C.; Chien, C.L. Manipulation of Nanowires in Suspension by Ac Electric Fields. *Appl. Phys. Lett.* **2004**, *85*, 4175–4177. [CrossRef]
189. Jeon, S.; Malyarchuk, V.; White, J.O.; Rogere, J.A. Optically Fabricated Three Dimensional Nanofluidic Mixers for Microfluidic Devices. *Nano Lett.* **2005**, *5*, 1351–1356. [CrossRef] [PubMed]
190. Ghosh, A.; Mandal, P.; Karmakar, S.; Ghosh, A. Analytical Theory and Stability Analysis of an Elongated Nanoscale Object under External Torque. *Phys. Chem. Chem. Phys.* **2013**, *15*, 10817–10823. [CrossRef] [PubMed]
191. Koplik, J.; Banavar, J.R. Physics of Fluids at Low Reynolds Numbers–A Molecular Approach. *Comput. Phys.* **1998**, *12*, 424–431. [CrossRef]
192. Cavalcantil, A.; Hogg, T.; Shirinzadeh, B. Nanorobotics System Simulation in 3D Workspaces with Low Reynolds Number. In Proceedings of the 2006 IEEE International Symposium on Micro-Nano Mechanical and Human Science, MHS, Nagoya, Japan, 5–8 November 2006. [CrossRef]
193. Davis, H.; Kottas, H.; Moody, A.M.G. The Influence of Reynolds Number on the Performance of Turbomachinery. *J. Fluids Eng.* **1951**, *73*, 499–503. [CrossRef]
194. Strub, A.R.; Bonciani, L.; Borer, C.J.; Casey, M.V.; Cole, S.L.; Cook, B.B.; Kotzur, J.; Simon, H.; Strite, M.A. Influence of the Reynolds Number on the Performance of Centrifugal Compressors. *J. Turbomach.* **1987**, *109*, 541–544. [CrossRef]
195. Wiesner, F.J. A New Appraisal of Reynolds Number Effects on Centrifugal Compressor Performance. *J. Eng. Power* **1979**, *101*, 384–392. [CrossRef]
196. Quin, D.; Grimes, R. The Effect of Reynolds Number on Microaxial Flow Fan Performance. *J. Fluids Eng.* **2008**, *130*, 1011011. [CrossRef]
197. Carpenter, C. Reynolds Number Is Important in Understanding Wax Deposition. *J. Pet. Technol.* **2018**, *70*, 94–95. [CrossRef]
198. Hirsch, C.; Tartinville, B. Reynolds-Averaged Navier-Stokes Modelling for Industrial Applications and Some Challenging Issues. *Int. J. Comput. Fluid Dyn.* **2009**, *23*, 295–303. [CrossRef]
199. Jeong, G.S.; Chung, S.; Kim, C.B.; Lee, S.H. Applications of Micromixing Technology. *Analyst* **2010**, *135*, 460–473. [CrossRef] [PubMed]
200. Heinz, S.; Roekaerts, D. Modelling of Turbulent Reacting Flows: Reynolds and Damkoehler Number Effects. In Proceedings of the First Symposium on Turbulence and Shear Flow Phenomena, Santa Barbara, CA, USA, 12–15 September 1999; Begel House Inc.: Santa Barbara, CA, USA, 1999; pp. 339–344. [CrossRef]
201. Heidarinejad, G.; Ghoniem, A. Vortex Simulation of the Reacting Shear Layer—Effects of Reynolds and Damkohler Numbers. In Proceedings of the 27th Aerospace Sciences Meeting, Reno, NV, USA, 9–12 January 1989. [CrossRef]
202. Patterson, G.K. Application of Turbulence Fundamentals to Reactor Modelling and Scaleup. *Chem. Eng. Commun.* **1981**, *8*, 25–52. [CrossRef]
203. Gladden, L.F.; Sederman, A.J. Magnetic Resonance Imaging and Velocity Mapping in Chemical Engineering Applications. *Annu. Rev. Chem. Biomol. Eng.* **2017**, *8*, 227–247. [CrossRef] [PubMed]

204. Mishra, P.; Ein-Mozaffari, F. Using Statistical and Experimental Methods to Investigate the Mixing of Dense Slurries with Coaxial Mixers: Effects of Design Parameters and Novel Equations for Power and Reynolds Numbers. *Ind. Eng. Chem. Res.* **2021**, *60*, 6306–6326. [CrossRef]
205. Zhou, Z.; Rujin, M. Numerical Simulation Study of the Reynolds Number Effect on Two Bridge Decks Based on the Deterministic Vortex Method. *Wind Struct.* **2010**, *13*, 347–362. [CrossRef]
206. Liu, Q.; Ma, W.; Liu, X. Reynolds Number Effect on Bridge Wind Engineering. *Wind Eng. JAWE* **2014**, *39*, 340–343. [CrossRef]
207. Matsuda, K.; Cooper, K.R.; Tanaka, H.; Tokushige, M.; Iwasaki, T. An Investigation of Reynolds Number Effects on the Steady and Unsteady Aerodynamic Forces on a 1:10 Scale Bridge Deck Section Model. *J. Wind Eng. Ind. Aerodyn.* **2001**, *89*, 619–632. [CrossRef]
208. Schewe, G.; Larsen, A. Reynolds Number Effects in the Flow around a Bluff Bridge Deck Cross Section. *J. Wind Eng. Ind. Aerodyn.* **1998**, *74–76*, 829–838. [CrossRef]
209. Lee, S.; Kwon, S.D.; Yoon, J. Reynolds Number Sensitivity to Aerodynamic Forces of Twin Box Bridge Girder. *J. Wind Eng. Ind. Aerodyn.* **2014**, *127*, 59–68. [CrossRef]
210. Schewe, G. Influence of the Reynolds-Number on Flow Induced Vibrations of Generic Bridge Sections. In Proceedings of the International Conference on Bridges, Dubrovnik, Croatia, 21–24 May 2006; SECON HDGK: Göttingen, Germay, 2006; pp. 351–358.
211. Kazutoshi, M.; Masafumi, T.; Tooru, I. Reynolds Number Effects on the Steady and Unsteady Aerodynamic Forces Acting on the Bridge Deck Sections of Long-Span Suspension Bridge. *IHI Eng. Rev.* **2007**, *40*, 12–26.
212. Hui, M.C.H.; Zhou, Z.Y.; Chen, A.R.; Xiang, H.F. The Effect of Reynolds Numbers on the Steady State Aerodynamic Force Coefficients of the Stonecutters Bridge Deck Section. *Wind Struct.* **2008**, *11*, 179–192. [CrossRef]
213. Kargarmoakhar, R.; Chowdhury, A.G.; Irwin, P.A. Reynolds Number Effects on Twin Box Girder Long Span Bridge Aerodynamics. *Wind Struct.* **2015**, *20*, 327–347. [CrossRef]
214. Zwicky, F. Reynolds' Numbers for Extragalactic Nebulae. *Astrophys. J.* **1941**, *93*, 411. [CrossRef]
215. Canuto, V.M.; Christensen-Dalsgaard, J. Turbulence in Astrophysics: Stars[1]. *Annu. Rev. Fluid Mech.* **1998**, *30*, 167–198. [CrossRef]
216. Lazarian, A.; Kowal, G.; Takamoto, M.; de Gouveia Dal Pino, E.M.; Cho, J. Theory and Applications of Non-Relativistic and Relativistic Turbulent Reconnection. *Magn. Reconnect. Concepts Appl.* **2016**, *427*, 409–471. [CrossRef]
217. Mason, J.; Perez, J.C.; Cattaneo, F.; Boldyrev, S. Extended Scaling Laws in Numerical Simulations of Magnetohydrodynamic Turbulence. *Astrophys. J. Lett.* **2011**, *735*, L26. [CrossRef]
218. Zhou, Y. Unification and Extension of the Similarity Scaling Criteria and Mixing Transition for Studying Astrophysics Using High Energy Density Laboratory Experiments or Numerical Simulations. *Phys. Plasmas* **2007**, *14*, 082701. [CrossRef]
219. Ryutov, D.D.; Remington, B.A. A "Perfect" Hydrodynamic Similarity and Effect of the Reynolds Number on the Global Scale Motion. *Phys. Plasmas* **2003**, *10*, 2629–2632. [CrossRef]
220. Degrave, J.; Felici, F.; Buchli, J.; Neunert, M.; Tracey, B.; Carpanese, F.; Ewalds, T.; Hafner, R.; Abdolmaleki, A.; de las Casas, D.; et al. Magnetic Control of Tokamak Plasmas through Deep Reinforcement Learning. *Nature* **2022**, *602*, 414–419. [CrossRef]
221. Beaux, M.F.; Hass, J.; Bridges, N.; Kwon, N.H.; McIlroy, D.N. Reynolds Number Manipulation of Mean Nanowire Lengths and Nanowire Suspension Quantification. *J. Appl. Phys.* **2011**, *110*, 024313. [CrossRef]
222. Kim, Y.; Lee Chung, B.; Ma, M.; Mulder, W.J.M.; Fayad, Z.A.; Farokhzad, O.C.; Langer, R. Mass Production and Size Control of Lipid-Polymer Hybrid Nanoparticles through Controlled Microvortices. *Nano Lett.* **2012**, *12*, 3587–3591. [CrossRef] [PubMed]
223. Deepak, K.; Rathore, J.S.; Sharma, N.N. Nanorobot Propulsion Using Helical Elastic Filaments at Low Reynolds Numbers. *J. Nanotechnol. Eng. Med.* **2011**, *2*, 011009. [CrossRef]
224. Venu Vinod, A.; Sadasiva Rao, P. Microfluidics of Nanodrug Delivery: Effect of Reynolds Number Ratio and Particle Size. *Chem. Prod. Process Model.* **2011**, *6*, 32. [CrossRef]
225. Driss, Z.; Karray, S.; Damak, A.; Abid, M.S. Effect of the Reynolds Number on the Aerodynamic Characteristics of a Horizontal Axis Wind Turbine. In Proceedings of the 2012 1st International Conference on Renewable Energies and Vehicular Technology, REVET 2012, Nabeul, Tunisia, 26–28 March 2012; pp. 317–323. [CrossRef]
226. Bourhis, M.; Pereira, M.; Ravelet, F. Experimental Investigation of the Effects of the Reynolds Number on the Performance and Near Wake of a Wind Turbine. *Renew. Energy* **2023**, *209*, 63–70. [CrossRef]
227. Bachant, P.; Wosnik, M. Effects of Reynolds Number on the Energy Conversion and Near-Wake Dynamics of a High Solidity Vertical-Axis Cross-Flow Turbine. *Energies* **2016**, *9*, 73. [CrossRef]
228. Cunningham, J.B.; Lemieux, P. Study of Low Reynolds Number Effects on Small Wind Turbine Performance. In Proceedings of the AIAA SciTech Forum, San Diego, CA, USA, 3–7 January 2022. [CrossRef]
229. Burdett, T.A.; Gregg, J.A.; Van Treuren, K.W. A Numerical Study on Improving Airfoil Performance at Low Reynolds Numbers for Small-Scale Wind Turbines Using Intentional Roughness. In Proceedings of the 50th AIAA Aerospace Sciences Meeting Including the New Horizons Forum and Aerospace Exposition, Nashville, TN, USA, 9–12 January 2012. [CrossRef]
230. Mohan, M.; Talukdar, P.K.; Saha, U.K. Influence of Reynolds Number on Aerodynamic and Performance Coefficients of a Novel Parabolic-Bladed Savonius Wind Rotor. *J. Energy Resour. Technol.* **2024**, *146*, 111801. [CrossRef]
231. Javed, A.; Djidjeli, K.; Naveed, A.; Xing, J.T. Low Reynolds Number Effect on Energy Extraction Performance of Semi-Passive Flapping Foil. *J. Appl. Fluid Mech.* **2018**, *11*, 1613–1627. [CrossRef]
232. Gurka, R.; Nafi, A.S.; Weihs, D. On an Adaptation of the Reynolds Number, Applicable to Body-Caudal-Fin Aquatic Locomotion. *Front. Mar. Sci.* **2022**, *9*, 914214. [CrossRef]

233. Cohen, N.; Boyle, J.H. Swimming at Low Reynolds Number: A Beginners Guide to Undulatory Locomotion. *Contemp. Phys.* **2010**, *51*, 103–123. [CrossRef]
234. McHenry, M.J.; Azizi, E.; Strother, J.A. The Hydrodynamics of Locomotion at Intermediate Reynolds Numbers:Undulatory Swimming in Ascidian Larvae (*Botrylloides* sp.). *J. Exp. Biol.* **2003**, *206*, 327–343. [CrossRef] [PubMed]
235. Gazzola, M.; Argentina, M.; Mahadevan, L. Scaling Macroscopic Aquatic Locomotion. *Nat. Phys.* **2014**, *10*, 758–761. [CrossRef]
236. Govardhan, R.N.; Arakeri, J.H. Fluid Mechanics of Aquatic Locomotion at Large Reynolds Numbers. *J. Indian Inst. Sci.* **2011**, *91*, 329–352.
237. Webb, P.W. Simple Physical Principles and Vertebrate Aquatic Locomotion. *Integr. Comp. Biol.* **1988**, *28*, 709–725. [CrossRef]
238. Techet, A.H. Propulsive Performance of Biologically Inspired Flapping Foils at High Reynolds Numbers. *J. Exp. Biol.* **2008**, *211*, 274–279. [CrossRef] [PubMed]
239. Moslemi, A.A.; Krueger, P.S. The Effect of Reynolds Number on the Propulsive Efficiency of a Biomorphic Pulsed-Jet Underwater Vehicle. *Bioinspir. Biomim.* **2011**, *6*, 026001. [CrossRef]
240. Wu, T.Y. *Introduction to the Scaling of Aquatic Animal Locomotion*; California Institute of Technology: Pasadena, CA, USA, 1977.
241. Channouf, S.; Jami, M. Impact Dynamics of Droplets on Circular Bodies: Exploring the Influence of Wettability, Viscosity, and Body Dimensions Using the Lattice Boltzmann Method. *Phys. Fluids* **2024**, *36*, 053316. [CrossRef]
242. Lv, Q.; Li, J.; Guo, P.; Zhang, B.; Tang, P. Effect of Reynolds Number on Impact Force and Collision Process of a Low-Velocity Droplet Colliding with a Wall Carrying an Equal-Mass Deposited Droplet. *Int. J. Multiph. Flow* **2023**, *163*, 104432. [CrossRef]
243. Gordillo, L.; Sun, T.P.; Cheng, X. Dynamics of Drop Impact on Solid Surfaces: Evolution of Impact Force and Self-Similar Spreading. *J. Fluid Mech.* **2018**, *840*, 190–214. [CrossRef]
244. Tretola, G.; Vogiatzaki, K. Unveiling the Dynamics of Ultra High Velocity Droplet Impact on Solid Surfaces. *Sci. Rep.* **2022**, *12*, 7416. [CrossRef]
245. Marmanis, H.; Thoroddsen, S.T. Scaling of the Fingering Pattern of an Impacting Drop. *Phys. Fluids* **1996**, *8*, 1344–1346. [CrossRef]
246. Bolleddula, D.A.; Berchielli, A.; Aliseda, A. Impact of a Heterogeneous Liquid Droplet on a Dry Surface: Application to the Pharmaceutical Industry. *Adv. Colloid Interface Sci.* **2010**, *159*, 144–159. [CrossRef] [PubMed]
247. Chen, H.; Hugo, R.; Park, S. In-Situ Modal Response Characterization of Pipe Structures Through Reynolds Number Variation. In Proceedings of the Biennial International Pipeline Conference, IPC, Calgary, AB, Canada, 24–28 September 2018; Volume 3, p. 9. [CrossRef]
248. Bakshi, S.; Roisman, I.V.; Tropea, C. Investigations on the Impact of a Drop onto a Small Spherical Target. *Phys. Fluids* **2007**, *19*, 032102. [CrossRef]
249. Anyoji, M.; Hamada, D. High-Performance Airfoil with Low Reynolds-Number Dependence on Aerodynamic Characteristics. *Fluid Mech. Res. Int. J.* **2019**, *3*, 76–80. [CrossRef]
250. Baljé, O.E. A Study on Reynolds Number Effects in Turbomachines. *J. Eng. Power* **1964**, *86*, 227–235. [CrossRef]
251. Ivanco, T.G.; Keller, D.F.; Pinkerton, J.L. Effect of Reynolds Number and Aeroelastic Scaling Upon Launch-Vehicle Ground-Wind Loads. *J. Spacecr. Rocket.* **2024**, *61*, 1305–1312. [CrossRef]
252. Penland, J.A.; Romeo, D.J. Advances in Hypersonic Exploration Capability- Wind Tunnel to Flight Reynolds Number. *J. Aircr.* **1971**, *8*, 881–884. [CrossRef]
253. Bensignor, I.; Seth, D.; McCrink, M. Rotor Propulsion Modeling for Low Reynolds Number Flow (Re < 10^5) for Martian Rotorcraft Flight. In Proceedings of the AIAA AVIATION 2022 Forum, Chicago, IL, USA, 27 June–1 July 2022; American Institute of Aeronautics and Astronautics: Reston, VA, USA, 2022.
254. Schneider, S.P. Laminar-Turbulent Transition on Reentry Capsules and Planetary Probes. *J. Spacecr. Rocket.* **2006**, *43*, 1153–1173. [CrossRef]
255. Purcell, E.M. Life at Low Reynolds Number. In *Physics and Our World*; World Scientific Publishing Company: London, UK, 2013; pp. 47–67. [CrossRef]
256. Gad-El Hak, M.; Bandyopadhyay, P.R. Reynolds Number Effects in Wall-Bounded Turbulent Flows. *Appl. Mech. Rev.* **1994**, *47*, 307–365. [CrossRef]
257. Donnelly, R.; Smith, M.; Swanson, C. Superfluid Wind Tunnels. *Phys. World* **1990**, *3*, 39. [CrossRef]
258. Salac, D.; Miksis, M.J. Reynolds Number Effects on Lipid Vesicles. *J. Fluid Mech.* **2012**, *711*, 122–146. [CrossRef]
259. Varela, P.; Suárez, P.; Alcántara-Ávila, F.; Miró, A.; Rabault, J.; Font, B.; García-Cuevas, L.M.; Lehmkuhl, O.; Vinuesa, R. Deep Reinforcement Learning for Flow Control Exploits Different Physics for Increasing Reynolds Number Regimes. *Actuators* **2022**, *11*, 359. [CrossRef]
260. Heuveline, V.; Wittwer, P. Exterior Flows at Low Reynolds Numbers: Concepts, Solutions, and Applications. In *Fundamental Trends in Fluid-Structure Interaction*; World Scientific Publishing Company: London, UK, 2010; pp. 77–169. [CrossRef]
261. Akhter, M.Z.; Riyadh Ali, A.; Omar, F.K. Effect of Flow Reynolds Number on the Aerodynamics of a Novel Bionic Morphing Flap. In Proceedings of the 2022 Advances in Science and Engineering Technology International Conferences, ASET 2022, Dubai, United Arab Emirates, 21–24 February 2022. [CrossRef]
262. Zheng, C.; Zhou, P.; Zhong, S.; Zhang, X. On the Cylinder Noise and Drag Reductions in Different Reynolds Number Ranges Using Surface Pattern Fabrics. *Phys. Fluids* **2023**, *35*, 035111. [CrossRef]

263. Ajayi, O.E.; Lawal, K.A.; Ukaonu, C.E.; Alabi, T.; Okoh, O.; Igbokwe, O. On the Characterisation of the Flow Regimes of Drilling Fluids. In Proceedings of the Society of Petroleum Engineers—SPE Nigeria Annual International Conference and Exhibition 2019, NAIC 2019, Lagos, Nigeria, 5–7 August 2019. [CrossRef]
264. Pavlíček, P. Local Reynolds Number. *AIP Conf. Proc.* **2019**, *2118*, 030035. [CrossRef]
265. Ghosh, B.; Kozarević, E. Identifying Explosive Behavioral Trace in the CNX Nifty Index: A Quantum Finance Approach. *Investig. Manag. Financ. Innov.* **2018**, *15*, 208–223. [CrossRef]
266. Ghosh, B.; Krishna, M.C. Econophysics Reviews. *SSRN Electron. J.* **2019**. [CrossRef]
267. Hildebrand, N.; Choudhari, M.M.; Duan, L. Effect of the Reynolds Number on the Freestream Disturbance Environment in a Mach 6 Nozzle. In Proceedings of the AIAA AVIATION 2022 Forum, Chicago, IL, USA, 27 June–1 July 2022. [CrossRef]
268. Silva, T.S.; Zecchetto, M.; Da Silva, C.B. The Scaling of the Turbulent/Non-Turbulent Interface at High Reynolds Numbers. *J. Fluid. Mech.* **2018**, *843*, 156–179. [CrossRef]
269. Yeung, P.K.; Donzis, D.A.; Sreenivasan, K.R. High-Reynolds-Number Simulation of Turbulent Mixing. *Phys. Fluids* **2005**, *17*, 081703. [CrossRef]
270. Ishihara, T.; Gotoh, T.; Kaneda, Y. Study of High-Reynolds Number Isotropic Turbulence by Direct Numerical Simulation. *Annu. Rev. Fluid Mech.* **2009**, *41*, 165–180. [CrossRef]
271. Alfonsi, G. Reynolds-Averaged Navier-Stokes Equations for Turbulence Modeling. *Appl. Mech. Rev.* **2009**, *62*, 040802. [CrossRef]
272. Tang, L. Reynolds-Averaged Navier-Stokes Simulation of Low-Reynolds-Number Airfoil Aerodynamics. *J. Aircr.* **2008**, *45*, 848–856. [CrossRef]
273. Piomelli, U. High Reynolds Number Calculations Using the Dynamic Subgrid-Scale Stress Model. *Phys. Fluids A Fluid Dyn.* **1993**, *5*, 1484–1490. [CrossRef]
274. Voke, P.R. Subgrid-Scale Modelling at Low Mesh Reynolds Number. *Theor. Comput. Fluid Dyn.* **1996**, *8*, 131–143. [CrossRef]
275. Xu, C.X.; Wang, Z.Y.; Cui, G.X.; Zhang, Z.S. Multiscale Large Eddy Simulation of Scalar Transport in Turbulent Channel Flow. In *New Trends in Fluid Mechanics Research*; Springer: Berlin/Heidelberg, Germany, 2007; pp. 111–114. [CrossRef]
276. Madlener, K.; Frey, B.; Ciezki, H.K. Generalized Reynolds Number for Non-Newtonian Fluids. *Prog. Propuls. Phys.* **2009**, *1*, 237–250. [CrossRef]
277. Foong, L.K.; Shirani, N.; Toghraie, D.; Zarringhalam, M.; Afrand, M. Numerical Simulation of Blood Flow inside an Artery under Applying Constant Heat Flux Using Newtonian and Non-Newtonian Approaches for Biomedical Engineering. *Comput. Methods Programs Biomed.* **2020**, *190*, 105375. [CrossRef] [PubMed]
278. Giesekus, H. Stoff- Und Wärmeübertragung Beim Strömen Schwach Viskoelastischer Flüssigkeiten Um Eine Rotierende Kugel. *Rheol. Acta* **1970**, *9*, 30–38. [CrossRef]
279. Garduño, I.E.; Tamaddon-Jahromi, H.R.; Webster, M.F. Oldroyd-B Numerical Solutions about a Rotating Sphere at Low Reynolds Number. *Rheol. Acta* **2015**, *54*, 235–251. [CrossRef]
280. Garduño, I.E.; Tamaddon-Jahromi, H.R.; Walters, K.; Webster, M.F. The Interpretation of a Long-Standing Rheological Flow Problem Using Computational Rheology and a PTT Constitutive Model. *J. Nonnewton Fluid Mech.* **2016**, *233*, 27–36. [CrossRef]
281. Chen, L.; Yang, C.; Xiao, Y.; Yan, X.; Hu, L.; Eggersdorfer, M.; Chen, D.; Weitz, D.A.; Ye, F. Millifluidics, Microfluidics, and Nanofluidics: Manipulating Fluids at Varying Length Scales. *Mater. Today Nano* **2021**, *16*, 100136. [CrossRef]
282. Hu, G.; Li, D. Multiscale Phenomena in Microfluidics and Nanofluidics. *Chem. Eng. Sci.* **2007**, *62*, 3443–3454. [CrossRef]
283. Wornom, S.; Ouvrard, H.; Salvetti, M.V.; Koobus, B.; Dervieux, A. Variational Multiscale Large-Eddy Simulations of the Flow Past a Circular Cylinder: Reynolds Number Effects. *Comput. Fluids* **2011**, *47*, 44–50. [CrossRef]
284. Chen, S.; Wang, M.; Xia, Z. Multiscale Fluid Mechanics and Modeling. *Procedia IUTAM* **2014**, *10*, 100–114. [CrossRef]
285. Ferziger, J.H.; Koseff, J.R.; Monismith, S.G. Numerical Simulation of Geophysical Turbulence. *Comput. Fluids* **2002**, *31*, 557–568. [CrossRef]
286. Chapman, D. *An Approximate Analytical Method for Studying Entry into Planetary Atmospheres*; Moffett Field: Mountain View, CA, USA, 1958.
287. Intrieri, P.F.; De Rose, C.E.; Kirk, D.B. Flight Characteristics of Probes in the Atmospheres of Mars, Venus and the Outer Planets. *Acta Astronaut.* **1977**, *4*, 789–799. [CrossRef]
288. Reeves, M.T.; Billam, T.P.; Anderson, B.P.; Bradley, A.S. Identifying a Superfluid Reynolds Number via Dynamical Similarity. *Phys. Rev. Lett.* **2015**, *114*, 155302. [CrossRef]
289. Takeuchi, H. Quantum Viscosity and the Reynolds Similitude of a Pure Superfluid. *Phys. Rev. B* **2024**, *109*, L020502. [CrossRef]
290. Barenghi, C.F. Is the Reynolds Number Infinite in Superfluid Turbulence? *Phys. D* **2008**, *237*, 2195–2202. [CrossRef]
291. Volovik, G.E. Classical and Quantum Regimes of Superfluid Turbulence. *JETP Lett.* **2003**, *78*, 533–537. [CrossRef]
292. Marusic, I.; McKeon, B.J.; Monkewitz, P.A.; Nagib, H.M.; Smits, A.J.; Sreenivasan, K.R. Wall-Bounded Turbulent Flows at High Reynolds Numbers: Recent Advances and Key Issues. *Phys. Fluids* **2010**, *22*, 065103. [CrossRef]
293. Dmitrenko, A.V.; Ovsyannikov, V.M. Theoretical Estimates of the Critical Reynolds Number in the Flow around the Sphere on the Basis of Theory of Stochastic Equations and Equivalence of Measures. *Fluids* **2023**, *8*, 81. [CrossRef]
294. Liebendorfer, A. New Theoretical Model for Effects of Reynolds Number on Laminar Boundary Layer along Ribbed Surfaces. *Scilight* **2019**, *2019*, 180004. [CrossRef]
295. Sanfins, G.; Anbarlooei, H.R.; Cruz, D.O.A.; Ramos, F. Complete and Incomplete Similarity for the Mean Velocity Profile of Turbulent Pipe and Channel Flows at Extreme Reynolds Number. *Phys. Fluids* **2021**, *33*, 085118. [CrossRef]

296. Anbarlooei, H.R.; Cruz, D.O.A.; Ramos, F. New Power-Law Scaling for Friction Factor of Extreme Reynolds Number Pipe Flows. *Phys. Fluids* **2020**, *32*, 095121. [CrossRef]
297. Moriconi, L.; Pereira, R.M. Statistics of Extreme Turbulent Circulation Events from Multifractality Breaking. *Phys. Rev. E* **2022**, *106*, 054121. [CrossRef] [PubMed]
298. Chini, G.P. Exact Coherent Structures at Extreme Reynolds Number. *J. Fluid Mech.* **2016**, *794*, 1–4. [CrossRef]
299. Klewicki, J.C.; Chini, G.P.; Gibson, J.F. Prospectus: Towards the Development of High-Fidelity Models of Wall Turbulence at Large Reynolds Number. *Philos. Trans. R. Soc. A Math. Phys. Eng. Sci.* **2017**, *375*, 20160092. [CrossRef] [PubMed]
300. Laguarda, L.; Hickel, S.; Schrijer, F.F.J.; van Oudheusden, B.W. Reynolds Number Effects in Shock-Wave/Turbulent Boundary-Layer Interactions. *J. Fluid Mech.* **2024**, *989*, A20. [CrossRef]
301. Ashoor, M.; Khorshidi, A.; Pirouzi, A.; Abdollahi, A.; Mohsenzadeh, M.; Barzi, S.M.Z. Estimation of Reynolds Number on Microvasculature Capillary Bed Using Diffusion and Perfusion MRI: The Theoretical and Experimental Investigations. *Eur. Phys. J. Plus* **2021**, *136*, 152. [CrossRef]
302. Martínez, I.A.; Roldán, E.; Dinis, L.; Petrov, D.; Parrondo, J.M.R.; Rica, R.A. Brownian Carnot Engine. *Nat. Phys.* **2015**, *12*, 67–70. [CrossRef] [PubMed]
303. Jayapregasham, S. *Effect of Varying Reynolds Number On The Aerodynamic Design of Lifting Surfaces*; Embry-Riddle Aeronautical University: Daytona Beach, FL, USA, 2021.
304. Dimotakis, P.E. Turbulent Mixing. *Annu. Rev. Fluid Mech.* **2005**, *37*, 329–356. [CrossRef]
305. Hanjalić, K. Will RANS Survive LES? A View of Perspectives. In Proceedings of the ASME Heat Transfer/Fluids Engineering Summer Conference 2004, HT/FED 2004, Charlotte, NC, USA, 1–15 July 2004; Volume 2A, pp. 715–725. [CrossRef]
306. Ericsson, L.; Reding, J. Reynolds Number Criticality in Dynamic Tests. In Proceedings of the 16th Aerospace Sciences Meeting, Huntsville, AL, USA, 16–18 January 1978; American Institute of Aeronautics and Astronautics: Reston, VA, USA, 1978.
307. Quinn, H.M. A Fluid Dynamic Development Like None Other. *Eur. J. Appl. Sci.* **2023**, *11*, 407–429. [CrossRef]
308. Das, R.; Girimaji, S.S. On the Reynolds Number Dependence of Velocity-Gradient Structure and Dynamics. *J. Fluid Mech.* **2019**, *861*, 163–179. [CrossRef]
309. Jones, W.P.; Launder, B.E. The Calculation of Low-Reynolds-Number Phenomena with a Two-Equation Model of Turbulence. *Int. J. Heat Mass. Transf.* **1973**, *16*, 1119–1130. [CrossRef]
310. Brodkey, R.S. Limitations on a Generalized Velocity Distribution. *AIChE J.* **1963**, *9*, 448–451. [CrossRef]
311. Heller, V. Scale effects in physical hydraulic engineering models. *J. Hydraul. Res.* **2012**, *50*, 244–246. [CrossRef]
312. Patel, V.C. Perspective: Flow at High Reynolds Number and Over Rough Surfaces—Achilles Heel of CFD. *J. Fluids Eng.* **1998**, *120*, 434–444. [CrossRef]
313. Heller, V. Self-Similarity and Reynolds Number Invariance in Froude Modelling. *J. Hydraul. Res.* **2017**, *55*, 293–309. [CrossRef]
314. Avila, M.; Barkley, D.; Hof, B. Transition to Turbulence in Pipe Flow. *Annu. Rev. Fluid Mech.* **2023**, *55*, 575–602. [CrossRef]
315. Chhabra, R.P.; Richardson, J.F. *Non-Newtonian Flow in the Process Industries: Fundamentals and Engineering Applications*; Butterworth-Heinemann: Oxford, UK, 2004; ISBN 9780750637701.
316. Yalin, M.S. *Theory of Hydraulic Models*; Macmillan: London, UK, 1971. [CrossRef]
317. Brennen, C.E. *Fundamentals of Multiphase Flow*; Cambridge University Press: London, UK, 2005. [CrossRef]
318. Anderson, J.D.; Cadou, C. *Fundamentals of Aerodynamics*, 7th ed.; McGraw Hill LLC: New York, NY, USA, 2024; p. 1142. ISBN 9781264151929.
319. Poelma, C. Ultrasound Imaging Velocimetry: A Review. *Exp. Fluids* **2016**, *58*, 3. [CrossRef]
320. Karniadakis, G.; Beskok, A.; Aluru, N. *Microflows and Nanoflows*; Antman, S.S., Marsden, J.E., Sirovich, L., Eds.; Interdisciplinary Applied Mathematics; Springer: New York, NY, USA, 2005; Volume 29, ISBN 0-387-22197-2.
321. Ewoldt, R.H.; Johnston, M.T.; Caretta, L.M. *Experimental Challenges of Shear Rheology: How to Avoid Bad Data*; Springer: New York, NY, USA, 2015; pp. 207–241.
322. Sreenivasan, K.R. Turbulent Mixing: A Perspective. *Proc. Natl. Acad. Sci. USA* **2019**, *116*, 18175–18183. [CrossRef] [PubMed]
323. Lee, M.; Moser, R.D. Direct Numerical Simulation of Turbulent Channel Flow up To. *J. Fluid Mech.* **2015**, *774*, 395–415. [CrossRef]
324. Balachandar, S.; Eaton, J.K. Turbulent Dispersed Multiphase Flow. *Annu. Rev. Fluid Mech.* **2010**, *42*, 111–133. [CrossRef]
325. Chabannes, V.; Prud'Homme, C.; Szopos, M.; Tarabay, R. High Order Finite Element Simulations for Fluid Dynamics Validated by Experimental Data from the Fda Benchmark Nozzle Model. *Comput. Methods Biomech. Biomed. Engin* **2017**, *20*, 139–140.
326. Marchetti, M.C.; Joanny, J.F.; Ramaswamy, S.; Liverpool, T.B.; Prost, J.; Rao, M.; Simha, R.A. Hydrodynamics of Soft Active Matter. *Rev. Mod. Phys.* **2013**, *85*, 1143–1189. [CrossRef]
327. Westerweel, J.; Elsinga, G.E.; Adrian, R.J. Particle Image Velocimetry for Complex and Turbulent Flows. *Annu. Rev. Fluid Mech.* **2013**, *45*, 409–436. [CrossRef]
328. Wang, T. A Brief Review on Wind Turbine Aerodynamics. *Theor. Appl. Mech. Lett.* **2012**, *2*, 062001. [CrossRef]
329. Snel, H. Review of Aerodynamics for Wind Turbines. *Wind Energy* **2003**, *6*, 203–211. [CrossRef]
330. Trivedi, C.; Gandhi, B.; Michel, C.J. Effect of Transients on Francis Turbine Runner Life: A Review. *J. Hydraul. Res.* **2013**, *51*, 121–132. [CrossRef]
331. Bahaj, A.S. Generating Electricity from the Oceans. *Renew. Sustain. Energy Rev.* **2011**, *15*, 3399–3416. [CrossRef]
332. Nalla, G.; Shook, G.M.; Mines, G.L.; Bloomfield, K.K. Parametric Sensitivity Study of Operating and Design Variables in Wellbore Heat Exchangers. *Geothermics* **2005**, *34*, 330–346. [CrossRef]

333. Bridgwater, A.V. Review of Fast Pyrolysis of Biomass and Product Upgrading. *Biomass Bioenergy* **2012**, *38*, 68–94. [CrossRef]
334. Falcão, A.F.d.O. Wave Energy Utilization: A Review of the Technologies. *Renew. Sustain. Energy Rev.* **2010**, *14*, 899–918. [CrossRef]
335. Williamson, S.J.; Stark, B.H.; Booker, J.D. Low Head Pico Hydro Turbine Selection Using a Multi-Criteria Analysis. *Renew. Energy* **2014**, *61*, 43–50. [CrossRef]
336. Anderson, J.D. *Hypersonic and High-Temperature Gas Dynamics*, 3rd ed.; AIAA Education Series: Las Vegas, NV, USA, 2019. [CrossRef]
337. Gnoffo, P.A. Planetary-Entry Gas Dynamics. *Annu. Rev. Fluid Mech.* **1999**, *31*, 459–494. [CrossRef]
338. Sutton, G.P.; Biblarz, O. *Rocket Propulsion Elements*, 9th ed.; Wiley: Hoboken, NJ, USA, 2016; Volume 59, ISBN 978-1-118-75365-1.
339. Braun, R.D.; Manning, R.M. Mars Exploration Entry, Descent, and Landing Challenges. *J. Spacecr. Rocket.* **2007**, *44*, 310–323. [CrossRef]
340. Goebel, D.M.; Katz, I. *Fundamentals of Electric Propulsion: Ion and Hall Thrusters*; John Wiley & Sons: Hoboken, NJ, USA, 2008; pp. 1–507. [CrossRef]
341. Marren, D.; Lu, F. *Advanced Hypersonic Test Facilities*; American Institute of Aeronautics and Astronautics: Reston, VA, USA, 2002.
342. McNamara, J.J.; Friedmann, P.P. Aeroelastic and Aerothermoelastic Analysis in Hypersonic Flow: Past, Present, and Future. *AIAA J.* **2011**, *49*, 1089–1122. [CrossRef]
343. Boyd, I.D.; Schwartzentruber, T.E. *Nonequilibrium Gas Dynamics and Molecular Simulation*; Cambridge University Press: Cambridge, UK, 2017; pp. 1–360. [CrossRef]
344. Read, P.L.; Lewis, S.R. *The Martian Climate Revisited: Atmosphere and Environment of a Desert Planet*; Springer: Berlin/Heidelberg, Germany, 2004; ISBN 9783540407430.
345. Laub, B.; Venkatapathy, E.; Laub, B.; Venkatapathy, E. Thermal Protection System Technology and Facility Needs for Demanding Future Planetary Missions. *ESASP* **2004**, *544*, 239–247.
346. Lani, A.; Sharma, V.; Giangaspero, V.F.; Poedts, S.; Viladegut, A.; Chazot, O.; Giacomelli, J.; Oswald, J.; Behnke, A.; Pagan, A.S.; et al. A Magnetohydrodynamic Enhanced Entry System for Space Transportation: MEESST. *J. Space Saf. Eng.* **2023**, *10*, 27–34. [CrossRef]
347. Samaha, M.A.; Tafreshi, H.V.; Gad-el-Hak, M. Superhydrophobic Surfaces: From the Lotus Leaf to the Submarine. *Comptes Rendus Méc.* **2012**, *340*, 18–34. [CrossRef]
348. Mahian, O.; Kolsi, L.; Amani, M.; Estellé, P.; Ahmadi, G.; Kleinstreuer, C.; Marshall, J.S.; Siavashi, M.; Taylor, R.A.; Niazmand, H.; et al. Recent Advances in Modeling and Simulation of Nanofluid Flows-Part I: Fundamentals and Theory. *Phys. Rep.* **2019**, *790*, 1–48. [CrossRef]
349. Cushman, J.H. *The Physics of Fluids in Hierarchical Porous Media: Angstroms to Miles*; Theory and Applications of Transport in Porous Media; Springer: Dordrecht, The Netherlands, 1997; Volume 10, ISBN 978-90-481-4909-4.
350. Lejars, M.; Margaillan, A.; Bressy, C. Fouling Release Coatings: A Nontoxic Alternative to Biocidal Antifouling Coatings. *Chem. Rev.* **2012**, *112*, 4347–4390. [CrossRef]
351. Thévenot, J.; Oliveira, H.; Sandre, O.; Lecommandoux, S. Magnetic Responsive Polymer Composite Materials. *Chem. Soc. Rev.* **2013**, *42*, 7099–7116. [CrossRef]
352. Ceccio, S.L. Friction Drag Reduction of External Flows with Bubble and Gas Injection. *Annu. Rev. Fluid Mech.* **2010**, *42*, 183–203. [CrossRef]
353. Saidur, R.; Leong, K.Y.; Mohammed, H.A. A Review on Applications and Challenges of Nanofluids. *Renew. Sustain. Energy Rev.* **2011**, *15*, 1646–1668. [CrossRef]
354. Yim, E.K.F.; Sheetz, M.P. Force-Dependent Cell Signaling in Stem Cell Differentiation. *Stem Cell Res. Ther.* **2012**, *3*, 41. [CrossRef]
355. Bhushan, B.; Jung, Y.C. Natural and Biomimetic Artificial Surfaces for Superhydrophobicity, Self-Cleaning, Low Adhesion, and Drag Reduction. *Prog. Mater. Sci.* **2011**, *56*, 1–108. [CrossRef]
356. Cattafesta, L.N.; Sheplak, M. Actuators for Active Flow Control. *Annu. Rev. Fluid Mech.* **2011**, *43*, 247–272. [CrossRef]
357. D'Alessandro, D.M.; Smit, B.; Long, J.R. Carbon Dioxide Capture: Prospects for New Materials. *Angew. Chem. Int. Ed.* **2010**, *49*, 6058–6082. [CrossRef] [PubMed]

Disclaimer/Publisher's Note: The statements, opinions and data contained in all publications are solely those of the individual author(s) and contributor(s) and not of MDPI and/or the editor(s). MDPI and/or the editor(s) disclaim responsibility for any injury to people or property resulting from any ideas, methods, instructions or products referred to in the content.

fluids

MDPI

Review

The Universal Presence of the Reynolds Number

Aldo Tamburrino * and Yarko Niño *

Department of Civil Engineering and Advanced Mining Technology Center (AMTC), University of Chile, Santiago 8370449, Chile
* Correspondence: atamburr@ing.uchile.cl (A.T.); ynino@ing.uchile.cl (Y.N.)

Abstract: The Reynolds number is a fundamental parameter in fluid dynamics, initially introduced by O. Reynolds in 1883 to characterize the transition between laminar and turbulent flow in fluids and necessary in the scaling of viscous resistance. Over time, its application has expanded significantly, becoming essential for studying a vast range of fluid phenomena—from microscopic scales such as cellular motion to macroscopic scales like turbulent flows and even intergalactic dynamics. The article highlights the universal relevance of the Reynolds number across various fields, including its adaptation to non-Newtonian fluids and granular flows. It emphasizes how the Reynolds number has evolved from a simple dimensionless group to a critical tool for understanding complex physical processes across different scales and environments.

Keywords: Reynolds number; similarity; scaling

1. Introduction

Osborne Reynolds (1842–1912) was a British engineer, physicist, and educator renowned for his pioneering contributions to fluid mechanics and hydraulics. Born in Belfast and educated at Queens' College, Cambridge, he became the first professor of engineering at Owens College, today the University of Manchester, a position he held for nearly four decades. Reynolds' early work spanned magnetism, electricity, and heat transfer, but he soon focused on fluid mechanics, making significant advancements in turbine pumps, lubrication theory (1886), and heat transfer. His 1883 paper on the law of resistance in parallel channels remains a classic, as does his 1895 work, which presents the criterion for flow stability—a seminal contribution to fluid dynamics. The "Reynolds number" and "Reynolds stress" are named after him due to his foundational work in turbulence and fluid flow modeling. A Fellow of the Royal Society since 1877, Reynolds contributed to developing the Whitworth laboratories at Owens College and received numerous accolades, including the Royal Society's gold medal in 1888.

It is interesting to note that Reynolds' 1895 article is not mentioned in the obituaries written after his death. These obituaries highlight that "he wrote more than 70 papers on mechanics and physics, including 'The Law of Resistance in Parallel Channels', 'Theory of Lubrication', and 'Flow of Gases', all of which are familiar to engineers". However, his work *Sub-Mechanics of the Universe* is praised as "one of the most remarkable scientific works of the 19th century" [1,2]. *Sub-Mechanics* represents a more speculative exploration of fundamental concepts, aiming to develop a unified theory for various physical phenomena. Unlike his other seminal papers, it lacks the experimental rigor and mathematical precision that defined his contributions to mechanics and physics. In this work, Reynolds [3,4] hypothesized the existence of a foundational element of the universe, which he called the "purely mechanical medium" (comparable to the ether), to explain the mechanics of fluid motion, the behavior of forces, and the interaction of matter at a fundamental level. However, this line of reasoning was soon overshadowed by developments in physics, notably the Michelson–Morley experiment (1887) [5] and Einstein's theory of relativity (1905) [6], both of which eliminated the need for an ether in explaining the behavior of light and gravity.

Rott presents a vivid history of the Reynolds number [7]. He mentions that Arnold Sommerfeld introduced the term "Reynolds number" in hydrodynamics in 1908 at the Fourth International Congress of Mathematicians held in Rome [8], using "R" to represent the dimensionless quantity $\rho UH/\mu$, which became widely adopted across fluid mechanics. U is the flow velocity, H is the pipe diameter, and ρ and μ are the density and dynamic viscosity of the fluid, respectively. However, this term was often mistakenly attributed to Blasius, who used it in his 1912 and 1913 publications [9,10]. Prandtl also played a significant role in the history of the Reynolds number, introducing it in his 1910 work on hydrodynamic similarity, where it was denoted by ξ [11]. Lord Rayleigh's 1892 work first referenced Reynolds' similarity law, which became fundamental in resolving the sphere drag problem [12]. Using dimensional analysis, he concluded that the pressure gradient in a pipe is a power function of the dimensionless parameter (cw/ν), i.e., the (dimensionless) pressure drop should be proportional to $(cw/\nu)^n$, where c is the pipe diameter, w is the velocity, and ν is the kinematic viscosity, respectively. Experimental data indicated that for large cw/ν values, n must be equal to 2, after which Raleigh shows that the pressure gradient is independent of the viscosity. In a note of less than one page published in 1913 [13], Rayleigh discusses results obtained by Eiffel regarding the resistance and fall velocity of spheres. He mentions (without providing references) that, according to the analyses by Stokes and Reynolds, the drag coefficient must depend on a single parameter (ν/VL), where V is the fall velocity, and L is a linear dimension proportional to $S^{1/2}$, with S being the "diametral surface" of the sphere. According to Rott [7], Rayleigh refers to Stokes' 1850 work [14], where, in studying the similarity of pendulum motion in a viscous medium, Stokes concludes that for systems to be similar, in addition to geometric similarity, they must be dynamically similar if they satisfy the proportion $a^2/T \propto \mu'$, where a corresponds to a line, and T is the time it takes the pendulum to travel along the line. μ' is defined as μ/ρ, which Stokes refers to as the "index of friction of the fluid" and endorses its use, noting that its dimensions are "the square of a line divided by a time, whence it will be easy to adapt the numerical value of μ' to a new unit of length or of time". In this way, Stokes would be the first to establish the importance of the Reynolds number (without naming it as such) in preserving the dynamic similarity of systems where viscosity is the cause of the resistive force.

Reynolds' 1883 pioneering work on flow visualization and critical Reynolds numbers was crucial in understanding turbulent flow. His experiments demonstrated that turbulent flow occurs above a critical dimensionless number, which varies with factors like pipe diameter and fluid properties [15]. In his paper, using dimensional analysis and taking advantage of Stokes' observation that μ/ρ or μ' "is a quantity of the nature of the product of a distance and a velocity", he established that the "birth of eddies [depends] on some definite value of $\rho\ c\ U/\mu$", where c is a "single linear parameter, say the radius of the tube", and U is a "single velocity parameter, say the mean velocity along a tube". To determine the cause of the "eddies", he used a result obtained by Stokes in 1843, which states that "under certain circumstances, the steady motion becomes unstable, so that an indefinitely small disturbance may lead to a change to sinuous motion". To establish the instability condition, Reynolds did not use the dimensionless parameter he derived. As was customary at the time, viscosity was not directly used in the computations. Instead, he used the Poiseuille relation that connects relative viscosity $P = \nu/\nu_0$ with temperature

(ν_0 is the viscosity of water at 0 °C), expressing numerically the condition of instability not in terms of the parameter $\rho\, c\, U/\mu$, as he had initially proposed in his paper, but as $P/(UD) = B_s$, with B_s in s/m^2. Reynolds found two extreme values for B_s. The first, corresponding to "at which steady motion broke down", is 43.73 s/m^2, and the second, "at which previously existing eddies would die out", is 278 s/m^2. In terms of the Reynolds number $Re = UD/\nu = 1/(\nu_0 B_s)$, using $\nu_0 = 1.793 \times 10^{-6}$ m^2/s [16], these threshold values correspond to Re equal to 12,736 and 2006, respectively. Reynolds later refined these ideas in his 1895 paper, using the Reynolds number explicitly to explore the transition from laminar to turbulent flow, establishing that the flow is stable for $Re < 1900$ and unstable for $Re > 2000$ [17]. Reynolds' application of dimensional analysis and power laws laid the foundation for modern turbulence theory, with his findings influencing both theoretical and practical aspects of fluid mechanics. The Reynolds number became crucial in understanding flow behavior and pressure losses in pipes, with significant contributions from Prandtl and von Kármán in later years [7]. By 1910, the Reynolds number had gained broader acceptance in engineering, particularly in aerodynamics and hydraulics, as seen in Blasius' influential work on pressure loss in pipes and turbulence theory [10]. At the 50th General Assembly of the Association of German Engineers in Mainz in 1909, Prandtl mentioned, referring to the scaling of airplanes and airships for testing in wind tunnels, that "in the absence of gravity effects, another similarity rule can be specified (it was probably first specified by O. Reynolds), to which the friction processes also strictly conform" [18]. At the beginning of the second decade of the 20th century, the concept of similarity, scaling, and the use of dimensionless parameters in fluid mechanics and, particularly, in hydraulics, was already well established, with notable works by von Mises (1914) [19], Forchheimer (1914) [20], Weber (1918) [21], among others. In the 1918 Annual General Meeting of the Shipbuilding Technical Society, held in Berlin, Weber presented an extensive paper (119 pages) in which he definitively laid the foundations of the laws of similarity and their use in physical models, "with special consideration for applications in shipbuilding". In it, Weber states that to ensure the dominance of gravitational forces, the Froude number, which he defines as $\varphi = V^2/(L\, g)$, must be preserved in the model and prototype. V and L correspond to characteristic velocity and length, respectively, and g is the acceleration due to gravity. To preserve the effect of viscous frictional forces, the Reynolds number must be the same in both the model and the prototype, denoted as $\psi = VL/\nu$. The preservation of capillary effects is achieved through the parameter $\xi = V^2L/\omega$, where ω is a "measure of dynamic capillarity", defined as $\omega = S/\rho$, with S being the surface tension and ρ the density. Weber refers to the dimensionless parameter ξ as "the unnamed number". It is now known as the Weber number. Additionally, in his work, Weber analyzes the deformation of an elastic solid, determining the Cauchy number (which he denotes as χ), and the case of phenomena subject to "general gravity". The dimensionless parameter he introduces is denoted η and corresponds to the ratio of centrifugal inertial force to the resultant force of Newton's law of universal gravitation. Sommerfeld's notation for the Reynolds number had also become popular in turbulence studies, at least among German authors. As an example, R is used by Noether (1921) [22], Schiller (1922) [23], and Heisenberg (1924) [24], among other authors.

The reader can find information about Osborne Reynolds' biography, his works, and the history of the dimensionless parameter named after him in the articles by Rott (1990), Jackson (1995 and 2004), and Jackson and Launder (2007) [7,25–27]. The purpose of this article is not to recount the history of the number of Reynolds but to examine the concept that the Reynolds number has evolved beyond its origins in fluid dynamics, demonstrating its diverse applications in systems involving motion. Initially defined as the ratio of inertial to viscous forces in a fluid, the Reynolds number has undergone significant reinterpretation. Today, it is a crucial analytical tool for understanding phenomena across various scales—from cellular motion at the microscopic level to turbulent flows and even intergalactic dynamics. This article highlights the Reynolds number's universal relevance across multiple disciplines, including its innovative adaptations for non-Newtonian fluids and granular flows.

Recently, a review article titled "The Reynolds number: A journey from its origin to modern applications" by Saldana et al. [28] was published in *Fluids*. Although it addresses the same topic as the present article, both covering the Reynolds number from its origin to more recent applications across a wide variety of fields, there are notable differences between the two. The article by Saldana et al. includes a bibliometric analysis and is more reference-oriented, featuring an extensive list of citations. It is more focused on the practical history of the Reynolds number, with a strong emphasis on its role in engineering and the development of numerical and experimental techniques. Although the present manuscript also provides a broad exploration of the Reynolds number applications across various disciplines, its central research focus lies in examining whether the Reynolds number, as a concept, has evolved beyond its original definition in fluid dynamics. Specifically, we investigate how its reinterpretation and adaptation have enabled its application to diverse systems, such as non-Newtonian fluids, granular flows, plasma physics, and relativistic fluid dynamics. This work hypothesizes that the Reynolds number's universal relevance stems from its ability to adapt to different physical contexts while retaining its fundamental meaning as a ratio of inertial to viscous forces. By addressing this hypothesis, the article aims to demonstrate that the Reynolds number is not merely a historical artifact of fluid mechanics but a versatile analytical tool with far-reaching implications.

2. Derivation of the Reynolds Number

2.1. The Reynolds Number as the Ratio Between Inertia and Viscous Forces

The derivation of the Reynolds number as a dimensionless parameter that relates inertia to viscous forces can be done in several ways. Probably the simplest is through dimensional analysis, as Reynolds did [15]. Generating dimensionless parameters is quite simple. The tricky part is the ability to recognize the variables that genuinely govern the physical process being studied and a deep understanding of the phenomenon. In his derivation, Reynolds acknowledges the effect of viscosity on the flow characteristics ("*the more viscous a fluid is, the less prone it is to eddying or sinuous motion*", in cursive in Reynolds' paper), as well as the usefulness of characterizing this effect through μ', the "index of friction of fluid" defined by Stokes [14] and now called kinematic viscosity, v, which is "a quantity of the nature of the product of a distance and a velocity". Assuming the motion depends "on a single velocity parameter" $\mathcal{U}$, and "on a single linear parameter" $\mathcal{L}$, we obtain the following:

$$Re = \frac{\mathcal{U}\,\mathcal{L}}{\nu} \tag{1}$$

Another way to obtain the Reynolds number is by taking the ratio between the inertia and the viscous force in the flow. Considering, following Reynolds [15], a unique length and velocity scales, $\mathcal{L}$ and $\mathcal{U}$, and no external temporal forcing acting on the flow, so that the only possible time scale is $\mathcal{T} = \mathcal{L}/\mathcal{U}$, the inertia of the flow, I, scales as follows:

$$I = \frac{d}{dt}(m\,\mathcal{U}) \sim \frac{\rho\,\mathcal{V}\,\mathcal{U}}{\mathcal{T}} \sim \frac{\rho\,\mathcal{L}^3\,\mathcal{U}}{\mathcal{L}/\mathcal{U}} = \rho\,\mathcal{L}^2\,\mathcal{U}^2 \tag{2}$$

In the previous relation, m is the mass, ρ is the density, and $\mathcal{V}$ is the volume of the fluid element. The resistive force of viscous origin scales with the flow characteristics in the following way:

$$F_V = \tau\, A_w \;\sim\; \mu\, \dot{\gamma}\, A_w \;\sim\; \mu\, \frac{\mathcal{U}}{\mathcal{L}}\, \mathcal{L}^2 = \mu\, \mathcal{U}\, \mathcal{L} \tag{3}$$

where $\dot{\gamma}$ is the shear rate, and A_w is the area element where the viscous stress acts. Defining the Reynolds number as the ratio of inertia to viscous force, $Re = I/F_V$, we obtain $Re = \mathcal{U}\,\mathcal{L}/\nu$.

The Reynolds number naturally arises as one of the control parameters governing the Navier–Stokes equations, which for body forces resulting from a gravitational field ($\vec{g}$) is written as follows:

$$\rho\left(\frac{\partial \vec{V}}{\partial t} + \vec{V}\cdot\nabla\vec{V}\right) = -\nabla p + \mu\,\nabla^2\vec{V} + \rho\,\vec{g} \tag{4}$$

Always assuming only one velocity and one characteristic length of the flow, denoting dimensionless variables with an asterisk, the velocity $\vec{V}$ and the coordinate vector $\vec{x}$ are $\vec{V} = \mathcal{U}\,\vec{V}^*$ and $\vec{x} = \mathcal{L}\,\vec{x}^*$. Note that the scaling of time, pressure, and gravity are not given; they result from the prescribed length $\mathcal{L}$ and velocity $\mathcal{U}$. Thus, the non-dimensional form of the Navier–Stokes equation becomes

$$\rho\left(\mathcal{U}\,\frac{\partial \vec{V}^*}{\partial t} + \frac{\mathcal{U}^2}{\mathcal{L}}\,\vec{V}^*\cdot\nabla^*\vec{V}^*\right) = -\frac{1}{\mathcal{L}}\,\nabla^* p + \mu\frac{\mathcal{U}}{\mathcal{L}^2}\,\nabla^{*2}\vec{V}^* + \rho\, g\,\hat{e}_g \tag{5}$$

$\hat{e}_g$ is the unit vector of $\vec{g}$.

$$\rho\frac{\mathcal{U}^2}{\mathcal{L}}\left(\frac{\mathcal{L}}{\mathcal{U}}\,\frac{\partial \vec{V}^*}{\partial t} + \vec{V}^*\cdot\nabla^*\vec{V}^*\right) = -\frac{1}{\mathcal{L}}\nabla^* p + \mu\frac{\mathcal{U}}{\mathcal{L}^2}\,\nabla^{*2}\vec{V}^* + \rho\, g\,\hat{e}_g \tag{6}$$

Thus, the time scale arises naturally as $\mathcal{T} = \mathcal{L}/\mathcal{U}$. The dimensionless time and pressure are $t^* = t\,\mathcal{U}/\mathcal{L}$ and $p^* = p/(\rho\mathcal{U}^2)$, and Equation (6) is written as follows:

$$\frac{\partial \vec{V}^*}{\partial t^*} + \vec{V}^*\cdot\nabla^*\vec{V}^* = -\nabla^* p^* + \frac{\mu}{\rho\mathcal{U}\mathcal{L}}\,\nabla^{*2}\vec{V}^* + \frac{g\,\mathcal{L}}{\mathcal{U}^2}\,\hat{e}_g \tag{7}$$

In this non-dimensional equation, two dimensionless control parameters appear: the inverse of the Reynolds number and the inverse of the Froude number, as defined by Weber [21]:

$$Fr = \frac{\mathcal{U}^2}{g\mathcal{L}} \tag{8}$$

The dimensionless Navier–Stokes equation becomes

$$\frac{\partial \vec{V}^*}{\partial t^*} + \vec{V}^*\cdot\nabla^*\vec{V}^* = -\nabla^* p^* + \frac{1}{Re}\nabla^{*2}\vec{V}^* + \frac{1}{Fr}\hat{e}_g \tag{9}$$

It is customary to join pressure and gravity in one term. Defining h as a vertical coordinate, positive upwards, $g\,\hat{e}_g = \nabla\, g\, h$. Defining a modified pressure as $\hat{p} = p + \rho\, g\, h$ [29], Equation (9) is reduced to

$$\frac{\partial \vec{V}^*}{\partial t^*} + \vec{V}^* \cdot \nabla^* \vec{V}^* = -\nabla^* \hat{p}^* + \frac{1}{Re} \nabla^{*2} \vec{V}^* \tag{10}$$

As $\hat{p} = p + \rho\, g\, h = p + \rho\, g\, \mathcal{L}\, h^*$ and $\nabla h = \nabla^* h^*$, the Froude number is included in $\nabla^* \hat{p}^*$.

2.2. The Reynolds Number from the Point of View of Similarity

In his articles related to the law of similarity of flows with friction, Blasius [9,10] scales the fluid and flow variables in system 1 for those in system 2, such that he defines $f_l = l_2/l_1$ as the ratio of lengths, $f_u = u_2/u_1$ as the ratio of velocities, $f_\nu = \nu_2/\nu_1$ as the ratio of kinematic viscosities, and other similar ratios. Instead of working with pressure, he works with the piezometric head, $h = p/\gamma$, where γ is the specific weight of the fluid. Considering that the inertial, pressure, and viscous friction terms are of the following form

$$\frac{\gamma}{g}\, u \frac{\partial u}{\partial x}, \quad \gamma \frac{\partial h}{\partial x}, \quad \frac{\gamma}{g}\, \nu \frac{\partial^2 u}{\partial x^2}, \tag{11}$$

their scales must be preserved in both systems, resulting in the following [20]:

$$\frac{f_\gamma\, f_u^2}{f_g} = \frac{f_\gamma\, f_h}{f_l} = \frac{f_\gamma\, f_\nu\, f_u}{f_g\, f_l^2} \tag{12}$$

from which it follows:

$$\frac{f_l\, f_u}{f_\nu} = 1, \tag{13a}$$

$$f_h = \frac{f_u^2}{f_g} \tag{13b}$$

From the Equation (13a), it is deduced that the processes are similar only if

$$\frac{u_1 l_1}{v_1} = \frac{u_2 l_2}{v_2} \tag{14}$$

that is, the Reynolds number in system 1 must be equal to the Reynolds number in system 2, $Re_1 = Re_2$. From the second scaling condition (Equation (13b)), the dimensionless parameter gh/u^2 is generated, corresponding to the Euler number in terms of the pressure head.

2.3. The Reynolds Number as the Ratio of Characteristic Times

The Navier–Stokes equation (Equation (4)) represents the transport of momentum in the flow. When presented divided by the density, it is easy to see that the kinematic viscosity ν gives the momentum diffusion coefficient. Thus, two-time scales are distinguished: one associated with the flow, corresponding to $\mathcal{T} = \mathcal{L}/\mathcal{U}$, and another associated with the time scale related to momentum transport by viscous diffusion over the length scale $\mathcal{L}$, denoted as $\mathcal{T}_V$. By dimensional analysis, $\mathcal{T}_V = \mathcal{L}^2/\nu$. Hence, the Reynolds number can also be interpreted as the ratio between the time scale associated with viscous diffusion and the time scale of the flow:

$$Re = \frac{\mathcal{T}_V}{\mathcal{T}} = \frac{\mathcal{U}\,\mathcal{L}}{\nu} \tag{15}$$

2.4. Generalization of the Reynolds Number for Non-Cylindrical Ducts

Equation (4) can be written as follows:

$$\rho\frac{d\vec{V}}{dt} = -\nabla\hat{p} + \nabla\cdot\tau_{ij} \tag{16}$$

where the stress tensor for a Newtonian fluid is given by $\tau_{ij} = \mu\,\dot{\gamma}_{ij}$, with $\dot{\gamma}_{ij}$ the shear rate given by $\dot{\gamma}_{ij} = \left(\partial u_i/\partial x_j + \partial u_j/\partial x_i\right)$. Calling $|\dot{\gamma}|$ to the norm of $\dot{\gamma}_{ij}$, a scale of the shear stress tensor is $\mu\,|\dot{\gamma}|$. From the definition of $\dot{\gamma}_{ij}$, the time scale associated with the shear rate is $\mathcal{T}_{shear} \sim 1/|\dot{\gamma}|$. Thus, the dimensionless stress tensor can be written as $\tau_{ij} = \left(\mu/\mathcal{T}_{shear}\right)\tau_{ij}^*$. Replacing the scaled variables in Equation (16), it becomes

$$\rho\,\frac{\mathcal{U}}{\mathcal{T}}\,\frac{d\vec{V}^*}{dt^*} = -\nabla\hat{p} + \frac{\mu}{\mathcal{L}\,\mathcal{T}_{shear}}\,\nabla^*\cdot\tau_{ij}^* \tag{17}$$

or,

$$\frac{d\vec{V}^*}{dt^*} = -\frac{\mathcal{T}}{\rho\,\mathcal{L}\,\mathcal{U}}\nabla\hat{p} + \frac{\mu}{\rho\,\mathcal{L}\,\mathcal{U}}\,\frac{\mathcal{T}}{\mathcal{T}_{shear}}\nabla^*\cdot\tau_{ij}^* \tag{18}$$

To recover Equation (10), the time scale associated with the flow must match the time scale associated with the shear rate, i.e., $\mathcal{T}_{shear} = \mathcal{T}$.

So far, only one length scale has been considered in the analysis, which is appropriate for the flow in a cylindrical tube, where $\mathcal{L}$ can be taken as the diameter, as Reynolds did. The question is as follows: What is the proper scale $\mathcal{L}$ when the geometry involves more than one longitude? Equations (2) and (3) allow us to consider a duct of arbitrary cross-section. Let P_w the "wetted perimeter", i.e., the perimeter of the duct wall where the shear stress is acting, and A the cross-section area of the tube, and Δx the length of a portion of the duct in the direction of the main flow. The fluid volume in Δx is $\mathcal{V} = A\,\Delta x$, and the area where the shear stress acts is $A_w = P_w\,\Delta x$. Thus, inertia and viscous force are given by

$$I = \frac{d}{dt}(m\,\mathcal{U}) \sim \frac{\rho\,\mathcal{V}\,\mathcal{U}}{\mathcal{T}} \sim \frac{\rho\,A\,\Delta x\,\mathcal{U}}{\mathcal{T}} \tag{19}$$

$$F_V = \tau\,A_w \sim \mu\,|\dot{\gamma}|\,A_w \sim \frac{\mu}{\mathcal{T}_{shear}}\,P_w\,\Delta x \tag{20}$$

Using the result that the flow and shear rate time scales are equal, the Reynolds number, defined as the ratio between inertia and viscous forces, becomes

$$Re = \frac{I}{F_V} = \frac{\mathcal{U}}{\nu}\frac{A}{P_w} \tag{21}$$

The hydraulic radius is defined as $R_H = A/P_w$ so that an appropriate length scale for a tube of arbitrary cross section is $\mathcal{L} = R_H$, and the proper Reynolds number is as follows:

$$Re_{R_H} = \frac{\mathcal{U}R_H}{\nu} \tag{22}$$

The hydraulic radius for the flow in a cylindrical tube of diameter D is $R_H = D/4$, for a rectangular duct of sides a and b is $R_H = ab/(2a+2b)$, and for the flow between two concentrical cylinders of diameters D_{outer} and D_{inner} is $R_H = (D_{outer} - D_{inner})/4$. For a flow with a free surface in a rectangular channel, with width b and flow depth H, the hydraulic radius is $R_H = bH/(b+2\mathrm{H})$.

The standard values of Re that define flow regime are as follows: for Re (Re_{R_H}) < 2000 (500), the regime is laminar; for Re (Re_{R_H}) > 4000 (1000), the regime is turbulent. Between those limits, the regime is named the laminar-turbulent transition.

2.5. The Reynolds Number as a Control Parameter in the Mathematical Analysis of the Navier–Stokes Equations

In the dimensionless form of the Navier–Stokes equation (Equation (10)), the Reynolds number is its only control parameter, and mathematical analysis and the derivation of analytical solutions may be possible by considering limiting values of Re. The most obvious cases are the extremes. One of them is when $Re \rightarrow \infty$, that is, a fluid with zero viscosity in which the second-order differential terms of the equations disappear (reducing the spatial boundary conditions), and the Euler equation is recovered. The other extreme case is when $Re \rightarrow 0$, a situation in which the flow's inertia is negligible, simplifying the analysis of the Navier–Stokes equation since, in this case, the nonlinear terms disappear (Stokes regime). The book by Drazin and Riley [30] presents a wide variety of problems whose analytical solutions feature the Reynolds number as a control parameter. Analytical solutions based on asymptotic expansions or through perturbation theory are parameterized in terms of the Reynolds number. The solutions are expressed as series, which are usually expanded in terms of powers (integer or fractional) of Re or $1/Re$, or functions of $\log(Re)$ [31,32]. The Reynolds number also naturally appears as a parameter that defines the stability of flows described by the Navier–Stokes equations [33–35].

3. The Universality of the Reynolds Number

As seen, the physical meaning of the Reynolds number is a quantification of the relevance of fluid inertia compared to the viscous force opposing the motion. At high Re values, inertia dominates, while at low values, viscous forces prevail. Thus, the Reynolds number predicts the upper limit of laminar flow and the lower boundary of turbulent flow. Furthermore, it is a key parameter in scaling resistance to motion. Although its application began with pressurized flows of Newtonian fluids in pipes, its use quickly extended to other fluid mechanics phenomena involving a length scale, a velocity variable, and a resistive force. Without aiming to present an exhaustive compilation, this document outlines various fields of fluid mechanics where a dimensionless parameter has been defined and identified as a Reynolds number. It must be noted that what specific authors call the Reynolds number, or the modified Reynolds number, others provide a different name.

3.1. Reynolds Number and Biofluid Mechanics

Broadly, biofluid mechanics refers to fluid flows influenced by the surfaces of living beings. These flows can be external or internal. Examples of the former include the flow around a bird or insect in flight or around bacteria, fish, or swimmers moving in a liquid medium. Internal biological flows pertain to phenomena such as blood flow and respiration, always associated with mass and heat transfer. A characteristic of biological flows is that the motion is typically affected by highly flexible and structured surfaces [36]. The relationship between fluid mechanics and biology is not new. It is known that Poiseuille's interest in pipe flow arose from his studies on blood flow in the circulatory system [37,38]. However, this association dates even further back to the Greeks Herophilus (before 300 B.C.) and Erasistratus (310–250 B.C.) of the Alexandria medical school, who were likely the first to study blood flow and liken the heart to a pump [39].

The motion of living organisms in a fluid medium spans a vast range of Reynolds numbers. In these cases, the characteristic length is associated with a body dimension, and the velocity corresponds to its displacement velocity in a stationary fluid or the relative velocity in a current. The medium is typically water or air. Water is considered the fluid medium for bacteria: $\mathcal{L} \sim 10\ \mu\text{m}$ and $\mathcal{U} \sim 10\ \mu\text{m/s}$ yielding $Re \sim 10^{-4}$. On the other extreme of the length scales, for a whale $\mathcal{L} \sim 10$ m and $\mathcal{U} \sim 10$ m/s, resulting in

$Re \sim 10^7$ [40]. Diverse dominant forces in motion can be expected in this wide range of Reynolds numbers. For example, at very low Reynolds numbers, the fluid behaves as if it has no inertia (Stokes flow), and Equation (10) reduces to

$$0 = -\nabla^* \hat{p}^{**} + \nabla^{*2} \vec{V}^* \tag{23}$$

In this very low Re limit, the appropriate scaling for the pressure is the viscous component rather than inertia, so the dimensionless pressure in this case is

$$\hat{p}^{**} = \frac{\hat{p}\,\mathcal{L}}{\mu\,\mathcal{U}} \tag{24}$$

Written with dimensions, Equation (23) is $\nabla \hat{p} = \mu \nabla^2 \vec{V}$, indicating a balance between the driving force (due to the pressure gradient) and the resistive viscous force. The linearity of this equation, along with the continuity equation and the boundary conditions, imply flow reversibility. Organisms with motion in low Reynolds number regimes have developed unique locomotion mechanisms distinct from those outside the Stokes range. This mechanism is associated with the existence of flagella, thin organelles whose periodic motion in a fluid environment enables movement [41].

Some organisms have developed locomotion based on undulatory and oscillatory body motion [42]. In this case, a temporal forcing of motion arises, associated with a certain frequency ω. Thus, the time scale associated with the flow is no longer $\mathcal{L}/\mathcal{U}$ but $\mathcal{T} = 1/\omega$, i.e., $t = t^{**}/\omega$, so the dimensionless Navier–Stokes equation is rewritten as follows:

$$Re_\omega \frac{\partial \vec{V}^*}{\partial t^{**}} + Re\, \vec{V}^* \cdot \nabla \vec{V}^* = -\nabla^* \hat{p}^{**} + \nabla^{*2} \vec{V}^* \tag{25}$$

A new Reynolds number emerges:

$$Re_\omega = \frac{\mathcal{L}^2\,\omega}{\nu} \tag{26}$$

Re_ω is named the oscillatory Reynolds number [40]. Taylor presented it in his pioneering article on the swimming of microscopic organisms [43]. Its velocity scale is given by the characteristic velocity of the fluid induced by the oscillatory motion, $\mathcal{L}\omega$. When a body vibrates, inertial stresses are generated by the interaction between the vibrating surface and the surrounding fluid. Re and Re_ω do not necessarily have the same order of magnitude. For example, an insect hovering in the air in a fixed position has a small Re but a very large Re_ω due to the high-frequency movement of its wings [40]. The oscillatory Reynolds number is frequently expressed in terms of another dimensionless parameter, the Strouhal number, defined as $St = \mathcal{L}\,\omega/\mathcal{U}$, such that $Re_\omega = St\, Re$.

To explain how a propulsive tail can move a body through a viscous fluid without relying on the reaction caused by the fluid's inertia, Taylor [43] simplified the three-dimensional problem of flagellar propulsion in spermatozoa and large arrays of cilia by modeling as transversely wavy sheets. If k is the wavenumber, ω is the oscillation frequency, the wave propagation velocity is ω/k, and pressure scales with $\mu\,\omega$. It is possible to define a new Reynolds number $Re_{\omega k} = \omega/(k^2\,\nu)$ considering the length scale as $1/k$. The effect of $Re_{\omega k}$ on the displacement velocity of the transversely wavy sheets is of second order [40].

The locomotion of microorganisms possessing dense arrays of motile cilia has been modeled as a spherical squirmer of radius a that moves through small, axisymmetric deformations of its surface. The radial and tangential velocities on its surface in a co-moving frame of reference are expressed in terms of series of Legendre and associated Legendre polynomial [44]. B_1 is the amplitude of the first coefficient of the series describing the surface tangential velocity, which is associated with the velocity of a squirmer in potential flow, $u_{spf} = 2B_1/3$. The only length scale for a squirmer in an unbounded Newtonian fluid is the radius a. Thus, considering $\mathcal{L} \sim a$ and $\mathcal{U} \sim u_{spf}$, Chisholm et al. define the Reynolds number associated with the motion of the squirmer as $Re = 2B_1\, a/(3\nu)$ [44].

A simplified model for analyzing experimental data regarding fish locomotion was presented by Gurka et al. [45]. It considers that the flow around the fish can be assimilated to the motion of a harmonic oscillatory motion parallel to a plate (Stokes second problem). Taking the frequency of the oscillatory motion of the fish's body as the frequency of the oscillating plate ω, the length scale as the depth penetration of the viscous wave $\mathcal{L} = 5\sqrt{2\,\nu/\omega}$, and the fish speed U as the velocity scale $\mathcal{U}$, a Reynolds number is formed: $Re = 10\, U/\sqrt{2\,\omega\,\nu}$. Gurka et al. call this Reynolds number "periodic swimming number" and denote it as P [45].

3.2. Reynolds Number in Nano and Microfluidics

Among all the dimensionless numbers, the Reynolds number is the most referenced in the context of microfluidics. Interestingly, it may also be one of the least significant for microfluidics. In nearly all cases, the fluids used in microfluidic devices have Reynolds numbers low enough that inertial effects become negligible [46]. The usual forms of the Reynolds number are those presented in Equations (1) or (2), with $\mathcal{U}$ being the flow average velocity and $\mathcal{L}$ equal to the diameter of the micro pipe or, in the case of rectangular ducts, using the corresponding hydraulic radius. It is also common to find a Reynolds number based on the flow rate Q and the diameter D [47,48]:

$$Re = \frac{Q}{vD} \tag{27}$$

Studies in microfluidics have shown that a slip velocity exists on the boundary. Although the boundary condition considering a slip velocity goes back to Navier, it was shadowed by Stokes' experiments on the non-slip condition [49]. Although non-slip is a valid condition in common large-scale flows, it cannot be overlooked in nano and microfluidics. Consider a plane Couette flow between two plates, driven by the motion of the upper plate, which does not allow slip, but in the lower wall, a slip condition exists. If λ_s is the slip (or penetration) length, Niavarani and Priezjev proposed the following Reynolds number [50]:

$$Re = \frac{U_{upp}\, h}{\nu}\left(1 + \frac{\lambda_s}{h}\right) \tag{28}$$

where U_{upp} is the velocity of the upper wall, and h the plates' separation. The previous result is obtained directly from $Re \sim I/F_V$. The shear stress remains constant across the fluid in the case of a two-dimensional pure Couette flow and from the Navier condition: $\tau_{wall} = \mu u_s/\lambda_s$, where u_s is the slip velocity at the wall. The velocity distribution indicates $U_{upp}/u_s = (h + \lambda_s)/\lambda_s$. Additionally, $\mathcal{V} = h\Delta x$ and $A = \Delta x$, where Δx is a length element of the wall in the direction of flow. Considering $\mathcal{L} = h$ and $\mathcal{U} = U_{upp}$, we have $I \sim \rho U_{upp}^2 \Delta x$. and $F_V = 2\,\tau_{wall}\,\Delta x$, which follows that the ratio I/F_V corresponds to the Reynolds number defined in Equation (28).

3.3. Reynolds Number in Boundary Layers

For simplicity, two-dimensional or axisymmetric flow will be considered next. The key of Prandtl's boundary layer concept introduced in 1904 [51] is the existence of two length scales, one in the flow direction, say $\mathcal{L}$, and another normal to it, the boundary layer thickness δ, such that $\delta \ll \mathcal{L}$. Navier–Stokes equations are greatly simplified with this assumption, and the so-called Prandtl equations are obtained [52,53]. The flow is rotational within the boundary layer, and the potential flow equations are valid outside it. Usually, $\mathcal{L}$ is associated with a coordinate attached to the body or the main flow direction, x. Thus, two Reynolds numbers arise, Re_x and Re_δ:

$$Re_x = \frac{U_0\ x}{\nu}, \tag{29a}$$

$$Re_\delta = \frac{U_0\ \delta}{\nu} \tag{29b}$$

where U_0 is the velocity of the irrotational flow (outside the boundary layer); generally, it depends on the x coordinate. For a boundary layer over a plate or solid body, Prandtl defined δ as the distance from the wall to the location where the flow velocity is $0.99U_0$. Other definitions of boundary layer thickness exist, like displacement thickness, momentum thickness, and energy thickness [52,53], which originate from Reynolds numbers based upon them. In Pohlhausen's method for determining the boundary layer separation point, the parameter $\Lambda_p = -(\delta^2/(\mu\ U_0))(dp/dx)$ appears. The pressure gradient can be related to the velocity gradient in the irrotational region, where Bernoulli's equation holds. This gradient expresses the previous term as $\Lambda = (\delta^2/\nu)(dU_0/dx)$. The parameter Λ_p can be interpreted as the ratio between the force associated with the pressure gradient and the viscous forces [54]. However, when the pressure gradient is related to the velocity gradient, the dimensionless Λ can be interpreted as the ratio between the momentum change due to the acceleration of the velocity in space and the viscous forces. Thus, Λ represents an "accelerating Reynolds number". If the momentum thickness is used instead of the boundary layer thickness δ, the accelerating Reynolds number is denoted as λ [52,53,55].

The no-slip condition and viscosity govern the flow near a flat disk rotating with a constant angular velocity Ω about an axis perpendicular to its plane in a fluid initially at rest. As a result, the fluid layer in direct contact with the disk is dragged along and pushed outward by centrifugal force. New fluid particles are continuously drawn toward the disk in the axial direction and then ejected centrifugally, creating a fully three-dimensional flow that behaves like a pump. The thickness δ of the layer "carried" with the disk is given by $\delta \sim \sqrt{\nu/\Omega}$. For a circular disk of radius R, the Reynolds number is defined as follows [52,53]:

$$Re = \frac{R^2\ \Omega}{\nu} \tag{30}$$

Equation (26) is similar to Equation (30), but there is a difference. In the latter, Ω is the disk's angular velocity, and in Equation (26), ω is the surface's oscillation frequency.

3.4. Reynolds Number in Rotating Flows and Hydraulic Machines

A glance at a rotating flow was given in the previous subsection. Another kind of rotating flow is generated in the gap between two concentric cylinders when one is rotating with an angular velocity Ω. A Reynolds number used for this flow is the following [56]:

$$Re = \frac{r_o\ \Omega\ d}{\nu} \tag{31}$$

where r_o is the radial distance to the outer cylinder, $d = D_o - D_i$ is the annular gap, with D_0 and D_i the diameter of the outer and inner cylinders, respectively. Singh and Rajvanshi [57], in their study with eccentric cylinders of inner and outer radii R_i and R_o rotating with angular velocities Ω_i and Ω_o worked with a Reynolds number defined as follows:

$$Re = \frac{R_o\sqrt{R_i^2\Omega_i^2 + R_0^2\Omega_o^2}}{v} \tag{32}$$

Ludwieg (cited in reference [58]) analyzed his data obtained from concentric cylinder experiments with the Reynolds number given by

$$Re = \frac{\Omega_o \, (R_o - R_i)^2}{\nu} \tag{33}$$

Joseph and Munson [58] studied the stability of the flow between two concentric cylinders with the inner moving along its axis with velocity U_c. They defined a generalized Reynolds number as

$$Re = \frac{\sqrt{R_i^2 \, (R_o - R_i)^2 + U_c^2} \, (R_o - R_i)}{\nu} \tag{34}$$

Bishop et al. [59], in their research concerning the characteristics of a natural convective oscillatory flow in horizontal, cylindrical annuli, defined three Reynolds numbers that they named amplitude Reynolds number Re_A, wavelength Reynolds number Re_W, and period Reynolds number Re_T:

$$Re_A = \frac{R_o \, A_o (R_o - R_i)}{\nu \, T} \tag{35}$$

$$Re_W = \frac{R_i \, A_i \, W}{\nu \, T} \tag{36}$$

$$Re_T = \frac{R_i \, A_i \, (R_o - R_i)}{\nu \, T} \tag{37}$$

In the above equations, A_o and A_i represent the amplitude of the oscillatory flow at the inner and outer cylinder radius, respectively, and W and T are the wavelength and the period of the oscillatory flow, respectively. Weidman [60] investigated experimentally the spin-up and spin-down of a fluid in the gap between two concentric cylinders, where the inner cylinder either accelerated to a constant angular velocity or decelerated until stopped. He defined a Reynolds number based on the fluid's local geostrophic velocity, $V_g = |u - \Omega \, r|$, and the Eckman layer thickness, $\delta = \sqrt{\nu / \Omega}$, i.e., $Re = V_g \, \delta / \nu$, resulting in the following:

$$Re = \frac{|u - \Omega \, r|}{\sqrt{\nu \, \Omega}} \tag{38}$$

It is easy to conclude that the appropriate Reynolds number for a sphere of diameter D rotating with angular velocity Ω in an infinite medium with kinematic viscosity ν is as follows [61]:

$$Re = \frac{D^2 \, \Omega}{\nu} \tag{39}$$

In the analysis of the effect of a wall located a distance d from the center of a sphere with radius R, the following modified Reynolds number appears [62]:

$$Re = \frac{R^2 \, \Omega}{\nu} \left(\frac{d}{R} - 1 \right) \tag{40}$$

In general, the flow of a fluid located in the space between two concentric spheres rotating around an axis does not include the gap thickness or the difference in the radii of the spheres in the definition of the Reynolds number, as is the case for flow between two cylinders. A relationship of the following type is usually used:

$$Re = \frac{\Omega_{ref} \, R_{ref}^2}{\nu} \tag{41}$$

where R_{ref} and Ω_{ref} are the reference variables used for the radius and angular velocity, which depend on the author. Typically, the analysis is accompanied by other dimensionless parameters to account for the gap and the difference in angular velocities of the spheres, such as $(\Omega_o - \Omega_i)/\Omega_i$ and $(R_o - R_i)/R_i$ [63], or the angular velocity ratio Ω_o/Ω_i and the radius ratio R_o/R_i [64,65]. $R_{ref} = R_o$ is used in references [64–68], among others. In contrast, $R_{ref} = R_i$ is used in [63,69–71]. The angular velocity of the outer sphere, $\Omega_{ref} = \Omega_o$, is employed in references [64,66,67], while $\Omega_{ref} = \Omega_i$ is used in [63,71]. Gulwadi and Elkouh [72] use two Reynolds numbers in their analysis: one associated with R_o and Ω_o and another with R_i. Nakabayashi et al. [70] also use two Reynolds numbers, with Ω_i or Ω_o as the reference angular velocity, but keeping $R_{ref} = R_i$ fixed. Bartels uses $\Omega_{ref} = 800\nu/R_o^2$ [68].

In turbomachines, the Reynolds number is presented in terms of the rotational speed N and the characteristic diameter D, such that

$$Re = \frac{\rho\, N\, D^2}{\mu} \tag{42}$$

Additionally, other Reynolds numbers are defined based on some characteristic of the blades, $Re_i = u_i \ell_i/\nu$. For example, u_i can be the impeller tip speed, and ℓ_i the impeller tip width at the impeller exit; or u_i can be the inlet relative velocity at mid-span and ℓ_i the chord length (the length of the flow path) [73]. In gas turbines, the Reynolds number also depends on the exhaust pressure, decreasing with decreasing exhaust pressures as indicated by the definition of the Reynolds number (e.g., peripheral Reynolds number) [74]:

$$Re = \frac{\rho\, u\, D}{\mu} = \frac{u\, D\, p_s}{\mu_s\, R\, T_s} \tag{43}$$

where u is the rotor tip speed, p is the pressure, μ is the dynamic viscosity, T is the temperature, and R is the gas constant. The sub-index s denotes the turbine's exit section or exhaust.

In performance simulations of radial compressors, instead of the actual Reynolds number, the Reynolds Number Index RNI is used, which is the ratio of the actual Reynolds number Re, and a reference Reynolds number Re_{ref}, for the same Mach number [75]:

$$RNI = \frac{Re}{Re_{ref}} \tag{44}$$

Use $\rho_s = p_s/(R\, T_s)$ and the fact that commonly there is no change in length scale, and preserve the Mach number in the actual and reference turbine ($Ma = Ma_{ref}$, i.e., $u/\sqrt{\gamma\, R\, T_s} = u_{ref}/\sqrt{\gamma_{ref}\, R_{ref}\, T_{s,ref}}$), where γ is the exponent of the isentropic process. Considering that $\gamma \approx \gamma_{ref}$, Equation (44) becomes

$$RNI = \frac{p_s}{p_{ref}} \sqrt{\frac{R_{ref}\, T_{ref}}{R\, T}}\, \frac{\mu_{ref}}{\mu} \tag{45}$$

3.5. Reynolds Number in Turbulence

Since its origin, the Reynolds number and turbulence have been inseparably linked, and it is impossible to refer to turbulence without referencing some version of the Reynolds number. As seen in the Introduction, this dimensionless parameter allows us to establish when the flow transitions from laminar to turbulent. We can say that the Reynolds number is at the origin of turbulence, not only because of its importance in experimentally discerning when "the steady motion becomes unstable" and "changes to sinuous motion". As Sommerfeld emphasized analytically, the Reynolds number is a key control parameter in the theoretical approach for determining when an infinitesimal disturbance in the flow grows exponentially, making it unstable. The equation that describes the behavior of disturbances is called the Orr–Sommerfeld equation [52,53], and it allows the generation of a neutral stability curve from which critical values for the Reynolds number, velocity, and wave number of the disturbance can be determined, beyond which any flow disturbance is amplified. For boundary layers over a flat plate, such curves can be found in the books by Schlichting and Gersten [52] and White [53].

In its origin, the length scale $\mathcal{L}$ was associated with the diameter D of the pipe, and the velocity scale $\mathcal{U}$ with the mean flow velocity U. In other words, they correspond to "large" length and velocity scales. The concept of eddies in turbulent flow had already been developed before Reynolds' publications [49]. The size of these eddies is limited by the size of the installation or scenario where the flow occurs, L. Considering that the energy dissipation rate ϵ is due to viscous friction, a dimensional analysis allows us to obtain the size of the smallest vortices in the flow, as well as their velocity and associated times, resulting in

$$\eta = \left(\frac{\nu^3}{\epsilon}\right)^{1/4}, \quad \mathcal{v} = (\nu\,\epsilon)^{1/4}, \quad t_K = \left(\frac{\nu}{\epsilon}\right)^{1/2} \tag{46}$$

The above length, velocity, and time scales are known as Kolmogorov scales, and the Reynolds number formed from them is 1 [76]:

$$Re_K = \frac{\mathcal{v}\,\eta}{\nu} = 1 \tag{47}$$

The energy from the largest vortices passes through the smaller ones until it is dissipated in the smallest vortices in a cascade process. Suppose ℓ corresponds to the size of the largest vortices in the flow, which is associated with a velocity $\mathcal{u}$, and the kinetic energy per unit mass of these vortices is proportional to $\mathcal{u}^2$, the rate of kinetic energy transfer per unit mass to the smaller vortices is $\mathcal{u}^3/\ell$. This energy is entirely dissipated at the Kolmogorov scale, so $\mathcal{u}^3/\ell \sim \epsilon$. Using Equation (46), the scale ratio is

$$\frac{\eta}{\ell} = Re_\ell^{-3/4}, \quad Re_\ell = \frac{\mathcal{u}\;\ell}{\nu} \tag{48}$$

The scale ℓ is related to the integral scales of turbulence, which are determined from the velocity correlation functions [76]. The Reynolds number from Equation (48) can be expressed in terms of turbulent kinetic energy k and the dissipation rate ϵ. In fact, $k \sim \mathcal{u}^2$ and $\ell \sim \mathcal{u}^3/\epsilon$, resulting in

$$Re_\ell = \frac{k^2}{\nu\,\epsilon} \tag{49}$$

Between the integral length scale ℓ and the Kolmogorov scale η, another important scale in turbulence, known as the Taylor microscale λ, is defined by the turbulent fluctuations of the strain rate. For isotropic turbulence, the following holds:

$$\overline{\left(\frac{\partial u'}{\partial x}\right)^2} \equiv \frac{\overline{u'^2}}{\lambda^2} = \frac{\mathcal{u}^2}{\lambda^2} \tag{50}$$

In Equation (50), u' is the turbulent component along the x-direction of the velocity vector, resulting from the Reynolds decomposition, where the instantaneous velocity is separated into a time-averaged component and a fluctuation around the mean value: $\vec{V} = \overline{\vec{V}} + \vec{V}'$. In homogeneous turbulence, it holds that $\overline{u'^2} = \overline{v'^2} = \overline{w'^2}$, so it is possible to consider $\mathcal{u}^2 = \overline{u'^2}$ in Equation (50) unambiguously. The Taylor microscale measures the scale of the turbulence at which velocity fluctuations are highly correlated over small distances. In simpler terms, it is the scale at which the flow begins to exhibit significant local velocity gradients, but before it reaches the Kolmogorov scales where dissipation occurs. The Taylor microscale is a key intermediate scale in the cascade process, often seen as a boundary between energy-containing and dissipative eddies.

The Reynolds number based on the Taylor microscale is defined as

$$Re_{\lambda} = \frac{u\,\lambda}{\nu} \tag{51}$$

The ratio of the Taylor microscale and integral length scale is as follows [76]:

$$\frac{\lambda}{\ell} = \left(\frac{15}{A}\right)^{1/2} Re_{\ell}^{-1/2} \tag{52}$$

where A is a constant of the order of 1. It is easy to see that $Re_K \ll Re_{\lambda} \ll Re_{\ell}$ from the above relationships, which implies $\eta \ll \lambda \ll \ell$.

Near absolute zero, quantum effects significantly influence fluid properties, leading to behaviors so distinct from ordinary fluids that they are termed superfluids. A superfluid behaves as a mixture of two components: a frictionless superfluid and a normal viscous fluid. These components move independently without momentum transfer. Superfluid flow exhibits no viscosity, no heat transfer, and is always potential flow, while normal flow exerts drag forces. This dual-flow concept explains the macroscopic dynamical properties of superfluid and was proposed by Landau in 1941, as mentioned in reference [77]. The Euler equation governs the superfluid component, but it includes an additional term corresponding to the gradient of the chemical potential.

Additionally, a parameter q, which governs mutual friction and dissipation, appears to multiply the term involving the cross product of the superfluid vorticity and velocity [78]. Since the superfluid component has no viscosity, the term $1/q$ plays a similar role in the generation of turbulence. However, this parameter alone is not sufficient to describe the characteristics of turbulence. A circulation coefficient κ, which characterizes the quantized circulation around a quantum vortex, also arises. Based on this, a Reynolds number associated with the superfluid component is defined as $Re_s = \mathcal{U}\mathcal{L}/\kappa$. For fully developed turbulence, $q \ll 1$ and $Re_s \gg 1$ must be satisfied simultaneously [78]. It should be noted that Landau's model disagrees with more recent experimental observations, which indicate that energy dissipation in superfluids occurs not through viscosity, as in ordinary fluids, but through the emission of acoustic waves [79]. A theoretical explanation emerges when a term accounting for the quantum stress tensor $\overline{\overline{\Sigma}}$ is added to the Euler equation. This term is given by $\nabla \cdot \overline{\overline{\Sigma}}$, with $\overline{\overline{\Sigma}} = (\hbar/(2m))^2 \rho_s \nabla\nabla\rho_s$, with $\hbar$ being the reduced Planck's constant, m the mass of one bosom, and ρ_s the superfluid density. Noting the similitude between $\nabla \cdot \overline{\overline{\Sigma}}$ and the stress tensor gradient $\nabla \cdot \overline{\overline{\tau}}$ in the Navier–Stokes equation, the quantum circulation is defined as $\kappa = h/m$, where is h the Planck's constant. For helium II, $\kappa \sim 10^{-3}\ \mathrm{cm^2/s}$ [79].

3.6. Reynolds Number in Porous Media

The porous medium consists of a matrix with interconnected void spaces, typically exhibiting irregular characteristics, through which a liquid or gas flows driven by a pressure gradient. The quantification of these spaces is often simplified by defining a characteristic length and a measure of the porosity or void fraction in the solid matrix. Porous media are frequently heterogeneous and anisotropic, and the flow within the pores can be either laminar or turbulent. The transition between these two regimes, typically characterized by a Reynolds number, is difficult to predict. Additionally, interactions between the fluid and the pore surfaces, such as the liquid's viscosity and capillary forces, can influence the flow in complex ways, particularly in microscopic pores, where intermolecular and capillary forces play a significant role.

There are numerous definitions of the Reynolds number in the context of flow through porous media. Discrepancies observed in the analysis and conclusions of some experimental data have been attributed to differences in the definitions of this dimensionless parameter [80]. The characteristic velocity $\mathcal{U}$ and the length scale $\mathcal{L}$ are challenging to determine, contributing to the variety of Reynolds number definitions used in this field.

In their work on mass transfer between particles and fluid in fluidized beds, Dwivedi and Upadhyay [81] use the variable G (superficial mass flow rate) to define four Reynolds numbers. In three of them, the length scale used is the diameter of the equivalent particle forming the porous medium, $\mathcal{L} = D_p$, and in the fourth, $\mathcal{L} = \sqrt{A_p}$, where A_p is the geometric surface area of the particle. In terms of the Darcy velocity U_D (defined as the flow rate per unit total surface area, including both the solid phase and the pores), instead of G and denoting ε as the void fraction, they are

$$Re_p = \frac{\rho\, U_D\, D_p}{\mu} \tag{53}$$

$$Re_{p,\varepsilon 1} = \frac{\rho\, U_D\, D_p}{\mu\, \varepsilon} \tag{54}$$

$$Re_{p,\varepsilon 2} = \frac{\rho\, U_D\, D_p}{\mu\, (1-\varepsilon)} \tag{55}$$

$$Re_{A_p} = \frac{\rho\, U_D\, \sqrt{A_p}}{\mu} \tag{56}$$

Ziołkowska and Ziołkowski [80] add a Reynolds number based on the specific surface area a as the characteristic length:

$$Re_a = \frac{4\, U_D}{\nu\, a} \tag{57}$$

The specific surface area is computed from

$$a = \frac{6\,(1-\varepsilon)}{\psi\, D_z} \tag{58}$$

$D_z = (6V/\pi)^{1/3}$ is the grain (or pellet) equivalent diameter, with V the grain (or pellet) volume. $\psi = F/(\pi D_z^2)$ is the shape factor of bed element, with F the grain (or pellet) surface area. Similar to Equations (53) and (55), in reference [80], Reynolds numbers based on D_z instead of D_p are also used.

Prieur du Plessis and Woudberg [82] use the following Reynolds number:

$$Re_{D_h} = \frac{\rho U_D D_h}{\mu} \tag{59}$$

where D_h is the hydraulic diameter, which for porous media can be related to the solid surface A_p as $D_h = 6V/A_p$.

Boleve et al. [83] studied electrical field's controlling parameters associated with water seepage through a porous material. When fluid flows through a porous medium, the shear forces on the fluid near the surface of the pores cause ions in the fluid to move, leading to a separation of charge. This separation creates an electrical potential across the material. In their definition of the Reynolds number, they used the seepage velocity, U, and $\mathcal{L} = R$, where R is the radius of the capillaries of the porous medium or $\mathcal{L} = \Lambda$, with Λ a characteristic flow length. For a granular medium formed by a unimodal particle size distribution, $\Lambda = d_0/(2m(F-1))$. d_0 is the mean particle diameter, and $F = \varepsilon^{-m}$ is the electrical formation factor. The exponent m is often called the cementation exponent. The Reynolds number neglecting electroosmotic effects is obtained from

$$Re_{\Lambda} = \frac{\rho^2 g\, K\, \Lambda}{\mu^2\, (1 + Re_{\Lambda})} \frac{h}{L} \tag{60}$$

where K is the permeability, g is the acceleration due to gravity, h is the hydraulic head, and L is the length of the porous pack. From Equation (60), a quadratic equation for Re_{Λ} is obtained, from which the positive root must be considered. Note that in this equation, the dependence on velocity has been replaced by the hydraulic gradient, h/L, which is often easier to determine.

Some authors [84,85] have used a characteristic length based on the permeability K of the porous media, $\mathcal{L} = \sqrt{K}$, such as

$$Re_K = \frac{U_D\, \sqrt{K}}{\nu} \tag{61}$$

K can be estimated using the Carman–Kozeny equation, $K = D_p^2\, \varepsilon^3 / \left(180\, (1 - \varepsilon)^2\right)$.

Jin and Kuznetsov [85] in their study on turbulent flows in porous media defined a Reynolds number based on the macroscopic strain tensor, $s_{Dij} = \left(\partial u_{Di}/\partial x_j + \partial u_{Dj}/\partial x_i\right)/2$, such that $\mathcal{U} = \sqrt{K}\, \left|s_{Dij}\right|$ with $\left|s_{Dij}\right| = \sqrt{2 s_{Dij} s_{Dij}}$, and $\mathcal{L} = \sqrt{K}$, resulting in:

$$Re_{s_{Dij}} = \frac{K\, \left|s_{Dij}\right|}{\nu} \tag{62}$$

Versions of the previous relations have been used for two-phase flows, such that a Reynolds number is defined for the solid phase and another for the liquid phase [86,87].

3.7. Reynolds Number in Multiphase Flows

Multiphase flows are those in which two or more phases of different nature (liquid, gas, or solid) coexist and move simultaneously through a medium. These phases can be found in dispersed or stratified mixtures, and the flows can have various configurations depending on how the phases are distributed. The phases can interact with each other through their interfaces, which influences the overall flow dynamics. These interactions include mass, heat, and momentum transfer processes. The Reynolds number is crucial for characterizing multiphase flows because it determines their interactions in addition to defining the flow regime in each phase. The most important interaction between the phases is the drag force, which depends on the Reynolds number and the concentration of the dispersed phase in the carrier phase.

The drag force $\vec{F}_D$ for an isolated particle in an infinite fluid is given by

$$\vec{F}_D = \frac{1}{2}\, \rho\, C_D\, A_p \left|\vec{V}_r\right| \vec{V}_r \tag{63}$$

where ρ is the fluid density, A_p is the particle's cross-sectional area, and $\vec{V}_r = \vec{V}_p - \vec{V}_f$ is the relative velocity of the particle ($\vec{V}_p$) to the fluid velocity ($\vec{V}_f$). C_D is the drag coefficient, which depends on the particle's Reynolds number, geometry, and orientation relative to the approaching flow. The previous relationship is valid for solid particles, drops, or bubbles. However, in the latter two cases, their interaction with the fluid can deform them, altering their surface area and drag coefficient as they move. The Reynolds number of a spherical particle (or bubble or drop) is given by

$$Re_P = \frac{\left|\vec{V}_r\right| D_P}{\nu} \tag{64}$$

If the particle is not spherical, a suitable characteristic length is used instead of D_P.

When the particle is not isolated, but there is suspension, the particle's concentration affects the drag coefficient, and the definition of the Reynolds number needs to consider the concentration of the particles in the mixture, c. For dilute suspensions ($c \lesssim 0.05 - 0.1$), the following Reynolds number can be used [88]:

$$Re_s = \frac{(1-c)\left|\vec{V}_r\right| D_P}{\nu} \tag{65}$$

The above expression considers the kinematic viscosity of the fluid phase. For higher concentrations, the particle "feels" a higher drag effect that can be considered using a dynamic viscosity associated with the suspension or mixture, μ_m, used to define the Reynolds number Re_m [89,90]:

$$Re_m = \frac{\rho \left|\vec{V}_r\right| D_P}{\mu_m} \tag{66}$$

The viscosity μ_m can be computed from the following [89–91]:

$$\mu_m = \mu \left(1 - \frac{c}{c_{pm}}\right)^{-2.5 c_{pm} \mu^*} \tag{67}$$

Equations (66) and (67) are valid for solid particles, drops, or bubbles. For solid particles, $c_{pm} \approx 0.62$ is the concentration for maximum packing, and $\mu^* = 1$. For bubbles and drops, it depends on the viscosity of the carrier phase, μ, and the viscosity of the fluid within the bubble or drop, μ_p, according to

$$\mu^* = \frac{\mu_p + 0.4\,\mu}{\mu_p + \mu} \tag{68}$$

Many other expressions for μ_m can be found in the literature [92].

In the design of spray systems, the flow characteristics at the nozzle exit are crucial for liquid atomization and the generation of fluid particles with a relatively homogeneous size distribution. One of the atomization methods involves tangentially injecting the liquid into the nozzle, creating a strong swirling motion that forms an air core at the center of the nozzle. This results in the liquid exiting as thin, axisymmetric sheet, then breaking into tiny fluid particles. The formation of the air core depends on a Reynolds number associated with the liquid flow rate and the geometry of the tangential entry ports, Re_t. The discharge coefficient depends on a Reynolds number, $Re_{\Delta p}$, which is based on the pressure gradient between the nozzle inlet and outlet, defined as

$$Re_t = \frac{\rho\, Q\, D_{tp}}{\mu\, A_{tp}}, \tag{69a}$$

$$Re_{\Delta p} = \frac{\sqrt{\rho\, \Delta p}\, D_0}{\mu} \tag{69b}$$

where Q is the discharge, D_{tp} is the diameter of the tangential entry ports of the nozzle, A_{tp} is the total area of entry ports, and D_0 is the diameter of outlet's nozzle [93]. The final size of the droplets depends on other variables like surface tension, gas properties (density and viscosity), and gas-to-liquid momentum ratio.

The mathematical modeling of multiphase flows involves coupling the continuity, momentum, and energy equations for the different phases involved. One of the most significant challenges is the interaction between phases, where one component exerts a drag on the others. Modeling can be quite complex depending on the conditions and the phases involved. For example, in gas–liquid mixtures within a closed pipe, a liquid film may form along the perimeter of the pipe, with a gas core in the center, which also drags liquid drops. Without delving into the various modeling approaches developed in this paper, Reynolds numbers—local or global—are frequently encountered in these models to calculate the pressure losses associated with the different phases or to determine the interphase momentum transfer. Below, as a sample, we present some of the Reynolds numbers found in multiphase flow modeling.

Brennen [94] considers a Reynolds number for the effective gas flow given by

$$Re_{eff} = \frac{U_C\, L_C}{\nu_C}\,(1+\xi) \tag{70}$$

where the subindex C refers to the "continuous" (carrier) phase, and ξ is the loading parameter, defined as follows:

$$\xi = \frac{c_D\,\rho_D}{c_C\,\rho_C} \tag{71}$$

The subindex D designs the "discrete" or disperse phase. The density of the mixture is $\rho = c_D\,\rho_D + c_C\,\rho_C$.

It has been observed in laminar flow that when small particles move through a pipe, they migrate away from the pipe wall and toward the center of the flow (Segre and Silberberg effect). This migration contrasts with the intuitive expectation that particles should follow the fluid's motion and tend to settle. This phenomenon occurs because the particles experience shear forces in the fluid and move through a region with a velocity gradient. Thus, a shear Reynolds number ($Re_{\dot{\gamma}}$) is generated [93], defined as follows:

$$Re_{\dot{\gamma}} = \frac{|\dot{\gamma}|\, D_p^2}{\nu} = \left|\frac{\partial u}{\partial r}\right| \frac{D_p^2}{\nu} \tag{72}$$

where u is the axial velocity, and r is the radial coordinate. It must be noted that $Re_{\dot{\gamma}}$ is a local Reynolds number as it is a function of r.

In the study of the role of the ambient fluid in gravitational granular flow dynamics [95], the authors derived two Reynolds numbers while nondimensionalizing the equations for the solid and liquid phases. One Reynolds number is associated with grain collisions (Re_{gc}), and the other with the fluid (Re_f). They are

$$Re_{gc} = \frac{c_{LP}\,\rho_s\,h_0\,\sqrt{g' h_0}}{\mu_{sc}}, \tag{73a}$$

$$Re_f = \frac{(1-c)\,\rho\,h_0\,\sqrt{g'\,h_0}}{\mu+\mu_T} \tag{73b}$$

c_{LP} is the loose packing concentration of the solids, ρ_s is the solid density, h_0 is the length scale of the liquid–solid mixture, $g' = g(\rho_s - \rho)/\rho_s$ is a reduced gravity, and μ_{sc} is the grain collisional viscosity, which depends on the diameter and density of the grains, solid concentration, elastic restitution coefficient, and the granular temperature, computed from the grain velocity fluctuations. The turbulent viscosity μ_T was determined using a k-epsilon model of the turbulence.

3.8. Reynolds Number in Fluvial Mechanics

Like the flow in pipes and channels, the Reynolds number plays a fundamental role in river mechanics. In addition to the Reynolds number from Equation (22), which determines whether the flow will be laminar or turbulent, other dimensionless numbers related to flow resistance arise. If turbulent flow develops over a smooth wall, a region in contact with the wall is generated, called the viscous sublayer, where viscous forces dominate over turbulent forces. The thickness of this sublayer (δ_V) is unknown a priori, but the variables it scales with can be determined through dimensional analysis. In this region, a shear stress acts, which can be characterized as the one acting on the wall, τ_0. The relevant fluid properties are its density ρ and viscosity μ. Thus, it is possible to generate a dimensionless parameter, $\Pi = (\delta_V \, \rho/\mu)\,\sqrt{\tau_0/\rho}$. The value of Π is determined experimentally, yielding $\Pi \approx 5$. The term $\sqrt{\tau_0/\rho}$ has the dimensions of velocity, is denoted u_*, and is called friction velocity. Therefore, $\Pi = \delta_V u_* \,/\nu$ corresponds to a Reynolds number whose length scale is the thickness of the viscous sublayer $\mathcal{L} = \delta_V$, and the velocity scale is the friction velocity $\mathcal{U} = u_*$, so we have the following:

$$Re_{\delta_V} = \frac{\delta_V \, u_*}{\nu} \tag{74}$$

The notation $Re_{\delta_V} = \delta_V^+$ is also used. Equation (74) indicates the presence of another length scale, ℓ_V, associated with the relevant variables near the wall: τ_0, ρ and μ or, equivalently, u_* and ν. These variables are called wall variables or inner variables and $\ell_V = \nu/u_*$. Then, Re_{δ_V} corresponds to the ratio of two lengths, the thickness of the viscous sublayer, and the inner length: $Re_{\delta_V} = \delta_V/\ell_V$.

We have referred to a smooth wall without specifying what exactly defines a smooth wall. Strictly speaking, every surface has roughness; however, from a hydrodynamic perspective, the wall is defined as hydrodynamically smooth if the roughness is entirely within the viscous sublayer. If the size of the roughness is k_s, this means $k_s < \delta_V$. Thus, a Reynolds number naturally arises by comparing k_s with ℓ_V, $Re_{k_s} = k_s\,/\ell_V$, named the roughness Reynolds number:

$$Re_{k_s} = \frac{k_s \, u_*}{\nu} \tag{75}$$

The notation $Re_{k_s} = k_s^+$ is also used. Thus, a wall is called hydrodynamically smooth if $Re_{k_s} = k_s^+ < 5$. If the size of the roughness is such that it completely inhibits the development of a viscous sublayer, the wall is called hydrodynamically rough; this occurs if $Re_{k_s} = k_s^+ > 70$. For Re_{k_s}'s intermediate values, the wall is named a smooth-to-rough transition.

The type of wall conditions the resistance to flow. The shear stress at the bottom (τ_0) is related to the average flow velocity (U), for example, through the Darcy friction factor (f) using the following relationship:

$$\tau_0 = \frac{1}{8}\, f\, U^2, \quad \frac{u_*}{U} = \sqrt{\frac{f}{8}} \tag{76}$$

In general, f is a function of the Reynolds number and the relative roughness, i.e., $f = f(Re_{R_H},\ k_s/R_H)$. For a laminar regime, it depends only on the Reynolds number, $f = 16/Re_{R_H}$. There are relationships for turbulent flows in open channels based on the semi-analytical Prandtl–Nikuradse expression obtained for pipes where the numerical constants have been adjusted [96]. In natural, non-cohesive channels, the roughness k_s is determined by a representative size of the sediment that forms the bed.

When the material that forms the bed can move due to the flow, a complex and continuous relationship exists between bed deformation, flow pattern alteration, and sediment movement. This deformation requires determining the condition under which the flow can move the sediment (incipient motion condition) and the amount of sediment transported (solid discharge). Shields [97] determined that the critical shear stress at which non-cohesive particle movement occurs depends on what he called the particle Reynolds number, Re_*:

$$Re_* = \frac{u_* \, D_s}{\nu} \tag{77}$$

where D_s is the characteristic size of the bed's sediment. As first noticed by Vanoni [98], the Shield diagram (i.e., the dimensionless critical shear stress, $\tau_* = u_*^2/(g\, R\, D_s)$, as a function of Re_*, where $R = \rho_s/\rho - 1$, and ρ_s and ρ are the sediment and water densities) is not practical because in order to find the critical shear stress for incipient motion, one must know the critical shear velocity u_*. The relationship can be cast in an explicit form if another Reynolds number exists: $Re_p = \sqrt{g\, R\, D_s^3}/\upsilon$, such that $Re_* = \tau_*^{1/2}\, Re_p$ [99]. Other relationships based on the critical shear stress have been developed by various authors [100].

When the particles of the bed are set in motion, change the bed configuration forming bedforms that depend on the flow characteristics and sediment properties. Thus, for example, ripples are present when a viscous sublayer exists [100,101] and antidunes in supercritical flows [100]. The flow resistance when there are bedforms has two components: one associated with the friction induced by the grains (roughness) and the other with the geometry. The first is computed using a friction coefficient, as mentioned above. The second demands ad hoc modeling.

Vortex sand ripples are bedforms created by oscillatory flow and are commonly generated beneath waves [102,103]. The behavior of vortex ripples in an oscillatory flow is controlled by the wave Reynolds number, which is defined $Re = A^2\, \omega/\nu$, where A is the wave semi-orbital excursion, ω is the radian frequency of the wave, given by $\omega = 2\pi/T$, T is the flow oscillation's period, and ν is the kinematic viscosity of the water [103,104].

The sediment discharge in rivers is divided into bedload and suspension load. Einstein's approach to determining the bedload depends on the ratio $Re_{k_s}/Re_{\delta_V} = k_s/\delta_V$ [100,105]. Other bed-load formulas require computing a critical shear stress, which can depend on Re_*.

Bagnold's approach to the sediment transport problem must be mentioned [106]. He defines a Reynolds number for a layer formed by solids sheared in a fluid and denoted by G:

$$G = \frac{D_s}{\mu}\sqrt{\frac{\rho_s}{\lambda}\,\tau_0} \tag{78}$$

where λ is the linear concentration. The above equation can be written as

$$G = \frac{D_s}{\mu/\rho_s\sqrt{\lambda}}\sqrt{\frac{\tau_0}{\rho_s}} \tag{79}$$

Bagnold expresses that G is closely analogous to the fluid Reynolds number $(D_s/\nu)\, \sqrt{\tau_0/\rho} = D_s\, u_*/\nu$. At the bed, the linear concentration remains constant at 14, the limiting concentration at which a mixture of water and a sheared array of solid grains was found to cease to behave as a Newtonian fluid and begin to behave as a rheological "paste" [106].

3.9. Reynolds Number in Granular Flows

The distinction between granular and multiphase flows is somewhat arbitrary, and some references from one subsection could easily belong to the other or both. Granular flows refer to "dry" flows of solid particles, meaning there is no liquid between the particles, only air, which remains stationary. Due to the significant difference between the density of air and that of liquids, it is common to assume that air does not play an important role in the dynamics of dry granular flow. However, in the reference [95], it is shown that while buoyancy can be neglected, the same does not apply to the effect of the viscosity of interstitial air (the kinematic viscosity of air is an order of magnitude higher than that of water), and the movement of the solid phase induces air movement.

Granular flows are classified into three regimes: quasi-static, with slow deformations and frictional particle interactions; gaseous, with rapid, dilute flow where particles interact via collisions; and liquid-like, where the material is dense but flows, with both collisional and frictional interactions [107].

The initial definition of the Reynolds number as the ratio of inertia to viscous forces can be extended to granular flows if it is considered as the ratio between inertia and the dissipative term in the equation of motion. In granular flows, the dissipative term is associated with frictional (tangential) stresses. For a dense material flow (liquid-like), Schaeffer [108] deduces that the following equation gives the velocity:

$$\rho_s \left(\frac{\partial \vec{V}}{\partial t} + \vec{V} \cdot \nabla \vec{V} \right) = -k \nabla \left(T_\tau \mid \dot{\gamma} \mid^{-1} \dot{\gamma}_{ij} \right) - \nabla T_\tau + \rho \vec{g} \tag{80}$$

where $T_\tau = tr(\tau_{ij})/3$ is the trace of the stress tensor divided by 3, and k is a material-dependent constant related to the internal friction angle. Thus, the inertia-related term scales as $\rho_s \mathcal{U}^2/\mathcal{L}$, while the dissipative term $-k \nabla \left(\tau \mid \dot{\gamma} \mid^{-1} \dot{\gamma}_{ij} \right)$ scales as $k\,T/\mathcal{L}$, thus *inertia/dissipative term* $\sim \rho_s \mathcal{U}^2/(k\,T)$. If, for example, a silo is considered, $T \sim \rho_s\, g\, \mathcal{L}$, then

$$Re_{gran} = \frac{\mathcal{U}^2}{kg\mathcal{L}} \tag{81}$$

Some authors define an effective viscosity for the granular medium as [108,109]

$$\mu_{eff} = \frac{\eta\, p_s}{|\dot{\gamma}|} \tag{82}$$

where η is the effective friction coefficient given by

$$\eta = \eta_1 + \frac{\eta_2 - \eta_1}{I_0/I + 1} \tag{83}$$

Here, η_1, η_2, and I_0 are constants with typical $\tan 21^\circ$, $\tan 33^\circ$, and 0.3, values, respectively. I is the inertial number defined as

$$I = \frac{\dot{\gamma}\, D_s}{\sqrt{p_s/\rho_s}} \tag{84}$$

where D_s is the particle diameter, ρ_s its density, p_s is the pressure, and $\dot{\gamma}$ is the second invariant of the shear rate. As can be seen, this definition of viscosity is not an intrinsic property of the material but depends on local flow characteristics.

For dilute granular flows, expressions for effective viscosity have been derived based on the kinetic theory of gases, for example [110].

$$\mu_{eff} = \frac{5\pi}{96}\rho_s\, D_p\, \sqrt{T_g} \tag{85}$$

T_g is the so-called granular temperature, corresponding to the root mean square value of grain velocity fluctuations.

3.10. Reynolds Number in Non-Newtonian Fluid Flows

Non-Newtonian fluids are those in which the relationship between stress and strain rate is not linear. In these cases, no single definition exists for viscosity since it depends on the strain rate. Essentially, two types of viscosity can be defined: apparent viscosity (μ_A) and local (or tangent) viscosity (μ_L). They are usually determined from a rheogram, which represents the relationship between shear stress and strain rate, obtained experimentally for each specific fluid. The measurement results in a relationship between τ_{ij} and $\dot{\gamma}_{ij}$, or, when using a rotational rheometer, $\tau_{r\theta}$ and $\dot{\gamma}_{r\theta}$. For simplicity, in the following discussion, stress and strain rate will be denoted simply as τ and $\dot{\gamma}$. The apparent and local viscosities are defined as follows:

$$\mu_A = \frac{\tau}{\dot{\gamma}}, \qquad \mu_L = \frac{d\tau}{d\dot{\gamma}} \tag{86}$$

For an Ostwald–de Waele fluid, defined as $\tau = K\,\dot{\gamma}^n$, the two viscosities are $\mu_A = K\,\dot{\gamma}^{n-1}$ and $\mu_L = n\,K\,\dot{\gamma}^{n-1}$.

One way to define the Reynolds number is to express the shear stress on the wall of a pipe in terms of the friction factor ($\tau_0 = \frac{1}{8}\,\rho\, f\, U^2$) and ensure that the friction factor in the laminar regime in a pipe for a non-Newtonian fluid has the same form as that of a Newtonian fluid, that is, $f = 64/Re$. The equivalent Reynolds number Re_E is frequently written as $F\, Re_{eq}$, where F is a function that depends on the yield stress, τ_y. The way to obtain Re_{eq} is as follows: in a cylindrical pipe with a radius R_0, the discharge is given by $Q = \pi\, R_0^2\, U = 2\pi\, \int_0^{R_0} u\, r\, dr$. We can write the strain rate $\dot{\gamma}$ as a function of τ, $\dot{\gamma} = g(\tau)$. As $\tau/\tau_0 = r/R_0$ give the shear stress distribution, changing variables from u to τ, the equation for the discharge can be integrated in terms of the shear stress and relate the velocity with the shear stress on the wall τ_0. The following relation is obtained: $2U/D = -(1/\tau_0^3)\, \int_0^{\tau_0} \tau^2 g(\tau)\, d\tau$. Thus, a relation among U, τ_0, and the parameters of the specific rheological model can be used to determine $f = 8\tau_0/(\rho\, U^2)$ and obtain Re_{eq}. The results of applying of the aforementioned procedure to three common types of rheology are presented in Table 1.

Table 1. Equivalent Reynolds for common rheological models. $\xi = \tau_y/\tau_0$.

MODEL	F	Re_{eq}
Ostwald–de Waele $\tau = K\,\dot{\gamma}^n$	1	$8\left(\frac{n}{3n+1}\right)^n \frac{\rho\, U^2}{K\,(2U/D)^n}$
Bingham's plastic $\tau = \tau_y + \mu_B\,\dot{\gamma}$	$1 - \frac{4}{3}\,\xi + \frac{1}{3}\,\xi^4$	$\frac{\rho\, U\, D}{\mu_B}$
Herschel and Bulkley $\tau = \tau_y + K\,\dot{\gamma}^n$	$\left[(3n+1)^n(1-\xi)^{n+1}\right] \times \left[\frac{(1-\xi)^2}{3n+1} + \frac{2\xi(1-\xi)}{2n+1} + \frac{\xi^2}{n+1}\right]^n$	$8\left(\frac{n}{3n+1}\right)^n \frac{\rho\, U^2}{K\,(2U/D)^n}$

The equivalent Reynolds number for Ostwald–de Waele or Herschel and Bukley models is often referred as "Metzner and Reed Reynolds number", denoted as Re_{MR}, was introduced in the analysis performed by those authors in their study on power law fluid flows [111]. As with Newtonian fluids, when inertia overcomes the stabilizing effect of viscous forces, non-Newtonian fluid flow becomes unstable and turbulence arises. Metzner and Reed studied the limits of laminar, transitional, and turbulent-flow regions for power-law (Ostwald–de Waele) fluids and their associated friction factors. Hanks determined the criterion for the laminar-turbulent transition in the flow of fluids presenting a yield stress, modeled as Bingham plastics or Powell–Eyring fluids. In these cases, the critical Reynolds number depends on the Hedström number, defined for a Bingham plastic fluid as $He = \rho \tau_y D^2 / \mu_B^2$. For the Powell–Eyring model (given by $\tau = \mu_\infty \dot{\gamma} + B^{-1} \sinh^{-1}(\dot{\gamma}/A)$) the Reynolds and Hedström numbers are defined as $Re = \rho\ U\ D/\mu_\infty$ and $He = \rho D^2 / (\mu_B^2 B)$, respectively [112,113]. A review by Yusufi et al. on modeling the flow of Herschel–Bulkley fluids in pipes [114] presents an analysis of the laminar-turbulent transition in these fluids. They use a Reynolds number defined as $Re_w = \rho U D / \mu_w$, where $\mu_w = K^{1/n} \tau_0 / (\tau_0 - \tau_y)^{1/n}$. Haldenwang et al. studied the impact of different Reynolds number definitions on the pressure drop prediction in sludge flows in pipes [115]. The laminar-turbulent transition of non-Newtonian fluids in open channel flows has been found to depend on both the Reynolds and Froude numbers [116].

3.11. Reynolds Number in Plasma and Magnetohydrodynamics

A plasma is an ionized gas in which a significant fraction of its atoms or molecules have lost or gained electrons, forming ions and free electrons. It has unique properties, such as conducting electricity and responding to electromagnetic fields. The equations of Vlasov–Maxwell govern plasma. The Vlasov equation describes the time evolution of the distribution function of a collisionless plasma consisting of charged particles with long-range interaction. Maxwell's equations are a set of equations that describe entirely electromagnetic phenomena. On the other hand, magnetohydrodynamics (MHD) is a macroscopic approximation in which the plasma is treated as a continuous conducting fluid without resolving the individual dynamics of electrons and ions. Assuming quasi-neutral equilibrium and frequent collisions, MHD is described by the following:

(i) The Navier–Stokes equations, including the electromagnetic Lorentz force $\vec{J} \times \vec{B}$, where $\vec{J}$ is the current density and $\vec{B}$ the magnetic field

$$\rho \left(\frac{\partial \vec{V}}{\partial t} + \vec{V} \cdot \nabla \vec{V} \right) = -\nabla \hat{p} + \mu \nabla^2 \vec{V} + \vec{J} \times \vec{B} \tag{87}$$

(ii) The following induction equation

$$\frac{\partial \vec{B}}{\partial t} = \nabla \times \nabla \times \vec{B} + \frac{1}{\sigma \mu_0} \nabla^2 \vec{B} \tag{88}$$

where σ is the electrical conductivity, and μ_0 is the permeability of free space. The product $\sigma\ \mu_0$ has dimensions of (lengh^2/time)$^{-1}$. Thus, the magnetic diffusivity or magnetic viscosity is defined as $\nu_m = 1/(\sigma\ \mu_0)$.

(iii) The following Maxwell's equations simplified for MHD

$$\nabla \cdot \vec{B} = 0, \tag{89a}$$

$$\nabla \times \vec{B} = \mu_0\ \vec{J} \tag{89b}$$

Making dimensionless the equation of momentum and induction with a characteristic fluid density ρ_0, magnetic field value, B_0, and using Equation (89b) ($J_0 = B_0/(\mu_0\,\mathcal{L})$), Equation (87) becomes

$$\rho^* \left(\frac{\partial \vec{V}^*}{\partial t^*} + \vec{V}^* \cdot \nabla^* \vec{V}^* \right) = -\nabla^* \hat{p}^* + \frac{\mu}{\rho_0\,\mathcal{U}\,\mathcal{L}} \nabla^{*2} \vec{V}^* + \frac{B_0^2}{\mu_0\,\rho_0\,\mathcal{U}^2} \vec{J}^* \times \vec{B}^* \tag{90}$$

Another dimensionless parameter appears in addition to the Reynolds number. It is named Alfvén number, $Al = B_0^2/(\mu_0\rho_0\mathcal{U}^2)$.

(iv) Similarly, the induction equation becomes

$$\frac{\partial \vec{B}^*}{\partial t^*} = \nabla^* \times \nabla^* \times \vec{B}^* + \frac{\nu_m}{\mathcal{U}\,\mathcal{L}} \nabla^{*2} \vec{B}^* \tag{91}$$

The new dimensionless parameter in front of $\nabla^{*2}\vec{B}^*$ is the inverse of a magnetic Reynolds number [117,118]:

$$Re_m = \frac{\mathcal{U}\,\mathcal{L}}{\nu_m} = \mathcal{U}\,\mathcal{L}\,\sigma\,\mu_0 \tag{92}$$

The magnetic Reynolds number is the ratio of advection of the magnetic field to magnetic diffusion. It compares the rate of macroscopic motion of the fluid with the rate of magnetic diffusion. If $Re_m >> 1$, magnetic induction dominates, and the magnetic field is "frozen" into the fluid, i.e., the hydrodynamic and electromagnetic fields are coupled rigidly. If $Re_m << 1$, magnetic diffusion dominates, and the magnetic field decouples from the fluid motion. In this case, the electromagnetic field's diffusion rate through the hydrodynamic field is appreciable and there is leakage flow or free decay [118,119].

The magnetic Reynolds number can be interpreted in terms of time scales. A characteristic time for free decay of the magnetic field is $\mathcal{T}_{mDecay} \sim \mathcal{L}^2/\nu_m$. A characteristic time of the flow field is $\mathcal{T} = \mathcal{L}/\mathcal{U}$. Thus, $Re_m = \mathcal{T}_{mDecay}/\mathcal{T}$.

In plasmas, the magnetic Reynolds number (Re_m) is often more important than the hydrodynamic Reynolds number (Re) because plasmas are highly conducting, and magnetic effects dominate. Examples are the solar corona and interstellar medium. Depending on the specific problem, the hydrodynamic Reynolds number (Re) and the magnetic Reynolds number (Re_m) are important in MHD. Examples of MHD are found in industrial applications (liquid metals) and geophysics (Earth core).

The matter of the initial stages of the universe is a conducting plasma, and the need of a primordial turbulence in the formation of galaxies and stars has been emphasized since the works of von Weizsäcker [120] and Gamow [121]. In the early universe, in addition to the Re_m, the so-called kinetic Reynolds number Re_{kin} is also important [122], where the coefficient of thermal diffusivity replaces the kinematic viscosity, ν_{th}, as they are directly related according to the kinetic theory of gases [123]. Thermal diffusivity can be estimated as $\nu_{th} \sim (\alpha_{em}T)^{-1}$, where α_{em} is the electromagnetic (or fine-structure) coupling constant [124], and T is the temperature. In both Re_m and Re_{kin}, the velocity scale $\mathcal{U}$ is a function of the electron and ion velocities, and the length scale $\mathcal{L}$ is considered to be on the order of the Hubble radius (the distance at which the universe's expansion rate equals the speed of light) at the corresponding epoch. As the early universe evolves, both the kinetic and magnetic Reynolds numbers decrease, with Re_m being approximately twenty orders of magnitude larger than Re_{kin} which is, in turn, smaller than one.

3.12. Reynolds Number in Relativistic Fluid Mechanics

Navier–Stokes theory is well established for describing dissipative fluid dynamics in non-relativistic settings. However, extending it to relativistic systems is problematic because the naive relativistic generalization leads to acausal and unstable behavior. This issue arises because dissipative currents in Navier–Stokes theory respond instantaneously to dissipative forces. This behavior is unphysical, according to this theory. Efforts have been made to address this by introducing a finite response time for dissipative currents, even in non-relativistic fluid dynamics. Israel and Stewart were among the first to address this issue in their 1979 article by developing a second-order theory of relativistic dissipative fluid dynamics [125]. Their approach introduces a finite relaxation time for dissipative currents, ensuring they do not respond instantaneously to dissipative forces. The term "second-order" includes terms proportional to the square of the Knudsen number. These terms are necessary to describe this relaxation process [126].

The fluid-dynamical limit is typically assessed using the Knudsen number ($Kn = \ell_{micro}/\mathcal{L}_{macro}$), defined as the ratio of microscopic to macroscopic length or time scales. Macroscopic scales are estimated from fluid-dynamical gradients, while microscopic scales correspond to the mean-free path or collision time. When $Kn \ll 1$, a clear separation exists between microscopic and macroscopic scales. This separation allows the system's dynamics to be described using only a few macroscopic fields as microscopic details can be effectively integrated out [127].

Fluid dynamics is expected to be valid near local thermal equilibrium. Deviations from equilibrium can be quantified using macroscopic variables by defining a set of ratios of dissipative quantities to equilibrium pressure or density, which serve as generalized forms of the inverse Reynolds number [127].

Denicol et al. proposed three inverse Reynolds numbers in their 2012 paper. By 2024, this set had expanded to six [126]. Following the notation of Denicol et al. [127], they are

$$Re_{\Pi}^{-1} = \frac{|\Pi|}{P_0}, \quad Re_{n}^{-1} = \frac{|n^{\mu}|}{n_0}, \quad Re_{\pi}^{-1} = \frac{|\pi^{\mu\nu}|}{P_0} \tag{93}$$

where Π is the bulk viscous pressure, n^{μ} is the particle-diffusion current, and $\pi^{\mu\nu}$ is the shear stress tensor. Additionally, P_0 and n_0 represent the pressure and particle density in local equilibrium, respectively. The term "inverse Reynolds number" comes from Re_{π}^{-1} as it relates to the usual Reynolds number under non-relativistic conditions.

4. Reynolds Number Presence: From Nanostructures to the Cosmic Horizon

As shown in the previous section, when there is fluid flow or the movement of a body within a fluid, it is possible to find Reynolds numbers associated with the phenomenon across virtually the entire range of measurable length scales. These scales can range from the order of a micrometer (as in the case of the size of *E. coli* [41] or, if the characteristic length is considered to be the diameter of the flagellum, on the order of 100 Å (~0.01 μm) [41]) to the size of the observable universe or cosmic horizon [122]. Similarly, the Reynolds number spans a range from approximately 10^{-5} for the movement of bacteria [41] to 10^{8} for a typhoon [128]. For stars, using Equation (48), Canuto and Christensen-Dalsgaard estimate that $Re > 10^{10}$ [129]. Since its conception and definition in the 19th century, the meaning of the Reynolds number has remained unchanged: it is the ratio of inertial forces to viscous forces in a flow, regardless of how it is derived. The velocity or length scales must be chosen appropriately depending on the application. Similarly, viscosity—originally considered a fluid property—can sometimes also depend on the flow, as seen in non-Newtonian fluids. In contrast, in others, such as plasmas, it remains primarily a fluid property.

By analogy, dimensionless parameters have been named "Reynolds numbers" in various fields, such as the magnetic Reynolds number defined by Elsasser in 1954 [117], inverse Reynolds numbers in relativistic fluid mechanics [126], the kinetic Reynolds number in cosmology [122], and Reynolds numbers in superfluids [78,79]. In all cases, viscosity is replaced by a coefficient that represents a resistive effect on the flow or arises from a term in the equation of motion that plays a role analogous to the stress tensor gradient in the Navier–Stokes equation.

As has been shown, the Reynolds number originated as a criterion for determining the transition of flow from laminar to turbulent [15,17] and as a criterion of similarity [10,18–21]. It has also been extensively used in the search for analytical solutions to the Navier–Stokes equations as it constitutes a control parameter for the equations, defining two extreme cases: $Re \ll 1$ (Stokes regime) and $Re \to \infty$ (potential flow). However, the Reynolds number alone is insufficient to represent all possible flow dynamics or maintain similarity in physical models. For example, the following are two situations where this occurs: laminar flow at very low Reynolds numbers and the scaling of flow dynamics in wind tunnels to full-scale flow.

4.1. When the Number of Reynolds Is Not Enough: Flows at Milli, Micro, and Nano Length Scales

Millifluidics refers to a field that deals with fluids in channels with dimensions on the scale of 1 mm to 10 mm. It is not simply a flow or a flow regime but rather a field of study and application focused on manipulating and understanding fluid behavior at this specific scale. The flows within millifluidic systems are typically laminar due to their low Reynolds number, which is usually in the range of $Re = 10\text{–}10^2$. When the characteristic dimensions of the channels are on the order of 100 nm to 1 mm, the associated Reynolds numbers span from 10^{-6} to 10. Nanofluidics, on the other hand, involves length scales smaller than 100 nm, with $Re < 10^{-6}$ [130].

Fluid bulk properties dominate millifluidics and microfluidics. In addition to inertia and viscous force, those derived from surface and interfacial tension are relevant [131]. Capillary force can drive fluids forward or lift them against gravity when confined within a narrow tube with a hydrophilic surface. This phenomenon enables passive fluid pumping in microfluidic devices [130]. Here, in addition to the Reynolds number, other dimensionless parameters are relevant as indicated below: Capillary number, corresponding to the ratio between viscous and surface tension forces, given by $Ca = \mu\mathcal{U}/\sigma$ (σ is the surface tension), and Péclet number, relating convection to diffusion, $Pe = \mathcal{U}\mathcal{L}/D$. D is the molecular diffusion coefficient of a specific component, and Pe determines the channel length required for that component to diffuse across the channel width [46]. When polymers are present in a solvent, two other important dimensionless parameters arise, both associated with the polymer's relaxation time T_R, which stem from the viscoelastic behavior of the solution. The first is the Deborah number, which measures the polymer's relaxation time relative to the deformation process' characteristic time, or the flow time scale, T_P. It is defined as $De = T_R/T_P$ [46,132]. The value of T_P depends on the process. For example, in a flow with a characteristic velocity scale U_0 in a channel that contracts over a length L_0, the geometric time scale is given by $T_P = L_0/U_0$, which is the time required for a polymer to traverse the channel. Similarly, in an oscillatory flow with frequency ω, the oscillation time scale is $T_P = 1/\omega$. The second dimensionless number arising in viscoelastic fluids is the Weissenberg number, given by $Wi = T_R\dot{\gamma}$, where $\dot{\gamma}$ is the shear rate. It can be shown that Wi represents the ratio of elastic to viscous forces [132,133].

In nanofluidics, channel dimensions are smaller than 100 nm, comparable to several molecules' size. As a result, intermolecular forces such as electrostatic forces, van der Waals attractions, hydrogen bonding, and steric repulsion become significant [130]. For instance, when a surface is charged, it attracts counterions (ions with a charge opposite to that of the charged surface, they are attracted to the surface due to electrostatic forces) and repels coions (ions with a charge of the same sign as the charged surface, they are repelled from the surface due to electrostatic forces) in the fluid, creating a characteristic length known as the Debye length (λ_D). Within this length, the electric potential decreases exponentially to zero within this length as the distance from the charged surface increases. The Debye length typically ranges from 1 to 100 nm. If a is the characteristic size of the particle, the Debye number is defined as $Db = a/\lambda_D$. This parameter indicates whether electrostatic interactions are long range (Db << 1) or short range (Db >> 1). In millifluidics or microfluidics, where channel dimensions are much larger than the Debye length, surface charges are generally shielded by counterions and can be considered negligible. However, in nanofluidics, where channel dimensions are comparable to or smaller than the Debye length, the electric field generated by the charged surface can penetrate the nanochannels and influence phenomena such as ion movement.

Overall, interactions in nanofluidics are highly complex and involve a combination of various intermolecular forces like those resulting in van der Waals attraction, hydrogen bonding, steric repulsion, electrostatic force, etc. The van der Waals force is related to the Hamaker constant (A), a material-dependent constant that quantifies the strength of van der Waals forces between particles, surfaces, or molecules. A is typically on the order of 10^{-20} to 10^{-19} joules for many common materials [134]. The Hamaker number is defined as $Ha = A/(k_B T)$, where k_B is the Boltzmann constant and T the temperature.

The Derjaguin number arises from the comparison between the van der Waals force to viscous force in a system: $De = F_{vdW}/(6\pi\mu RU)$, where F_{vdW} is the van der Waals force, μ the fluid dynamic viscosity, R is the particle effective radius, and U is the relative velocity. The force F_{vdW} depends on the geometry of the particles and their separation (h). For two spheres with radii R_1 and R_2, $F_{vdW} = AR_{eff}/(6h^2)$, where $R_{eff} = R_1R_2/(R_1 + R_2)$. For two flat parallel surfaces of area S, $F_{vdW} = AS/(6\pi h^3)$ [135,136].

There is not a specific name for the dimensionless parameter involving hydrogen bonding. However, the ratio of hydrogen bond energy (E_{HB}) to thermal energy: $Hy = E_{HB}/(k_B T)$ is commonly used in studies of hydrogen bonding to determine whether thermal fluctuations can disrupt hydrogen bonds. This ratio is often discussed in the context of molecular interactions.

Coulomb's law governs electrostatic interactions. A standard dimensionless parameter for electrostatic forces is the Debye–Hückel number or the ratio of electrostatic energy to thermal energy: $DH = q^2/(4\pi\epsilon k_B T)$, where q is the charge of the particle, ϵ is the permittivity of the medium, and $k_B T$ is the thermal energy. This parameter determines the strength of electrostatic interactions relative to thermal fluctuations.

As can be seen, for flows with $Re \ll 1$, the Reynolds number is insufficient to characterize the flow adequately, and other relevant phenomena begin to emerge, described by sets of new dimensionless parameters. Notably, the small dimensions that suppress inertial nonlinearity bring attention to other physical phenomena that are less familiar at the macroscale [41,46]. The article by Squires and Quake [46] provides a comprehensive list of applications within the range of very low Reynolds numbers for Newtonian fluid flows (specifically, laminar regime flows), which are now commonly referred to as milli-, micro-, and nanofluidics.

It should also be noted that the laminar flow condition for $Re \ll 1$ does not apply to all types of rheologies. A surprising phenomenon occurs in flows with low Reynolds numbers of viscoelastic fluids (typically polymeric solutions) as they can exhibit turbulence-like behavior despite their negligible inertia. This behavior is similar to the turbulence found in Newtonian fluid flows that exceed the critical Reynolds number [137]. This phenomenon, known as "elastic turbulence", arises from the elastic properties of the solution, which stem from the elastic characteristics of viscoelastic fluids, even at very low Reynolds numbers ($Re \ll 1$). While traditional turbulence is driven by inertia, elastic turbulence arises from the elastic stresses caused by the stretching and deformation of polymers in the flow. The Weissenberg number characterizes the onset of elastic turbulence. Schiamberg et al. reported measurements in torsional, parallel-plate flow of a polymer solution that showed elastic turbulence in experiments with Reynolds number $Re = R^2\Omega/\nu$ (Equation (30)) in the range 0.023–0.59 [138]. Weissenberg numbers were in the range $Wi = 0.30$–4.97. It must be emphasized that elastic turbulence is not the same as hydrodynamic turbulence as the destabilizing mechanisms are fundamentally different.

Even though the Reynolds number is a crucial dimensionless parameter for characterizing fluid flow and enabling scaling between different systems, it has inherent limitations. Scaling issues are not limited to flow conditions where $Re \ll 1$, and it is not an exaggeration to say that they occur across the entire range of Reynolds numbers. Preserving Re alone does not guarantee perfect similarity as it only accounts for the ratio of inertial to viscous forces. Other dimensionless numbers, such as the Froude number (for gravitational effects), the Capillary or Weber numbers (for surface tension effects), the Weissenberg number (for viscoelastic effects), and the Mach number (for compressibility effects), may also be significant depending on the flow conditions. Neglecting these other parameters can lead to inaccurate predictions, especially when dealing with complex fluids, high-speed flows, or systems where surface tension or gravitational forces play a significant role. Therefore, while matching Reynolds number is a valuable starting point, a comprehensive scaling analysis requires consideration of all relevant dimensionless groups to capture the underlying physics of the flow accurately.

4.2. When the Number of Reynolds Is Not Enough: Limitations of the Reynolds Number in Flow Scaling

While the concept of the Reynolds number has not changed since its origin, the range of values it takes in practical applications has significantly expanded, as shown in the review by Saldana et al. [28]. These authors present a wide array of technical applications in which the range of Reynolds number values continues to grow. These applications include the design of aircraft, automobiles, ships, and submarines, as well as industrial processes such as heat exchangers, mixers, and more—alongside the classical applications involving flow in pipes and channels. As noted in Section 2, the Reynolds number is fundamental for the proper scaling of flows with friction and represents a key advancement in the development of studies using reduced-scale models. Wind tunnel experiments have been crucial to the advancement of aviation. It is worth noting that wind tunnels were already being used for studies on aerial locomotion long before the scaling laws were fully established. Although the first wind tunnel was built by Francis H. Wenham in 1871 to study the lift and drag of flat plates, and in 1901 the Wright brothers constructed a wind tunnel to investigate the lift of wing models (their first flight took place in 1903) [139], it was not until 1908, with Prandtl's wind tunnel in Göttingen, that these facilities began to be fully exploited. It was then that the application of Reynolds similarity relations in experiments and the analysis of results became standard practice [18].

It is interesting to note that, in the study of the aerodynamic characteristics of vehicles, the current challenges—such as shape optimization to reduce drag, stability, control effectiveness, leading-edge separation, pitching, and others ([28] and references therein)—are virtually the same as those outlined by Prandtl in his presentation at the 50th General Assembly of the Association of German Engineers [18]. Prandtl emphasized the importance of stability, controllability, and accurate scaling in aerodynamic testing, highlighting the need for Reynolds number similarity between models and full-scale aircraft. He pointed out the scaling limitations due to the inverse relationship between velocity and length, which can lead to impractical test conditions. While using water as a working fluid offers lower kinematic viscosity, its high density makes it unsuitable due to excessive power demands and forces on the tested model. Pressurized air emerged as a viable alternative as its dynamic viscosity remains nearly constant with pressure while density increases linearly. In 1926, Munk and Miller reported the pressurized wind tunnel developed by NACA (now NASA), capable of reaching 20 atmospheres, achieving a Reynolds number of 2.7×10^6 for a Fokker D.VII model, compared to 1.35×10^5 at atmospheric pressure [140].

It is common to express the Reynolds number of a wind tunnel as Re per unit length ($Re/\mathcal{L}$) since the characteristic length corresponds to that of the tested object (for example, for an aircraft wing, it is usually the chord length). The maximum $Re/\mathcal{L}$ values reached by the early wind tunnels were 6.5×10^5 m^{-1} (Göttingen, 1908), 1.4×10^6 m^{-1} (Eiffel, 1909), 2.6×10^6 m^{-1} (Eiffel, 1914), 3.6×10^6 m^{-1} (Göttingen, 1917), etc., (values calculated from Reference [141]). A significant leap in the maximum $Re/\mathcal{L}$ value came with the pressurized wind tunnel of the NACA, which increased the values of earlier wind tunnels by an order of magnitude, reaching 3×10^7 m^{-1} [140].

As Prandtl pointed out [18], higher flow velocities make air compressibility increasingly important, requiring that pressure waves be considered in the scaling process. These waves must be scaled to preserve the Mach number between the model and the full-scale prototype. The Mach number is defined as $Ma = U/c$, where U is the flow velocity, and c is the speed of sound in air. As the Reynolds number increases, the Mach number also increases. The flow is considered subsonic for $Ma < 0.8$, but mild compressibility corrections are required when Ma approaches 0.8 locally. When $0.8 < Ma < 1.3$, the flow is transonic. It should be noted that even if an aircraft is flying, for example, at $Ma = 0.85$, the flow over the wings can exceed $Ma = 1$, and shock waves may be generated. For $1.2 < Ma < 3$, the flow is called supersonic. For $3 < Ma < 5$, the flow is high supersonic. In this range, the temperature at stagnation points can exceed the melting point of certain metals, and near $Ma = 5$, air temperature can reach combustion levels. The flow is called hypersonic for $Ma > 5$, and it is in this regime that spacecrafts reentering the atmosphere operate [142].

Pressurized wind tunnels must be used to physically model flows in the transonic or supersonic regime. However, due to the high pressures required to achieve the working velocities, there is an increase in air temperature, which affects the flow properties, dynamic similarity, and the integrity of materials. To address this, cryogenic wind tunnels have been designed and built. These tunnels operate at ambient pressure but use air (or another gas, such as gaseous nitrogen) at low temperatures to achieve high Reynolds numbers, compatible with the condition imposed by the Mach number [143]. This approach decouples aerodynamic effects due to Reynolds number, Mach number, and aeroelasticity. The largest high-pressure, cryogenic, transonic wind tunnels are the National Transonic Facility from NASA ($0.1 < Ma < 1.2$, $Re/\mathcal{L}$ up to 4.4×10^8 m^{-1}, working pressures between 1 to 8 atm, temperature range: -157 °C to 66 °C) and the European Transonic Wind Tunnel ($0.13 < Ma < 1.3$, $Re/\mathcal{L}$ up to 2.3×10^8 m^{-1}, working pressures between 1.1 to 4.4 atm, temperature range -160 °C to 40 °C) [144,145].

Low-temperature wind tunnels are not restricted to aerodynamic studies as they have also been used to study the effects of the Reynolds number on high-speed trains [146].

The requirements that resulted from the development of aviation after World War II, as well as in aerospace engineering, have demanded the need to design and build supersonic and hypersonic wind tunnels. In these, the Mach number is large enough that the compression and temperature increase in the air (or gas used) must be taken into account, along with the resulting changes in density and viscosity, complicating scaling [147,148]. In high-speed flows, the transition from laminar flow to turbulent flow is important because turbulent boundary layers significantly increase the heat transfer to the surface, which is a major concern for thermal protection systems [149]. Hypersonic air plasma wind tunnels, such as the SCIROCCO arc jet facility (Italian Aerospace Research Center), are designed to simulate the extreme conditions experienced by spacecraft during atmospheric re-entry. These facilities generate high-enthalpy flows, which include high temperatures and pressures, to replicate the thermal and aerodynamic loads on spacecraft materials and components. The gas used in the wind tunnel must faithfully replicate the atmospheric composition relevant to the mission. For re-entry to Earth, this usually means air (a mixture of nitrogen and oxygen). For other planetary missions, such as Mars, the gas composition may need to simulate atmospheres rich in CO_2. The Reynolds numbers achieved in arc-jet facilities are often much lower than those encountered during actual re-entry. These high Reynolds numbers are due to the plasma flow's low density and high viscosity [150–152].

High Reynolds water tunnels are used frequently in cavitation studies like those from the U.S. Navy Large Cavitation Channel (LCC) or the Garfield Thomas Water Tunnel. The LCC is the largest water tunnel in the world reaching a Reynolds number based on the momentum thickness θ ($Re_\theta = \mathcal{U}\theta/\nu$) of about 10^5 (corresponding to a Reynolds number based on the boundary layer thickness (δ) of $Re_\delta = \mathcal{U}\delta/\nu \sim 10^6$) for turbulent boundary layers. The withdrawal of LCC is its high operating costs. Less expensive to operate are water tunnels with lower maximum Reynolds numbers, like that of Oklahoma State University, with maximum Reynolds number of $Re_\theta \sim 10^4$ ($Re_\delta \sim 10^5$) [153,154].

In conclusion, while the Reynolds number remains a cornerstone of fluid dynamics and scaling analysis, its limitations necessitate the consideration of additional dimensionless parameters to fully capture the complexities of fluid behavior. The evolution of experimental facilities, from early wind tunnels to modern cryogenic and hypersonic wind tunnels, reflects the ongoing effort to address these challenges. These advancements have enabled more accurate modeling of high-speed, compressible, and extreme flow conditions, critical for aerospace engineering applications to cavitation studies. However, achieving perfect scaling remains an elusive goal due to the interplay of various forces and the practical constraints of experimental setups. Developing specialized facilities, such as high-pressure and cryogenic wind tunnels, has provided valuable tools for decoupling the effects of Reynolds number, Mach number, and aeroelasticity.

5. Discussion and Conclusions

The Reynolds number, introduced initially to describe the transition between laminar and turbulent flows and as a scaling parameter, has evolved into a fundamental dimensionless parameter applicable across various disciplines. This paper has demonstrated its universal relevance by highlighting its use in various physical systems, including classical fluid mechanics, biofluid dynamics, nano- and microfluidics, boundary layer theory, rotating flows, multiphase systems, porous media, and even relativistic fluid mechanics. The adaptability of the Reynolds number underscores its role as a unifying principle in the study of fluid motion and transport phenomena.

One key takeaway from this study is that while the traditional definition of the Reynolds number is based on the ratio of inertial to viscous forces, its interpretation varies significantly depending on the specific system under consideration. For example, in microfluidics, where inertial effects are negligible, the Reynolds number primarily confirms the dominance of viscous forces. In turbulence studies, it governs the energy cascade process from large to small eddies, while in biofluid mechanics, it dictates locomotion strategies at different scales, from bacteria to whales. In relativistic fluid dynamics, inverse

Reynolds numbers are used to quantify deviations from equilibrium, demonstrating that even in extreme conditions, the concept remains indispensable.

Furthermore, this study has shown that modifications and generalizations of the Reynolds number naturally arise when applied to complex systems. For non-Newtonian fluids, effective viscosity-dependent Reynolds numbers better describe flow behavior. In rotating flows, alternative formulations account for centrifugal effects, in magnetohydrodynamics, the magnetic Reynolds number governs the interaction between fluid motion and magnetic fields, and the kinetic Reynolds number appears as an important parameter in the initial stages of the Universe. These variations highlight the flexibility of the Reynolds number as an analytical tool capable of adapting to different governing forces and constraints.

A notable observation is that different scientific communities sometimes define Reynolds-like numbers differently based on their needs. While this flexibility is beneficial, it also underscores the importance of clearly specifying definitions when comparing results across fields. Standardizing certain modifications could improve interdisciplinary communication and facilitate a broader understanding of flow phenomena.

The Reynolds number remains fundamentally defined as the ratio of inertial to viscous forces, a concept without significant changes in its physical interpretation. Its core meaning—capturing the balance between these forces—continues to be universally applicable across fluid mechanics. Attempts to redefine or extend Re beyond this fundamental ratio often result in parameters that no longer represent the original concept. Thus, while the physical meaning of Re has remained consistent, its application has evolved to address increasingly complex systems.

By analogy, "Reynolds numbers" have been defined in different fields, corresponding to dimensionless parameters that compare the effect of inertia with some resistive force to motion. The analogous Reynolds number has emerged as a critical ability for describing the interaction between fluid motion and magnetic fields, highlighting the adaptability of Re to new governing forces. Similarly, inverse Reynolds numbers are used in relativistic fluid dynamics to quantify deviations from equilibrium in extreme conditions, such as those in astrophysical or high-energy systems. These developments demonstrate that while the concept of Re has been adapted to analogous situations, it has expanded into new and challenging domains, offering valuable insights into diverse physical phenomena.

In conclusion, the Reynolds number remains a cornerstone of fluid mechanics and beyond. Its widespread applicability across multiple disciplines demonstrates its fundamental importance in understanding and predicting flow behavior in diverse physical systems. Future research may continue to refine and expand its use, spanning from the smallest to the largest scales of the universe. The Reynolds number solidifies its status as an essential tool for scientific and engineering advancements by continually adapting to new challenges.

Author Contributions: Conceptualization: A.T.; investigation: A.T. and Y.N.; writing—original draft preparation: A.T.; writing—review and editing: Y.N. In general, both authors contributed to the general development of the document. All authors have read and agreed to the published version of the manuscript.

Funding: This research received no external funding.

Data Availability Statement: No new data were generated for this article.

Acknowledgments: The authors acknowledge ANID Project AFB230001: Advanced Mining Technology Center.

Conflicts of Interest: The authors declare no conflicts of interest.

Abbreviations

The following abbreviations are used in this manuscript:

LCC	U.S. Navy Large Cavitation Channel
MHD	Magnetohydrodynamics
NACA	U.S. National Advisory Committee for Aeronautics (1915–1958)
NASA	U.S. National Aeronautics and Space Administration (is the successor to NACA)

References

1. *Death of Prof. Reynolds*; Supplement; Manchester Courier: Manchester, UK, 1912; p. 8.
2. *Funeral of a Distinguished Scientist*; The West Somerset Free Press: Taunton, UK, 1912; p. 3.
3. Reynolds, O. On the Sub-Mechanics of the Universe, communicated to the Royal Society on February 3, 1902, and read on February 13, 1902, for publication in the Philosophical Transactions. *Proc. R. Soc. Lond.* **1902**, *69*, 425–433.
4. Reynolds, O. *The Sub-Mechanics of the Universe*; The University Press: Cambridge, UK, 1903.
5. Michelson, A.A.; Morley, E.W. On the Relative Motion of the Earth and the Luminiferous Ether. *Am. J. Sci.* **1887**, *34*, 333–345. [CrossRef]
6. Einstein, A. Zur Elektrodynamik bewegter Körper. *Ann. Phys.* **1905**, *17*, 891–921. [CrossRef]
7. Rott, N. Note on the history of the Reynolds number. *Annu. Rev. Fluid Mech.* **1990**, *22*, 1–11. [CrossRef]
8. Sommerfeld, A. Ein Beitrag zur hydrodynamischen Erklaerung der turbulenten Fluessigkeitsbewegüngen. In Proceedings of the Atti del IV Congresso Internazionale dei Matematici, Rome, Italy, 6–11 April 1908; Tipografia della R. Accademia dei Lincei: Rome, Italy, 1909; Volume III, pp. 116–124.
9. Blasius, H. Das Aehnlichkeitsgesetz bei Reibungsvorgängen. *Z. Des Vereinesdeutscher Ingenieure. (VDI-Z)* **1912**, *56*, 639–643.
10. Blasius, H. Das Aehnlichkeitsgesetz bei Reibungsvorgängen in Flüssigkeiten. In *Mitteilungen Über Forschungsarbeiten Auf Dem Gebiete Des Ingenieurwesens*; Springer: Berlin/Heidelberg, Germany, 1913; pp. 1–41. [CrossRef]
11. Prandtl, L. Eine Beziehung zwischen Wärmeaustausch und Strömungswiderstand der Flüssigkeiten. *Phys. Z.* **1910**, *11*, 1072–1078.
12. Rayleigh, L. On the question of the stability of the flow of fluids. *Philos. Mag. J. Sci.* **1892**, *34*, 59–70. [CrossRef]
13. Rayleigh, L. Sur la résistance des sphères dans l'air en mouvement. *Comptes Rendus De l'Acad. Des Sci.* **1913**, *154*, 109–110.
14. Stokes, G.G. On the Effect of the Internal Friction of Fluids on the Motion of Pendulums. In *Transactions of the Cambridge Philosophical Society*; The Pitt Press: Cambridge, UK, 1851; Volume IX, Part II; pp. 1–99. (Read on December 9, 1850).
15. Reynolds, O. An Experimental Investigation of the Circumstances Which Determine Whether the Motion of Water Shall Be Direct or Sinuous, and of the Law of Resistance in Parallel Channels. *Phil. Trans. R. Soc. Lond.* **1883**, *174*, 935–982.
16. Popiel, C.O.; Wojtkowiak, J. Simple Formulas for Thermophysical Properties of Liquid Water for Heat Transfer Calculations (from 0 °C to 150 °C). *Heat Transf. Eng.* **1988**, *19*, 87–101. [CrossRef]
17. Reynolds, O. On the Dynamical Theory of Incompressible Viscous Fluids and the Determination of the Criterion. *Phil. Trans. R. Soc. Lond. A* **1895**, *186*, 123–164.
18. Prandtl, L. Die Bedeutung von Modellversuchen für die Luftschiffahrt und Flugtechnik und die Einrichtungen für solche Versuche in Göttingen (The Importance of Model Experiments for Airship Navigation and Aviation, and the Facilities for Such Experiments in Göttingen). *Z. Des Vereines Dtsch. Ingenieure* **1909**, *53*, 1711–1719.
19. von Mises, R. *Elemente der Technischen Hydromechanik*; B. G. Teubner: Leipzig, Germany, 1914.
20. Forchheimer, P. *Hydraulik*; B. G. Teubner: Leipzig, Germany, 1914.
21. Weber, M. Die Grundlagen der Ähnlichkeitsmechanik und ihre Verwertung bei Modellversuchen, unter besonderer Berücksichtigung schiffbautechnischer Anwendungen. Schiffbautechnische Gesellschaft. In Proceedings of the XX Ordentliche Hauptversammlung, Berlin, Germany, 21–23 November 1918; Springer: Berlin/Heidelberg, Germany, 1918. [CrossRef]
22. Noether, F. Das Turbulenzproblem. *Z. Für Angew. Math. Und Mech.* **1921**, *1*, 125–138. [CrossRef]
23. Schiller, L. *Untersuchungen Über Laminare und Turbulente Strömung (Leipziger Habilitationsarbeit)*; Forschungsarbeiten auf dem Gebiete des Ingenieurwesens Herausgegeben vom Verein Deutscher Ingenieure; Meyer, D., Seyffert, M., Eds.; Verlag des Vereines Deutscher Ingenieure: Berlin, Germany, 1922.
24. Heisenberg, W. Über Stabilität und Turbulenz von Flüssigkeitsströmen. *Ann. Der Phys.* **1924**, *379*, 577–627. [CrossRef]
25. Jackson, J.D. Osborne Reynolds: Scientist, Engineer and Pioneer. *Proc. R. Soc. Lond. Ser. A Math. Phys. Sci.* **1995**, *451*, 49–86.
26. Jackson, J.D. The Reynolds number—A fundamental unifying parameter in fluid mechanics based on physical insight and experiment. In Proceedings of the Opening Lecture of the 4th European Thermal Sciences Conference, Birmingham, UK, 29–31 March 2004; National Exhibition Centre: Birmingham, UK, 2004.
27. Jackson, D.; Launder, B. Osborne Reynolds and the Publication of His Papers on Turbulent Flow. *Annu. Rev. Fluid Mech.* **2007**, *39*, 19–35. [CrossRef]

28. Saldana, M.; Gallegos, S.; Gálvez, E.; Castillo, J.; Salinas-Rodríguez, E.; Cerecedo-Sáenz, E.; Fernández-Ávila, J.; Navarra, A.; Toro, N. The Reynolds Number: A Journey from Its Origin to Modern Applications. *Fluids* **2024**, *9*, 299. [CrossRef]
29. Batchelor, G.K. *An Introduction to Fluid Dynamics*; Second Reprint of the First Cambridge Mathematica Library Edition; Cambridge University Press: New York, NY, USA, 2002.
30. Drazin, P.G.; Riley, N. *The Navier–Stokes Equations: A Classification of Flows and Exact Solutions*; Cambridge University Press: New York, NY, USA, 2007.
31. Van Dyke, M. *Perturbation Methods in Fluid Dynamics, Annotated ed.*; The Parabolic Press: Stanford, CA, USA, 1975.
32. Nayfeh, A.H. *Perturbation Methods, Wiley Classics Library ed.*; Wiley-VCH Verlag GmbH & Co. KGaA: Weinheim, Germany, 2004.
33. Lin, C.C. *The Theory of Hydrodynamic Stability*; Cambridge University Press: Cambridge, UK, 1955.
34. Chandrasekhar, S. *Hydrodynamic and Hydromagnetic Stability*; Dover Publications, Inc.: New York, NY, USA, 1981.
35. Drazin, P.; Reid, W. *Hydrodynamic Stability*, 2nd ed.; Cambridge University Press: Cambridge, UK, 2004.
36. Oertel, H. (Ed.) *Prandtl–Essentials of Fluid Mechanics*, 3rd ed.; Springer Science+Business: New York, NY, USA, 2010.
37. Goldsmith, H.L.; Shalak, R. Hemodynamics. *Annu. Rev. Fluid Mech.* **1975**, *7*, 213–247. [CrossRef]
38. Sutera, S.P.; Shalak, R. The History of Poiseuille's Law. *Annu. Rev. Fluid Mech.* **1993**, *25*, 1–19. [CrossRef]
39. Copley, A.L. Fluid Mechanics and Biorheology. *Clin. Hemorheol.* **1990**, *10*, 3–19. [CrossRef]
40. Lauga, E. *The Fluid Dynamics of Cell Motility*; Cambridge University Press: Singapore, 2020.
41. Purcell, E.M. Life at Low Reynolds number. *Am. J. Phys.* **1977**, *45*, 3–11. [CrossRef]
42. Smits, A.J. Undulatory and oscillatory swimming. *J. Fluid Mech.* **2019**, *874*, P1-1–P1-70. [CrossRef]
43. Taylor, G.I. Analysis of the swimming of microscopic organisms. *Proc. R. Soc. Lond. Ser. A. Math. Phys. Sci.* **1951**, *209*, 447–461.
44. Chisholm, N.G.; Legendre, D.; Lauga, E.; Khair, A.S. A squirmer across Reynolds numbers. *J. Fluid Mech.* **2016**, *796*, 233–256. [CrossRef]
45. Gurka, R.; Nafi, A.-S.; Weihs, D. On an adaptation of the Reynolds number, applicable to body-caudal-fin aquatic locomotion. *Front. Mar. Sci.* **2022**, *9*, 914214. [CrossRef]
46. Squires, T.M.; Quake, S.R. Microfluidics: Fluid physics at the nanoliter scale. *Rev. Mod. Phys.* **2005**, *77*, 977–1026. [CrossRef]
47. Gañán-Calvo, A.M.; Gordillo, J.M. Perfectly Monodisperse Microbubbling by Capillary Flow Focusing. *Phys. Rev. Lett.* **2001**, *87*, 274501. [CrossRef]
48. Lin, W.; Shi, R.; Lin, J. Distribution and Deposition of Cylindrical Nanoparticles in a Turbulent Pipe Flow. *Appl. Sci.* **2021**, *11*, 962. [CrossRef]
49. Tamburrino, A. From Navier to Stokes: Commemorating the Bicentenary of Navier's Equation on the Lay of Fluid Motion. *Fluids* **2024**, *9*, 15. [CrossRef]
50. Niavarani, A.; Priezjev, N.V. The effective slip length and vortex formation in laminar flow over a rough surface. *Phys. Fluids* **2009**, *21*, 052105. [CrossRef]
51. Prandtl, L. Über Flüssigkeitsbewegung bei sehr kleiner Reibung. Verhandlungen des III. In Proceedings of the Internationalen Mathematiker-Kongresses, Heidelberg, Germany, 8–13 August 1904.
52. Schlichting, H.; Gersten, K. *Boundary-Layer Theory*, 9th ed.; Springer: Berlin/Heidelberg, Germany, 2017.
53. White, F.M. *Viscous Fluid Flow*, 3rd ed.; McGraw-Hill: Singapore, 2006.
54. Cebeci, T.; Bradshaw, P. *Momentum Transfer in Boundary Layers*; Hemisphere Publishing Corporation: Washington, DC, USA, 1977.
55. Rosenhead, L. *Laminar Boundary Layers*; Clarendon Press: Oxford, UK, 1966.
56. Chithrakumar, V.K.; Venugopal, G.; Rajkumar, M.R. Convection in vertical annular gap formed by stationary heated inner cylinder and rotating unheated outer cylinder. *Heat Mass Transf.* **2019**, *55*, 2873–2888. [CrossRef]
57. Singh, M.; Rajvanshi, S.C. Heat transfer between rotating eccentric cylinders with different radii. *Int. J. Heat Mass Transf.* **1982**, *25*, 1719–1724. [CrossRef]
58. Joseph, D.D.; Munson, B.R. Global stability of spiral flow. *J. Fluid Mech.* **1970**, *43*, 545–575. [CrossRef]
59. Bishop, E.H.; Carley, C.T.; Powes, R.E. Natural Convective Oscillatory Flow in Cylindrical Annuli. *Int. J. Heat Mass Transf.* **1968**, *11*, 1741–1752. [CrossRef]
60. Weidman, P.D. On the spin-up and spin-down of a rotating fluid. Part 2. Measurements and stability. *J. Fluid Mech.* **1976**, *77*, 709–735. [CrossRef]
61. Kreith, J.; Roberts, L.G.; Sullivan, J.A.; Sinha, S.N. Convection Heat Transfer and Flow Phenomena of Rotating Spheres. *Int. J. Heat Mass Transf.* **1963**, *6*, 881–885. [CrossRef]
62. Liu, Q.; Prosperetti, A. Wall effects on a rotating sphere. *J. Fluid Mech.* **2010**, *657*, 1–21. [CrossRef]
63. Yavorskaya, I.M.; Belyaev, Y.M. Hydrodynamical Stability in Rotating Spherical Layers: Application to Dynamics of Planetary Atmospheres. *Acta Astronaut.* **1986**, *13*, 433–440. [CrossRef]
64. Douglass, R.W.; Shaughnessy, E.J.; Munson, B.R. Small Reynolds Number Convection in Rotating Spherical Annuli. *J. Heat Transf.* **1979**, *101*, 427–433. [CrossRef]

65. Munson, B.R.; Joseph, D.D. Viscous incompressible flow between concentric rotating spheres. Part 1. Basic flow. *J. Fluid Mech.* **1971**, *49*, 289–303. [CrossRef]
66. Pearson, C.E. A numerical study of the time-dependent viscous flow between two rotating spheres. *J. Fluid Mech.* **1967**, *28*, 323–336. [CrossRef]
67. Riley, T.A.; Mack, L.R. Thermal effects on slow viscous flow between rotating concentric spheres. *Int. J. Non-Linear Mech.* **1972**, *7*, 275–288. [CrossRef]
68. Bartels, F. Taylor vortices between two concentric rotating spheres. *J. Fluid Mech.* **1982**, *119*, 1–25. [CrossRef]
69. Yakushin, V.I. Motion of a fluid between two rotating concentric spheres. *Fluid Dyn.* **1968**, *3*, 35–36. [CrossRef]
70. Nakabayashi, K.; Tsuchida, Y.; Honda, A. Flow between Two Independently Rotating Concentric Spheres. *Trans. Jpn. Soc. Mech. Eng. Part B* **2002**, *68*, 1008–1015. [CrossRef]
71. Marcus, P.S.; Tuckerman, L.S. Simulation of flow between concentric rotating spheres. Part 1. Steady states. *J. Fluid Mech.* **1987**, *185*, 1–30. [CrossRef]
72. Gulwadi, S.D.; Elkouh, A.F. Unsteady flow in an annulus between two concentric rotating spheres. *Appl. Sci. Res.* **1992**, *49*, 147–159. [CrossRef]
73. Valdés, M.; Sebastián, A.; Abbas, R. Reynolds-number-dependent efficiency characterization of a micro-scale centrifugal compressor using non-conventional working fluids. *Energy Convers. Manag.* **2018**, *177*, 224–232. [CrossRef]
74. Baljé, O.E. A Study on Reynolds Number Effects in Turbomachines. *J. Eng. Power* **1964**, *86*, 227–235. [CrossRef]
75. Kurzke, J.; Halliwell, I. *Propulsion and Power*; Springer: Cham, Switzerland, 2018.
76. Tennekes, H.; Lumley, J.L. *A First Course in Turbulence*; The MIT Press: Cambridge, UK, 1972.
77. Landau, L.D.; Lifshitz, E.M. *Fluid Mechanics*; Third Impression; Pergamon Press: Oxford, UK, 1966.
78. Volovik, G.E. Classical and Quantum Regimes of Superfluid Turbulence. *JETP Lett.* **2003**, *78*, 533–537. [CrossRef]
79. Barenghi, C.F. Is the Reynolds Number Infinite in Superfluid Turbulence? *Phys. D Nonlinear Phenom.* **2008**, *237*, 2195–2202. [CrossRef]
80. Ziołkowska, I.; Ziołkowski, D. Fluid Flow Inside Packed Beds. *Chem. Eng. Process.* **1988**, *23*, 137–164. [CrossRef]
81. Dwivedi, P.N.; Upadhyay, S.N. Particle-Fluid Mass Transfer in Fixed and Fluidized Beds. *Ind. Eng. Chem. Process Des. Dev.* **1977**, *16*, 157–165. [CrossRef]
82. Prieur du Plessis, J.; Woudberg, S. Pore-scale derivation of the Ergun equation to enhance its adaptability and generalization. *Chem. Eng. Sci.* **2008**, *63*, 2576–2586. [CrossRef]
83. Boleve, A.; Crespy, A.; Revil, A.; Janod, F.; Mattiuzzo, J.-L. Streaming potentials of granular media: Influence of the Dukhin and Reynolds numbers. *J. Geoph. Res. Solid Earth* **2007**, *112*, B08204. [CrossRef]
84. Ahmed, N. Discussion to Friction Factor and Reynolds Number in Porous Media Flow by by A. Arbhabhirama and A. A. Dinoy (Proc. Paper 9784). *J. Hydraul. Div.* **1974**, *100*, 701–703. [CrossRef]
85. Jin, Y.; Kuznetsov, A.V. Multiscale modeling and simulation of turbulent flows in porous media. *Int. J. Fluid Eng.* **2024**, *1*, 010601. [CrossRef]
86. Clavier, R.; Chikhi, N.; Fichot, F.; Quintard, M. Modeling of inertial multi-phase flows through high permeability porous media: Friction closure laws. *Int. J. Multiph. Flow* **2017**, *91*, 243–261. [CrossRef]
87. Davit, Y.; Quintard, M. One-Phase and Two-Phase Flow in Highly Permeable Porous Media. *Heat Transf. Eng.* **2018**, *40*, 391–409. [CrossRef]
88. Shook, C.A.; Rocco, M.C. *Slurry Flow*; Butterworth-Heinemann: Boston, MA, USA, 1991.
89. Ishii, M.; Mishima, K. Two-Fluid Model and Hydrodynamic Constitutive Relations. *Nucl. Eng. Des.* **1984**, *82*, 107–126. [CrossRef]
90. Manninen, M.; Taivassalo, V.; Kallio, S. *On the Mixture Model for Multiphase Flow*; VTT Publications 288; Technical Research Center of Finland: Espoo, Finland, 1996.
91. Ishii, M.; Zuber, N. Drag Coefficient and Relative Velocity in Bubbly, Droplet or Particulate Flows. *AIChE J.* **1979**, *25*, 843–855. [CrossRef]
92. Coussot, P. *Mudflow Rheology and Dynamics*; A.A. Balkema: Rotterdam, The Netherlands, 1997.
93. Michaelides, E.E.; Crowe, C.T.; Schwarzkopf, J.D. *Multiphase Flow Handbook*, 2nd ed.; Taylor & Francis: Abingdon, UK; CRC Press: Boca Raton, FL, USA, 2017.
94. Brennen, C.E. *Fundamentals of Multiphase Flows*; Cambridge University Press: Cambridge, UK, 2005.
95. Meruane, C.; Tamburrino, A.; Roche, O. On the role of the ambient fluid on gravitational granular flow dynamics. *J. Fluid Mech.* **2010**, *648*, 381–404. [CrossRef]
96. Yen, B.C. Open Channel Flow Resistance. *J. Hydraul. Eng.* **2002**, *128*, 20–39. [CrossRef]
97. Shields, A. Anwendung der Ähnlichkeitsmechanik und der Turbulenzforschung auf die Geschiebebewegung. In *Mitteilungen der Preussischen Versuchsanstalt fur Wasserbau and Schiffbau*; No. 26; Elgenverlage der Preußischen Versuchsanstalt für Wasserbau und Schiffbau: Berlin, Germany, 1936.

98. Vanoni, V.A. *Measurements of Critical Shear Stress for Entraining Fine Sediments in a Boundary Layer*; Report No. KH-R-7; California Institute of Technology: Pasadena, CA, USA, 1964.
99. Garcia, M. *Sedimentation Engineering (Manual 110): Processes, Management, Modeling, and Practice*; American Society of Civil Engineers: Reston, VA, USA, 2008; 1132p.
100. Dey, S. *Fluvial Hydrodynamics*; Springer: Berlin/Heidelberg, Germany, 2014.
101. Yalin, M.S. On the Determination of Ripple Length. *J. Hyd. Div.* **1977**, *103*, 439–442. [CrossRef]
102. Ayrton, H. The Origin and Growth of Ripple-mark. *Proc. R. Soc. Lond. Ser. A Contain. Pap. A Math. Phys. Character* **1910**, *A84*, 285–310. [CrossRef]
103. Admiraal, D.; Musalem-Jara, R.; García, M.; Niño, Y. Vortex trajectory hysteresis above self-formed vortex ripples. *J. Hydraul. Res.* **2006**, *44*, 437–450. [CrossRef]
104. Yue, L.; Hsu, T.; Horner-Devine, A. Selection of vortex ripple dimensions in sinusoidal oscillatory flows. Part 1. Ripple dimensions and fluid kinematics. *J. Fluid Mech.* **2023**, *960*, A40. [CrossRef]
105. Einstein, H.A. *The Bed-Load Function for Sediment Transportation in Open Channel Flows*; Technical bulletin number 1026; United States Department of Agriculture, Soil Conservation Service: Washington, DC, USA, 1950.
106. Bagnold, R.A. *An Approach to the Sediment Transport Problem from General Physics*; Geological Survey Professional Paper 422-1; United States Government Printing Office: Washington, DC, USA, 1966.
107. Forterre, Y.; Pouliquen, O. Flows of Dense Granular Media. *Annu. Rev. Fluid Mech.* **2008**, *40*, 1–24. [CrossRef]
108. Schaeffer, D.G. Instability in the Evolution Equations Describing Incompressible Granular Flow. *J. Diff. Eqs.* **1987**, *66*, 19–50. [CrossRef]
109. Jop, P.; Forterre, Y.; Pouliquen, O. A constitutive law for dense granular flows. *Nature* **2006**, *441*, 727–730. [CrossRef]
110. Rao, K.K.; Nott, P.R. *An Introduction to Granular Flow*; Cambridge University Press: Cambridge, UK, 2008.
111. Metzner, A.B.; Reed, J.C. Flow of Non-Newtonian Fluids-Correlation of the Laminar, Transition, and Turbulent-flow Regions. *AIChE J.* **1955**, *1*, 434–440. [CrossRef]
112. Hanks, R.W. *A Generalized Criterion for the Laminar-Turbulent Transition in the Flow of Fluids*; Report K-153; Union Carbide Nuclear Company: Oak Ridge, TN, USA, 1962.
113. Hanks, R.W. The Laminar-Turbulent Transition for Fluids with a Yield Stress. *AIChE J.* **1963**, *9*, 306–309. [CrossRef]
114. Yusufi, B.K.; Kapelan, Z.; Mehta, D. Advances in modeling the flow of Herschel–Bulkley fluids in pipes: A review. *Phys. Fluids* **2025**, *37*, 021302. [CrossRef]
115. Haldenwang, R.; Sutherland, A.P.N.; Fester, V.G.; Holm, R.; Chhabra, R.P. Sludge pipe flow pressure drop prediction using composite power-law friction factor-Reynolds number correlations based on different non-Newtonian Reynolds numbers. *Water SA* **2012**, *38*, 615–622. [CrossRef]
116. Haldenwang, R.; Slatter, P.T.; Chhabra, R.P. An experimental study of non-Newtonian fluid flow in rectangular flumes in laminar, transition and turbulent flow regimes. *J. S. Afr. Inst. Civil Eng.* **2010**, *52*, 11–19.
117. Elsasser, W.M. Dimensional Relations in Magnetohydrodynamics. *Phys. Rev.* **1954**, *95*, 1–5. [CrossRef]
118. Cambel, A.B. *Plasma Physics and Magnetofluidmechanics*; McGraw-Hill Book Company, Inc.: New York, NY, USA, 1963.
119. Kulsrud, R.M. *Plasma Physics for Astrophysics*; Princeton University Press: Princeton, NJ, USA, 2005.
120. von Weizsäcker, C.F. The Evolution of Galaxies and Stars. *Astrophys. J.* **1951**, *114*, 165–186. [CrossRef]
121. Gamow, G. The Role of Turbulence in the Evolution of the Universe. *Phys. Rev.* **1952**, *86*, 251. [CrossRef]
122. Giovannini, M. Reynolds Numbers in the Early Universe. *Phys. Lett. B* **2012**, *711*, 327–331. [CrossRef]
123. Chapman, S.; Cowling, T.G. *The Mathematical Theory of Non-Uniform Gases: An Account of the Kinetic Theory of Viscosity, Thermal Conduction and Diffusion in Gases*, 3rd ed.; Cambridge University Press: Cambridge, UK, 1970.
124. Jackson, J.D. *Classical Electrodynamics*, 3rd ed.; John Wiley & Sons, Inc.: Hoboken, NJ, USA, 1999.
125. Israel, W.; Stewart, J.M. Transient relativistic thermodynamics and kinetic theory. *Ann. Phys.* **1979**, *118*, 341–372. [CrossRef]
126. Rocha, G.S.; Wagner, D.; Denicol, G.S.; Noronha, J.; Rischke, D.H. Theories of Relativistic Dissipative Fluid Dynamics. *Entropy* **2024**, *26*, 189. [CrossRef] [PubMed]
127. Denicol, D.S.; Niemi, H.; Molnár, E.; Rischke, D.H. Derivation of transient relativistic fluid dynamics from the Boltzmann equation. *Phys. Rev. D* **2012**, *85*, 114047. [CrossRef]
128. Wu, H.; Ke, S.; Wang, F.; Wang, W. Typhoon-Induced Failure Process and Collapse Mechanism of Super-Large Cooling Tower Based on WRF-CFD-LS/DYNA Nesting Technology. *Appl. Sci.* **2022**, *12*, 4178. [CrossRef]
129. Canuto, V.M.; Christensen-Dalsgaard, J. Turbulence in Astrophysics: Stars. *Annu. Rev. Fluid Mech.* **1998**, *30*, 167–198. [CrossRef]
130. Chen, L.; Yang, C.; Xiao, Y.; Yan, X.; Hu, L.; Eggersdorfer, M.; Chen, D.; Weitz, D.A.; Ye, F. Millifluidics, Microfluidics, and Nanofluidics: Manipulating Fluids at Varying Length Scales. *Mater. Today Nano* **2021**, *16*, 100136. [CrossRef]
131. Sackmann, E.K.; Fulton, A.L.; Beebe, D.J. The present and future role of microfluidics in biomedical research. *Nature* **2014**, *507*, 181–189. [CrossRef]
132. Poole, R.J. The Deborah and Weissenberg numbers. *Br. Soc. Rheol. Rheol. Bull.* **2012**, *53*, 32–39.

133. Thompson, R.L.; Oishi, C.M. Reynolds and Weissenberg numbers in viscoelastic flows. *J. Non-Newt. Fluid Mech.* **2021**, *292*, 104550. [CrossRef]
134. Das, S.; Sreeram, P.A.; Raychaudhuri, A.K. A method to quantitatively evaluate the Hamaker constant using the jump-into-contact effect in atomic force microscopy. *Nanotechnology* **2007**, *18*, 035501. [CrossRef]
135. Derjaguin, B.; Landau, L. Theory of the stability of strongly charged lyophobic sols and of the adhesion of strongly charged particles in solutions of electrolytes. *Prog. Surf. Sci.* **1941**, *43*, 30–59. [CrossRef]
136. Israelachvili, J.N. *Intermolecular and Surface Forces*, 3rd ed.; Elsevier Inc.: San Diego, CA, USA, 2011.
137. Li, F.-C.; Kinoshita, H.; Li, X.B.; Oishi, M.; Fujii, T.; Oshima, M. Creation of very-low-Reynolds-number chaotic fluid motions in microchannels using viscoelastic surfactant solution. *Exp. Therm. Fluid Sci.* **2010**, *34*, 20–27. [CrossRef]
138. Schiamberg, B.A.; Shereda, L.T.; Hu, H.; Larson, R.G. Transitional pathway to elastic turbulence in torsional, parallel-plate flow of a polymer solution. *J. Fluid Mech.* **2006**, *554*, 191–216. [CrossRef]
139. Randers-Pehrson, N.H. Pioneer Wind Tunnels. In *Smithsonian Miscellaneous Collections*; The Smithsonian Institution: Washington, DC, USA, 1935; Volume 93, Publication 3294.
140. Munk, M.M.; Miller, E.W. *The Variable Density Wind Tunnel of the National Advisory Committee for Aeronautics*; NACA Report No. 227; Government Printing Office: Washington, DC, USA, 1926.
141. Prandtl, L.; Tietjens, O.G. *Applied Hydro- and Aeromechanics*; McGraw-Hill Book Company, Inc.: York, PA, USA, 1934.
142. Katz, J. *Automotive Aerodynamics*; John Wiley & Sons, Ltd.: Chichester, UK, 2016.
143. Lawing, P.L.; Kilgore, R.A.; McGuire, P.G. *Cryogenic Wind Tunnels for High Reynolds Number Testing*; NASA Technical Memorandum 87743; NASA: Washington, DC, USA, 1986.
144. NASA. National Transonic Facility Fact Sheet. Available online: https://www.nasa.gov/directorates/armd/armd-aetc/national-transonic-facility-ntf-quick-facts/ (accessed on 10 April 2025).
145. ETW (European Transonic Windtunnel). *User Guide*; Report Number: ETW/D/95001/A; European Transonic Windtunnel: Köln, Germany, 2004.
146. Han, Y.; Chen, D.; Liu, S.; Xu, G. An Investigation into the Effects of the Reynolds Number on High-Speed Trains Using a Low Temperature Wind Tunnel Test Facility. *Fluid Dyn. Mater. Process.* **2020**, *16*, 1–19. [CrossRef]
147. Kendall, J.M. Wind Tunnel Experiments Relating to Supersonic and Hypersonic Boundary-Layer Transition. *AIAA J.* **1975**, *13*, 290–299. [CrossRef]
148. Steelant, J. *Hypersonic Technology Developments with EU Co-Funded Projects*; AVT-185 RTO AVT/VKI Lecture Series held at the von Karman Institute; NATO-OTAN: Brussels, Belgium, 2010.
149. Beckwith, I.E.; Miller, C.G. Aerothermodynamics and Transition in High-Speed Wind Tunnels at NASA Langley. *Annu. Rev. Fluid Mech.* **1990**, *22*, 419–439. [CrossRef]
150. De Filippis, F.; Del Vecchio, A.; Caristia, S.; Graps, E. SCIROCCO arc-jet facility for large scale spacecraft TPS verification. In Proceedings of the 55th International Astronautical Congress 2004, Vancouver, BC, Canada, 4–8 October 2004. Paper IAC-04-I.1.06.
151. Purpura, C.; De Filippisa, F.; Barrera, P.; Mandanici, D. Experimental characterisation of the CIRA plasma wind tunnel SCIROCCO test section. *Acta Astronaut.* **2008**, *62*, 410–421. [CrossRef]
152. Cipullo, A.; Savino, L.; Marenna, E.; De Filippis, F. Thermodynamic state investigation of the hypersonic air plasma flow produced by the arc-jet facility SCIROCCO. *Aerosp. Sci. Tech.* **2012**, *23*, 358–362. [CrossRef]
153. Farsiani, Y.; Elbing, B.R. Characterization of the Custom-Designed, High Reynolds Number Water Tunnel. In Proceedings of the ASME 2016 Fluids Engineering Division Summer Meeting, FEDSM2016, Washington, DC, USA, 10–14 July 2016.
154. Elbing, B.R.; Daniel, L.; Farsiani, Y.; Petrin, C.E. Design and Validation of a Recirculating, High-Reynolds Number Water Tunnel. *J. Fluids Eng.* **2018**, *140*, 081102. [CrossRef]

Disclaimer/Publisher's Note: The statements, opinions and data contained in all publications are solely those of the individual author(s) and contributor(s) and not of MDPI and/or the editor(s). MDPI and/or the editor(s) disclaim responsibility for any injury to people or property resulting from any ideas, methods, instructions or products referred to in the content.

fluids

MDPI

Review

Convergence towards High-Speed Steady States Using High-Order Accurate Shock-Capturing Schemes

Juan C. Assis [1], Ricardo D. Santos [1], Mateus S. Schuabb [2], Carlos E. G. Falcão [3], Rômulo B. Freitas [4] and Leonardo S. de B. Alves [3,*]

[1] Programa de Pós-Graduação em Engenharia Mecânica, Universidade Federal Fluminense, Niterói 24210-240, RJ, Brazil
[2] Department of Mechanical and Aerospace Engineering, The Ohio State University, Columbus, OH 43210, USA
[3] Departamento de Engenharia Mecânica, Universidade Federal de Santa Maria, Santa Maria 97105-900, RS, Brazil
[4] Centro Federal de Educação Tecnológica Celso Suckow da Fonseca, Nova Iguaçu 26041-271, RJ, Brazil
* Correspondence: lsbalves@id.uff.br

Abstract: Creating time-marching unsteady governing equations for a steady state in high-speed flows is not a trivial task. Residue convergence in time cannot be achieved when using most low- and high-order spatial discretization schemes. Recently, high-order, weighted, essentially non-oscillatory schemes have been specially designed for steady-state simulations. They have been shown to be capable of achieving machine precision residues when simulating the Euler equations under canonical coordinates. In the present work, we review these schemes and show that they can also achieve machine residues when simulating the Navier–Stokes equations under generalized coordinates. This is carried out by considering three supersonic flows of perfect fluids, namely the flow upstream a cylinder, the flow over a blunt wedge, and the flow over a compression ramp.

Keywords: TVD schemes; WENO schemes; residue convergence; increment convergence

1. Introduction

The high-speed flows under discussion in this work are modeled by unsteady and compressible Navier–Stokes equations for perfect fluids. These governing equations are solved using the method of lines. This represents a two-step simulation process that first generates a system of ordinary differential equations from the partial differential equations composing the original model governing equations and then marches it forward in time. A wide range of methods can be applied to generate this system. A review of them is beyond the scope of this work, but several relevant books can be used as a starting point [1–5]. This system is obtained here by applying finite-difference schemes under a generalized coordinate framework to the spatial operators of the model. The code developed to run all the simulations presented here is called 3D4S [6,7]. It stands for one-, two-, and three-dimensional structured steady-state solvers and uses the C++ programming language [8].

1.1. Literature Review

1.1.1. Spatial Discretization

Apart from unconventional source terms that are not employed in the present work, all spatial operators in the unsteady and compressible Navier–Stokes equations for perfect fluids are applied to inviscid and viscous fluxes. Their discretizations are usually treated using different methodologies, whose implementations are briefly discussed below.

On the one hand, the discretization of viscous flux spatial gradients employs the traditional second-, fourth-, and sixth-order central-difference schemes in conservative form. It is important to note that the third one usually renders the resulting code too numerically unstable [9], and hence, it is not used here.

On the other hand, an accurate discretization of inviscid fluxes that can capture shocks is significantly more complex. This is due to the nonlinear stability properties that are required to prevent the Gibbs phenomenon near discontinuities. Monotone schemes are capable of doing so. This property states that (i) new minima and maxima cannot be created, (ii) a local minimum cannot be reduced, and (iii) a local maximum cannot be increased. They are, however, first-order accurate. High-order accuracy requires the use of special techniques to control the oscillations induced by discretizing across the shock. One possibility is to relax the monotonicity requirement for something else less restrictive. Following this path [10] leads to second-order accurate schemes that are total-variation diminishing (TVD). This property states that the solution's total variation at a given time step is always smaller than or equal to the solution's total variation at the previous time step. These schemes also satisfy the entropy condition, which guarantees the convergence of weak solutions of hyperbolic conservation laws. Furthermore, all monotone schemes are TVD and all TVD schemes are monotonicity-preserving [11]. This property states that a solution that is monotone at a given time step will remain monotone at the next time step. Furthermore, TVD schemes work by employing flux-limiters, which are weights placed on the forward and backward fluxes. They are designed to locally reduce the scheme to first-order accurate near discontinuities [12]. There are many flux limiters. Superbee [13] and Minmod [14–16] are the ones that introduce the least and most numerical diffusion, respectively. All others fall in between these two, such as the Monotonized Central Difference [17], van Leer [18], Koren [19], Sweby [20], Osher [21], Ospre [22] and van Albada [23]. 3D4S can use either the first-order accurate upwind/downwind discretization or the second-order accurate TVD scheme developed by [10], coupled with any of these flux limiters.

Unfortunately, it is well known that TVD schemes suffer from an accuracy order reduction near discontinuities [18,24,25]. Furthermore, they are first-order accurate for two-dimensional problems [26]. A path towards higher-order accuracy is to relax the TVD requirement for something that is less restrictive. This was performed by introducing the class of essentially non-oscillatory (ENO) schemes [27]. They represent higher-order generalizations of the first-order Godunov scheme [28] and its second-order extension known as the Monotonic Upstream-centered Scheme for Conservation Laws (MUSCL) [29]. ENO schemes choose between different candidate stencils in order to avoid including discontinuity in the discretization. The selection is performed by comparing the local smoothness of the discretization on these different stencils, measured using divided differences [30,31]. A variation of these schemes, known as weighted ENO (WENO) schemes, was then developed to achieve higher-order accuracy and smoother numerical fluxes [32]. Instead of using only one of the candidate stencils, as in ENO, WENO uses a convex combination of all candidate stencils based on nonlinear weights determined by smoothness indicators. 3D4S can use several different fifth-order WENO schemes for unsteady simulations, such as the original WENO5-JS [33], as well as WENO5-M [34], WENO5-Z [35] and WENO5-Z+ [36], and steady simulations, such as WENO5-ZQ [37] and its variants [38–41].

A distinguishing feature of the aforementioned upwind/downwind schemes is that their discretization must be performed according to the propagation direction of the inviscid flux characteristics. There are essentially two different ways to identify these directions, which are known as the flux-difference and vector-splitting methods [42]. The former are known as Riemann solvers whereas the latter are known as Boltzmann solvers, although they can also be considered Riemann solvers in a wider sense. On one hand, flux-difference splitting first discretizes the inviscid fluxes using their cell-faced values and only then separates their propagation directions to evaluate these values. Flux-vector splitting, on the other hand, first separates the inviscid fluxes according to their propagation directions and only then discretizes each part accordingly. It is important to note that they can also be categorized as either complete or incomplete solvers. The former resolves all waves that are present, whereas the latter neglects some of these waves. 3D4S can currently employ many of these methods, such as Roe [43], HLL [44] and Roe-HLL [45] based on flux-difference splitting as well as Steger–Warming [46] and Lax–Friedrichs [33] based on flux-vector splitting, which are applied in either conservative or characteristic form.

1.1.2. Temporal Discretization

Having applied the spatial discretization of choice on a given grid to the unsteady and compressible Navier–Stokes equations for perfect fluids, the resulting system can then be marched in time. This is performed in 3D4S using different methodologies to generate either time-accurate unsteady or steady states. Since physically stable flows are studied in this work, explicit time-marching schemes can be employed to reach their steady states.

The explicit Euler scheme is a good choice for time-marching the unsteady governing equations towards time-asymptotic stable stead states not only due to its simplicity but also due to its parallelization efficiency. This is the approach used when employing 3D4S in this scenario. When time accuracy is required for shock-capturing, however, high-order temporal integration should be used. Furthermore, this should be performed in such a way as to preserve the nonlinear numerical stability properties of the chosen spatial discretization schemes. All time-marching schemes with this property are known today as Strong-Stability-Preserving (SSP), although they were originally called TVD as well because this property does not allow the solution's total variation to increase in time. High-order time-marching schemes with this property preserve the nonlinear numerical stability properties of the Euler scheme, as long as their maximum time step is restricted. The SSP property was first proposed for an explicit Runge–Kutta (RK) scheme [30]. Although this SSP-RK scheme was originally presented in a format that did not belong to the traditional Butcher formulation [5], it can be written in this way using appropriate algebraic procedures [47,48]. The nonlinear upper bound imposed on the maximum time step of high-order SSP marching schemes, which is smaller than its linear counterpart, led to the optimization of these schemes so they can achieve the largest possible maximum time steps. Optimal explicit SSP-RK schemes have been derived with up to fourth-order accuracy and five stages [49]. They are the ones used in 3D4S for time-accurate simulations.

It is important to mention here that implicit SSP-RK schemes exist as well [50], where optimal diagonally implicit schemes have been developed with up to fourth-order accuracy and eight-stages [51,52] and with up to sixth-order accuracy and eleven stages [53]. Second- and third-order accurate implicit–explicit (IMEX) RK schemes also exist [54], but only the explicit part is SSP. However, there is strong evidence that the nonlinear upper bound imposed on the time steps of implicit and IMEX SSP-RK schemes renders them inefficient when compared to their explicit counterparts [55]. Hence, they are not used in 3D4S for time-accurate simulations.

1.1.3. Convergence towards Steady States

As far as the authors are aware, the effect that high-order WENO schemes have on convergence towards steady-state was first evaluated almost two decades ago [56]. This study shed light on the existence of small-amplitude post-shock oscillations, which prevented residue convergence to machine precision. It also led to the development of WENO schemes specially designed for steady-state simulations [37–40]. The use of these schemes did lead to residue convergence towards machine precision zero by reducing the amplitude of post-shock oscillations.

1.2. Present Contributions

Nevertheless, there are still issues that must be addressed on this front. For instance, steady-state convergence in the aforementioned studies was evaluated when considering the Euler equations, which do not include viscous effects. Furthermore, the test cases simulated in these studies considered quite simple geometries; i.e., they could be simulated using standard coordinate systems. Both issues are addressed in the present paper, namely the ability to achieve steady-state convergence to machine precision residues when using

WENO schemes (i) in the presence of viscous diffusion and (ii) when performing numerical simulations in generalized coordinates.

In order to do so, a comparison of several shock-capturing schemes is performed using the Navier–Stokes equations over a framework with generalized coordinates. In Section 2, the methodology used to implement the numerical discretization of both spatial and temporal terms of the governing equations is discussed. It also contains a brief review of the grid-generation techniques employed here. In Section 3, the results obtained for three different test cases are presented, namely the two-dimensional supersonic flow over a compression ramp, over a planar blunt body and upstream of a cylinder. Finally, the major conclusions derived from this work are highlighted in Section 4.

2. Methodology

2.1. Governing Equations

Using a Cartesian coordinate system, it is possible to write the Navier–Stokes equations in dimensionless form as

$$\frac{\partial \boldsymbol{Q}}{\partial t}+\frac{\partial \boldsymbol{E}^i}{\partial x}+\frac{\partial \boldsymbol{F}^i}{\partial y}+\frac{\partial \boldsymbol{G}^i}{\partial z}=\frac{\partial \boldsymbol{E}^v}{\partial x}+\frac{\partial \boldsymbol{F}^v}{\partial y}+\frac{\partial \boldsymbol{G}^v}{\partial z}\quad, \tag{1}$$

where $\boldsymbol{Q}$ is the vector of conservative variables and $\boldsymbol{E}$, $\boldsymbol{F}$ and $\boldsymbol{G}$ are the conservative flux vectors in the x, y, and z directions, respectively. The superscripts i and v stand for inviscid and viscous, respectively. These vectors are given by

$$\boldsymbol{Q}=\begin{bmatrix}\rho\\ \rho u\\ \rho v\\ \rho w\\ \rho E\end{bmatrix}\ ,\ \boldsymbol{E}^i=\begin{bmatrix}\rho u\\ \rho u^2+p\\ \rho v u\\ \rho w u\\ (\rho E+p)u\end{bmatrix}\ ,\ \boldsymbol{F}^i=\begin{bmatrix}\rho v\\ \rho v u\\ \rho v^2+p\\ \rho w v\\ (\rho E+p)v\end{bmatrix}\ ,\ \boldsymbol{G}^i=\begin{bmatrix}\rho w\\ \rho u w\\ \rho v w\\ \rho w^2+p\\ (\rho E+p)w\end{bmatrix}\ ,$$

$$\boldsymbol{E}^v=\begin{bmatrix}0\\ \tau_{xx}\\ \tau_{xy}\\ \tau_{xz}\\ u\,\tau_{xx}+v\,\tau_{xy}+w\,\tau_{xz}-q_x\end{bmatrix}\ ,\ \boldsymbol{F}^v=\begin{bmatrix}0\\ \tau_{yx}\\ \tau_{yy}\\ \tau_{yz}\\ u\,\tau_{yx}+v\,\tau_{yy}+w\,\tau_{yz}-q_y\end{bmatrix}\ \text{and} \tag{2}$$

$$\boldsymbol{G}^v=\begin{bmatrix}0\\ \tau_{zx}\\ \tau_{zy}\\ \tau_{zz}\\ u\,\tau_{zx}+v\,\tau_{zy}+w\,\tau_{zz}-q_z\end{bmatrix}\ .$$

where the viscous stress tensor is defined by

$$\begin{aligned}
\tau_{xx}&=\Pi_1\left[2\mu\frac{\partial u}{\partial x}+\lambda\left(\frac{\partial u}{\partial x}+\frac{\partial v}{\partial y}+\frac{\partial w}{\partial z}\right)\right]\ ,\quad \tau_{xy}=\tau_{yx}=\Pi_1\left[\mu\left(\frac{\partial u}{\partial y}+\frac{\partial v}{\partial x}\right)\right]\ ,\\
\tau_{yy}&=\Pi_1\left[2\mu\frac{\partial v}{\partial y}+\lambda\left(\frac{\partial u}{\partial x}+\frac{\partial v}{\partial y}+\frac{\partial w}{\partial z}\right)\right]\ ,\quad \tau_{xz}=\tau_{zx}=\Pi_1\left[\mu\left(\frac{\partial u}{\partial z}+\frac{\partial w}{\partial x}\right)\right]\ \text{and}\\
\tau_{zz}&=\Pi_1\left[2\mu\frac{\partial z}{\partial x}+\lambda\left(\frac{\partial u}{\partial x}+\frac{\partial v}{\partial y}+\frac{\partial w}{\partial z}\right)\right]\ \text{and}\quad \tau_{yz}=\tau_{zy}=\Pi_1\left[\mu\left(\frac{\partial v}{\partial z}+\frac{\partial w}{\partial y}\right)\right]\ ,
\end{aligned} \tag{3}$$

and the heat flux is defined by

$$q_x=-\Pi_2 k\frac{\partial T}{\partial x}\ ,\quad q_y=-\Pi_2 k\frac{\partial T}{\partial y}\ \text{and}\ q_z=-\Pi_2 k\frac{\partial T}{\partial z}\ . \tag{4}$$

The aforementioned system of equations has more variables than equations. However, it is possible to employ the equation of state for an ideal gas, i.e.,

$$p = -\Pi_3 \rho T \quad , \tag{5}$$

where Π_k represents the dimensionless parameters, calculated with respect to the reference quantities. They can be written in generic form as follows:

$$\Pi_1 = \frac{\mu_R}{\rho_R U_R L_R} \quad , \quad \Pi_2 = \frac{k_R T_R}{\rho_R U_R^3 L_R} \quad \text{and} \quad \Pi_3 = \frac{R T_R}{U_R^2} \quad . \tag{6}$$

The reference quantities, on the other hand, are selected based on the free stream conditions. However, an exception is made for the reference velocity, which is derived from the free stream's speed of sound. Consequently, the dimensionless parameters become

$$\Pi_1 = \frac{Ma_\infty}{Re_\infty} \quad , \quad \Pi_2 = \frac{Ma_\infty}{(\gamma - 1) Re_\infty \, \mathrm{Pr}_\infty} \quad \text{and} \quad \Pi_3 = \frac{1}{\gamma} \quad , \tag{7}$$

where Re is the Reynolds number, Pr is the Prandtl number, Ma is the Mach number, and the subscript ∞ indicates free stream quantities. Having selected these three dimensionless parameters, one must still select the dynamic viscosity and thermal conductivity. The dimensionless dynamic viscosity is assumed to follow Sutherland's law, i.e.,

$$\mu(T) = \frac{1 + T_0}{T + T_0} T^{3/2} \quad , \tag{8}$$

and the dimensionless thermal conductivity is computed assuming that the Prandtl number is constant $Pr = Pr_\infty$. Hence, one obtains

$$\mu(T) = k(T) \quad . \tag{9}$$

since both heat capacities are constant in ideal gases.

The numerical methods utilized here were originally designed for rectangular domains. In order to use them in complex geometries, a coordinate transformation is applied to the governing equations to map the generalized physical space (x, y, z) to the rectangular computational space (ξ, η, ζ). It must be noted that this transformation ensures that the transformed equations retain their conservative behavior. Within the computational space, the governing equations are reformulated as

$$\frac{\partial q}{\partial t} + \frac{\partial e^i}{\partial \xi} + \frac{\partial f^i}{\partial \eta} + \frac{\partial g^i}{\partial \zeta} = \frac{\partial e^v}{\partial \xi} + \frac{\partial f^v}{\partial \eta} + \frac{\partial g^v}{\partial \zeta} \quad , \tag{10}$$

noting that the new coordinate system depends on the old one, i.e.,

$$\xi = \xi(x, y, z) \quad , \quad \eta = \eta(x, y, z) \quad \text{and} \quad \zeta = \zeta(x, y, z) \quad , \tag{11}$$

and, hence, the transformation Jacobian

$$\mathbf{J} = \frac{\partial(\xi, \eta, \zeta)}{\partial(x, y, z)} = \begin{bmatrix} \xi_x & \xi_y & \xi_z \\ \eta_x & \eta_y & \eta_z \\ \zeta_x & \zeta_y & \zeta_z \end{bmatrix} \quad , \tag{12}$$

has a nonzero determinant, i.e.,

$$J = \frac{1}{J^{-1}} = 1 \Big/ \frac{\partial(x, y, z)}{\partial(\xi, \eta, \zeta)} = 1 \Big/ \begin{vmatrix} x_\xi & x_\eta & x_\zeta \\ y_\xi & y_\eta & y_\zeta \\ z_\xi & z_\eta & z_\zeta \end{vmatrix} \quad . \tag{13}$$

and the necessary metrics are computed using

$$\begin{bmatrix} \xi_x & \xi_y & \xi_z \\ \eta_x & \eta_y & \eta_z \\ \zeta_x & \zeta_y & \zeta_z \end{bmatrix} = J \begin{bmatrix} y_\eta z_\zeta - y_\zeta z_\eta & -(x_\eta z_\zeta - x_\zeta z_\eta) & x_\eta y_\zeta - x_\zeta y_\eta \\ -(y_\xi z_\zeta - y_\zeta z_\xi) & x_\xi z_\zeta - x_\zeta z_\xi & -(x_\xi y_\zeta - x_\zeta y_\xi) \\ y_\xi z_\eta - y_\eta z_\xi & -(x_\xi z_\eta - x_\eta z_\xi) & x_\xi y_\eta - x_\eta y_\xi \end{bmatrix} , \tag{14}$$

leading to the transformed conservative variables and fluxes

$$\begin{gathered} q = J^{-1} Q \ , \\ e^i = J^{-1}\left(E^i \xi_x + F^i \xi_y + G^i \xi_z\right) \ , \quad f^i = J^{-1}\left(E^i \eta_x + F^i \eta_y + G^i \eta_z\right) \ , \\ g^i = J^{-1}\left(E^i \zeta_x + F^i \zeta_y + G^i \zeta_z\right) \ , \quad e^v = J^{-1}(E^v \xi_x + F^v \xi_y + G^v \xi_z) \ , \\ f^v = J^{-1}(E^v \eta_x + F^v \eta_y + G^v \eta_z) \ \text{ and } \ g^v = J^{-1}(E^v \zeta_x + F^v \zeta_y + G^v \zeta_z) \ . \end{gathered} \tag{15}$$

2.2. Grid Generation

All spatial discretization schemes described so far are implemented on a generalized coordinate framework using uniformly distributed points between 0 and 1 in all directions. Hence, the physical domain of interest must be mapped into this computational domain for all simulations. Analytical transformations are employed whenever they are available. If this is not the case, grid-generation tools are required. This is an important issue because grid quality is directly related to solution accuracy. For example, lack of convergence can be a consequence of poor grid quality [2].

There are two fundamental types of grids for multidimensional regions: structured and unstructured. They differ by the way in which the grid points are locally organized. Structured grids have regular connectivity, which is implicitly taken into account. They have quadrilateral elements in two-dimensional domains or hexahedron elements in three-dimensional ones. On the other hand, unstructured grids can be identified by irregular connectivity that must be explicitly described and stored. This type of grid quite often employs triangles and tetrahedral elements, respectively, in two- and three-dimensional problems [57]. Besides that, they also differ in the type of method that can use them. While finite difference methods require structured grids, finite volume and finite element methods allow both types of grids. Since 3D4S is based on finite differences schemes, this brief review focuses only on structured grid generation tools.

Since elliptic partial differential equations were introduced for grid generation [58], this approach has established itself as the one that arguably produces the best grids in terms of smoothness and grid point distribution. When these elliptic differential equations are homogeneous, the smoothness of the Laplace operator makes the grid evenly spaced throughout the domain. This makes grid refinement at specific locations, such as the wall for boundary-layer problems, more difficult. Hence, grids based solely on Laplace equations are unusable in practice. However, Poisson equations can be used instead so that control functions can be introduced through their source terms. The quality of an elliptic grid now depends on how well these control functions can be tailored [59]. They are usually employed to control grid clustering [60] as well as enforce orthogonality at boundaries [61]. The latter is essential to avoid coupling between wall grid points when wall-normal derivatives are required. The particular grid-generation procedure used in 3D4S [62] introduces an intermediate step between physical and computational spaces [63] so that both grid clustering and orthogonality can be enforced. Figure 1 shows an illustration of its mapping procedure.

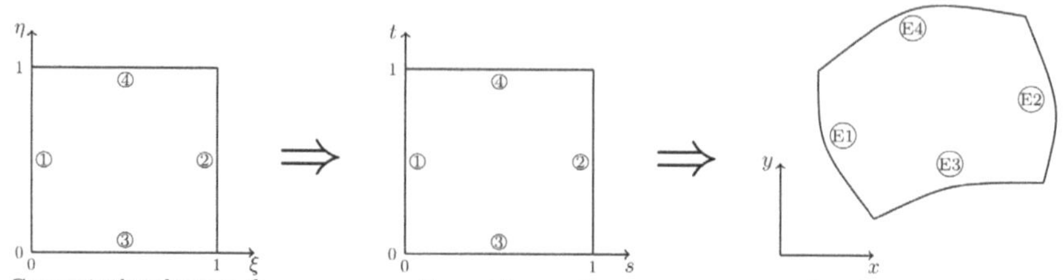

Figure 1. Transformation from computational (ξ, η) space to a physical domain in Cartesian (x, y) space going through a parametric (s, t) space.

2.3. Numerical Methods

2.3.1. Viscous Flux Discretization

In the 3D4S implementation, standard central differences are applied to discretize the viscous fluxes. To illustrate this approach, consider the scalar viscous flux

$$\frac{\partial f}{\partial x} = \frac{\partial}{\partial x}\left(\mu \frac{\partial u}{\partial x}\right) \quad , \tag{16}$$

which can be approximated using a generic central difference scheme,

$$\left.\frac{\partial f}{\partial x}\right|_i = \bar{\delta}_x^{(p)} f_i + \mathbf{O}(\Delta x^p) \quad , \tag{17}$$

also applied to its scalar flux, i.e.,

$$f_i = \mu_i \bar{\delta}_x^{(p)} u_i + \mathbf{O}(\Delta x^p) \quad , \tag{18}$$

where p is the scheme accuracy order and δ_x is the central-difference operator, which can be developed for any order p. For instance, the second-order $(p = 2)$ operator is

$$\bar{\delta}_x^{(2)} f_i = \frac{f_{i+1} - f_{i-1}}{2\,\Delta x} \quad , \tag{19}$$

the fourth-order $(p = 4)$ operator is

$$\bar{\delta}_x^{(4)} f_i = \frac{f_{i-2} - 8f_{i-1} + 8f_{i+1} - f_{i+2}}{12\,\Delta x} \quad , \tag{20}$$

and the sixth-order $(p = 6)$ operator is

$$\bar{\delta}_x^{(6)} f_i = \frac{-f_{i-3} + 9f_{i-2} - 45f_{i-1} + 45f_{i+1} - 9f_{i+2} + f_{i+3}}{12\,\Delta x} \quad . \tag{21}$$

Introducing Equation (18) into Equation (17) yields

$$\left.\frac{\partial f}{\partial x}\right|_i = \delta_x(\mu_i \delta_x u_i + \mathbf{O}(\Delta x^q)) + \mathbf{O}(\Delta x^p) = \delta_x(\mu_i \delta_x u_i) + \mathbf{O}(\Delta x^p, \Delta x^{q-1}) \quad , \tag{22}$$

which means that the latter operation in Equation (22) can lead to odd–even decoupling as well as order loss whenever μ is not a smooth enough function [9]. Hence, high-order operators, such as the one in Equation (21), must be used with care.

Figure 2 illustrates the stencil dependence associated with the central-difference schemes discussed previously. It is important to recognize that, as the order of the scheme increases, so does the number of points required for the approximation. Near boundary schemes must be biased upwind near the inlet and downwind near the outlet due to the lack of external grid points. Generally, the scheme's accuracy near boundaries is reduced to maintain numerical stability. Nevertheless, the overall accuracy can be preserved by ensuring that the boundary order is one less than the inner domain order [64].

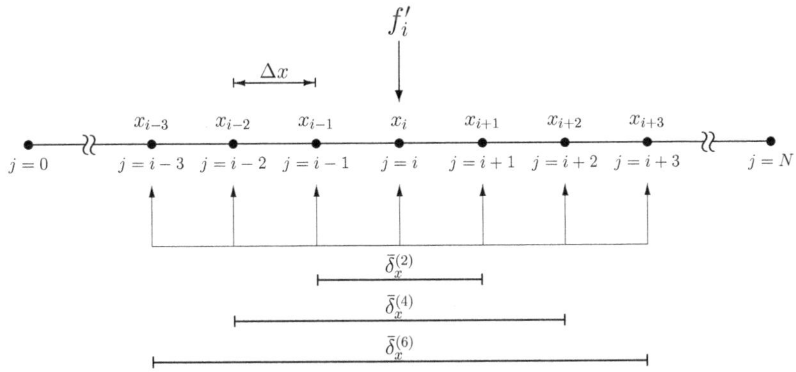

Figure 2. Stencil dependence of central-difference operators.

2.3.2. Inviscid Flux Discretization

The discretization of inviscid fluxes in 3D4S is significantly more complex. It is described using the scalar hyperbolic conservation law

$$\frac{\partial u}{\partial t} + \frac{\partial f}{\partial x} = 0 \quad , \tag{23}$$

where x is the position in a one-dimensional space, t is time, and $u = u(x,t)$ and $f = f(u(x,t))$ are the conservative variable and the conservative flux function, respectively. The goal is to describe the evolution of u within $t \in [0,\infty]$ and $x \in [x_0, x_N]$.

Conservative Schemes

A numerical scheme used to solve Equation (23) is called conservative if it ensures that u is conserved across all cells. Thus, numerical conservation is achieved using a single flux function that describes the flow of u between neighboring cells. Using a semi-discretized form of Equation (23), namely

$$\left.\frac{du}{dt}\right|_i = -\left.\frac{\partial f}{\partial x}\right|_i \quad , \tag{24}$$

a conservative finite-difference scheme used to approximate the right-hand-side term of Equation (24) must be written as

$$\left.\frac{df}{dx}\right|_i = \frac{h_{i+\frac{1}{2}} - h_{i-\frac{1}{2}}}{\Delta x} \quad , \tag{25}$$

where $h_{i+1/2} = h(x = x_{i+1/2})$ is a numerical flux function describing the flux crossing the interface $x_{i+1/2}$ between x_i and x_{i+1}. This can be visualized in Figure 3, which shows an illustration of a typical computational grid. The numerical flux function is defined such that Equation (25) involves no truncation error. Traditionally [31], the numerical flux function has been implicitly defined as

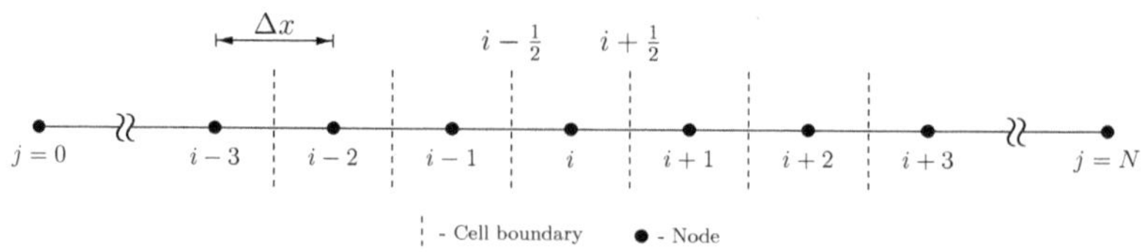

Figure 3. Illustration of a typical computational grid.

$$f(x) = \frac{1}{\Delta x} \int_{x-\frac{\Delta x}{2}}^{x+\frac{\Delta x}{2}} h(\xi) d\xi \quad , \tag{26}$$

which leads, after integration, to

$$f(x) = \frac{H(x + \frac{\Delta x}{2}) - H(x + \frac{\Delta x}{2})}{\Delta x} \quad , \tag{27}$$

where $H(x)$ is the primitive of $h(\xi)$, i.e., $H(x) = \int_{-\infty}^{x} h(\xi)\, d\xi$. Furthermore, taking the derivative of Equation (27) and evaluating it at x_i results in Equation (25). One must note that the numerical flux function is not a numerical approximation of the flux function, even though the flux function derivative is exactly equal to a finite difference approximation of the numerical flux function. Unfortunately, it is not possible to obtain the exact numerical flux function, and, hence, a numerical approximation must be used. High-order conservative numerical schemes can be constructed by employing high-order approximations of the numerical flux function. Defining $\widehat{f}(x)$ as this arbitrary polynomial approximation of $h(x)$, its coefficients can be determined from the flux function by replacing $h(\xi)$ by $\widehat{f}(\xi)$ in Equation (26). Hence, a numerical approximation of Equation (25) can be written as

$$\left. \frac{df}{dx} \right|_i \approx \frac{\widehat{f}_{i+\frac{1}{2}} - \widehat{f}_{i-\frac{1}{2}}}{\Delta x} \quad . \tag{28}$$

Flux Function Splitting

The previous discussion did not take into account the propagation direction, i.e., $\partial f / \partial u$, of variable u. In the particular case of a nonlinear flux function, the variable u can propagate both downstream, where $\partial f / \partial u > 0$, and upstream, where $\partial f / \partial u < 0$, simultaneously. Hence, the flux function must be split into positive and negative parts, since they require different spatial discretization techniques. In other words,

$$f = f^+ + f^- \quad \text{with} \quad \frac{\partial f^+}{\partial u} \geq 0 \quad \text{and} \quad \frac{\partial f^-}{\partial u} \leq 0 \quad , \tag{29}$$

which allows the numerical flux function to be computed as

$$\widehat{f} = \widehat{f}^+ + \widehat{f}^- \quad . \tag{30}$$

The Lax–Friedrichs flux splitting is a simple and inexpensive example of flux splitting. It splits the fluxes according to

$$f^{\pm} = \frac{1}{2}(f \pm \alpha u) \quad \text{with} \quad \alpha = \max_u \left| \frac{\partial f}{\partial u} \right| \quad . \tag{31}$$

WENO Schemes

In the vicinity of discontinuities, high-order polynomial interpolations inherently exhibit oscillatory behavior. This is illustrated in Figure 4, which shows spurious oscillations from high-order polynomial fits across a step function discontinuity. Shock waves present in high-speed compressible flows are an example of such a discontinuity, which leads to spurious oscillations in simulations employing high-order schemes. They not only lead to accuracy loss but also negatively impact the numerical stability of said schemes.

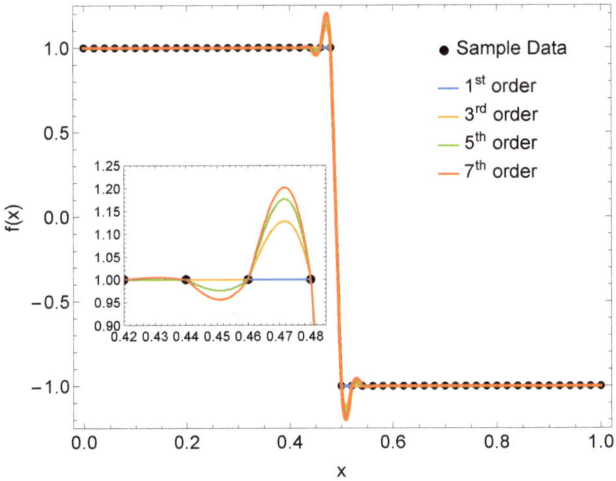

Figure 4. Polynomial interpolations of data taken from a step function.

Weighted Essentially Non-Oscillatory (WENO) schemes were developed to prevent such oscillations while preserving the high-order accuracy of the approximation. The foundational principle of a WENO scheme is its use of a convex combination of numerical flux approximations from lower-order polynomials to construct a high-order approximation. In smooth regions, WENO schemes become upstream central schemes. Near discontinuities, on the other hand, the stencil that includes the discontinuity is assigned a lower weight in the convex combination, minimizing its impact.

The WENO reconstruction [33] can be written as

$$\widehat{f}_{i+1/2} = \sum_{k=0}^{r-1} \omega_k \widehat{f}^k_{i+\frac{1}{2}} \quad , \tag{32}$$

where r is the number of stencils candidates, $\widehat{f}^k_{i+1/2}$ is the numerical flux approximation using the kth stencil, and ω_k is the nonlinear weight given to the kth stencil. Figure 5 illustrates the classical fifth-order WENO-JS [33], where each $\widehat{f}^k_{i+1/2}$ is formulated as a third-order polynomial approximation of the numerical flux function using different stencils. Following this approach, the numerical flux approximations based on upstream central schemes are given by

$$\begin{aligned} \widehat{f}^0_{i+\frac{1}{2}} &= \frac{1}{6}(2f_{i-2} - 7f_{i-1} + 11f_i) \quad , \\ \widehat{f}^1_{i+\frac{1}{2}} &= \frac{1}{6}(-f_{i-1} + 5f_i + 2f_{i+1}) \quad \text{and} \\ \widehat{f}^2_{i+\frac{1}{2}} &= \frac{1}{6}(2f_i + 5f_{i+1} - 1f_{i+2}) \quad , \end{aligned} \tag{33}$$

and the weights ω_k in WENO-JS are given by

$$\omega_k^{(JS)} = \frac{\alpha_k}{\sum_{n=0}^{r-1} \alpha_n} \quad \text{with} \quad \alpha_k = \frac{\bar{\omega}_k}{(\varepsilon + IS_k)^p} \quad , \tag{34}$$

where $\epsilon = 10^{-6}$ is used to prevent division by zero, IS_k is the indicator of smoothness for the kth stencil, $\bar{\omega}_k$ is the linear weight, and p is a constant used to accelerate the rate at which $\omega_k^{(JS)}$ goes to zero in non-smooth regions. On one hand, the linear weighted combination of numerical flux approximations must yield a fifth-order upstream central scheme, i.e.,

$$\bar{\omega}_0 \widehat{f}^0_{i+\frac{1}{2}} + \bar{\omega}_1 \widehat{f}^1_{i+\frac{1}{2}} + \bar{\omega}_2 \widehat{f}^2_{i+\frac{1}{2}} = \frac{1}{60}(2f_{i-2} - 13f_{i-1} + 47f_i + 27f_{i+1} - 3f_{i+2}) \quad , \tag{35}$$

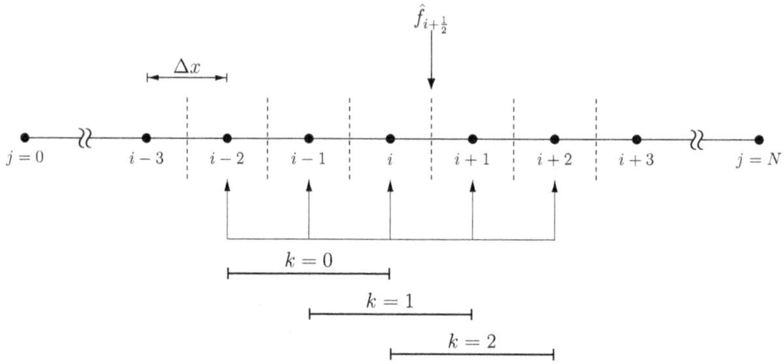

Figure 5. Fifth-Order WENO-JS scheme stencil dependence.

and, hence, the linear weights must be

$$\bar{\omega}_0 = \frac{1}{10} \quad , \quad \bar{\omega}_1 = \frac{6}{10} \quad \text{and} \quad \bar{\omega}_2 = \frac{3}{10} \quad . \tag{36}$$

On the other hand, the smoothness indicator should converge to one, making the linear and nonlinear weights equivalent. This can be achieved using the following definition [33]:

$$IS_k = \sum_{j=1}^{2} \Delta x^{2j-1} \int_{x_{i-\frac{1}{2}}}^{x_{i+\frac{1}{2}}} \left(\frac{d^j \hat{f}^k}{dx^j} \right)^2 dx \quad , \tag{37}$$

which depends on the polynomial approximations employed. In the particular case of the WENO-JS scheme, these indicators of smoothness become

$$\begin{aligned} IS_0 &= \frac{13}{12}(f_{j-2} - 2f_{j-1} + f_j)^2 + \frac{1}{4}(f_{j-2} - 4f_{j-1} + 3f_j)^2 \quad , \\ IS_1 &= \frac{13}{12}(f_{j-1} - 2f_j + f_{j+1})^2 + \frac{1}{4}(-f_{j-1} + f_{j+1})^2 \quad \text{and} \\ IS_2 &= \frac{13}{12}(f_j - 2f_{j+1} + f_{j+2})^2 + \frac{1}{4}(3f_j - 4f_{j+1} + f_{j+2})^2 \quad . \end{aligned} \tag{38}$$

After the early seminal contributions [32,33], a diverse spectrum of WENO schemes has emerged. For instance, the WENO-Z scheme [35] introduces a new procedure for the calculation of the nonlinear weights, i.e.,

$$\omega_k^{(Z)} = \frac{\alpha_k^z}{\sum_{n=0}^{r-1} \alpha_n^z} \quad \text{with} \quad \alpha_k^z = \bar{\omega}_k \left[1 + \left(\frac{\tau}{\varepsilon + IS_k} \right)^p \right] \quad , \tag{39}$$

where $\tau = |IS_0 - IS_2|$. Another example is the WENO-ZP [36], which proposes a different formulation for the weights, namely

$$\alpha_k^{zp} = \bar{\omega}_k \left[1 + \left(\frac{\tau + \varepsilon}{\varepsilon + IS_k} \right)^p + \lambda \left(\frac{IS_k, +\varepsilon}{\tau + \varepsilon} \right) \right] \quad , \tag{40}$$

noting that it reduces to the WENO-Z scheme when $\lambda = 0$, although $\lambda = \Delta x^{2/3}$ is suggested instead. For the sake of simplicity, the variants that modify the reconstruction polynomials are omitted. Nevertheless, their derivation can be found elsewhere [37,39,56].

Extension to a System of Equations

In the case of a system of hyperbolic equations, i.e.,

$$\frac{\partial \boldsymbol{q}}{\partial t} + \frac{\partial \boldsymbol{f}}{\partial x} = 0 \quad , \tag{41}$$

$\boldsymbol{q} = \boldsymbol{q}(x,t)$ is the vector of conservative variables and $\boldsymbol{f} = \boldsymbol{f}(\boldsymbol{q}(x,t))$ is the vector of conservative fluxes. It is possible rewrite this equation as

$$\frac{\partial \boldsymbol{q}}{\partial t} + \frac{\partial \boldsymbol{f}}{\partial \boldsymbol{q}}\frac{\partial \boldsymbol{q}}{\partial x} = 0 \quad , \tag{42}$$

using the chain rule. Solving Equation (42) is known as the component-wise approach. A different approach, known as the characteristic-wise approach, can be pursued by taking advantage of Equation (41)'s hyperbolicity. As a consequence, $\partial \boldsymbol{f}/\partial \boldsymbol{q}$ has a complete set of real eigenvalues, and also, Equation (42) can be decoupled.

When using a high-order discretization, the characteristic-wise approach has provided more accurate results than its component-wise counterpart. It can be implemented by first defining $\boldsymbol{R}(\boldsymbol{q})$ and $\boldsymbol{L}(\boldsymbol{q}) = \boldsymbol{R}^{-1}$, which are the right and left eigenvector matrixes of $\partial \boldsymbol{f}/\partial \boldsymbol{q}$, respectively. Hence, the Lax–Friedrichs flux split for systems of equations is written as

$$\boldsymbol{f}_j^{\pm} = \frac{1}{2}\left(\boldsymbol{f}_j \pm \boldsymbol{R}_{i+\frac{1}{2}}\boldsymbol{\Lambda}_{i+\frac{1}{2}}\boldsymbol{L}_{i+\frac{1}{2}}\boldsymbol{q}_j\right) \quad \text{with} \quad j \in [i-l, i+l] \quad , \tag{43}$$

where l is the stencil width. Furthermore, $\boldsymbol{R}_{i+1/2} = \boldsymbol{R}(\boldsymbol{q}_{i+1/2})$ and $\boldsymbol{L}_{i+1/2} = \boldsymbol{L}(\boldsymbol{q}_{i+1/2})$, where $\boldsymbol{q}_{i+1/2}$ is an average state between i and $i+1$, such as the Roe average [12], and

$$\boldsymbol{\Lambda}_{i+\frac{1}{2}} = Diag\left[\lambda^{(0)}, \lambda^{(1)}, \ldots, \lambda^{(m)}\right] \quad \text{with} \quad \lambda^{(p)} = \max_{\boldsymbol{q}}\left|\lambda^{(p)}(\boldsymbol{q}_{i+\frac{1}{2}})\right| \quad , \tag{44}$$

where m is the number of equations and $\lambda^{(p)}$ is the pth eigenvalue of $\partial \boldsymbol{f}/\partial \boldsymbol{q}$. Multiplying Equation (43) by $\boldsymbol{L}_{i+\frac{1}{2}}$ as well as denoting $\bar{\boldsymbol{f}}_j = \boldsymbol{L}_{i+\frac{1}{2}}\boldsymbol{f}_j$ and $\boldsymbol{w}_j = \boldsymbol{L}_{i+\frac{1}{2}}\boldsymbol{q}_j$, yields

$$\bar{\boldsymbol{f}}_j^{\pm} = \frac{1}{2}\left(\bar{\boldsymbol{f}}_j \pm \boldsymbol{\Lambda}_{i+\frac{1}{2}}\boldsymbol{w}_j\right) \quad , \tag{45}$$

which is the projection of positive and negative fluxes into the characteristic space. A WENO reconstruction of $\bar{\boldsymbol{f}}_j^{\pm}$ generates the numerical flux function approximation $\widehat{\bar{\boldsymbol{f}}}{}^{\pm}_{i+\frac{1}{2}}$. Projecting back to the physical space with $\widehat{\boldsymbol{f}}{}^{\pm}_{i+\frac{1}{2}} = \boldsymbol{R}_{i+\frac{1}{2}}\widehat{\bar{\boldsymbol{f}}}{}^{\pm}_{i+\frac{1}{2}}$, it is possible to build the scheme using Equation (28). Multi-dimensional problems are built in the same way, one dimension at a time. The eigenvalues and eigenvectors can be found elsewhere [65].

2.3.3. Temporal Integration

The spatial discretization of both viscous and inviscid terms present in Equation (10) leads to the system of ordinary differential equations

$$\frac{d\boldsymbol{q}}{dt} = \mathcal{F}(\boldsymbol{q}) \quad , \tag{46}$$

where $\mathcal{F}(\boldsymbol{q})$ represents discrete residue obtained after spatial discretization. Following the method of lines, the time integration of Equation (46) can be now discussed. The simplest way of marching it forward in time is

$$\boldsymbol{q}^{n+1} = \boldsymbol{q}^n + \Delta t \mathcal{F}(\boldsymbol{q}^n) + \mathbf{O}(\Delta t) \quad , \tag{47}$$

known as the explicit Euler method, which is first-order-accurate in time. Higher-order time integration schemes also exist. They are designed to increase scheme accuracy in time, reducing the CPU time required to generate highly accurate solutions. Runge–Kutta (RK) schemes are arguably the most used multi-stage high-order marching schemes. However, classical schemes [5] that do not satisfy the nonlinear proprieties required by a WENO discretization lead to spurious oscillations near discontinuities.

Strong-Stability-Preserving (SSP) RK schemes [47] ensure the preservation of these nonlinear stability proprieties. They can be written as

$$\begin{aligned} \boldsymbol{k}^{(0)} &= \boldsymbol{q}^n \quad , \\ \boldsymbol{k}^{(i)} &= \sum_{k=0}^{i-1} \left(\alpha_{i,k}\, \boldsymbol{k}^{(k)} + \beta_{i,k}\, \Delta t\, \mathcal{F}(\boldsymbol{k}^{(k)}) \right) \quad \text{with} \quad 1 \leq i \leq s \quad , \\ \boldsymbol{q}^{n+1} &= \boldsymbol{k}^{(s)} \quad , \end{aligned} \tag{48}$$

where s is the number of stages and $\boldsymbol{k}^{(i)}$ is the intermediate stage variable vector. It is important to note that SSPRK schemes are subject to a nonlinear time-step restriction that is more stringent than its linear counterpart. In other words, the aforementioned nonlinear stability properties will only be preserved when $\Delta t \leq C \Delta t_{EE_{max}}$, for a given C constant.

Both the optimal second-order SSPRK scheme, given by

$$\begin{aligned} \boldsymbol{k}^{(1)} &= \boldsymbol{q}^n + \Delta t\, \mathcal{F}(\boldsymbol{q}^n) \quad \text{and} \\ \boldsymbol{q}^{n+1} &= \frac{1}{2} \boldsymbol{q}^n + \frac{1}{2} \boldsymbol{k}^{(1)} + \frac{\Delta t}{2}\, \mathcal{F}(\boldsymbol{k}^{(1)}) \quad , \end{aligned} \tag{49}$$

and the optimal third-order SSPRK scheme, given by

$$\begin{aligned} \boldsymbol{k}^{(1)} &= \boldsymbol{q}^n + \Delta t\, \mathcal{F}(\boldsymbol{q}^n) \quad , \\ \boldsymbol{k}^{(2)} &= \frac{3}{4} \boldsymbol{q}^n + \frac{1}{4} \boldsymbol{k}^{(1)} + \frac{\Delta t}{4}\, \mathcal{F}(\boldsymbol{k}^{(1)}) \quad \text{and} \\ \boldsymbol{q}^{n+1} &= \frac{1}{3} \boldsymbol{q}^n + \frac{2}{3} \boldsymbol{k}^{(2)} + \frac{2\,\Delta t}{3}\, \mathcal{F}(\boldsymbol{k}^{(2)}) \quad , \end{aligned} \tag{50}$$

are currently implemented in 3D4S [49].

3. Results

Three different test cases are presented to analyze the ability of high-order schemes to achieve accurate steady states when simulating the Navier–Stokes equations. First, a Mach 3 flow over a compression ramp is discussed. The importance of such problem comes from the preponderance of corners in high-speed flows. The adverse (favorable) pressure gradient upstream (downstream) of the corner separates (reattaches) the boundary layer. At these same regions of the flow, compression fans appear and lead to the formation of separation and reattachment shocks. Hence, the main goal here is to evaluate steady-state convergence in the presence of shock boundary layer interactions (SBLIs). Subsequently, a Mach 6 flow over a blunt body is discussed. The body geometry is constructed by connecting a cylinder to a wedge through a C^1 curve. A bow shock appears near the leading edge and propagates away from the body. The relevance of this geometry arises from the effects that blunted noses have in the laminar–turbulent boundary layer transition. Finally, a Mach 6 flow upstream of a cylinder is discussed. This case is employed here to highlight the difference between different WENO schemes regarding steady-state convergence. 3D4S code verification can be found elsewhere in the literature [66–68].

3.1. Compression Ramp

The first attempts to obtain such steady states used low-order schemes, which are here based on the explicit Euler scheme for time integration and a second-order accurate spatial discretization, which employed a TVD scheme with the Roe flux-difference splitting and the SuperBee fluxn limiter for inviscid fluxes as well as a conservative central-difference scheme for the viscous fluxes. No slip and isothermal walls are considered with $T_w = T_\infty = 216.67$. Figure 6 presents a sketch of the analytic grid used in these simulations, and a complete discussion of grid generation can be found elsewhere [6]. The density wall-normal profile at the corner (left) and density-maximum residue convergence in time (right) obtained for three different free stream Reynolds numbers, namely $Re_\infty = 2 \times 10^3$ (top), 5×10^3 (middle), and 10×10^3 (bottom), are show in Figure 7 for the isotherm wall under different grid sizes. Simulations require an increasingly larger number of grid points for their spatial profiles to converge towards the same tolerances as the free stream Reynolds number increases. Nevertheless, the number of grid points used in this figure shows spatial profiles that are grid-converged for graphical accuracy. On the other hand, residue convergence in time has a much stronger dependence on the free-stream Reynolds number. Residue converges in time for all grids employed when $Re_\infty = 2 \times 10^3$, for only the four largest grids employed when $Re_\infty = 5 \times 10^3$, and for none of the grids employed when $Re_\infty = 10 \times 10^3$. Convergence is quantified here by how much the maximum density residue decreases in time. It does so by approximately 10 orders of magnitude in the former two cases. In the latter case, however, it does so by only approximately four orders of magnitude, suggesting a six-digit accuracy loss. These low-order trends were also observed when using OpenFOAM, even though its second-order accurate finite-volume solver used approximately 16 million grid points to simulate the $Re_\infty = 10^4$ case [66].

In order to emphasize the impact of higher-order schemes in such a steady-state simulation, Figure 8 (left) shows the convergence history for $Re_\infty = 10 \times 10^3$ considering an adiabatic wall when a fifth-order WENO-JS and a fourth-order central difference scheme are applied to inviscid and viscous flux discretizations, respectively. Unlike the low-order simulations, a 10-orders-of-magnitude residue decrease in time is observed when using a $N_x = 2401$, $N_y = 801$ grid, while temporal content is still present in the residue when using a $N_x = 1801$, $N_y = 601$ grid. Figure 8 (right) shows the residue convergence in time for both grids. One can note in the latter case that numerical oscillations are generated at the leading-edge shock and propagate downstream, preventing residue convergence in the time-to-machine precision. Furthermore, wall-properties of case $Re_\infty = 10 \times 10^3$ are presented in Appendix A.1.

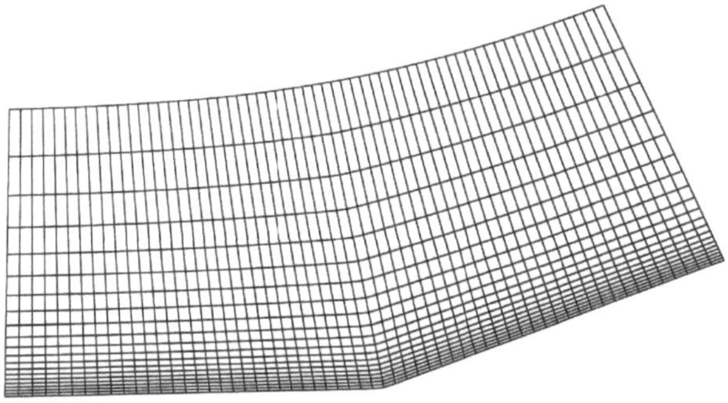

Figure 6. Compression ramp grid using $N_x = 61$ and $N_y = 21$ points.

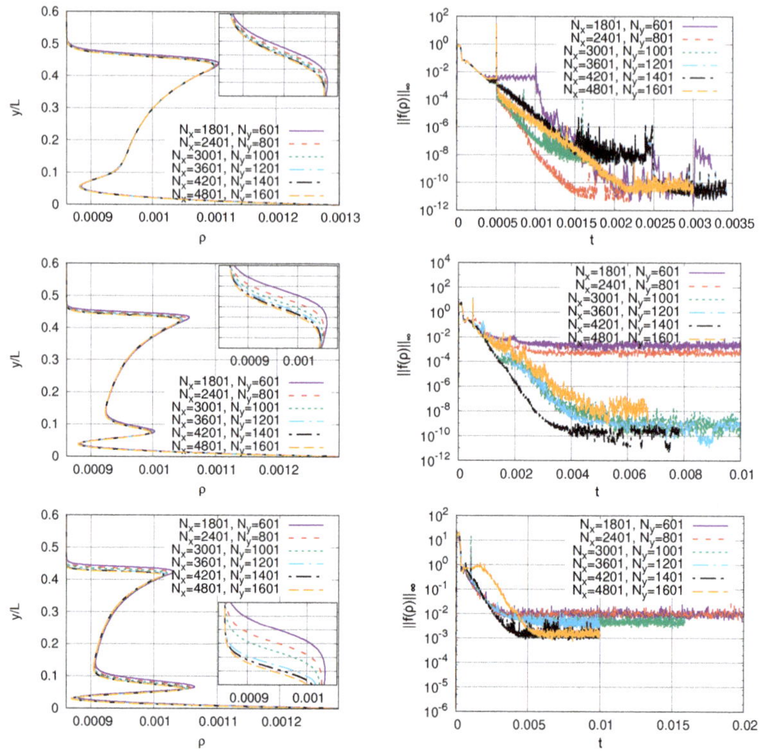

Figure 7. Density wall-normal profile at the corner (**left**) and density maximum residue convergence in time (**right**) obtained with $Re_\infty = 2 \times 10^3$ (top), 5×10^3 (**middle**), and 10×10^3 (**bottom**) from low-order 3D4S for the isothermal wall case under different grid sizes.

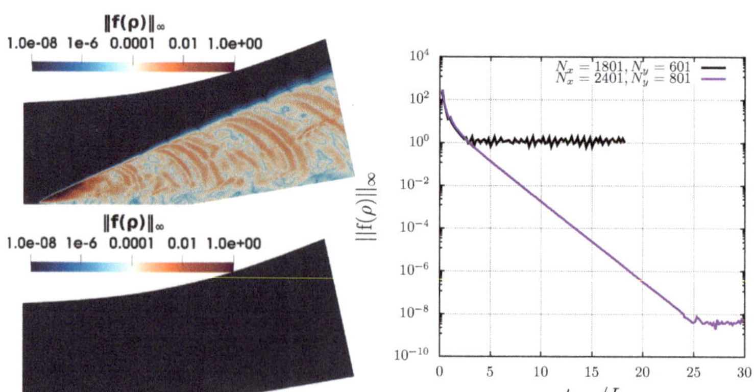

Figure 8. Convergence history (**right**) for the compression ramp simulation using different grid sizes, as well as the steady-state residue of mass conservation equation using (**left top**) $N_x = 1801$, $N_y = 601$ and (**left bottom**) $N_x = 2401$, $N_y = 801$.

3.2. Blunt Wedge

WENO schemes developed for unsteady simulations are known to introduce post-shock oscillations, with a magnitude proportional to the local truncation error, that prevent residue convergence in time toward machine zero [38]. This issue was overcome with the development of WENO5-ZQ for steady simulations [37,39–41]. Its ability to do so is verified here by simulating the hypersonic flow over a blunt wedge.

Flow conditions are set by imposing the free stream Mach number $Ma_\infty = 6$, the Reynolds number based on the nose radius $Re_R = 1.5 \times 10^3$, the free stream Prandtl number $Pr_\infty = 0.72$, the ratio between constant specific heats $\gamma = 1.4$, the free stream temperature $T_\infty = 200$ K, the dynamic viscosity defined by Sutherland's law, and the thermal conductivity defined by the constant Prandtl assumption. Furthermore, a no-slip and isothermal wall with $T_w = T_0$ is considered. Grid-converged simulations were achieved with $(N_x, N_y) = (1201, 801)$ using a time step of $\Delta t = 5 \times 10^{-6}$. Time integration was performed using the third-order accurate and three-stage ERK scheme, and the viscous fluxes were discretized with the fourth-order accurate conservative central-difference scheme. Figure 9 shows a sketch of the elliptic grid used in these simulations. A complete discussion of grid generation can be found elsewhere [63].

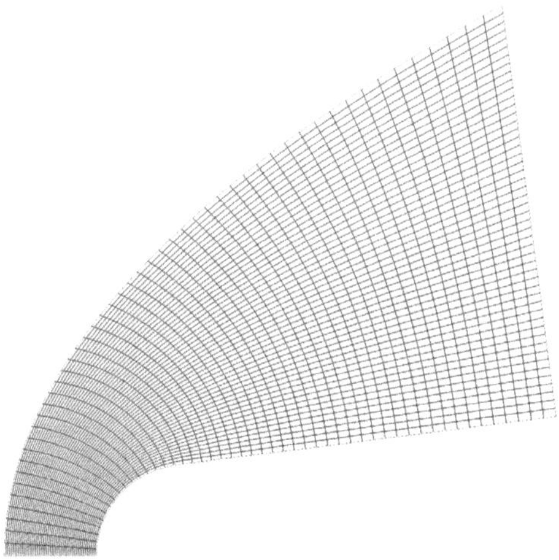

Figure 9. Blunt-body grid using $N_x = 51$ and $N_y = 51$ points.

Figure 10 (right) shows the convergence in time of the L_∞ norm of the residue for the mass, momentum and energy conservation equations using WENO5-JS (dashed) and WENO5-ZQ (solid) schemes. As already expected, post-shock oscillations prevent the maximum residue calculated with the former from converging in time to machine zero. Maximum residues calculated with the latter, on the other hand, decrease by approximately 10 orders of magnitude in time. Figure 10 (left) shows the spatial distribution of the steady mass conservation equation residue obtained while using WENO-JS (top) and WENO-ZQ (bottom) schemes. One can note that the maximum residues are, in fact, located at the shock. Elsewhere, these residues are two to five orders of magnitude smaller. There was no need to either align the grid and the shock or cluster grid points around the shock whenever using WENO. This is in contrast with said literature, which simulated this flow using a second-order TVD-type scheme instead. They spent a large amount of effort trying to improve the alignment between the grid and steady shock, as well as clustering grid points around this shock due to the strong gradients in the stagnation region. Moreover, wall-properties of case $Re_\infty = 1.5 \times 10^3$ are presented in Appendix A.2.

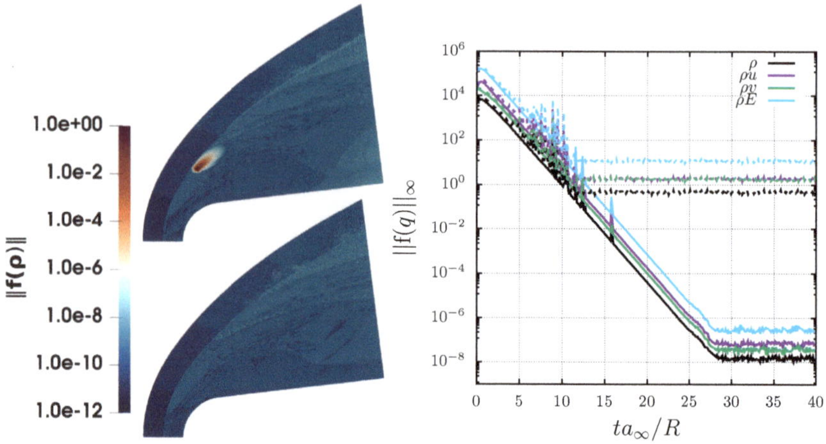

Figure 10. Instantaneous spatial distribution of the mass conservation residue obtained while using WENO-JS (**left top**) and WENO-ZQ (**left bottom**). Residue time convergence to the steady state (**right**) for the blunt-body simulation using WENO5-JS (dashed) and WENO5-ZQ (solid) schemes.

3.3. Quarter Cylinder

WENO-ZQ, which is especially designed for steady simulations, has shown better results than the classical WENO-JS scheme so far. This is also true for other WENO schemes as well; i.e., they also have a worse performance for steady simulations when compared to WENO-ZQ. In order illustrate this, a set of simulation results is presented for a Mach 6 flow upstream of a cylinder using four different WENO schemes, namely WENO-JS [33], WENO-ZP [36], WENO-ZPs [36,56] and WENO-ZQ [37]. The reference Reynolds number is $Re_D = 1 \times 10^4$ and the free-stream temperature $T_\infty = 271.155$ K. Furthermore, a no-slip and adiabatic wall is considered. Figure 11 shows a sketch of the analytic grid used in these simulations, where the grid generation formulas can be found elsewhere [33].

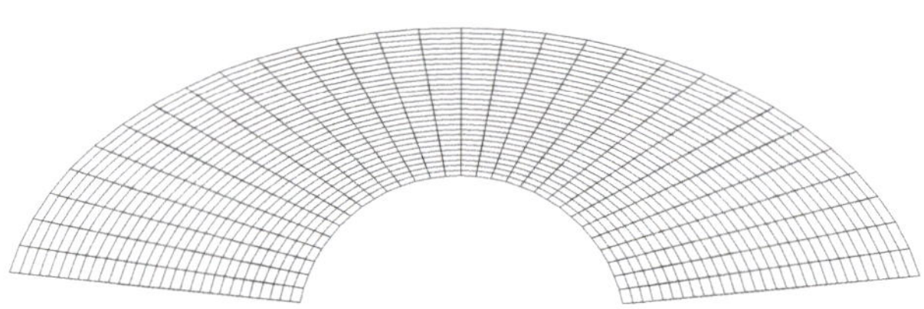

Figure 11. Cylinder grid using $N_x = 31$ and $N_y = 31$ points.

Figure 12 shows the mass conservation equation's maximum residue convergence in time obtained while employing WENO-JS (top left), WENO-ZP (top right), WENO-ZPs (bottom left), and WENO-ZQ (bottom right). For each case, a grid-independence study was performed. While the former three schemes were only able to reduce this residue in time by approximately 3 orders of magnitude, WENO-ZQ was able to do so for 12 orders of magnitude, essentially reaching machine precision. Figure 13 presents the spatial distribution of the steady density field obtained for these four cases. One can note that slight post-shock oscillations prevent the residue convergence to machine precision in the three leftmost plots, as occurred in the compression ramp and blunt-wedge cases. Furthermore, wall-properties of case $Re_\infty = 10 \times 10^3$ are presented in Appendix A.3.

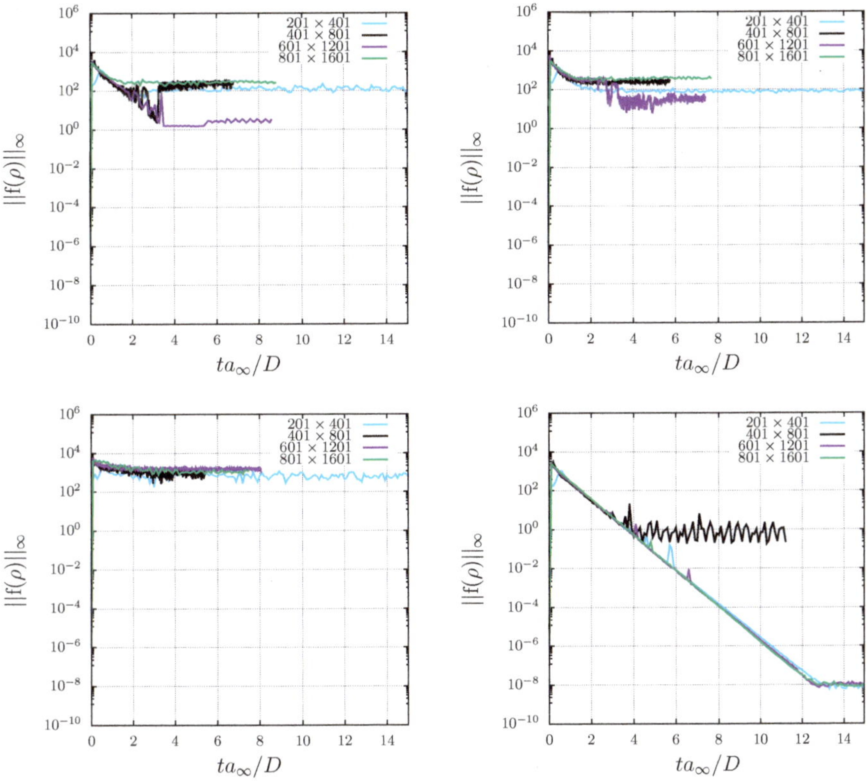

Figure 12. Density maximum residue convergence in time obtained with WENO-JS (**top left**), WENO-ZP (**top right**), WENO-ZPs (**bottom left**), and WENO-ZQ (**bottom right**) for different grid sizes.

Figure 13. Steady-state density residue spatial distribution obtained using WENO-JS (first column), WENO-ZP (second colomn), WENO-ZPs (third column), and WENO-ZQ (fourth column). The results were obtained considering the $(N_x, N_y) = (801, 1201)$ grid.

4. Conclusions

The ability of high-order shock-capturing schemes to achieve residue convergence in time and, hence, generate accurate steady-states was discussed in the present work. A carefully analysis of three different supersonic flows showed the benefits of using higher-order schemes. They are more efficient when compared to their lower-order counterparts. As the Reynolds number increases, the high grid resolution required to achieve machine precision steady-states when using lower-order schemes makes them unfeasible. A considerably smaller grid resolution is required to do so when using high-order schemes. However, this is not true of all high-order schemes. Only the WENO-ZQ scheme, which is specially designed for steady-state simulations, was able to achieve machine precision residues in all test cases explored in this study. It is important to highlight that, even though a double-precision floating point was used, machine precision for the computation of residue $\mathcal{F}$ was greater than $O(10^{-16})$. This is expected since the increment per time step Δt for the conservative variables is $O(\Delta t\,\mathcal{F})$, which leads to an increment of $O(10^{-16})$.

Funding: This research was funded by the US grants FA9550-18-1-0419, FA9550-22-1-0033 and FA8655-23-1-7031 under SOARD/AFOSR as well as the Brazilian agencies CAPES (graduate fellowships), FAPERJ (CNE and Bolsa Nota 10 SEI E-26/204.043/2022) and CNPq (PQ1D).

Acknowledgments: The authors would like to acknowledge Pedro Paredes (NIA/NASA LaRC) and Vassilis Theofilis (Technion) for their support during the development of this work.

Conflicts of Interest: The authors declare no conflicts of interest.

Appendix A. Wall Properties

Appendix A.1. Compression Ramp

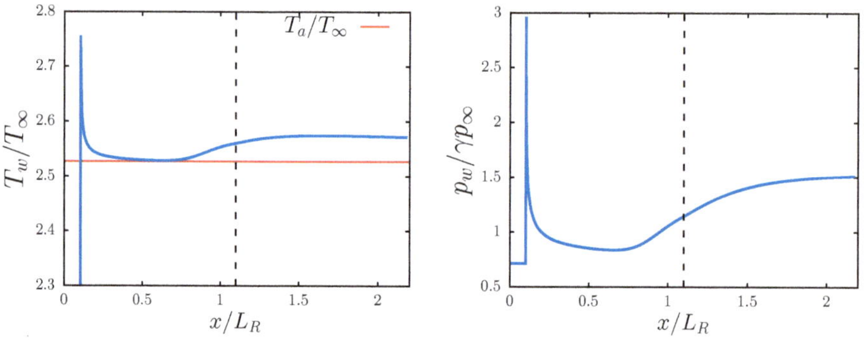

Figure A1. Dimensionless wall temperature (**left**) and pressure (**right**). In both figures, the vertical dashed black line marks the corner position. The solid red line represents the adiabatic wall temperature using the theoretical recovery temperature for laminar boundary layers. The leading edge is placed at $x/L_R = 0.1$, and the reference length is the distance from the leading edge to the corner.

Appendix A.2. Blunt Body

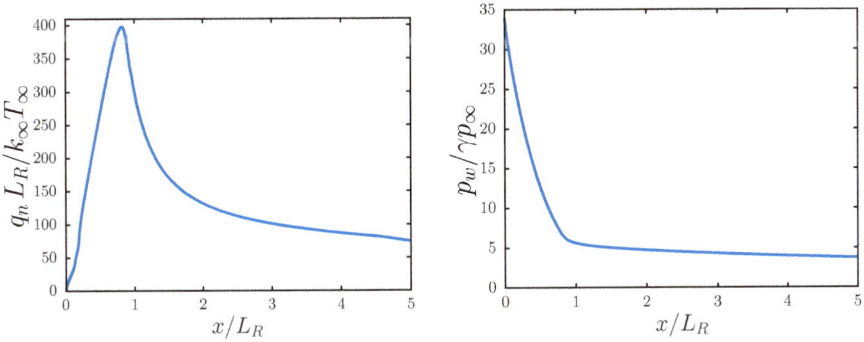

Figure A2. Dimensionless wall heat flux (**left**) and pressure (**right**). The reference length L_R is the nose radius and x/L_R measures the distance along the wall.

Appendix A.3. Cylinder

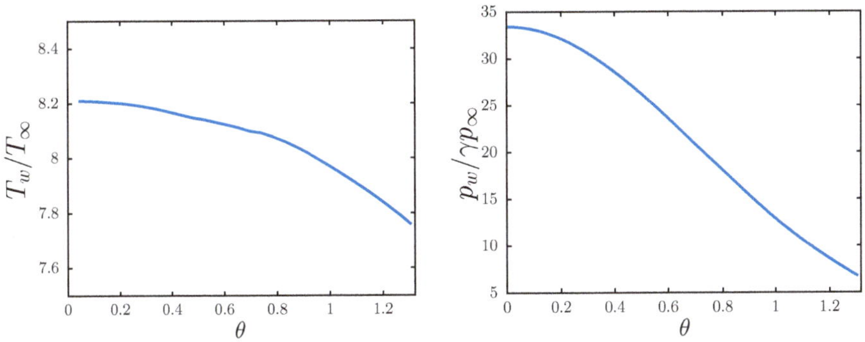

Figure A3. Dimensionless wall temperature (**left**) and pressure (**right**). $\theta = 0$ corresponds to the stagnation point. Due to symmetry, only the upper half of the solution is presented.

References

1. Versteeg, H.K.; Malalasekera, W. *An Introduction to Computational Fluid Dynamics. The Finite Volume Method*; Prentice Hall: London, UK, 1995.
2. Tannehill, J.C.; Anderson, D.A.; Pletcher, R.H. *Computational Fluid Mechanics and Heat Transfer*; Taylot & Francis: Philadelphia, PA, USA, 1997.
3. Lomax, H.; Pulliam, T.H.; Zingg, D.W. *Fudamentals of Computational Fluid Dynamics*; Scientific Computation, Springer & Verlag: Berlin/Heidelberg, Germany, 2001.
4. Hirsch, C. *Numerical Computation of Internal and External Flows*; Fundamentals of Computational Fluid Dynamics; Elsevier: Amsterdam, The Netherlands, 2007; Volume 1.
5. Butcher, J.C. *Numerical Methods for Ordinary Differential Equations*; John Wiley & Sons, Inc.: Hoboken, NJ, USA, 2008.
6. Santos, R.D. A Study of Runge–Kutta Schemes with Stronge Linear and Nonlinear Numerical Stability. Ph.D. Thesis, Fluminense Federal University, Niterói, Brazil, 2020.
7. Assis, J.C. Grid Shock Alignment, Boundary Conditions Axial Symmetry and 3D simulations of Hypersonic Flows over Blunt Bodies. Master's Thesis, Universidade Federal Fluminense, Niterói, Brazil, 2024.
8. Harbison, S.P.; Steele, G.L., Jr. *C—A Reference Manual*, 4th ed.; Tartan, Inc.: Monroeville, PA, USA, 1995.
9. Santiago, K.F.S.; Alves, L.S.B. Conservative and nondispersive schemes for diffusion terms with strong property variations. *Numer. Heat Transf. Part B Fundam.* **2017**, *71*, 133–145. [CrossRef]
10. Harten, A. High resolution schemes for hyperbolic conservation laws. *J. Comput. Phys.* **1983**, *49*, 357–393. [CrossRef]
11. Hirsch, C. *Numerical Computation of Internal and External Flows, Computational Methods for Inviscid and Viscous Flows*; Wiley: Hoboken, NJ, USA, 1991.
12. Laney, C.B. (Ed.) *Computational Gas Dynamics*; Cambridge University Press: Cambridge, UK, 1998.

13. Roe, P.L. Some contributions to the modelling of discontinuous flows. In *Large-Scale Computations in Fluid Mechanics*; Engquist, B.E., Osher, S., Somerville, R.C.J., Eds.; AMS-SIAM: Providence, RI, USA, 1985; Volume 22, pp. 163–193.
14. Roe, P.L. *Numerical Algorithms for the Linear Wave Equation*; Royal Aircraft Establishment: Farnborough, UK, 1981.
15. Roe, P.L.; Baines, M.J. Algorithms for advection and shock problems. In Proceedings of the 4th Conference on Numerical Methods in Fluid Mechanics, Paris, France, 7–9 October 1981.
16. Sweby, P.K.; Baines, M.J. On convergence of Roe's scheme for the general non-linear scalar wave equation. *J. Comput. Phys.* **1984**, *56*, 135–148. [CrossRef]
17. van Leer, B. Towards the ultimate conservative difference scheme. IV. A new approach to numerical convection. *J. Comput. Phys.* **1977**, *23*, 276–299. [CrossRef]
18. van Leer, B. Towards the ultimate conservative difference scheme. II. Monotonicity and conservation combined in a second-order scheme. *J. Comput. Phys.* **1974**, *14*, 361–370. [CrossRef]
19. Koren, B. Chapter A Robust Upwind Discretization Method for Advection, Diffusion and Source Terms. In *Numerical Methods for Advection-Diffusion Problems*; Centrum voor Wiskunde en Informatica: Amsterdam, The Netherlands, 1993.
20. Sweby, P.K. High Resolution Schemes Using FLux Limiters for Hyperbolic Conservation Laws. *SIAM J. Numer. Anal.* **1984**, *21*, 995–1011. [CrossRef]
21. Chakravarthy, S.; Osher, S. High resolution applications of the Osher upwind scheme for the Euler equations. In Proceedings of the 6th Computational Fluid Dynamics Conference Danvers, Danvers, MA, USA, 13–15 July 1983.
22. Waterson, N.P.; Deconinck, H. *A Unified Approach to the Design and Applications of Bounded Higher-Order Convection Schemes*; Preprint/Von Karman Institute for Fluid Dynamics, VKI: Sint-Genesius-Rode, Belgium, 1995; pp. 182–207.
23. van Albada, G.D.; van Leer, B.; Roberts, W.W. A Comparative Study of Computational Methods in Cosmic Gas Dynamics. *Astron. Astrophys.* **1982**, *108*, 76–84.
24. Boris, J.P.; Book, D.L. Flux-Corrected Transport. I. SHASTA, A Fluid Transport Algorithm That Works. *J. Comput. Phys.* **1973**, *11*, 38–69. [CrossRef]
25. van Leer, B. Towards the ultimate conservative difference scheme I. The quest of monotonicity. In *Proceedings of the Third International Conference on Numerical Methods in Fluid Mechanics*; Lecture Notes in Physics; Springer: Berlin/Heidelberg, Germany, 1973; Volume Fundamental Numerical Techniques; pp. 163–168.
26. Goodman, J.B.; LeVeque, R.J. On the Accuracy of Stable Schemes for 2D Scalar Conservation Laws. *Math. Comput.* **1985**, *45*, 15–21. [CrossRef]
27. Harten, A.; Engquist, B.; Osher, S.; Chakravarthyy, S.R. Uniformly High Order Accurate Essentially Non-Oscillatory Schemes, III. *J. Comput. Phys.* **1987**, *131*, 3–47. [CrossRef]
28. Godunov, S.K. A difference scheme for numerical solution of discontinuous solution of hydrodynamic equations. *Math. Sb.* **1959**, *47*, 271–306.
29. van Leer, B. Towards the ultimate conservative difference scheme. V. A second-order sequel to Godunov's method. *J. Comput. Phys.* **1979**, *32*, 101–136. [CrossRef]
30. Shu, C.W.; Osher, S. Efficient Implementation of Essentially Non-Oscillatory Shock-Capturing Schemes. *J. Comput. Phys.* **1988**, *77*, 439–471. [CrossRef]
31. Shu, C.W.; Osher, S. Efficient Implementation of Essentially Non-Oscillatory Shock-Capturing Schemes II. *J. Comput. Phys.* **1989**, *83*, 32–78. [CrossRef]
32. Liu, X.D.; Osher, S.; Chan, T. Weighted Essentially Non-Oscillatory Schemes. *J. Comput. Phys.* **1994**, *115*, 200–212. [CrossRef]
33. Jiang, G.S.; Shu, C.W. Efficient Implementation of Weighted ENO Schemes. *J. Comput. Phys.* **1996**, *126*, 202–228. [CrossRef]
34. Henrick, A.K.; Aslam, T.D.; Powers, J.M. Mapped weighted essentially non-oscillatory schemes: Achieving optimal order near critical points. *J. Comput. Phys.* **2005**, *207*, 542–567. [CrossRef]
35. Borges, R.; Carmona, M.; Costa, B.; Don, W.S. An improved weighted essentially non-oscillatory scheme for hyperbolic conservation laws. *J. Comput. Phys.* **2008**, *227*, 3191–3211. [CrossRef]
36. Acker, F.; Borges, R.; Costa, B. An improved WENO-Z scheme. *J. Comput. Phys.* **2016**, *313*, 726–753. [CrossRef]
37. Zhu, J.; Qiu, J. A new fifth order finite difference WENO scheme for solving hyperbolic conservation laws. *J. Comput. Phys.* **2016**, *318*, 110–121. [CrossRef]
38. Zhu, J.; Shu, C.W. Numerical study on the convergence to steady state solutions of a new class of high order WENO schemes. *J. Comput. Phys.* **2017**, *349*, 80–96. [CrossRef]
39. Zhu, J.; Shu, C.W. A new type of multi-resolution WENO schemes with increasingly higher order of accuracy. *J. Comput. Phys.* **2018**, *375*, 659–683. [CrossRef]
40. Zhu, J.; Shu, C.W. Convergence to Steady-State Solutions of the New Type of High-Order Multi-resolution WENO Schemes: A Numerical Study. *Commun. Appl. Math. Comput.* **2019**, *2*, 429–460. [CrossRef]
41. Zhang, S.; Zhu, J.; Shu, C.W. A brief review on the convergence to steady state solutions of Euler equations with high-order WENO schemes. *Adv. Aerodyn.* **2019**, *1*, 25. [CrossRef]
42. Toro, E.F. *Riemann Solvers and Numerical Methods for Fluid Dynamics: A Practical Introduction*; Springer-Verlag: Berlin/Heidelberg, Germany, 2009.
43. Roe, P.L. Approximate Reimann Solvers, Parameter Vectors and Difference Schemes. *J. Comput. Phys.* **1981**, *43*, 357–372. [CrossRef]

44. Harten, A.; Lax, P.D.; van Leer, B. On Upstream Differencing and Godunov-Type Schemes for Hyperbolic Conservation Laws. *SIAM Rev.* **1983**, *25*, 35–61. [CrossRef]
45. Einfeldt, B. On Godunov-Type Methods for Gas Dynamics. *SIAM J. Numer. Anal.* **1988**, *25*, 294–318. [CrossRef]
46. Steger, J.L.; Warming, R.F. Flux vector splitting of the inviscid gas dynamic equations with application to finite-difference methods. *J. Comput. Phys.* **1981**, *40*, 263–293. [CrossRef]
47. Gottlieb, S.; Shu, C.W. Total-Variation-Diminishing Runge–Kutta Schemes. *Math. Comput.* **1998**, *67*, 73–85. [CrossRef]
48. Spiteri, R.J.; Ruuth, S.J. A new class of optimal high-order strong-stability- preserving time discretization methods. *SIAM J. Numer. Anal.* **2002**, *40*, 469–491. [CrossRef]
49. Gottlieb, S.; Ketcheson, D.; Shu, C.W. High-order strong stability preserving time discretizations. *J. Sci. Comput.* **2009**, *38*, 251–289. [CrossRef]
50. Ferracina, L.; Spijker, M.N. An Extension and Analysis of the Shu-Osher Representation of Runge–Kutta Methods. *Math. Comput.* **2004**, *74*, 201–219. [CrossRef]
51. Ferracina, L.; Spijker, M.N. *Computing Optimal Monotonicity-Preserving Runge–Kutta Methods*; Technical Report MI2005-07; Leiden University: Leiden, The Netherlands, 2005.
52. Ferracina, L.; Spijker, M.N. Strong Stability of Singly-Diagonally-Implicit Runge–Kutta Methods. *Appl. Numer. Math.* **2008**, *58*, 1675–1686. [CrossRef]
53. Ketcheson, D.I.; Macdonald, C.B.; Gottlieb, S. Optimal Implicit Strong Stability Preserving Runge–Kutta Methods. *Appl. Numer. Math.* **2009**, *59*, 373–392. [CrossRef]
54. Pareschi, L.; Russo, G. implicit–explicit Runge–Kutta Schemes and Applications to Hyperbolic Systems with Relaxation. *J. Sci. Comput.* **2005**, *25*, 129–155.
55. Santos, R.D.; Alves, L.S. A comparative analysis of explicit, IMEX and implicit strong stability preserving Runge–Kutta schemes. *Appl. Numer. Math.* **2021**, *159*, 204–220. [CrossRef]
56. Zhang, S.; Shu, C.W. A new smoothness indicator for the WENO schemes and its effect on the convergence to steady state solutions. *J. Sci. Comput.* **2007**, *31*, 273–305. [CrossRef]
57. Liseikin, V. *Grid Generation Methods*; Scientific Computation; Springer: Amsterdam, The Netherlands, 2009.
58. Thompson, J.F. Elliptic grid generation. *Appl. Math. Comput.* **1982**, *10–11*, 79–105. [CrossRef]
59. Thompson, J.F.; Soni, B.K.; Weatherill, N.P. *Handbook of Grid Generation*; CRC Press: Boca Raton, FL, USA, 1998.
60. Steger, J.L.; Sorenson, R.L. Automatic mesh-point clustering near a boundary in grid generation with elliptic partial differential equations. *J. Comput. Phys.* **1979**, *33*, 405–410. [CrossRef]
61. Hsu, K.; Lee, S.L. A numerical technique for two-dimensional grid generation with grid control at all of the boundaries. *J. Comput. Phys.* **1991**, *96*, 451–469. [CrossRef]
62. Spekreijse, S.P. Elliptic Grid Generation Based on Laplace Equations and Algebraic Transformations. *J. Comput. Phys.* **1995**, *118*, 38–61. [CrossRef]
63. Nunes, M.S.S. Generation of Steady-States for Stationary and Convectively Unstable Flows Using the Frequency Displacement Procedure. Master's Thesis, Fluminense Federal University, Niterói, Brazil, 2021.
64. Poinsot, T.J.; Lele, S.K. Boundary conditions for direct simulations of compressible viscous flows. *J. Comput. Phys.* **1992**, *101*, 104–129. [CrossRef]
65. Masatsuka, K. *I Do Like CFD*; Lulu. com: Morrisville, NC, USA, 2009; Volume 1.
66. Santos, R.D.; Alves, L.S.; Cerulus, N.; Theofilis, V. On Two-Dimensional Steady-States of Supersonic Flows over Compression Ramps and their Global Linear Instability. In Proceedings of the AIAA Scitech Forum, San Diego, CA, USA, 7–11 January 2019; pp. 2019–2321.
67. Burtsev, A.; Quintanilha, H.R.D.; Theofilis, V.; Santos, R.D.; Alves, L.S. Linear Instability Mechanisms of Supersonic Flow Past Blunt Bodies. In Proceedings of the AIAA Scitech Forum, Virtual, 11–15 and 19–21 January 2021; AIAA 2021-0050.
68. Santos, R.D.; Alves, L.S.; Cerulus, N.; Quintanilha, H.; Theofilis, V. Accurate Two Dimensional Steady-State for the Supersonic Flow over a Compression Corner. In Proceedings of the 33th ICAS Congress, Stockholm, Sweden, 4–9 September 2022.

Disclaimer/Publisher's Note: The statements, opinions and data contained in all publications are solely those of the individual author(s) and contributor(s) and not of MDPI and/or the editor(s). MDPI and/or the editor(s) disclaim responsibility for any injury to people or property resulting from any ideas, methods, instructions or products referred to in the content.

fluids

MDPI

Review

A Nonlinear Approach in the Quantification of Numerical Uncertainty by High-Order Methods for Compressible Turbulence with Shocks

H. C. Yee [1,*], P. K. Sweby [2], Björn Sjögreen [3] and D. V. Kotov [4,†]

1 NASA Ames Research Center, Moffett Field, CA 94035, USA
2 Department of Mathematics and Statistics, University of Reading, Whiteknights, Reading RG6 6AX, UK; p.k.sweby@reading.ac.uk
3 MultiD Analyses AB, 421 65 Göteborg, Sweden; bjorn.sjogreen@multid.se
4 Independent Researcher, Canada; dmitry.kotov84@gmail.com
* Correspondence: helen.m.yee@nasa.gov
† Self (Work done while associated with BARI, NASA Ames Research Center).

Abstract: This is a comprehensive overview on our research work to link interdisciplinary modeling and simulation techniques to improve the predictability and reliability simulations (PARs) of compressible turbulence with shock waves for general audiences who are not familiar with our nonlinear approach. This focused nonlinear approach is to integrate our **"nonlinear dynamical approach"** with our **"newly developed high order entropy-conserving, momentum-conserving and kinetic energy-preserving methods"** in the quantification of numerical uncertainty in highly nonlinear flow simulations. The central issue is that the solution space of discrete genuinely nonlinear systems is much larger than that of the corresponding **genuinely nonlinear continuous systems**, thus obtaining numerical solutions that might not be solutions of the continuous systems. Traditional uncertainty quantification (UQ) approaches in numerical simulations commonly employ linearized analysis that might not provide the true behavior of genuinely nonlinear physical fluid flows. Due to the rapid development of high-performance computing, the last two decades have been an era when computation is ahead of analysis and when very large-scale practical computations are increasingly used in poorly understood multiscale data-limited complex nonlinear physical problems and non-traditional fields. This is compounded by the fact that the numerical schemes used in production computational fluid dynamics (CFD) computer codes often do not take into consideration the genuinely nonlinear behavior of numerical methods for more realistic modeling and simulations. Often, the numerical methods used might have been developed for weakly nonlinear flow or different flow types other than the flow being investigated. In addition, some of these methods are not discretely physics-preserving (structure-preserving); this includes but is not limited to entropy-conserving, momentum-conserving and kinetic energy-preserving methods. Employing theories of nonlinear dynamics to guide the construction of more appropriate, stable and accurate numerical methods could help, e.g., (a) delineate solutions of the discretized counterparts but not solutions of the governing equations; (b) prevent numerical chaos or numerical "turbulence" leading to FALSE predication of transition to turbulence; (c) provide more reliable numerical simulations of nonlinear fluid dynamical systems, especially by direct numerical simulations (DNS), large eddy simulations (LES) and implicit large eddy simulations (ILES) simulations; and (d) prevent incorrect computed shock speeds for problems containing stiff nonlinear source terms, if present. For computation intensive turbulent flows, the desirable methods should also be efficient and exhibit scalable parallelism for current high-performance computing. Selected numerical examples to illustrate the genuinely nonlinear behavior of numerical methods and our integrated approach to improve PARs are included.

Keywords: nonlinear dynamics; discrete dynamics; dynamics of numerics; numerics of dynamics; uncertainty quantification; chaotic transients; direct numerical simulations; nonlinear approach to uncertainty quantification; high-order shock-capturing methods; entropy-conserving methods; physics-preserving (structure-preserving) methods; hybrid methods; nonlinear filter methods; momentum-conserving methods; kinetic-energy-preserving methods; compressible turbulence; gas dynamics; MHD

1. Introduction

This paper presents an overview on our continuing long-term effort in reliable and accurate high-order numerical method development by integrating our earlier work on the application of nonlinear dynamical systems theory to improve the understanding of global nonlinear behavior of finite discretization and their connection to numerical uncertainties in CFD and computational physics. See [1–3] and the references cited therein for more details. Our approach is to use interdisciplinary knowledge gained from nonlinear studies to aid the construction of more appropriate and to improve high-order numerical methods for the simulation of compressible turbulent flows. The areas of application are the modeling and simulation of turbulence combustion, supersonic turbulence and computational astrophysics, and aiding the design of the next-generation spacecrafts and high-speed civil transports. Supersonic turbulence occurs in, e.g., scramjet engines, re-entry spacecrafts, black hole accretions, star formations, stellar winds and volcanic eruptions. We also employ these guiding principles as a nonlinear approach for uncertainty quantification (UQ) of multiscale CFD simulations. Due to the fact that our integrated multi-disciplinary nonlinear approach is not well known in the CFD community, the intent of this comprehensive overview is for general audiences who are not familiar with our nonlinear approach. It is hoped that the long overview will be helpful.

This nonlinear approach is particularly vital to improve the predictability and reliability for numerical simulation (PARs) of time-accurate, long-time integration of compressible turbulence with shock waves by DNS, LES and ILES computations. One of the many key contributing factors is that even though the numerical methods were designed to be stable and accurate for rapidly developing flows, they are either too dissipative, encounter nonlinear instability or exhibit aliasing error and other anomalies that drastically deviate from linear or linearized analyses in long-time integration of the aforementioned genuinely nonlinear multiscale fluid flows. For compressible turbulence with shock waves, designing methods with the minimal amount of numerical dissipation while maintaining numerical stability and accuracy is an intricate balancing act. Proper adaptive numerical dissipation control in high-order methods would provide accurate and stable solutions and play an integral role in the success of PARs. See our numerical method development for a variety of flow types and characteristics presented in [1,2,4–30]. See also a recent article by Glimm et al. concerning the nonlinear phenomena in turbulence simulations [31].

For definitions of linear, weakly nonlinear, moderately nonlinear and strongly nonlinear/genuinely nonlinear continuum systems and their discretized counterparts, see [31–39] for some definitions. Oversimplified definitions for weakly, moderately and strongly nonlinear systems are as follows: Consider a nonlinear system. Loosely speaking, a weakly nonlinear theory keeping only the first term of the Taylor series expansion as a function of the considered system variable/parameter would provide the behavior of a weakly nonlinear system as a function of the chosen variable/parameter. A moderately nonlinear system keeps the first two terms or the first three terms of the Taylor series expansion. A strongly nonlinear system keeps the high-order terms or all terms of the Taylor series expansion.

1.1. Central Issues

Often, numerical methods based on solution performance behavior for model linear or weakly nonlinear systems are employed by some practitioners to simulate highly coupled genuinely nonlinear systems. For example, we found in the open literature that some claims of computed turbulence are NOT turbulence of the chosen governing equations. In addition, some claims of transition to turbulence for a considered Reynolds number are **numerical transitions** to turbulence. Some of the causes are as follows: (a) employing diffusive methods, (b) lacking sufficient grid points, (c) trapping into a long numerical chaotic transient phase and mistakenly identifying as turbulence transitions that have already occurred, (d) introducing spatial nonlinearity by nonlinear shock-capturing schemes, (e) introducing nonlinearity by time discretizations, (f) introducing nonlinearity by grid adaptations, (g) insufficient amount of numerical dissipation by low-dissipative high-order methods, (h) exhibiting wrong propagation speed of discontinuities by standard conservative shock-capturing methods for flows containing stiff source terms and discontinuities, (i) employing wrong numerical methods for flows with source terms and (j) employing entropy violating methods for external hypersonic blunt-body flows. Additional causes also exist but are beyond the scope of this overview. Examples to illustrate some of the causes or combined causes can be found in [1,2,4,6–8,18–24,40–42] and the references cited therein.

The central issue is that the solution space of discrete genuinely nonlinear systems is much **LARGER** than that of the corresponding **genuinely nonlinear continuous systems**, thus resulting in numerical solutions that are not solutions of the physical systems. A nonlinear method can introduce incorrect solutions when solving a linear system. Different numerical methods resulting from discretizing the same governing equations are separate discrete nonlinear dynamical systems as a function of the finite (non-zero) time step, finite grid spacing and tunable scheme parameter aside from the inherent physical parameters of the continuum governing equations. The traditional uncertainty quantification (UQ) approaches in numerical simulations often employ linearized analysis that might not provide the true behavior of the genuinely nonlinear physical flows. Linearized analysis relies on local behavior, and the convergence criterion is based on "**as the time step and grid spacing approach to zero**". In all computational physics practices, finite (not extremely small) time steps and grid spacings are used. Unlike the linear model equations used for conventional stability and accuracy considerations in time-dependent partial different equations (PDEs), there is no equivalent unique nonlinear model equation for nonlinear hyperbolic and parabolic PDEs for fluid dynamics. A numerical method behaving in a certain way for a particular nonlinear differential equation (DE) (PDE or ordinary differential equation (ODE)) might exhibit a different behavior for a different nonlinear DE even though the DEs are of the same type. The need to guarantee PARs of genuinely nonlinear systems becomes acute when computations offer the **ONLY** way of generating this type of data-limited simulations; the lack of reliable observable data and/or experimental means is unfeasible for any of a number of possible reasons.

At the same time, due to the rapid advancement of high-performance computing, the last two decades have been an era when computation is far ahead of analysis and when very large-scale practical computations are increasingly used in poorly understood multiscale data-limited complex nonlinear physical problems and non-traditional fields. This is compounded by the fact that some of the numerical schemes used in production that applied CFD computer codes for modeling and simulations do not take into account the genuine nonlinear behavior of numerical methods. In addition, most of the numerical methods in use are not discretely physics-preserving (structure-preserving) methods. This includes but is not limited to entropy-conserving, momentum-conserving, and kinetic energy-preserving methods. For example, a well known non-physical solution exhibited by entropy-violating numerical methods raises concerns with the computation on grid-aligned blunt-body hypersonic external flow. Some key points are explained below.

Entropy-Violating Methods Relating to Carbuncle Effect in Hypersonic Blunt Body Simulations: Take, for example, in hypersonic modeling and simulations, the issue of "the carbuncle effect and hypersonic CFD simulations" has attracted numerical method developers for forty years, even before the term "carbuncle" was born [9,43–46]. There were many heated debates at some of the CFD conferences. This is due partially to some of the developers not being aware of earlier related work or due to the disconnect between entropy-violating numerical method development and applied CFD physical insights, and much more. Actually, these numerical anomalies are related to entropy-violating methods for one dimension (1D) flows, multidimensional (multi-D) flows and much more. They are also due to the added nonlinearity introduced by flux limiters by shock-capturing methods, the computation of external flow involving dimension-by-dimension approaches and the scaling of the eigenvector if Roe's Riemann solver or the local Lax–Friedrichs flux is used in the construction of the resulting numerical method. Some proposed cures to multi-D high-speed blunt body simulation were presented as grid-aligned shock instability at the stagnation point. Some of these proposed cures did not provide the full picture or were not aware of earlier published work, e.g., [21,22,47]. In our work, we include integrated cures on the inaccuracy, stability and convergence issues aside from a multi-D entropy fix to cure the grid-aligned instability by entropy-violating methods. Our works preceded the heavily cited work of R. Sanders and J.J. Quirk [48,49]. See samples of later published works [43,50–52] on the subject. Private communication with D.K. Prabhu of NASA Ames Research Center in 2018 indicated that some of the CFD hypersonic production computer codes, e.g., DPLR and their unstructured version, continue to use a non-grid-aligned grid to avoid grid-aligned shock instability. Section VII will include excerpts from our 1980s published work [21,22,47,53], illustrating our enhancement of reliable numerical simulation for both steady and unsteady hypersonic blunt-body flows using entropy-violating methods.

Below, we expand our introductory discussion on the fundamentals of nonlinear behavior of numerical simulations further.

1.2. Treating Fluid Dynamics Equations as Dynamical Systems

Traditional fluid dynamics and CFD-related texts and reference books consider the Euler and Navier–Stokes equations in differential forms as coupled systems of nonlinear PDEs consisting of physical parameters. These equations are rarely classified as dynamical systems. Basically, fluid flow occurs in three possible regimes: turbulent, laminar and a transition region. The conditions that lead to each type of flow behavior are system-specific. Fluid dynamicists are interested in how the flow behaves as a function of one or more physical parameters. Of particular importance to fluid dynamicists is locating the critical value of the physical parameter where the fluid undergoes drastic changes in the flow behavior. Some examples are the prediction of transition to turbulence or laminar instability as a function of the Reynolds number; flow separation and stall as a function of Reynolds number and angle of attack; rotorcraft vibration as a function of rotation speed and flight speed; the occurrence of shock waves as a function of the body shape and/or Mach number; and the formation of vortices, flutter and other flow phenomena as a function of the angle of attack or other physical parameters.

One the other hand, one also can recast the study of admissible shock wave solutions of hyperbolic conservation laws as the study of the dynamics of heteroclitic orbits of a system of nonlinear ODEs [54] Another application is in the area of aiding the understanding of the topology of flow patterns (flow visualizations) of laboratory experiments, observable physical phenomena and numerical data. An additional important topic for CFD is the control and optimization of dynamical systems. This involves the application of optimization and control theory to dynamical systems. Some researchers use these interdisciplinary ideas to study, for example, the control of turbulence, the control of vortex generation and/or shock waves, the control of vibration in rotorcraft and the control of aerodynamic noise such as sonic boom and jet noise.

One historical note, the application of dynamical system theory to the study of spatiotemporal instabilities of aerodynamic and hydrodynamic flows and chaotic systems in fluid dynamics was discussed, respectively, in the 1994 and 1996 von Karman Institute for Fluid Dynamics Lecture Series. How the solution behaves as one or more of the system parameters is varied precisely belongs to the discipline of dynamical systems and bifurcation theory in applied mathematics. According to Ian Stewart (1990) [38],

"Bifurcation theory is a method for finding interesting solutions to nonlinear equations by tracking dull ones and waiting for them to lose stability".

Figure 1 shows a schematic diagram of stability of steady-state solutions arising through three types of bifurcation phenomena as a function of the system parameter α. The phenomenon of generating spurious steady-state numerical solutions (or other spurious asymptotes) by certain numerical schemes is often confused with the non uniqueness (or multiple steady states) of the governing equation. Actually, the existence of non-unique steady-state solutions of the continuum can complicate the numerics tremendously and is independent of the occurrence of spurious asymptotes of the associated scheme. For example, assuming the governing equation has a unique steady-state solution as a function of the system parameter α for a bracketed interval, a numerical method can introduce non-unique spurious steady-state (and asymptotes) numerical solutions as time step Δt increases within the bracketed interval. Referring to the same figure for the discretized counterpart, the system parameter α of Figure 1 in this case would be $\alpha\Delta t$, indicating the possible spurious steady-state phenomena portion of the bifurcation diagram with a reduced bracketed interval of α than the allowable unique steady state for the continuum. The full bifurcation diagram as a function of system parameter will be illustrated later for some of the commonly used time discretizations. The occurrence of possible spurious steady-state solutions, e.g., spurious zero wave speed standing wave phenomena by some numerical method, can influence the reliability of unsteady (time-accurate) simulations. See Yee et al., Yee and Sweby, Lafon and Yee, Griffiths et al. and Sullivan et al. [6,23,24,35,55] for some examples and discussion.

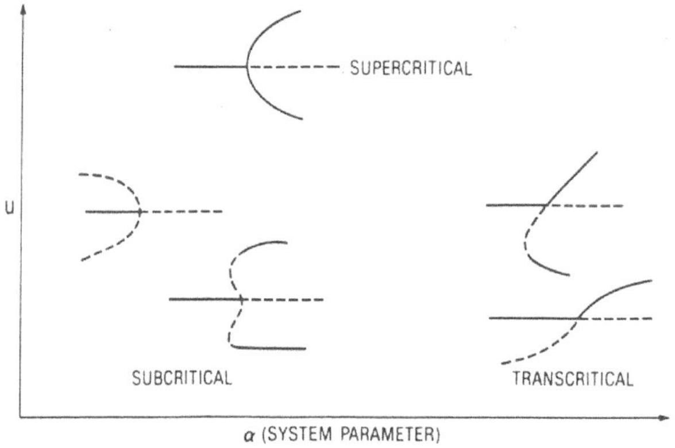

Figure 1. Stability of steady-state solutions arising through three types of bifurcation phenomena as a function of the system parameter α (- stable, - - unstable).

1.3. *Isolation of the Sources of Numerical Uncertainties*

In order to isolate the sources of numerical uncertainties, we need to set aside the possible discrepancy between the chosen model and the real physics and/or experimental data. Therefore, to gain first-hand insight into the "dynamics of numerics", we concentrate only on how well numerical schemes can mimic the solution behavior of the underlying governing PDEs for finite time steps and grid spacings. Even with this restriction, the study of the PARs of genuinely nonlinear compressible turbulence with shock waves encompasses elements and factors far beyond what is discussed here. The basic idea is to have a very clear distinction of numerical uncertainties from each source. These include but are not limited to the following:

- Stability and well-posedness of the governing PDEs;
- Type, order of accuracy, nonlinear stability and convergence of finite discretizations as functions of finite grid spacing, time steps and system parameters;
- Limits and barriers of existing finite discretizations for highly nonlinear stiff problems with source terms and forcing, and/or for long-time wave propagation phenomena;
- Numerical boundary condition (BC) treatments;
- Finite representation of infinite domains;
- Solution strategies in solving the nonlinear discretized equations;

- Procedures for obtaining the steady-state numerical solutions;
- Grid quality and grid adaptations;
- Multigrids;
- Domain decomposition (zonal or multicomponent approach) in solving large problems.

Although some of the numerical uncertainties can be explained and minimized by traditional numerical analysis and standard CFD practices, especially for weakly nonlinear and non-stiff governing PDEs, such practices might not be sufficient for strongly nonlinear and/or stiff problems. The need of a better understanding of the nonlinear behavior of numerical schemes being used as an integral part of computer code verification, validation and certification is needed.

1.4. Lack of Unique Nonlinear Model for Nonlinear Hyperbolic and Parabolic PDEs

As briefly mentioned earlier, a major stumbling block in genuinely nonlinear studies is that unlike the linear model equations used for conventional stability and accuracy considerations in time-dependent PDEs, there is no equivalent unique nonlinear model equation for nonlinear hyperbolic and parabolic PDEs for fluid dynamics. On one hand, a numerical method behaving in a certain way for a particular nonlinear continuum equation might exhibit a different behavior for a different nonlinear continuum equation even though they are of the same PDE type. On the other hand, even for simple nonlinear model continuum equations with known solutions, the discretized counterparts can be extremely complex to analyze, depending on the numerical methods. Except in special cases, there is no general theory at the present time to characterize the various nonlinear behaviors of the underlying discretized counterparts. Our strategy is based on the knowledge gained for nonlinear model problems with known analytical solutions to identify and explain the possible sources and remedies of numerical uncertainties in practical computations.

Throughout the discussion, the term **"discretized counterparts"** is used to mean the finite difference equations resulting from finite discretizations of the underlying DEs. The term **"dynamics"** is used loosely to mean the dynamical behavior of nonlinear dynamical systems (continuum or discrete), and **"numerics"** is used loosely to mean the numerical methods and procedures in solving dynamical systems. We would like to emphasize that in our study of the dynamics of numerics, unless otherwise stated, we always assume the continuum (governing equations) is nonlinear.

Although this overview paper is intended primarily for computational fluid dynamicists, the approaches should be useful for computational scientists, physicists, engineers and computer scientists who have a need for reliable numerical simulations.

1.5. Objective and Outline

DNS, LES, ILES and turbulence modelings are traditional methods for the simulation of compressible turbulence. The DNS, LES and ILES computations of compressible turbulence require time-accurate, nonlinearly stable methods suitable for long-time integration with minimal use of numerical dissipation, as such flows consist of the disparity of space and time scales during different stages of turbulence development. They are usually characterized by very large Reynolds numbers and even larger magnetic Reynolds numbers for magnetized turbulence in magnetohydrodynamics (MHD). In order to improve reliable compressible turbulence simulations, our interdisciplinary design of new numerical methods (a) employs quantification of numerical uncertainty via a nonlinear dynamical approach; (b) mimics and preserves physical properties of the chosen governing equations as much as possible; (c) is high order, low dissipative, low dispersive and suitable for a wide range of flow speeds within the same computational domain; (d) is suitable for flows consisting of disparity of space and time scales during different stages of turbulence development; (e) possesses high-order stable summation-by-parts (SBP) numerical boundary operators for non-periodic boundaries [56–58]; (f) is applicable for 3D spatial and time-varying deforming grids with geometric conservation law property (GCL); and (g) is nonlinear, stable, efficient and exhibits scalable parallelism for current high-performance computing.

The specific background and motivation are presented in Section 2. Sources of nonlinearities and knowledge gained from nonlinear model problems are discussed in Section 3. Section 4 discusses spurious numerics by time discretizations. Section 5 discusses the nonlinear behavior of adaptive numerical methods. Section 6 gives a brief description on source term treatments in reacting flows to obtain the correct propagation speed of discontinuities for flows containing stiff source terms, shocks and shear layers. Section 7 devotes a large section to illustrate spurious numerics relevant to turbulent flow computations, especially relevant to false turbulent flow transitions (numerical transitions). Due to a large volume of disjoint proposed carbuncle cures devoted to the simulations of hypersonic blunt-body flows, we devote a complete Section 8 to include excerpts from [21,22] for the cures. We believe our early work presents an integrated cure related to the simulation of multi-D hypersonic blunt body and its relations with the carbuncle phenomena than some of the last thirty years of studies reported in the open literature. In particular, nonlinearity introduced by the nonlinear shock-capturing method that creates more spurious numeric than mere carbuncle cures will be discussed. Section 9 illustrates some test cases in spurious numerics related to problems with stiff source terms and our methods to minimize them. The test cases include simple scalar, simple system cases to a very complex 13 species of non-equilibrium reacting flow simulations. Section 10 gives a brief description of our high-order physical-preserving nonlinear filter methods. It discusses how numerical dissipation affects the predictability and reliability of turbulence simulation of compressible turbulence with shock waves. Several numerical examples illustrating our physical-preserving seventh-order nonlinear filter methods compared with standalone standard high-order shock-capturing methods will be included. In Section 11, we illustrate a comparison of eighth-order physical-preserving methods for several gas dynamics simulations, while Section 12 presents some concluding remarks.

2. Background, Motivation and Integrated Approach

The period of the 1970s and 1980s was an era of explosion of new numerical method developments in CFD, new theories of nonlinear dynamical systems and the beginning of increasing computer speed to aid numerical simulations of increases in complexity for modeling and simulation of fluid flows. Routinely, numerical methods were used to understand the dynamics of highly nonlinear continuum dynamical systems without analyzing the global nonlinear behavior of the numerical methods as the discrete dynamical system being used. Fast forward to three decades later, with more sophisticated developments in CFD, many new theories in nonlinear dynamical systems and quantum jumps in the computation speed of high-performance computing appeared, yet these two disciplines of **"Dynamics of Numerics"** and **"Numerics of Dynamics"** are still lacking close interactions.

This work consisted of a culmination of over 35 years of continuing research work by the authors and collaborators in bridging some of the gaps of interaction between these disciplines as a pairing to aid the development of more reliable and accurate simulations of genuinely nonlinear fluid flows. Some of the key concepts and our earlier efforts in bridging the gaps are reported in the lecture notes/long overview paper by Yee and Sweby, Yee and Sjögreen [1–3], and the references cited therein.

The dynamics of numerics work of Yee and Sweby [1,2] summarized the global nonlinear behavior of finite discretizations for constant or variable time steps and fixed or adaptive grid spacings using tools from dynamical systems theory. Detailed analysis of commonly used temporal and spatial discretizations for simple model problems was presented. The role of dynamics in the understanding of long-time behavior of numerical integration and the nonlinear stability, convergence and reliability of using time-marching approaches for obtaining steady-state numerical solutions in CFD was explored. Their study is complemented with examples of spurious behavior observed in steady and unsteady CFD computations. Numerically induced chaos and numerically induced chaotic transient can be mistaken as turbulence transitions. The CFD examples were chosen to illustrate non-apparent spurious behavior that was difficult to detect without extensive grid and temporal refinement studies, and some knowledge from dynamical systems theory. These studies revealed the various possible dangers of misinterpreting the numerical simulation of realistic complex flows. The details of published work related to Yee and Sweby [1,2] encompass several dozens of their peer-reviewed publications and conference proceedings papers. See, e.g., [1,6–8,23,35,59–63].

The work of Yee and Sjögreen [3] presents their recent adaptive numerical dissipation control development for long-time integration of DNS, LES and ILES computations of compressible turbulence for gas dynamics and MHD. Their approach uses **adaptive blending of high-order physics-preserving non-dissipative methods (classical central, Padé (compact) and dispersion relation preserving (DRP))** with high-order shock-capturing methods in such a way that high-order shock-capturing methods are active only in the vicinity of shock/shear waves, and high gradient and spurious high-frequency oscillation regions guided by flow sensors. Typically, the adaptive blending of more than one method falls under two categories: **hybrid methods and nonlinear filter methods**. By examining the construction of these two approaches, the authors' nonlinear filter approach is more efficient and less CPU intensive while obtaining similar accuracy. The different stages of our high-order nonlinear filter method's development are also encompassed in several dozens of peer-reviewed publications and conference proceedings. See [3] and the references cited therein for more details. For some of the details, see [4,5,9–11,13–16,26–30,64,65].

Here, the key concepts of our work [1–3,6,7,23,24,35,66] are discussed first before the discussion of our integrated nonlinear approach in the quantification of numerical uncertainty and our high-order physics-preserving methods for more reliable simulation of compressible turbulence with shock waves.

2.1. *Quantification of Numerical Uncertainty in the 1980s and 1990s*

Since the late 1980s, many CFD-related journals imposed an editorial policy statement on numerical uncertainty, which pertained mainly to the accuracy, computer code and solution verification under the linearized analysis guidelines. See the book of Oberkampf and Roy [67] for a broader development and Glimm et al. [31] concerning the nonlinear phenomena in turbulence simulations.

However, the study of numerical uncertainties in practical modeling and simulation in computational physics and scientific computing encompasses very broad subject areas. At present, some of the numerical uncertainties can be explained and minimized by traditional numerical analysis and standard CFD practices. However, such practices might not be sufficient for strongly nonlinear and/or stiff problems. Examples of this type of problem are combustion, turbulent combustion, high speed and reacting flows, and certain turbulence models in Navier–Stokes computations. A good understanding of the nonlinear behavior of numerical schemes being used is a conduit to the quantification of numerical uncertainty and should be an integral part of code verification and validation. See [36] for the definition of highly (or genuinely) nonlinear problems.

2.2. *Interaction of Dynamical Systems and Numerical Methods*

During the late 1970s and 1980s, a new area of applied mathematics emerged from the interaction of dynamical system theory and numerical analysis. These developments addressed mainly mathematical principles and their applications of numerics in the understanding of the dynamics of differential equations (DEs) without discussing the connection between dynamics and numerics for initial value problems (IVPs) and initial boundary value problems (IBVPs). There was, however, some discussion on this connection for boundary value problems (BVPs) [32,68–74]. Studies of BVPs of the elliptic type continue to the present day. See, for example, the early four SIAM Conferences on Dynamical Systems (1990, 1992, 1995, 1997) and the Proceedings of IMA Conference on Dynamics of Numerics and Numerics of Dynamics, 31 July–2 August 1990, Bristol, England.

In the late 1980s, developments concerned with the connection between dynamics and numerics for IVPs and IBVPs slowly emerged. See, for example, [37,75–81]. These developments raised many interesting and important issues of concern that are useful to practitioners in computational sciences. Some of the issues are as follows: (a) Can advances in dynamical systems provide new insights into better understanding of numerical algorithms and the construction of new ones? (b) Can these advances aid in the determination of a more reliable criterion on the use of existing numerical schemes for strongly nonlinear problems? (c) What is the influence of finite time steps and grid spacings rather than time steps and grid spacings approaching zero on the overall nonlinear behavior and stability of the scheme in terms of allowable initial data and discretized parameters? In the early 1990s, the use of dynamics to address the long-time behavior of numerical schemes for IVPs and IBVPs began to flourish. The 1990s work includes the Conference on Dynamics of Numerics and Numerics of Dynamics (University of Bristol, 31 July–2 August 1990), the Chaotic Numerics Workshop (Deakin University, Geelong, Australia, 12–16 July 1993), the Conference on Dynamical Numerical Analysis (Georgia Institute of Technology, Atlanta, Georgia, 14–16 December 1995) and the "Innovative Time Integrators Workshop" (Center for Mathematics and computer Science, Amsterdam, 6–8 November 1996, the Netherlands). These conferences were devoted almost entirely to dynamical numerical analysis. See the proceedings and references cited therein. See also [39,44,82–88]. The majority of the later developments concentrated on the long-time behavior of ODE solvers using variable step size based on local error controls [89]. This type of local error control in time enjoyed much success in controlling accuracy and stability for transient computations for ODEs. For time-dependent PDEs, the caveat is that regardless of whether finite difference (and finite volume) or finite element spatial discretization methods are employed, when time-marching approaches are used to obtain steady-state numerical solutions, local error controls similar to those used in ODE solvers that were designed for accuracy purposes are neither practical nor appropriate to use since such local step size error control methods might prevent the solution from reaching the correct steady-state solutions within a reasonable number of iterations. We note that the standard practice of using "local time step" (varied from grid point to grid point with the same CFL) in time-marching to the steady state is not the same as the variable step size based on local error controls.

Our studies indicated that the understanding of the dynamics of numerics for constant step size is necessary as many CFD practitioners employ the constant step size approach. Besides, the study of the dynamics of ODE solvers using variable step size based on local error control requires knowledge of the constant step size case [82]. In a series of papers, Yee et al. [23], Yee and Sweby [24,61,62], Sweby et al. [60], Sweby and Yee [59], and Lafon and Yee [6,7] studied the dynamics of finite discretization for constant time steps. The examples used in these papers were deliberately kept simple to permit explicit analysis. The approach was to take nonlinear model ODEs and PDEs with known explicit solutions (the most straightforward way of being sure of what is 'really' happening), discretize them according to various standard numerical methods and apply techniques from discrete dynamics to analyze the behavior of the discretized counterparts. To set the stage for later discussion, next, we discuss the connection of dynamical systems with CFD. Yee et al., Yee and Sweby, and Lafon and Yee [1,6,7,23,62,90] divide their studies into two categories, steady-state and time-accurate computations. Within each category, they further divide the governing PDEs into homogeneous and inhomogeneous (i.e., with or without source terms), and rapidly/slowly developing and long-time integration problems. The following subsections list outstanding issues of numerical uncertainties in CFD in which the tools of dynamical systems theory can play a role.

3. Sources of Nonlinearities and Knowledge Gained from Nonlinear Model Problems

One of the primary building blocks for the PAR of numerical simulations is to identify sources of nonlinearities (from the continuum governing equations and their discretized counterparts) and to isolate the elements and issues of numerical uncertainties due to these nonlinearities.

3.1. Nonlinearities Due to the Governing Equations vs. Nonlinearity Due to the Numerics

Typical sources of nonlinearities in CFD due to the physics are convection, diffusion, forcing, turbulence source terms, reacting flows, turbulence combustion, gravity in black holes, turbulence in magnetic clouds related problems or any combination of the above. Nonlinearities due to the numerics are less familiar. There are generally three major sources:

- Nonlinearities due to time discretizations: The discretized counterpart is nonlinear in the time step. Examples of this type are Runge–Kutta methods. If fixed time steps are used, spurious steady-state or spurious asymptotic numerical solutions can occur, depending on the initial condition (IC). Linear multistep methods (LMMs) [89] are linear in the time step; they do not exhibit spurious steady states, but implicit LMMs can stabilize unstable steady states. See Yee and Sweby [1,6,7,23,62,90] and the references cited therein for the dynamics of numerics of standard time discretizations. However, unconditionally stable LMMs can stabilize unstable steady states and much more.
- Nonlinearities due to spatial discretizations: The discretized counterpart can be nonlinear in the grid spacing and/or the scheme. Examples of nonlinear schemes are the total variation diminishing (TVD), essentially nonoscillatory (ENO) and weighted ENO (WENO) schemes and spatial schemes consisting of flux limiters. The resulting discretized counterparts are nonlinear (in the dependent variables) even though the governing equation is linear. See Yee [21] and the references cited therein for some forms of these nonlinear schemes.
- Nonlinearities due to complex geometries, boundary interfaces, grid generation, grid refinements and grid adaptations [59]: Each of these procedures can introduce nonlinearities even though the governing equation is linear.

3.2. Discrete Dynamical Systems and CFD

When we try to use numerical methods to gain insight into the fluid physics, there is an added new dimension to the overall problem. Even though we freeze the physical parameters of the governing equations, the resulting discretized counterparts are not just a nonlinear system of difference equations but also a nonlinear but discrete dynamical system as functions of the discretized parameter on their own. Depending on the scheme, the discretized counterparts usually preserve the steady states of the continuum. In addition, the discretized counterparts possess their own dynamics, which is usually richer than the continuum [23,24,37,76]. These resulting discrete dynamical systems are a function of all of the discretized parameters that are not present in the governing equations. This is one of the key factors in influencing the numerical solution to depart from the physical one if the governing equations are strongly nonlinear and stiff. It is important that before analyzing the dynamics of numerics, it is necessary to analyze (or understand) as much as possible the dynamical behavior of the governing equations and/or the physical problems using theories of DEs (ODEs and PDEs), dynamical systems of DEs and also physical guidelines. See, e.g., [91,92] for two specialized studies.

3.2.1. Nonlinear Numerical Methods

We define nonlinear numerical methods for time-dependent PDEs in the sense that the final algorithm is nonlinear even for the constant-coefficient linear PDE. It is well known that all of the TVD, total variation bounded (TVB), ENO and WENO schemes and related later shock-capturing methods development are nonlinear schemes. These types of schemes are known to have a slower convergence rate than classical shock-capturing methods and can occasionally produce nonphysical solutions for certain combinations of entropy-satisfying parameters and flux limiters (in spite of the fact that entropy-satisfying TVD, TVB and ENO schemes can suppress nonphysical solutions). See Yee [21] for a summary of the subject in the late 1980s. The second aspect of these nonlinear schemes is that even if the numerical method is formally of more than first order and if the approximation converges, the rate may still be only first order behind the shock (not just around the shock). This can happen for systems where one characteristic may propagate part of the error at a shock into the smooth domain. See, e.g., Engquist and Sjögreen [93], which illustrated this phenomena with examples. Different designs of flux limiters to improve the time-marching to the steady states have been ongoing for the last 40 years, e.g., see two recent works [94,95].

Numerical Methods that are Linear vs. Nonlinear in Δt: In CFD, the usual classification of time-accurate schemes for time-marching approaches to the steady state are explicit, implicit and hybrid explicit and implicit methods. A less commonly known classification of numerical schemes for time-marching approaches is the identification of schemes that are linear or nonlinear in the time step (Δt) parameter space when applied to nonlinear DEs. All linear multistep methods (LMMs) of explicit or implicit type are linear in Δt, and all multistage Runge–Kutta methods are nonlinear in Δt. Lax–Wendroff/MacCormick and ADER [96,97] type of non-separable full discretizations also are nonlinear in Δt. For non-separable discretization methods and multistage Runge–Kutta time discretizations, the accuracy of the steady-state numerical solutions depends on Δt. A desirable property for a scheme that is linear in Δt is that, if the numerical solution converges, its steady-state numerical solutions are independent of the time step. Unconditional stable implicit LMMs' time integrators can stabilize unstable steady states. See [1,2,66] for some examples. For a representative fluid simulation example on stabilizing unstable steady states with implicit time integrators, see, e.g., Poliashenko and Yee 1995, an unpublished joint work. This unpublished work was presented at the 10th International Conference in Finite Element Methods, 5–8 January 1998, Tucson, Arizona. A summary is included in [1,2].

3.2.2. Dynamical Behavior of Numerical Approximations of ODEs vs. Time-Dependent PDEs

From our analyses and studies, we have shown that spurious numerical solutions can be independently introduced by time and spatial discretizations. Take the case where the ODEs are obtained from semi-discrete approximations of PDEs. The resulting system of ODEs contains more parameters (due to spatial discretizations) than those in the physical problems governed by ODEs. The parameters due to spatial discretizations for the semi-discrete approximation become the system parameter (instead of the discretized parameter) of the resulting system of ODEs. Depending on the differencing scheme, the resulting discretized counterparts of a PDE can be nonlinear in Δt, the grid spacing and the numerical dissipation parameters, even though the PDEs have only one parameter or none. One major consideration is that one might be able to choose a "safe" numerical method to solve the resulting system of ODEs to avoid spurious stable numerical solutions due to time discretizations. However, spurious numerical solutions, especially spatially varying spurious steady states introduced by spatial discretizations in nonlinear hyperbolic and parabolic PDEs for CFD applications, appear to be more difficult to avoid due to the use of a fixed mesh. In the case of the semi-discrete approach, such as methods of lines or finite element methods, if spurious numerical solutions due to spatial discretizations exist, the resulting ODE system has already inherited this spurious feature as part of the exact solution of the semi-discrete case. Thus, care must be taken in using the ODE solver computer packages for PDE applications. See Lafon and Yee [6,7] for a discussion.

3.2.3. Dynamics of Time-Marching Approaches to Obtain Steady-State Solutions

The use of time-marching approaches to obtain steady-state numerical solutions has been considered the method of choice in CFD for five decades since the pioneering work of Crocco [98] and Moretti and Abbett [99]. Moretti and Abbett used this approach to solve the inviscid supersonic flow over a blunt body without resorting to solving the steady form of PDEs of the mixed type. The introduction of efficient CFD algorithms of MacCormack [100], Beam and Warming [101], Briley and McDonald [102], and Steger [103] marked the beginning of numerical simulations of 2D and 3D Navier–Stokes equations for complex configurations. It enjoyed much success in computing a variety of weakly and moderately nonlinear fluid flow problems. For strongly nonlinear problems, the situation is more complicated. In addition to the understanding of the sources of nonlinearities, it is necessary to isolate all elements and issues of numerical uncertainties due to these nonlinearities in time-marching to the steady state.

The following isolates some of the key elements and issues of numerical uncertainties in time-marching to the steady state.

Solving an IBVP with Unknown Initial Data: When time-marching approaches are employed to obtain steady-state numerical solutions, a BVP is transformed into an IBVP with unknown initial data. The time differencing in this case acts as a pseudo time. Linearized stability analysis indicates that a subset of the numerical solutions for certain ranges of the discretized parameters and numerical boundary conditions mimic the true solution behavior of the governing equation. However, it is less known that there exist asymptotic numerical solutions (including spurious steady states) that are not solutions of the continuum inside as well as outside the safe regions [23,61,62,104], depending on the initial data. Unlike nonlinear problems, the numerical solutions of linear or nearly linear problems are "independent" of the discretized parameters and initial data as long as the discretized parameters are inside the stability limit (or the Courant–Friedrich–Lewy (CFL) condition)—that is, the topological shapes of these solutions remain the same within the stability limit and accuracy of the scheme for linear behavior. It turns out that if constant step sizes are used, the stability, convergence rate and occurrence of spurious numerical solutions are intimately related to the choice of initial data (or start up solution).
Reliability of Residual Test: It is a common practice in time-marching approaches to obtaining the steady-state numerical solutions by the "local time step" (using the same CFL) via, e.g., time relaxation and multigrid. The criterion or guideline to measure the speed of convergence to the steady state is to monitor the residual of the computed solution if the quantity approaches (computer) machine zero. The danger is that if the computed solution is a spurious steady state, the residual would still approach machine zero as well.
Influence of Spurious Steady-State Numerical Solutions and Time-Accurate Simulations: The occurrence of spurious steady-state solutions by some numerical method can influence the true behavior of unsteady (time-accurate) simulations. For example, the spurious zero wave speed standing wave phenomena can interfere with the reliability of unsteady simulations. See Yee et al. and Griffiths et al. [23,35] for some discussion. Another issue is on the simulation of turbulence flows riding on top of a steady flow. See Sullivan et al. [55] for a practical physical example.

3.2.4. Non-Unique Steady-State Solutions of Nonlinear DEs vs. Spurious Asymptotes of the Numerical Methods

In the literature, the phenomenon of generating spurious steady-state numerical solutions (or other spurious asymptotes) by certain numerical schemes is often confused with the non-uniqueness (or multiple steady states) of the governing equation. In fact, the existence of non-unique steady-state solutions of the continuum can complicate the numerics tremendously and is independent of the occurrence of spurious asymptotes of the associated scheme. If possible, an analysis of the chosen nonlinear ODEs and PDEs and their dynamical behavior would help the study of the dynamics of numerical methods for nonlinear PDEs. A full understanding of the subject can shed some light on the controversy about the "true" existence of multiple steady-state solutions through numerical experiments for certain flow types of the Euler and/or Navier–Stokes equations. See, for example, Gimperlein et al. [105] for some discussion on the non-uniqueness in fluid flows.

3.3. Dynamics of Numerics via Tools from Bifurcation Theory

In many fluid problems, the solution behavior is well known for certain values of the physical parameters but unknown for other values. For these other values of the parameters, the problem might become very stiff and/or strongly nonlinear, making the available numerical schemes (or the scheme in use) intractable. In this situation, continuation methods in bifurcation theory can become very useful. If possible, one should start with the physical parameter of a known or reliable steady state (e.g., flow behavior is usually known for low angles of attack but not for high angles of attack). One can then use a continuation method such as the improved pseudo arclength continuation method of Keller [106] (or the recent developments in this area) to solve for the bifurcation curve as a function of the physical parameter. See, e.g., early development by Doedel [32,69], Shroff and Keller [107], and Davidson [108]. The equations used are the discretized counterpart of the steady PDEs or the time-dependent PDEs. See Stephens and Shubin [74,109] for earlier related work.

If time-marching approaches are used, a reliable steady-state numerical solution (as a starting value on the correct branch of the bifurcation curve for a particular value of the physical parameter) is assumed. This starting steady-state numerical solution is assumed to have the proper time step and initial data combination and to have a grid spacing fine enough to resolve the flow feature. The continuation method will produce a continuous spectrum of the numerical solutions as the underlying physical parameter is varied until it arrives at a critical value p_c such that it either experiences a bifurcation point or fails to converge. Since we started on the correct branch of the bifurcation curve, the solution obtained before that p_c should be more reliable than if one starts with the physical parameter in question with unknown initial data and tries to stretch the limitation of the scheme. Note that by starting a reliable solution on the correct branch of the bifurcation curve, the dependence of the numerical solution on the initial data associated with time-marching methods can be avoided before a spurious bifurcation occurs.

Finally, when one is not sure of the numerical solution, the continuation method can be used to double check it. This approach can even reveal the true limitations of the existing scheme. In other words, the approach can reveal the critical physical parameter for which the numerical method breaks down. On the other hand, if one wants to find out the largest possible time step and/or grid spacing that one can use for a particular problem and physical parameter, one can also use continuation methods to trace out the bifurcation curve as a function of the time step and/or grid spacing. In this case, one can start with a small time step and/or grid spacing with the correct steady state and observe the critical discretized parameter as it undergoes instability or spurious bifurcation. Of course, this method for minimizing spurious steady states can still suffer from spurious behavior due to an under-resolved grid because of limited computer resources for complex practical problems. Practical guidelines to avoid under-resolved grids are yet another important building block toward reliable numerical simulations. The efficient treatment of solving the extremely large set of eigenvalue problems to study the type and stability of bifurcation points is yet another challenge. See, e.g., Fortin et al. [110], Davidson [108], and Shroff and Keller [107] for some discussions. Consequently, further development in numerical bifurcation analysis and new concepts in adaptive methods for time-marching to steady state hold a key to the minimization of spurious numerics.

3.4. A Primary Step to Minimize Spurious Numerics via Knowledge Gained from Nonlinear Model Problems

A straightforward first step of being sure of what is "really" happening with the nonlinear behavior of numerics is with the aid of elementary examples. Yee et al., Yee and Sweby (1991–1997), Sweby and Yee (1994–1995) and Griffiths et al. (1992a,b) [1,2,23, 35,61,62,66,66,90] discuss the fundamentals of spurious behavior of commonly used time and spatial discretizations in CFD. Details of these examples can be found in their earlier papers. These examples consist of nonlinear model ODEs and PDEs with known analytical solutions. They illustrate the danger of employing fixed (constant) time steps and grid spacings. They were selected to illustrate the following different nonlinear behavior of numerical methods:

- Strong dependence of numerical solutions on initial data (numerical basins of attraction);
- Global asymptotic behavior of super-stable time discretizations;

- Occurrence of stable and unstable spurious asymptotes **above** the linearized stability limit of the scheme (for constant time steps);
- Occurrence of stable and unstable spurious steady states **below** the linearized stability limit of the scheme (for constant time steps);
- **Stabilization** of unstable physical steady states by implicit and semi-implicit methods;
- Interplay of initial data and time steps on the occurrence of spurious numerical asymptotes;
- Interference with the dynamics of the underlying implicit scheme by procedures in solving the nonlinear algebraic equations (resulting from implicit discretizations of the continuum equations);
- Dynamics of the linearized implicit Euler scheme solving the time-dependent equations to obtain steady states vs. Newton's method solving the steady equation;
- Spurious numerics independently introduced by spatial and time discretizations;
- Convergence problems and spurious behavior of high-resolution shock-capturing methods;
- Numerically induced and suppressed (spurious) chaos, and numerically induced chaotic transients;
- Spurious numerics generated by grid adaptations;
- Effect of spurious numerical standing wave in unsteady computations;
- Wrong propagation speed of discontinuities for problems containing stiff source terms.

We denote "spurious numerical solutions (and asymptotes)" to mean numerical solutions (asymptotes) that are solutions (asymptotes) of the discretized counterparts but are not solutions (asymptotes) of the underlying DEs. We also include steady-state solutions, periodic solutions, limit cycles, chaos and strange attractors as part of asymptotic solutions of the DEs and the corresponding discretized counterparts. See [111,112] for the definition of chaos and strange attractors. Super-stable scheme here refers to the region of numerical stability enclosing the physical instability of the true solution of the governing equation.

Aside from earlier studies by the authors and us, the aforementioned investigator, some later work carried out by Alicia Serfaty de Markus and Ronald E. Mickens [113,114] discussed the suppression of numerically induced chaos and reduction of computational errors. Their work was independently studied in 1999 and 2005 for different applications.

4. Dynamics of Numerics of Time Discretizations and Representative Examples

In this section, we briefly include the procedures we used for the study and select basic numerical examples on the dynamics of numerics by time discretizations.

4.1. Basin of Attraction of the Continuum vs. Basin of Attractions of the Discretized Counterparts

Strong dependence of solutions on initial data by the continuum and its discretized counterparts can be best analyzed by comparing the basin of attraction of the continuum and its discretized counterparts. The definition of basin of attraction will be defined shortly. It is one of the key tools to the understanding of spurious numerics. Unconditional stable time discretizations might have the property of a "numerical basin of attraction" that is larger than the underlying "exact basin of attraction" of the continuum as it can stabilize unstable steady states. Multistage Runge–Kutta methods can contribute to extra spurious steady states and asymptotes. In order to have a basic understanding of dynamical behavior of numerical methods for constant time steps, we take continuum models with known explicit solutions, discretize them according to various standard numerical methods commonly used in CFD and apply techniques from discrete dynamics to analyze the behavior of the resulting systems and to compare the nonlinear behavior of these schemes. Consequently, some of the schemes used for illustration might obviously not be suitable for that particular nonlinear model or spurious dynamics could have been avoided easily under that environment.

In our series of published work [1,2,23,35,61,62,66,90], we performed studies to gain a better understanding of the global asymptotic nonlinear behavior of (a) multistage Runge–Kutta type time discretizations and (b) explicit and implicit linear multipstep (LMM) time discretizations, including super-stable implicit LMMs. Examples of nonlinear effect caused by grid adaptation and super-stable implicit total variation diminishing (TVD) schemes on the overall performance of the numerical procedures were given. Some methods to minimize spurious steady-state numerical solutions were discussed. Here, only the fundamental concepts of strong dependence of computed solutions on initial data, basins of attraction of the continuum vs. basins of attraction of the corresponding numerical basins of attractions for two simple nonlinear scalar ODEs are shown. Here, "Exact" Basin of Attraction is a set of all initial data asymptotically approaching that asymptote for the continuum-governing equations. We use the terms "exact" basin of attraction and "numerical" basin of attraction to mean the basin of attraction of the DE and basin of attraction of the underlying discretized counterpart for the considered numerical method.

Consider the simple scalar autonomous ODE,

$$du/dt = S(u). \tag{1}$$

Here, autonomous ODEs are for $S(u)$ not explicitly dependent on t, i.e., not $S(u,t)$. The two scalar first-order autonomous nonlinear ODEs are

$$du/dt = S(u) = \alpha u(1-u) \tag{2}$$

$$du/dt = S(u) = \alpha u(1-u)(b-u), 0 < b < 1. \tag{3}$$

Equation (2) is the simplified form of the heavily studied logistic ODE with an application in biology and applied mechanics. The fixed points (steady states) u^* for (2) with $\alpha > 0$ are when $S(u^*) = 0$—that is, $u^* = 0$ (unstable) and $u^* = 1$ (stable), and no additional periodic solutions or asymptotes exist. The basin of attraction for the stable fixed point u is the entire positive half-plan for all values of $\alpha > 0$.

The fixed points for Equation (3) with $\alpha > 0$ are $u^* = 0$ (unstable), $u^* = 1$ (unstable) and $u^* = b$ (stable), and no additional periodic solutions or asymptotes exist. The basin of attraction for the stable fixed point $u^* = b$ is $0 < u < 1$ for all $\alpha > 0$.

Denoting Δt as the time step, and considering $\alpha > 0$, Figure 2 shows the full bifurcation diagram of the explicit Euler method for the logistic ODE. Note that the explicit Euler applied to the logistic ODE resulted in the famous logistic map $u^{n+1} = \alpha u^n(1-u^n)$. Unlike the underlying logistic ODE, it is well known that the logistic map possesses very rich dynamical behavior such as period-doubling (of period 2^n for any positive integer n) cascades resulting in chaos (Feigenbaum [34]). One can find Figure 2 in most of the elementary dynamical systems text books. The exact values of $r = \alpha\Delta t$ for all of the period-doubling bifurcation points and chaotic windows (intervals of r) were discovered by Feigenbaum in the late 1970s. Interested readers should consult these elementary textbooks for details. In other words, one can obtain the analytical (exact) behavior of the spurious asymptotes and numerical (spurious) chaos of the logistic map. Next, we look at more time discretizations used to solve these two different ODEs, where we present full bifurcation diagrams overlaid on top of their corresponding numerical basin of attractions.

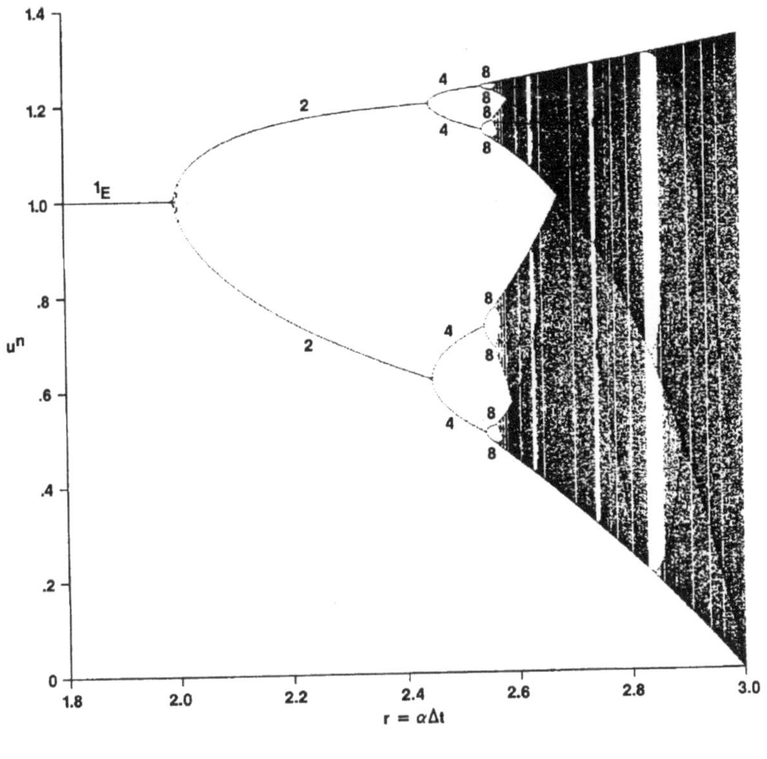

Figure 2. Full bifurcation diagram of the explicit Euler method for the logistic ODE with $\alpha > 0$.

4.2. *Strong Dependence of Solutions on Initial Data by Examining Their Corresponding Basin of Attraction*

It is possible that for the same Δt but two different initial data or vice versa, the scheme can converge to two different distinct numerical solutions of which one or neither of them is the true solution of the underlying ODE. Thus, in a situation where there is no prior information about the exact steady-state solution, and where a time-marching approach is used to obtain the steady-state numerical solution when initial data are not known, a stable spurious steady-state could be computed and mistaken for the correct steady-state solution. We need the corresponding "full" bifurcation diagram using this numerical approach, and we must over-plot all of the individual bifurcation diagrams of existing asymptotes of any period and chaotic attractors obtained by using the entire domain of u values as starting initial data. Thus, a better method in numerically approximating the full bifurcation diagram is dividing the domain of interest of the u axis into equal increments and using these u values as initial data. The "full" bifurcation diagram is obtained by simply over-plotting all of these individual diagrams on one. Figure 2 shows the "full" bifurcation diagrams for the corresponding fixed point diagrams of the logistic ODE using the explicit Euler method for $\alpha > 0$. Note that the full bifurcation diagram computed this way might miss some of the windows of bifurcations that occur inside the intervals of the adjacent r and/or the initial data values.

The commonly used time discretizations in CFD are multistage Runge–Kutta methods, implicit Euler/linearized implicit Euler, trapezoidal/linearized trapezoidal, three-level backward differentiation formula (BDF) and their linearized versions. Figures 3–12 illustrate selected full bifurcation diagrams with the corresponding basins of attraction overlaid on top for the two scalar ODE models. One can see how the different methods exhibit spurious steady states and asymptotes. The complexity of the different basins of attraction and their differences in comparing with the two ODE models are also shown.

Figure 3 shows the bifurcation diagrams and basins of attraction for the logistic ODE using four different Runge–Kutta methods (Modified Euler, Improved Euler, Kutta and the classical fourth-order Runge–Kutta (R-K 4)).

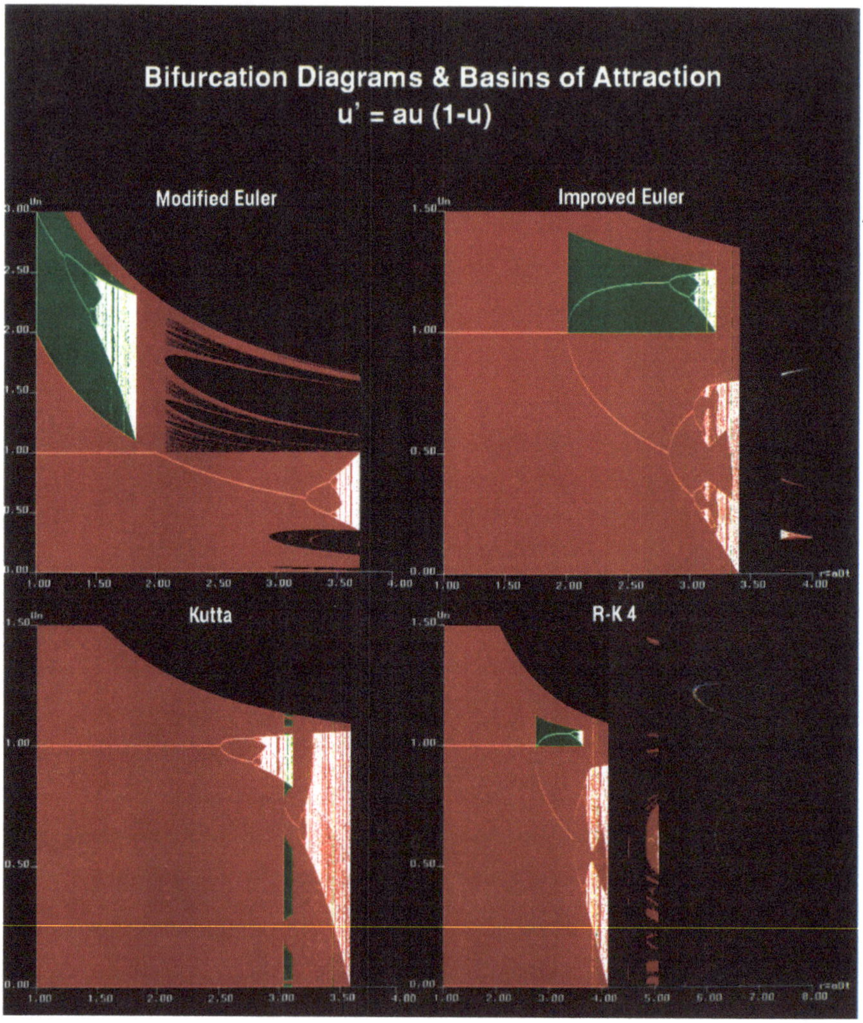

Figure 3. Full bifurcation diagram and basin of attraction for the logistic ODE using four different Runge–Kutta methods (Modified Euler, Improved Euler, Kutta and the classical fourth-order Runge–Kutta (R-K 4)), $\alpha > 0$.

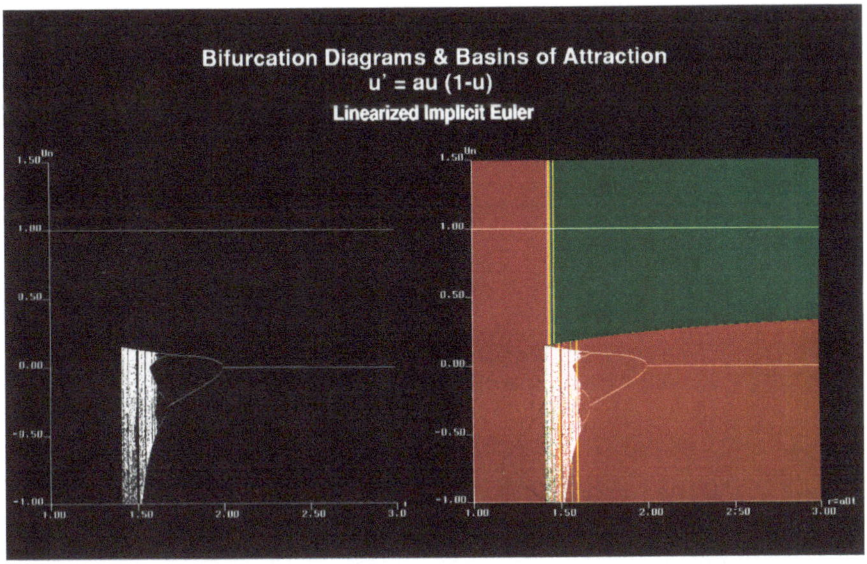

Figure 4. Full bifurcation diagram and basin of attraction for the logistic ODE using linearized implicit Euler, $\alpha > 0$.

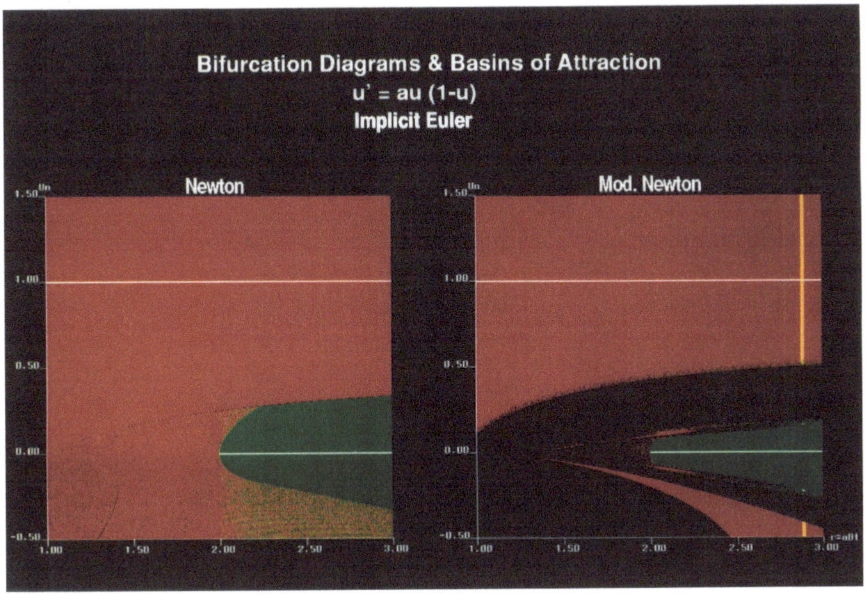

Figure 5. Full bifurcation diagram and basin of attraction for the logistic ODE using implicit Euler, $\alpha > 0$.

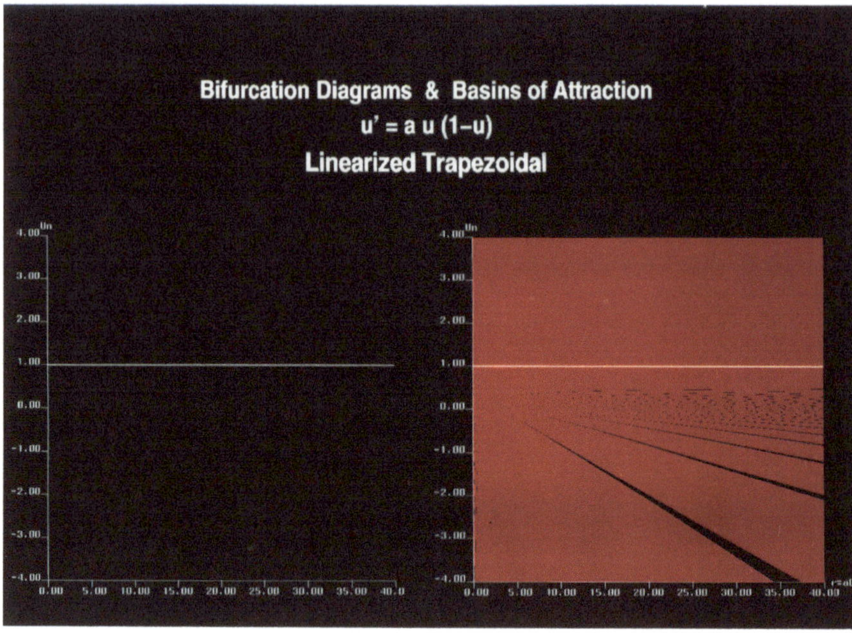

Figure 6. Full bifurcation diagram and basin of attraction for the logistic ODE using linearized trapezoidal, $\alpha > 0$.

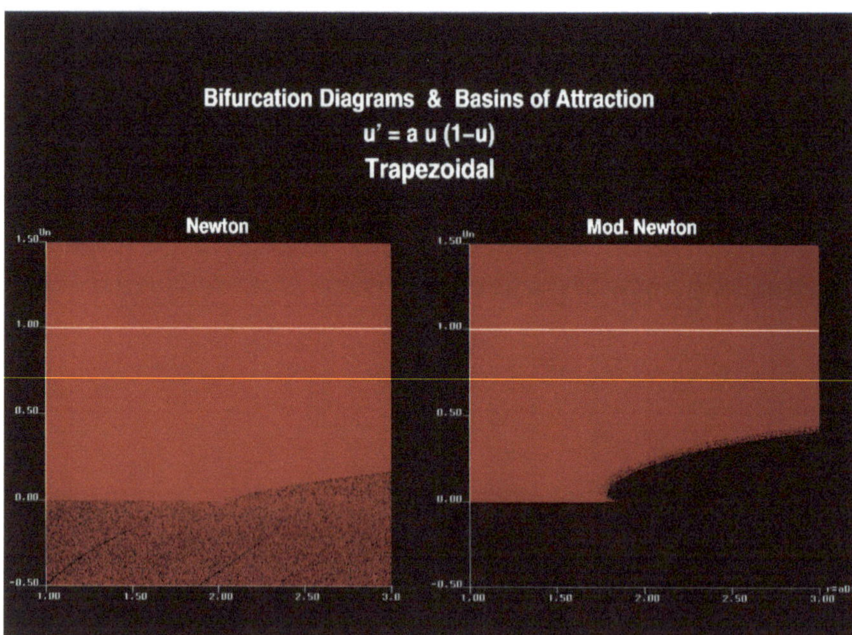

Figure 7. Full bifurcation diagram and basin of attraction for the logistic ODE using trapezoidal, $\alpha > 0$.

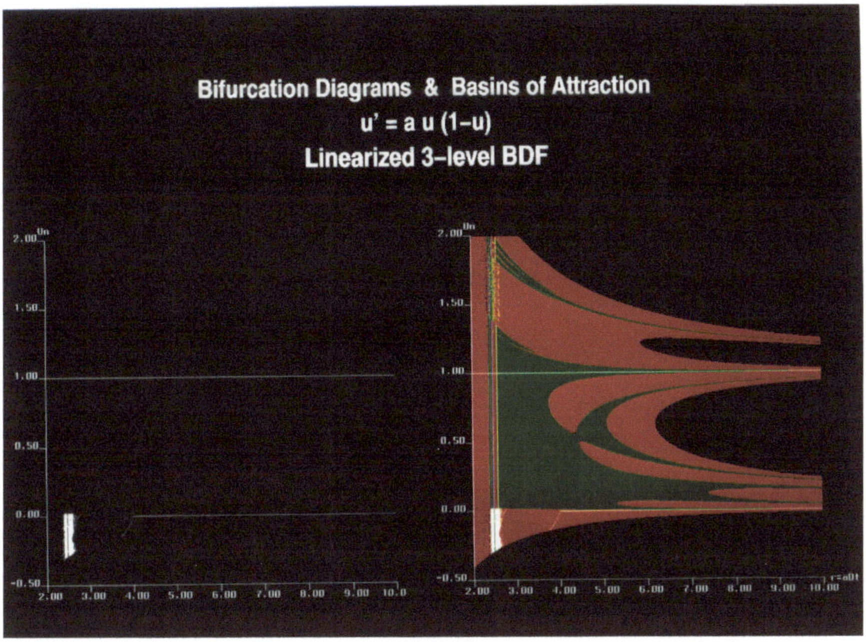

Figure 8. Full bifurcation diagram and basin of attraction for the logistic ODE using linearized 3-level BDF, $\alpha > 0$.

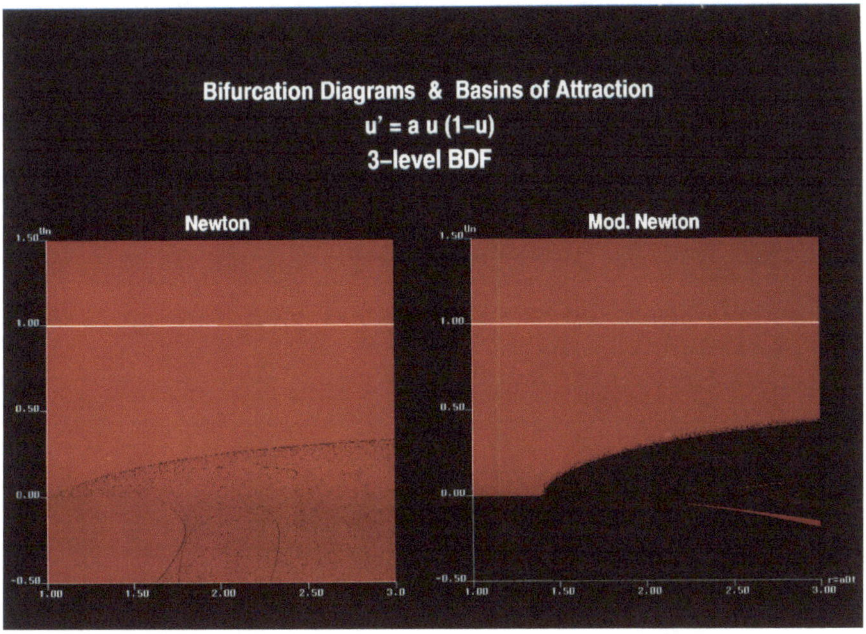

Figure 9. Full bifurcation diagram and basin of attraction for the logistic ODE using 3-level BDF, $\alpha > 0$.

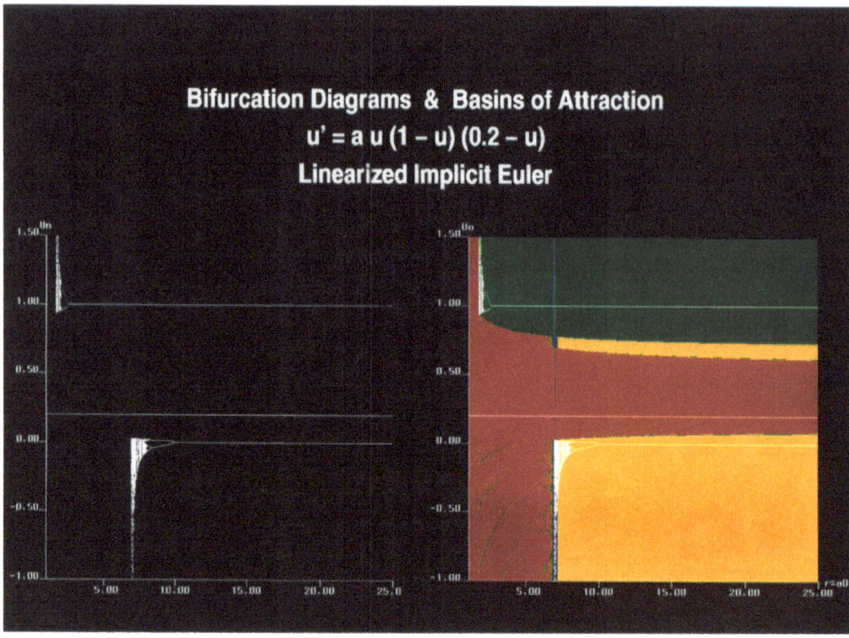

Figure 10. Full bifurcation diagram and basin of attraction for ODE (3) using linearized implicit Euler, $\alpha > 0$.

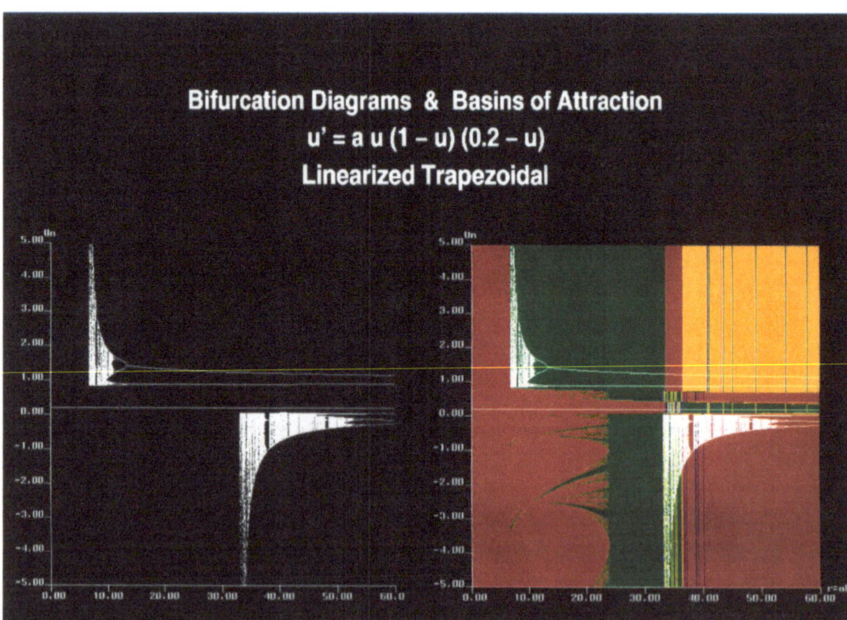

Figure 11. Full bifurcation diagram and basin of attraction for ODE (3) using linearized trapezoidal, $\alpha > 0$.

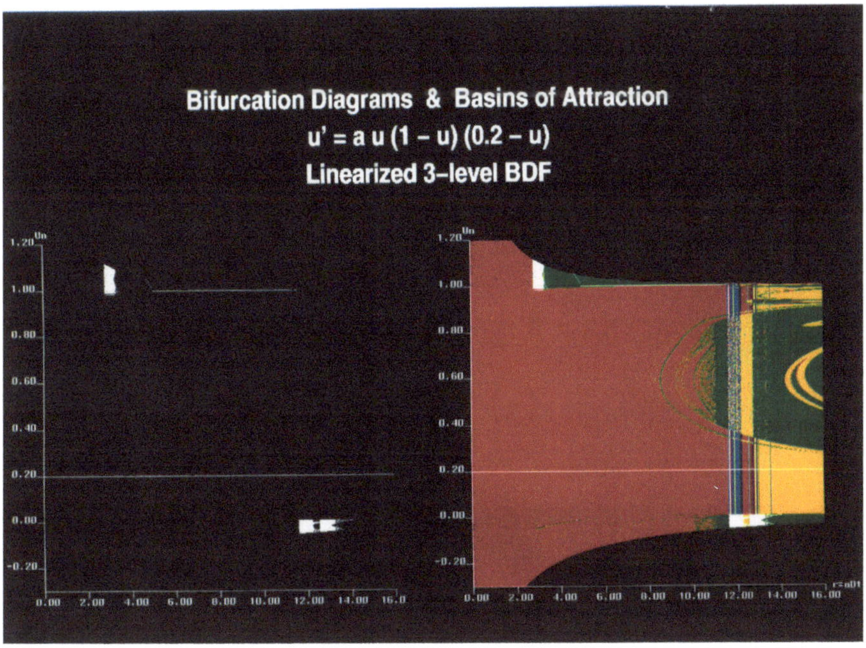

Figure 12. Full bifurcation diagram and basin of attraction for ODE (3) using linearized 3-level BDF, $\alpha > 0$.

It is interesting to note that the corresponding bifurcation diagrams of the respective discrete maps produced by the studied schemes consist of unions of "logistic-map-like" bifurcations and/or "inverted logistic-map-like" bifurcations with similar yet slightly complicated period-doubling cascades resulting in chaos. In all of the fixed point diagrams shown, the majority of the bifurcation phenomena can be divided into three kinds: flip, supercritical and transcritical bifurcations (Seydel [115]). The global asymptotic behavior of super-stable implicit LMMs indicates that unconditional stable LMMs can stabilize unstable physical steady states. From these illustrations, one can conclude that all of the studied explicit methods eventually undergo "period doubling bifurcations" leading to the "logistic-map-type bifurcations". The birth of the logistic-map-type bifurcations can occur below or beyond the linearized stability limit of the true steady state of the governing equation. Besides the regular logistic-map-type bifurcations, some of these methods undergo the so-called "inverted logistic-map-type" of period-doubling bifurcations. The above example explains the role of initial data in the generation of spurious steady-state numerical solutions, stable and unstable spurious numerical chaos and other asymptotes. Readers are referred to our work [1,66] for the forms of these studied ODE methods.

Although the basin of attraction for the stable fixed point u is the entire positive half-plan for all values of $\alpha > 0$ for the logistic ODE, the role of initial data in the occurrence of stabilizing unstable steady states of the governing equation and the introduction of stable and unstable spurious numerical chaos and other asymptotes by implicit LMMs are illustrated. See our previous papers [1,66,90] for more discussions.

It is noted that the computed bifurcation diagram cannot distinguish between the types of bifurcation and the periodicity of the spurious fixed points of any order. With the numerical basins of attraction and their respective bifurcation diagrams superimposed on the same plot, the type of bifurcation and the determination of which initial data lead to which stable asymptotes become apparent. In order to obtain the corresponding numerical basins of attraction for the schemes discussed above, one immediately realizes that, in most cases, a numerical approach is the only recourse until more theoretical tools for searching for the basin boundaries of general discrete maps become available. We would like to add that there are isolated theories or approximate methods to locate some basin boundaries for simple discrete maps or special classes of discrete maps. Even in this case, these methods are neither practical nor are there fixed guidelines for the actual implementation of discrete maps for more complex ones of similar type. See early work by Hsu [116] for an approximate method and Friedman [117] for numerical algorithms that compute connecting orbits.

Local Error Control vs. Suppression of Spurious Numerics The previous sections discussed mainly the spurious behavior of long-time integrations of initial value problems of nonlinear ODE solvers for constant step sizes. The use of adaptive step size based on local error control for implicit methods was studied by Dieci and Estep [85]. Dieci and Estep concluded that for super-stable LMMs with local step size error control and depending on the procedure for solving the resulting nonlinear algebraic equations, spurious behavior can occur. Our preliminary study on the two variable step size control methods indicated that one shortcoming is that the size of Δt needed to avoid spurious dynamics is impractical (too small) to use, especially for needing more than one ODE method for error control. Theoretical studies on the adaptive explicit Runge–Kutta method for long-time integration have been gaining more attention. Early work by Stuart [39,88], Humphries [44], Higham and Stuart [118] and Aves et al. [82] showed that local error control offers benefits for long-term computations with certain problems and methods. Aves et al. addressed the heart of the question of whether local error control confers global properties of steady states of the IVP of autonomous ODEs using adaptive Runge–Kutta type methods. Aves et al.'s work is concerned with long-time behavior and global quantities of general explicit Runge–Kutta methods with step size control for autonomous ODEs. They studied spurious fixed points that persist for arbitrarily small error tolerances. This type of adaptive Runge–Kutta method usually consists of primary and secondary Runge–Kutta methods of different orders. Their main result is positive. When standard local error control is used, the chance of encountering spuriosity is extremely small. For general systems of ODEs, the constraints imposed by the error control criterion make spuriosity extremely unlikely. Recent development on this subject is beyond the scope of this overview paper.
Spurious Steady States and Nonphysical Wave Speeds: The possible connection of the numerical phenomenon of incorrect propagation speeds of discontinuities with the existence of some stable spurious steady states introduced by the spatial discretization was discussed in [1,2,8,25,42]. The topic will be discussed in more detail in Sections 6 and 9.

5. Adaptive Numerical Methods in Space and Time

Another important building block for PARs is adaptive numerical methods. This includes adaptive temporal and spatial schemes and grid adaptation as an integral part of the numerical solution process. For adaptive temporal discretization, it includes using more than one time integrator for error control and/or time step control, adaptive grid refinement (AMR) and, most of all, adaptive numerical dissipation controls by blending of more than one scheme based on smart flow sensors. Nearly three decades ago, using tools from dynamical systems, Yee et al. (1991–1997), Yee and Sweby (1993–1997), Griffiths et al. (1992a,b) and Lafon and Yee (1991, 1992) ([1,6,7,23,62,90]) showed that adaptive temporal and adaptive spatial schemes are important in minimizing numerically induced chaos, numerically induced chaotic transients and the false prediction of flow instability by direct numerical simulation (DNS). Their studies further indicated the need in the development of practical adaptive temporal schemes based on error controls to minimize spurious numerics due to the **full discretizations**. In addition, the development of adaptive temporal and spatial schemes based on error controls to minimize numerical artifacts due to the full discretizations is also needed. This is due to the fact that adaptive temporal or adaptive spatial schemes **alone** will not be able to provide an accurate and reliable process to minimize numerical artifacts for time-accurate computations. Guided by the theory of nonlinear dynamics, Yee et al. [63] and Yee and Sweby [119] presented practical examples that illustrated the danger of using nonadaptive temporal and spatial schemes for studying flow instability.

Another important numerical method development is accurate, stable and reliable simulations of compressible turbulence with shock waves by adaptive numerical dissipation controls. It is well known that reliable, accurate and efficient direct numerical simulation (DNS) of turbulence in the presence of shock waves represents a significant challenge for numerical methods. Standard TVD, ENO, WENO and discontinuous Galerkin (DG) types of shock-capturing methods for the Euler equations has been routinely used for over three decades in high-speed blast wave simulations with virtually non-oscillatory, crisp resolution discontinuities. For the unaveraged unsteady compressible Navier–Stokes equations, it was observed that these schemes are still too dissipative for turbulence and transition predictions. On the other hand, hybrid schemes, where spectral and/or higher-order compact (Padé) schemes are switched to higher-order WENO or ENO schemes when shock waves are detected, have their deficiencies for complex turbulence and shock interaction. One shortcoming of this type of hybridization is that the numerical solution might experience a non-smooth transition at the switch to a different type of scheme. For 2D and 3D complex shock wave and shear surface interactions, the switch mechanism can become non-trivial and frequent activation of shock-capturing schemes is possible.

The early work of Yee et al. [26,63], Sjögreen and Yee [10,120], and Yee and Sjögreen [121,122] indicates that appropriate adaptive numerical dissipation control is essential to control nonlinear instability in general and particularly for long-time integration. A more recent integrated design approach on the construction of adaptive numerical dissipation controls can be found in Yee and Sjögreen [3] and the references cited therein. An earlier, excellent overview paper on hybrid methods for high-speed flows was presented by Pirozzoli in 2011 [123]. The nonlinear filter methods of Yee et al., which blended more than one method using a high-order entropy-conserving method [9,10,26–28], were not included in [123].

6. Numerical Source Term Treatments in Reacting Flows

We devote this section to source term treatments in reacting flows as there is a wide area of CFD, computational astrophysics and the next-generation spacecrafts and high-speed civil transport design applications. Examples of application areas are combustion, turbulence combustion, scramjet engine designs and black hole accretion. In the modeling of problems containing finite-rate chemistry or combustion, a wide range of space and time scales is often present due to the reacting terms, over and above the different scales associated with turbulent flows, leading to additional numerical difficulties. This stems mainly from the fact that the majority of widely used numerical algorithms in reacting flows were originally designed to solve non-reacting fluid flow problems. Fundamental studies on the behavior of these schemes for reacting model problems by the author and collaborators were reported in Yee and Sweby [119] and the references cited therein. In the majority of these studies, theories from dynamical systems were used to gain a better understanding of the nonlinear effects on the performance of these schemes. The main findings of this report and our recent development are as follows:

- It was shown in LeVeque and Yee [8] that, for stiff reactions containing shock waves, it is possible to obtain stable solutions that look reasonable and yet are completely wrong because the discontinuities are in the wrong locations. Stiff reaction waves move at nonphysical wave speeds, often at the rate of one grid cell per time step, regardless of their proper speed. There exist several methods that can overcome this difficulty for a single reaction term. For more than a single reaction term in fully coupled nonlinear systems, more research is needed. One impractical way of minimizing the wrong speed of propagation of discontinuities is to demand orders of magnitude grid size reduction compared with what appears to be a reasonable grid spacing in practice.

- It was shown in Lafon and Yee [6,7] that the numerical phenomenon of incorrect propagation speeds of discontinuities may be linked to the existence of some stable spurious steady-state numerical solutions.
- It was also shown in Lafon and Yee[6,7] that various ways of discretizing the reaction term can affect the stability and convergence of the spurious numerical steady states and/or the exact steady states. Pointwise evaluation of the source terms appears to be the least stable.
- It was shown in Yee et al. [23] and Griffiths et al. [35,124] that spurious discrete traveling waves can exist, depending on the method of discretizing the source term. When physical diffusion is added, it is not known what type of numerical difficulties will surface.
- Our later work with Chi-Wang Shu, Wei Wang and Dmitry V. Kotov [18–20,40–42] developed new methods to minimize spurious numerics for combustion and nonequilibrium reacting flows.

Reaction-Convection Model: Based on the work of [6–8,23,35,119], in addition to the incorrect propagation speed of discontinuities, other spurious numerics that are directly tied to the amount of numerical dissipation contained in the chosen scheme and the numerical treatment of source terms may result in possible spurious steady-state numerical solutions and spurious standing waves [6,7,23,35]. It was shown in Lafon and Yee [6,7] and in Griffiths et al. [35] that various ways of discretizing the nonlinear reaction terms can affect the stability of, and convergence to, the spurious numerical steady states and/or the exact steady states. Pointwise evaluation of the source terms appears to be the least stable. The studies of Lafon and Yee [6,7] indicated that numerical phenomena of incorrect propagation speeds of discontinuities may be linked to the existence of some stable spurious steady-state numerical solutions. More importantly, the different combination of time step, grid spacing and initial condition plays a major role in obtaining the correct solution. In addition, it was shown in Yee et al. [23] and Griffiths et al. [35] that spurious discrete traveling waves can exist, depending on the method of discretizing the source term.

The effect of spatial as well as time discretizations on the existence and stability of spurious steady-state solutions was illustrated with examples in LeVeque and Yee and in Lafon and Yee (1991, 1992) [6–8,18–20,40,42]. This phenomenon is due to the smearing of the discontinuity caused by the spatial discretization of the advection term. This introduces a nonequilibrium state into the calculation. Studies linking spurious numerical standing waves for a nonlinear reaction model by first- and second-order spatial and temporal discretizations can be found in Lafon and Yee [6,7] and in Griffiths, Stuart and Yee [2,35]. The possible wrong prediction of transition point Reynolds number by DNS due to spurious bifurcation that created a false transition point was discussed in [2]. Inaccuracy of the scheme or insufficient grid points might lead to possible spurious bifurcation as well as creating wrong propagation speed of discontinuities and smearing of turbulent fluctuations. See [2] for a discussion. For various project methods to minimize wrong propagation speed of discontinuities, see [125–128] and the references cited therein.

Section IX is devoted to illustrating selected test cases on spurious numerics related to problems with stiff source terms and our recent methods to minimize wrong computed solutions.

7. Spurious Numerics Relevant to Turbulent Flow Computations [1,2]

Four numerical examples that exhibit spurious numerics relevant to turbulent flow computations were illustrated together with other CFD examples and can be found in [1,2]. This section illustrates two examples: (a) Chaotic transients near the onset of turbulence in DNS of channel flow (Keefe 1992, Keefe and Yee 1997, [129,130]); (b) false flow transition simulation on a 2D incompressible flow over a backward-facing step as a function of Reynolds number. The numerical transition is due to a lack of grid refinement, not long enough time integration and/or lack of sufficient numerical dissipation for high-order methods. If care is not taken, spurious bifurcation of the discretized counterpart and/or a numerically induced chaotic transient can be mistaken for the onset of physical turbulence of the governing equation. These simple examples are useful in modeling and simulations of turbulence combustion, supersonic turbulence, computational astrophysics and the next-generation spacecrafts and high-speed civil transport design. See Figure 13 for a schematic of the flow pattern of a scramjet engine. The figure source is from https://en.m.wikipedia.org/wiki/File:ScramjetDiagram.gif. This figure is in the public domain in the United States because it was solely created by NASA. See also Moura et al. [131] on the characterization of supersonic turbulent combustion in a Mach-10 scramjet combustor.

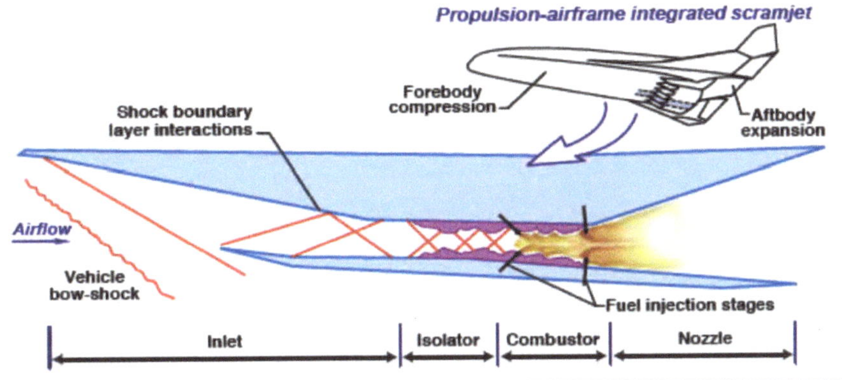

Figure 13. Schematic of flow pattern of a scramjet engine. Source: https://en.m.wikipedia.org/wiki/File:ScramjetDiagram.gif. This figure is in the public domain in the United States because it was solely created by NASA.

7.1. Chaotic Transients Near the Onset of Turbulence in Direct Numerical Simulations of Channel Flow (Keefe 1988, Keefe and Yee 1997 (Unpublished))

In addition to the inherent chaotic and chaotic transient behavior in some physical systems, numerics can independently introduce and suppress chaos as well as chaotic transients. Loosely speaking, a chaotic transient behaves like a chaotic solution ([132]). A chaotic transient can occur in a continuum or a discrete dynamical system. One of the major characteristics of a numerically induced chaotic transient is that if one does not integrate the discretized equations long enough, the numerical solution has all the characteristics of a chaotic solution. The required number of integration steps might be far beyond those found in standard CFD simulation practice before the numerical solution can get out of the chaotic transient mode. Furthermore, standard numerical methods, depending on the initial data, usually experience drastic reductions in step size and convergence rate near a bifurcation point of the continuum in addition to the bifurcation points due solely to the discretized parameters. See Yee and Sweby [1,62,90] for a discussion. Consequently, a possible numerically induced chaotic transient is especially worrisome in direct numerical simulations of the transition from laminar to turbulent flows. Except for special situations, it is extremely difficult to bracket closely the physical transition point by mere DNS of the Navier–Stokes equations. Even away from the transition point, this type of numerical simulation is already very CPU intensive and the convergence rate is usually rather slow. Due to limited computer resources, the numerical simulation can result in chaotic transients indistinguishable from sustained turbulence, yielding a spurious picture of the flow for a given Reynolds number. Consequently, it casts some doubt on the reliability of numerically predicted transition points and chaotic flows. It also influences the true connection between chaos and turbulence. See also Moore et al. [133].

Assuming a known physical bifurcation or transition point, Figure 14 illustrates the schematic of four possible spurious bifurcations due to constant time steps and constant grid spacings. Different examples to illustrate the occurrence of these scenarios can be found in Sections 6.3 and 6.4 of our work [2]. Section 6.4 in [2] discusses the stability of the steady state (as a function of the Reynolds number) of a 2D backward-facing step problem using direct simulations. The present section is the computation by Laurence Keefe performed in the late 1980s. In 1996, we made use of the knowledge from continuum and discrete dynamical systems theory to interpret his result. We identified some of the aforementioned numerical uncertainties in his computations. The result is reported in [1].

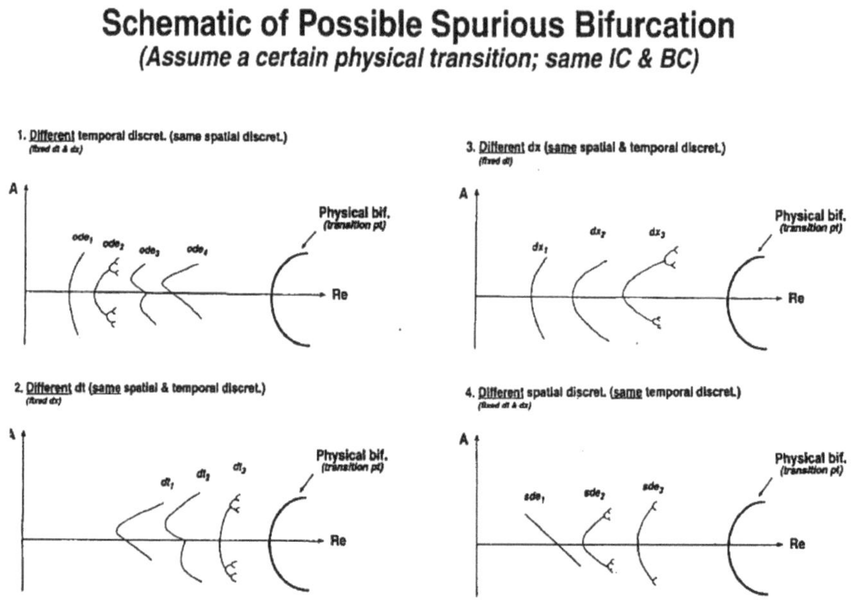

Figure 14. Schematic of possible spurious bifurcation for constant time steps and grid spacings. (**1**) Different temporal discretizations ode_1, ode_2, ode_3 and ode_4 (same spatial discretization and the same constant dt and dx). (**2**) Different constant time steps dt_1, dt_2, dt_3 and dt_4 (same temporal and spatial discretizations, and the same constant dx). (**3**) Different constant grid spacings dx_1, dx_2, dx_3 and dx_4 (same spatial and temporal discretizations, and the same constant dt). (**4**) Different spatial discretizations sde_1, sde_2 and sde_3 (same temporal discretization and the same constant dt).

The physical problem that Keefe considered is depicted in Figure 15, where a flow is confined between planes at $y = \pm 1$ and is driven in the x-direction by a mean pressure gradient dp/dx. The flow is characterized by a Reynolds number $Re = U_\infty L/\nu$, where U_∞ is the mean centerline velocity, L is the channel half-height and ν is the kinematic viscosity. Within the channel, the flow satisfies the incompressible Navier–Stokes equations and no-slip boundary conditions are applied at the walls. In the particular calculations shown here, these equations have been manipulated into velocity–vorticity form, where one integrates equations for the wall-normal velocity v and normal vorticity η, and recovers the other two velocity components from the incompressibility condition and the definition of η.

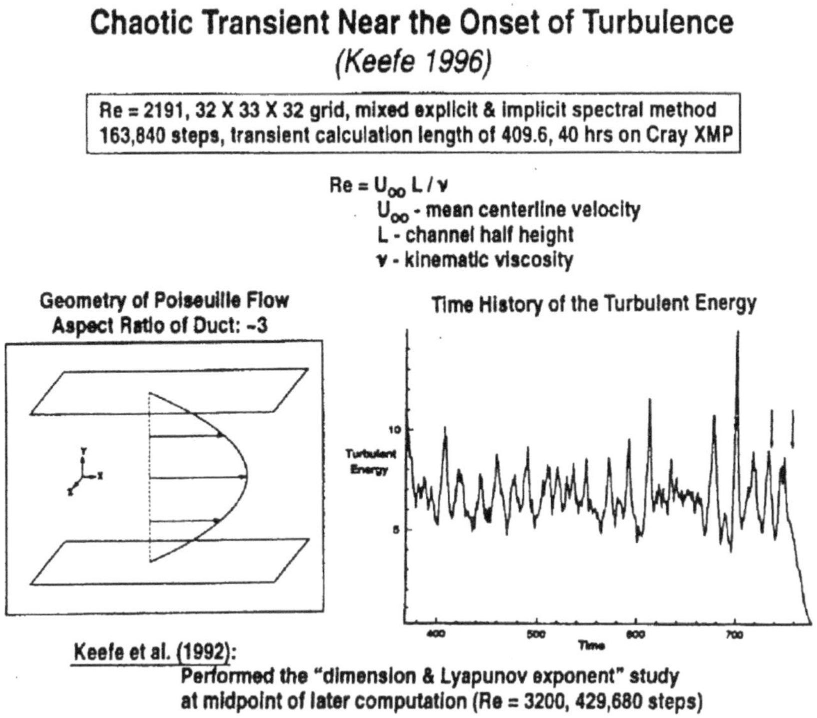

Figure 15. Three-dimensional channel flow computation by Keefe (1997) [130].

$$\frac{\partial}{\partial t}\Delta^2 v = h_v + \frac{1}{Re}\Delta^4 v, \tag{4}$$

$$\frac{\partial}{\partial t}\eta = h_g + \frac{1}{Re}\Delta^2 \eta, \tag{5}$$

$$f + \frac{\partial v}{\partial y} = 0, \tag{6}$$

where

$$f = \frac{\partial u}{\partial x} + \frac{\partial w}{\partial z}, \qquad \eta = \frac{\partial u}{\partial z} - \frac{\partial w}{\partial x}, \tag{7}$$

$$h_v = -\frac{\partial}{\partial y}\left(\frac{\partial H_1}{\partial x} + \frac{\partial H_3}{\partial z}\right) + \left(\frac{\partial^2}{\partial x^2} + \frac{\partial^2}{\partial z^2}\right)H_2, \tag{8}$$

$$h_g = \frac{\partial H_1}{\partial z} - \frac{\partial H_3}{\partial x}. \tag{9}$$

Here, the H_i contains the nonlinear terms in the primitive form of the Navier–Stokes equations and the mean pressure gradient.

The velocity increases extremely rapidly normal to the wall, and turbulent channel flows are essentially homogeneous in planes parallel to the wall. The first requires a concentration of grid points near the wall, and the second suggests use of a doubly periodic domain in planes parallel to the wall. A spectral representation of the velocity field (u, v, w) is

$$\vec{u} = \sum_l \sum_m \sum_n \vec{A}_{lmn}(t) T_l(y) e^{im\alpha x + in\beta z}, \tag{10}$$

where the $T_l(y)$ are Chebyshev polynomials used for the spatial discretization. The numerical problem then becomes dependent on α and β in addition to Re. For the time discretization, mixed explicit–implicit methods are used. The nonlinear terms in the equations are advanced using second-order Adams–Bashforth or a low-storage, third-order Runge–Kutta scheme (Spalart et al. [134]), while the viscous terms are advanced by Crank–Nicholson.

One of the central problems in studies of wall-bounded shear flows is the determination of when a steady laminar flow becomes unstable and transitions to turbulence. In dynamical systems terms, the Navier–Stokes equations always have a fixed point solution for low enough Reynolds numbers; but for each flow geometry, the Reynolds number at which this fixed point bifurcates needs to be determined. In channel flow, the fixed point solution (a parabolic velocity profile across the channel, $u(y) = (1 - y^2)$) becomes linearly unstable at $Re = 5772$ (Orszag [135]). However, since turbulence appears in experiments at much lower Reynolds numbers, it was conjectured that this bifurcation must be subcritical. A subsequent numerical solution of the nonlinear stability equations (Herbert 1976, Ehrenstein and Koch 1991) [33,136] demonstrated this to be true, showing that limit cycle solutions with amplitude ϵ branch back to lower Reynolds numbers before subsequently passing through a turning point and curving back toward higher Reynolds numbers. Thus, for Reynolds numbers just above the turning point, the flow equations have at least four solutions: the fixed point, two unstable limit cycles and a chaotic solution (experimentally observed turbulence). Determining the location of the turning point in $(\alpha, \beta, \epsilon, Re)$ space is known as the minimum-critical-Reynolds-number problem, and its solution is by no means complete.

One way to investigate the turning point problem is to perform DNS of channel flow for conditions believed to be near this critical condition. Beginning with a known turbulent initial condition from higher Reynolds number, one integrates in time at the target Reynolds number to determine whether the flow decays back to the fixed point or sustains itself as turbulence. Although this may not be the most efficient way to bracket the turning point, it has the advantage that the peculiar dynamics of the flow near the turning point, whether in decay or sustained turbulence, are observable. This yields information about the path along which flows become turbulent at these low Reynolds numbers.

Unfortunately, the flow dynamics are very peculiar near the turning point, and extremely long chaotic transients are observed in the computations that make a fine determination of that point all but impossible by this method. This can be seen in Figure 15, where a time history of the turbulent energy in a channel flow (energy above that in the laminar flow) is plotted for a Reynolds number of 2191. To understand the time scale of the phenomenon, some experimental facts need to be recalled. In typical experimental investigations of channel flow, the infinite transverse and streamwise extent of the ideal flow are approximated by studying flow in high aspect ratio (10–40) rectangular ducts that are typically 50–100 duct heights long. If times are non-dimensionalized by the centerline mean velocity U_∞ and the duct half height L, then statistics on turbulence are gathered by averaging hot-wire data over intervals $\Delta t U_\infty / L \sim 200$. In the simulations and figure, the time scale is based on the friction velocity u_τ and L, where typically 15–20 $u_\tau \sim U_\infty$. Thus, averaging over intervals $\Delta t u_\tau / L \sim 10$ should and does yield stable flow statistics that compare well with experiments. The near-wall velocity profile, cross-channel turbulence intensities, and Reynolds and shear stress distribution for the $\Delta t u_\tau / L \sim 10$ interval near the end of the transient, delineated by the arrows in Figure 15, indicate the good comparison. In each case, they correspond well to available experimental data. However, consider the time scale of the transient; it spans $\Delta t u_\tau / L \sim 300$, thirty times longer than the time needed to obtain stable statistics that would convince most experimentalists that they are viewing a fully developed turbulent channel flow. This is further complicated by the wide variation of the transient length, depending upon both the grid resolution (number of modes in the spectral representation) and the linearly stable time step of the integration. In fact, for fixed (α, β, Re), it is possible to obtain sustained turbulence for one time step but see it rapidly decay to laminar flow for another, lower value of the step.

Extended chaotic transients near bifurcation points are not an unknown phenomenon; the "meta-chaos" of the Lorenz system is but one of many known examples. However, the practicalities of numerical computation in fluid dynamics usually interfere with one's ability to discern whether transient, or sustained turbulence, is being calculated. The computations required to obtain the transient plot in Figure 15 needed 40 h of single processor time on a Cray XMP, some thirty years ago. Such a small amount of expended time was only possible because the spatial resolution of the calculation was relatively coarse ($32 \times 33 \times 32$), in keeping with the large scales of the phenomena expected at these flow conditions. Higher resolution calculations ($192 \times 129 \times 160$) (Kim et al. [137]) at greater Reynolds numbers typically have taken hundreds of hours ($\sim$250) to barely obtain the $\Delta t u_\tau / L = 10$ averaging interval that is so inadequate for detecting transients. Because such calculations are so time consuming, one typically chooses an integration time step that is a substantial fraction of the linear stability limit of the algorithm so as to maximize the calculated "flow time" for expended CPU time. However, it is clear from these transient results that this practice has some dangers when close to critical points of the underlying continuous dynamical system. Thus, it appears that just as pseudo-time integration to obtain steady solutions can result in spurious results, genuine time integration can result in chaotic transients indistinguishable from sustained turbulence and yield a spurious picture of the flow for a given Reynolds number.

7.2. Temporal and Spatial Refinement Studies of 2D Incompressible Flow over a Backward-Facing Step

The 2D incompressible flow over a backward-facing step has been addressed by many authors using a wide variety of numerical methods. Figure 16 shows the flow geometry. Fluid with constant density ρ and viscosity μ enters the upstream channel of height h with a prescribed velocity profile (usually parabolic). After traveling a distance l, the fluid passes over a backward-facing step of height s and enters the downstream channel of height $H = h + s$. After traveling a distance L downstream of the step, the fluid exits the region of interest. For Reynolds numbers considered here, the flow separates at the corner and forms a recirculating region behind the step. Additional recirculating regions form on the upper and, subsequently, lower walls of the downstream channel as the Reynolds number is increased.

Results of sustained unsteady flow from various numerical simulations have been reported for Reynolds numbers (*Re*) ranging from 250 up to 2500. The formulations included the vortex method, unsteady equations in stream function form, steady equations and the associated linear-stability problem, and the unsteady equations in primitive variable form. The numerical methods used cover almost all of the existing schemes in the literature. The majority of the numerical results are summarized in Gresho et al. [138]. The work of Gresho et al. was an answer to a controversy concerning the stability of the stationary solution at $Re = 800$. It was concluded by Kaiktsis et al. [139] computed by a high-order spectral element method that transition to turbulent flow has occurred at $Re = 800$. See [139] for the description of the numerical method. Kaiktsis et al. examined the long-time temporal behavior of the flow and found that the flow is steady at $Re = 500$, time-periodic at $Re = 700$ and chaotic at $Re = 800$. Gresho et al. conducted a detailed grid refinement study using four different numerical methods and concluded that the backward-facing step at $Re = 800$ is a stable steady flow.

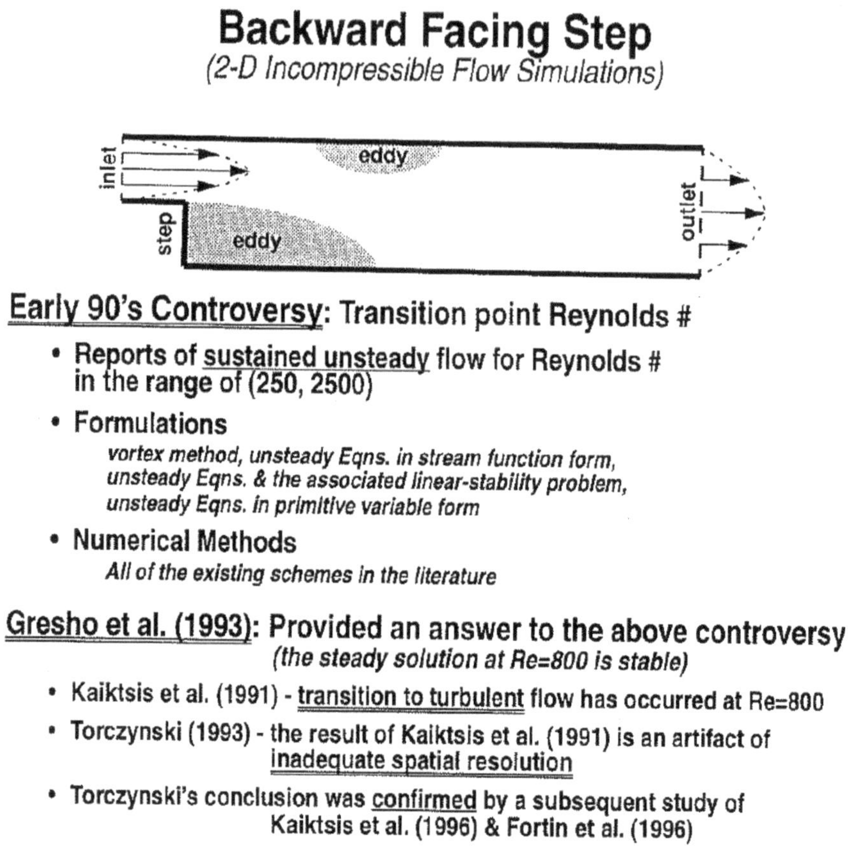

Figure 16. Schematic of the backward-facing step problem.

In addition to the study of Gresho et al., an extensive grid refinement study of this flow using a spectral element method was conducted in Torczynski [140]. The simulated geometry and the numerical method correspond to those of Kaiktsis et al. Flow was examined at Reynolds numbers of 500 and 800. His systematic grid refinement study was performed by varying both the element size and the order of the polynomial representation within the elements. For both Reynolds number values with the transient computations stopped at $t = 800$, it was observed that low-resolution grid cases exhibit chaotic-like temporal behavior whereas high-resolution grid cases evolve toward asymptotically steady flow by a monotonic decay of the transient. The resolution required to obtain asymptotically steady behavior is seen to increase with Reynolds number. These results suggest that the reported transition to sustained chaotic flow by Kaiktsis et al. at Reynolds numbers around 700 is an artifact of inadequate spatial resolution. Torczynski's conclusion was further confirmed by a subsequent study of Kaiktsis et al. [141] and Fortin et al. [110]. Fortin et al. employed tools from dynamical systems theory to search for the Hopf bifurcation point (transition point). They showed that the flow remains steady at least up to $Re = 1600$.

7.2.1. Grid Refinement Study of Torczynski (1993) [140]

In Torczynski (1993) [140], $Re = \rho\overline{u}2h/\mu$ is based on upstream conditions. The variable $\overline{u}$ is the spatial average of the horizontal velocity u over h. The geometry is specified to match that of Kaiktsis et al. [139]. The upstream channel height h and step height s have values of $h = 1$ and $s = 0.94231$, yielding a downstream channel height of $H = 1.94231$. The corner of the step is at $(x, y) = (1, 0)$. The channel extends a distance $L = 1$ upstream from the step and a distance $L = 34$ downstream from the step to preclude undue influence of the finite channel length on the flow at $Re = 800$. The following conditions are applied on the boundaries of the computational domain: $u = v = 0$ on the upper and lower channel walls, $-p + \mu \partial u/\partial n = 0$ and $\partial v/\partial n = 0$ on the outflow boundary, and $u = [\tanh(t/16)]u_B(y) + [1 - \tanh(t/16)]u_P(y)$ and $v = 0$ on the inflow boundary and the step surface. Here, $u_B(y) = max[0, 3y(1-y)]$ is the correct boundary condition for flow over a backward-facing step and $u_P(y) = 3(1-y)(s+y)/(1+s)^3$ is the Poiseuille flow observed infinitely far downstream whenever steady flow is asymptotically obtained. The initial velocity field is set equal to $u = u_P(y)$ and $v = 0$ throughout the domain. Here, v is the vertical velocity and p is the pressure. Thus, the above combination of boundary and initial conditions initially allows flow through the step surface so that the simulations can be initialized using an exact divergence-free solution of the Navier–Stokes equations. Furthermore, since the inflow boundary condition is varied smoothly in time from Poiseuille flow to flow over a backward-facing step, the flow experiences an order-unity transient that is probably strong enough to excite sustained unsteady behavior, if that is the appropriate asymptotic state for the numerical solution.

The simulations were performed using the commercial code NEKTON v2.8, which employs a time-accurate spectral-element method with the Uzawa formulation (NEKTON, 1991). Let D be the dimensionality. Each element has N^D velocity nodes located at Gauss–Lobatto–Legendre collocation points, some of which are on the element boundaries, and $(N-2)^D$ pressure nodes located at Gauss–Legendre collocation points, all of which are internal. Within each element, the velocity components and the pressure are represented by sums of D-dimensional products of Lagrangian-interpolant polynomials based on nodal values. This representation results in continuous velocity components but discontinuous pressure at element boundaries. Henceforth, the quantity N is referred to as the element order, even though the order of the polynomials used to represent the velocity is $N - 1$. NEKTON employs mixed explicit and implicit temporal discretizations. To avoid solving a nonlinear, nonsymmetric system of equations at each time step, the convective term is advanced explicitly in time using a third-order Adams–Bashforth scheme. All other terms are treated implicitly (implicit Euler for the pressure and for the viscous terms).

In our further investigation [63], Torczynski and the first author used three spectral-element grids of differing resolution, denoted L (low), M (medium) and H (high). Detailed investigation can be found in [2,63]. Here, we include selected results of those studies. Figure 17 shows the computational domain and the grid distribution of the three spectral element grids in which the **distribution of nodes within each spectral element is not shown**. The L grid with $N = 9$ is identical to the grid of Kaiktsis et al. (1991) [139]. Four general classes of behavior are observed for the numerical solutions. First, "steady monotonic" denotes evolution of the numerical solution toward an asymptotically steady state. Second, "steady oscillatory" denotes evolution toward an asymptotically steady state with a decaying oscillation superimposed on the monotonic decay. Third, "unsteady chaotic" denotes irregular transient behavior of the numerical solution that shows no indication of evolving toward steady behavior. Fourth, "diverge" denotes a numerical solution terminated by a floating-point exception. In Figure 17, the first character denotes the grid resolution L, M or H; the first digit indicates the Reynolds number 500 or 800; and the last two digits indicate the order of the spectral element being used. For example, $L807$ means $Re = 800$ using the L grid with $N = 7$.

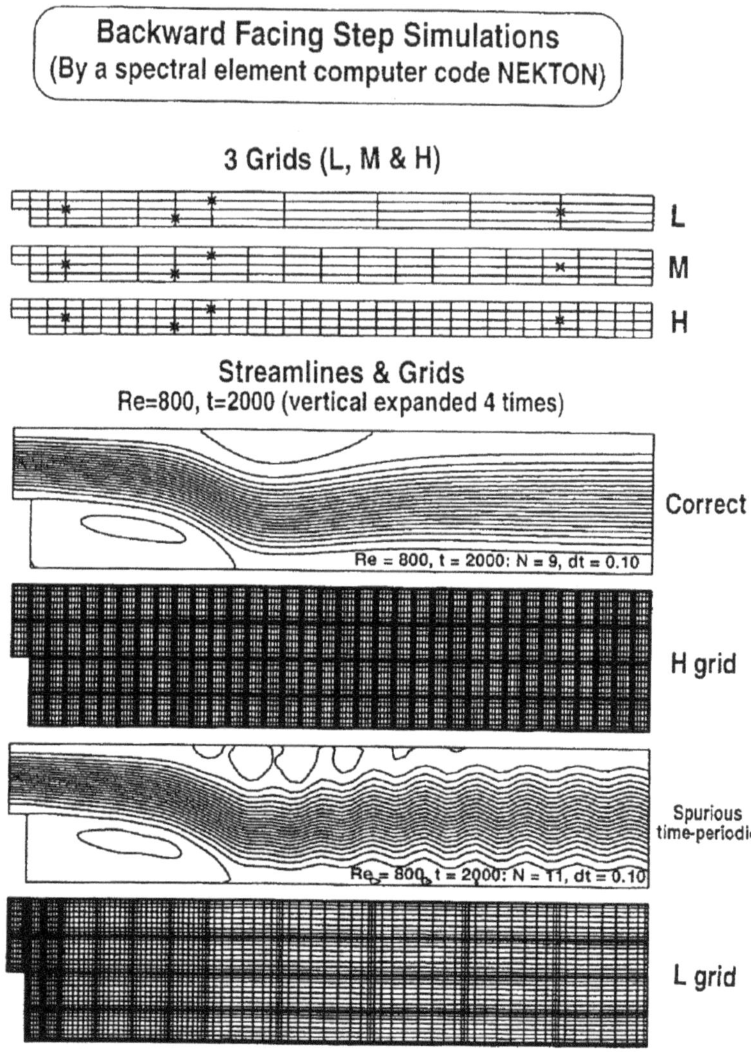

Figure 17. Three different grids, and streamlines of *H*809 and *L*811 and their corresponding grids, by the spectral element method.

Our extensive grid refinement study resulted in grid-independent steady-state numerical solutions for both $Re = 500$ and $Re = 800$. As the grid resolution is reduced below the level required to obtain grid independent solutions, chaotic-like temporal behavior occurred. The degree of grid resolution required to obtain a grid-independent solution was observed to increase as the Reynolds number is increased. Figure 17 shows the streamlines for *H*809 (steady solution) and *L*811 (spurious time-periodic solution) and the corresponding grids with the distribution of the nodes of the spectral elements shown.

7.2.2. Temporal Refinement Studies Using Knowledge from Dynamical Systems Theory

All of Torczynski's numerical solutions integrate to $t = 800$. With the knowledge of possible nonlinear behavior of numerical schemes such as long-time transients before a steady state is reached; numerically induced chaotic transients; numerically induced or suppressed chaos; existence of spurious steady states and asymptotes; and the intimate relationship among initial data, time step and grid spacing observed in discrete dynamical systems theory, Yee et al. [63] examined the Torczynski cases in more detail.

In the Yee et al. (1997) study ([2,63]), in addition to grid refinement, temporal refinements are made on all of the under-resolved grid cases to determine if these cases sustain the same temporal behavior at a much later time or evolve into a different type of spurious behavior. At $t = 800$, cases $L506, L507, L508, L509, L811$, $M807$ and $M808$ either exhibit "unsteady chaotic" or "steady oscillatory" behavior. We integrate these cases to $t = 2000$ to determine if a change in solution behavior occurs. From the phenomena observed in Keefe's 3D channel flow computation and others, $t = 2000$ might not be long enough for a long-time transient or long chaotic transient to die out. There is also the potential of evolving into a different type of spurious or divergent behavior at a much later time. However, for this study, it appears that $t = 2000$ is sufficient. For $Re = 500$, we also recomputed some of these cases with a sequence of Δt that bracketed the benchmark study of Torczynski. The Δt values are $0.02, 0.05, 0.10, 0.125, 0.2, 0.3, 0.4$ and 0.5 for $Re = 500$. The CFL number for all of these cases is above 1 for $\Delta t > 0.10$. The reason for the investigation of $\Delta t = 0.3, 0.4$ and 0.5 is to find out, after the transients have died out, if the solution converges to the correct steady state for Δt that are a few times larger than 0.10.

For $Re = 800$, we integrate $L811$ and $M808$ with $\Delta t = 0.10$ and $M807$ with $\Delta t = 0.02, 0.05$ and 0.10 to $t = 2000$. Aside from integrating to $t = 2000$, five different initial data were examined for cases $M807$, $M809$ and $M811$ for $\Delta t = 0.10$ to determine the influence of the initial data and the grid resolution on the final numerical solution. The five initial data are as follows:

- Uniform: $u, v = 0$;
- Shear layer: $u = u_B(y) = max[0, 3y(1-y)], v = 0$;
- Solution from solving the steady Stokes equation (with no convection terms);
- Torczynski (1993) [140]: $u = u_P(y) = 3(1-y)(s+y)/(1+s)^3, v = 0$;
- Channel flow both upstream and downstream of step: Same as previous initial data except the boundary conditions.

The boundary conditions for the first three and last initial condition were parabolic inflow and no-slip at walls, whereas the boundary conditions for the fourth initial condition were those of Torczynski [140]:

$$u = [\tanh(t/16)]u_B(y) + [1 - \tanh(t/16)]u_P(y);\ v = 0.$$

The CPU time required to run the above cases ranged from less than a day to several days on a Sparc Center 2000 using one processor.

The chaotic-like behavior evolves into a time-periodic solution beyond $t = 800$ for $L506$ and $L507$, whereas the chaotic-like behavior evolves into a time-periodic solution beyond $t = 800$ for $L811$ and a divergent solution for $M807$. The "steady oscillatory" case $L508$ slowly evolves to the correct steady state with an amplitude of oscillation of 10^{-5}. The oscillation is not detectable within the plotting accuracy. The "steady oscillatory" time evolution of $M808$ is similar to that of $L508$. The numerical solutions with "steady oscillatory" and "steady monotonic" behavior at early stages of the time integration are almost identical at later stages of the time integration. They all converge to the correct steady state. The initial data study at $Re = 800$ with $\Delta t = 0.10$ is summarized in Table 5.5 of Yee et al. [63], which illustrates the intimate relationship between initial data and grid resolution.

Figure 18 shows the vertical velocity time histories at $(x, y) = (30, 0)$ advanced to a time of $t = 2000$ for $M807$ with $\Delta t = 0.02, 0.05$ and 0.10, and for $L811$ with $\Delta t = 0.10$. Case $M807$ diverges at $t = 1909.2$ with $\Delta t = 0.02$, at $t = 972.4$ with $\Delta t = 0.05$ and at $t = 827.77$ with $\Delta t = 0.10$. The time histories for these three time steps appear to show chaotic-like behavior if one stops the computations at $t = 800$. The bottom plot of Figure 18 shows the vertical velocity time histories advanced to a time of $t = 2000$ for $L811$ with $\Delta t = 0.10$. It shows the definite time-periodic spurious solution pattern. On the other hand, the time history for this case appears to show an aperiodic-like pattern if one stops the computation at $t = 800$. Note that the $L809$ grid case was used by Kaiktsis et al. [139], and they concluded that "2-D transition" has already occurred at $Re = 800$.

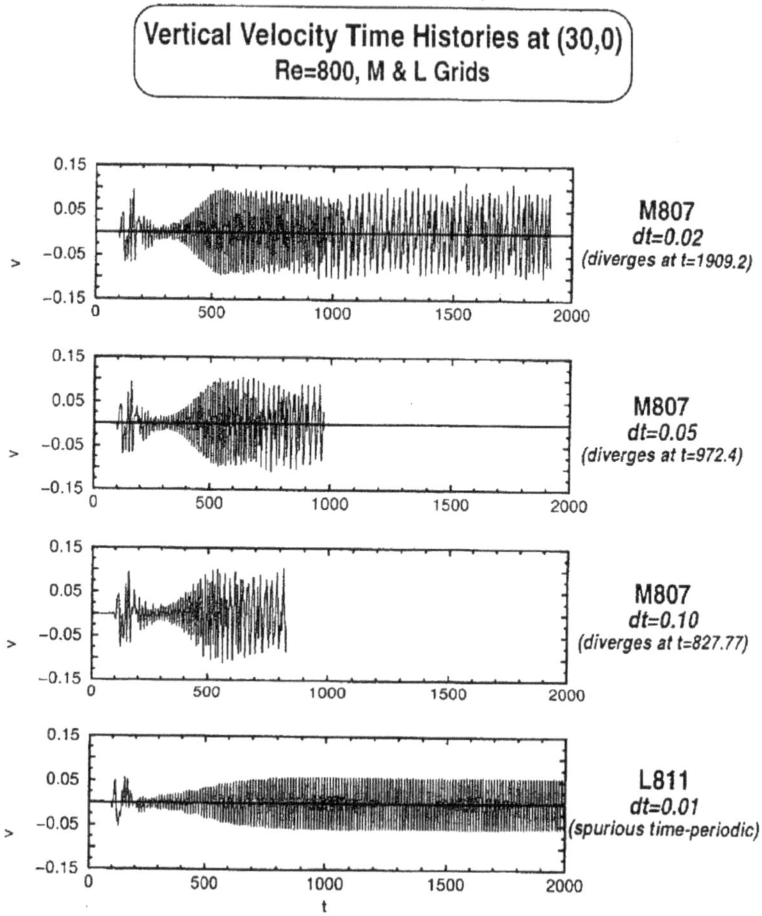

Figure 18. Vertical velocity time histories for $M807$ with time steps 0.02, 0.05, 0.10 and $L811$ with time step 0.01 for $t = 2000$.

In summary, without the temporal refinement study (longer time integration), the $L506$, $L507$, $L811$ and $M807$ cases can be mistaken to be chaotic-like (or aperiodic-like) flow. Although the time history up to $t = 800$ appears chaotic-like, one cannot conclude it is chaotic without longer transient computations. One can conclude that with transient computations that are 2.5 times longer than Torczynski's original computations, what appeared to be aperiodic-like or chaotic-like behavior at earlier times evolved toward either a time-periodic or divergent solution at later times. These temporal behaviors appear to be long-time aperiodic-like transients or numerically induced chaotic-like transients. For $Re = 800$, five different initial data were examined to determine if the flow exhibits strong dependence on initial data and grid resolution. Results showed that the numerical solutions are sensitive to these five initial data. Note that the results presented pertain to the characteristic of the studied scheme and the direct simulations. However, if one is certain that $Re = 800$ is a stable steady flow, a non-time-accurate method such as time-marching to obtaining the steady-state numerical solution would be a more efficient numerical procedure.

7.2.3. Spurious Bifurcation by Different Time Integrators (Henderson and Yee 1998, Unpublished)

This was a joint work with Ronald Henderson in 1998. The unpublished work was presented at the 10th International Conference in Finite Element Methods, 5–8 January 1998, Tucson, Arizona, and has also been presented at various invited lectures during the last four years. Our joint work illustrates the situation where solving the nonlinear terms of the Navier–Stokes equations by two different explicit time integrators (same implicit time integrator for the linear terms) results in spurious bifurcation. This spurious bifurcation is shown in Figure 19 as a function of the Reynolds number. These computations use the implicit Euler time integrator for the linear terms. Also, the same spatial discretization $L809$ is used with a fixed time step of $t = 0.10$. The two explicit time integrators are the third-order Adams–Bashforth (AB3) and a second-order explicit stiffly stable method (SS2) (Henderson [142]). The AB3 method experiences a spurious bifurcation near $Re = 720$, whereas the SS2 method experiences a spurious bifurcation at a larger Reynolds number near $Re = 800$. The method and the scaling for this figure can be found in Henderson [142]. Finding the exact location of these spurious bifurcation points requires more complicated computation, which is not performed here. In addition, the exact representation of this bifurcation plot is rather complicated to explain and is not important for the current discussion as it is not the main illustration for this study. When an adaptive version of the spectral element method (Henderson [142]) is used, the problem remains laminar up to $Re = 1800$. Future work, which is indicated in Figure 19, is planned, but the case has not been revisited thus far.

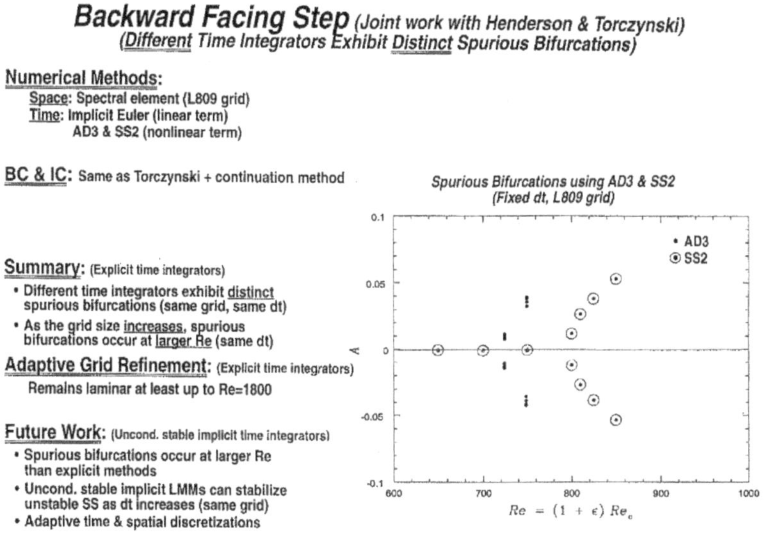

Figure 19. Different explicit time integrators (for the nonlinear terms of the Navier–Stokes equations) exhibit distinct spurious bifurcation using the same spatial discretization $L809$ and a time step of 0.10.

7.2.4. Minimization of Spurious Bifurcation by a Suitable Filter (Fischer 2001, Unpublished)

A few year later, Fischer (2001) [143] computed the same $L811$ spectral element grid using a time integrator based on the operator integration-factor splitting (OIFS) method developed by Maday, Patera and Rønquist [144]. This scheme decouples the convective step from the Stokes update, thereby allowing CFL numbers in excess of unity. At the end of each step, Fischer applies a filter to the velocity that effectively scales the Nth-order Legendre modes within each element by $(1-\alpha)$, where, typically, $0.05 \leq \alpha \leq 0.30$ (Fischer and Mullen [143]. Because the filter is applied on each step, its strength is a function of Δt as well as α. The spurious behavior observed by Kaiktsis et al. [139] is cured by the filter, and a stable steady-state numerical solution is obtained without further grid refinement. Figure 20 illustrates the velocity time histories at $(30,0)$ by the filtered and unfiltered spectral element methods with $\Delta t = 0.10$.

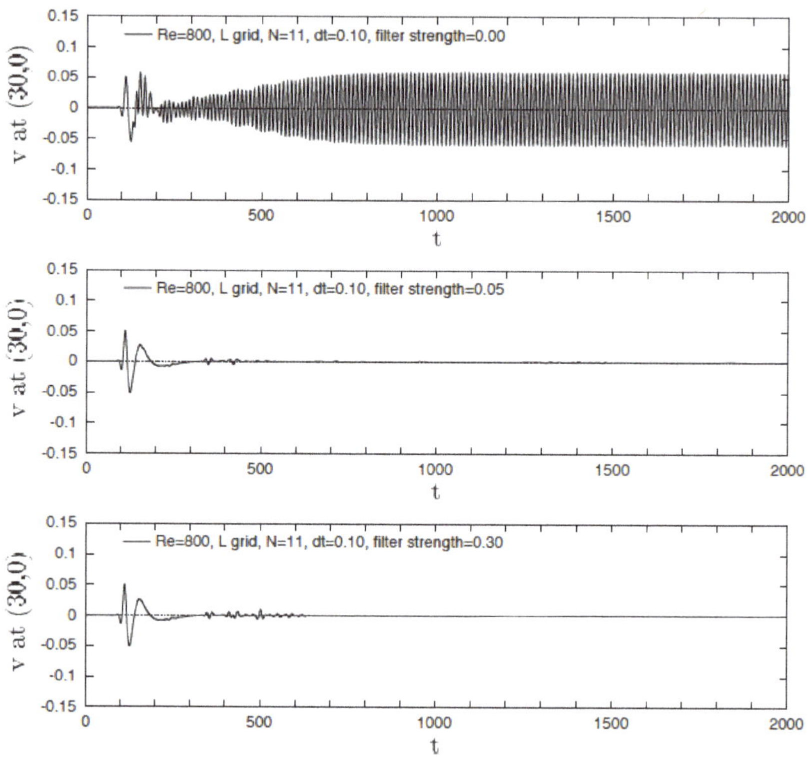

Figure 20. Comparison of the filtered with the unfiltered solutions of $L811$ with time step $dt = 0.10$. (**Top**) no filter; (**Middle**) filter (filter strength 0.05); (**Bottom**) filter (filter strength 0.3).

In summary, the above test cases illustrate all of the possible scenarios of spurious bifurcations indicated on the schematic bifurcation diagram as a function of Reynolds number by different temporal and spatial discretizations. See Gresho et al. [138] and the references cited therein for some more examples.

8. Entropy-Violating Numerical Methods and Nonlinearity-Induced Spurious Numerics by Shock-Capturing Methods

Among other spurious numerics discussed earlier, here, we would like to concentrate on the topic of "the carbuncle effect and hypersonic blunt body CFD simulations". Aside from issues in employing entropy-violating numerical methods in 1D and multi-D flows, there are added complications for the computation of external flow involving dimension-by-dimension numerical approaches on high-speed blunt body simulations. The topic has attracted numerical method developers for forty years, even before the term "carbuncle" phenomena was born. Since the mid 1990s until the present time, there have been many explanations and proposed similar cures in the open literature. See, e.g., these later published works [43,48,49,51,52].

Below, we summarize our earlier work in the late 1980s with extensive numerical studies using our multi-D entropy fixes and other cures associated with this issue. Our key work related to this topic can be found in [21,22,47]. Since the 1983 Harten and Hyman entropy fixed by entropy violating methods [145], we have been using the various form of entropy fixes since 1983 and continuing to the present time, including high-order ENO, WENO and variants of ENO and WENO schemes. Our cure exhibits no grid-aligned shock instability issue for the hypersonic blunt body simulation for both time-marching to steady states or unsteady flows, including moving impinging shock of Types I–VI relating to the Edney experiment [146].

8.1. Avoidance of Spurious Numerics, and Enhancement of Stability and Convergence Rate for Hypersonic Flows Simulations

In [21,22,53], some elements and parameters that can affect the stability and convergence rate in high Mach number cases but have a negligible effect in low Mach number cases for steady-state inviscid blunt-body flows were identified. The current study indicated that the same elements and parameters can affect the stability and convergence rate at hypersonic speeds for viscous computations as well. They are as follows: (1) the choice of the entropy correction parameter δ_1 for non-entropy conserving methods, (2) the choice of the dependent variables on which the limiters are applied (related to proper scaling of the eigenvectors for high-speed flows) and (3) the prevention of nonphysical solutions during the initial transient stage. Our study indicated that these elements can also improve the stability of unsteady as well as steady hypersonic flows. Here, we use the same notation as in [21,22] for the TVD and MUSCL schemes entropy fix parameter. Their finding by the TVD and MUSCL schemes using Roe's approximate Riemann solver and implicit LMM time discretization are discussed as follows. It is remarked that for all of our later computations using high-order ENO and WENO methods, an analog entropy fix is needed if the local Lax–Friedrich flux formulation is used.

1. For Mach numbers ranging from 1.2 to 15, numerical experiments for one and higher dimensional unsteady flows containing unsteady shocks show that the second-order explicit TVD schemes are insensitive to the entropy correction for $0 \leq \delta_1 \leq 0.1$. In most cases, $\delta_1 = 0$ was used. For $0.1 \leq \delta_1 \leq 0.25$, there is a possibility of improving stability in the sense of allowing a higher CFL number at the expense of a slight smearing of the discontinuities. However, for unsteady complex shock wave interactions, a small positive δ_1 or a variable δ_1 can help stabilize the time-accurate implicit algorithm.

For subsonic to low supersonic steady-state NACA 0012 airfoil computations, the resolution of the shock waves was found to be quite insensitive to $0.1 \leq \delta_1 \leq 0.25$ and a constant value seems to be sufficient. However, for hypersonic flows, especially for blunt-body flows, a constant δ_1 or a variable δ suggested by Harten and Hyman [145] was found to be insufficient, but a variable δ_1 depending on the spectral radius of the Jacobian matrices of the fluxes is very helpful in terms of stability and convergence rate. For multi-D, a multi-D entropy fix is needed. In fact, a proper choice of the entropy parameter δ_1 for higher Mach number flows not only helps in preventing nonphysical solutions but can act as a control of the convergence rate and of the sharpness of shocks and slip surfaces (or shear layer in viscous flows). The smaller the δ_1 that is used, the slower the convergence rate. The larger the $\delta_1 \leq 1$ that is used, the larger the numerical dissipation being added. For the present blunt-body steady-state calculations with Mach numbers $M > 4$, the initial flow conditions at the wall are obtained using the known wall temperature in conjunction with pressures computed from a modified Newtonian expression. Also, for implicit methods, a slow start-up procedure from initial conditions is necessary. See [21,22] for additional details.

2. Higher order TVD-type schemes in general involve limiter functions. However, there are options in choosing the types of dependent variables when applying limiters for systems of hyperbolic conservation law, particularly for systems in generalized coordinates. The choice of the dependent variables on which limiters are applied can affect the stability and convergence process. Due to the non-uniqueness of the eigenvectors, the choice of the characteristic variables on which the limiters are applied play an important role in the stability and convergence rate as the Mach number increases. For example, the path to arrive to the steady states might get trapped into a spurious limit cycle or spurious oscillations. This is directly related to proper scaling of the eigenvectors for high-speed flows. For low and moderate Mach numbers, the different choices of the eigenvectors have less effect on the stability and convergence rate. However, for large Mach number cases, the magnitudes of all the variables at the jump of the bow shock are not the same. In general, the jumps are much larger for the pressures than for the densities or total energy. Studies have indicated that employing the form of the eigenvectors such that the variation of the jump of the characteristic variables in transformed coordinates are of the same order of magnitude as the jump in the pressure would be a good choice for hypersonic flows.

3. The entropy fix for multi-D case for curvilinear grids should use the transformed variables, and the entropy variable parameter should include all dimensions' effects. All of our simulations for blunt body hypersonic simulations [21,22] use the multi-D entropy fix in the transformed variables.

4. A less known property of Roe's average state is that it allows the square of the average sound speed $c^2{}_{j+1/2}$ to lie outside the interval between $c^2{}_j$ and $c^2{}_{j+1}$ for equilibrium real gases. Roe's average state $c_{j+1/2}$ might be negative even though c_j and c_{j+1} are positive during the transient stage when the initial conditions are far from the steady-state physical solution. In this case, we replace $c^2{}_{j+1/2}$ by $\max\left(c^2{}_{j+1/2}, \min\left(c^2{}_j, c^2{}_{j+1}\right)\right)$. This latter safety check is particularly helpful for the symmetric TVD algorithm but not necessary for the upwind algorithm. Here, $c^2{}_{j+1/2}$ is the square of the sound speed evaluated at Roe's average state. In addition, due to the large gradients and to the fact that the initial conditions are far from the steady-state physical solution, the path used by the implicit method can go through states with negative pressures if a large time step is employed. A convenient way to overcome this difficulty is to fix a minimum non-negative allowed value for the density and the pressure. With this safety check, the scheme allows a much larger time step and converges several times faster.

Choice of Limiters: Unlike flows with transonic and low supersonic shock waves, problems containing strong hypersonic shock waves are more sensitive to the treatment of limiters. Using the more diffusive limiter turns out to be more stable than other more compressive limiters. The more compressive limiters have a very low stability and slow convergence rate for steady flows. The same conclusion applies for unsteady flows where the more compressive limiters have a very restricted time step limit. Using a different limiter that is more appropriate for different linear and nonlinear characteristic fields would help. See [21,22] for more details. For over four decades, different designs of flux limiters to improve the time-marching to the steady states have been ongoing; see, e.g., two recent works [94,95,147].

Below, hypersonic blunt body simulations from [21,22] have been selected to illustrate our work in the 1980s to avoid spurious numerics and to enhance stability and convergence rates for hypersonic blunt body simulations. In [21,22], many types of blunt body test cases were conducted and all types of Edney shock-impingement on the blunt body [146] were considered. They include inviscid, perfect and real gas, non-interfering blunt-body steady and unsteady flows. In addition, test cases also include steady and unsteady inviscid, perfect gas blunt-body flow with an impinging shock.

8.2. Inviscid Blunt Body Steady Computations for a Real Gas

The freestream conditions for the current illustration are $M_\infty = 25$, $p_\infty = 1.22 \times 10^3$ N/m^2, $\rho_\infty = 1.88 \times 10^{-2}$ kg/m^3 and $T_\infty = 226$ K. Figure 21 shows half of the 61×33 grid used for the blunt-body problem. The shock standoff distance is at approximately 14 points from the wall on the symmetry axis. See [21,22] for details of the numerical method and simulations.

Figure 22 shows the Mach number, density, pressure and κ contours for Mach number $M_\infty = 25$. Here, $\kappa = (\partial p/\partial \bar{\epsilon})_\rho$, where $\bar{\epsilon} = \rho\epsilon$ and ϵ is the specific internal energy.

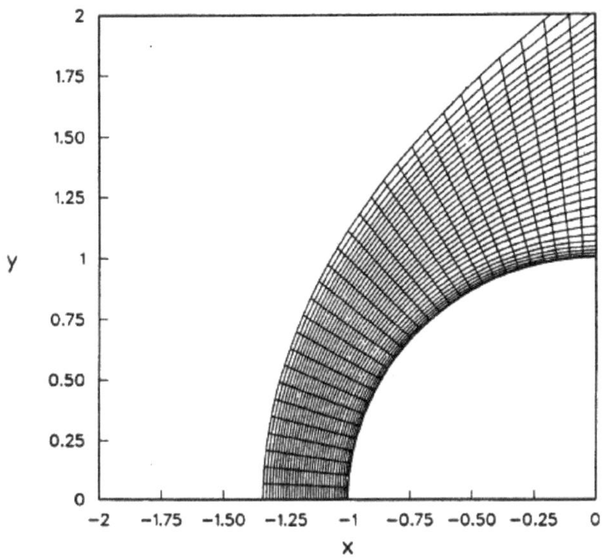

Figure 21. Blunt body grid: 31 × 33 grid points.

8.3. Inviscid Blunt Body with an Impinging Shock Steady Computations for a Perfect Gas

Take an example of a 2D inviscid steady blunt-body flow with an impinging shock for a perfect gas with $M_\infty = 4.6$ computation. The freestream conditions for this flow field are the freestream temperature $T_\infty = 167$K and $\gamma = 1.4$ for a perfect gas. This flow field is typical of what may be experienced by the inlet cowl of a hypersonic aerodynamics vehicle. Figures 23 and 24 show the schematic of the computational domain, the Mach contours and the L2-norm residual of an inviscid shock-on-shock interaction on a blunt body with radius R_l and thickness $D = 2R_l$. An oblique shock with an angle of 20.9° relative to the free stream impinges on the bow shock. Various types of interactions occur depending on where the impingement point is located on the bow shock. As shown by the Mach contours ranging from 0 to 4.55 in increments of 0.05, the impinging shock has caused the stagnation point to move away from its undisturbed location at the symmetry line. The surface pressures at the new stagnation point can be several times larger than those at the undisturbed location of the stagnation point. In addition, a slip surface emanates from the bow shock and impinging shock intersection point and is intercepted by a shock wave that starts at the upper kink of the bow shock. The interacting shock waves and slip surfaces are confined to a very small region and must be captured accurately by the numerical scheme if the proper surface pressures are to be predicted correctly. The 77 × 77 grid used and the convergence rate computed by the implicit TVD scheme are shown in Figure 24. Though the pattern of the flow is significantly more complicated than for the previous cases, the convergence rate remains quite satisfactory.

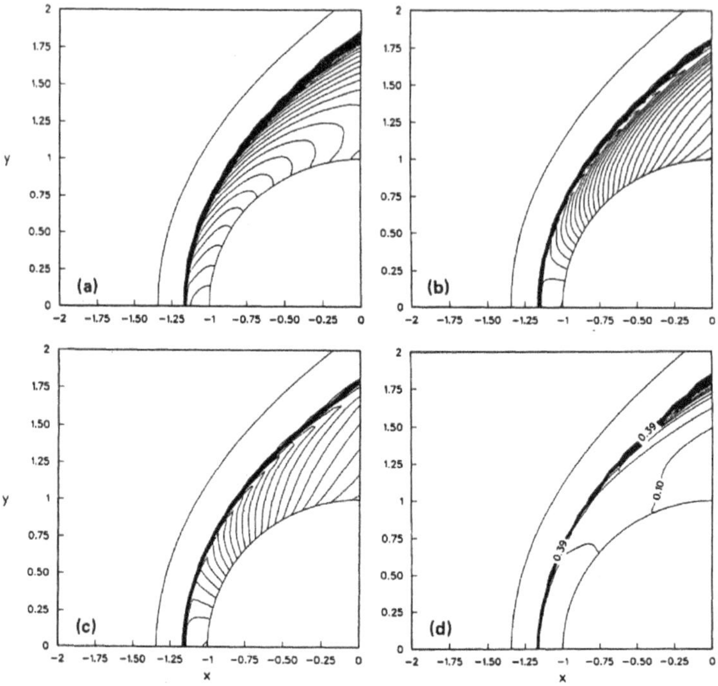

Figure 22. Blunt body steady computation with $M_\infty = 25$: the Mach contours (**a**), density contours (**b**), pressure contours (**c**) and (**d**) κ computed by an implicit scheme for an equilibrium real gas.

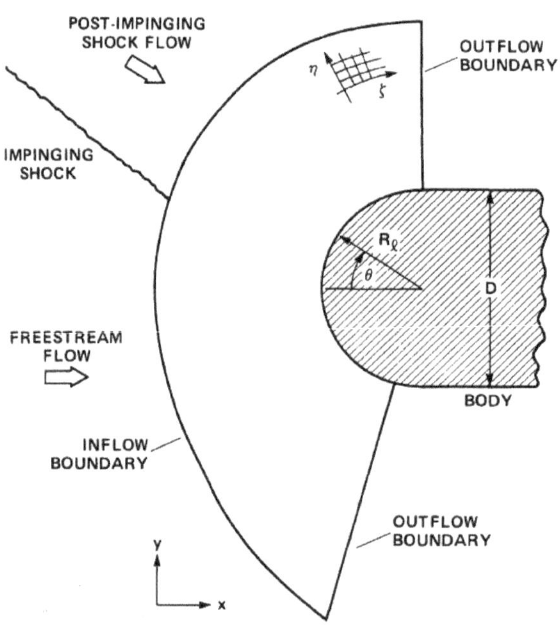

Figure 23. Schematic of the computational domain for a blunt-body flow with an impinging shock.

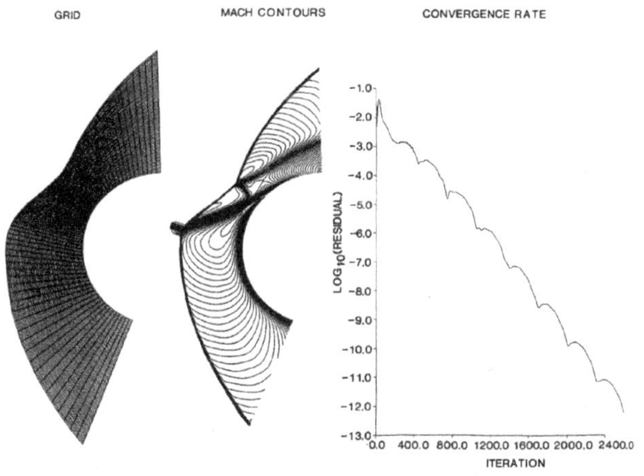

Figure 24. Two-dimensional inviscid steady blunt-body flow with an impinging shock for a perfect gas with $M_\infty = 4.6$.

8.4. Viscous Steady and Unsteady $M_\infty = 15$ Hypersonic Perfect Gas Computations with Impinging Shock

Figure 25 illustrates the shock resolution of unsteady and steady thin-layer Navier–Stokes computations by a second-order implicit time-accurate TVD scheme. The steady test case is similar to the previous impinging shock study except that the freestream Mach number is 15, the impingement shock angle is 22.75°, the free stream temperature is $T_\infty = 255.6$ K and the Reynolds number based on the diameter is 186,000. Shown are the Mach contours from 0 to 15 in increments of 0.1. For the unsteady computation, the impingement shock at the same angle relative to the freestream moves downward across the bow-shock of the blunt body. The impingement shock velocity is 10% of the freestream velocity ($M_\infty = 15$). Although the impingement shock locations for the unsteady and steady computations are similar, the shock patterns are very different. A 241 × 141 non-adaptive grid is used for both computations. A time step of 0.002 (equivalent to a maximum CFL of 48) is used for steady-state computations, whereas a time step of 0.0005 (equivalent to a maximum CFL of 10–12 at the vicinity of the boundary layer and a CFL of 1 for the rest of the flow field) is used for the unsteady computations. The steady-state solution can be reached in 1200 steps with a three order of magnitude drop in the L2-norm residual. Extra iterations are needed to bring the residual to a lower level; however, no change in the contour plots or surface pressures at least to within 3–4 digits of accuracy is observed.

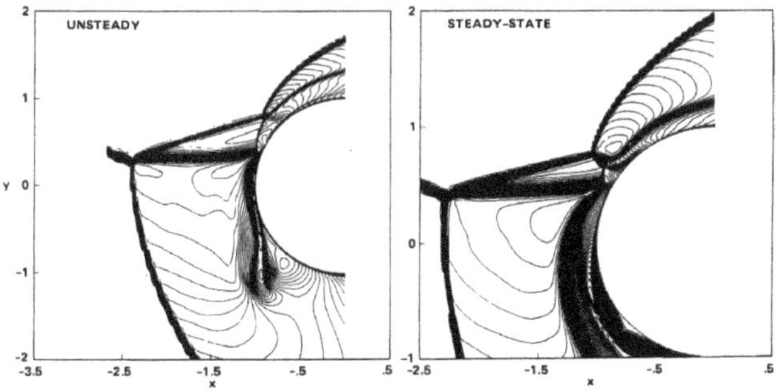

Figure 25. The Mach contours of a 2D viscous steady and unsteady hypersonic perfect gas with $M_\infty = 15$ and Re_0 = 186,000.

9. Numerical Examples: Spurious Numerics Related to Problems with Stiff Source Terms and Methods to Overcome the Issues

For nearly four decades, the wrong speed phenomenon has attracted a large volume of research work; see, e.g., [2,6–8,18–20,40,42,148–155]. In combustion and reacting flows, various strategies have been proposed to overcome this wrong speed difficulty for one to two species cases with a single reaction. Since numerical dissipation that spreads the discontinuity front is the cause of the wrong propagation speed of discontinuities, a natural strategy is to avoid any numerical dissipation in the scheme. In combustion, level set and front tracking methods were used to track the wave front to minimize this spurious behavior [149,152,153]; see Wang et al. [20] for a comprehensive overview of the last two decades of development. Wang et al. also proposed a new high-order finite difference method with subcell resolution for advection equations with stiff source terms for a single species to overcome the difficulty. For a subcell resolution method for multi-species, see Wang et al. [41]. See [18–20,40–42] for various examples in our effort to minimize the wrong propagation speed of discontinuities. For various project methods to minimize the wrong propagation speed of discontinuities, see [125–128] and the references cited therein.

9.1. One-Dimensional Scalar Reaction-Convection Test Case

This numerical phenomenon was first observed by Colella et al. [156] in 1986, who considered both the reactive Euler equations and a simplified system obtained by coupling the inviscid Burgers equation with a single convection/reaction equation. LeVeque and Yee (1990) [8] showed that a similar spurious propagation phenomenon can be observed even with scalar equations, by properly defining a model problem with a stiff source term. They introduced and studied the simple one-dimensional (1D) scalar conservation law with an added inhomogeneous parameter-dependent source term

$$u_t + u_x = S(u), \tag{11}$$

$$S(u) = -\mu u(u - \frac{1}{2})(u - 1). \tag{12}$$

Between the time frame of 2010–2013, the authors revisited the LeVeque and Yee test case, and combustion and nonequilibrium reacting flows coupled system cases [18–20,40,42]. Here, we include the study of the LeVeque–Yee scalar test case using the subcell-resolution method reported in [42]. Figure 26 shows the shock location error as a function of the CFL number using 128 discrete CFL values between $(0.001, 8)$ with $6.291338583 \times 10^{-3}$ equal increment by WENO5 and WENO5/SR. WENO5/SR is the original Shu and Jiang fifth-order WENO using Wang et al.'s subcell-resolution method. Three grid refinements using $50, 150, 300$ uniform grid points and for three stiffness coefficients of the source term $\mu = 1, 100, 1000$ are compared. "Err" represents number of grid points away from the reference solution on the figure. Figure 26 indicates a general trend of the scheme behavior by WENO5/SR for the 1D scalar test case conducted by LeVeque and Yee (1990) [8]. In this case, WENO5 behaves differently from the system test case. In this case, all the nonlinearity and stiffness contained in the governing equation are due to the source term as the convection term in the LeVeque and Yee's scalar model PDE is linear. It appears that the nonlinearity due to the convection terms does not alter the general spurious behavior pattern.

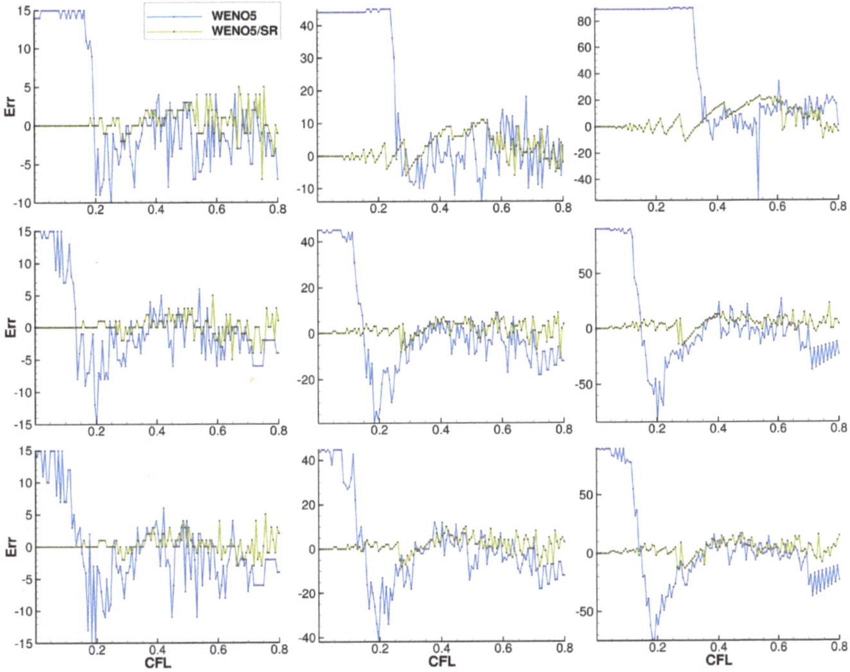

Figure 26. LeVeque and Yee linear advection and nonlinear stiff source term test case [8]. Number of grid points away from the reference shock solution (Err) as a function of the CFL number (128 discrete CFL values between (0.001, 8) with $6.291338583 \times 10^{-3}$ equal increment) by WENO5 and WENO5/SR using 50, 150, 300 uniform grid points (across) and for stiffness coefficients $\mu = 1, 100, 1000$ (top to bottom).

When the parameter μ is very large, a wrong propagation speed of discontinuity phenomenon by dissipative numerical methods is observed in coarse grids. In reacting flows, $\frac{1}{\mu}$ can be described as the reaction time. In order to isolate the problem, LeVeque and Yee solved (11) and (12) by the fractional step method using Strang splitting [157]. For this particular source term, the reaction (ODE) step of the fractional step method can be solved exactly. In their study using pointwise evaluation of the source term ($S(u)$ is evaluated at the j grid point index, i.e., $S(u_j)$ for each time evolution, the phenomenon of wrong propagation speed of discontinuities is connected with the smearing of the discontinuity caused by the spatial discretization of the advection term. They found that the propagation error is due to the numerical dissipation contained in the scheme, which smears the discontinuity front and activates the source term in a nonphysical manner. The smearing introduces a nonequilibrium state into the calculation. Thus, as soon as a nonequilibrium value is introduced in this manner, the source term turns on and immediately restores equilibrium, while at the same time shifting the discontinuity to a cell boundary. By increasing the spatial resolution by an order of magnitude, they were able to improve towards the correct propagation speed. It is remarked here that in a general stiff source term problem, a sufficient spatial resolution is as important as temporal resolution when the reaction step of the fractional step method cannot be solved exactly. Moreover, employing finite time steps and grid spacings that are below the standard Courant–Friedrich–Levy (CFL) limit on shock-capturing methods for compressible Euler and Navier–Stokes equations containing stiff reacting source terms and discontinuities reveals surprising counter-intuitive results.

9.2. Minimization of Wrong Propagation Speed of Discontinuities for Problems Containing Stiff Source and Discontinuities

Our studies in [42] found that using the Strang form of the operator splitting between the homogeneous part of the system of governing equations (without the source term) and the nonlinear stiff source terms is more stable than solving the fully coupled fluid-reaction equations. Here, Strang operator splitting is different from the skew-symmetric splitting of the inviscid flux derivatives. Strang operator splitting is commonly used in the combustion community to solve reacting flow problems. Figure 27 shows the schematic of Strang operator splitting. For the subcell resolution method, the subsequent two Figures 28 and 29 show the schematic of subcell-resolution methods. For problems with source terms employing any higher than first-order methods, we need to use methods that are well-balanced. See [18,19] for a well-balanced WENO method for chemical reacting flows.

On the Strang operator splitting approach, all of the aforementioned numerical methods (physics-preserving methods, shock-capturing methods and blending of more than one method) can be used to solve the reacting equations portion without the source terms first. Then, the source term is solved using an ODE (ordinary differential equations) solver, as indicated in Figures 28 and 29.

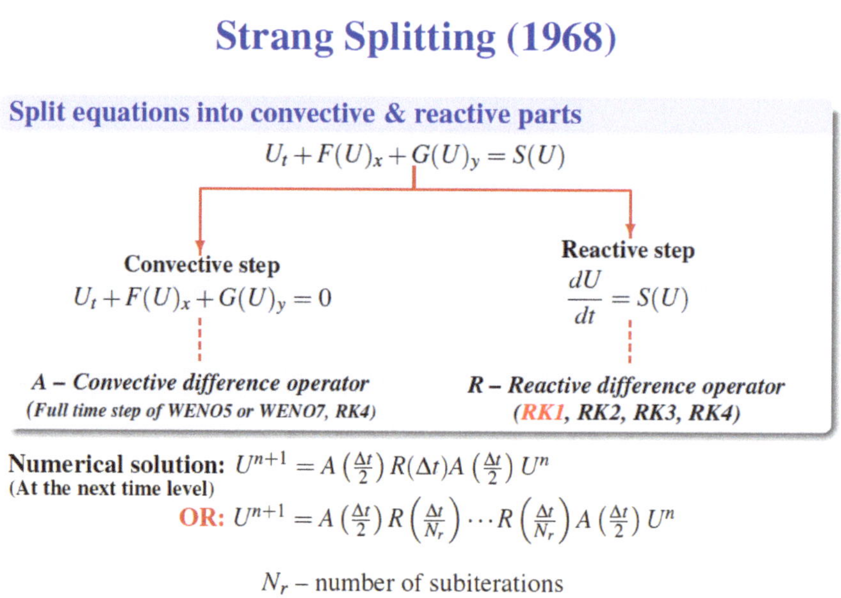

Figure 27. Schematic of Strang operator splitting on the homogeneous portion and the source term portion of the governing equations.

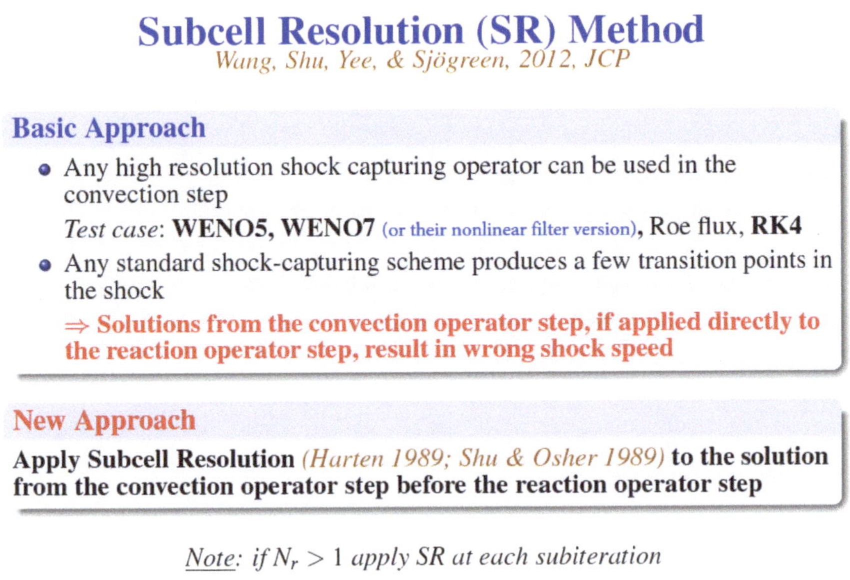

Figure 28. Subcell- resolution method in solving equations containing nonlinear source terms.

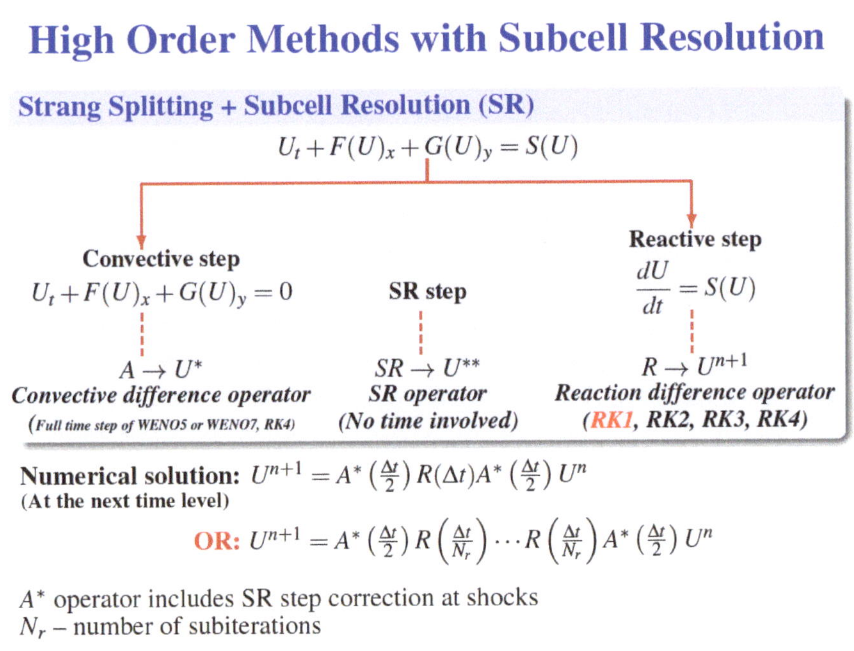

Figure 29. Subcell resolution using three staggered steps.

9.3. One-Dimensional Chapman–Jogùet (C-J) Detonation Containing Stiff Source Terms and Discontinuities

The second example illustrates the effect of numerical dissipation on the propagation speed of a discontinuity for a 1D system in the presence of a stiff source term. We compute a 1D

C-J detonation for a one-reaction Arrhenius source term. This example, which was considered in [20], has been extensively studied, e.g., in [20,40,42,151,155]. See [20,42,151,155] for the initial flow conditions and additional results.

Figure 30 shows the 1D problem setup. Figure 31 shows a comparison of four high-order accurate shock-capturing schemes WENO5, WENO5fi, WENO5fi+split and our high-order subcell resolution version of WENO5 [20,42] (WENO5/SR). The new WENO5/SR method uses the Strang form of the splitting between the homogeneous part of the system of governing equations (without the source term) and the nonlinear stiff source terms. For each time integration, the first step is to solve the homogeneous system of governing equations. Then, a subcell resolution detector is used to correct the wrong numerical speed of propagation of discontinuities before solving the next step of the nonlinear stiff source term. It is remarked that the Strang splitting is not to be confused with the skew-symmetric splitting of the inviscid flux derivative as in Ducros et al.'s splitting [158]. The reference solution is computed by the WENO5 scheme on a uniform grid with 10,000 points. Here, WENO5fi+split denotes the use of sixth-order classical central spatial discretization together with the dissipative portion of WENO5. The term "split" here denotes the use of the Strang form of the splitting between the homogeneous part of the system of governing equations (without the source term) and the nonlinear stiff source terms in solving the governing equations.

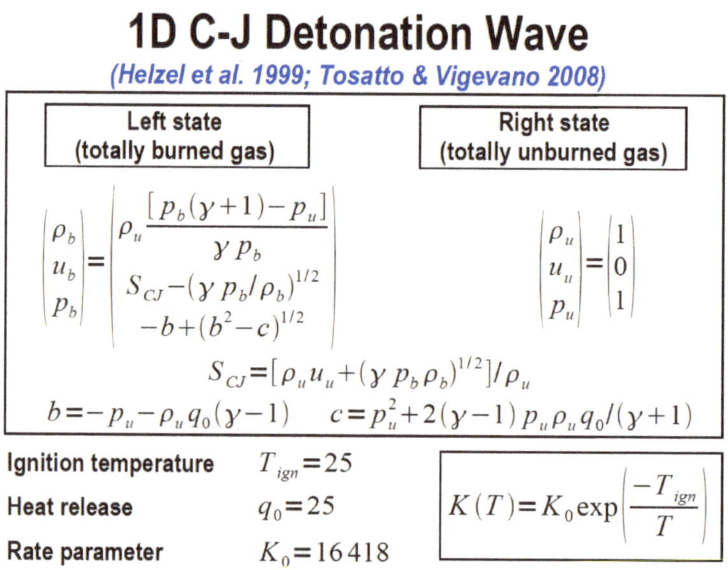

Figure 30. One-dimensional C-J detonation problem setup.

Figure 32 shows the number of grid points away from the reference shock solution (Err) as a function of the CFL number (128 discrete CFL values between (0.001, 8) with $6.291338583 \times 10^{-3}$ equal increment) for three low dissipative shock-capturing methods WENO5/SR, WENO5fi+split and WENO5fi/SR+split using three 50, 150 and 300 uniform grid points and for three stiffness $K_0, 100K_0, 1000K_0$. Err is highly dependent on the grid, method and stiffness coefficient. The two subcell-resolution methods exhibit the least Err error.

Figure 33 shows the number of grid points away from the reference shock solution (Err) as a function of the CFL number (128 discrete CFL values between (0.001, 8) with $6.291338583 \times 10^{-3}$ equal increment) for three low dissipative shock-capturing methods

WENO5/SR, WENO5fi+split and WENO5fi/SR+split, RK3 and RK4 time discretizations using two 150 and 300 uniform grid points, and stiffness K_0.

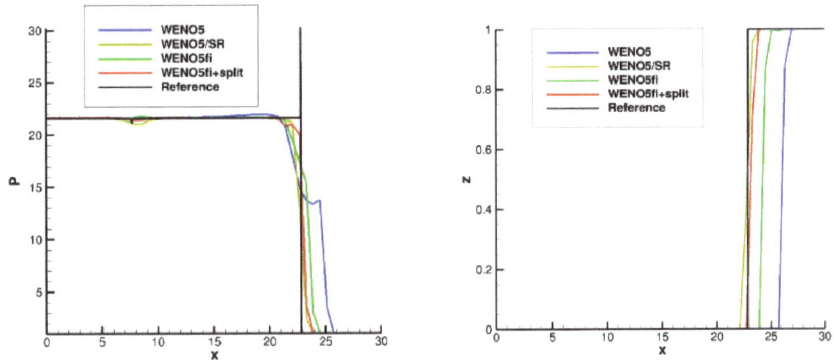

Figure 31. Pressure (**left**) and mass fraction of unburnt gas (**right**). Comparison among four high-order shock-capturing methods for a 1D C-J detonation problem. One reaction, Arrhenius model, using 50 uniformly distributed grid points.

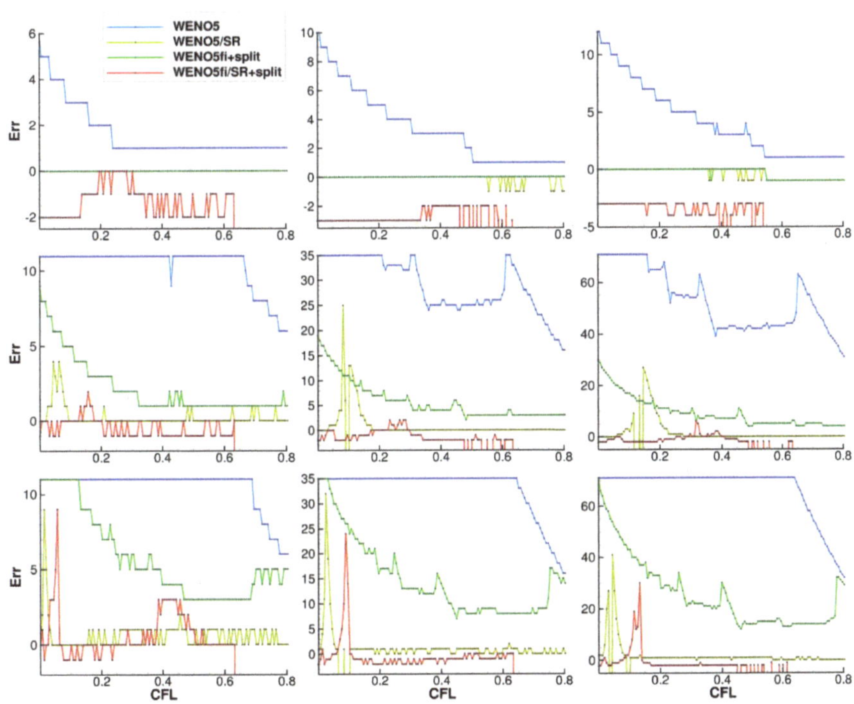

Figure 32. One-dimensional C-J detonation problem: Number of grid points away from the reference shock solution (Err) as a function of the CFL number (128 discrete CFL values between (0.001, 8) with $6.291338583 \times 10^{-3}$ equal increment) for three low dissipative shock-capturing methods using 50, 150 and 300 uniform grid points (across) and for stiffness $K_0, 100K_0, 1000K_0$ (top to bottom).

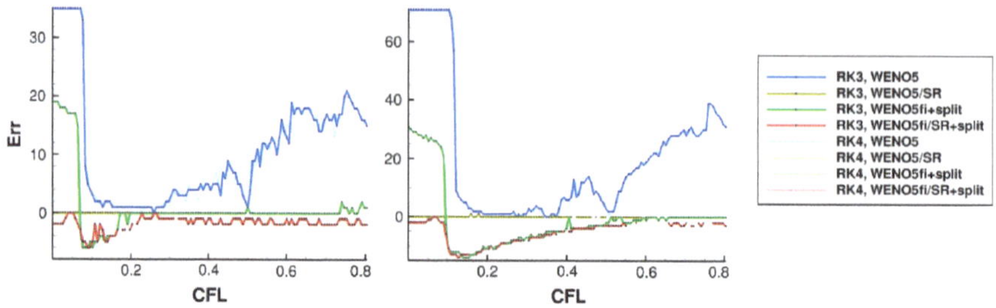

Figure 33. One-dimensional C-J detonation problem, Arrhenius case at $t = 1.8$. Comparison of the same three spatial discretizations with RK4 and RK3 temporal discretization using two 150, 300 uniform grid points (across) and stiffness K_0. Number of grid points away from the reference shock solution (Err) as a function of the CFL number (128 discrete CFL values between $(0.001, 8)$ with $6.291338583 \times 10^{-3}$ equal increment).

9.4. Two-Dimensional Chapman–Joguet (C-J) Detonation Containing Stiff Source Terms and Discontinuities [42]

Figure 34 shows a 2D C-J detonation problem setup. Figure 35 shows the density comparison by three methods comparing with the reference solution by WENO5 using 4000×800 uniform grid points. One important feature of this solution is the appearance of triple points, which travel in the transverse direction and reflect from the upper and lower walls. A discussion of the mechanisms driving this solution is given in [159]. Again, a pointwise evaluation of the source is employed for the 2D test case. Scheme behavior as a function of CFL, grid refinement and stiffness of the source terms was performed in Yee et al. (2013) [42]. WENO5/SR and WENO5fi+split are able to obtain the correct shock speed with similar accuracy compared to the reference solution. WENO5fi gives a slightly oscillatory solution near $x = 0.004$. WENO5 and WENO5/SR produce no oscillations at the same location. Further improvement of the flow sensor of the filter scheme is needed in order to remove the spurious oscillations. Furthermore, for the 500×100 grid, WENO5fi also obtained the correct shock speed. For $CFL = 0.05$, however, WENO5fi/SR+split is not able to obtain the correct shock speed for the stiff coefficient K_0. The WENO5fi+split method is the nonlinear filter approach using WENO5 on the entropy split form of the classical fourth-order central spatial discretization. Section X includes a brief introduction on the WENO5fi+split method and for any WENO order. Details can be found in [11–16,26,29,30].

Figure 36 shows a 1D cross-section of density at $t = 1.7 \times 10^{-7}$ by the same four methods as the 1D case on a uniform coarse grid of 200×40. The CFL=0.05 and $K_0 = 0.5825 \times 10^{10}$. The right figure is a close up of the vicinity of the discontinuity. The reference solution is by WENO5 using 4000×800 uniform grid points. WENO5 is more dissipative and also gives the largest error in the location of the discontinuity. WENO5fi performs better and is less dissipative because the WENO5 dissipation is not used everywhere in the computational domain. WENO5/SR gives the most accurate propagation speed but WENO5fi+split compares very well with WENO5/SR for this particular problem and grid size. The probable reason is that because of the stabilizing effect of the splitting, less dissipation is needed to keep WENO5fi+split stable. WENO5/SR and WENO5fi+split capture the correct structure using fewer grid points than in Helzel et al. [151] and in Tosatto and Vigevano [155].

Figure 37 shows the 2D detonation problem at $t = 1.7 \times 10^{-7}$ and $K_0 = 0.5825 \times 10^{10}$. The figure indicates the number of grid points away from the reference shock solution (Err) as a function of the CFL number (128 discrete CFL values between $(0.001, 8)$ with $6.291338583 \times 10^{-3}$ equal increment) for three low dissipative shock-capturing methods using 200×40 and 500×100 uniform grid points (across) and for stiffness $K_0, 100K_0, 1000K_0$ (top to bottom).

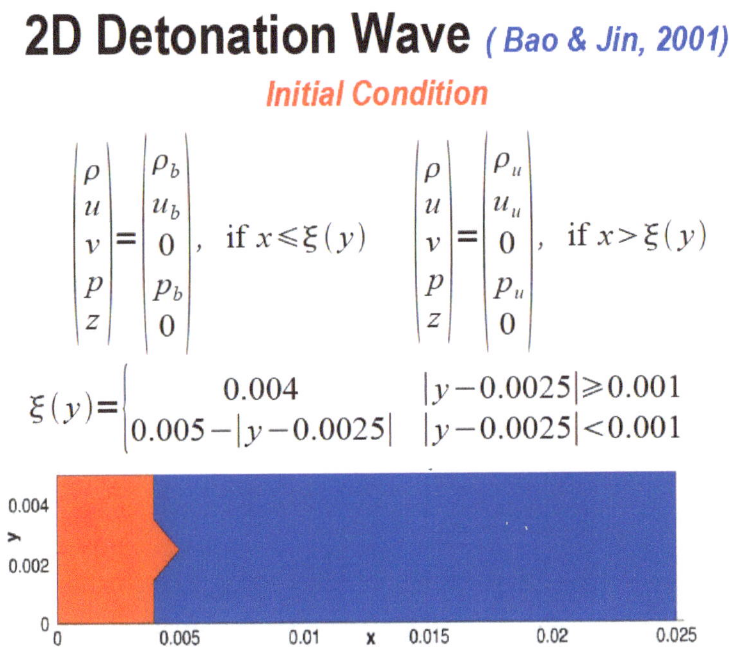

Figure 34. Two-dimensional C-J detonation problem setup.

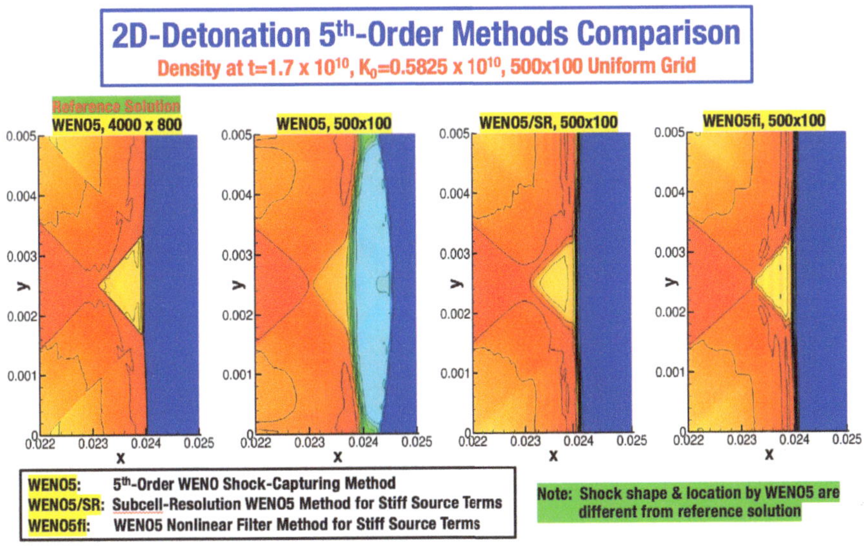

Figure 35. Two-dimensional detonation problem at $t = 1.7 \times 10^{-7}$ and $K_0 = 0.5825 \times 10^{10}$: Density computed by different methods. From left to right: reference solution by the standard WENO5 method using 4000×800 uniform grid points; WENO5, WENO5/SR and WENO5fi+split using 500×100 uniform grid points with $CFL = 0.05$.

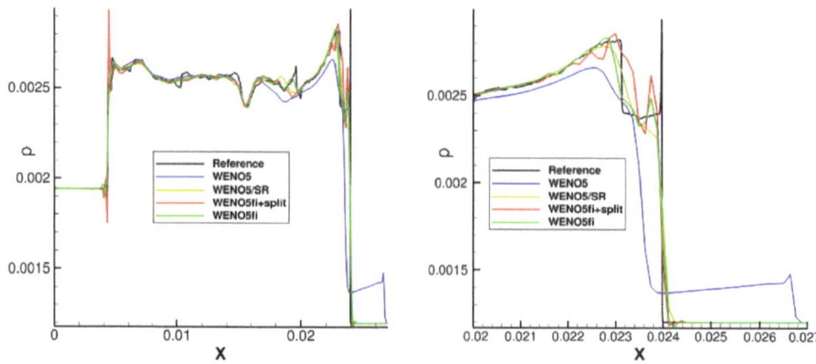

Figure 36. Two-dimensional C-J detonation comparison: 1D cross-section of density at $t = 1.7 \times 10^{-7}$ by the same four methods as the 1D case on a uniform coarse grid of 200×40. The CFL=0.05 and $K_0 = 0.5825 \times 10^{10}$. The right figure is a close up of the vicinity of the discontinuity. Note that there is an enlargement of the x domain in order to illustrate the wrong shock location. The reference solution is by WENO5 using 4000×800 uniform grid points.

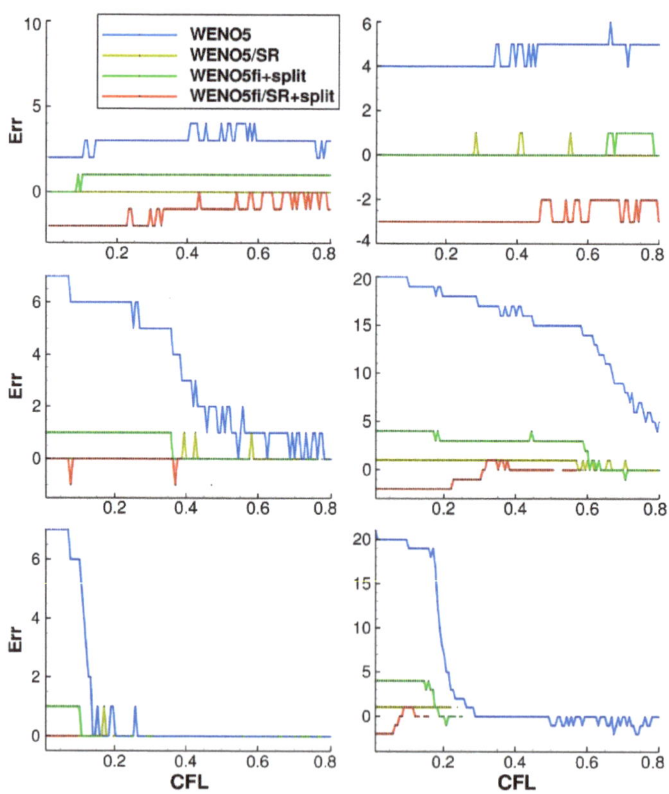

Figure 37. Two-dimensional detonation problem at $t = 1.7 \times 10^{-7}$ and $K_0 = 0.5825 \times 10^{10}$: Number of grid points away from the reference shock solution (Err) as a function of the CFL number (128 discrete CFL values between $(0.001, 8)$ with $6.291338583 \times 10^{-3}$ equal increment) for three low dissipative shock-capturing methods using 200×40 and 500×100 uniform grid points (across) and for stiffness $K_0, 100K_0, 1000K_0$ (top to bottom).

9.5. One-Dimensional and Two-Dimensional 13 Species EAST Simulations

The next two examples show the computations of a 1D and 2D 13-species chemical reacting flow by Kotov et al. and Yee et al. [4,42].

These 1D and 2D simulations consist of a 13-species chemical reacting flow, which is a simplified problem related to the NASA Electric Arc Shock Tube (EAST) experiment. Figure 38 shows the schematic of the EAST experiment. Figure 39 shows the EAST model governing equations in 1D, 2D and 3D. See [4,42] for flow notations. Figures 40 and 41 show the 1D and 2D problem setup. Figure 42 shows the comparison of three grids using the same WENO5fi method (left) and the comparison of five methods with the reference solution (right). The left subfigure of Figure 42 indicates how the shock/shear locations are dependent on the grid size using the same high-order method. The relative width between the shock and shear are the function of the grid spacing compared with the reference solution. The right subfigure indicates how the shock/shear locations are dependent on five different numerical methods (TVD, TVDafi+split, WENO5-llf, WENO5Pafi+split and TVD/SR). See [4] for the description of the five methods.

Figure 43 shows the temperature and pressure evolution of the 2D 13-species test case. Figure 44 shows the comparison of methods indicating how the shock/shear locations are dependent on the three methods. Different boundary layers are predicted by the three different methods as well. Here, WENO5-LLF denotes WENO5 using the local Lax–Friedrichs flux (LLF), and WENO5P-LLF denotes a positivity-preserving WENO5 of Zhang and Shu [160] using the LLF. The relative width between the shock and shear is a function of the methods.

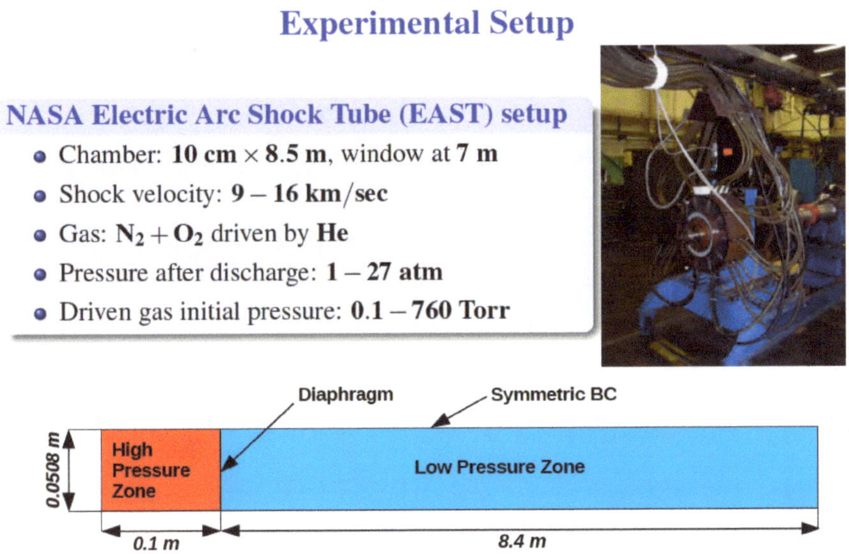

Figure 38. The EAST experiment.

EAST Problem. Governing equations

NS equations for 2D (i=1,2) or 3D (i=1,2,3) chemically non-equilibrium flow:

$$\frac{\partial \rho_s}{\partial t}+\frac{\partial}{\partial x_j}(\rho_s u_j+\rho_s d_{sj}) = \Omega_s$$

$$\frac{\partial}{\partial t}(\rho u_i)+\frac{\partial}{\partial x_j}(\rho u_i u_j+p\delta_{ij}-\tau_{ij}) = 0$$

$$\frac{\partial}{\partial t}(\rho E)+\frac{\partial}{\partial x_j}\left(u_j(E+p)+q_j+\sum_s \rho_s d_{sj} h_s-u_i\tau_{ij}\right) = 0$$

$$\rho=\sum_s \rho_s \qquad p=RT\sum_{s=1}^{N_s}\frac{\rho_s}{M_s} \qquad \rho E=\sum_{s=1}^{N_s}\rho_s\left(e_s(T)+h_s^0\right)+\frac{1}{2}\rho v^2$$

$$\tau_{ij}=\mu\left(\frac{\partial u_i}{\partial x_j}+\frac{\partial u_j}{\partial x_i}\right)-\mu\frac{2}{3}\frac{\partial u_k}{\partial x_k}\delta_{ij} \qquad d_{sj}=-D_s\frac{\partial X_s}{\partial x_j} \qquad q_j=-\lambda\frac{\partial T}{\partial x_j}$$

$$\Omega_s=M_s\sum_{r=1}^{N_r}(b_{s,r}-a_{s,r})\left[k_{f,r}\prod_{m=1}^{N_s}\left(\frac{\rho_m}{M_m}\right)^{a_{m,r}}-k_{b,r}\prod_{m=1}^{N_s}\left(\frac{\rho_m}{M_m}\right)^{b_{m,r}}\right]$$

Figure 39. The EAST model governing equations in 1D, 2D and 3D. See [4,42] for flow notations.

1D EAST Problem Setup

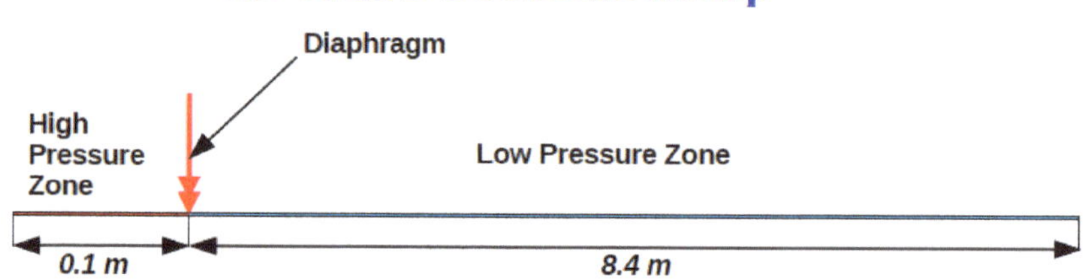

13 species mixture:

$e^-, He, N, O, N_2, NO, O_2, N_2^+, NO^+, N^+, O_2^+, O^+, He^+$.

High Pressure Zone

ρ	$1.10546\ kg/m^3$
T	$6000\ K$
p	$12.7116\ MPa$
Y_{He}	0.9856
Y_{N_2}	0.0144

Low Pressure Zone

ρ	$3.0964\times10^{-4} kg/m^3$
T	$300\ K$
p	$26.771\ Pa$
Y_{O_2}	0.21
Y_{N_2}	0.79

Figure 40. One-dimensional 13-species problem setup related to the EAST experiment.

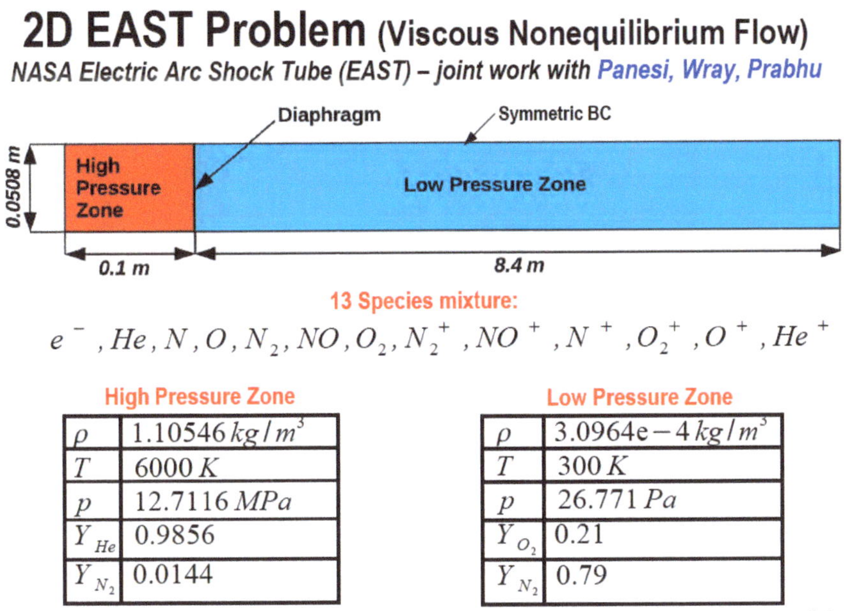

Figure 41. Two-dimensional 13-species problem setup related to the EAST experiment.

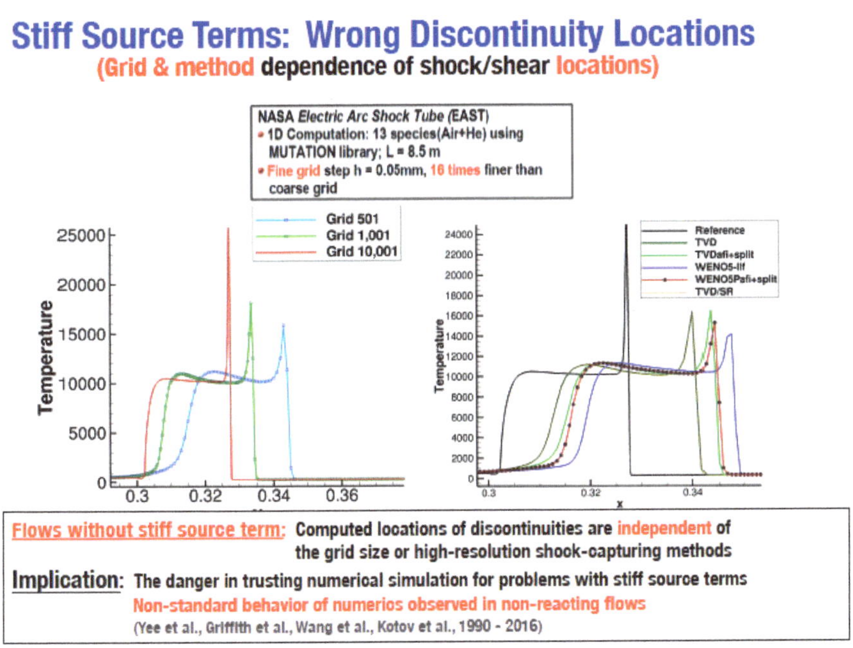

Figure 42. Temperature comparison using three grids and comparison among five high-order shock-capturing methods for 1D 13-species chemical reacting flows.

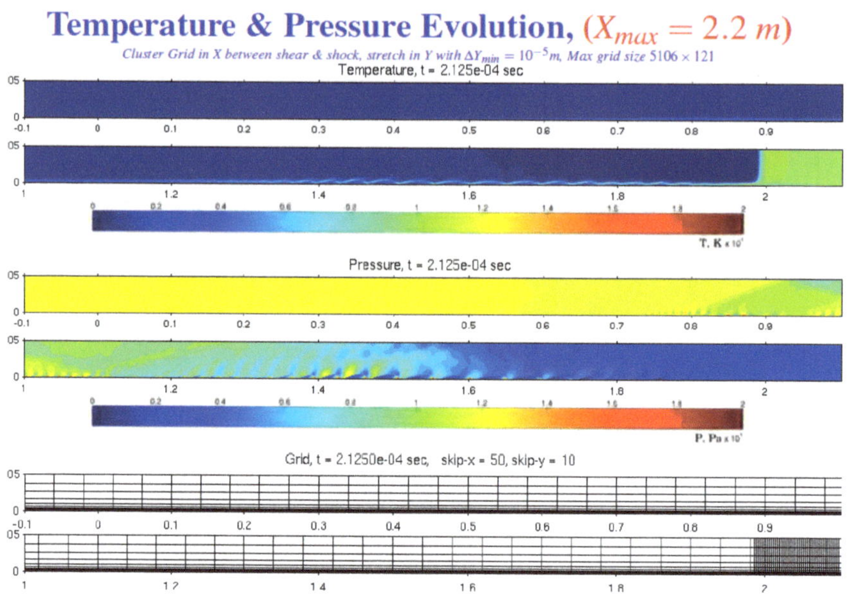

Figure 43. Temperature and pressure evolution of the 2D 13 species chemical reacting flows.

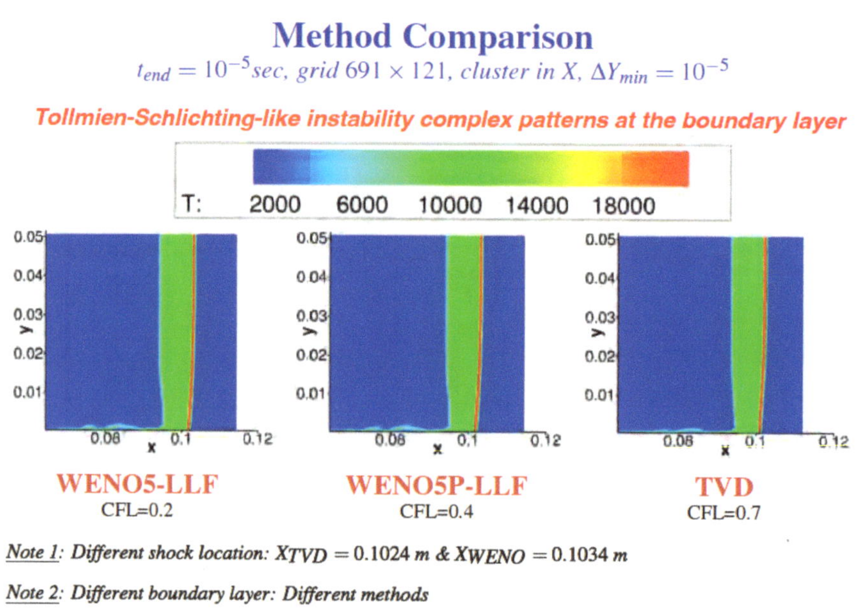

Figure 44. Temperature comparison among three high-order shock-capturing methods for the same 2D 13-species chemical reacting flows.

10. How Numerical Dissipation Affects the Predictability and Reliability of Simulation of Compressible Turbulence with Shock Waves

This section show how numerical dissipation affects PAR of compressible turbulence with shock waves. A short historical perspective and a brief description of the high-order methods being heavily used are included in this section.

ENO and WENO methods utilizing the shock resolving capability of TVD schemes together with higher order accuracy were developed in the late 1980s and early 1990s. In the past two decades, many variants of ENO and WENO methods were derived, yet they were still too dissipative away from shock waves for long-time integration of turbulent flows. For this reason, straightforward application of these original and recently developed high-order variants of ENO and WENO methods alone appear to not be a preferred choice for highly resolved turbulence simulations that push the resolution to the limit without resorting to using an extremely fine grid. From now on, the terms TVD, ENO and WENO refer to their original formulations and their variants. In other words, using one single method during the entire flow evolution usually posted shortcomings for complex turbulence flow development. A blending of more than one method to cater to the particular flow regions or different stages of turbulence development guided by a smart flow sensor at each time step on the type and amount of numerical dissipation is needed and would help to improve the overall stability, accuracy and reliability of simulations.

Since the late 1990s, blending of more than one method basically concentrated on two categories of such development: (a) the hybrid method, which switches between high-order non-dissipative methods and high-order shock-capturing methods (e.g., high-order WENO or ENO) guided by a flow sensor on the computed data; (b) the high-order nonlinear filter method of Yee et al., Sjögreen and Yee, and Yee and Sjögreen [4,10,13,25,27,28,40,161]. For our nonlinear filter methods, after the completion of a full time step of the primary high-order non-dissipative linear scheme, a smart local flow sensor would examine the computed data by flagging the locations and the amount of numerical dissipations that are needed. Then, the computed data would be filtered by a dissipative portion of high-order shock-capturing methods according to the flagged locations and the percentage amount of shock-capturing dissipation. More sophisticated forms with more than one type of numerical dissipation and local sensors are reported in [4,5,27,28]. For the performance of the nonlinear filter methods using high-order non-dissipative Padé (compact) spatial methods as the primary (base) methods, see [11,14,162]. For performance of the nonlinear filter methods using the high-order dispersion relation preserving (DRP) methods for aeroacoustic application, see Many of the recent developers of more complex variants of ENO and WENO are considering the aforementioned hybridized approach with their own flow sensors.

The majority of existing high-order methods, including hybrid or high-order nonlinear filter methods, might not discretely preserve the desirable physics-preserving properties (e.g., entropy conserving, momentum conserving and or kinetic energy preserving with added mechanism). To further minimize the use of numerical dissipation and to preserve some of the key physical properties, the current trend in turbulence/shock numerical simulations uses adaptive blending of a high-order **skew-symmetric form of physics-preserving non-dissipative methods** with the desired high-order shock-capturing methods guided by a smart flow sensor in a manner to further minimize the aid of added numerical dissipation. Here, the non-dissipative classical central, Padé (compact) or dispersion relation preserving (DRP) spatial discretization is applied to the skew-symmetric form of the inviscid flux derivatives. High-order skew-symmetric forms of physics-preserving methods blended with high-order shock-capturing methods can be found in [11–16,26,29,30]. Our studies indicated that the nonlinear filter approach provides similar accuracy but is more efficient than the hybrid approach in conjunction with higher than one stage of time discretization, e.g., higher than the first order of Runge–Kutta methods.

10.1. Short Overview of Hybrid Methods

Consider a 1D hyperbolic conservation law, where $\mathbf{U}$ is the vector of flow variables, $\mathbf{F}_x$ is the inviscid flux derivative, and t and x represent time and the computational space in x in a computational domain

$$\mathbf{U}_t + \mathbf{F}_x = 0.$$

For simplicity, we assume the semi-discrete form uses a uniform finite difference method (FDM) grid $x_j = (j-1)\Delta x, j = 1, \ldots, N$, where Δx is the grid spacing. Denote $\mathbf{U}_j$ and $\mathbf{F}_j$ as the approximation of $\mathbf{U}$ and $\mathbf{F}$ at grid spacing j. A semi-discrete form of a chosen method is

$$\frac{d\mathbf{U}_j}{dt} + \frac{\mathbf{H}_{j+1/2} - \mathbf{H}_{j-1/2}}{\Delta x} = \mathbf{0}, \tag{13}$$

where $\mathbf{H}_{j+1/2}$ denotes numerical fluxes consistent with $\mathbf{F}$. A simple hybrid method switching between two methods can be written as

$$\mathbf{H}_{j+1/2} = (1 - \nu_{j+1/2})\mathbf{H}^C_{j+1/2} + \nu_{j+1/2}\mathbf{H}^S_{j+1/2},$$

where $\mathbf{H}^C_{j+1/2}$ is a high-order non-dissipative numerical flux and $\mathbf{H}^S_{j+1/2}$ is a high-order shock-capturing numerical flux. $\nu_{j+1/2}$ is a local flow sensor. For a 1D system of hyperbolic conservation laws using a complete set of characteristic waves, the high-order shock-capturing methods are preferred to discretize locally each of the characteristic waves by a complete set of Riemann solvers [17,21,25,26,163]. If a multidimensional method is not used, it is very common to apply a 1D Riemann solver method dimension-by-dimension for a multi-dimensional system of hyperbolic conservation laws [4,5,10,17,27,28,163]. See some recent related development [164–166].

10.2. Overview of Nonlinear Filter

The nonlinear filter approach is less known in the aerodynamics CFD application community as it has not been described in CFD references or textbooks. This section gives a longer overview than the hybrid method approach, including test cases from our extensive work [10,12,13,15,16,27,28,65,161,167].

In a nutshell, a nonlinear filter approach consists of a base scheme step and a post-processing nonlinear filter step.

10.2.1. Base Scheme Step

A full time step is advanced using a high-order non-dissipative spatially central scheme (classical central, Padé or DRP method). A summation-by-parts (SBP) boundary operator [56,57] and a matching order conservative high-order freestream metric evaluation for curvilinear grids [17] are used. A high-order temporal discretization such as the third-order TVD Runge–Kutta or fourth-order Runge–Kutta (RK3 or RK4) method is used.

10.2.2. Nonlinear Filter Step—A Post-Processing Step

Instead of the hybrid approaches, to further improve nonlinear stability and accuracy from the non-dissipative spatial base scheme, the computed data are nonlinearly filtered by a dissipative portion of a high-order shock-capturing scheme with a local smart flow sensor. At each grid point, the local flow sensor(s) is employed to analyze the regularity of the computed flow field from the base scheme step. Only strong discontinuity locations would receive the full amount of shock-capturing dissipation. In smooth regions, no dissipation is added unless high-frequency oscillations are detected. In regions with strong turbulence, if needed, a small fraction of the shock-capturing dissipation can be added to improve stability. Note that the nonlinear filter numerical fluxes only involve the inviscid flux derivatives regardless of whether the flow is viscous or inviscid. If viscous terms are present, for ease of the SBP boundary closure implementation for the viscous flux derivatives, the same inviscid central difference operator used for the first derivative is employed twice for the viscous flux derivatives. For a variety of local flow sensors with automatic selection of different flow types, see [28].

The nonlinear filter idea was first introduced and tested by Yee et al. [25,26], using an artificial compression method (ACM) of Harten [168] as the flow sensor. Later, smart flow sensors were developed by Sjögreen and Yee, Yee and Sjögreen, and Kotov et al. [4,5,10,27,28]. Unlike the hybrid approach, in the presence of physical viscosity, the primary high-order non-dissipative scheme includes the discretizations on both the inviscid and viscous portion of the governing equation before the nonlinear filtering post-processing step. This is opposed to the hybrid method. It examines the computed flow data from the previous time step information. In addition, switching between two methods that can create instability during frequent switching between two methods is opposed to the nonlinear filter methods that do not switch among methods. The high-order base scheme again can be a standard central, Padé or DRP method, depending on the flow type in question.

Denote the computed solution by the base scheme step by $U^*_{j,k,l}$ of 3D Euler equations of gas dynamics. After the completion of a full time step of the spatial base scheme step at time n, the final update of the solution after the filter step at time $n+1$ (with the numerical fluxes in the y- and z-directions suppressed as well as their corresponding y- and z-direction indices on the x inviscid flux suppressed) is

$$U^{n+1}_{j,k,l} = U^*_{j,k,l} - \frac{\Delta t}{\Delta x}[H^*_{j+1/2} - H^*_{j-1/2}], \tag{14}$$

$$H^*_{j+1/2} = R_{j+1/2}\overline{H}_{j+1/2}. \tag{15}$$

The nonlinear filter numerical fluxes usually involve the use of field-by-field approximate Riemann solvers. If the Roe type of approximate Riemann solver [169] is employed, for example, the x-filter numerical flux vector $H^*_{j+1/2}$ is evaluated at the $U^*_{j,k,l}$ solution from the base scheme step. $R_{j+1/2}$ is the matrix of right eigenvectors of the Jacobian of the inviscid flux vector in terms of Roe's average states based on U^*. $H^*_{j+1/2}$ and $H^*_{j-1/2}$ are "filter" numerical fluxes in terms of Roe's average states based on $U^*_{j,k,l}$. Denote the elements of the filter numerical flux vector $\overline{H}_{j+1/2}$ by $\overline{h}^m_{j+1/2}, m = 1, 2, ..., 5$. The element of the filter numerical flux $\overline{h}^m_{j+1/2}$ has the form

$$\overline{h}^m_{j+1/2} = \frac{\kappa^m_{j+1/2}}{2} w^m_{j+1/2}\phi^m_{j+1/2}. \tag{16}$$

where $w^m_{j+1/2}$ is a flow sensor to activate the nonlinear numerical dissipation portion of a high-order shock-capturing scheme $\frac{1}{2}\phi^m_{j+1/2}$. The term $\kappa^m_{j+1/2}$ represents a locally determined positive parameter that is less than or equal to one, based on the regularity of the computed data. The choice of the parameter κ can be different for different flow types and is automatically chosen by using the local $\kappa^m_{j+1/2}$ described in [28]. However, if the computation used the Ducros et al. flow sensor, we set $\kappa^m_{j+1/2} = 1$ [170].

Figure 45 shows the nonlinear filter procedure for a system of 1D hyperbolic conservation laws. Figure 46 shows the nonlinear filter procedure using a combination of more than one flow sensor. Figure 46 also shows how to obtain, e.g., the dissipative portion of a seventh-order WENO (WENO7) denoted by $\phi^m_{j+1/2} = g^m_{j+1/2} - b^m_{j+1/2}$. The term $g^m_{j+1/2}$ represents the mth characteristic term of WENO7 and $b^m_{j+1/2}$ represents the mth characteristic term of the eighth-order central method. The nonlinear filter approach using WENO7 is denoted by WENO7fi on some of the later numerical test result presentations.

10.2.3. A Historical Note on Nonlinear Filter Approaches

The original idea of the nonlinear filter approach of Yee et al. (2000) [25] was based on shock-capturing methods that are written into a central discretization portion and a nonlinear shock-capturing portion (dissipative portion). All variants of TVD schemes are already written in this form by design. In order to obtain the dissipative portion of other high-order shock-capturing methods, the dissipative portion of the considered shock-capturing $\phi^m_{j+1/2} = g^m_{j+1/2} - b^m_{j+1/2}$ can be obtained in a similar manner as WENO7fi.

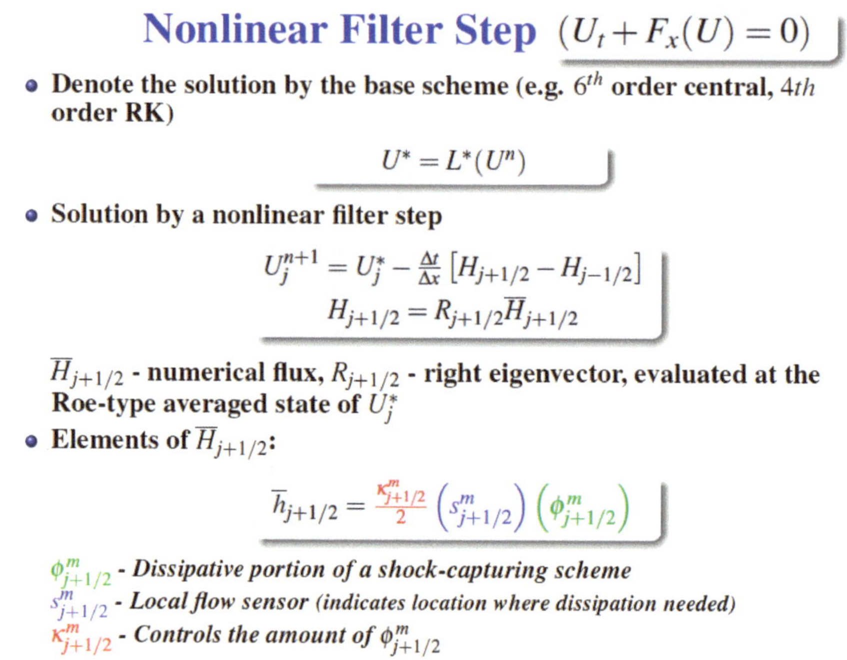

Figure 45. Nonlinear filter procedure.

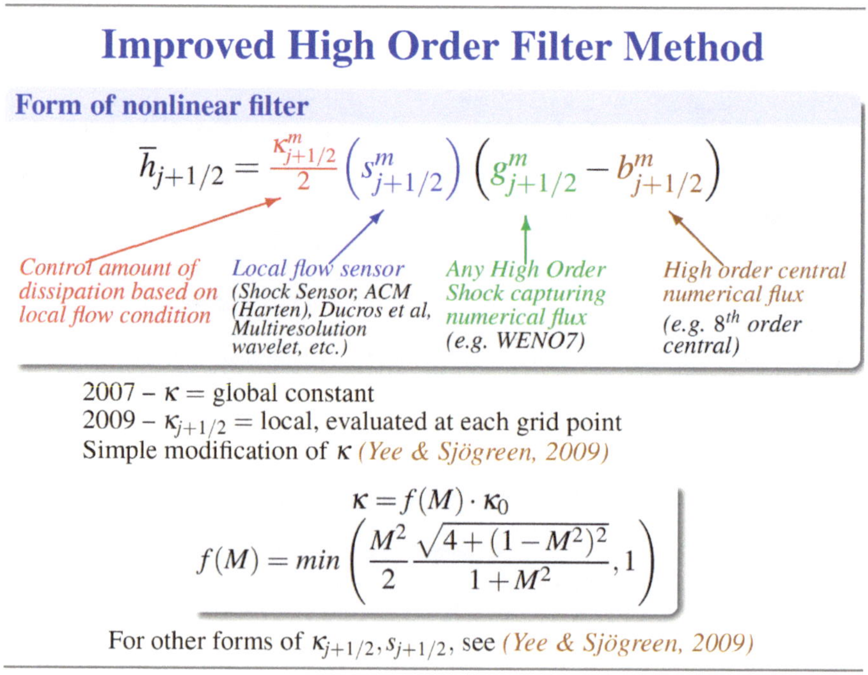

Figure 46. Nonlinear filter procedure with local flow sensors.

It is noted that the nonlinear filter step described above should not be confused with the LES filtering operation. For extension of the blending of more than one numerical method to the ideal MHD, see [12,13,15,16,27,28,65,161,167].

As mentioned briefly above, the nonlinear filter schemes are efficient by construction. The total computational cost for a given error tolerance is significantly lower than for standard shock-capturing schemes or their hybrid cousins of the same order. One important reason for their efficiency is that the nonlinear shock-capturing filter dissipation is applied after each full time step of the base scheme, whereas a standard shock-capturing/hybrid method evaluates the shock-capturing method at each stage of the, e.g, m stage Runge–Kutta (R-K) time stepping scheme. Hence, the nonlinear filter approach requires only one shock-capturing method evaluation per time step per grid point per dimension, independent of the time discretization involved. Hybrid schemes, which switch between the high-order non-dissipative method and high-order shock-capturing method, require m Riemann solvers for each m stages of the R-K time discretization. Therefore, the hybrid methods are less efficient than the nonlinear filter methods. The hybrid method also can encounter numerical instability as well as conservation issues related to switching between schemes at the switch locations.

In addition, if the computation considers the dimension-by-dimension and a complete set of characteristic waves Riemann solver approach, only **ONE** CPU intensive evaluation of a dissipative portion of a chosen high-order shock-capturing method in local characteristic variables is need per dimension, per time step and per grid spacing for the nonlinear filter approach, regardless of the time integrator. Contrary to the nonlinear filter method, for the hybrid method, the number of high-order shock-capturing methods using the same Riemann solver approach evaluations per dimension, per time step and per grid spacing would be four times by a four-stage Runge–Kutta time integrator.

Figure 47 shows a 3D isotropic turbulence with shocklets problem setup. Figure 48 shows the accuracy of our nonlinear filter method for the isotropic turbulence simulation with the smart flow sensor built-in. Comparing the resolution by the hybrid method reported in Johnson et al. [171], one can see that they produce the same accuracy. For the same 3D isotropic problem setup as in Johnson et al. [171] and performance of our nonlinear filter method, see [5].

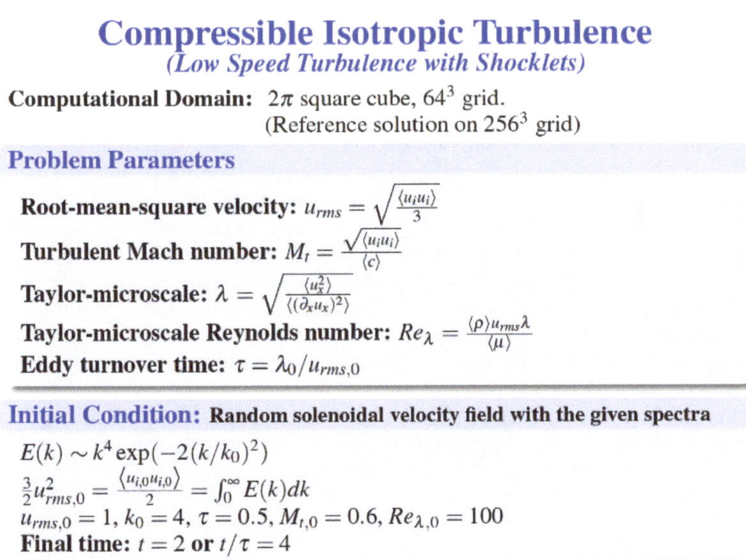

Figure 47. Three-dimensional isotropic turbulence with shocklets problem setup.

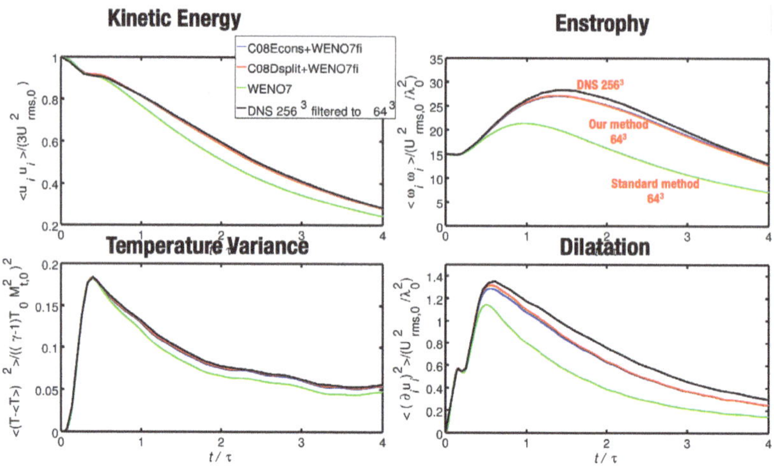

Figure 48. Three-dimensional Isotropic turbulence with shocklets: Comparison among standard 7th-order WENO method (WENO7) and our two 7th-order nonlinear filter methods (C08Econs+WENO7fi and C08Dsplit+WENO7fi) using a very coarse 64^3 grid with the filtered DNS computation on a 256^3 grid. Kinetic energy (**top left**), Enstrophy (**top right**), Temperature Variance (**bottom left**) and Dilatation (**bottom right**). C08Econs denotes the 8th-order classical central applied to the entropy split form, and C08Dsplit denotes the 8th-order classical central applied to the Ducros et al. split form of the Euler inviscid flux derivative.

10.3. The Need for Efficient High-Order Physics-Preserving Methods Blending with High-Order Shock-Capturing Methods

In classical mechanics, temporal, spatial or both finite difference discretizations that conserve a certain physical property (or combination of physical properties) of the governing equations are most often referred to, in a broad sense, as structure-preserving numerical methods. See e.g., [172]. In the literature, structure-preserving methods are also denoted as **physics-preserving methods**. Here, we use these terms interchangeably. Temporal and spatially finite difference discretizations that discretely conserve fundamental properties of the chosen governing equations are hereafter are denoted as **structure-preserving methods (SPMs)**. SPMs have been gaining more attention in long-time integration of turbulent flows. A combination of these SPMs can further increase the reliability, stability and accurate simulation of fluid flows, especially for turbulent flows. It is noted that the majority of existing high-order methods, including hybrid or high-order nonlinear filter methods, might not discretely preserve the desirable physical-preserving properties (NOT SPM).

In the applied mathematics CFD community, entropy-conserving and entropy-stable methods have been flourishing for the last two decades [3,12,13,15,16,26,173–175]. Symplectic time integrators are used to prevent phase distortion, e.g., Hamiltonian systems or the Korteweg–deVries equations [176]. In the applied mechanics and CFD community, the importance of developing SPMs that conserve one or more fundamental properties of the governing equations (mass, minimizing phase distortion, momentum and physical entropy conserving, positivity of density and pressure, and kinetic energy preserving, etc.) has been ongoing for the last four decades. See e.g., [3,12,13,15,16,26,65,158,174,177,178] for some of the developments. Combinations of more than one SPM is also be possible. See [14,30] or the next section for formulations. Formulations using the DRP and Padé high-order SPM non-dissipative linear methods are presented in [11,14].

10.4. SPM Finite Difference Formulation in Split Form: Skew-Symmetric Splittings of the Compressible Euler Flux Derivatives

When the high-order non-dissipative spatial method discussed previously discretized a class of the skew-symmetric form of the inviscid Euler flux derivative (instead of the unsplit flux derivative), it was shown that the appearance of instabilities or aliasing errors in long-time integration of turbulence flows can be delayed. The splitting can be performed for certain flow variables or the entire convective flux derivative system. They are written in a special split form which is equivalent to the unsplit version, i.e., the split and unsplit governing equations are equivalent. However, after finite difference discretizations, certain different split forms can maintain a discrete entropy conservation, momentum conservation, kinetic energy preservation property or can provide a stable L2-like energy norm estimate for smooth solutions. See articles [3,12,13,15,16,26,30,45,65,158,174,177,179,180] for discussions of the performance of various skew-symmetric splitting approaches in DNS and LES applications. See Sjögreen et al. [13] for the extension of these skew-symmetric splittings to any even-order split approximation formulation for the Euler and MHD equations.

In the simplest form, a split approximation starts from rewriting the derivative of the product $(ab)_x$ as

$$(ab)_x = \alpha(ab)_x + \gamma a b_x + \beta a_x b, \tag{17}$$

before discretization, where a and b are functions of x. The parameters α, γ and β are so chosen to be still equivalent to the original $(ab)_x$ before discretization. A common split derivative is found by setting $\alpha = \gamma = \beta = 1/2$, which results in the form

$$(ab)_x = \frac{1}{2}(ab)_x + \frac{1}{2}ab_x + \frac{1}{2}a_x b. \tag{18}$$

The Ducros et al. splitting [158] starts with this particular non-conservative split form using the classical central discretization operator on the derivative terms. They then apply mathematical manipulations on the difference operators to make the final discretization into a conservative difference method. Their construction will be discussed shortly.

A split approximation for the derivative of the product of three functions $(abc)_x$ starts with rewriting the derivative as

$$\begin{aligned}(abc)_x \;&= \alpha(abc)_x + \gamma[a(bc)_x + bca_x] + \beta[b(ac)_x + acb_x] \\ &+\kappa[c(ab)_x + abc_x] + \delta[bca_x + acb_x + abc_x]\end{aligned} \tag{19}$$

before discretization, where a, b and c are functions of x. The parameters $\alpha, \gamma, \beta, \kappa$ and δ are chosen to be still equivalent to the original $(abc)_x$ before discretization. A well-known split derivative that can be written in conservative form and can improve nonlinear stability is the Kennedy and Gruber [45] splitting, which is kinetic energy preserving for the Euler equation by setting the first four parameters to be $1/4$ and $\delta = 0$ for the triple product derivative. See [45,179,180] for details.

For the nonlinear Euler equation of gas dynamics, a less known splitting of the inviscid flux derivative is the entropy splitting method of [26,178,181]. The key mathematical idea of the entropy splitting is that their split formulas can be used to estimate the L^2-norm-like or the energy norm of the computed solution for periodic and non-periodic boundary conditions. In the recent terminology, it is entropy stable (in the sense of possessing an energy estimate) for the Euler equations using high-order classical central or DRP central schemes in conjunction with summation-by-parts boundary operators [11,26,64,178,181–183]. See [64] for a follow on study of the subject. The next two subsections give a short overview of the Ducros et al. and entropy splitting methods.

10.5. Ducros et al. Type Conservative Splitting

The Ducros et al. split approximation [158] for the Euler equations of gas dynamics starts with (18) the terms of the split form approximated by

$$\frac{1}{2}D(ab) + \frac{1}{2}D(a)b + \frac{1}{2}aD(b), \tag{20}$$

where D is a centered finite difference operator, and a and b are functions of x. For this split approximation by a central, Padé or DRP discretization, after mathematical manipulations, the resulting method conserves momentum (see below). However, it is not obvious how to obtain a norm estimate for nonlinear systems.

The key step in the Ducros et al. [158] split approximation is to rewrite (20) in conservation form. For the second-order operator $Du_j = (u_{j+1} - u_{j-1})/(2\Delta x)$, we have

$$\frac{1}{2}D(ab) + \frac{1}{2}D(a)b + \frac{1}{2}aD(b) = \frac{1}{4\Delta x}\Delta_+[(a_j + a_{j-1})(b_j + b_{j-1})], \tag{21}$$

where $\Delta_+ q_j = (q_{j+1} - q_j)$.

The beauty is that Equation (21) can be generalized to standard centered difference operators of $2p$th-order of accuracy,

$$D_p u_j = \frac{1}{\Delta x}\sum_{k=1}^{p} \alpha_k^{(p)}(u_{j+k} - u_{j-k}). \tag{22}$$

The coefficients $\alpha_k^{(p)}$ satisfy

$$\sum_{k=1}^{p} k\alpha_k^{(p)} = \frac{1}{2} \qquad \sum_{k=1}^{p} \alpha_k^{(p)} k^{2n+1} = 0, \;\; n = 1, \dots, p-1. \tag{23}$$

The right-hand side of the algebraic identity

$$a_{j+k}b_{j+k} - a_{j-k}b_{j-k} + (a_{j+k} - a_{j-k})b_j + a_j(b_{j+k} - b_{j-k}) = \\ (a_{j+k} + a_j)(b_{j+k} + b_j) - (a_j + a_{j-k})(b_j + b_{j-k}) \tag{24}$$

is written on conservative form by

$$(a_{j+k} + a_j)(b_{j+k} + b_j) - (a_j + a_{j-k})(b_j + b_{j-k}) = \\ \sum_{m=0}^{k-1}(a_{j-m} + a_{j+k-m})(b_{j-m} + b_{j+k-m}) - \sum_{m=0}^{k-1}(a_{j-1-m} + a_{j-1+k-m})(b_{j-1-m} + b_{j-1+k-m}). \tag{25}$$

The conservative form of the split approximation becomes

$$\begin{aligned} &\frac{1}{2}D_p(ab) + \frac{1}{2}D_p(a)b + \frac{1}{2}aD_p(b) = \\ &\quad \frac{1}{\Delta x}\sum_{k=1}^{p}\frac{1}{2}\alpha_k^{(p)}\Big((a_{j+k}b_{j+k} - a_{j-k}b_{j-k}) + a_j(b_{j+k} - b_{j-k}) + (a_{j+k} - a_{j-k})b_j\Big) \\ &\quad = \frac{1}{\Delta x}\sum_{k=1}^{p}\frac{\alpha_k^{(p)}}{2}\left(\sum_{m=0}^{k-1}(a_{j-m} + a_{j+k-m})(b_{j-m} + b_{j+k-m})\right. \\ &\quad \left. - \sum_{m=0}^{k-1}(a_{j-1-m} + a_{j-1+k-m})(b_{j-1-m} + b_{j-1+k-m})\right) = \frac{1}{\Delta x}(h_{j+1/2} - h_{j-1/2}), \end{aligned} \tag{26}$$

where the numerical flux is defined by

$$h_{j+1/2} = \sum_{k=1}^{p} \frac{1}{2}\alpha_k^{(p)} \sum_{m=0}^{k-1} (a_{j-m} + a_{j+k-m})(b_{j-m} + b_{j+k-m}). \tag{27}$$

Define

$$\Theta_{j+1/2}^{(p)}(a,b) = \sum_{k=1}^{p} \frac{1}{2}\alpha_k^{(p)} \sum_{m=0}^{k-1} (a_{j-m} + a_{j+k-m})(b_{j-m} + b_{j+k-m}). \tag{28}$$

Consider the x-direction inviscid flux for 3D Euler equations of gas dynamics equations

$$\mathbf{F} = [\rho u, \rho u^2 + p, \rho uv, \rho uw, (e+p)u]^T,$$

where the inviscid flux $\mathbf{F} = \mathbf{F}(\mathbf{U})$; the velocity vector $\mathbf{u} = [u(x,y,z), v(x,y,z), w(x,y,z)]^T$ in the x-, y- and z-directions; ρ denotes the density; p is the pressure; and e is the total energy. Denote $\mathbf{u}_j = (u_j, v_j, w_j)^T$ as the discretization at the j grid location with the y and z discretizations indices suppressed for simplicity. The flux components can be written as products of two factors in many different ways, leading to different split approximations. One Ducros et al. split-type approximation of the gas dynamics flux derivative that will be used in this study is given by

$$\mathbf{F}_x|_{x=x_j} \approx \begin{pmatrix} \frac{1}{2}D\rho_j u_j + \frac{1}{2}\rho_j D u_j + \frac{1}{2}u_j D\rho_j \\ \frac{1}{2}D\rho_j u_j^2 + \frac{1}{2}\rho_j u_j D u_j + \frac{1}{2}u_j D\rho_j u_j + Dp_j \\ \frac{1}{2}D\rho_j u_j v_j + \frac{1}{2}\rho_j v_j D u_j + \frac{1}{2}u_j D\rho_j v_j \\ \frac{1}{2}D\rho_j u_j w_j + \frac{1}{2}\rho_j w_j D u_j + \frac{1}{2}u_j D\rho_j w_j \\ \frac{1}{2}Du_j(e_j + p_j) + \frac{1}{2}u_j D(e_j + p_j) + \frac{1}{2}(e_j + p_j)Du_j \end{pmatrix}, \tag{29}$$

which, by (27), can be written in conservative form with numerical flux function

$$\mathbf{H}_{j+1/2} = \frac{1}{2}\sum_{k=1}^{p} \alpha_k^{(p)} \sum_{m=1}^{k-1} \begin{pmatrix} (\rho_{j-m} + \rho_{j+k-m})(u_{j-m} + u_{j+k-m}) \\ (\rho_{j-m}u_{j-m} + \rho_{j+k-m}u_{j+k-m})(u_{j-m} + u_{j+k-m}) + p_{j-m} + p_{j+k-m} \\ (\rho_{j-m}v_{j-m} + \rho_{j+k-m}v_{j+k-m})(u_{j-m} + u_{j+k-m}) \\ (\rho_{j-m}w_{j-m} + \rho_{j+k-m}w_{j+k-m})(u_{j-m} + u_{j+k-m}) \\ (e_{j-m} + p_{j-m} + e_{j+k-m} + p_{j+k-m})(u_{j-m} + u_{j+k-m}) \end{pmatrix}. \tag{30}$$

The more compact notation introduced in (28) allows (30) to be rewritten as

$$\mathbf{H}_{j+1/2} = \begin{pmatrix} \Theta_{j+1/2}^{(p)}(\rho, u) \\ \Theta_{j+1/2}^{(p)}(\rho u, u) + \Theta_{j+1/2}^{(p)}(p, 1) \\ \Theta_{j+1/2}^{(p)}(\rho v, u) \\ \Theta_{j+1/2}^{(p)}(\rho w, u) \\ \Theta_{j+1/2}^{(p)}(e + p, u) \end{pmatrix}. \tag{31}$$

Unlike the linearized and symmetrized system of Euler equations, where the split approximations lead to a stability estimate, there is no such stability estimate for the $2p$th-order accurate conservative Ducros et al. splitting. However, since the Ducros et al. splitting results in a conservative scheme, it is applicable for problems containing discontinuities.

10.6. *Special Class of Entropy Conserving SPM—Entropy Split Method (Entropy Splitting Approach)*

One class of SPM that the authors have been developing is the entropy split method for the compressible Euler flux derivatives. The entropy split method was originally denoted as the **"Entropy Spitting Approach"** [17,26]. This is a less-known skew-symmetric splitting for the Euler equation of gas dynamics [26,178,181]. This skew-symmetric splitting made use of Harten's symmetrizable form of the Euler equations in terms of the entropy variables [174] to obtain a semi-discrete splitting of the Euler equations with a discrete entropy stability (in space) by the summation-by-parts approach. The entropy splitting is written in terms of the sum of a conservative portion and a nonconservative portion. If central (classical central, Padé or DRP) discretizations are used for the interior scheme (interior grid points) and a summation-by-parts boundary scheme (boundary points), it can be proved that the resulting split method is entropy conservative and entropy stable. See the recent result by Sjögreen and Yee [3,11,12,14,15,29,30,64,65]. During the entropy splitting development, the Harten [174] and the Gerritsen and Olsson entropy splitting forms incorrectly selected the un-physical branch of the inequality and were later corrected by Yee et al., hereafter referred to as the entropy splitting of the Euler equations. It is considered to be a semi-conservative splitting except at the boundary grid points. The entropy splitting of Olsson and Oliger, Gerritsen and Olsson, and Yee et al. [26,178,181] is a splitting of a form that is more suitable for the discrete stable energy norm estimate technique, including boundary scheme estimate for arbitrary order of central spatial schemes. See Yee et al. [26] for the formulation.

Consider the 1D Euler equations with inviscid flux derivative $\mathbf{F}(\mathbf{U})_x$ for a perfect gas. For extension to multi-D in a time-varying deforming grid, see [17,26,184]. The inviscid flux derivative is split into the following via the entropy variables W discussed in Harten [174].

$$\mathbf{F}_x = \frac{\beta}{\beta+1}\mathbf{F}_x + \frac{1}{\beta+1}\mathbf{F}_W W_x, \quad \beta \neq -1 \tag{32}$$

$$W = [w_1,\ w_2,\ w_3,\ w_4,\ w_5]^T = \frac{p^*}{p}[e + \frac{\alpha-1}{\gamma-1}p,\ -\rho u,\ -\rho v,\ -\rho w,\ \rho]^T, \tag{33}$$

where

$$p^* = -(p\rho^{-\gamma})^{\frac{1}{\alpha+\gamma}} \tag{34}$$

and

$$\beta = \frac{\alpha+\gamma}{1-\gamma}, \quad \alpha > 0 \text{ or } \alpha < -\gamma. \tag{35}$$

See Yee et al. [9,25,26] for the formulation, the choice for β and early numerical examples.

In the original entropy splitting method, the (**Entropy Splitting Approach**), the Euler flux derivatives approximation as a sum of a conservative portion and a non-conservative portion in conjunction with summation-by-parts (SBP) difference boundary closure [58] of Olsson and Oliger, Gerritsen and Olsson, and Yee et al. [26,178,181] are not written in the usual numerical flux form. Yee at al. and Vinokur and Yee [17,26] re-examined the split form of the Euler flux derivative by Harten, Olsson and Oliger, and Gerritsen and Olsson [174,178,181], to select the physical relevant branch of the split parameter β. They extended the entropy split form to include a thermally perfect gas for moving curvilinear grids [17,26,184]. In addition, they performed a detailed study that applied high-order spatial central discretization on the entropy splitting form of the Euler flux derivatives and compared it with the un-split form. It was shown that the split form of the Euler flux derivatives is more stable than the unsplit form for longer time integration without the need of added numerical dissipation. In [11], DRP (dispersion relation-preserving) finite discretizations [182,185–188] were also applied to the entropy split form of the Euler flux derivatives and with a similar gain in numerical stability. Here, the various high-order methods resulting from applying classical spatial central, DRP and Padé methods to the split form of the Euler flux derivative are referred to as entropy split methods as a function of the splitting parameter β. These entropy split methods are entropy conserving and stable but they are usually not conservative numerical methods without additional reformulation, as proposed in, e.g., Sjögreen and Yee [3,12,13,15,16,30,189].

The entropy split methods and other SPM high-order skew-symmetric splitting of the inviscid Euler or ideal MHD flux derivative formulation of any even order can be found in [12,13,15]. Most of the skew symmetric splittings are not entropy conserving numerically.

Some of them are momentum conserving or kinetic energy preserving numerically, see [45, 158,179,180]. See [3,12,13,15,16,30,189] for additional development for the compressible Euler and ideal MHD equations.

Other Entropy Conserving Methods for the Compressible Euler and Ideal MHD Equations

The entropy-conserving method by Tadmor [175,190] is the method of choice for rapidly developing unsteady flows, especially in development of unstructured grid methods (e.g., DG method). Our work on typical test cases using the FDM formulation found that the Tadmor entropy conserving method [175,190] of the same order requires twice the arithmetic operations than the entropy split methods. In addition, Harten's entropy function can be part of a Tadmor-type entropy conserving method family [12,65] with similar accuracy and stability. Comparisons among the methods can be found in [3,11,12,30,65].

10.7. Newer Class of Entropy Split Methods in Numerical Flux Form: Not Relying on Homogeneous Property of the Inviscid Flux and Symmetrizable Inviscid Flux Derivatives

In 2019, Sjögreen and Yee [15] derived a high-order conservative numerical flux for the non-conservative portion of the entropy splitting of the Euler flux derivatives by taking advantage of the homogeneity property of the inviscid flux and symmetrizable inviscid flux derivatives. Due to the construction, this conservative numerical flux requires more arithmetic operations and is less stable than the original entropy split method (not fully conservative). An alternative simple approach for problems containing shock waves is to use a shock detector to switch the entropy split discretization to a standard centered approximation in the neighborhood of shock waves to avoid wrong shock speed.

Instead of taking advantage of the homogeneity property of the inviscid flux and symmetrizable inviscid flux derivatives, a wider class of entropy split methods that do not require the homogeneous property were developed, see our 2022 paper [16]. The formulation is written in standard numerical flux form. This wider class of entropy split method was extended for the equations of ideal MHD by the authors in [16].

It is our assessment that employing high-order SPM non-dissipative methods in conjunction with the nonlinear filter approach should be a method of choice for high-resolution simulations involving compressible turbulence up to mildly supersonic regimes. For turbulence with shocks, there is an even larger gain both in the accuracy and CPU time of the nonlinear filter schemes over their standard WENO counterparts.

Forms of the structure-preserving (SPM) numerical fluxes that belong to the class of skew-symmetric splitting of the inviscid flux derivatives in curvilinear grids are given in [16,30]. Depending on the type of flow physics, high-order central, DRP or Padé methods are applied to the skew-symmetric split form of the inviscid flux derivative as the baseline method before the post-processing nonlinear filter step, if needed, especially for problems with shocks. Selected methods with multiple structure-preserving properties are given with numerical examples.

10.8. A Complex 3D Supersonic Shock-Turbulence Interaction by a Physics-Preserving Nonlinear Filter Method [4]

The 3D supersonic shock-turbulence interaction is a more challenging test case and is a very CPU intensive computation with long transient simulation before obtaining meaningful turbulent statistics. The 3D test case considered here concerns an initial turbulence disturbance at the inflow boundary interacting with a stationary supersonic planar shock wave. The problem has been studied by previous investigators, mainly related to DNS computations; see e.g., [191–193]. Here, we choose the configuration considered in the DNS study of [192]. Figure 49 shows a schematic of the problem setup. The computational domain is $-2 \leq x \leq -2+4\pi$, $0 \leq y \leq 2\pi$ and $0 \leq z \leq 2\pi$. The grid is uniform in all directions with the spacing in x being several times finer than in y and z (see [192] for explanation). We solve the filtered governing equations in a non-dimensional form. The Yee et al. nonlinear filter scheme with the Ducros et al. flow sensor [170] is used for integration of the Ducros et al. [158] split form of the governing equations. The spatial base scheme is the eighth-order central differencing and the nonlinear filter scheme is the dissipative portion of the seventh-order WENO scheme (WENO7). This nonlinear filter scheme, including the Ducros et al. splitting of the governing equation, is denoted by WENO7fi + split (or WENO7fi for ease of discussion). Since the initial data consist of a planar shock in the x-direction, numerical dissipation should be mainly needed in the x-direction. In order to obtain more accurate results, WENO dissipation is employed only in the x-direction at the postprocessing stage of the Yee et al. nonlinear filter. The inflow and outflow boundary conditions are applied in the streamwise direction and periodic boundary conditions are applied in the transverse directions.

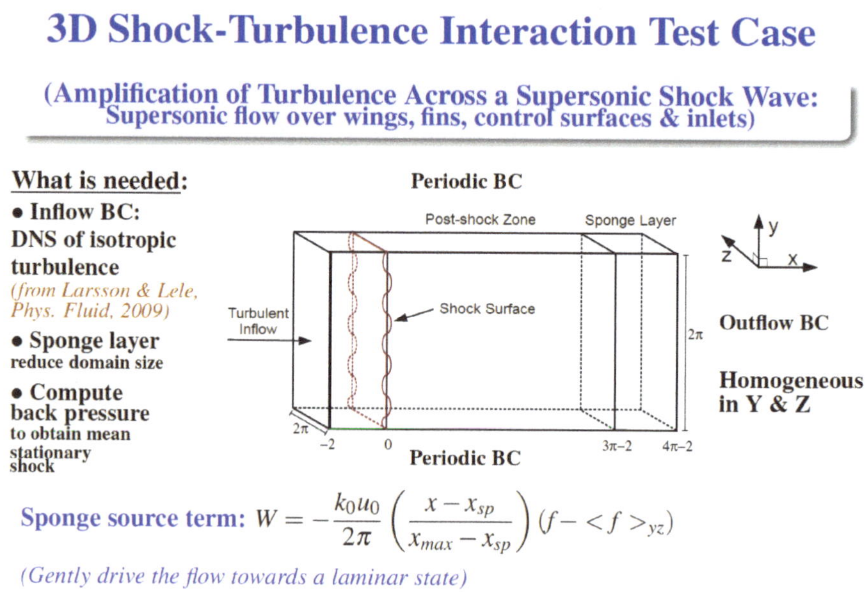

Figure 49. Three-dimensional supersonic shock-turbulence interaction test case.

Inflow boundary condition. A fully developed turbulent inflow condition is applied using a turbulent database. This database is generated as follows. First, an initial isotropic turbulent field with the energy spectrum $E(k) \sim k^4 \exp(-2k^2/k_0^2)$ and microscale Reynolds number $Re_\lambda = \rho\lambda u'_{rms}/\mu = 140$ is generated using the methodology described in Ristorcelli and Blaisdell [194]. Here, the energy peak wavenumber $k_0 = 4$ is used. Next, the decay of this field in a periodic box is simulated for approximately three eddy turnover times $\tau = \lambda/u'_{rms}$ to ensure fully developed turbulence. After the decay, the Reynolds number $Re_\lambda = 40$ and the turbulent Mach number $M_t = \overline{u'_i u'_i}^{1/2}/c_0 = 0.16$. Here, c_0 is the mean speed of sound. The generated isotropic turbulence is introduced at the inflow boundary with constant mean velocity u_0. We consider two cases with mean flow Mach numbers $M = 1.5$ and $M = 3.0$. In order to compare the DNS results, we use the inflow database from [192].

Outflow boundary condition. In order to avoid acoustic reflections of subsonic flow from the outflow boundary, a non-reflective sponge layer is employed on the region near the outflow. The length of this layer is $x_{max} - x_{sp} = \pi$. The sponge layer is implemented by introducing the following source term into the equations:

$$\Omega = -\frac{k_0 u_0}{2\pi}\left(\frac{x - x_{sp}}{x_{max} - x_{sp}}\right)(f - <f>_{yz}), \tag{36}$$

where $f = \rho, \rho u_i, \rho E$ and $\langle \cdot \rangle_{yz}$ denotes averaging in the y- and z-directions.

The outflow pressure p_∞ is chosen such that the mean shock location is stationary. For laminar flow, Rankine–Hugoniot conditions give

$$\frac{p_\infty}{p_0} = 1 + \frac{2\gamma}{\gamma+1}\left[\frac{(u_0 - U_s)^2}{c_0^2} - 1\right], \tag{37}$$

where p_0 is the inflow mean pressure, u_0 is the mean inflow velocity, c_0 is the mean inflow speed of sound and U_s is the shock velocity. As the inflow condition is turbulent, the Rankine–Hugoniot conditions are valid only instantaneously but not on average. After an initial guess based on (37), the outflow pressure is refined by an iteration procedure, integrating the governing equations on a coarse grid and updating the pressure according to the formula

$$p'_\infty = p_\infty + 4U_s\rho_u u_0/(\gamma+1). \tag{38}$$

See [192] for more details.

To illustrate an instantaneous turbulent pattern, Figure 50 shows the instantaneous velocity field u_x (top) and u_y (bottom) at a slice where z = const, obtained with DNS on a grid of 1553×256^2 points. A turbulent flow enters over the left boundary at Mach 1.5. The turbulence interacts with a quasi-steady shock wave near the inflow boundary. The figure shows the x- and y-directions velocity field in a $z = const.$ slice of the three-dimensional computational domain. See [4] for details of the problem set up and method comparisons. The numerical method is an eighth-order accurate non-dissipative method together with an adaptive nonlinear filter by a dissipation portion of the seventh-order positivity preserving WENO7 of Zhang and Shu [160] (WENO7Pfi). The Ducros et al. skew-symmetric splitting is the structure-preserving eighth-order central spatial discretization. The turbulent Mach number is 0.16. The incoming turbulence causes wrinkles in the shock front. The classical fourth-order Runge–Kutta method (RK4) is used as temporal discretization. Grid refinement and comparison of standard high-order shock-capturing methods with the high-order structure-preserving nonlinear filter method were performed in [4].

The comparison between the filtered DNS data and the results obtained with different numerical schemes on the coarse grid with 389×64^2 points and $M = 3$ is shown in Figure 51. Here, we compare the results obtained by the fifth-order WENO (WENO5) and WENO7 with the results obtained by the filter counterpart of WENO7 with Ducros et al. splitting of the governing equations (WENO7fi+split). The comparison of WENO5, WENO7 and WENO7fi+split is shown in Figure 52. For Figure 52, each of the considered schemes is employed in all three directions. The solution by WENO7fi is closer to the filtered DNS solution than WENO5 and WENO7. WENO5 is far too dissipative for DNS computation with this coarse grid. Figure 53 shows the velocity spectrum of DNS by WENO7fi with $M = 3$, sliced at $x = 0$: four-level DNS grid refinement study on grids with 389×64^2, 777×128^2, 1553×256^2 and 2329×256^2 points.

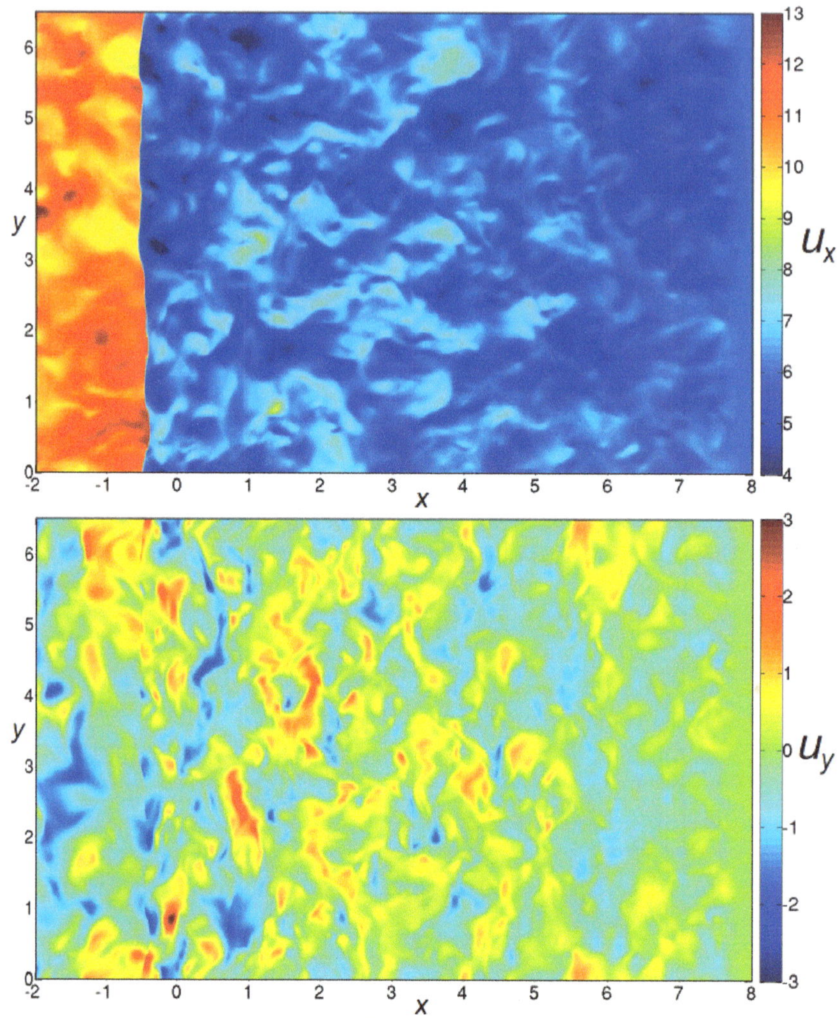

Figure 50. Instantaneous velocity field u_x (top) and u_y (bottom) obtained with DNS on grid of 1553×256^2 points. Slice $z = \text{const}$.

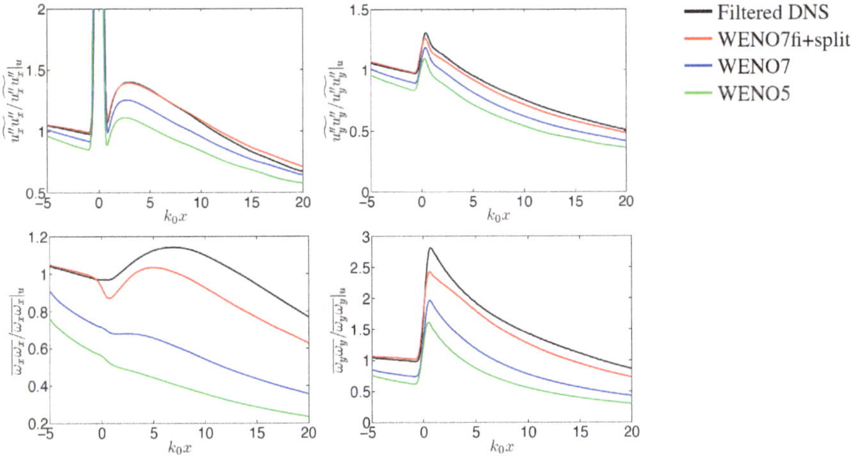

Figure 51. DNS, WENO5, WENO7 and WENO7fi method comparison: Comparison to filtered DNS data of the statistics obtained by the different numerical schemes on grid 389×64^2, $M = 1.5$. Top row: streamwise (**left**) and transverse (**right**) Reynolds stress components. Bottom row: streamwise (**left**) and transverse (**right**) vorticity components.

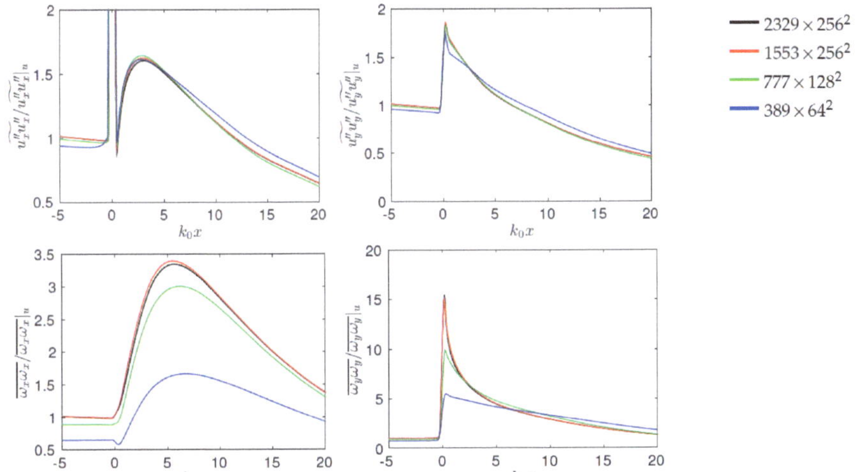

Figure 52. DNS, WENO7fi, M = 3: Four-level DNS grid refinement study on grids with 389×64^2, 777×128^2, 1553×256^2 and 2329×256^2 points. Top row: streamwise (**left**) and transverse (**right**) Reynolds stress components. Bottom row: streamwise (**left**) and transverse (**right**) vorticity components.

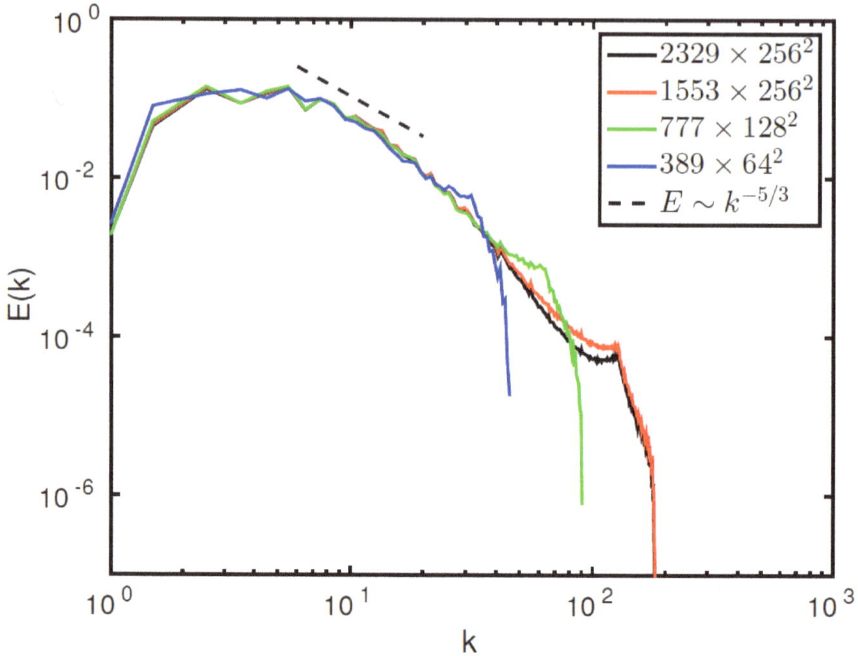

Figure 53. DNS, WENO7fi, $M = 3$, velocity spectrum, slice at $x = 0$: four-level DNS grid refinement study on grids with 389×64^2, 777×128^2, 1553×256^2 and 2329×256^2 points.

11. High-Order Skew-Symmetric Physical-Preserving Methods (SPM) Comparison for Gas Dynamics

For MHD computation and using the extension of our new entropy split method for the equation of ideal MHD, see [16,30]. Here, some test cases comparing a set of high-order structure-preserving methods for compressible gas dynamics will be given in this section. Nomenclature for these method comparisons are defined below.

11.1. Various Structure-Preserving Properties Nomenclature

For test cases comparison, the high-order spatial structure-preserving methods considered are as follows:

- ECHKP: Entropy conserving using the Harten class of entropy functions $E_H = -\frac{\gamma+\alpha}{\gamma-1}\rho(p\rho^{-\gamma})^{\frac{1}{\alpha+\gamma}}$ [174,175]. It turned out that this method in its base form also satisfies Ranocha's kinetic energy preservation condition (KEP), so there is only one variant for this method [46,65].
- ECLOG: Tadmor-type entropy-conserving method using the Tadmor entropy function $E_L = -\rho\log(p\rho^{-\gamma})$.
- ES: Skew-symmetric splitting of the inviscid flux derivative that is entropy conserving and stable using the Harten entropy function [174] and the generalized energy norm with summation-by-parts (SBP) [12,26,64].
- DS: Momentum-conserving Ducros et al. skew-symmetric split of the inviscid flux derivative [158].
- KGP: Kennedy–Gruber–Pirozzoli (KGP) skew-symmetric splitting of the inviscid flux derivative that is kinetic energy preserving [45,179,180].
- ESDS: Entropy split with Ducros et al. splitting [12].
- ESSW: Entropy split with Ducros et al. splitting but switch to regular central near discontinuities [15].

- ECLOGKP: Tadmor-type entropy conserving method using the Tadmor entropy function with Ranocha's kinetic energy preserving modification [46].
- DSKP: Ducros split with KEP.

From our previous studies [12,13,15], the Tadmor-type entropy conserving methods EC, ECLOG and ECLOGKP are the most CPU intensive methods among the nine methods. For results by EC, see [12,13,15]. It is approximately twice the CPU per time step than the ES, ESDS and ESSW methods. DS is the least CPU intensive. Comparisons of execution times were given in previous published works.

See [12,13,15] for more test case comparisons, including strong shock waves. Here, only the 2D isotropic vortex convection, 3D Taylor–Green vortex shock free turbulence and the Brio-Wu MHD shock tube test cases are selected for the nine method comparison. The comparison includes the maximum-norm error, mass conservation errors, entropy and kinetic energy vs. time for the eighth-order methods. It is noted that the performance comparison discussed here pertains to the chosen flow type, governing equation set and uniform grid spacings without any grid adaptation. The performance of the entropy split methods as a function of the split parameter β for other test cases can be found in our previous published works indicated above. Just like other current high-order method developments in the literature, the performance of the entropy split method is highly dependent on the grid size, flow type, flow condition, shock-free turbulence, turbulence with shocklets and turbulence with strong shocks. Below, only results by the eighth-order classical spatial discretization are shown. Studies by DRP and compact spatial discretizations were also performed but are not shown here. See [12–15,29] for additional investigations.

11.2. A Smooth Flow Gas Dynamics Test Case: 2D Isotropic Vortex Convection

A standard test case to examine the stability and accuracy of long-time integration for a smooth flow is the 2D isentropic vortex convection with the initial data indicated in Figure 54.

2D Isentropic Vortex Convection

(Inviscid, pure convection of the IC)

Computational Domain: Period BC, $0 \le x \le 18, 0 \le y \le 18$

Initial Condition:

$$\rho(x,y) = ((1 - \frac{(\gamma-1)\beta^2}{8\gamma\pi^2} e^{1-r^2})^{\frac{1}{\gamma-1}}$$

$$u(x,y) = u_\infty - \frac{\beta(y-y_0)}{2\pi} e^{(1-r^2)/2}$$

$$v(x,y) = v_\infty + \frac{\beta(x-x_0)}{2\pi} e^{(1-r^2)/2}$$

$$p(x,y) = \rho(x,y)^\gamma,$$

$$r^2 = x^2 + y^2, \beta = 5, \gamma = 1.4, u_\infty = 1, v_\infty = 0.$$

Figure 54. Two-dimensional isentropic vortex convection problem setup.

The exact solution is the initial data translated, $\mathbf{u}(x,t) = \mathbf{u}_0(x - u_\infty t, y - v_\infty t)$. The computational domain was of size $0 \leq x \leq 18$, $0 \leq y \leq 18$ with periodic boundary conditions. The center of the vortex is $(x_0, y_0) = (9,9)$. We integrate this test case to a final time $T = 800$ for a majority of the comparisons. Two grids 101^2 and 201^2 are considered for the comparison. For $\beta = 2$ and the coarse grid 101^2, we integrate to a final time $T = 1440$, which is more than 100 times longer than reported in the literature and twice as long as most of our previous studies with end time 720. To obtain a better time accuracy, the CFL number is set to 0.4, leading to a fairly small time step.

Investigating the level of maximum-norm error by the eighth-order ES and ESSW methods as a function of β vs. time using the two grids, Figure 55 shows that a lower maximum-norm error can be prolonged for longer time integration using $\beta = 1$ than $\beta = 2$ by the finer grid for both ES and ESSW methods. For the same grid, the maximum-norm error is highly dependent on the value of β. The error remains the same for $\beta = 1$ by both ES and ESSW methods. However, for other $\beta = 2$ and 2.5, the error using the ES method is slightly different from the ESSW method.

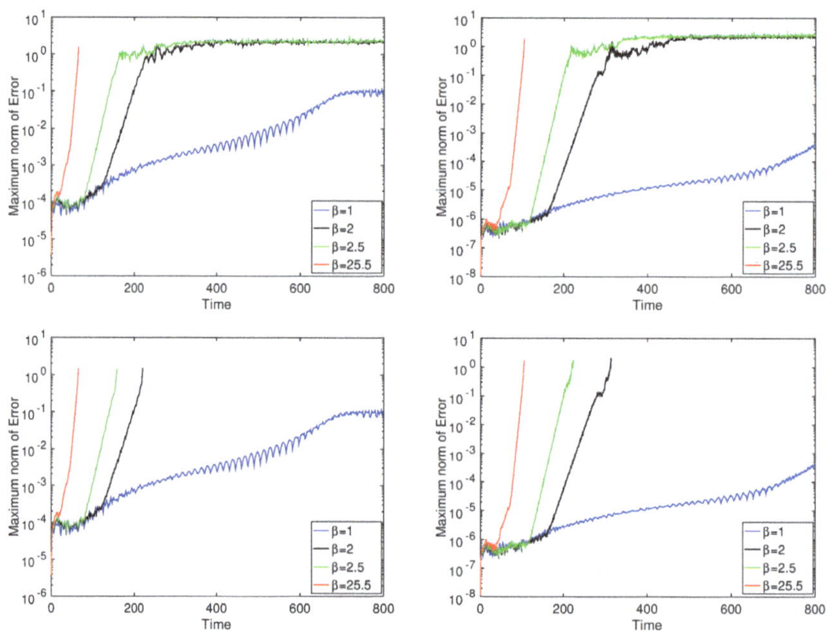

Figure 55. Two-dimensional inviscid isentropic vortex convection: Comparison of maximum-norm error vs. time for different β by eighth-order ES (**top**) and ESSW (**bottom**) using 101^2 (**left**) and 201^2 (**right**) grid points.

Figure 56 shows the maximum-norm error, mass conservation errors, entropy and kinetic energy for the nine eighth-order methods using $\beta = 1$ on the fine grid. Overall, ES performs the best, except the conservation of mass breaks down at around time 680. The kinetic energy and entropy results show the quantity with its value at time zero subtracted, e.g., the kinetic energy ($E_{kin(t)}$) shown is $E_{kin}(t) - E_{kin}(0)$.

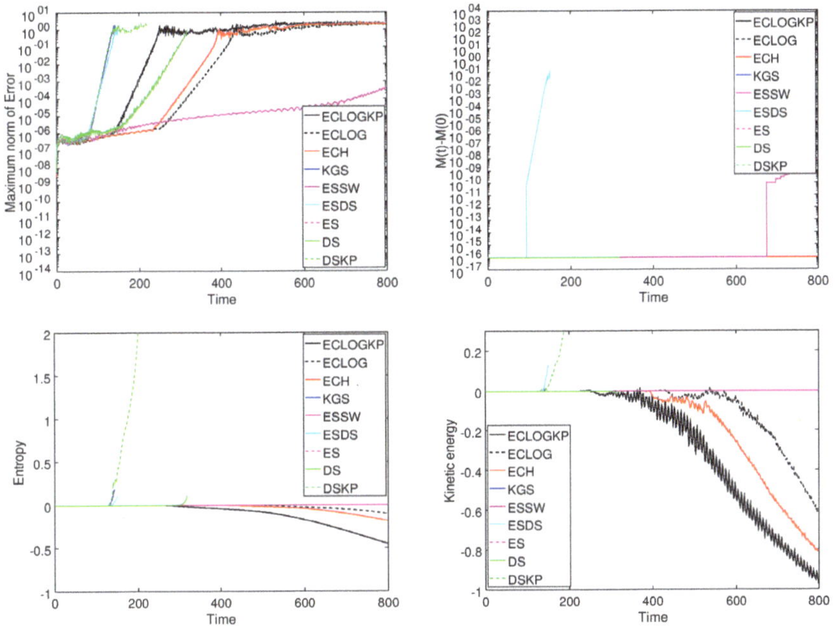

Figure 56. Two-dimensional inviscid isentropic vortex convection: comparison of maximum-norm error, mass conservation error, entropy errors and kinetic energy by E_H vs. time for the nine eighth-order methods using a fine 201^2 grid points and $\beta = 1$.

To show the performance of $\beta = 2$, Figure 57 shows the time-evolution to the final end time of $T = 1440$ using the coarse grid 100^2 in terms of maximum-norm error, the two entropies and the kinetic energy as function of time for eight methods using $\beta = 2$ in the entropy split methods. The performance by the ES method using $\beta = 2$ is very different to using $\beta = 1$ using the same grid.

The entropy E_L shows the quantity with its value at time zero subtracted. The methods KGS, DS and ESDS blow up before time 300, while the four methods that conserve entropy all run to the final time. However, the methods that do not blow up have very large errors after time 300.

From the computed results (not all results are shown here), the entropies are not perfectly conserved for longer times. The likely explanation is that the poor accuracy for long times makes time discretization error significant. This, in turn, destroys the conservation when entropy conservation holds only for the semi-discrete problem. However, as seen from the zoomed-in plots, the entropy and kinetic energy are very close to being constant, up to the time when the errors become large. This is an improvement that we expected to see with the more accurate time integration.

Extensive method comparisons were conducted in [12,13,15,30]. Results by ES and ESDS are highly dependent on the split parameter β. Only comparisons using $\beta = 1$ and $\beta = 2$ based on the study in [12] are shown here. Using $\beta = 1$ with grid refinement, the performance by ES is stable and more accurate for a longer time integration. In summary, for the 2D vortex convection case, ESDS, ESSW and KGP methods conserve entropy in a shorter time integration and are less accurate for longer time integration than the other considered methods. For ESDS and ESSW, modification from the original ES method might interfere with the entropy conservation property.

For the splitting parameter β, even for shock-free long-time integration flows, for physical argument, it is preferred to use $\beta = 1$ or greater as for $\beta < 1$, more than 50% non-conservative portion of the split flux derivative is used.

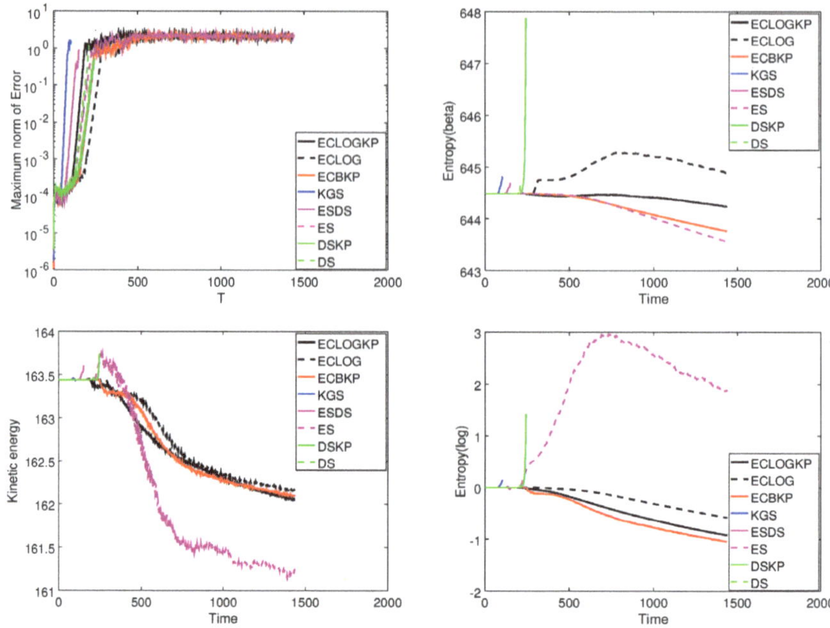

Figure 57. Two-dimensional inviscid isentropic vortex convection: Final end time of $T = 1440$ comparison of maximum-norm error, entropy E_H, entropy E_L and kinetic energy vs. time for the eight eighth-order methods using a 100^2 grid points and $\beta = 2$.

11.3. A Well-Known 3D Shock-Free Compressible Turbulence Gas Dynamics Test Case—3D Taylor–Green Vortex

The well-known shock-free compressible turbulence test case to evaluate the stability and accuracy for gas dynamics is the Taylor–Green vortex [195]. The 3D Euler equations of compressible gas dynamics are solved with $\gamma = 5/3$. The computational domain is a cube with sides of length 2π and with periodic boundary conditions in all three directions. The initial conditions are

$$\rho = 1 \quad p = 100 + \{[\cos(2z) + 2)(\cos(2x) + \cos(2y)] - 2\}/16 \tag{39}$$

$$u = \sin x \cos y \cos z, \quad v = -\cos x \sin y \cos z, \quad w = 0. \tag{40}$$

The problem is solved to time 20. In our previous studies, the uniform coarse grid DNS of 64^3 grid points solutions were compared with the filtered fine grid DNS solutions using a uniform 256^3 grid points. Here, we use the same uniform coarse grid to examine the nonlinear stability and accuracy of the three new eighth-order NEW, ESSW and ESDS methods. The total kinetic energy of the exact solution is constant in time.

Although grid refinement studies were performed and reported in [12,13,15,30], here, we only report the numerical results for a coarse grid DNS comparison among methods. It is noted that for this Taylor–Green inviscid problem, small scales are generated that eventually cause large errors in the solution due to inadequate resolution. This probably occurs around $T = 5$. Another issue is that for very low dissipative or non-dissipative numerical methods for the simulation of turbulent flows, even with extreme grid refinement, grid convergence cannot be obtained as the original inviscid Euler equations are chaotic in nature. With sufficient but not excessive numerical dissipations, one is solving the equivalent of very high Reynolds number Navier–Stokes equations. See Yee and Sjögreen [196] for a study. The end time is 20 instead of the standard end time 10 to observe the solution behavior twice as long using the same RK4 time discretization and CFL number 0.4. Harten's entropy was used for all schemes except ECLOG and ECLOGKP, where we used the log-entropy. Furthermore, the entropy $E(t)$ is normalized to show deviation from the initial entropy. Thus, $E(t) - E(0)$ is shown in the figures.

Figure 58 shows the comparison of kinetic energy, maximum-norm error and entropy vs. time for the nine eighth-order methods. It is interesting to see the behavior of doubling the time integration duration for such coarse grid DNS computations.

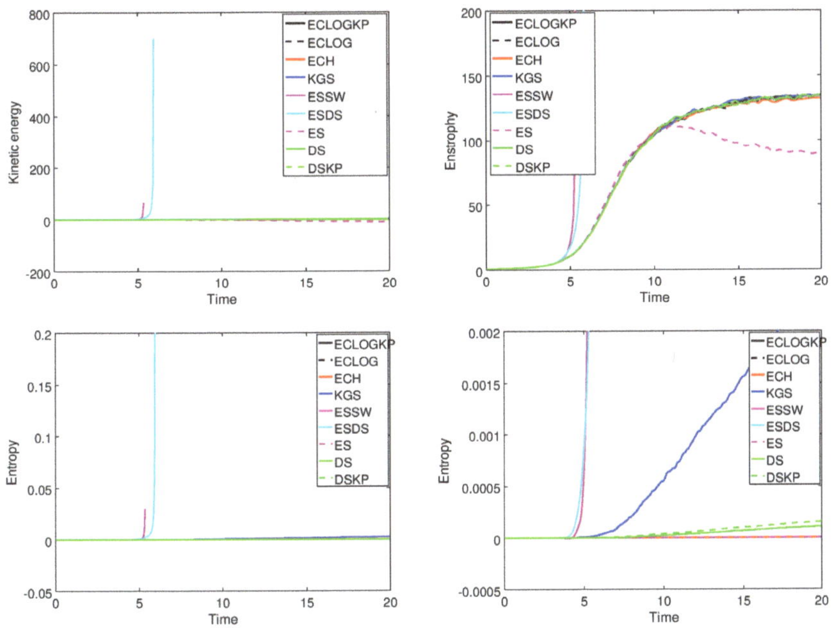

Figure 58. Three-dimensional inviscid Taylor–Green vortex using 64^3 grid points: Comparison of kinetic energy (**top left**), enstrophy (**top right**), entropy (**bottom left**) and entropy (closed up, **bottom right**) vs. time for the nine eighth-order methods using $\beta = 2$.

Method ESDS becomes unstable at around time 6, and method DSKP becomes unstable around time 7. All other schemes ran to completion. Again, the kinetic energy and entropy results show the quantity with its value at time zero subtracted, e.g., the kinetic energy ($E_{kin(t)}$) shown is $E_{kin}(t) - E_{kin}(0)$. The ES method starts to lose some energy at a later time. Otherwise, the stable results are similar. ECLOGKP and KGS are indistinguishable and fall on top of each other in the zoomed-in figure. One surprising result is that method ECBKP is expected to preserve kinetic energy in the same manner but does not. The other stable methods are a little off, but this is only visible in the closeup. Methods ES and ECBKP fall on top of each other, as we would expect, since these two schemes conserve the entropy in a discretized sense. Method ECLOGKP is also on top of ES and ECBKP, making it hard to visualize the differences in the results.

Not included here are studies illustrating that the logarithmic entropy function conserved Harten's entropy almost perfectly. Methods ECLOGKP and ECLOG conserve entropy, as illustrated in the figure, where their solutions are on top of each other. Overall, ECLOGKP, ECLOG, ECBKP, KGS and DS are very similar. One has to zoom in very much in this region to see any differences for this test case. However, differences might be larger for other flow problems. As can be seen for this test case, method ES behaves somewhat different. DSKP and ESDS are not performing well. It is noted again that results by ES and ESDS are highly dependent on β. Only results with $\beta = 2$ are shown based on the study in [12].

Figure 59 compares how well these eighth-order methods conserve mass and momentum in the x-, y- and z-directions. The figures show total mass (or momentum) as a function of time, normalized so that the initial mass (or momentum) is zero. For the mass-conservation plot, all schemes except ES, ESSW and ESDS are perfectly conservative. The normalized $M(t) - M(0)$ is zero as long as the method is stable for all methods except the ES scheme, which is around e^{-9} for most of the time. In addition, these figures are shown in the log-scale, where we artificially set the schemes with zero conservation error to e^{-16}.

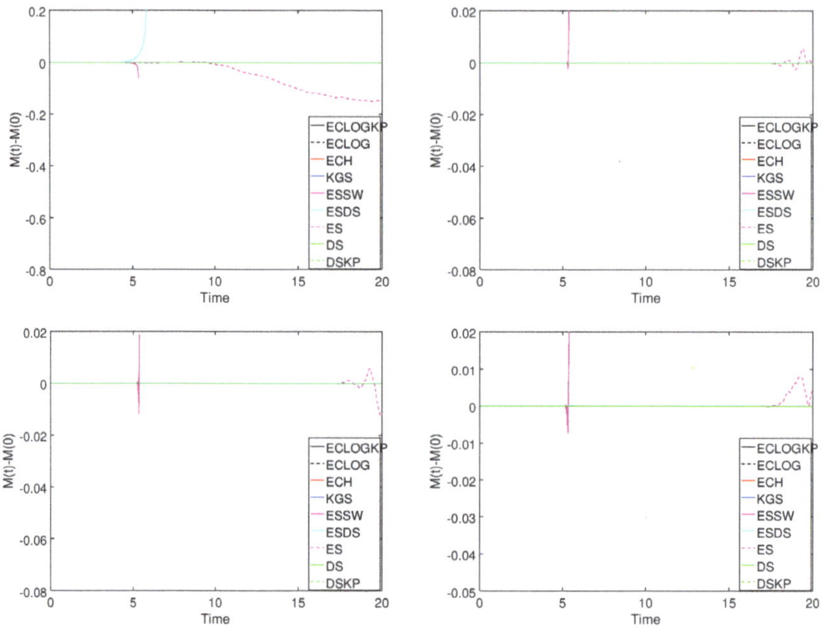

Figure 59. Three-dimensional inviscid Taylor–Green vortex using 64^3 grid points: comparison of mass conservation (**top left**), x-momentum conservation (**top right**), y-momentum conservation (**bottom left**) and z-momentum conservation (**bottom right**) vs. time for the nine eighth-order methods using $\beta = 2$.

The Taylor–Green vortex test case has a similar method comparison conclusion as the vortex convection test case. ESDS, ESSW and KGP methods conserve entropy in a shorter time integration and are less accurate for longer time integration than the other considered methods. For ESDS and ESSW, their modification from the original ES method might interfere with the entropy conservation property.

For the extension of these high-order methods for the ideal MHD in curvilinear grids with method comparisons, see [30].

Only the key numerical results of a very difficult Brio–Wu MHD shock tube problem is briefly discussed here. The Brio–Wu MHD shock tube problem involves two fast rarefaction waves, a slow compound wave, a contact discontinuity and a slow shock wave. According to the Flash code [197,198], standard high-order shock-capturing methods exhibit oscillatory solutions. Examples are PPM (piecewise parabolic method of Collela and Woodward [199]) and third- to fifth-order WENO (WENO3–WENO5). Most researchers resorted to use first order or very diffusive second-order methods with local Lax–Freidrichs (LLF) or variants of the HLL-type numerical fluxes [200,201]. The oscillations increase with an increase in B_y, with the reason being that stronger B_y introduces a more transverse effect that resists shock propagation in the x-direction, causing the shock to move slowly. Our simulation using our high-order method provides higher accuracy with decreased oscillation at the discontinuity regions than the Flash code results. Details of the test case can be found in [16,30].

11.4. A Short Summary of Comparative Performance of High-Order Skew-Symmetric SPM

In summary, for the chosen test cases, the long-time integration performance of the entropy split methods as a function of the split parameter β is dependent on the flow type, flow condition and the grid spacing. For finer grids, the stability of long-time integration for the same β value can be different. The general guideline is that for very smooth pure convection flows, $\beta = 1$ (50% conservative and 50% non-conservative of the split term) or β between 1 and 2 can improve the stability tremendously without added numerical dissipation. A finer grid in general prolongs stability for longer time integration. The split method conserves mass and entropy. For problems with shock waves, β between 20 and 26 performs well with the NEW and ESSW methods in conjunction with the Yee et al. and Yee and Sjögreen nonlinear filter approach.

The relative stability and accuracy are comparable among the KGP-type, DS-type split methods and the ES and ESSW methods. In certain test cases, ES and ESSW are more stable for longer time integration. KGP-type and DS-type methods are the least CPU intensive for both the gas dynamics and MHD test cases. However, they do not conserve entropy. For the difficult Brio–Wu test case, unlike most published work reported in the literature, the current nonlinear filter approach in conjunction with the various splitting methods indicated that stability and high accuracy were obtained using very high-order spatial discretization and the Roe-type approximate Riemann solver. The numerical instability and highly spurious oscillatory solutions by standard high-order shock-capturing methods are drastically reduced by the ES and ESSW methods. With the current studies and from our previous studies [12,13,15,202], EC, ECLOG and ECLOGKP are the most CPU intensive methods among the nine methods. They have at least twice the CPU usage per time step than the ES, ESDS and ESSW methods and yet exhibit similar resolutions. DS is the least CPU intensive.

For a generalization to a wider class of entropy split methods for compressible ideal MHD using the two-point numerical flux with numerical examples, see [16],

12. Concluding Remark

Some building blocks to ensure a higher level of confidence in PAR of numerical simulations have been discussed. The discussion concentrates only on how well numerical schemes can mimic the solution behavior of the underlying PDEs. The possible discrepancy between the chosen model and the real physics and/or experimental data is set aside. These building blocks are based largely on the author's view, background and integrated experience in computational physics, numerical analysis and the dynamics of numerics. They also represent the end result of the various studies with the author's collaborators indicated in the acknowledgment section. Among many other important building blocks for the PAR of numerical simulations, the author believes the following five building blocks are essential. The first building block is to analyze as much as possible the dynamical behavior of the governing equation. For stability and well-posedness considerations, whenever it is possible, it is also necessary to condition (not pre-condition) the governing PDEs before the application of the appropriate scheme (Yee and Sjögreen [121,122]). The second building block is to understand the nonlinear behavior, limits and barriers and to isolate the spurious behavior of existing numerical schemes. The third building block is to include nonlinear dynamics and bifurcation theories as an integral part of the numerical process whenever it is possible. The fourth building block is to construct appropriate adaptive spatial and temporal discretizations that are suitable for the underlying governing equation. The last building block is to construct appropriate adaptive numerical dissipation/filter controls for long-time integrations, and complex high-speed turbulent and combustion simulations (Sjögreen and Yee 2001 [10], Yee and Sjögreen 2001 [120]).

We have revealed some of the causes of spurious phenomena due to the numerics in an attempt to improve the understanding of the effects of numerical uncertainties in CFD. We have shown that guidelines developed using linearization methods are not always valid for nonlinear problems. We have gained an improved understanding of long-time behavior of nonlinear problems and the nonlinear stability, convergence and reliability of time-marching approaches. We have learned that numerics can introduce and suppress chaos and can also introduce chaotic transients. The danger of relying on DNS to closely bracket the onset of turbulence and chaos is evident.

We have shown that employing theories of nonlinear dynamics to guide the construction of more appropriate, stable and accurate numerical methods could help (a) delineate solutions of the discretized counterparts but not solutions of the governing equations; (b) prevent numerical chaos/turbulence leading to FALSE predication of transition to turbulence; (c) provide reliable numerical simulations of nonlinear fluid dynamical systems, especially by direction numerical simulations (DNS), large eddy simulations (LES) and implicit large eddy simulations (ILES) simulations; and (d) prevent incorrect computed shock speeds for problems containing stiff nonlinear source terms, if present.

From the wide varieties of test cases for compressible gas dynamics and MHD performed in [1–3], the performance of the blending of more than one method is highly dependent on the flow physics. Regardless of the choice of these blendings of high-order methods, these test cases illustrate the improved stability and accuracy over the standalone single high-order method approach. Nine eighth-order structure-preserving methods were compared for a wide range of test cases in order to illustrate the effectiveness of the blending of selected numerical method approaches. These current trends of adaptive blending of high-order, structure-preserving non-dissipative methods with high-resolution, high-order shock-capturing methods are also applicable to discontinuous Galerkin method, and finite element methods and spectral element methods. In addition, our studies indicated that these current approaches can also improved stability and accuracy in rapidly developed unsteady flows containing steep gradients, and shear and shock waves with the benefit of minimizing added numerical dissipations.

Our future work is planned to develop newer methods catering to some of the aforementioned disciplines with the knowledge gained from these nonlinear studies.

Funding: Financial support from the DOE/SciDAC program and the NASA TTT/RCA program for the first author is gratefully acknowledged. The authors are grateful to Murray Tobak, Marcel Vinokur, Karim Shariff, Dinesh Prabhu, Jasim Ahmad and Alan Wray of NASA Ames Research Center for the numerous invaluable discussions throughout the course of this work. Special thanks to W.L. Oberkampf of his support of our work in the early 2000s and all of the collaborators—A. Harten, R. F. Warming, D.V. Kotov, C.-W. Shu, P.L. Roe, A. Lafon, W. Wang, R.J. LeVeque, N.D. Sandham, J.M. Djomehri, D.F. Griffiths, A.M. Stuart, M. Vinokur, L. Keefe, J.R. Torczynski, P. Fischer, R.D. Henderson, M. Poliashenko, J.-L. Montagn'e, A.A. Wray, A. Hadjadj, A. Kritsuk, T. Magin, and M. Panesi, D.K. Prabhu, G.H. Klopfer, M.F. Barone, S.A. Morton and M.R. Visbal—during the various stages of research development. The 2nd and 3rd of this research received no external funding.

Conflicts of Interest: Author Björn Sjogreen was employed by the company MultiD Analyses AB. The work done on this work is not supported by MultiD Analyses AB, but on his own time. All authors declare that the research was conducted in the absence of any commercial or financial relationships that could be construed as a potential conflict of interest.

References

1. Yee, H.C.; Sweby, P.K. *Dynamics of Numerics & Spurious Behaviors in CFD Computations*; RIACS Technical Report 97.06; NASA AMES Research Center: Mountain View, CA, USA, 1997.
2. Yee, H.C. Building Blocks for Reliable Complex Nonlinear Numerical Simulations. In *Turbulent Flow Computation*; Drikakis, D., Geurts, B., Eds.; Springer: New York, NY, USA, 2002.
3. Yee, H.C.; Sjögreen, B.; Kotov, D.V. Numerical Dissipation Control in High Order Methods for Compressible Turbulence: Recent Development. *Fluids* **2024**, *9*, 127. [CrossRef]

4. Kotov, D.V.; Yee, H.C.; Wray, A.A.; Hadjadj, A.; Sjögreen, B. High Order Numerical Methods for the Dynamic SGS Model of Turbulent Flows with Shocks. *Commun. Comput. Phys.* **2016**, *19*, 273–300. [CrossRef]
5. Kotov, D.V.; Yee, H.C.; Wray, A.A.; Sjögreen, B.; Kritsuk, A.G. Numerical dissipation control in high order shock-capturing schemes for LES of low speed flows. *J. Comput. Phys.* **2016**, *307*, 189–202. [CrossRef]
6. Lafon, A.; Yee, H.C. Dynamical Approach Study of Spurious Steady-State Numerical Solutions for Nonlinear Differential Equations, Part III: The Effects of Nonlinear Source Terms in Reaction-Convection Equations. *Comp. Fluid Dyn.* **1996**, *6*, 1–36. [CrossRef]
7. Lafon, A.; Yee, H.C. Dynamical Approach Study of Spurious Steady-State Numerical Solutions of Nonlinear Differential Equations, Part IV: Stability vs. Numerical Treatment of Nonlinear Source Terms. *Comput. Fluid Dyn.* **1996**, *6*, 89–123. [CrossRef]
8. LeVeque, R.J.; Yee, H.C. A Study of Numerical Methods for Hyperbolic Conservation Laws with Stiff Source Terms. *J. Comp. Phys.* **1990**, *86*, 187–210. [CrossRef]
9. Sandham, N.D.; Li, Q.; Yee, H.C. Entropy Splitting for High-Order Numerical Simulation of Compressible Turbulence. *J. Comput. Phys.* **2002**, *23*, 307–322.
10. Sjögreen, B.; Yee, H.C. Multiresolution Wavelet Based Adaptive Numerical Dissipation Control for Shock-Turbulence Computation. *J. Sci. Comput.* **2004**, *20*, 211–255. [CrossRef]
11. Sjögreen, B.; Yee, H.C. Accuracy Consideration by DRP Schemes for DNS and LES of Compressible Flow Computations. *Comput. Fluids* **2017**, *159*, 123–136. [CrossRef]
12. Sjögreen, B.; Yee, H.C. Entropy Stable Method for the Euler Equations Revisited: Central Differencing via Entropy Splitting and SBP. *Sci. Comput.* **2019**, *81*, 1359–1385. [CrossRef]
13. Sjögreen, B.; Yee, H.C.; Kotov, D.V.; AKritsuk, A.G. Skew-Symmetric Splitting for Multiscale Gas Dynamics and MHD Turbulence Flows. *Sci. Comput.* **2020**, *83*, 43. [CrossRef]
14. Sjögreen, B.; Yee, H.C. High Order Compact Central Spatial Discretization Under the Framework of Entropy Split Methods. In Proceedings of the ICOSAHOM21, Vienna, Austria, 12–16 July 2021.
15. Sjögreen, B.; Yee, H.C. Construction of Conservative Numerical Fluxes for the Entropy Split Method. *Comm. Appl. Math. Comput.* **2021**, *5*, 653–678. [CrossRef]
16. Sjögreen, B.; Yee, H.C. Generalization to a Wider Class of Entropy Split Methods for Compressible Ideal MHD. *Comput. Fluids* **2024**, *268*, 106087. [CrossRef]
17. Vinokur, M.; Yee, H.C. Extension of Efficient Low Dissipation High-Order Schemes for 3D Curvilinear Moving Grids. In *Frontiers of Computational Fluid Dynamics, Proceedings of the Robert MacCormack 60th Birthday Conference, 26–28 June 2000*; NASA/TM-2000-209598; World Scientific: Half Moon Bay, CA, USA, 2002.
18. Wang, W.; Yee, H.C.; Shu, C.-W.; Yee, H.C.; Sjögreen, B. High-Order well-balanced schemes and applications to non-equilibrium flow. *J. Comput. Phys.* **2009**, *228*, 6682–6702 [CrossRef]
19. Wang, W.; Yee, H.C.; Sjögreen, B.; Magin, T.; Shu, C.W. Construction of Low Dissipative High-Order Well-Balanced Filter Schemes for Nonequilibrium Flows. *J. Comput. Phys.* **2011**, *230*, 4316–4335. [CrossRef]
20. Wang, W.; Shu, C.W.; Yee, H.C.; Sjögreen, B. High Order Finite Difference Methods with Subcell Resolution for Advection Equations with Stiff Source Terms. *J. Comput. Phys.* **2012**, *231*, 190–214. [CrossRef]
21. Yee, H.C. *A Class of High-Resolution Explicit and Implicit Shock-Capturing Methods*; VKI Lecture Series 1989-04; NASA: Washington, DC, USA, 1989.
22. Yee, H.C.; Klopfer, G.H.; Mongtagné, J.-L. High-Resolution Shock-Capturing Schemes for Inviscid and Viscous Hypersonic Flows. *J. Comput. Phys.* **1990**, *88*, 31–61. [CrossRef]
23. Yee, H.C.; Sweby, P.K.; Griffiths, D.F. Dynamical Approach Study of Spurious Steady-State Numerical Solutions for Nonlinear Differential Equations, Part I: The Dynamics of Time Discretizations and Its Implications for Algorithm Development in Computational Fluid Dynamics, NASA TM-102820, April 1990. *J. Comput. Phys.* **1991**, *97*, 249–310.
24. Yee, H.C.; Sweby, P.K. Dynamical Approach Study of Spurious Steady-State Numerical Solutions for Nonlinear Differential Equations, Part II: Global Asymptotic Behavior of Time Discretizations. *Int. J. Fluid Dyn.* **1995**, *4*, 219–283. [CrossRef]
25. Yee, H.C.; Sandham, N.D.; Djomehri, M.J. Low-Dissipative High Order Shock-Capturing Methods Using Characteristic-Based Filters. *J. Comput. Phys.* **1999**, *150*, 199–238. [CrossRef]
26. Yee, H.C.; Vinokur, M.; Djomehri, M.J. Entropy Splitting and Numerical Dissipation. *J. Comp. Phys.* **2000**, *162*, 33–81. [CrossRef]
27. Yee, H.C.; Sjögreen, B. Development of Low Dissipative High Order Filter Schemes for Multiscale Navier-Stokes and MHD Systems. *J. Comput. Phys.* **2007**, *225*, 910–934. [CrossRef]
28. Yee, H.C.; Sjögreen, B. High Order Filter Methods for Wide Range of Compressible Flow Speeds. In Proceedings of the ICOSAHOM09, Trondheim, Norway, 22–26 June 2009.
29. Yee, H.C.; Sjögreen, B. Comparative Study on a Variety of Structure-Preserving High Order Spatial Discretizations with the Entropy Split Methods for MHD. In Proceedings of the ICOSAHOM21, Virtual, 12–16 July 2021.
30. Yee, H.C.; Sjögreen, B. Recent Advancement of Entropy Split Methods for Compressible Gas Dynamics and MHD. *J. Appl. Math. Comput.* **2023**, *463*, 127545. [CrossRef]
31. Glimm, J.; Cheng, B.; Sharp, D.H.; Kaman, T. A Crisis for the Verification and Validation of Turbulence Simulations. *Phys. D Nonlinear Phenom.* **2020** *404*, 132346. [CrossRef]

32. Doedel, E.J.; Beyn, W.-J. Stability and Multiplicity of Solutions to Discretizations of Nonlinear Ordinary Differential Equations. *SIAM J. Sci. Stat. Comput.* **1981**, *2*, 107–120.
33. Ehrenstein, U.; Koch, W. Nonlinear bifurcation study of plane Poiseuille flow. *J. Fluid Mech.* **1991**, *228*, 111–148.
34. Feigenbaum, M.J. Quantitative Universality for a Class of Nonlinear Transformations. *J. Stat. Phys.* **1978**, *19*, 25–52. [CrossRef]
35. Griffiths, D.G.; Stuart, A.M.; Yee, H.C. Numerical Wave Propagation in Hyperbolic Problems with Nonlinear Source Terms. *SIAM J. Numer. Anal.* **1992**, *29*, 1244–1260. [CrossRef]
36. Jackson, E.A. *Perspectives of Nonlinear Dynamics*; Cambridge University Press: New York, NY, USA, 1989; Volume I.
37. Mitchell, A.R.; Griffiths, D.F. *Beyond the Linearized Stability Limit in Non Linear Problems*; Report NA/88; Department of Mathematical Sciences, University of Dundee: Scotland, UK, 1985.
38. Stewart, I. Bifurcation Theory Old and New. In Proceedings of the IMA Conferences on Dynamics of Numerics and Numerics of Dynamics, Bristol, UK, 31 July–2 August 1990.
39. Stuart, A.M. *Numerical Analysis of Dynamical Systems*; Cambridge University Press: Cambridge, UK, 1994.
40. Kotov, D.V.; Yee, H.C.; Pansei, M.; Prabhu, D.K.; Wray, A.A. Computational Challenges for Simulations Related to the NASA Electric Arc Shock Tube (EAST) Experiments. *J. Comput. Phys.* **2014**, *269*, 215–233. [CrossRef]
41. Wang, W.; Shu, C.W.; Yee, H.C.; Kotov, D.V.; Sjögreen, B. High order finite difference methods with subcell resolution for stiff multispecies detonation capturing. *Comm. Comput. Phys.* **2015**, *17*, 317–336. [CrossRef]
42. Yee, H.C.; Kotov, D.V.; Wang, W.; Shu, C.H. Spurious behavior of shock-capturing methods by the fractional step approach: Problems containing stiff source terms and discontinuities. *J. Comput. Phys.* **2013**, *241*, 266–291. [CrossRef]
43. Azevedo, J.; Ramalho, M.; Azevedo, J. An Investigation into the Origin of the Carbuncle Phenomenon in High Speed CFD Calculations. In Proceedings of the COBEM, Natal, RN, Brazil, 24–28 October 2011.
44. Humphries, A.R. Numerical Analysis of Dynamical Systems. Ph.D. Thesis, University of Bath, Bath, UK, 1993.
45. Kennedy, C.A.; Gruber, A. Reduced Aliasing Formulations of the Convective Terms Within the Navier-Stokes Equations. *J. Comput. Phys.* **2008**, *227*, 1676–1700. [CrossRef]
46. Ranocha, H. Entropy Conserving and Kinetic Energy Preserving Numerical Methods for the Euler Equations Using Summation-by-Parts Operators. In Proceedings of the ICOSAHOM-2018, Imperial College, London, UK, 9–13 July 2018.
47. Yee, H.C.; Shinn, J.L. Semi-Implicit and Fully Implicit Shock-Capturing Methods for Nonequilibrium Flows. In Proceedings of the A/AA 8th Computational Fluid Dynamics Conference, Honolulu, HI, USA, 9–11 June 1987; McCroskey, W.J., Ed.; AIAA-87-1116-CP; AIAA: Washington, DC, USA, 1987; p. 159.
48. Quirk, J.J. A contribution to the great Riemann solver debate, Intern. *J. Num. Meth. Fluids* **1994**, *18*, 55–574.
49. Sanders, R.; Morano, E.; Druguet, M.-C. Multidimensional Dissipation for Upwind Schemes: Stability and Applications to Gas Dynamics. *J. Comput. Phys.* **1998**, *145*, 511–537. [CrossRef]
50. Fleischmann, N.; Adami, S.; Hu, X.Y.; Adams, N.A. A low dissipation method to cure the grid-aligned shock instability. *J. Comput. Phys.* **2020**, *401*, 109004. [CrossRef]
51. Hu, L.; Feng, Z. Development of a Carbuncle-free and low-dissipation Roe-type Scheme: Applications to Multidimensional Euler Flows. *Commun. Nonlinear Sci. Num. Simul.* **2023**, *116*, 106798. [CrossRef]
52. Kemm, F. Heuristical and Numerical Considerations for the Carbuncle Phenomenon. *arXiv* **2017**, arXiv:1507.00666v2. [CrossRef]
53. Montagne, J.-L.; Yee, H.C.; Klopfer, G.H.; Vinokur, M. Hypersonic blunt body computations including real gas effects. In *Nonlinear Hyperbolic Equations—Theory, Computation Methods, and Applications, Proceedings of the 2nd International Conference in Hyperbolic Problems, Aachen, Germany, 14–18 March 1988*; Ballmann, J., Jeltsch, R., Eds.; Vieweg+Teubner Verlag: Wiesbaden, Germany, 1989.
54. Shearer, M.; Schaeffer, D.G.; Marchesin, D.; Paes-Leme, P. Solution of the Riemann Problem for a Prototype 2×2 System of Non-Strictly Hyperbolic Conservation Laws. *Arch. Rat. Mech. Anal.* **1987**,*97*, 299–320. [CrossRef]
55. Sullivan, P.P.; Banner, M.L.; Morison, R.P. Turbulent Flow over Steep Steady and Unseady Waves under Strong Wind Forcing. *J. Phys. Ocean.* **2018**, *48*, 3–27. [CrossRef]
56. Olsson, P. Summation by Parts, Projections, and Stability. I. *Math. Comp.* **1995**, *64*, 1035–1065. [CrossRef]
57. Sjögreen, B.; Yee, H.C. On Tenth-Order Central Spatial Schemes. In Proceedings of the TSFP-5, Munich, Germany, 27–29 August 2007.
58. Strand, B. Summation by Parts for Finite Difference Approximations for d/dx. *J. Comput. Phys.* **1996**, 110, 47–97. [CrossRef]
59. Sweby, P.K.; Yee, H.C. On the Dynamics of Some Grid Adaptation Schemes. In Proceedings of the 4th International Conference on Numerical Grid Generation in CFD and Related Fields, Swansea, UK, 6–8 April 1994; RIACS Technical Report 94.02; University College of Swansea: Swansea, UK, 1994.
60. Sweby, P.K.; Lafon, A.; Yee, H.C. On the Dynamics of Computing a Chemically Relaxed Nonequilibrium Flow. In Proceedings of the ICFD Conference on Numerical Methods for Fluid Dynamics, Oxford, UK, 3–6 April 1995.
61. Yee, H.C.; Sweby, P.K. *Global Asymptotic Behavior of Iterative Implicit Schemes*; RIACS Technical Report 93.11; NASA AMES Research Center: Mountain View, CA, USA, 1994.
62. Yee, H.C.; Sweby, P.K. *Dynamical Approach Study of Spurious Steady-State Numerical Solutions for Nonlinear Differential Equations, II. The Dynamics of Numerics of Systems of 2 × 2 ODEs and Its Connection to Finite Discretizations of Nonlinear PDEs*; RNR-92-008; NASA Ames Research Center: Mountain View, CA, USA, 1992.

63. Yee, H.C.; Torczynski, J.R.; Morton, S.A.; Visbal, M.R. ; Sweby, P.K. On Spurious Behavior of CFD Simulations, AIAA 97-1869. In Proceedings of the 13th AIAA Computational Fluid Dynamics Conference, Snowmass, CO, USA, 29 June–2 July 1997. *Int. J. Num. Meth. Fluids* **1999**, *30*, 675–711. [CrossRef]
64. Sjögreen, B.; Yee, H.C. An entropy stable method revisited: Central differencing via entropy splitting and SBP. In Proceedings of the ICOSAHOM-2018, London, UK, 9–13 July 2018.
65. Sjögreen, B.; Yee, H.C. High order entropy conservative central schemes for wide ranges of compressible gas dynamics and MHD flows. *J. Comput. Phys.* **2018**, *364*, 153–185. [CrossRef]
66. Yee, H.C.; Sweby, P.K. On Spurious Behavior of Super-Stable Implicit Methods. *Intern. J. CFD* **1997**, *8*, 265–286. [CrossRef]
67. Oberkampf, W.L.; Roy, C.J. *Verification and Validation in Scientific Computing*; Cambridge University Press: Cambridge, UK, 2010.
68. Beyn, W.-J.; Lorenz, J. Spurious Solutions for Discrete Superlinear Boundary Value Problems. *Computing* **1982**, *28*, 43–51. [CrossRef]
69. Doedel, E. *AUTO: Software for Continuation and Bifurcation Problems in Ordinary Differential Equations*; California Institute of Technology: Pasadena, CA, USA, 1986.
70. Doedel, E. *AUTO: Software for Continuation and Bifurcation Problems in Ordinary Differential Equations*; Concordia University: Montreal, QC, Canada; California Institute of Technology: Pasadena, CA, USA, 2000.
71. Kellogg, R.B.; Shubin, G.R.; Stephens, A.B. Uniqueness and the Cell Reynolds Number. *SIAM J. Numer. Anal.* **1980**, *17*, 733–739. [CrossRef]
72. Peitgen, H.-O.; Saupe, D.; Schmitt, K. Nonlinear Elliptic Boundary Value Problems versus Their Finite Difference Approximations: Numerically Irrelevant Solutions. *J. Reine Angew. Mathematik* **1981**, *322*, 74–117.
73. Schreiber, R.; Keller, H.B. Spurious Solution in Driven Cavity Calculations. *J. Comput. Phys.* **1983**, *49*, 310–333. [CrossRef]
74. Shubin, G.R.; Stephens, A.B.; Glaz, H.M. Steady Shock Tracking and Newton's Method Applied to One-Dimensional Duct Flow. *J. Comput. Phys.* **1981**, *39*, 364–374. [CrossRef]
75. Griffiths, D.F.; Mitchell, A.R. *Stable Periodic Solutions of a Nonlinear Partial Difference Equation in Reaction Diffusion*; Report NA/113; Department of Mathematics and Computer Science, University of Dundee: Scotland, UK, 1988.
76. Iserles, A. *Stability and Dynamics of Numerical Methods for Nonlinear Ordinary Differential Equations*; DAMTP 1988/NAI; University of Cambridge: Cambridge, UK, 1988.
77. Lorenz, E.N. Computational Chaos—A Prelude to Computational Instability. *Physica D* **1989**, *35*, 299–317. [CrossRef]
78. Prüfer, M. Turbulence in Multistep Methods for Initial Value Problems. *SIAM J. Appl. Math.*, **1985**, *45*, 32–69. [CrossRef]
79. Sanz-Serna, J.M.; Vadillo, F. *Nonlinear Instability, the Dynamic Approach*; Watson, G.A., Griffiths, D.F., Eds.; Pitman: London, UK, 1985.
80. Sanz-Serna, J.M. Studies in Numerical Nonlinear Instability I: Why do leapfrog schemes go unstable? *SIAM J. Sci. Stat. Comput.* **1985**, *6*, 923–938. [CrossRef]
81. Sanz-Serna, J.M. *Numerical Ordinary Differential Equations vs. Dynamical Systems*; Applied Mathematics and Computation Report 1990/3; Universidad de Valladolid: Valladolid, Spain, 1990.
82. Aves, M.A.; Griffiths, D.F.; Higham, D.J. Does Error Control Suppress Spuriosity? *SIAM J. Num. Anal.* **1995**, *34*, 756–778. [CrossRef]
83. Corless, R.M. Error Backward. *Contemp. Math.* **1994**, *172*, 31–62.
84. Corless, R.M. What Good Are Numerical Simulations of Chaotic Dynamical Systems? *Comput. Math. Appl.* **1994**, *28*, 107–121. [CrossRef]
85. Dieci, L.; Estep, D. Some Stability Aspects of Schemes for the Adaptive Integration of Stiff Initial Value Problems. *SIAM J. Sci. Stat. Comput.* **1991**, *12*, 1284–1303. [CrossRef]
86. Hairer, E.; Iserles, A.; Sanz-Serna, J.M. Equilibria of Runge-Kutta Methods. *Numer. Math.* **1989**, *58*, 243–254. [CrossRef]
87. Poloashenko, M.; Aidun, C.K. A direct method for computation of simple bifurcations. *J. Comput. Phys.* **1995**, *121*, 246–260. [CrossRef]
88. Stuart, A.M. Probabilistic and Deterministic Convergence Proofs for Software for Initial Value Problems. *Numer. Algorithms* **1997**, *14*, 227–260. [CrossRef]
89. Butcher, J.C. *Numerical Analysis of Ordinary Differential Equations*; John Wiley & Son: Chichester, UK, 1987.
90. Yee, H.C.; Sweby, P.K. *Nonlinear Dynamics & Numerical Uncertainties in CFD*; NASA Technical Memorandum; NASA Ames Research Center: Mountain View, CA, USA, 1996; p. 110398.
91. Bhattacharyya, B. Uncertainty qantification of dynamical systems by a POD-Kriging surrogate model. *J. Comput. Sci.* **2022**, *60*, 101602. [CrossRef]
92. Frohlich, B.; Hose, D.; Dieterich, O.; Hanss, M.; Eberhard, P. Uncertainty quantification of large-scale dynamical systems using parametric model order reducion. *Mech. Syst. Signal Process.* **2022**, *171*, 108855. [CrossRef]
93. Engquist, B.; Sjögreen, B. High Order Shock Capturing Methods. In *CFD Reviews*; Hafez, M., Oshima, K.; Eds.; John Wiley: New York, NY, USA, 1995; pp. 210–233.
94. Moujaes, P.; Kuzmin, D. Monolithic convex limiting and implicit pseudo-time stepping for calculating stead-state solutions of the Euler Equations. *arXiv* **2024**, arXiv:2407.03746v1.
95. Zhu, J.; Shu, C.-W.; Qiu, J. RKDG Methods with Multi-resolution WENO Limiters for Solving Steady-State Problems on Triangular Meshes. *Comm. Appl. Math. Comput.* **2023**, *6*, 1575-1599. [CrossRef]

96. Toro, E.F.; Hidalgo, A. ADER finite volume schemes for nonlinear reaction-diffusion equations. *Appl. Numer. Math.* **2009**, *59*, 73–100. [CrossRef]
97. Toro, E.F. *Riemann Solver & Numerical Methods for Fluid Dynamics*; Springer: Berlin/Heidelberg, Germany, 1997; ISBN 978-3-662-03492-7.
98. Crocco, L. A Suggestion for the Numerical Solution of the Steady Navier-Stokes Equations. *AIAA J.* **1965**, *3*, 1824–1832. [CrossRef]
99. Moretti, G.; Abbett, M. A Time-Dependent Computational Method for Blunt Body Flows. *AIAA J.* **1966**, *4*, 2136–2141. [CrossRef]
100. MacCormack, R.W. *The Effect of Viscosity in Hypervelocity Impact Cratering*; AIAA Paper No. 69–354; American Institute of Aeronautics and Astronautics (AIAA): Cincinnati, OH, USA, 1969.
101. Beam, R.M.; Warming, R.F. An implicit Factored Scheme for the Compressible Navier-Stokes Equations. *AIAA J.* **1978**, *16*, 293–402. [CrossRef]
102. Briley, W.R.; McDonald, H. Solution of the multidimensional compressible Navier-Stokes equations by a generalized implicit method. *J. Comp. Phys.* **1977**, *24*, 372–397. [CrossRef]
103. Steger, J. Implicit Finite-Difference Simulation of Flow about Arbitrary Two Dimensional Geometries. *AIAA J.* **1978**, *16*, 679–686. [CrossRef]
104. Yee, H.C.; Sweby, P.K. On Super-Stable Implicit Methods and Time-Marching Approaches; RIACS Technical Report 95.12; NASA AMES Research Center: Mountain View, CA, USA, 1995; also, Proceedings of the Conference on Numerical Methods for Euler and Navier-Stokes Equations, Montreal, QC, Canada, 14–16 September 1995. *Int. J. Comput. Fluid Dyn.* **1997**, *8*, 265–286.
105. Gimperlain, H.; Grinfeld, M.; Knops, R.J.; Slemrod, M. Non-Uniqueness in Plan Fluid Flows. *arXiv* **2023**, arXiv:2301.09122v2.
106. Keller, H.B. Numerical Solution of Bifurcation and Nonlinear Eigenvalue Problems. In *Applications of Bifurcation Theory*; Rabinowitz, P.H., Ed.; Academic Press: Cambridge, MA, USA, 1977; pp. 359–384.
107. Shroff, G.M.; Keller, H.B. Stabilisation of unstable procedures: The RPM. *SIAM J. Numer. Analy.* **1993**, *30*, 1099–1120. [CrossRef]
108. Davidson, B. Large Scale Continuation and Numerical Bifurcation for PDE's. *SIAM J. Numer. Anal.* **1997**, *34*, 2008–2027. [CrossRef]
109. Stephens, A.B.; Shubin, G.R. Multiple Solutions and Bifurcation of Finite Difference Approximations to Some Steady Problems of Fluid Dynamics. *SIAM J. Sci. Stat. Comput.* **1981**, *2*, 404–415. [CrossRef]
110. Fortin, A.; Jardak, M.; Gervais, J.J.; Pierre, R. Localization of Hopf Bifurcations in Fluid Flow Problems. *Intern. J. Numer. Meth. Fluids* **1997**, *24*, 1185–1210. [CrossRef]
111. Thompson, J.M.T. and Stewart, H.B. *Nonlinear Dynamics and Chaos*; John Wiley: Hoboken, NJ, USA, 1986.
112. Hoppensteadt, F.C. *Analysis and Simulation of Chaotic Systems*; Springer: New York, NY, USA, 1993.
113. De Markus, A.S.; Mickens, R.E. Suppression of Numerically Induced Chaos with Nonstandard Finite Difference Schemes. *J. Comput. Appl. Math.* **1999**, *106*, 317–324. [CrossRef]
114. de Markus, A.S.; Mickens, R.E. Reduction of Computational Errors Using Nonstandard Finite Difference Models. *Differ. Eqn. Appl.* **2005**, *1*, 83–91. [CrossRef]
115. Seydel, R. *From Equilibrium to Chaos*; Elsevier: New York, NY, USA, 1988.
116. Hsu, C.S. *Cell-to-Cell Mapping*; Springer: New York, NY, USA, 1987.
117. Friedman, M. On Computing Connecting Orbits. In Proceedings of the Conference on Dynamical Numerical Analysis, Atlanta, GA, USA, 14–16 December 1995; Georgia Institute of Technology: Atlanta, GA, USA, 1995.
118. Higham, D.J.; Stuart, A.M. Analysis of the Dynamics of Local Error Control via a Piecewise Continuous Residual. *BIT Numer. Math.* **1998**, *38*, 44–57. [CrossRef]
119. Yee, H.C.; Sweby, P.K. *Dynamics of Numerics & Spurious Behaviors in CFD Computations, Proceedings of the 7th ISCFD Conference, Beijing, China, 15–19 September 1997*; RIACS Technical Report 97.06; NASA: Washington, DC, USA, 1997.
120. Sjögreen, B.; Yee, H.C. On Entropy Splitting, Linear and Nonlinear Numerical Dissipations and Long-Time Integrations. In Proceedings of the 5th Internattional Conference on Spectral and High Order Methods, Uppsala, Sweden, 11–15 June 2001.
121. Yee, H.C.; Sjögreen, B. Adaptive Numerical-Dissipation/Filter Controls for High Order Numerical Methods. In Proceedings of the 3rd International Conference on DNS/LES, Arlington, TX, USA, 4–9 August 2001.
122. Yee, H.C.; Sjögreen, B. Designing Adaptive Low Dissipative High Order Schemes for Long-Time Integrations. In *Turbulent Flow Computation*; Drikakis, D., Geurts, B., Eds.; Springer: New York, NY, USA, 2001; also RIACS Technical Report, Dec. (2001).
123. Pirozzoli, S. Numerical Methods for High-Speed Flows. *Annu. Rev. Fluid Mech.* **2011**, *43*, 163–194. [CrossRef]
124. Griffiths, D.F.; Sweby, P.K.; Yee, H.C. On Spurious Asymptotes Numerical Solutions of Explicit Runge-Kutta Schemes. *IMA J. Numer. Anal.* **1992**, *12*, 319–338. [CrossRef]
125. Bao, W.; Jin, S. The random projection method for hyperbolic conservation laws with stiff reaction terms. *J. Comput. Phys.* **2000**, *163*, 216–248. [CrossRef]
126. Bao, W.; Jin, S. The random projection method for stiff detonation capturing. *SIAM J. Sci. Comput.* **2001**, *23*, 1000–1026. [CrossRef]
127. Bao, W.; Jin, S. The random projection method for stiff multispecies detonation capturing. *J. Comput. Phys.* **2002**, *178*, 37–57. [CrossRef]
128. Chertock, A.; Chu, A.; Kurganov, A. Accurate Deterministic Projection Methods for Stiff Detonation Waves. *Commun. Math. Sci.* **2024**, *22*, 871–910. [CrossRef]
129. Keefe, L.; Moin, P.; Kim, J. The Dimension of Attractors Underlying Periodic Turbulent Poiseuille Flow. *J. Fluid Mech.* **1992**, *242*, 1–29. [CrossRef]
130. Keefe, L. Private communication, 1997. *unpublished*.

131. Moura, A.F.; Gibbons, N.; Wheatley, V.; McIntyre, T.; Jahn, I. Characterization of Supersonic Turbulent Combustion in a Mach-10 Scramjet Combustor. *AIAA J.* **2020**, *58*, 2180–2196. [CrossRef]
132. Grebogi, C.; Ott, E.; Yorke, J.A. Crises, Sudden Changes in Chaotic Attractors, and Transient Chaos. *Physica D* **1983**, *7*, 181–200. [CrossRef]
133. Moore, D.R.; Weiss, N.O.; Wilkins, J.M. The Reliability of Numerical Experiments: Transitions to Chaos in Thermo solutal Convection. *Nonlinearity* **1990**, *3*, 997–1014. [CrossRef]
134. Spalart, P.R.; Moser, R.D.; Rogers, M.M. Spectral methods for the Navier-Stokes equations with one infinite and two periodic directions. *J. Comput. Phys.* **1991**, *96*, 297. [CrossRef]
135. Orszag, S. Accurate solution of the Orr-Sommerfeld stability equation. *J. Fluid Mech.* **1971**, *50*, 689–703. [CrossRef]
136. Herbert, T. *Lecture Notes in Physics 59*; Springer: Berlin/Heidelberg, Germany, 1976; p. 235.
137. Kim, J.; Moin, P.; Moser, R. Turbulence statistics in fully developed channel flow at low Reynolds number. *J. Fluid Mech.* **1987**, *177*, 133–166. [CrossRef]
138. Gresho, P.M.; Gartling, D.K.; Torczynski, J.R.; Cliffe, K.A.; Winters, K.H.; Garratt, T.J.; Spence, A.; Goodrich, J.W. Is the Steady Viscous Incompressible Two?Dimensional Flow Over a Backward-Facing Step at Re = 800 Stable? *Intern. J. Numer. Meth. Fluids* **1993**, *17*, 501–541. [CrossRef]
139. Kaiktsis, L.; Karniadakis, G.E.; Orszag, S.A. Onset of Three-Dimensionality, Equilibria, and Early Transition in Flow Over a Backward-Facing Step. *J. Fluid Mech.* **1991**, *231*, 501–528. [CrossRef]
140. Torczynski, J.R. *A Grid Refinement Study of Two-Dimensional Transient Flow Over a Backward-Facing Step Using a Spectral-Element Method*; Separated Flows; Dutton, J.C., Purtell, L.P., Eds.; ASME: New York, NY, USA, 1993; Volume 149.
141. Kaiktsis, L.; Karniadakis, G.E.; Orszag, S.A. Unsteadiness and Convective Instabilities in Two-Dimensional Flow Over a Backward-Facing Step. *J. Fluid Mech.* **1996**, *321*, 157–187. [CrossRef]
142. Henderson, R.D. Adaptive Spectral Element Methods for Turbulence and Transition. In *High-Order Methods for Computational Physics*; Barth, T.J., Deconinck, H., Eds.; Springer: Berlin/Heidelberg, Germany, 1999.
143. Fischer, P.F.; Mullen, J.S. Filter-Based Stabilization of Spectral Element Methods. *C. R. L'AcadéMie Sci. Ser. I Math.* **2001**, *332*, 265–270. [CrossRef]
144. Maday, Y.; Patera, T.; Rønquist, E.M. An operator-integration-factor splitting method for time-dependent problems: Application to incompressible fluid flow. *J. Sci. Comput.* **1990**, *5*, 263–292. [CrossRef]
145. Harten, A.; Hyman, J.M. Gui-Qiang Chen1 and Philippe G. LeFloch2. *J. Comput. Phys.* **1983**, *50*, 235. [CrossRef]
146. Edney, B.E. *FFA Report 115*; The Aeronautical Research Institute of Sweden: Stockholm, Sweden, 1968; *unpublished*.
147. Wei, L.; Xia, Y. Steady-state simulations of Euler equations by the discontinuous Galerkin method with the hybrid limiter. *J. Comput. Phys.* **2024**, *515*, 113288. [CrossRef]
148. Berkenbosch, A.; Kaasschieter, E.; Klein, R. Detonation Capturing for Stiff Combustion Chemistry. *Combust. Theory Model* **1998**, *2*, 313–348. [CrossRef]
149. Bihari, B.; Schwendeman, D. Multiresolution Schemes for the Reactive Euler Equations. *J. Comput. Phys.* **1999**, *154*, 197–230. [CrossRef]
150. Bourlioux, A.; Majda, A.J.; Roytburd, V. Theoretical and Numerical Structure for Unstable One-Dimensional Detonations. *SIAM J. Appl. Math.* **1991**, *51*, 303–343. [CrossRef]
151. Helzel, C.; LeVeque, R.; Warneke, G. A Modified Fractional Step Method for the Accurate Approximation of Detonation Waves. *SIAM J. Sci. Stat. Comp.* **1999**, *22*, 1489–1510. [CrossRef]
152. Jeltsch, R.; Klingenstein, P. Error Estimators for the Position of Discontinuities in Hyperbolic Conservation Laws with Source Term Which are Solved Using Operator Splitting. *Comput. Vis. Sci.* **1999**, *1*, 231–249. [CrossRef]
153. LeVeque, R.J.; Shyue, K.-M. One-Dimensional Front Tracking Based on High Resolution Wave Propagation Methods. *SIAM J. Sci. Comput.* **1995**, *16*, 348–377. [CrossRef]
154. Pember, R. Numerical Methods for Hyperbolic Conservation Laws with Stiff Relaxation, I. Spurious Solutions. *SIAM J. Appl. Math.* **1993**, *53*, 1293–1330. [CrossRef]
155. Tosatto, L.; Vigevano, L. Numerical Solution of Under-Resolved Detonations. *J. Comp. Phys.* **2008**, *227*, 2317–2343. [CrossRef]
156. Colella, P.; Majda, A.; Roytburd, V. Theoretical and Numerical Structure for Numerical Reacting Waves. *SIAM J. Sci. Stat. Comput.* **1986**, *7*, 1059–1080. [CrossRef]
157. Strang, G. On the Construction and Comparison of Difference Schemes. *SIAM J. Numer. Anal.* **1968**, *5*, 506–517. [CrossRef]
158. Ducros, F.; Laporte, F.; Soulères, T.; Guinot, V.; Moinat, P.; Caruelle, B. High-Order Fluxes for Conservative Skew-Symmetric-Like Schemes in Structured Meshes: Application to Compressible. *Flows J. Comput. Phys.* **2000**, *161*, 114–139. [CrossRef]
159. Kailasanath, K.; Oran, E.; Boris, J.; Young, T. A computational method for determining detonation cell size. In Proceedings of the 23rd Aerospace Sciences Meeting, Reno, NV, USA, 14–17 January 1985.
160. Zhang, X.; Shu, C.-W. Positivity-Preserving High Order Finite Difference WENO Schemes for Compressible Euler Equations. *J. Comput. Phys.* **2012**, *231*, 2245–2258. [CrossRef]
161. Yee, H.C.; Sjögreen, B. Efficient Low Dissipative High Order Schemes for Multiscale MHD Flows, II: Minimization of div **B** Numerical Error. *J. Sci. Comput.* **2006**, *29*, 115–164. [CrossRef]
162. Yee, H.C.; Sjögreen, B. Adaptive Filtering and Limiting in Compact High Order Methods for Multiscale Gas Dynamics and MHD Systems. *Comput. Fluid* **2008**, *37*, 593–619. [CrossRef]

163. Yee, H.C.; Warming, R.F.; Harten, A. Implicit total variation diminishing (TVD) schemes for steady-state calculations. *J. Comput. Phys.* **1985**, *57*, 327–360. [CrossRef]
164. Ching, E.J.; Johnson, R.F.; Kercher, A.D. Postitivity-preserving and entropy-bounded discontinuous Galerkin method for the chemically reating, compressible Euler equations. Part I: The one-dimensional case. *J. Comput. Phys.* **2024**, *505*, 112881. [CrossRef]
165. Corcos, L.P.; Saye, R.I.; Sethian, J.A. A hybrid finite difference level set-implicit mesh discontinuous Galerkin method for multi-layer coating flows. *J. Comput. Phys.* **2024**, *507*, 112960. [CrossRef]
166. Maltsev, V.; Yuan, D.; Jenkins, K.W.; Skote, M.; Tsoutscanis, P. Hybrid discontinuous Galerkin-finite volume techniques for compressible flows on unstructured meshes. *J. Comput. Phys.* **2023**, *473*, 111755. [CrossRef]
167. Sjögreen, B.; Yee, H.C.; Kotov, D.V. Skew-symmetric Splitting and Stability of High Order Central Schemes. *J. Phys. Conf. Ser.* **2017**, *837*, 012019. [CrossRef]
168. Harten, A. The Artificial Compression Method for Computation of Shocks and Contact Discontinuities: (III). Self-Adjusting Hybrid Schemes. *Math. Comput.* **1978**, *32*, 363–389.
169. Roe, P.L. Approximate Riemann Solvers, Parameter Vectors, and Difference Schemes. *J. Comput. Phys.* **1981**, *43*, 357–372. [CrossRef]
170. Ducros, F.; Ferrand, V.; Nicoud, F.; Weber, C.; Darracq, D.; Gacherieu, C.; Poinsot, T. Large-Eddy Simulation of the Shock/Turbulence Interaction. *J. Comput. Phys.* **1999**, *152*, 517–549. [CrossRef]
171. Johnsen, E.; Larsson, J.; Bhagatwala, A.; Cabot, W.; Moin, P.; Olson, B.; Rawat, P.; Shankar, S.; Sjgreen, B.; Yee, H.; et al. Assessment of high-resolution methods for numerical simulations of compressible turbulence with shock waves. *J. Comput. Phys.* **2010**, *229*, 1213–1237. [CrossRef]
172. Palha, A.; Gerritsma, M. A Mass Energy, Enstrophy and Vorticity Conserving (MEEVC) Mimetic Spectral Element Discretization for the 2D Incompressible Navier-Stokes Equations. *J. Comput.* **2017**, *328*, 200–220. [CrossRef]
173. Gassner, G.; Winters, A.R. A Novel Robust Strategy for Discontinuous Galerkin Methods in Computational Fluid Mechanics: Why? When? What? Where? *Front. Phys.* **2021**, *8*, 500690. [CrossRef]
174. Harten, A. On the Symmetric Form of Systems for Conservation Laws with Entropy. *J. Comput. Phys.* **1983**, *49*, 151. [CrossRef]
175. Tadmor, E. Entropy Stability Theory for Difference Approximations of Nonlinear Conservation Laws and Related Time-Dependent Problems. *Acta Numer.* **2003**, 12, 451–512. [CrossRef]
176. Cano, B.; Sanz-Serna, M.J. Error growth in the numerical integration of periodic orbits, with application to Hamiltonian and reversible systems. *SIAM J. Numer. Anal.* **1997**, *34*, 1391–1417. [CrossRef]
177. Blaisdell, G.A.; Spyropoulos, E.T.; Qin, J.H. The Effect of the Formulation of Nonlinear Terms on Aliasing Errors in Spectral Methods. *Appl. Num. Math.* **1996**, *21*, 207–219. [CrossRef]
178. Gerritsen, M.; Olsson, P. Designing an Efficient Solution Strategy for Fluid Flows. I. A Stable High Order Finite Difference Scheme and Sharp Shock Resolution for the Euler Equations. *J. Comput. Phys.* **1996**, *129*, 245–262. [CrossRef]
179. Coppola, G.; Capuano, F.; Pirozzoli, S.; de Luca, L. Numerically Stable Formulations of Convective Terms for Turbulent Compressible Flows. *J. Comput. Phys.* **2019**, *382*, 86–104. [CrossRef]
180. Pirozzoli, S. Generalized conservative approximations of split convective derivative operators. *J. Comput. Phys.* **2010**, *219*, 7180–7190. [CrossRef]
181. Olsson, P.; Oliger, J. *Energy and Maximum Norm Estimates for Nonlinear Conservation Laws*; RIACS Technical Report 94.01; NASA: Washington, DC, USA, 1994.
182. Johansson, S. *High Order Summation by Parts Operator Based on a DRP Scheme Applied to 2D*; Technical Report 2004-050; Uppsala University: Uppsala, Sweden, 2004.
183. Yee, H.C.; Sjögreen, B. Recent Developments in Accuracy and Stability Improvement of Nonlinear filter Methods for DNS and LES of Compressible Flows. *Comput. Fluids* **2018**, *169*, 331–348. [CrossRef]
184. Sjögreen, B.; Yee, H.C.; Vinokur, M. On high order finite-difference metric discretizations satifying GCL on mmoving and deforming grids. *J. Comput. Phys.* **2014**, *265*, 211–220 [CrossRef]
185. Bogey, C.; Bailly, C. A Family of Low dispersive and Low Dissipative Explicit Schemes for Computing the Aerodynamic Noise. In Proceedings of the AIAA-Paper 2002-2509, 8th ARAA/CEAS Aeroacoustics Conference & Exhibit, Breckenridge, CO, USA, 17–19 June 2002.
186. Brambley, E.J. Optimized finite-difference (DRP) schemes perform poorly for decaying or growing oscillations. *J. Comput. Phys.* **2015**, *324*, 258–274. [CrossRef]
187. Tam, C.K.W. *A CAA Primer for Practicing Engineers*; AEDC-TR-08-2; Arnold Engineering Development Center: Arnold Air Force Base, TN, USA, 2008.
188. Tam, C.K.W. *Computational Aeroacoustics: A Wave Number Approach*; Cambridge Aerospace Series; Cambridge University Press: New York, NY, USA, 2012; Volume 33.
189. Sjögreen, B.; Yee, H.C. A New Approach for a Wider Class of Entropy Split Methods for Compressible Gas Dynamics and MHD. In Proceedings of the ICCFD11 Conference, Maui, HI, USA, 11–15 July 2022.
190. Tadmor, E. Numerical Viscosity and the Entropy Condition for Conservative Difference Schemes. *Math. Comput.* **1984**, 43, 369–381. [CrossRef]
191. Bermejo-Moreno, I.; Larsson, J.; Lele, S.K. LES of canonical shock-turbulence interaction. In *Annual Research Briefs, Center for Turbulence Research*; Stanford University: Stanford, CA, USA, 2010; pp. 209–222.

192. Larsson, J.; Lele, S.K. Direct numerical simulation of canonical shock/turbulence interaction. *Phys. Fluids* **2009**, *21*, 101–114. [CrossRef]
193. Lee, S.; Lele, S.K.; Moin, P. Interaction of isotropic turbulence with shock waves: Effect of shock strength. *J. Fluid Mech.* **1997**, *340*, 225–247 [CrossRef]
194. Ristorcelli, J.R.; Blaisdell, G.A. Consistent initial conditions for the DNS of compressible turbulence. *Phys. Fluids* **1997**, *9*, 4–6 [CrossRef]
195. Taylor, G.; Green, A. Mechanism of the Production of Small Eddies from Large Ones. *Proc. R. Soc. Lond. A* **1937**, 158, 499–521.
196. Yee, H.C.; Sjögreen, B. Simulation of Richtmyer-Meshkov Instability by Sixth-order Filter Methods. *Shock Waves J.* **2007**, *17*, 185–193. [CrossRef]
197. Castro, M.J.; Gallardo, J.M.; Marquina, A. Jacobian-Free Incomplete Riemann Solvers. In *Theory, Numerics and Application Problems I: Aachen Germany*; Springer: Berlin/Heidelberg, Germany, 2016; pp. 292–307.
198. *Flash Manual–UserManual.wiki*; University of Chicago: Chicago, IL, USA, 2021.
199. Colella, P.; Woodward, P.R. The Piecewise Parabolic Method (PPM) for Gas-Dynamical Simulations. *J. Comput. Phys.* **1984**, *54*, 174–201. [CrossRef]
200. Gurski, K.F. An HLLC-Type Approximate Riemann Solver for Ideal Magnetohydrodynamics. *SIAM J. Sci. Comput.* **2004**, *25*, 2165–2187. [CrossRef]
201. Li, S. An HLLC Riemann Solver for Magneto-Hydrodynamics. *J. Comput. Phys.* **2005**, *203*, 344–357. [CrossRef]
202. Yee, H.C.; Sjögreen, B. On Entropy Conservation and Kinetic Energy Preservation Methods. In Proceedings of the ICOSAHOM-2019, Paris, France, 1–5 July 2019.

Disclaimer/Publisher's Note: The statements, opinions and data contained in all publications are solely those of the individual author(s) and contributor(s) and not of MDPI and/or the editor(s). MDPI and/or the editor(s) disclaim responsibility for any injury to people or property resulting from any ideas, methods, instructions or products referred to in the content.

fluids

MDPI

Review

Review of the Color Gradient Lattice Boltzmann Method for Simulating Multi-Phase Flow in Porous Media: Viscosity, Gradient Calculation, and Fluid Acceleration

Fizza Zahid [1,*] and Jeffrey A. Cunningham [2]

1 Institute of Environmental Engineering and Research, University of Engineering and Technology, Lahore 54890, Pakistan
2 Department of Civil and Environmental Engineering, University of South Florida, Tampa, FL 33620, USA; cunning@usf.edu
* Correspondence: fzahid1813@gmail.com

Abstract: The lattice Boltzmann method (LBM) is widely applied to model the pore-scale two-phase flow of immiscible fluids through porous media, and one common variant of the LBM is the color gradient method (CGM). However, in the literature, many competing algorithms have been proposed for accomplishing different steps in the CGM. Therefore, this paper is the first in a series that aims to critically review and evaluate different algorithms and methodologies that have been proposed for use in the CGM. Specifically, in this paper, we (1) provide a brief introduction to the LBM and CGM that enables and facilitates consideration of more sophisticated topics subsequently; (2) compare three methods for modeling the behavior of fluids of moderately different viscosities; (3) compare two methods for calculating the color gradient; and (4) compare two methods for modeling external forces or accelerations acting upon the fluids of interest. These topics are selected for the first paper in the series because proper selection of these algorithms is necessary and sufficient to perform two common "benchmark" simulations, namely bubble tests and layered Poiseuille flow. Future papers in the series will build upon these topics, considering more challenging conditions or phenomena. By systematically reviewing key aspects, features, capabilities, and limitations of the CGM, this series of papers will extend our collective ability to apply the method to a variety of important fluid flow problems in geosciences and engineering.

Keywords: lattice Boltzmann; color gradient; multi-phase flow; porous media; viscosity; interfacial tension; bubble test; Poiseuille flow

1. Introduction

The flow of immiscible (multi-phase) fluids in porous media is important in many problems in geosciences and environmental engineering. Examples include water percolation through the vadose zone, the storage of supercritical carbon dioxide in deep saline aquifers, and the clean-up of groundwater contaminated by non-aqueous-phase liquids (e.g., petroleum and chlorinated solvents). To fully understand such systems, it is necessary to characterize fluid behavior at the pore scale, because macroscopic (bulk) fluid behavior depends on processes occurring at the pore scale [1]. Pore-scale factors that might be important include grain and pore morphology, inter-pore connectivity, interfacial tension, and fluid–fluid–solid interactions (i.e., wettability). However, quantifying the importance of such factors is challenging: the subsurface environment cannot be observed directly, and it is difficult to create physical (laboratory) models that are both realistic and controllable.

Because of the challenges of running controlled physical experiments with realistic porous media, pore-scale multi-phase fluid flow is often studied via numerical models that aim to accurately capture the most important physics of real systems. There are several different methods available that model pore-scale processes and systems, including pore network models, the lattice Boltzmann method (LBM), Monte Carlo methods, particle dynamics methods, and grid-based computational methods that track interfaces [2]. Many modern pore-scale models are able to sufficiently resolve multi-phase fluid interactions and adequately depict the behavior that is seen in laboratory experiments [3].

Of the methods developed heretofore, the LBM has proven to be a robust and reliable tool to simulate fluid flow (including multi-phase flow) in porous media [4–9]. A defining feature of the LBM is that it accounts for pore-scale fluid dynamics that satisfy the incompressible Navier–Stokes equations, without explicitly solving the partial differential equation [10]. Among the advantages of the LBM are that it is well suited to application in complex geometries such as porous media [11] and that computations are local and explicit, making the method amenable to parallel implementations [12–19].

Different types of lattice Boltzmann methods have been developed for describing multi-phase fluid flow. These include the pseudopotential model (sometimes called the "Shan-Chen" method) [20], free energy method [21], mean field theory [22], stabilized diffuse-interface model [23], and color gradient method (sometimes called the "Rothman-Keller" method) [24,25]. Each of these variants has its own strengths, weaknesses, and applications [14,16]. Over the years, the color gradient method (CGM) has matured into a tool that has characterized various flows in porous media [26].

The CGM was developed and implemented for immiscible fluids in porous media by Gunstensen et al. [25], based on the lattice gas model [24]. A feature of the CGM is its ability to keep the immiscible fluids sufficiently distinct and separate with the help of an interface thickness parameter. This minimizes the mixing of the fluids at the interface and reduces distortion [27]. This method also conserves momentum at the local and global scale [28], which makes it possible to implement different types of boundary conditions (free exit) that cannot be implemented with other LBM variants. However, the method's main strength is the ability to independently tune fluid and flow parameters (e.g., interfacial tension, contact angle, fluid densities, and interface thickness) with individual color gradient constants or parameters [29]. The CGM has been utilized to simulate various porous media fluid flow applications such as carbon dioxide transport in subsurface systems [30], water transport in proton exchange membrane fuel cells [31], temperature-driven fluid flow in heated microchannels [32], drainage and imbibition in oil-water systems [33], spontaneous imbibition in micromodels [34], the behavior of droplets on surfaces [35], the adsorption of fluids onto the surfaces of porous media [36,37], the flow of particle suspensions [38], the separation of gas mixtures [9], and the behavior of droplets during inkjet printing [39], to name just a few. Due to the CGM's popularity, some previous authors have compared its capabilities to those of other leading multi-phase LBM methods, establishing the CGM's capacity to capture the physics of multi-phase fluid flow phenomena [3,40–42].

The execution and application of the CGM have been described and reviewed previously by other authors, often as part of broader reviews of multi-phase LBM methods (e.g., [16,26,43]). However, certain important practical aspects of implementing the CGM (e.g., how to model the flow of fluids with widely differing densities or viscosities and how to implement "open" boundary conditions) have not yet been systematically reviewed in the previous literature. Also, improvements to the CGM continue to appear in the literature frequently, and some recent advancements to the CGM (e.g., how to best account for wettability and contact angles and how to extend the method from two spatial dimensions to three [35,42,44–46]) have not been considered by previous reviews of the method. Finally, the previous review papers generally do not guide the reader in terms of algorithm development and the practical implementation of the CGM. Therefore, a need exists to systematically review the capabilities and limitations of the CGM, starting with fundamental aspects for readers who are not yet experts in the method, then progressing to more sophisticated and "modern" topics, all while emphasizing the practical aspects of implementing the algorithms reviewed.

Therefore, the overall objective of this series of papers is to critically review the capabilities and implementation of the CGM as a method of modeling pore-scale multiphase flow in porous media. We aim to provide a review that is helpful both to new users of the LBM and to more experienced modelers with a higher level of expertise. For the current paper, the first in the series, the specific objectives are as follows:

1. Provide a brief introduction to the LBM and CGM that subsequently enables and facilitates consideration of more sophisticated topics;
2. Critically review different methods for modeling the flow of fluids of moderately different viscosities;
3. Critically compare two methods for calculating the color gradient;
4. Critically compare two methods for modeling external forces or accelerations acting upon the fluids of interest;
5. Summarize some of the challenges in applying the CGM with respect to the topics listed immediately above.

The rationale for focusing on these particular aspects of the CGM in the first paper of the series is that understanding these algorithms will allow the reader to run two of the most common "benchmark" tests for lattice Boltzmann modeling, namely bubble tests (droplet tests) and layered Poiseuille flow; examples of these tests will be provided at the end of this paper. Future papers in the series will build upon the topics covered here, considering phenomena or conditions such as fluids of widely differing viscosity and/or density, how to introduce the "wettability" of solid surface and specify a desired contact angle, how to simulate fluid flow in a domain with an open boundary condition (e.g., fluid flushing out of a porous medium), and how to extend the CGM from two spatial dimensions to three. By systematically reviewing these key aspects and features, this series of papers will extend our collective ability to apply the CGM to a variety of important fluid flow problems in geosciences and engineering.

2. Review of LBM Basics

The lattice Boltzmann method (LBM) for modeling fluid flow is based on a marriage of two sub-fields of physics: lattice gas cellular automata and Boltzmann's kinetic theory of gases (i.e., statistical mechanics).

A lattice gas cellular automaton is a system in which fictitious "gas" particles reside on a grid (lattice) and move between nodes of the grid. Typically, each particle has the same mass and moves with the same speed as each other. When two particles arrive at the same node at the same time, the particles collide, and the particles' velocities (momenta) may be altered by the collision. Frisch et al. [47] demonstrated the remarkable result that if the lattice is formulated properly, and if the collisions are modeled in a manner that conserves particle mass and momentum, then the lattice gas cellular automaton is able to simulate the flow of realistic fluids. That is, the aggregate or macroscopic behavior of the particles on the lattice can be shown to satisfy the Navier–Stokes equations. This discovery by Frisch et al. [47] enabled lattice gas cellular automata to be developed as tools for computational fluid dynamics.

An important advancement to the pioneering work of Frisch et al. [47] was modifying the lattice gas model to account for the spatio-temporal evolution of *distributions* of particles, or *probabilities* of particle velocities, rather than the movements of individual particles [48–50]. The evolution of the particle distributions can be described with the following equation, which, as pointed out by Wolfram [51], is similar to a Boltzmann transport equation:

$$f_i(\mathbf{x} + \mathbf{e}_i\,\delta t, t + \delta t) = f_i(\mathbf{x}, t) + \Omega_i(f(\mathbf{x}, t)) \quad (1)$$

where $f_i(\mathbf{x}, t)$ is the probability of a particle at location $\mathbf{x}$ having the velocity $\mathbf{e}_i$ at time t, δt is a small incremental time interval, and Ω_i is the collision operator that operates on f. Because particle movement is constrained to the lattice, there is a finite set of velocity vectors $\mathbf{e}_i$. Solving for the distribution function f allows for the simulation of macroscopic fluid behavior without the "noise" that is encountered when working with individual particles.

In practice, the time variable t is generally taken as a discrete integer-valued variable corresponding to the number of time steps taken; movement from one node on the lattice to an adjacent node requires one time step, and δt is always taken as one time step. Therefore, many papers in the literature re-write Equation (1) more compactly as the following:

$$f_i(\mathbf{x} + \mathbf{e}_i, t + 1) = f_i(\mathbf{x}, t) + \Omega_i(f(\mathbf{x}, t)) \quad (2)$$

In this form, the units on velocity $\mathbf{e}_i$ are lattice units per time step (l.u./t.s.); the reader must understand that it is implied that $\mathbf{e}_i$ is multiplied by one time step, so the addition $\mathbf{x} + \mathbf{e}_i$ is acceptable.

Figure 1 shows two possible lattice configurations that have been applied for modeling flow in two dimensions, along with the corresponding set of velocity vectors $\mathbf{e}_i$. These two configurations are commonly referred to as D2Q7 and D2Q9. The "D2" notation refers to the fact that the lattices are two-dimensional, and "Q7" or "Q9" refers to the number of neighbors for each node in the lattice (including the node itself as one of its neighbors, which allows for the possibility that a particle can remain stationary rather than moving to an adjoining node).

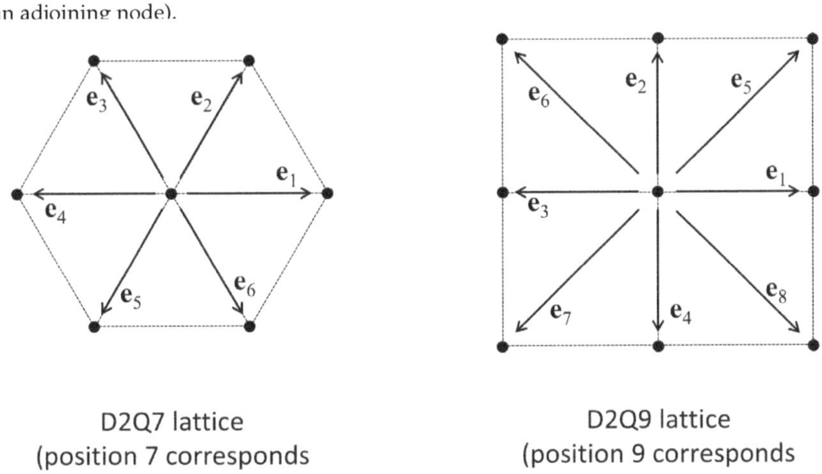

Figure 1. Two possible lattice configurations for simulating two-dimensional fluid flow.

In classical statistical mechanics, $f(\mathbf{x}, \mathbf{v}, t)$ is the probability density function for fluid particles at location $\mathbf{x}$ having velocity $\mathbf{v}$ at time t. The Boltzmann equation for the movement of the fluid particles (in the absence of an external force field) is

$$\frac{\partial f}{\partial t} + \mathbf{v} \cdot \nabla f = \Omega \tag{3}$$

where Ω represents the change in particle momentum due to collisions. It can be seen that Equation (2) is a form of Equation (3) in which the time derivative has been discretized and in which the particle velocities are constrained to the lattice. Therefore, Equation (2) is sometimes termed the "lattice Boltzmann equation".

A major advancement in the application of Equation (2) was developing a convenient way of estimating the collision operator. In classical statistical mechanics, Bhatnagar et al. [52] developed a method for estimating the collision operator in Equation (3); specifically, the collision operator is estimated as "relaxation" of the function f toward an equilibrium distribution f^{eq}. The method of Bhatnagar et al. [52] was applied to the lattice Boltzmann equation by Qian et al. [53]. Qian et al. [53] presented the following equations, which constitute the very popular Bhatnagar–Gross–Krook (or "BGK") formulation of the lattice Boltzmann method:

$$f_i(\mathbf{x} + \mathbf{e}_i\,\delta t, t + \delta t) = f_i(\mathbf{x}, t) + \frac{1}{\tau}\left[f_i^{\text{eq}}(\mathbf{x}, t) - f_i(\mathbf{x}, t)\right] \tag{4}$$

$$f_i^{\text{eq}}(\mathbf{x}, t) = w_i\,\rho\left[1 + 3(\mathbf{e}_i \cdot \mathbf{u}) + \frac{9}{2}(\mathbf{e}_i \cdot \mathbf{u})^2 - \frac{3}{2}(\mathbf{u} \cdot \mathbf{u})\right] \tag{5}$$

where τ is a relaxation time characterizing the particle collision, ρ is the fluid density (in the lattice domain), $\mathbf{u}$ is the macroscopic velocity vector of the fluid at location $\mathbf{x}$ and time t, and w_i is a weighting parameter. Equation (5) is the formulation of f^{eq} that enables the macroscopic fluid behavior to adhere to the Navier–Stokes equations [53]. For this formulation, the viscosity of the fluid (in the lattice domain, units l.u.^{2}/t.s.) is given by

$$\nu = \frac{1}{6}(2\tau - 1). \tag{6}$$

The weighting parameters, w_i, depend on which lattice geometry is employed. For a D2Q9 lattice (shown in Figure 1), w is given by the following [53–55]:

$$w = \begin{bmatrix} \frac{1}{9} & \frac{1}{9} & \frac{1}{9} & \frac{1}{9} & \frac{1}{36} & \frac{1}{36} & \frac{1}{36} & \frac{1}{36} & \frac{4}{9} \end{bmatrix}. \tag{7}$$

For the flow of a fluid in two dimensions, the solution of Equations (4)–(7) enables us to track the particle distribution function f over time and space. However, Equation (5) depends on the macroscopic fluid properties ρ and $\mathbf{u}$. These can be determined by calculating the moments of f, according to the following:

$$\rho(\mathbf{x}, t) = \sum_i f_i(\mathbf{x}, t) \tag{8}$$

$$\mathbf{u}(\mathbf{x}, t) = \frac{\sum_i f_i(\mathbf{x}, t)\,\mathbf{e}_i}{\sum_i f_i(\mathbf{x}, t)} \tag{9}$$

Equations (4)–(9) constitute the foundation for simulating two-dimensional fluid flow with the lattice Boltzmann method. In practice, the method is implemented through an alternating series of "streaming" steps and "collision" steps that together satisfy Equation (4). More information about these steps is provided below in Section 3, but some details of implementation are omitted here in the interest of space. The reader is referred to previous work by Sukop and Thorne [6], Guo and Shu [43], Huang et al. [26], Krüger et al. [56], or Mohamad [57] for additional details of implementation.

It is worth noting that the LBM can also be used to simulate transport as well as fluid flow—for instance, mass transport via diffusion [58] and/or heat transfer via conduction [59]. Using the LBM to model transport rather than flow requires modification to the selection of the equilibrium distribution function f^{eq} and the relaxation time τ (e.g., [58,59]). However, in the present paper, we are concerned only with the multi-phase flow of immiscible fluids in porous media, and we do not consider diffusive mass transport or conductive heat transfer.

3. CGM Framework

The original CGM [25] was developed as a combination of the immiscible cellular gas automaton [24] and the LBM [50]. The method considers two immiscible fluids, a "red" fluid and a "blue" fluid, and thus considers two particle distribution functions f^k, where the superscript k indicates the red or the blue fluid. Particles of both fluids collide and stream in the lattice domain following the regular LBM operation described above (Section 2), but with an added "perturbation" step that enables interfacial tension to be included in the model, and with a "recoloring" step to maintain the fluids as immiscible.

Over the past three decades, the overall model algorithm has remained essentially unchanged, but improvements to the equations have resulted in better accuracy and better representations of true multi-phase fluid physics. Some early developments include increased capability to model fluids of unequal viscosity and/or density [60], improvement in the recoloring step [27], and modification of the perturbation equation [55]. These advances have enabled the model to be used for a wider range of fluid applications and have paved the way for developing three-dimensional models [42,61,62]. This section provides the framework of a two-dimensional CGM that includes the aforementioned improvements of Grunau et al. [60], Latva-Kokko and Rothman [27], and Reis and Phillips [55]. More recent updates pertinent to the contrast in fluid viscosity, calculation of the color gradient, and application of external forces are considered subsequently in Sections 4–6. Also, the extension of the two-dimensional CGM to three dimensions will be discussed in a future paper in this series.

The essence of the CGM can be expressed as the following analog to Equation (2) [62]:

$$f_i^k(\mathbf{x}+\mathbf{e}_i, t+1) = f_i^k(\mathbf{x}, t) + \Omega_i^k(\mathbf{x}, t) \tag{10}$$

where the collision operator $\Omega_i^k(\mathbf{x}, t)$ now represents the aggregate of three steps: the single-phase collision $(\Omega_i^k)^1$, the perturbation operation $(\Omega_i^k)^2$, and a recoloring process. In practice, the order of the operations for implementing the CGM is as follows:

1. Calculate macroscopic variables ρ_{tot}, ρ^k, and $\mathbf{u}$;
2. Calculate equilibrium distributions for red and blue fluids, $f^{\text{eq},k}$;
3. Apply the single-phase collision operator, $(\Omega_i^k)^1$, for both fluids;
4. Calculate the "color gradient" between red and blue fluids, $\mathbf{C}(\mathbf{x}, t)$;
5. Apply the perturbation operator, $(\Omega_i^k)^2$, to account for interfacial tension;
6. Perform a "re-coloring" step to maintain the immiscibility of fluids;
7. Stream the particle distributions f^k.

These steps are discussed in more detail in the sub-sections following.

We note that there are variants in the CGM that do not exactly follow the sequence of steps listed above. For instance, one such variant [63–66] enables the whole algorithm to operate with any color distribution function. There can be advantages to such variants, including lower memory demands. However, in this review, we focus only on the more traditional CGM that relies on red and blue fluid populations, because of that method's ubiquity and relative ease of implementation.

3.1. Calculation of Macroscopic Variables

The first step in the algorithm is the calculation of the macroscopic variables: the total density of the combined (composite) fluid ρ_{tot}, the individual fluid densities ρ^k for the red and blue fluids, and the velocity $\mathbf{u}$ of the combined fluid. Analogous to Equations (8) and (9), these macroscopic variables can be calculated from the moments of the red and blue distribution functions:

$$\rho^k(\mathbf{x},t) = \sum_i f_i^k(\mathbf{x},t) \tag{11}$$

$$\rho_{\text{tot}}(\mathbf{x},t) = \sum_k \rho^k(\mathbf{x},t) \tag{12}$$

$$\mathbf{u}(\mathbf{x},t) = \frac{\sum_k \sum_i \mathbf{e}_i f_i^k(\mathbf{x},t)}{\sum_k \sum_i f_i^k(\mathbf{x},t)} \tag{13}$$

3.2. Equilibrium Distribution Functions

The second step involves computing the equilibrium distribution functions $f^{\text{eq},k}$ for the red and blue fluids, which are used subsequently (step 3) to calculate the single-phase collision operator [67,68]:

$$f_i^{\text{eq},k}(\mathbf{x},t) = \rho^k\left(C_i^k + w_i\left[3(\mathbf{e}_i\cdot\mathbf{u}) + \frac{9}{2}(\mathbf{e}_i\cdot\mathbf{u})^2 - \frac{3}{2}(\mathbf{u}\cdot\mathbf{u})\right]\right) \tag{14}$$

For a D2Q9 lattice, w_i in Equation (14) is given by Equation (7) and C_i^k is given by the following [55,67]:

$$C_i^k = \begin{cases} \frac{1-\alpha^k}{5} & \text{if } i = 1,2,3,4 \\ \frac{1-\alpha^k}{20} & \text{if } i = 5,6,7,8 \\ \alpha^k & \text{if } i = 9 \end{cases} \tag{15}$$

In Equation (15), α^k is a parameter that introduces a density contrast between the fluids while maintaining stable hydrodynamic pressures at the interface (cf. [55,60]). The selected values of α^r and α^b determine the density ratio between the two fluids, the speed of sound within each fluid (c_s^k), and the pressure within the system, as summarized in Table 1. In Table 1, the parameter ζ, which relates c_s^k to α^k, is lattice-dependent; values are provided by Leclaire et al. [69]. For a D2Q9 lattice, $\zeta = \frac{3}{5}$.

To simulate a stable density contrast, α^k must be selected within the range $0 < \alpha^k < 1$ for both fluids. In particular, for a D2Q9 lattice, if $\alpha^k = 4/9$, Equation (14) reduces to Equation (5) for fluid k, and the speed of sound within fluid k is $c_s^k = 1/\sqrt{3}$. For other values of α^k, the speed of sound in fluid k will be different, as given in Table 1. To model the flow of two fluids of equal density, α^k for both fluids may usually be set equal to $4/9$. An introduction of different values of α^k for the two fluids is able to achieve density ratios up to about 18.5 [55]; systems with greater density ratios require a different approach, which will be reviewed in a subsequent paper in this series.

Note that to achieve a desired density ratio γ, there is not one unique pair of α^r and α^b that must be selected; for instance, selecting $\alpha^b = 0.02$ and $\alpha^r = 0.51$ will yield a density ratio $\gamma = 2$, but selecting $\alpha^b = 0.4$ and $\alpha^r = 0.7$ will also yield $\gamma = 2$. In limited trials during which we tested different combinations of α^r and α^b, we found that setting both values of α^k close to $4/9$ led to slightly better results than allowing one value of α^k to be far from that value. The selection of α^r and α^b according to the equations shown in Table 1 ensures that the pressure in the two fluids is continuous at the fluid–fluid interface [55,67].

It has been demonstrated [66] that when using the aforementioned $f^{eq,k}$ relations for fluids of unequal density, the tangential momentum of the two fluids is conserved along the fluid–fluid interface, but the tangential velocity along the fluid–fluid interface then becomes discontinuous. It is not yet clear if there is a practical implication of conserving momentum rather than velocity during simulations, nor if a different method for modeling unequal densities would be able to ensure the continuity of tangential velocities [66].

Table 1. User-specified CGM constants and their relations to fluid properties.

Symbol	Parameter	Fluid Relation
τ^k	Relaxation time of the kth fluid	Kinematic viscosity $\nu^k = \frac{1}{3}(\tau^k - 0.5)$
α^k	Density parameter of the kth fluid	Density ratio of the two fluids $\gamma = \frac{\rho^r}{\rho^b} = \frac{1-\alpha^b}{1-\alpha^r}$; speed of sound $c_s^k = \sqrt{\zeta(1-\alpha^k)}$; pressure $P^k = \rho^k (c_s^k)^2 = \rho^k \zeta \, (1-\alpha^k)$
A^k	Interfacial tension constant of the kth fluid	Interfacial tension σ; see discussion in Section 3.5 regarding the relationship between A^k and σ
β	Interface thickness parameter	See discussion in Section 3.6; $\beta \leq 1$

3.3. Collision Operator

In the next step, the collision operator $(\Omega_i^k)^1$ is computed and applied. There are different ways to compute the collision operator $(\Omega_i^k)^1$ depending on the viscosity ratio of the two fluids. The simplest collision expression is the BGK formulation, which considers that collisions induce the "relaxation" of f^k toward the equilibrium population $f^{eq,k}$, as discussed previously in Section 2:

$$(\Omega_i^k)^1 = \frac{-1}{\tau}\left[f_i^k(\mathbf{x},t) - f_i^{eq,k}(\mathbf{x},t)\right] \tag{16}$$

The challenge with applying Equation (16) is that it depends on a relaxation time τ that is related to the fluid viscosity (as shown in Equation (6)), but the fluid viscosity might not be the same for the red and blue fluids. If the red and blue fluids have a viscosity ratio that is very close to one (say, within a factor of two), then they will have similar values of τ, and Equation (16) can be applied without difficulty. If the fluids have a viscosity contrast that is slightly less than $\sim$0.5 or slightly greater than $\sim$2, an "effective" viscosity can be computed by interpolation, as discussed subsequently in Section 4. For even greater viscosity contrasts, more sophisticated approaches are required; for instance, the BGK approach can be replaced by a multiple-relaxation-time (MRT) approach, which will be discussed in a subsequent paper in this series.

3.4. Computing the Color Gradient

Next, the "color gradient" (sometimes also called "color flux" or "color field"), $\mathbf{C}(\mathbf{x},t)$, is calculated. The color gradient $\mathbf{C}(\mathbf{x},t)$ is the vector perpendicular to the interface between the red and blue fluids, as shown in Figure 2.

In the literature, different approaches have been presented for computing the color gradient. The original approach, as proposed and presented by Rothman and Keller [24], Gunstensen et al. [25], and Grunau et al. [60], is the following:

$$\mathbf{C}(\mathbf{x},t) = \sum_i \mathbf{e}_i\left[\rho^r(\mathbf{x}+\mathbf{e}_i,t) - \rho^b(\mathbf{x}+\mathbf{e}_i,t)\right] \tag{17}$$

Two notable modifications to this "original" approach are the "isotropic" color gradient of Leclaire et al. [70] and the "modified" color gradient of Xu et al. [71]. Subsequently in this paper (Section 5), we will compare some results of simulations using the original color gradient to the results of simulations using the isotropic color gradient of Leclaire et al. [70].

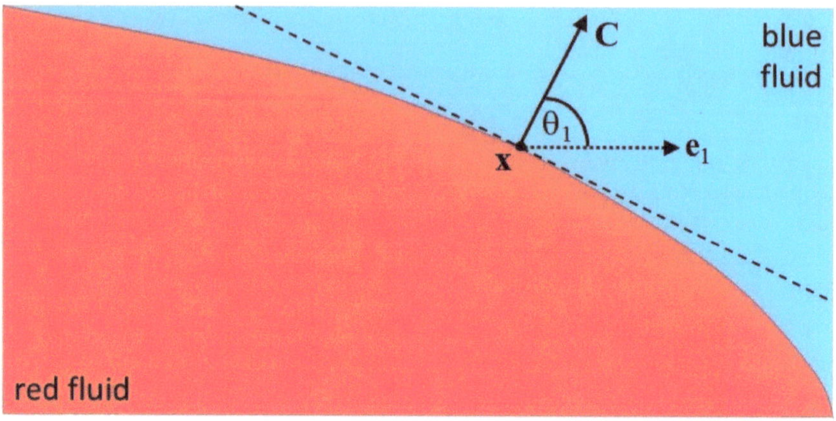

Figure 2. The color gradient **C**; figure modified from Reis and Phillips [55].

3.5. Perturbation

Gunstensen et al. [25] explain that the color gradient method proceeds by performing a collision step (as summarized above), then by "add[ing] a surface-tension-inducing perturbation, "then by recoloring the fluids and streaming the particles. In other words, the perturbation step of the CGM is the method of incorporating interfacial tension between the immiscible fluids; the color gradient **C** is utilized in this perturbation operation. In the paper of Gunstensen et al. [25], the perturbation step is represented by their Equation (17). Subsequently, perturbation strategies were proposed by Grunau et al. [60], Lishchuk et al. [72], Latva-Kokko and Rothman [27], and Reis and Phillips [55]. Ideally, the perturbation process should recover the correct interfacial force, reduce artificial velocities at the fluid–fluid interface (spurious velocities), retain the correct interface shape, and minimize interface thickness [62,72]. The approaches of Lishchuk et al. [72] and of Reis and Phillips [55] both fulfill these criteria. Furthermore, both methods have been applied successfully in both two-dimensional and three-dimensional models [42,62,72]. The approach of Lishchuk et al. [72] has been shown to model contact angles more accurately under certain conditions [45,73,74]. However, the approach of Reis and Phillips [55]—which is essentially an improvement upon the approaches of Grunau et al. [60] and Latva-Kokko and Rothman [27]—has been more extensively used because of its ability to include density ratios greater than one and because it is computationally less expensive. Here, we summarize the method of Reis and Phillips [55].

In this method, the perturbation step is accomplished through the application of the perturbation operator $(\Omega_i^k)^2$ that is applied to each f_i^k:

$$f_i^k(\mathbf{x},t) = f_i^k(\mathbf{x},t) + (\Omega_i^k)^2 \tag{18}$$

$$(\Omega_i^k)^2 = \frac{A^k}{2}|\mathbf{C}|\left[w_i\frac{(\mathbf{C}\cdot\mathbf{e}_i)^2}{\mathbf{C}\cdot\mathbf{C}} - B_i\right] \tag{19}$$

where the parameter A^k is a user-specified parameter (see Table 1) that controls the interfacial tension value. Equation (19) enables the conservation of mass and momentum, as explained by Reis and Phillips [55] and by Liu et al. [62]. In Equation (19), B is a vector of values that depends on the lattice employed; the proper selection of B enables the recovery of the correct interfacial force term in the Navier–Stokes equations [55,62]. Specifically, for a D2Q9 lattice:

$$B = \left[\frac{2}{27} \; \frac{2}{27} \; \frac{2}{27} \; \frac{2}{27} \; \frac{5}{108} \; \frac{5}{108} \; \frac{5}{108} \; \frac{5}{108} \; \frac{-4}{27} \right]. \tag{20}$$

This method of introducing interfacial tension between the fluids has been shown to recover correct macroscopic equations for two-phase flows; detailed mathematical derivations are given by Reis and Phillips [55] and Liu et al. [62].

The interfacial tension between the two fluids, σ, is related to the A^k parameters [27,60]. Either the user can specify the A^k parameters, and the interfacial tension σ will result from the choice of the A^k, or the user can specify the desired interfacial tension σ, and values of A^k can be calculated accordingly. The problem is that the relationship between A^k and interfacial tension has not been completely resolved; different papers give different derivations or approximations of the relationship. Also, the relation between A^k and σ is known to depend on the fluid viscosities and the associated collision relaxation times τ; if the fluids have unequal viscosities, then they also have unequal values of τ, which complicates the relation between A^k and σ.

To simplify the situation somewhat, we found that, if the two fluids have equal densities or if the density ratio is acceptably low ($0.5 \leq \gamma \leq 2.0$), a single value of A can often be used for both fluids, i.e., $A^r = A^b = A$. In this case, A can be computed from a desired value of σ by following Equations (36) and (37) of Leclaire et al. [70]. That is, if the color gradient $\mathbf{C}$ is computed according to Equation (17), then A and σ are related by

$$A = \frac{3}{4} \frac{\sigma}{(\rho^r + \rho^b)\tau} \tag{21}$$

in which τ is an interpolated or effective relaxation time, which will be explained subsequently in Section 4. If, instead of using Equation (17) to calculate $\mathbf{C}$, we instead use the isotropic color gradient of Leclaire et al. [70] (which will be discussed in more detail subsequently), then A must instead be computed by the following:

$$A = \frac{9}{2} \frac{\sigma}{(\rho^r + \rho^b)\tau} \tag{22}$$

in which, again, τ is an interpolated or effective relaxation time. Notably, because the interpolated relaxation time τ varies both spatially and temporally (as will be seen subsequently in this paper), the parameter A also varies spatially and temporally, so it must be computed at every lattice location at every time step.

If Equations (21) and (22) do not successfully produce the desired interfacial behavior, then alternatively, the interfacial tension σ for selected values of A^k can be verified or estimated by conducting a bubble test (static droplet test) and measuring the simulated pressures and droplet radius [62,68,75]. The perturbation term and its relationship to interfacial tension are also considered in more detail in Appendix A of Ginzburg and Steiner [63].

3.6. Recoloring

After the perturbation step, the fluids are still mixed and require separation. This is achieved by a "recoloring" step. The original recoloring procedure [25] suffered from numerical issues such as "lattice pinning" and fluctuations in velocity (also called "spurious currents" or "anomalous currents") near the interface [27,76]. Subsequent recoloring algorithms proposed in the literature (e.g., [27,55,61,67]) were designed largely to address these numerical issues. The algorithm proposed by Latva-Kokko and Rothman [27]—which is an improvement to that of D'Ortona et al. [77]—has been found to perform well for a variety of tests and is presented here:

$$f_i^r(\mathbf{x},t) = \frac{\rho^r}{\rho_{\text{tot}}} f_i^{\text{tot}} + \beta \frac{\rho^r \rho^b}{(\rho_{\text{tot}})^2} \cos(\Theta_i) f_i^{\text{eq},0} \tag{23}$$

$$f_i^b(\mathbf{x},t) = \frac{\rho^b}{\rho_{\text{tot}}} f_i^{\text{tot}} - \beta \frac{\rho^r \rho^b}{(\rho_{\text{tot}})^2} \cos(\Theta_i) f_i^{\text{eq},0} \tag{24}$$

where β is a user-specified parameter that controls the thickness of the interface (Table 1), and ρ_{tot} is the total fluid density as defined in Equation (12).

In Equations (23) and (24), $f_i^{\text{tot}}(\mathbf{x},t) = f_i^r(\mathbf{x},t) + f_i^b(\mathbf{x},t)$ and $f_i^{\text{eq},0}(\mathbf{x},t)$ are the equilibrium distribution function for the combined fluid under stationary conditions, i.e., evaluated at $\mathbf{u} = 0$. For the calculation of $f_i^{\text{eq},0}(\mathbf{x},t)$, we recommend using Equations (7), (14) and (15) given above, which differ slightly from Equations (1) and (2) of Latva-Kokko and Rothman [27]. Also, Θ_i represents the angle between the color gradient vector $\mathbf{C}$ and the lattice velocity vector $\mathbf{e}_i$, as shown in Figure 2 and given by the following equation:

$$\cos(\Theta) = \frac{\mathbf{e}_i \cdot \mathbf{C}}{|\mathbf{e}_i||\mathbf{C}|} \tag{25}$$

The parameter β in Equations (23) and (24) is specified by the user and is set between 0 and 1. A β value close to one gives minimum interface thickness but increases spurious velocities and reduces the accuracy of the interfacial tension value [62,68,78]. Conversely, if β is set too low, the spurious velocities at the interface reduce, but the interface thickness increases. Usually, a value of β between 0.70 and 0.99 is selected to maintain a thin interface without introducing significant spurious velocities (e.g., [79,80]), though Halliday et al. [78] argue that β should not exceed $\sqrt{2}/2 \approx 0.71$. Also, Leclaire et al. [81] offer a method for using a variable, rather than constant, value of β to improve performance. Despite these subtleties, we generally found that a constant value of $\beta \approx 0.8$ works well as a "default" value, unless it is determined that a more sophisticated approach is required.

In the literature, alternatives to the recoloring procedure have also been proposed. A few of these have been reviewed briefly by Ginzburg [66]. For instance, Kehrwald [65] proposes an "anisotropic diffusion" algorithm as an alternative to recoloring, and demonstrates that the algorithm is able to reduce the spurious currents (anomalous currents) mentioned above. However, most implementations of the CGM still rely on the recoloring algorithm described above, perhaps because it is relatively simpler to implement.

3.7. Summary of Parameters Specified by Model User

To summarize, the parameters that must be specified by the model user are those listed in Table 1.

- We generally recommend $\beta = 0.8$ as a default value for the interface thickness parameter β.
- For simulating fluids of equal density, we recommend $\alpha^r = \alpha^b = \frac{4}{9}$. For density ratios other than one, setting both values of α^k reasonably close to 4/9 is recommended over setting one of the values far from 4/9.
- For most applications, we recommend $A^r = A^b = A$, which can be calculated to produce a desired interfacial tension σ as given by Equation (21) or (22). However, A must be calculated at each lattice node at each time step because of the temporal and spatial variability of the effective relaxation time τ.
- Finally, the model user must specify either the relaxation time τ^k or the kinematic viscosity ν^k for each fluid, using Equation (6) to calculate whichever of the two was not specified explicitly.

4. Accounting for Fluids of (Moderately) Unequal Viscosity

The ratio of viscosity between the two fluids can strongly influence the overall behavior of the system. A classic example is "viscous fingering" that results from a less viscous fluid displacing a more viscous fluid in a porous medium [82,83]. The LBM is sometimes used as a tool to investigate and understand such mechanisms (e.g., [68,84–86]).

Although the CGM provides a robust framework to model multi-phase flow when the viscosity ratio is reasonably close to 1 (perhaps in the range 0.5–2), modifications to the methods of Section 3 are required if the viscosity ratio is far from 1. For instance, Rannou [87] demonstrated that the "standard" CGM could properly simulate two-phase flow in a channel for viscosity ratios as high as 10, but that simulated velocity profiles deviated from the correct values at higher viscosity ratios. This is because two fluids with unequal viscosities also have unequal values of τ, the relaxation time in the BGK collision operator, according to Equation (6). Therefore, at the interface between the two fluids, τ is not clearly defined, which can lead to incorrect calculations in the vicinity of the interface [68,88].

To address this problem, an effective value of τ near the interface can be interpolated from the adjacent known τ^k values before the first collision operator $(\Omega_i^k)^1$ is calculated. Methods available for interpolating τ have been presented by Grunau et al. [60], Tölke et al. [61], and Liu et al. [88], and are reviewed here.

In the method of Grunau et al. [60], an "order parameter" P (also sometimes called the "phase field" in the literature) can be calculated at any location $\mathbf{x}$ to indicate which of the two fluids, red or blue, dominantly occupies that location:

$$P(\mathbf{x},t) = \frac{\rho^r(\mathbf{x},t) - \rho^b(\mathbf{x},t)}{\rho^r(\mathbf{x},t) + \rho^b(\mathbf{x},t)} \tag{26}$$

Values of P range from -1 (region is occupied only by the blue fluid) to $+1$ (region is occupied only by the red fluid). The value of τ at any location $\mathbf{x}$ near the interface is determined from P and a user-specified interface parameter, δ, which must be ≤ 1:

$$\tau(\mathbf{x},t) = \begin{cases} \tau^r & P > \delta \\ g^r(P) & \delta \geq P > 0 \\ g^b(P) & 0 \geq P > -\delta \\ \tau^b & P < -\delta \end{cases} \tag{27}$$

$$\begin{aligned} g^r(P) &= s_1 + s_2 P + s_3 P^2 \\ g^b(P) &= t_1 + t_2 P + t_3 P^2 \\ s_1 = t_1 &= 2\frac{\tau^r \tau^b}{\tau^r + \tau^b} \\ s_2 &= 2\frac{\tau^r - s_1}{\delta} \\ s_3 &= -\frac{s_2}{2\delta} \\ t_2 &= 2\frac{t_1 - \tau^b}{\delta} \\ t_3 &= -\frac{t_2}{2\delta} \end{aligned} \tag{28}$$

Tölke et al. [61] use the same order parameter P as Grunau et al. [60], but the calculation of τ based on P is different. Specifically, Tölke et al. [61] use a simple linear interpolation to compute τ in the vicinity of the fluid–fluid interface:

$$\tau(\mathbf{x},t) = \frac{1}{2}(1 + P(\mathbf{x},t))\tau^r + \frac{1}{2}(1 - P(\mathbf{x},t))\tau^b \tag{29}$$

Finally, the interpolation scheme of Liu et al. [88] is based on a method devised by Zu and He [89] for use in the free-energy method. This interpolation scheme again uses the same order parameter P, but now τ is evaluated as a weighted harmonic mean of τ^r and τ^b:

$$\frac{1}{\tau(\mathbf{x},t) - 0.5} = \frac{1 + P(\mathbf{x},t)}{2(\tau^r - 0.5)} + \frac{1 - P(\mathbf{x},t)}{2(\tau^b - 0.5)} \tag{30}$$

Figure 3 demonstrates the effectiveness of these three popular interpolation schemes for the classic benchmark simulation of two-phase flow in a channel (layered Poiseuille flow). For Figure 3, the testing domain is a 10×200 lattice domain with periodic boundary conditions and with a force applied to the fluid in the middle of the channel (methods for applying a force to the fluid will be described subsequently in Section 6). Other details regarding the simulation parameters are provided by Liu et al. [88]. In this figure, we simulated the flow of two fluids with a viscosity ratio of 12.5, i.e., the fluid in the middle of the channel (from lattice units 50 to 150) has a higher viscosity than that of the fluid near the edges of the channel. The CGM is used according to Section 3, and the LBM results are compared to a known analytical solution provided in Chapter 2 of Huang et al. [26] (cf., [32,40,87]). In Figure 3, we generated the results in the top-left panel without viscosity interpolation, and we generated the results in the remaining panels using the viscosity interpolation methods of Grunau et al. [60], Tölke et al. [61], and Liu et al. [88], as indicated on the figure. The improvement afforded by each viscosity interpolation is obvious.

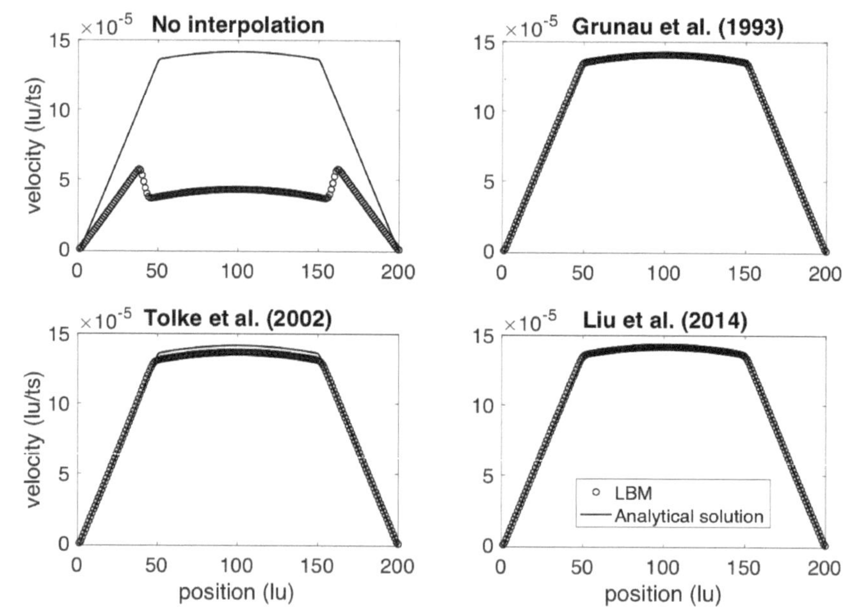

Figure 3. Predicted velocity profiles for two-phase flow in a channel, without and with the τ interpolation schemes [60,61,88]. The viscosity ratio of the two fluids is 12.5, i.e., the fluid in the center of the channel has a viscosity higher than that of the fluid at the edges of the channel.

All three interpolation schemes considerably reduce the error between the analytical and LBM velocities. Among these, the methods of Grunau et al. [60] and Liu et al. [88] perform similarly and accurately while the method of Tölke et al. [61] can not match the analytical solution completely. The methods of both Grunau et al. [60] and Liu et al. [88] have been successfully implemented and are popularly used [42,55,68,85,90]. The choice of which method to use is mostly up to the user's preference, but we note that the method of Liu et al. [88] requires fewer computations without any apparent degradation of performance. We therefore use the method of Liu et al. [88] in any simulations discussed subsequently in this paper (Sections 5 and 6).

In many applications, such as in simulations of drainage or imbibition, the fluid–fluid interface moves over time. Thus, the values of $P(\mathbf{x}, t)$ change over space and time; so too then do the values of the interpolated viscosity $\tau(\mathbf{x}, t)$. This means that any other variables that depend on τ—for instance, A in Equations (21) and (22)—also change over space and time. The values of P, τ, and A must therefore be recalculated at each lattice node at each time step.

As can be seen from Figure 3, the use of viscosity interpolation can, under some conditions, allow the CGM to perform well when the two fluids are of unequal viscosities. However, if the ratio of the fluid viscosities becomes extreme (e.g., the viscosity ratio is greater than $\sim$10), viscosity interpolation might not be sufficient to ensure that simulations are stable and accurate. Furthermore, the results observed in Figure 3 depend not only on the ratio of ν^r/ν_b but also on the specific values of ν^r and ν^b selected; choosing different values of ν^r and ν^b might alter the results even if the ratio ν^r/ν^b is held constant. BGK solutions are not necessarily fixed by non-dimensional numbers [66,91,92]. For either or both of these reasons, it might be necessary in some situations to employ more sophisticated approaches, such as replacing the BGK collision algorithm with a multiple relaxation time (MRT) algorithm. The application of the MRT to the color gradient method will be considered and reviewed in a future paper in this series.

When specifying values near an interface, as discussed immediately above, Ginzburg [66] distinguishes between an "implicit" interface approach (in which a blue fluid node is adjacent to a red fluid node, with an interface implied to exist halfway in between) and an "explicit" interface approach (in which a grid node is located at the interface, with blue fluid nodes on one side and red fluid nodes on the other side). For the case of an implicit interface, it is possible to select the CGM parameters in a way that the lattice Boltzmann method exactly reproduces the analytical solution for two-phase Poiseuille flow [66]. For the case of an explicit interface, Ginzburg [66] argues for using the harmonic mean value of viscosity, ν, rather than relaxation time, τ. This approach ensures the continuity of the tangential shear stress at the interface and the exact momentum solution inside each phase in two-phase Poiseuille flow, but needs to extrapolate it from the bulk to an explicit interface. Users may therefore wish to adopt the approach of Ginzburg [66] rather than interpolating values of τ. Nevertheless, for relatively simple problems, interpolating the value of τ near the interface is an expedient way of accounting for moderate differences in fluid viscosities, as demonstrated by Figure 3.

5. Comparison of Two Methods for Computing the Color Gradient

In Section 3.4, we presented the original method of calculating the color gradient as proposed by Rothman and Keller [24], Gunstensen et al. [25], and Grunau et al. [60]. This method of calculating the color gradient has been used successfully in many applications (e.g., [29,55]). However, Leclaire et al. [70] noted that standard color gradient models (as described heretofore in this paper) are plagued by problems such as spurious fluid velocities near fluid interfaces and the inability to model fluids of widely varying densitities, and that these problems are due partly to the fact that the calculation of the color gradient (Equation (17)) contains discretization errors that are directionally biased (anisotropic). Leclaire et al. [70] therefore presented alternative ways of calculating the color gradient that remove the anisotropy of the discretization errors, which improves the accuracy and robustness of the model. We here refer to these as "isotropic" color gradient calculations, in contrast to the "original" color gradient calculation presented in Equation (17).

For a D2Q9 lattice, as shown in Figure 1, the following modification to Equation (17) can be used [70]:

$$\mathbf{C}(\mathbf{x},t) = \sum_i \mathbf{e}_i \, \xi_i \left[\rho^r(\mathbf{x}+\mathbf{e}_i,t) - \rho^b(\mathbf{x}+\mathbf{e}_i,t)\right] \tag{31}$$

in which the weights ξ_i are given by the following:

$$\xi = \begin{bmatrix} \frac{1}{3} & \frac{1}{3} & \frac{1}{3} & \frac{1}{3} & \frac{1}{12} & \frac{1}{12} & \frac{1}{12} & \frac{1}{12} & 0 \end{bmatrix}. \tag{32}$$

For other lattice schemes (e.g., D3Q15 and D3Q27), the reader is referred to Leclaire et al. [70].

In Figure 4, we compare the original gradient approach to the isotropic gradient approach by applying both approaches to the problem of a "bubble test" or "droplet test". For this test, a red fluid droplet of radius R is emplaced in a field of blue fluid, the system is allowed to equilibrate, and then the pressure drop is measured across the interface of the droplet. In two dimensions, the theoretical pressure drop across the interface is $\Delta P = \sigma / R$, where σ is the interfacial tension specified by the user. The measured value of ΔP from the CGM simulation can be compared to the theoretical value of ΔP as a way to validate the CGM method. In Figure 4, the top row of panels shows the shape of the bubble after equilibration under different test conditions, and the bottom row of panels shows the agreement between measured and theoretical values of ΔP for those test conditions. For all simulations, the overall domain size was 128 × 128 l.u.; the specified interfacial tension was $\sigma = 0.0010$ (in LBM units); the interface thickness parameter $\beta = 0.85$; the viscosity $\nu^r = \nu^b = 0.16667$ l.u.^{2}/t.s.; the simulations were allowed to equilibrate for 80,000 time steps before ΔP was measured; and the radius of the red droplet was varied between 16 and 32 l.u. (the top row of panels shows the results for the cases in which $R = 32$ l.u.).

The left-most column in Figure 4 shows the results when $\rho^r = 2.0$, $\rho^b = 1.0$, and the original gradient is used. It can be seen that the droplet retains a mostly circular shape (panel (A)) and that the measured values of ΔP agree well with the theoretical relationship (panel (D)). A user might reasonably conclude that the original gradient calculation is acceptable for these test conditions. The center column of the figure shows the results when $\rho^r = \rho^b = 1.0$ and the original gradient is used. In this case, the measured values of ΔP are still close to the theoretical values (panel (E)), but the shape of the droplet is noticeably non-circular (anisotropic; panel (B)). The error in the gradient calculation is observable in the anisotropic shape of the droplet. Therefore, the right-most column of Figure 4 shows the results from the same conditions as the center column, but using the isotropic gradient calculation instead of the original gradient calculation. The measured values of ΔP are slightly improved (panel (F)), and the droplet shape attains the desired circular shape (panel (C)). A comparison of panel (B) to panel (C) shows that the isotropic gradient is clearly preferable to the original gradient under these test conditions.

Moreover, there is a distinct difference in the generated spurious velocities in the original color gradient and the isotropic color gradient methods. For the test case shown in Figure 4 when R = 32 l.u., the maximum spurious velocities are equal to 2.3×10^{-4} l.u./t.s. when using the original gradient calculation but are reduced to 7.1×10^{-6} l.u./t.s. when the isotropic color gradient is used. These results are in agreement not only with the observations of Leclaire et al. [70] but also with those of Lafarge et al. [93] and Mora et al. [94]. The subject of spurious velocities in these simulations will be revisited in Section 7.

Therefore, in general, we recommend the isotropic gradient calculation (Equations (31) and (32)) over the original gradient calculation (Equation (17)). The simulation results are improved without any meaningful increase in the complexity or run time of the computer codes. However, when employing the isotropic color gradient, the user must be certain to use Equation (22) rather than Equation (21) when calculating the A parameter to yield a desired value of σ.

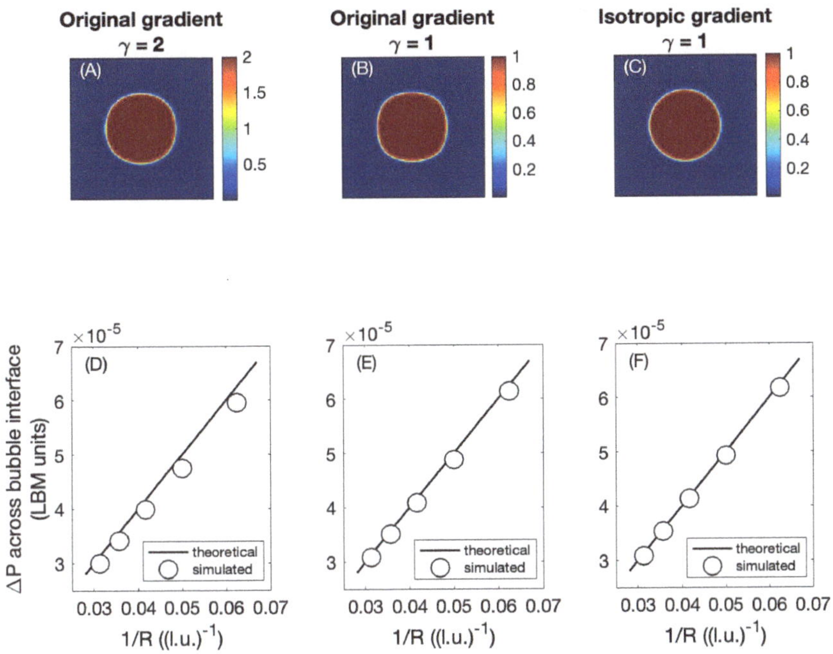

Figure 4. A comparison of the original and isotropic color gradient methods using a droplet test (bubble test) as a validation.

6. Application of a Force or Acceleration to the Fluids

A fluid flows in response to a pressure gradient or in response to a force acting on the fluid. Therefore, to simulate either single-phase or multi-phase flow, we must have methods for incorporating pressure gradients and/or external forces acting on the fluid(s). One way to do this is by specifying boundary conditions, e.g., specifying the fluid pressure or the fluid velocity on the boundaries of the domain [95]. Another approach is by including forces into the LBM simulations; methods that incorporate forces are often referred to as "forcing schemes" [56,96].

Many forcing schemes have been presented in the literature; a thorough review has been provided by Bawazeer et al. [96] for the flow of a single fluid. Here, we consider two of the most common forcing schemes and demonstrate how to apply them to a multi-phase flow simulation. In the first approach, originally proposed by Shan and Chen [20], external forces acting on the fluid are accounted for by including an acceleration term in the equilibrium distribution function f^{eq}. In the second approach, proposed by Luo [97,98], a source term is added to include the force or acceleration as part of the collision step of the CGM.

In the method of Shan and Chen [20], the macroscopic velocity $\mathbf{u}$ in the computation of $f^{\text{eq},k}$ is replaced by a new velocity $\mathbf{u}^{\text{eq},k}$ that includes the effects of an acceleration $\mathbf{a}^k$

acting on fluid k. The acceleration $\mathbf{a}^k$ need not necessarily be the same on the red fluid and the blue fluid. Thus, Equation (14) becomes

$$f_i^{\mathrm{eq},k}(\mathbf{x},t) = \rho^k\left(C_i^k + w_i\left[3\left(\mathbf{e}_i\cdot\mathbf{u}^{\mathrm{eq},k}\right) + \frac{9}{2}\left(\mathbf{e}_i\cdot\mathbf{u}^{\mathrm{eq},k}\right)^2 - \frac{3}{2}\left(\mathbf{u}^{\mathrm{eq},k}\cdot\mathbf{u}^{\mathrm{eq},k}\right)\right]\right) \tag{33}$$

in which

$$\mathbf{u}^{\mathrm{eq},k}(\mathbf{x},t) = \mathbf{u}(\mathbf{x},t) + \mathbf{a}^k\,\tau \tag{34}$$

and τ represents the interpolated relaxation time at location $\mathbf{x}$ and time t. The remainder of the color gradient algorithm remains unchanged. Thus, in practice, only the calculation of $f^{\mathrm{eq},k}$ needs to be changed to implement this forcing scheme.

The method of Luo [97,98] accounts for forces in a different manner than the method of Shan and Chen [20]. Here, the macroscopic velocity $\mathbf{u}$ and the calculation of $f^{\mathrm{eq},k}$ remain unchanged, i.e., Equation (14) is used as presented. To account for the acceleration $\mathbf{a}^k$ acting on the fluids, a source term S_i^k is calculated as:

$$S_i^k(\mathbf{x},t) = w_i\frac{\rho^k(\mathbf{x},t)}{(c_s^k)^2}\left(\mathbf{e}_i\cdot\mathbf{a}^k\right) \tag{35}$$

in which w_i is given by Equation (7). The source term S_i^k must be calculated for both fluids at each location $\mathbf{x}$, each time t, and for each of the nine directions. As with the method of Shan and Chen [20], the method of Luo [97,98] allows for a different acceleration acting on the red and blue fluids. The source term is then incorporated into the collision step as follows:

$$f_i^k(\mathbf{x}+\mathbf{e}_i,t+1) = f_i^k(\mathbf{x},t) + (\Omega_i^k)^1 + S_i^k(\mathbf{x},t) \tag{36}$$

in which $(\Omega_i^k)^1$ is given by Equation (16) assuming that the BGK algorithm is used.

We compared the two forcing schemes described above via a layered Poiseuille flow validation test. In this test, red and blue fluids both flow between two parallel plates. A layer of red fluid is "sandwiched" in the middle of the channel between two layers of blue fluid. An acceleration a is applied in the $+x$ direction to the red fluid only, which induces flow. The situation is depicted graphically in Figure 5. This is a classic validation test for simulations of multi-phase flow because there is an analytical solution for the resulting velocity profile in the channel; see, for instance, Liu et al. [88].

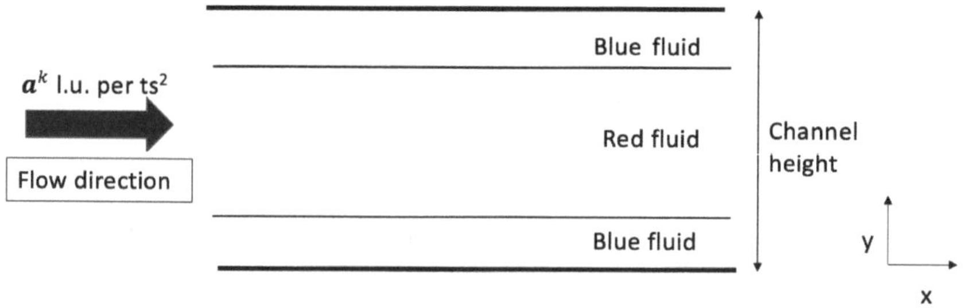

Figure 5. Two-phase Poiseuille flow simulation setup [88].

We performed the test using a domain of 10 nodes in the x-direction and 120 nodes in the y-direction. As shown in Figure 5, the top and bottom walls are solid boundaries at which a no-slip boundary condition (half-way bounceback) is imposed. Periodic boundary conditions are applied on the left and right boundaries, which is why only 10 nodes are required in the x-direction. We performed the test for both the Shan and Chen [20] forcing scheme, which we now call the "feq-accel" method following Bawazeer et al. [96], and the Luo [97,98] forcing scheme, which we now call the "source term" method. For both methods, we tested viscosity ratios (ν^r/ν^b) of 0.25 and 4.0; thus, a total of four simulations were performed. For all tests, the interfacial tension $\sigma = 0.010$, the applied acceleration on the red fluid was $a^r = 1.5 \times 10^{-8}$ l.u./(t.s.)^2, the interface thickness parameter $\beta = 0.9$, the fluid densities $\rho^r = \rho^b = 1.0$, and simulations were run for at least 80,000 time steps to ensure that steady state was reached.

The results of the simulations are shown in Figure 6. The figure shows the velocity profiles of the fluids between the two parallel plates. At the top and bottom of the domain (i.e., at the two plates), the fluid velocity is 0 because of the no-slip boundary condition. The velocity reaches a maximum in the middle of the channel. There are two key findings as illustrated in Figure 6: first, there is no noticeable difference between the the left two panels [20] and the right two panels [97,98] of the figure, indicating that the two forcing schemes yield indistinguishable results for this test problem; second, both methods yield results that agree nearly perfectly with the analytical solution, indicating that either method is acceptable for the conditions tested.

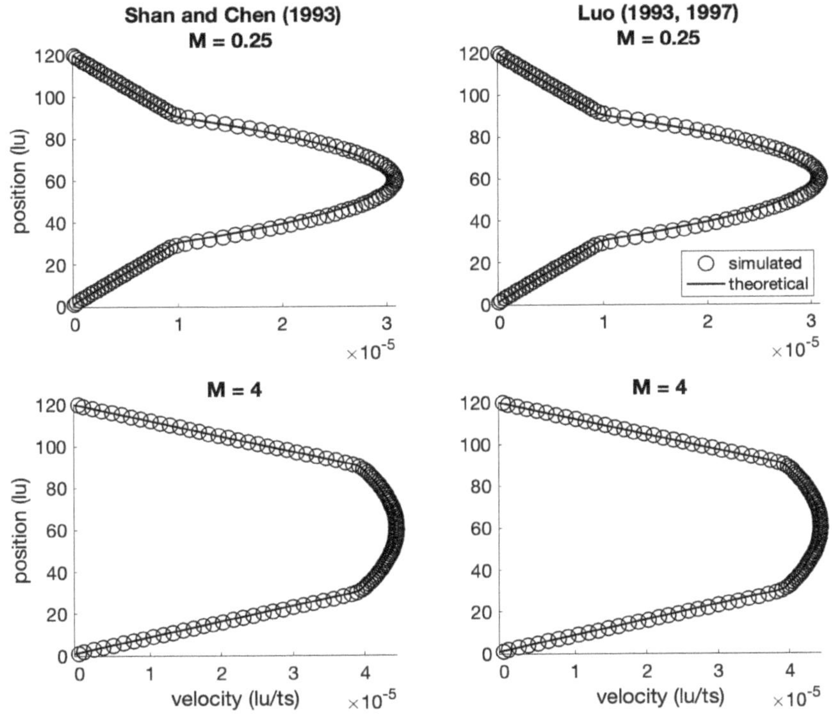

Figure 6. Comparison of layered Poiseuille flow results from the feq-accel forcing scheme of Shan and Chen [20] and the source term forcing scheme of Luo [97,98].

The equivalence of the two forcing schemes in Figure 6 is not surprising for the particular test problem considered. An examination of Equations (10), (14), (16) and (33)–(36) shows that the two methods are mathematically equivalent for cases in which the last two terms of the f^{eq} distribution functions can be ignored. This is the case for grid-aligned Poiseuille flow; see Ginzburg et al. [91]. Also, the relationship and agreement between these two methods for the case of single-fluid flow have been elucidated by Huang et al. [99].

Based on the agreement between the two methods, it is not the case that one method is generally "better" than the other for multi-phase flow simulations that use the BGK collision operator. Nevertheless, we find it useful to present and compare the two proposed approaches here, for a few reasons. First, some previous papers have indicated that using gravitational acceleration in multi-phase simulations is not preferred (e.g., [19,75]), which points toward the use of an "external" forcing scheme. Second, Silva [100] demonstrated that if the applied force is spatially varying, including the acceleration term in the f^{eq} expression leads to a discrepancy between the CGM and the incompressible Navier–Stokes equations with an external body force under steady-state conditions; this also argues in favor of the source term forcing scheme. Finally, we found that when deploying more advanced CGM techniques, such as multiple relaxation times instead of the BGK collision scheme, one or the other of these two proposed forcing schemes can sometimes be easier to implement or code in practice. It is therefore useful to the reader to be familiar with both approaches in case a particular problem favors one over the other from a practical standpoint.

7. Challenges of Simulating Multi-Phase Flow in CGM

A multi-phase simulation in the CGM must be free from artificial fluid effects, which can arise near fluid–fluid and fluid–solid boundaries. For example, a good multi-phase simulation involves distinct phases separated by a thin interface, maintains fluid continuity between the phases, provides minimal fictitious or spurious velocities, and provides minimal fluid distortion. Therefore, a main challenge in the CGM is accurately depicting fluid flow while eliminating any artificial fluid effects. There is no numerical multi-phase flow model that is 100% accurate and free from error. However, by understanding the principles and theoretical foundations of the CGM, such errors can be minimized, and the method can provide a reasonable depiction of fluid flow processes.

One common error in the CGM is the presence of spurious velocities, i.e., the presence of fictitious fluid velocities where the fluids are supposed to be stationary. Such spurious velocities can arise near the interface of a bubble or droplet, as in Section 5. In that section, we discussed how the calculation of the color gradient can influence the droplet shape, as shown in Figure 4. The calculation of the color gradient can also affect the magnitude of spurious velocities. The algorithms discussed heretofore in this paper have made efforts to reduce or eliminate those fictitious velocities, but they usually do not disappear completely [68,70,94]. Specifically, for the case shown in Figure 4, the maximum magnitude of these spurious velocities is around 2.3×10^{-4} l.u./t.s. when using the "original" color gradient and around 7.1×10^{-6} l.u./t.s. when using the isotropic color gradient; see Figure 7. By using a higher-order gradient discretization, artificial velocities can be reduced, which also accompanies a reduction in the distortion in fluid shape (as seen in Figure 4).

In a CGM simulation, the major challenges are minimizing the presence of spurious velocities, ensuring pressure continuity between phases, minimizing the interface thickness, and minimizing distortion in fluid shape (e.g., droplet shape). This paper focused on the BGK-CGM formulation, and errors were considered in the context of common CGM code-validation tests that can be implemented with the BGK collision algorithm—which means that we considered relatively "easy" test conditions. In cases where simulations must account for fluid wettability, a higher viscosity ratio, and/or a higher density ratio, errors are likely to become more prominent. In future papers in this series, we will review efforts that have been made to minimize errors in more sophisticated contexts [87,101,102].

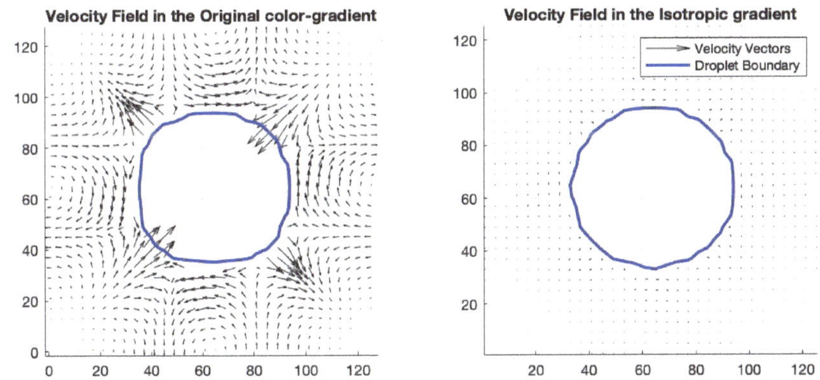

Figure 7. Spurious velocities generated by original and isotropic color gradient algorithms in a bubble test. Using the isotropic gradient reduces the magnitude of spurious velocities by a factor of about 30.

Another challenge in executing the LBM simulations—not limited to the color gradient method—is appropriately determining the relationship between "LBM variables" and "real-world variables". This is achieved by using appropriate unit conversions for length, time, and mass. For instance, what real-world length scale is represented by one lattice unit, and what real-world time scale is represented by one time step? The concepts of scaling the LBM units have been considered previously by Pan et al. [5], Krüger et al. [56], and Baakeem et al. [103], among others. A key point is that, in many instances, particularly if we want to model the flow of low-viscosity fluids, the LBM lattice unit and time step must correspond to very small values of real-world length and time—millimeters and milliseconds, or perhaps even smaller. Thus, a simulation that takes hours to run might only simulate, say, the fluid flow through a 10 cm $\times$ 10 cm square for a period of a few minutes. Therefore, modeling fluid flow through large domains (a scale of meters) over time scales of hours or days can be very computationally expensive. A way to address the challenge of computational expense is through the parallelization of the LBM algorithms, which has been considered elsewhere [12,15,18,19,104].

8. Summary and Conclusions

Over the past 30 years, the lattice Boltzmann method (LBM) has become increasingly popular for modeling multi-phase fluid flow in porous media, and one popular variant of the LBM is the color gradient method. However, along with the rising popularity of the color gradient method has come a proliferation of different proposed algorithms and strategies for implementing various aspects of the method. Selecting the "best" algorithms and subroutines is daunting, especially for relatively new users of the method. Therefore, we aim to critically review the capabilities and implementation of the CGM, particularly for modeling pore-scale multi-phase flow in porous media, through a series of papers, of which this is the first.

In this paper, we provided a brief introduction to the LBM and CGM that enables and facilitates consideration of the subsequent topics. Following that introduction, we reviewed three specific aspects of color gradient modeling: different methods for modeling the flow of fluids of moderately different viscosities, different methods for calculating the color gradient, and different methods for modeling external forces or accelerations acting upon the flowing fluids. Specific findings from this paper include the following:

1. Spatial interpolation of the collision relaxation times, τ, enables accurate modeling of the flow of fluids of different viscosities as long as the viscosity contrast is moderate;

2. Of the methods considered for τ interpolation, that of Liu et al. [88] is accurate and easy to implement;
3. Using the "original" color gradient calculation of Rothman and Keller [24], Gunstensen et al. [25], and Grunau et al. [60] is sufficient in some cases, but the "isotropic" color gradient calculation of Leclaire et al. [70] improves upon the original method without any significant increase in model complexity;
4. External accelerations acting on the flowing fluids can be modeled using a variety of forcing schemes, but for simple problems, there is no significant difference between the "feq-accel" forcing scheme of Shan and Chen [20] and the "source term" forcing scheme of Luo [97,98].

Understanding these aspects of the color gradient method is necessary and sufficient to allow the user to perform two common "benchmark" tests of lattice Boltzmann modeling, namely bubble tests and layered Poiseuille flow.

Future papers in the series will build upon these topics, considering phenomena or conditions such as fluids of widely differing viscosity and/or density, how to introduce the "wettability" of solid surfaces and specify a desired contact angle, how to simulate fluid flow in a domain with an open boundary condition (e.g., fluid flushing out of a porous medium), and how to extend the CGM from two spatial dimensions to three. By systematically reviewing these key aspects and features, this series of papers will extend our collective ability to apply the CGM to a variety of important problems in geosciences and engineering.

Author Contributions: Conceptualization, F.Z. and J.A.C.; methodology, F.Z. and J.A.C.; software, F.Z. and J.A.C.; validation, F.Z. and J.A.C.; formal analysis, F.Z. and J.A.C.; investigation, F.Z.; resources, J.A.C.; data curation, F.Z. and J.A.C.; writing—original draft preparation, F.Z.; writing—review and editing, F.Z. and J.A.C.; visualization, F.Z. and J.A.C.; supervision, J.A.C.; project administration, F.Z. and J.A.C.; funding acquisition, F.Z. and J.A.C. All authors have read and agreed to the published version of the manuscript.

Funding: This paper is based on the funding and support from the Foreign Fulbright PhD Fellowship (awarded to F.Z.) and from the Commonwealth Scientific and Industrial Research Organization (CSIRO, Australia, support of J.A.C. during 2013–2014). Any opinions, findings, conclusions, or recommendations expressed in this material are those of the authors and do not necessarily reflect the views of the Fulbright Association or of CSIRO.

Data Availability Statement: Code development data related to the CGM will be available upon request from the authors.

Acknowledgments: Some of the computations for this work were performed on the Research Computing Resources of the University of South Florida (USF). The authors thank Ethan Yan for his assistance with code development and Amy Stuart for her assistance with deploying the codes on USF's Research Computing Resources. The authors also thank Sebastian Leclaire for sharing his Laplace bubble test MATLAB Version R2020b codes, which helped us develop our own Laplace bubble test. Lastly, the authors thank Al Valocchi for his words of encouragement at various stages during the coding process.

Conflicts of Interest: The authors declare no conflicts of interest.

Abbreviations

The following abbreviations are used in this manuscript:

BGK	Bhatnagar, Gross, Krook
CGM	Color gradient method
LBM	Lattice Boltzmann method
MRT	Multiple relaxation time

References

1. Blunt, M.J. *Multiphase Flow in Permeable Media: A Pore-Scale Perspective*; Cambridge University Press: Cambridge, UK, 2017.
2. Meakin, P.; Tartakovsky, A.M. Modeling and simulation of pore-scale multiphase fluid flow and reactive transport in fractured and porous media. *Rev. Geophys.* **2009**, *47*, RG3002. [CrossRef]
3. Zhao, B.; MacMinn, C.W.; Primkulov, B.K.; Chen, Y.; Valocchi, A.J.; Zhao, J.; Kang, Q.; Bruning, K.; McClure, J.E.; Miller, C.T.; et al. Comprehensive comparison of pore-scale models for multiphase flow in porous media. *Proc. Natl. Acad. Sci. USA* **2019**, *116*, 13799–13806. [CrossRef] [PubMed]
4. Guo, Z.; Zhao, T.S. Lattice Boltzmann model for incompressible flows through porous media. *Phys. Rev. E* **2002**, *66*, 036304. [CrossRef] [PubMed]
5. Pan, C.; Hilpert, M.; Miller, C.T. Lattice-Boltzmann simulation of two-phase flow in porous media. *Water Resour. Res.* **2004**, *40*, W01501. [CrossRef]
6. Sukop, M.C.; Thorne, D.T.J. *Lattice Boltzmann Modeling: An Introduction for Scientists and Engineers*; Springer: New York, NY, USA, 2006.
7. Porter, M.L.; Schaap, M.G.; Wildenschild, D. Lattice-Boltzmann simulations of the capillary pressure–saturation–interfacial area relationship for porous media. *Adv. Water Resour.* **2009**, *32*, 1632–1640. [CrossRef]
8. Boek, E.S.; Venturolli, M. Lattice-Boltzmann studies of fluid flow in porous media with realistic rock geometries. *Comput. Math. Appl.* **2010**, *59*, 2305–2314. [CrossRef]
9. Ho, M.; Tucny, J.M.; Ammar, S.; Leclaire, S.; Reggio, M.; Trepanier, J.Y. Lattice Boltzmann model for rarefied gaseous mixture flows in three-dimensional porous media including Knudsen diffusion. *Fluids* **2024**, *9*, 237. [CrossRef]
10. Chen, L.; Kang, Q.; Mu, Y.; He, Y.L.; Tao, W.Q. A critical review of the pseudopotential multiphase lattice Boltzmann model: methods and applications. *Int. J. Heat Mass Transf.* **2014**, *76*, 210–236. [CrossRef]
11. Atykhan, M.; Kabdenova, B.; Monaco, E.; Rojas-Solorzano, L. Modeling immiscible fluid displacement in a porous medium using lattice Boltzmann method. *Fluids* **2021**, *6*, 89. [CrossRef]
12. Kandhai, D.; Koponen, A.; Hoekstra, A.G.; Kataja, M.; Timonen, J.; Sloot, P.M. Lattice Boltzmann hydrodynamics on parallel systems. *Comput. Phys. Commun.* **1998**, *111*, 14–26. [CrossRef]
13. Wang, M.; Pan, N. Prediction of effective physical properties of complex multiphase materials. *Mater. Sci. Eng. R Rep.* **2008**, *63*, 1–30. [CrossRef]
14. Aidun, C.K.; Clausen, J.R. Lattice-Boltzmann method for complex flows. *Annu. Rev. Fluid Mech.* **2010**, *42*, 439–472. [CrossRef]
15. Tölke, J. Implementation of a lattice Boltzmann kernel using the Compute Unified Device Architecture developed by nVIDIA. *Comput. Vis. Sci.* **2010**, *13*, 29–39. [CrossRef]
16. Liu, H.; Kang, Q.; Leonardi, C.R.; Schmieschek, S.; Narváez, A.; Jones, B.D.; Williams, J.; Valocchi, A.J.; Harting, J. Multiphase lattice Boltzmann simulations for porous media applications. *Comput. Geosci.* **2016**, *20*, 777–805. [CrossRef]
17. McClure, J.E.; Li, Z.; Berrill, M.A.; Ramstad, T. The LBPM software package for simulating multiphase flow on digital images of porous rocks. *Comput. Geosci.* **2021**, *25*, 871–895. [CrossRef]
18. Yang, J.; Xu, Y.; Yang, L. Taichi LBM3D: A single-phase and multiphase lattice Boltzmann solver on cross-platform multicore CPU/GPUs. *Fluids* **2022**, *7*, 270. [CrossRef]
19. Yang, G.; Chen, S.; Wang, M. Implementation of a direct-addressing based lattice Boltzmann GPU solver for multiphase flow in porous media. *Comput. Phys. Commun.* **2023**, *65*, 108828. [CrossRef]
20. Shan, X.; Chen, H. Lattice Boltzmann model for simulating flows with multiple phases and components. *Phys. Rev. E* **1993**, *47*, 1815–1819. [CrossRef]
21. Swift, M.R.; Orlandini, E.; Osborn, W.R.; Yeomans, J.M. Lattice Boltzmann simulations of liquid-gas and binary fluid systems. *Phys. Rev. E* **1996**, *54*, 5041–5052. [CrossRef]
22. He, X.; Chen, S.; Zhang, R. A lattice Boltzmann scheme for incompressible multiphase flow and its application in simulation of Rayleigh-Taylor instability. *J. Comput. Phys.* **1999**, *152*, 642–663. [CrossRef]
23. Lee, T.; Liu, L. Lattice Boltzmann simulations of micron-scale drop impact on dry surfaces. *J. Comput. Phys.* **2010**, *229*, 8045–8062. [CrossRef]

24. Rothman, D.H.; Keller, J.M. Immiscible Cellular-Automaton Fluids. *J. Stat. Phys.* **1988**, *52*, 1119–1127. [CrossRef]
25. Gunstensen, A.K.; Rothman, D.H.; Zaleski, S.; Zanetti, G. Lattice Boltzmann model of immiscible fluids. *Phys. Rev. A* **1991**, *43*, 4320–4327. [CrossRef] [PubMed]
26. Huang, H.; Sukop, M.C.; Lu, X.Y. *Multiphase Lattice Boltzmann Methods: Theory and Application*; John Wiley & Sons: Chichester, UK, 2015.
27. Latva-Kokko, M.; Rothman, D. Diffusion properties of gradient-based lattice Boltzmann models of immiscible fluids. *Phys. Rev. E* **2005**, *71*, 056702. [CrossRef] [PubMed]
28. McClure, J.E.; Berill, M.A.; Gray, W.G.; Miller, C.T. Tracking interface and common curve dynamics for two-fluid flow in porous media. *J. Fluid Mech.* **2016**, *796*, 211–232. [CrossRef]
29. Ahrenholz, B.; Tölke, J.; Lehmann, P.; Peters, A.; Kaestner, A.; Krafczyk, M.; Durner, W. Prediction of capillary hysteresis in a porous material using lattice-Boltzmann methods and comparison to experimental data and a morphological pore network model. *Adv. Water Resour.* **2008**, *31*, 1151–1173. [CrossRef]
30. Chen, Y.; Li, Y.; Valocchi, A.J.; Christensen, K.T. Lattice Boltzmann simulations of liquid CO_2 displacing water in a 2D heterogenous micromodel at reservoir pressure conditions. *J. Contam. Hydrol.* **2018**, *212*, 14–27. [CrossRef]
31. Sarkezi-Selsky, P.; Schmies, H.; Kube, A.; Latz, A.; Jahnke, T. Lattice Boltzmann simulation of liquid water transport in gas diffusion layers of proton exhange membrane fuel cells: Parametric studies on capillary hysteresis. *J. Power Sources* **2022**, *535*, 231381. [CrossRef]
32. Fu, X.; Sun, J. Improved three-dimensional multiple-relaxation-time color-gradient lattice Boltzmann finite-difference model for thermocapillary flows. *Phys. Fluids* **2023**, *35*, 072105.
33. Zahid, F.; Cunningham, J.A. Effect of Grain-Size Distribution on Temporal Evolution of Interfacial Area during Two-phase Flow in Porous Media. *Transp. Porous Med.* **2022**, *144*, 283–300. [CrossRef]
34. Liu, H.; Sun, S.; Wu, R.; Wei, B.; Hou, J. Pore-scale modeling of spontaneous imbibition in porous media using the lattice Boltzmann method. *Water Resour. Res.* **2021**, *57*, 2020WR029219. [CrossRef]
35. Sheikholeslam-Noori, S.M.; Taeibi-Rahni, M.; Shams-Taleghani, S.A. Multiple-relaxation time color-gradient lattice Boltzmann model for simulating contact angle in two phase flows with high density ratios. *Eur. Phys. J. Plus* **2019**, *134*, 399. [CrossRef]
36. Zakirov, T.R.; Zhuchkova1, O.S.; Khramchenkov, M.G. Mathematical Model for Dynamic Adsorption with Immiscible Multiphase Flows in Three-dimensional Porous Media. *Lobachevskii J. Math.* **2024**, *45*, 888–898. [CrossRef]
37. Zakirov, T.R.; Khayuzkin, A.S.; Khramchenkov, M.G.; Galeev, A.A.; Kosterina, E.A. Effect of gravity number on soil contamination during dynamic adsorption of heavy metal ions: A lattice Boltzmann study. *Int. Commun. Heat Mass Transf.* **2024**, *158*, 107852. [CrossRef]
38. Li, Q.; Yang, G.; Huang, Y.; Lu, X.; Min, J.; Wang, M. Lattice Boltzmann method for particulate multiphase flow system. *Int. J. Mech. Sci.* **2024**, *273*, 109217. [CrossRef]
39. Datadien, K.P.N.; Staso, G.D.; Diddens, C.; Wijshoff, H.M.A.; Toschi, F. Comparison of Lattice Boltzmann, Finite Element and Volume of Fluid Multicomponent Methods for Microfluidic Flow Problems and the Jetting of Microdroplets. *Commun. Comput. Phys.* **2023**, *33*, 912–936. [CrossRef]
40. Huang, H.; Wang, L.; Lu, X.Y. Evaluation of three lattice Boltzmann models for multiphase flows in porous media. *Comput. Math. Appl.* **2011**, *61*, 3606–3617. [CrossRef]
41. Yang, J.; Boek, E.S. A comparison study of multi-component lattice Boltzmann models for flow in porous media applications. *Comput. Math. Appl.* **2013**, *65*, 882–890. [CrossRef]
42. Leclaire, S.; Parmigiani, A.; Chopard, B.; Latt, J. Three-dimensional lattice Boltzmann method benchmarks between color-gradient and pseudo-potential immiscible multi-component models. *Int. J. Mod. Phys. C* **2017**, *28*, 1750085. [CrossRef]
43. Guo, Z.; Shu, C. *Lattice Boltzmann Method and Its Applications in Engineering*; World Scientific Publishing Co.: Hackensack, NJ, USA, 2013.
44. Ba, Y.; Liu, H.; Li, Q.; Kang, Q.; Sun, J. Multiple-relaxation-time color-gradient lattice Boltzmann model for simulating two-phase flows with high density ratio. *Phys. Rev. E.* **2016**, *94*, 023310. [CrossRef]
45. Akai, T.; Bijeljic, B.; Blunt, M.J. Wetting boundary condition for the color-gradient lattice Boltzmann method: Validation with analytical and experimental data. *Adv. Water Resour.* **2018**, *116*, 56–66. [CrossRef]
46. Wen, Z.X.; Li, Q.; Yu, Y.; Luo, K.H. Improved three-dimensional color-gradient lattice Boltzmann model for immiscible two-phase flows. *Phys. Rev. E* **2019**, *100*, 023301. [CrossRef] [PubMed]
47. Frisch, U.; Hasslacher, B.; Pomeau, Y. Lattice-gas automata for the Navier-Stokes equation. *Phys. Rev. Lett.* **1986**, *56*, 1505–1508. [CrossRef] [PubMed]
48. McNamara, G.R.; Zanetti, G. Use of the Boltzmann equation to simulate lattice-gas automata. *Phys. Rev. Lett.* **1988**, *61*, 2332–2335. [CrossRef]
49. Higuera, F.J.; Succi, S.; Benzi, R. Lattice gas dynamics with enhanced collisions. *Europhys. Lett.* **1989**, *9*, 345–349. [CrossRef]
50. Higuera, F.J.; Jiménez, J. Boltzmann approach to lattice gas simulations. *Europhys. Lett.* **1989**, *9*, 663–668. [CrossRef]

51. Wolfram, S. Cellular automaton fluids, 1: Basic Theory. *J. Stat. Phys.* **1986**, *45*, 471–526. [CrossRef]
52. Bhatnagar, P.L.; Gross, E.P.; Krook, M. A model for collision processes in gases. I. Small amplitude processes in charged and neutral one-component systems. *Phys. Rev. E.* **1954**, *94*, 511–525. [CrossRef]
53. Qian, Y.H.; d'Humières, D.; Lallemand, P. Lattice BGK models for Navier-Stokes equation. *Europhys. Lett.* **1992**, *17*, 479–484. [CrossRef]
54. Lallemand, P.; Luo, L.S. Theory of the lattice Boltzmann method: Dispersion, dissipitation, isotropy, Galilean invariance, and stability. *Phys. Rev. E* **2000**, *61*, 6546. [CrossRef]
55. Reis, T.; Phillips, T.N. Lattice Boltzmann Model for Simulating Immiscible Two-Phase Flows. *J. Phys. A Math.* **2007**, *40*, 4033–4053. [CrossRef]
56. Krüger, T.; Kusumaatmaja, H.; Kuzmin, A.; Shardt, O.; Silva, G.; Viggen, E.M. *The Lattice Boltzmann Method: Principles and Practice*; Springer International Publishing: Cham, Switzerland, 2017.
57. Mohamad, A.A. *Lattice Boltzmann Methods: Fundamentals and Engineering Applications with Computer Codes*, 2nd ed.; Springer: London, UK, 2019.
58. Chai, Z.; Huang, C.; Shi, B.; Guo, Z. A comparative study on the lattice Boltzmann models for predicting effective diffusivity of porous media. *Int. J. Heat Mass Trans.* **2016**, *98*, 687–696. [CrossRef]
59. Zhu, W.; Kan, A.; Chen, Z.; Zhang, Q.; Zhang, J. A modified lattice Boltzmann method for predicting the effective thermal conductivity of open-cell foam materials. *Int. J. Heat Mass Trans.* **2022**, *133*, 105957. [CrossRef]
60. Grunau, D.; Chen, S.; Eggert, K. A lattice Boltzmann model for multiphase fluid flows. *Phys. Fluids A* **1993**, *5*, 2557–2562. [CrossRef]
61. Tölke, J.; Krafczyk, M.; Schulz, M.; Rank, E. Lattice Boltzmann simulations of binary fluid flow through porous media. *Philos. Trans. A Math. Phys. Eng. Sci.* **2002**, *360*, 535–545. [CrossRef]
62. Liu, H.; Valocchi, A.J.; Kang, Q. Three-dimensional lattice Boltzmann model for immiscible two-phase flow simulations. *Phys. Rev. E* **2012**, *85*, 046309. [CrossRef]
63. Ginzburg, I.; Steiner, K. A free-surface lattice Boltzmann method for modelling the filling of expanding cavities by Bingham fluids. *Philos. Trans. R. Soc. Lond. A* **2002**, *360*, 453–466. [CrossRef]
64. Ginzburg, I.; Steiner, K. Lattice Boltzmann model for free-surface flow and its application to filling process in casting. *J. Comput. Phys.* **2003**, *185*, 61–99. [CrossRef]
65. Kehrwald, D. Numerical Analysis of Immiscible Lattice BGK. Ph.D. Thesis, Fraunhofer Institut für Techno- und Wirtschaftsmathematik, Kaiserslautern, Germany, 2002.
66. Ginzburg, I. Lattice Boltzmann modeling with discontinuous collision components: Hydrodynamic and advection-diffusion equations. *J. Stat. Phys.* **2007**, *126*, 157–206. [CrossRef]
67. Leclaire, S.; Reggio, M.; Trépanier, J.Y. Numerical evaluation of two recoloring operators for an immiscible two-phase flow lattice Boltzmann model. *Appl. Math. Model.* **2012**, *36*, 2237–2252. [CrossRef]
68. Huang, H.; Huang, J.J.; Lu, X.Y. Study of immiscible displacements in porous media using color-gradient based multiphase lattice Boltzmann method. *Comput. Fluids* **2014**, *93*, 164–172. [CrossRef]
69. Leclaire, S.; Parmigiani, A.; Malaspinas, O.; Chopard, B.; Latt, J. Generalized three-dimensional lattice Boltzmann color-gradient method for immiscible two-phase pore-scale imbibition and drainage in porous media. *Phys. Rev. E* **2017**, *95*, 033306. [CrossRef] [PubMed]
70. Leclaire, S.; Reggio, M.; Trépanier, J.Y. Isotropic color gradient for simulating very high density ratios with a two-phase flow lattice Boltzmann model. *Comput. Fluids* **2011**, *48*, 98–112. [CrossRef]
71. Xu, Z.; Liu, H.; Valocchi, A.J. Lattice Boltzmann simulation of immiscible two-phase flow capillary valve effect in porous media. *Water Resour. Res.* **2017**, *53*, 3770–3790. [CrossRef]
72. Lishchuk, S.V.; Care, C.M.; Halliday, I. Lattice Boltzmann algorithm for surface tension with greatly reduced microcurrents. *Phys. Rev. E* **2003**, *67*, 036701. [CrossRef]
73. Xu, M.; Liu, H. Prediction of immiscible two-phase flow properties in a two-dimensional Berea sandstone using the pore-scale lattice Boltzmann simulation. *Eur. Phys. J. E* **2018**, *41*, 124. [CrossRef]
74. Chen, Y.; Valocchi, A.J.; Kang, Q.; Visvanathan, H.S. Inertial effects during the process of supercritical CO_2 displacing brine in a sandstone: Lattice Boltzmann simulations based on the continuum-surface-force and geometrical wetting models. *Water Resour. Res.* **2019**, *55*, 144–165. [CrossRef]
75. Huang, J.; Xiao, F.; Yin, X. Lattice Boltzmann simulation of pressure-driven two-phase flows in capillary tube and porous medium. *Comput. Fluids* **2017**, *155*, 134–145. [CrossRef]
76. Ginzbourg, I.; Adler, P.M. Surface tension models with different viscosities. *Transp. Porous Media* **1995**, *20*, 37–76. [CrossRef]
77. D'Ortona, U.; Salin, D.; Cieplak, M.; Rybka, R.B. Two-color nonlinear Boltzmann cellular automata: Surface tension and wetting. *Phys. Rev. E* **1995**, *51*, 3718–3728. [CrossRef]

78. Halliday, I.; Hollis, A.P.; Care, C.M. Lattice Boltzmann algorithm for continuum multicomponent flow. *Phys. Rev. E* **2007**, *76*, 026078. [CrossRef] [PubMed]
79. Lishchuk, S.V.; Halliday, I.; Care, C.M. Multicomponent lattice Boltzmann method for fluids with density contrast. *Phys. Rev. E* **2008**, *77*, 036702. [CrossRef] [PubMed]
80. Liu, H.; Wu, L.; Ba, Y.; Xi, G.; Zhang, Y. A lattice Boltzmann method for axisymmetric multicomponent flows with high viscosity ratio. *J. Comput. Phys.* **2016**, *27*, 052103. [CrossRef]
81. Leclaire, S.; Pellerin, S.; Reggio, M.; Trépanier, J.Y. An approach to control the spurious currents in a multiphase lattice Boltzmann method and to improve the implementation of initial condition. *Int. J. Numer. Methods Fluids* **2015**, *77*, 732–746. [CrossRef]
82. Chen, J.D.; Wilkinson, D. Pore-scale viscous fingering in porous media. *Phys. Rev. Lett.* **1985**, *55*, 1892–1895. [CrossRef]
83. Homsy, G. Viscous fingering in porous media. *Annu. Rev. Fluid Mech.* **1987**, *19*, 271–311. [CrossRef]
84. Dong, B.; Yan, Y.Y.; Li, W.; Song, Y. Lattice Boltzmann simulation of viscous fingering phenomenon of immiscible fluids displacement in a channel. *Comput Fluids* **2010**, *39*, 768–779. [CrossRef]
85. Liu, H.; Zhang, Y.; Valocchi, A.J. Lattice Boltzmann simulation of immiscible fluid displacement in porous media: Homogenous versus heterogeneous pore network. *Phys. Fluids* **2015**, *27*, 052103. [CrossRef]
86. Bakhshian, S.; Hosseini, S.A.; Shokri, N. Pore-scale characteristics of multiphase flow in heterogenous porous media using the lattice Boltzmann method. *Sci. Rep.* **2019**, *9*, 3377. [CrossRef]
87. Rannou, G. Lattice Boltzmann Method and Immiscible Two-Phase Flow. Ph.D. Thesis, Georgia Institute of Technology, Atlanta, GA, USA, 2008.
88. Liu, H.; Valocchi, A.J.; Werth, C.; Kang, Q.; Oostrom, M. Pore-scale simulation of liquid CO_2 displacement of water using a two-phase lattice Boltzmann model. *Adv. Water Resour.* **2014**, *73*, 144–158. [CrossRef]
89. Zu, Y.Q.; He, S. Phase-field based lattice Boltzmann model for incompressible binary fluid systems with density and viscosity contrasts. *Phys. Rev. E* **2013**, *87*, 043301. [CrossRef] [PubMed]
90. Huang, J. Pore-Scale Simulation of Multiphase Flow Using Lattice Boltzmann Method: Developments and Applications. Ph.D. Thesis, Colorado School of Mines, Golden, CO, USA, 2017.
91. Ginzburg, I.; Verhaege, F.; d'Humières, D. Two-Relaxation-Time Lattice Boltzmann Scheme: About Parametrization, Velocity, Pressure and Mixed Boundary Conditions. *Commun. Comput. Phys.* **2008**, *3*, 427–478.
92. d'Humières, D.; Ginzburg, I. Viscosity independent numerical errors for lattice Boltzmann models: From recurrence equations to 'magic' collision numbers. *Comput. Math. Appl.* **2009**, *58*, 823–840. [CrossRef]
93. Lafarge, T.; Boivin, P.; Odier, N.; Cuenot, B. Improved color-gradient method for lattice Boltzmann modeling of two-phase flows. *Phys. Fluids* **2021**, *33*, 082110. [CrossRef]
94. Mora, P.; Morra, G.; Yuen, D.A. Optimal surface-tension isotropy in the Rothman-Keller color-gradient lattice Boltzmann method for multiphase flow. *Phys. Rev. E* **2021**, *103*, 033302. [CrossRef]
95. Zou, Q.; He, X. On pressure and velocity boundary conditions for the lattice Boltzmann BGK model. *Phys. Fluids* **1997**, *9*, 1591–1598. [CrossRef]
96. Bawazeer, S.H.; Baakeem, S.S.; Mohamad, A.A. A critical review of forcing schemes in lattice Boltzmann method: 1993–2019. *Arch. Comput. Methods Eng.* **2021**, *28*, 4405–4423. [CrossRef]
97. Luo, L.S. Lattice-Gas Automata and Lattice Boltzmann Equations for Two-Dimensional Hydrodynamics. Ph.D. Thesis, Georgia Institute of Technology, Atlanta, GA, USA, 1993.
98. Luo, L.S. Analytic solutions of linearized lattice Boltzmann equation for simple flows. *J. Stat. Phys.* **1997**, *88*, 913–926. [CrossRef]
99. Huang, H.; Krafczyk, M.; Lu, X. Forcing term in single-phase and Shan-Chen-type multiphase lattice Boltzmann models. *Phys. Rev. E* **2011**, *84*, 046710. [CrossRef]
100. Silva, G. Discrete effects on the forcing term for the lattice Boltzmann modeling of steady hydrodynamics. *Comput. Fluids* **2020**, *203*, 104537. [CrossRef]
101. Huang, H.; Huang, J.J.; Lu, X.; Sukop, M.C. On simulations of high-density ratio flows using color-gradient multi-phase lattice Boltzmann models. *Int. J. Mod. Phys. C* **2013**, *24*, 1350021. [CrossRef]
102. Leclaire, S.; Pellerin, N.; Reggio, M.; Trépanier, J.Y. A multiphase lattice Boltzmann method for simulating immiscible liquid-liquid interface dynamics. *Appl. Math. Model.* **2016**, *40*, 6376–6394. [CrossRef]
103. Baakeem, S.S.; Bawazeer, S.H.; Mohamad, A.A. A novel approach of unit conversion in the lattice Boltzmann method. *Appl. Sci.* **2021**, *11*, 6386. [CrossRef]
104. Namvar, M.; Leclaire, S. Simple lattice Boltzmann model with low memory usage. *J. Comput. Sci.* **2022**, *62*, 101723. [CrossRef]

Disclaimer/Publisher's Note: The statements, opinions and data contained in all publications are solely those of the individual author(s) and contributor(s) and not of MDPI and/or the editor(s). MDPI and/or the editor(s) disclaim responsibility for any injury to people or property resulting from any ideas, methods, instructions or products referred to in the content.

fluids

Review

Bridging Large Eddy Simulation and Reduced-Order Modeling of Convection-Dominated Flows through Spatial Filtering: Review and Perspectives

Annalisa Quaini [1,†], Omer San [2,†], Alessandro Veneziani [3,†] and Traian Iliescu [4,*,†]

1 Department of Mathematics, University of Houston, 3551 Cullen Blvd, Houston, TX 77204, USA; quaini@math.uh.edu
2 Department of Mechanical, Aerospace and Biomedical Engineering, University of Tennessee, 1512 Middle Drive, Knoxville, TN 37996, USA; osan@utk.edu
3 Department of Mathematics, Emory University, 400 Dowman Drive, Atlanta, GA 30322, USA; avenez2@emory.edu
4 Department of Mathematics, Virginia Tech, 225 Stanger Street, Blacksburg, VA 24061, USA
* Correspondence: iliescu@vt.edu
† These authors contributed equally to this work.

Abstract: Reduced-order models (ROMs) have achieved a lot of success in reducing the computational cost of traditional numerical methods across many disciplines. In fluid dynamics, ROMs have been successful in providing efficient and relatively accurate solutions for the numerical simulation of laminar flows. For convection-dominated (e.g., turbulent) flows, however, standard ROMs generally yield inaccurate results, usually affected by spurious oscillations. Thus, ROMs are usually equipped with numerical stabilization or closure models in order to account for the effect of the discarded modes. The literature on ROM closures and stabilizations is large and growing fast. In this paper, instead of reviewing all the ROM closures and stabilizations, we took a more modest step and focused on one particular type of ROM closure and stabilization that is inspired by large eddy simulation (LES), a classical strategy in computational fluid dynamics (CFD). These ROMs, which we call LES-ROMs, are extremely easy to implement, very efficient, and accurate. Indeed, LES-ROMs are modular and generally require minimal modifications to standard ("legacy") ROM formulations. Furthermore, the computational overhead of these modifications is minimal. Finally, carefully tuned LES-ROMs can accurately capture the average physical quantities of interest in challenging convection-dominated flows in science and engineering applications. LES-ROMs are constructed by leveraging spatial filtering, which is the same principle used to build classical LES models. This ensures a modeling consistency between LES-ROMs and the approaches that generated the data used to train them. It also "bridges" two distinct research fields (LES and ROMs) that have been disconnected until now. This paper is a review of LES-ROMs, with a particular focus on the LES concepts and models that enable the construction of LES-inspired ROMs and the bridging of LES and reduced-order modeling. This paper starts with a description of a versatile LES strategy called evolve–filter–relax (EFR) that has been successfully used as a full-order method for both incompressible and compressible convection-dominated flows. We present evidence of this success. We then show how the EFR strategy, and spatial filtering in general, can be leveraged to construct LES-ROMs (e.g., EFR-ROM). Several applications of LES-ROMs to the numerical simulation of incompressible and compressible convection-dominated flows are presented. Finally, we draw conclusions and outline several research directions and open questions in LES-ROM development. While we do not claim this review to be comprehensive, we certainly hope it serves as a brief and friendly introduction to this exciting research area, which we believe has a lot of potential in the practical numerical simulation of convection-dominated flows in science, engineering, and medicine.

Keywords: large eddy simulation; reduced-order modeling; spatial filtering; machine learning; incompressible fluids; compressible fluids; cardiovascular modeling; atmospheric modeling

1. Introduction

In a couple of seminal papers published in 1941 [1,2], Kolmogorov showed that the vortices and eddies in a fluid flow span an increasingly large range of scales as the inertial forces become dominant over the viscous forces. The aim of a direct numerical simulation (DNS) is to capture all these flow structures with a computational mesh sufficiently refined to resolve even the smallest scales. For a large number of important applications in science, engineering, and medicine, a DNS requires extremely fine meshes, leading to exorbitant computational costs for current computing technology. Examples of such applications include atmospheric flow for weather and climate predictions, flow around wind turbines for efficient energy production, and blood flow in larger arteries to predict the progression of a disease or plan for surgery.

The goal to reduce the computational cost of DNS while maintaining a reasonable level of accuracy has motivated a vast body of literature. Among the many proposed approaches, Reynolds averaged Navier–Stokes (RANS) and large eddy simulation (LES) have become popular and widespread across different sciences and industries. In broad terms, RANS models average the Navier–Stokes equations in different ways (quite often in time), while LES techniques filter them (usually in space). This considerably lowers the computational cost but requires closures to represent certain terms emerging from the averaging/filtering process. For closure, RANS models typically rely on empirical descriptions calibrated with available data. While one important line of research seeks to improve the accuracy of RANS models by systematically informing them with larger datasets [3], this review focuses on LES, which is a mathematically and physically justified approach [4–6].

LES techniques directly resolve a portion of the flow scales and require a model to account for the remaining (small) unresolved scales that are not directly captured due to mesh under-refinement. Much of the literature has been dedicated to the derivation and performance analysis of LES models, which inevitably feature parameters. The combination of properly tuned LES models and high mesh resolution has been shown to yield accurate computational results in fields ranging from blood flow studies [7–9] to weather and climate forecasts [10–19]. However, while feasible (unlike DNS), these highly accurate LES simulations still carry a high computational cost. These hefty computational costs could be acceptable in exploratory studies, but they are incompatible with pressing deadlines and/or multiquery contexts.

Multiquery contexts arise from needing to, e.g., assess the uncertainty in the computed solution, optimally control a system, or solve an inverse problem for parameter identification or optimization. They are as ubiquitous as tight deadlines in science, engineering, and medicine. Unable to be patient for the long computational times of accurate simulations, LES practitioners typically trade accuracy (high-resolution meshes and costly parameter tuning) for computational efficiency (coarse meshes and parameters set empirically). This trade-off limits the full potential of computational simulations and has possible detrimental consequences on the reliability of a prediction (e.g., weather and climate), the structural integrity of a physical asset (e.g., a wind turbine), or the life of an individual (e.g., surgical planning for cardiovascular disease). Drastically reducing the computational cost of highly accurate LES is a significant research goal that would enable tremendous advancements across sciences and industries.

A viable way to rectify the imbalance between computational efficiency and accuracy is by bridging two fields that have so far remained disconnected: LES and advanced reduced-order modeling (ROM) techniques. ROMs are methods specifically designed to speed up computations in multiquery settings using state-of-the-art linear algebra techniques and error estimation. The natural connections between these two fields emerge around the mathematical operations of *spatial (low-pass) filtering* and *approximate deconvolution*, both in the classical sense [4,20–22] and reinterpreted through a more modern, data-driven lens [23–28]. Filtering can be redefined through the integration of physics-based models and data-driven methodologies, fostering the evolution of hybrid solvers that respect the underlying physics while harnessing the robust structure and pattern recognition capabilities of machine learning. With a research effort on the complex computational mathematics behind spatial filtering and approximate deconvolution for LES and ROM, fast, accurate, and robust simulations of convection-dominated flows will be possible with minimal user set-up. The modular bidirectional interaction between ROM and LES will yield a converged computational twin usable for detailed flow analysis (LES counterpart) and swift decision making (ROM counterpart), dynamically updating itself with incoming measurements and synthetic data.

This review is structured as follows. Section 2 introduces general concepts in LES modeling and then, for clarity of presentation, focuses on one of the most popular and successful models, the EFR strategy, with details on the application of the EFR algorithm to the incompressible Navier–Stokes equations and the weakly compressible Euler equations. Section 3 presents results obtained with the EFR method as a full-order method for complex applications arising in hemodynamics and physics of the atmosphere and discusses related open problems. In Section 4, we show how key concepts in the construction of LES models, i.e., spatial filters and approximate deconvolution, can be leveraged to construct LES-ROMs. An important feature in Section 4 is the discussion of preliminary numerical analysis results obtained for LES-ROMs. Selected achievements and prospective applications of LES-ROMs for incompressible and compressible fluids are reported in Section 5. Section 6 draws conclusions from this review and summarizes the open perspectives.

2. LES as a Full-Order Model

In this section, we briefly outline some of the main ideas, concepts, and tools used in the development of LES models. We focus on spatial filtering and approximate deconvolution since these mathematical concepts have played an important role in connecting ("bridging") two research fields, LES and reduced-order modeling, that have been treated separately until recently. We also focus on one particular class of LES models, i.e., the evolve–filter–relax strategy, which has also been central in "bridging" LES and reduced-order modeling. Finally, we present these LES concepts and models for both incompressible and compressible flows. We made this choice to demonstrate that these concepts and models are truly universal and are not restricted to particular classes of flows or computational settings. Of course, our LES presentation is inherently brief and focused on spatial filtering, approximate deconvolution, and the evolve–filter–relax strategy. For a deeper and more comprehensive introduction to LES, the reader is referred to, e.g., the research monographs [4–6,29,30].

In order to explain the main ideas of LES, let us briefly summarize some key concepts in Kolmogorov's theory [1,2].

The kinetic energy (KE) of a fluid, which is the kinetic energy associated with eddies in the flow, defined as $\frac{1}{2}\rho \int_\Omega ||\mathbf{u}||^2 d\omega$, where ρ is the fluid density, $\mathbf{u}$ the fluid velocity, Ω the spatial domain where the fluid is flowing, and $||\cdot||_{L^2}$ denotes the L^2 norm, is injected in the system at large scales (low wave numbers). Since the large-scale eddies are unstable, they break down, transferring the energy to smaller and smaller eddies. Finally, the KE is dissipated by the viscous forces at small scales (high wave numbers). This process is usually referred to as *energy cascade*. See Figure 1. The scale at which the viscous forces dissipate energy takes the name of the *Kolmogorov scale*:

$$\eta = \left(\frac{\nu^3}{\varepsilon}\right)^{1/4}, \tag{1}$$

where ν is the kinematic viscosity of the fluid and ε is the time average of the rate at which the energy is dissipated (see, e.g., [31]). In order to formally define ε, let $[0, t_f]$ be the time interval of interest for the fluid flow. Then,

$$\varepsilon = \limsup_{t_f \to \infty} \frac{1}{t_f |\Omega|} \int_0^{t_f} \nu ||\nabla \mathbf{u}||_{L^2}^2 \, dt,$$

where $|\Omega|$ denotes the measure of the domain.

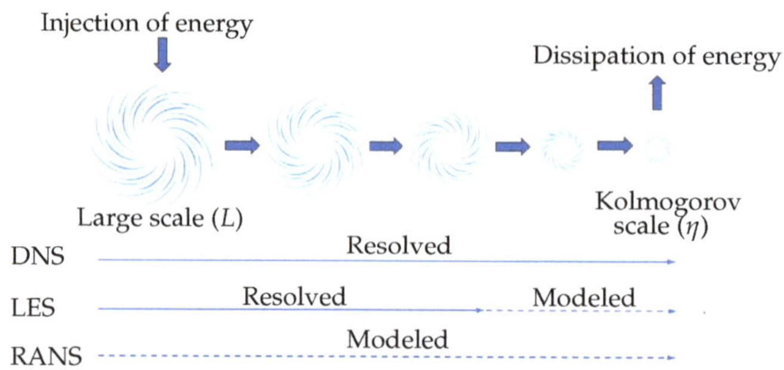

Figure 1. Overall view of the energy cascade, from injection to dissipation of energy, and associated types of modeling.

For a flow in a developed turbulent regime, the dissipation rate has to be of the same magnitude as the production rate, which is the rate at which the KE is supplied to the small scales. A common way to express ε is in terms of the macro-scale variables associated with the flow. For this, let U and L be characteristic macroscopic speed and length, respectively. Then, we can write $\varepsilon \sim U^3/L$ [32]. If we plug this into (1), we obtain

$$\eta \sim Re^{-3/4}L, \tag{2}$$

where Re is the Reynolds number

$$Re = \frac{UL}{\nu}. \tag{3}$$

The Reynolds number, which weights the inertial forces over the viscous forces, is often used as an indication of whether a flow is turbulent or not. Roughly speaking, at "high" Reynolds numbers, inertial forces dominate and flows tend to be turbulent.

Equation (2) explains the difficulty associated with the numerical solution of flows at high Reynolds numbers: in order to correctly capture the dissipated energy, one needs a grid with spacing $h \sim \eta$. This approach, called DNS, leads in general to a large number of unknowns and prohibitive computational costs as the Reynolds number increases. If one tries to reduce the computational cost simply by making the mesh coarser (i.e., using a mesh with size h, which fails to resolve the Kolmogorov scale), the energy physically dissipated at the small scales is not properly dissipated at the numerical level. The under-diffusion in the simulation leads to nonphysical computed velocities. In the best-case scenario, nonphysical oscillations in the velocity field eventually make the simulation crash, which points to an obvious problem in the simulation set-up. In the worst-case scenario, the simulation does not crash and one accepts a nonphysical solution as good.

A possible way to reduce the computational cost without sacrificing accuracy is to solve for the flow average using a mesh coarser than that required by DNS and introduce a model for the effects not captured by the flow average alone. Among such alternatives, RANS and LES are the most widely used for practical applications in science and engineering. In the RANS approach, none of the eddies in the flow are directly resolved from the flow equations, meaning that the effect of all eddies, from largest to smallest, is modeled. See Figure 1. In a sense, RANS could be considered the opposite extreme to DNS. RANS models are formally obtained from decomposing instantaneous quantities (velocity and pressure) into a time-averaged component and a fluctuating (over the average) component and applying an averaging procedure to the governing equations. Extra stresses and scalar fluxes emerging from this averaging procedure require closure, i.e., additional equations or assumptions to close the new system of equations. In the LES approach, one resolves only the larger scales by using a "coarse" mesh (again with respect to DNS) and models the effect of the smaller scales that are not directly solved. See Figure 1. LES models, which are a versatile compromise between DNS and RANS, are devised from applying a low-pass filter to the governing equations. Such low-pass filtering can be viewed as a space averaging.

Traditionally, LES models introduce dissipation via momentum fluxes that are linearly dependent upon the rate of strain of the resolved scales. To clarify this, let us consider the incompressible Navier–Stokes equations (NSEs): denoting the velocity by $\mathbf{u}$ and the pressure by p, the NSEs read

$$\rho\frac{\partial \mathbf{u}}{\partial t}+\rho\,\nabla\cdot(\mathbf{u}\otimes\mathbf{u})-\nabla\cdot(-p\mathbf{I}+2\mu\boldsymbol{\epsilon}(\mathbf{u}))=\mathbf{f}\quad \text{in}\,\Omega\times(0,t_f), \tag{4}$$

$$\nabla\cdot\mathbf{u}=0\quad \text{in}\,\Omega\times(0,t_f), \tag{5}$$

where $\boldsymbol{\epsilon}(\mathbf{u})=(\nabla\mathbf{u}+\nabla\mathbf{u}^T)/2$ is the strain-rate tensor, μ is the dynamic viscosity, and $\mathbf{f}$ is a forcing term. In real applications, appropriate initial and boundary conditions define the problem to solve, as we will see below. With a traditional LES model, one ultimately solves the following problem: find $\overline{u}$ and $\overline{p}$ such that

$$\rho\frac{\partial \overline{u}}{\partial t}+\rho\,\nabla\cdot(\overline{u}\otimes\overline{u})-\nabla\cdot(-\overline{p}\mathbf{I}+2(\mu+\mu_a)\boldsymbol{\epsilon}(\overline{u}))=\overline{\mathbf{f}}\quad \text{in}\,\Omega\times(0,t_f), \tag{6}$$

$$\nabla\cdot\overline{u}=0\quad \text{in}\,\Omega\times(0,t_f), \tag{7}$$

where $\overline{u}$ is the filtered velocity, $\overline{p}$ is the filtered pressure, $\overline{\mathbf{f}}$ is the filtered forcing, and μ_a is an artificial viscosity introduced by the LES model. This is known as *eddy viscosity closure* [4,5]. Different LES models arise for different definitions of μ_a.

One of the most widely used eddy viscosity models is the Smagorinsky model [33]. Introduced in the early 1960s, it defines the artificial viscosity as

$$\mu_a=\rho(C_s\alpha)^2\sqrt{2\boldsymbol{\epsilon}(\overline{u}):\boldsymbol{\epsilon}(\overline{u})},\quad C_s^2=C_k\sqrt{\frac{C_k}{C_\epsilon}}, \tag{8}$$

where α is a filter width (typically comparable with the mesh size) and C_k and C_ϵ are model parameters. The values of such parameters are not universal and need to be tuned for each specific application. The Smagorinsky model has enjoyed much success for several reasons. It is relatively simple and computationally inexpensive compared to other LES models. It is easy to implement and can be used with a wide range of numerical methods, also thanks to the tunable parameters. However, it has one important limitation: the assumption of local balance between the subgrid scale energy production and dissipation. Since such equilibrium conditions do not hold in many practical applications, the Smagorinsky model often results in over-diffusive simulations. A large body of research has been motivated by the aim of improving upon the Smagorinsky model.

Some alternative methods introduce artificial diffusion that can be solution-dependent (see, e.g., [34–38]) or residual-based (see, e.g., [39–44]) so that the artificial viscosity vanishes where the solution is smooth and/or decreases as the mesh is refined. Other methods add a set of equations to the discrete governing equations formulated on a coarse mesh. This extra problem can be devised in different ways, for example, by a functional splitting of the solved and unresolved scales as in variational multiscale methods (see, e.g., [45–48]) or by resorting to the concept of "suitability" of weak solutions [49]. All these methods entail adding one or more extra terms to the governing equations, in the spirit of (6), and have had great accomplishments. However, one practical downside, in this case, is the need to implement ad hoc extensions that are not part of standard solvers, requiring access to legacy source codes (if not rewriting the solver). An attractive option is, therefore, to add the extra problem sequentially to the fluid dynamics model, since in this case the implementation of the LES model would not require any major modification of a legacy solver. This can be achieved by applying a differential nonlinear low-pass filter to the (nonphysical) solution computed with the fluid dynamics model and a "coarse" mesh. Since this particular LES strategy, i.e., nonlinear spatial filtering, has already made an impact in ROMs, below we focus on this approach. This will improve the clarity of our presentation on how to establish a two-way connection between (i.e., "bridging") LES and ROMs.

2.1. Nonlinear Spatial Filtering for LES

Let us consider a generic fluid model (e.g., the momentum equation of the NSE (4))

$$\frac{\partial \mathbf{U}}{\partial t} + \mathcal{M}(\mathbf{U}) = \mathbf{0} \tag{9}$$

and discretize it in time with, e.g., the backward Euler scheme for simplicity:

$$\frac{\mathbf{U}^{n+1} - \mathbf{U}^{n}}{\Delta t} + \mathcal{M}(\mathbf{U}^{n+1}) = \mathbf{0}. \tag{10}$$

Here, $\mathbf{U}$ denotes a generic variable (e.g., $\mathbf{U} = \rho\mathbf{u}$ in (9) represents the variable in (4)) and $\mathcal{M}(\cdot)$ denotes nontrivial operators that characterize the fluid model. For simplicity, we do not include any forcing term in (9).

If we further apply space discretization to (10) and use a mesh of size $h \gg \eta$, the computed solution typically features nonphysical oscillations. Such oscillations can be attenuated or removed through a filtering operation. This is the idea at the core of a sequential algorithm called evolve–filter–relax (EFR). The EFR algorithm applied to (9) reads as follows:

- *Evolve*: Find intermediate variable $\mathbf{V}^{n+1}$ such that

$$\frac{\mathbf{V}^{n+1} - \mathbf{U}^{n}}{\Delta t} + \mathcal{M}(\mathbf{V}^{n+1}) = \mathbf{0}. \tag{11}$$

  For this step, one could adopt the same space discretization technique used for (10) and, hence, the same solver.

- *Filter*: Find filtered variable $\overline{\mathbf{V}}^{n+1}$ such that

$$\overline{\mathbf{V}}^{n+1} = \mathcal{F}\mathbf{V}^{n+1}, \tag{12}$$

  where $\mathcal{F}$ is a generic spatial filter. Different types of spatial filters will be described throughout this paper (e.g., in (16)). The solver for this step can be added as a module to the solver for (10).

- *Relax*: Set

$$\mathbf{U}^{n+1} = (1-\chi)\mathbf{V}^{n+1} + \chi\overline{\mathbf{V}}^{n+1}, \tag{13}$$

  with relaxation parameter $\chi \in [0,1]$. Here, $\mathbf{U}^{n+1}$ is the end-of-step approximation of $\mathbf{U}$ at time t^{n+1}, which "corrects" the intermediate approximation $\mathbf{V}^{n+1}$.

The appeal of the EFR algorithm lies not only in *modularity* but also in *flexibility*, since there exists a large variety of spatial filters that one could use (e.g., Gaussian, box, sharp cutoff) [5,6].

The main idea of using spatial filtering can be illustrated with a schematic like in Figure 2. The challenge is to represent a flow variable, e.g., the velocity, $\mathbf{u}$ (represented by the maroon curve in Figure 2), on a given coarse mesh of size h. As mentioned in Section 1, having to work with a coarse mesh is generally the case in realistic engineering, geophysical, and, often, medical applications of convection-dominated flows. As illustrated in Figure 2, if $\mathbf{u}$ has small-scale components of a size below the mesh size, h, then these components will not be accurately represented on the given mesh. More importantly, the inaccuracies in representing $\mathbf{u}$ on the given coarse mesh are generally propagated and amplified in convection-dominated, nonlinear problems (e.g., the NSEs), yielding spurious numerical oscillations. To eliminate/alleviate these spurious oscillations, the central idea in LES is extremely simple: instead of trying to approximate all the spatial scales in $\mathbf{u}$ on the given coarse mesh, try to approximate only the large scales in $\mathbf{u}$. This can be achieved by using a low pass spatial filter, $\mathcal{F}$, to filter $\mathbf{u}$. The filtered flow variable, $\overline{u} := \mathcal{F}\mathbf{u}$, which is represented by the orange curve in Figure 2, retains only the large spatial structures in $\mathbf{u}$. Thus, $\overline{u}$ can be represented accurately on the given coarse mesh. Furthermore, the spatial filtering eliminates the small scale, oscillatory component of $\mathbf{u}$. This, in a nutshell, is the guiding principle of LES [4,5].

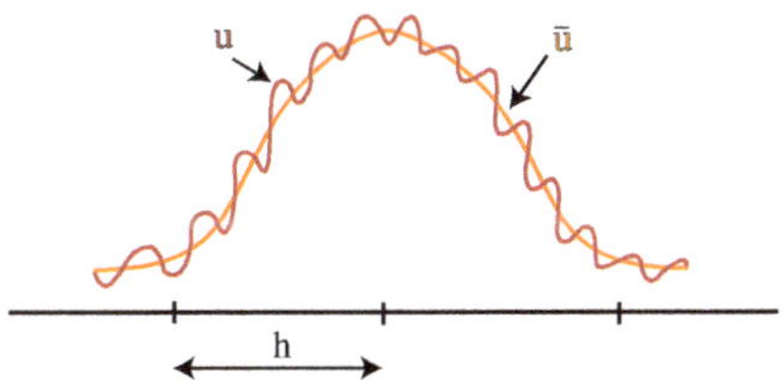

Figure 2. LES schematic showing the input flow variable, **u**, that cannot be represented on a given coarse mesh, and the filtered flow variable, $\overline{u}$, that can be accurately represented on the coarse mesh.

The EFR algorithms, which are also known as filter stabilizations (or regularizations [6]) because they reduce or eliminate unphysical fluctuations in the computed solution, can be categorized as eddy viscosity models for certain discretization choices [50,51]. The connection between the EFR algorithm and LES modeling is easily seen by shifting the index $n+1$ to n in (12) and (13) and plugging them into (11) to obtain

$$\frac{\mathbf{V}^{n+1}-\mathbf{V}^{n}}{\Delta t}+\mathcal{M}(\mathbf{V}^{n+1})+\frac{\chi}{\Delta t}(\mathcal{I}-\mathcal{F})\mathbf{V}^{n}=\mathbf{0}, \tag{14}$$

where $\mathcal{I}$ is the identity operator. Equation (14) gives us an implicit discretization of problem (9) with the backward Euler scheme and an additional, explicitly treated dissipation term (compare (14) with (10)). Let us assume that $\chi = \chi_0 \Delta t$, where χ_0 is a time-independent constant. Then, (14) can be seen as a time-stepping scheme for the following problem:

$$\frac{\partial \mathbf{V}}{\partial t}+\mathcal{M}(\mathbf{V})+\chi_0(\mathcal{I}-\mathcal{F})\mathbf{V}=\mathbf{0}. \tag{15}$$

Thus, the EFR algorithm (11)–(13) can be interpreted as a splitting scheme for problem (15). In turn, this can be read as problem (9) with an additional term designed to be dissipative.

Stabilization based on linear filters [52–54] has been widely studied (see, e.g., [55–57]). However, the breaking down of eddies into smaller ones until they get damped is a highly nonlinear process. Hence, it seems more appropriate to use nonlinear filters to select the eddies to be damped. A well-established [58–64] nonlinear filter for (12) is

$$\mathcal{F}=(\mathcal{I}+\mathcal{L})^{-1}, \quad \mathcal{L}=-\nabla\cdot(\delta\nabla), \quad \text{with } \delta=\alpha^2 a(\mathbf{V}^{n+1}), \tag{16}$$

where α can be interpreted as the *filtering radius* and $a(\cdot)\in(0,1]$ is the so-called *indicator function*. This is a critical function that determines where and how much eddy viscosity is added, and it is crucial to the success of the EFR method. We discuss possible definitions of $a(\cdot)$ in Section 2.1.3. Notice that $\mathcal{F}$ can be interpreted in view of the theory of maximal monotone operators and their Yosida regularized operators (see [65], Ch. 7). This provides a solid mathematical foundation to the methodology.

The rest of this section focuses on the application of the EFR algorithm to the incompressible NSE (in Section 2.1.1) and the weakly compressible Euler equations (in Section 2.1.2). However, it could be easily applied to other fluid dynamics models, like, e.g, the one-layer [66] and two-layer [67] quasi-geostrophic equations.

2.1.1. The EFR Method for the Incompressible Navier–Stokes Equations

For the sake of completeness, let us consider Equations (4) and (5) supplemented with initial conditions and boundary conditions. Precisely, we are given initial data $\mathbf{u}_0(\mathbf{x})$ such that

$$\mathbf{u}(\mathbf{x},0)=\mathbf{u}_0(\mathbf{x}), \quad \mathbf{x}\in\Omega,$$

and boundary data $\mathbf{u}_D$ and $\mathbf{g}$ such that

$$\mathbf{u}(\mathbf{x},t)=\mathbf{u}_D(\mathbf{x},t), \quad \mathbf{x}\in\Gamma_D, \tag{17}$$

$$-p\mathbf{n}+2\mu\epsilon(\mathbf{u})\cdot\mathbf{n}=\mathbf{g}(\mathbf{x},t), \quad \mathbf{x}\in\Gamma_N, \tag{18}$$

where $\Gamma_D \cup \Gamma_N = \partial\Omega$, $\Gamma_D \cap \Gamma_N = \emptyset$, and $\mathbf{n}$ is the outward unit vector normal to Γ_N. A mix of Dirichlet and Neumann boundary conditions as in (17) and (18) occurs in many practical problems. Other possible boundary conditions (Robin type) are addressed in Remark 1.

Algorithm (11)–(13), with a filter as in (16), applied to problem (4)–(5) reads as follows:

- *Evolve*: find intermediate velocity $\mathbf{v}^{n+1}$ and pressure q^{n+1} such that

$$\frac{\rho}{\Delta t}\mathbf{v}^{n+1} + \rho\,\nabla\cdot\left(\mathbf{u}^n \otimes \mathbf{v}^{n+1}\right) - 2\mu\nabla\cdot\boldsymbol{\epsilon}(\mathbf{v}^{n+1}) + \nabla q^{n+1} = \mathbf{f}^{n+1} + \frac{\rho}{\Delta t}\mathbf{v}^n, \quad (19)$$

$$\nabla\cdot\mathbf{v}^{n+1} = 0, \quad (20)$$

with boundary conditions

$$\mathbf{v}^{n+1}(\mathbf{x}) = \mathbf{u}_D(\mathbf{x}, t^{n+1}), \quad \mathbf{x} \in \Gamma_D, \quad (21)$$

$$-q^{n+1}\mathbf{n} + 2\mu\epsilon(\mathbf{v}^{n+1})\cdot\mathbf{n} = \mathbf{g}(\mathbf{x}, t^{n+1}), \quad \mathbf{x} \in \Gamma_N. \quad (22)$$

Again, for simplicity, we used the backward Euler scheme for time discretization. Of course, other time discretization schemes are possible. Notice that, consistent with the time-discretization order, we also linearized the problem with a first-order time extrapolation of the convective velocity. We also set $\mathbf{v}^0 = \mathbf{u}_0$.

- *Filter*: find filtered velocity $\overline{\mathbf{v}}^{n+1}$ and Lagrange multiplier λ^{n+1} such that

$$-\nabla\cdot\left(\delta\nabla\overline{\mathbf{v}}^{n+1}\right) + \overline{\mathbf{v}}^{n+1} + \nabla\lambda^{n+1} = \mathbf{v}^{n+1}, \quad (23)$$

$$\nabla\cdot\overline{\mathbf{v}}^{n+1} = 0, \quad (24)$$

with $\delta = \alpha^2 a(\mathbf{v}^{n+1})$ and $a(\cdot)$ is the indicator function. In this case, we apply boundary conditions

$$\overline{\mathbf{v}}^{n+1}(\mathbf{x}) = \mathbf{u}_D(\mathbf{x}, t^{n+1}), \ \mathbf{x} \in \Gamma_D, \quad (25)$$

$$-\lambda^{n+1}\mathbf{n} + 2\alpha^2 a(\mathbf{v}^{n+1})\epsilon(\overline{\mathbf{v}}^{n+1})\cdot\mathbf{n} = \mathbf{0}, \quad \mathbf{x} \in \Gamma_N. \quad (26)$$

The filtered velocity is introduced only in the time discrete problem, so we do not need initial conditions for it. The choice of the boundary conditions for the filtered velocity is driven by the need for consistency with the original conditions.

- *Relax*: set

$$\mathbf{u}^{n+1} = (1-\chi)\mathbf{v}^{n+1} + \chi\overline{\mathbf{v}}^{n+1}, \quad (27)$$

$$p^{n+1} = q^{n+1}. \quad (28)$$

This particular version of the EFR method for incompressible flows was first studied in [62].

Note that filter problem (23), (24) can be conveniently rewritten as a generalized Stokes problem. In fact, by multiplying Equation (23) by $\rho/\Delta t$ and rearranging the terms, we obtain

$$\frac{\rho}{\Delta t}\overline{\mathbf{v}}^{n+1} - \nabla\cdot\left(\mu_a\nabla\overline{\mathbf{v}}^{n+1}\right) + \nabla\overline{q}^{n+1} = \frac{\rho}{\Delta t}\mathbf{v}^{n+1}, \quad \mu_a = \rho\frac{\alpha^2}{\Delta t}a(\mathbf{v}^{n+1}), \quad (29)$$

where $\overline{q}^{n+1} = \rho\lambda^{n+1}/\Delta t$. Problem (29), (24) can be seen as a time-dependent Stokes problem with a nonconstant artificial viscosity μ_a, discretized by the backward Euler scheme. Unlike δ, which has the dimension of a length squared, μ_a is dimensionally a dynamic viscosity. The big advantage of this rewriting is that a solver for problem (29), (24) can then be obtained by adopting a standard linearized Navier–Stokes solver. Another advantage is that unlike λ^{n+1}, $\overline{q}^{n+1}$ has the same dimensional units as q^{n+1}. Hence, if one replaces (23) and (24) with (29) and (24), it is possible to relax the pressure too. In [59], it was proposed to replace (28) with $p^{n+1} = (1-\chi)q^{n+1} + \chi\overline{q}^{n+1}$.

Remark 1. *It can happen that a Robin boundary condition is enforced on part of the domain:*

$$-p\mathbf{n} + 2\mu\epsilon(\mathbf{u}) \cdot \mathbf{n} + \gamma\mathbf{u}(\mathbf{x},t) = \mathbf{g}_R(\mathbf{x},t), \quad \mathbf{x} \in \Gamma_R,$$

where $\mathbf{g}_R$ *and* γ *are given. This kind of boundary condition can arise in certain cases, such as domain decomposition algorithms (see, e.g., [68]) or the use of "transpiration" techniques for fluid–structure interaction problems (see, e.g., [69]). The corresponding boundary condition for the evolve step is*

$$-q^{n+1}\mathbf{n} + 2\mu\epsilon(\mathbf{v}^{n+1}) \cdot \mathbf{n} + \gamma\mathbf{v}^{n+1} = \mathbf{g}_R(\mathbf{x},t^{n+1}), \quad \mathbf{x} \in \Gamma_R,$$

while in the filter step—still driven by consistency arguments—one could impose

$$-\overline{q}^{n+1}\mathbf{n} + 2\mu\epsilon(\overline{\mathbf{v}}^{n+1}) \cdot \mathbf{n} + \gamma\overline{\mathbf{v}}^{n+1} = \gamma\mathbf{v}^{n+1}, \quad \mathbf{x} \in \Gamma_R.$$

Remark 2. *While Equation* (24) *is required to prove that the solution of the EFR algorithm* (19)–(28) *converges to the solution of the NSEs* (4) *and* (5) *[6], it can be disregarded in the implementation of the algorithms. Releasing constraint* (24) *leads to substantial simplification and computational time savings since there is one less variable (i.e., the Lagrange multiplier to enforce the incompressibility constraint). In fact, one would solve a simplified filter problem: find filtered velocity* $\overline{\mathbf{v}}^{n+1}$ *such that*

$$-\nabla \cdot \left(\delta\nabla\overline{\mathbf{v}}^{n+1}\right) + \overline{\mathbf{v}}^{n+1} = \mathbf{v}^{n+1}, \tag{30}$$

As noted in [70], the incompressibility is exactly preserved by the simplified differential filter (30) *only for periodic conditions. In all the other cases, the end-of-step velocity* $\mathbf{u}^{n+1}$ *does not strictly satisfy mass conservation. However, it can be shown numerically that at the discrete level, the mass conservation error is very low (see, e.g., [71]).*

EFR algorithms have been well studied for the incompressible Navier–Stokes equations with a finite element [50,59–63,70], spectral element [52–54], or finite volume method [58]. They can be seen as splitting schemes for a nonlinear version of the Leray model. The original Leray model was introduced in 1934 as a theoretical tool to prove the existence of weak solutions of the NSEs [72]. As a computational tool, Leray regularization was first used in [73] as a stabilization strategy for under-resolved simulations of turbulent flows with classical numerical discretizations [6]. It was also shown [39,74] that when a differential filter is used, the Leray model is similar to the NS-α model of Foias, Holm, and Titi [75]. The EFR algorithm with nonlinear differential filters was shown to provide numerical results that are more precise in localizing where the eddy viscosity is needed and are overall more accurate than results obtained with Smagorinsky-type models while featuring comparable computational efficiency [60–63,70]. In [58,59], the results given by the EFR algorithm for certain filter choices were shown to be very accurate when compared to measurements for complex 3D incompressible flow problems.

2.1.2. The EFR Method for the Weakly Compressible Euler Equations

Despite the encouraging results for incompressible flows, EFR algorithms have been used to simulate compressible flows [51] only recently. While ideas similar to filter stabilization have been applied to the Euler equations [76–79], its full potential for compressible flows remains to be tapped into.

Below, we apply the EFR algorithm (11)–(13) to the weakly compressible Euler equations in the formulation most used in the literature of atmospheric simulations (see, e.g, [80–82] and references therein). Let us state such a formulation first.

Let ρ be the air density, $\mathbf{u}$ the wind velocity, p the atmospheric pressure, and θ the potential temperature. Temperature θ is a quantity of interest in meteorology and is defined as $\theta = T/\pi$, where T is the absolute temperature and π is the Exner pressure. Since one is typically interested in fluctuations of ρ and p with respect to a constant (in time) and hydrostatically balanced background state ρ_0 and p_0, we write $p = p_0 + p'$ and $\rho = \rho_0 + \rho'$, with $\nabla p_0 + \rho_0 g\widehat{\mathbf{k}} = \mathbf{0}$, where g is the gravitational constant and $\widehat{\mathbf{k}}$ is the unit vector aligned with the vertical axis z. See, e.g., [83] for more details on this. Then, the conservation of mass, momentum, and potential temperature in spatial domain Ω and over a time interval $(0, t_f]$ can be written as

$$\frac{\partial \rho}{\partial t} + \nabla \cdot (\rho \mathbf{u}) = 0 \qquad \text{in } \Omega \times (0, t_f], \tag{31}$$

$$\frac{\partial (\rho \mathbf{u})}{\partial t} + \nabla \cdot (\rho \mathbf{u} \otimes \mathbf{u}) + \nabla p' + \rho' g \widehat{\mathbf{k}} = \boldsymbol{f_u} \qquad \text{in } \Omega \times (0, t_f]. \tag{32}$$

$$\frac{\partial (\rho \theta)}{\partial t} + \nabla \cdot (\rho \mathbf{u} \theta) = f_\theta \qquad \text{in } \Omega \times (0, t_f], \tag{33}$$

where $\boldsymbol{f_u}$ and f_θ are possible forcing terms. We close system (31)–(33) with a thermodynamics equation of state for p:

$$p = p_g (\rho R \theta / p_g)^{c_p / c_v}, \tag{34}$$

where $p_g = 10^5$ Pa is a standard pressure, R is the specific gas constant of dry air, and c_p (respectively, c_v) is the specific heat capacity at constant pressure (respectively, volume). Obviously, system (31)–(34) have to be supplemented with proper initial and boundary conditions. However, we do not specify them, as they would be treated as explained in Section 2.1.1. Other perspectives on this are included in Section Open Problems.

EFR algorithm (11)–(13), with filtering as in (16), applied to problem (31)–(34) after time discretization by the backward Euler scheme reads as follows:

- *Evolve*: find density ρ^{n+1}, density fluctuation $\rho'^{,n+1}$, and intermediate variables $\mathbf{v}^{n+1}$, q^{n+1}, $q'^{,n+1}$, and β^{n+1} such that

$$\frac{\rho^{n+1} - \rho^n}{\Delta t} + \nabla \cdot (\rho^{n+1} \mathbf{v}^{n+1}) = 0, \tag{35}$$

$$\frac{\rho^{n+1} \mathbf{v}^{n+1} - \rho^n \mathbf{u}^n}{\Delta t} + \nabla \cdot (\rho^{n+1} \mathbf{v}^{n+1} \otimes \mathbf{v}^{n+1}) + \nabla q'^{,n+1} + \rho'^{,n+1} g \nabla \rho^{n+1} = f_u^{n+1}, \tag{36}$$

$$\frac{\rho^{n+1} \beta^{n+1} - \rho^n \theta^n}{\Delta t} + \nabla \cdot (\rho^{n+1} \mathbf{v}^{n+1} \beta^{n+1}) = f_\theta^{n+1}, \tag{37}$$

$$q^{n+1} = p_g \left(\rho^{n+1} R \beta^{n+1} / p_g \right)^{c_p / c_v}, \tag{38}$$

$$q^{n+1} = p_0 + q'^{,n+1}, \tag{39}$$

$$\rho^{n+1} = \rho_0 + \rho'^{,n+1}. \tag{40}$$

- *Filter*: find filtered variables $\overline{\mathbf{v}}^{n+1}$ and $\overline{\beta}^{n+1}$ such that

$$-\nabla \cdot (\delta_v \nabla (\overline{\mathbf{v}}^{n+1})) + \overline{\mathbf{v}}^{n+1} = \mathbf{v}^{n+1}, \quad \delta_v = \alpha^2 a(\mathbf{v}^{n+1}), \tag{41}$$

$$-\nabla \cdot (\delta_\beta \nabla (\overline{\beta}^{n+1})) + \overline{\beta}^{n+1} = \beta^{n+1}, \quad \delta_\beta = \alpha^2 a(\beta^{n+1}). \tag{42}$$

- *Relax*: find end-of-step $\mathbf{u}^{n+1}$, p^{n+1}, $p'^{,n+1}$, and θ^{n+1} such that

$$\mathbf{u}^{n+1} = (1 - \chi) \mathbf{v}^{n+1} + \chi \overline{\mathbf{v}}^{n+1}, \tag{43}$$

$$\theta^{n+1} = (1 - \xi) \beta^{n+1} + \xi \overline{\beta}^{n+1}, \tag{44}$$

$$p^{n+1} = p_g \left(\rho^{n+1} R \theta^{n+1} / p_g \right)^{c_p / c_v}, \tag{45}$$

$$p^{n+1} = p_0 + p'^{,n+1}, \tag{46}$$

with relaxation parameters χ, $\xi \in [0, 1]$.

This particular version of the EFR method for compressible flows was introduced in [51]. Therein, the same filter was applied to the intermediate velocity and intermediate potential temperature. However, different variables have different physical properties and, thus, different types of filtering may be a better choice.

Similar to what was performed for (23), one can multiply (41) by $\rho^{n+1}/\Delta t$ to obtain a Stokes-like problem:

$$\frac{\rho^{n+1}}{\Delta t}(\overline{\mathbf{v}}^{n+1} - \mathbf{v}^{n+1}) - \nabla \cdot (\mu_a \nabla \overline{\mathbf{v}}^{n+1}) = \mathbf{0}, \quad \mu_a = \rho^{n+1}\frac{\alpha^2}{\Delta t}a(\mathbf{v}^{n+1}), \tag{47}$$

where μ_a is dimensionally a dynamic viscosity, i.e., a proper eddy viscosity.

2.1.3. Indicator Function

Once the filter is selected, the success of the EFR algorithm depends on the reliability of the indicator function. This function should take values close to zero where its argument (i.e., velocity or potential temperature) does not need regularization and close to one where the argument needs to be regularized. Linear filters, which are equivalent to setting

$$a(\mathbf{v}) = 1, \tag{48}$$

have limited efficacy, since they introduce the same amount of regularization everywhere in the domain and likely lead to overdiffusion. Much attention has been devoted to devising accurate indicator functions.

Before presenting a brief survey of the indicator functions available in the literature, we remark that one can recover a Smagorinsky-like model from the EFR algorithm by taking the following indicator function [84]:

$$a_S(\mathbf{v}) = \frac{|\nabla \mathbf{v}|}{||\nabla \mathbf{v}||_\infty}. \tag{49}$$

This indicator function is convenient because of its strong monotonicity properties. However, it is known to introduce excessive diffusion, just like (48). In fact, notice that a_S takes large values in the regions of the domain with shear flow, which is not turbulent.

Below, we group indicator functions into two categories (physical phenomenology-based and mathematics-based) and present their strengths and limitations.

Physical Phenomenology-Based Indicator Functions

In [62], the authors propose indicator functions based on physical quantities that are known to vanish for coherent flow structures, i.e., structures that persist in their form for "long" periods of time. One of the most popular methods to identify coherent vortices is the Q criterion [85], which classifies as persistent and coherent vortices those regions where

$$Q(\mathbf{u},\mathbf{u}) = \frac{1}{2}(\nabla^s \mathbf{u} : \nabla^s \mathbf{u} - \epsilon(\mathbf{u}) : \epsilon(\mathbf{u})) > 0,$$

where $\nabla^s \mathbf{u}$ is the spin tensor (skew-symmetric part of $\nabla \mathbf{u}$):

$$\nabla^s \mathbf{u} = \nabla \mathbf{u} - \epsilon(\mathbf{u}) = \frac{1}{2}(\nabla \mathbf{u} - \nabla \mathbf{u}^T). \tag{50}$$

It is easy to see that $Q > 0$ in the regions of the domain where spin dominates deformation. An indicator is obtained by rescaling $Q(\mathbf{u},\mathbf{u})$ so that the condition $Q(\mathbf{u},\mathbf{u}) > 0$ implies $a(\mathbf{u}) \simeq 0$, which means that regularization is not needed. A possibility is the following:

$$a_Q(\mathbf{u}) = \frac{1}{2} - \frac{1}{\pi}\arctan\left(\epsilon^{-1}\frac{Q(\mathbf{u},\mathbf{u})}{|Q(\mathbf{u},\mathbf{u})| + \epsilon^2}\right),$$

where ϵ is a parameter introduced to avoid the denominator vanishing.

A second indicator uses an eddy viscosity coefficient formula proposed by Vreman [86] that vanishes for 320 types of flow structures known to be coherent. The Vreman-based indicator function reads

$$a_V(\mathbf{u}) = \sqrt{\frac{B(\mathbf{u})}{|\nabla \mathbf{u}|_F^4}},$$

where the subindex F refers to the Frobenius norm and $B(\mathbf{u})$ is defined as

$$B = \beta_{11}\beta_{22} - \beta_{12}^2 + \beta_{11}\beta_{33} - \beta_{13}^2 + \beta_{22}\beta_{33} - \beta_{23}^2, \quad \beta_{ij}(\mathbf{u}) = \sum_{m=1,2,3} \frac{\partial u_i}{\partial x_m}\frac{\partial u_j}{\partial x_m}.$$

Since $0 \leq B(\mathbf{u})/|\nabla \mathbf{u}|_F^4 \leq 1$, $a_V(\mathbf{u}) \in [0,1]$. The Vreman-based indicator function was shown to be successful in [60].

Another physics-based indicator function uses the relative helicity density RH, which is a local quantity related to the macroscopic helicity H:

$$RH = \frac{\mathbf{u}\cdot\mathbf{w}}{|\mathbf{u}||\mathbf{w}|}, \quad H = \frac{1}{|\Omega|}\int_\Omega \mathbf{u}\cdot\mathbf{w}\, d\Omega,$$

where $\mathbf{w} = \nabla \times \mathbf{u}$ denotes vorticity. From the NSE in rotational form, it is possible to see that local high helicity suppresses local turbulent dissipation caused by the breakdown of eddies into smaller ones. A helicity-based indicator can be created by ensuring that values of RH near one imply $a(\mathbf{u}) \simeq 0$, i.e.,

$$a_H(\mathbf{u}) = 1 - \left|\frac{\mathbf{u}\cdot\mathbf{w}}{|\mathbf{u}||\mathbf{w}| + \epsilon^2}\right|,$$

where again ϵ is a parameter that prevents the denominator from vanishing.

Notice that other (more selective) indicator functions can also be obtained by taking the geometric average of two (or more) indicator functions.

All these physical phenomenology-based indicator functions have the advantage of requiring only algebraic operations on $\mathbf{u}$ and its derivatives. Hence, their implementation may be quite straightforward. However, the major drawback is that they do not allow for a rigorous convergence theory to verify the robustness of the associated filtering method. This limitation is overcome by the mathematics-based (rather than physics-based) indicators discussed next.

Mathematics-Based Indicator Functions

In order to define a class of mathematics-based indicator functions, let us recall a basic result about linear filters $\mathcal{F}$, which are invertible, self-adjoint, compact operators from a Hilbert space V to itself. From the spectral theorem (see, e.g., [65]), we have

$$\mathcal{F}\mathbf{x} = \sum_{i=0}^{\infty} \lambda_i \langle \mathbf{x}, \mathbf{e}_i\rangle \mathbf{e}_i, \qquad \mathcal{F}^{-1}\mathbf{y} = \sum_{i=0}^{\infty} \frac{1}{\lambda_i}\langle \mathbf{y}, \mathbf{e}_i\rangle \mathbf{e}_i,$$

where λ_i are the eigenvalues of $\mathcal{F}$ and $\mathbf{e}_i$ are the corresponding eigenfunctions, which form an orthonormal basis for V. Since $\mathcal{F}$ is compact, its eigenvalues accumulate to 0 and the inverse operator $\mathcal{F}^{-1}$ is unbounded.

Let $\mathcal{D}$ be a bounded, regularized approximation of $\mathcal{F}^{-1}$, whose action on $\mathbf{y}$ is given by

$$\mathcal{D}\mathbf{y} = \sum_{i=0}^{\infty} \phi\left(\frac{1}{\lambda_i}\right)\langle \mathbf{y}, \mathbf{e}_i\rangle \mathbf{e}_i$$

with

$$\phi\left(\frac{1}{\lambda_i}\right) \simeq \begin{cases} \dfrac{1}{\lambda_i} & \text{if } i \text{ is "small"}, \\ 0 & \text{if } i \text{ is "large"}. \end{cases}$$

Then, by composing $\mathcal{F}$ and $\mathcal{D}$, we can obtain a low-pass filter, so that

$$\|\mathbf{x} - \mathcal{D}\mathcal{F}\mathbf{x}\| \text{ is } \begin{cases} \text{small} & \text{if } \mathbf{x} \text{ is "smooth"}, \\ \text{large} & \text{if } \mathbf{x} \text{ is not "smooth"}, \end{cases}$$

where "smooth" is intended with respect to the eigenfunction of the operator $\mathcal{F}$. This means that $\mathbf{x}$ is smooth if $\langle \mathbf{x}, \mathbf{e}_i\rangle$ is significantly different from zero only for small values of

i. The composition of the two operators $\mathcal{F}$ and $\mathcal{D}$ can be interpreted as a low-pass filter. This motivates the definition of indicator function

$$a_{\mathcal{D}}(\mathbf{u}) = \frac{|\mathbf{u} - \mathcal{D}(\mathcal{F}(\mathbf{u}))|}{||\mathbf{u} - \mathcal{D}(\mathcal{F}(\mathbf{u}))||_\infty}. \tag{51}$$

A well-studied choice for the linear filter $\mathcal{F}$ in (51) is the linear Helmholtz filter, i.e., (16) with δ constant in space and time. Let us denote it with $\mathcal{F}_H$. A popular choice for $\mathcal{D}$ is the Van Cittert deconvolution operator $\mathcal{D}_N$, defined as

$$\mathcal{D}_N = \sum_{n=0}^{N} (\mathcal{I} - \mathcal{F})^n.$$

Note that the evaluation of $a_\mathcal{D}$ with $\mathcal{D} = \mathcal{D}_N$ requires the application of the linear filter $\mathcal{F}$ a total of $N + 1$ times. This is an argument in favor of limiting the value of N, since large N would increase the computational cost. Another argument is provided by interpreting $\mathcal{D}_N$ as the N-th iteration of a Richardson scheme to solve the filter problem. See [59] for more details. In practice, one takes $N = 0, 1$, corresponding to $\mathcal{D}_0 = \mathcal{I}$ and $\mathcal{D}_1 = 2\mathcal{I} - \mathcal{F}$, respectively. For these choices of N, the indicator function (51) becomes

$$a_{\mathcal{D}_0}(\mathbf{u}) = \frac{|\mathbf{u} - \mathcal{F}(\mathbf{u})|}{\max(1, ||\mathbf{u} - \mathcal{F}(\mathbf{u})||_\infty)}, \quad a_{\mathcal{D}_1}(\mathbf{u}) = \frac{|\mathbf{u} - 2\mathcal{F}(\mathbf{u}) + \mathcal{F}(\mathcal{F}(\mathbf{u}))|}{\max(1, ||\mathbf{u} - 2\mathcal{F}(\mathbf{u}) + \mathcal{F}(\mathcal{F}(\mathbf{u}))||_\infty)}. \tag{52}$$

Indicator function (51) with $\mathcal{D} = \mathcal{D}_N$ and $\mathcal{F} = \mathcal{F}_H$, which we will refer to as the *approximate deconvolution* (AD) indicator function, was proposed in [61]. However, the idea of using van Cittert approximate deconvolution in fluid models to increase accuracy is well established and has strong mathematical foundations [87–89]. It is possible to prove [89] that

$$\phi - D_N(F_H(\phi)) = (-1)^{N+1}\delta^{2N+2}\Delta^{N+1}\mathcal{F}_H^{N+1}\phi. \tag{53}$$

From this, it follows that, indeed, $a_{\mathcal{D}_N}(\mathbf{u})$ takes values close to zero in the regions of the domain where $\mathbf{u}$ is smooth.

In addition to allowing for rigorous mathematical analysis, the AD indicator function has been shown to be more selective than indicator $a_S(\cdot)$ (49) in identifying the regions of the domain where diffusion is needed for both incompressible [61] and compressible flows [51]. Hence, the EFR algorithm with indicator function $a_\mathcal{D}(\cdot)$ corrects the overdiffusivity of $a_S(\cdot)$. For incompressible flows, $a_\mathcal{D}(\cdot)$ has also been shown to be particularly accurate for realistic 3D problems [8,58,59,64].

2.2. Machine Learning for LES

In this section, we briefly outline several machine learning strategies that have been recently used in the construction of LES models. While not as well developed as classical LES models, the machine learning strategies are a dynamic research area, with a lot of potential for LES, reduced-order modeling, and CFD in general. Thus, in this section, we outline several machine learning approaches used in LES and reduced-order modeling.

Recent years have seen an escalating interest in leveraging machine learning techniques for CFD, as elucidated in [3,90]. This subsection is devoted to highlighting some opportunities offered by machine learning for LES modeling, starting with a literature review. Noteworthy endeavors within various research groups have demonstrated the feasibility of training deep learning-based closure models [23,91–93], while others have focused on the acquisition of closure models through symbolic regression techniques [94–96]. It is imperative to underscore that a relatively underexplored domain is the integration of blending and classification methodologies, particularly utilizing indicator functions as presented in the previous subsection.

There is an existing body of knowledge detailing the efficacy of various closure models for specific types of flows. Furthermore, current efforts are marked by a growing emphasis on generating extensive turbulent flow datasets [97]. However, a critical gap remains in establishing comprehensive maps between underlying patterns and flow structures, along with the identification of the best-suited, verified, tested, and mathematically analyzed closure models. On a more granular level, once trained, the same maps can also be used to find the most suitable filters or filtering parameters. This approach allows the utilization of indicator functions within certain latent spaces to accurately identify a low-pass filter or its associated shape and parameters, which can then be applied to mainstream closure approaches. In essence, machine learning tools can be leveraged not only to identify a filter but also to build a hybrid closure model atop it. Furthermore, it is important to note that references [98,99] demonstrate how a single hypothesis or map can be used to create a simple selection mechanism. This mechanism proves valuable for choosing between different structural and functional closures or even various numerical methods.

A hypothesis segregation concept proposed in [98] uses functional and structural closure models and in [99] tests upwind and central numerical methods. The key idea is to train a modular neural network for structure identification offline and use that model to define an expert system to adjust underlying closure models and numerical approaches in a weighted manner temporally or spatially online as needed. A key aspect is the modular nature of offline–online split approaches, where the same trained model defines filter strengths, radii, and shapes for use in a structural model. See Figure 3 for a schematic representation.

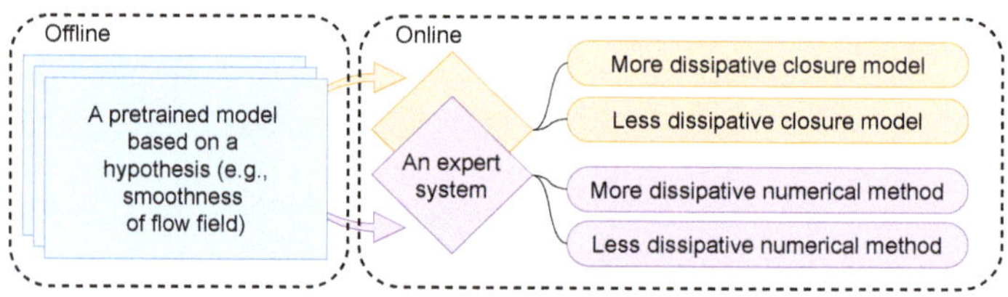

Figure 3. Schematic of the concept proposed in [99].

We also advocate for the judicious use of machine learning tools to facilitate the generation of diverse maps within the latent space [100–102]. Such learned maps serve as foundational elements for constructing an expert system capable of discerning flow regimes and their associated closure models from an array of existing models. This approach offers a departure from the conventional approach of relying purely on black-box closure models or resource-intensive symbolic regression practices.

By harnessing the pattern and structure recognition capabilities inherent in machine learning, one can effectively employ existing structural or functional closure models in a dynamic and adaptive manner. For example, an approach with two types of artificial neural networks for ML has been proposed [103]: a classifier to identify the contribution of each building block in the flow and a predictor to estimate wall shear stress by combining these building-block flows. Such approaches with structure recognition capabilities are poised to enhance the efficiency and accuracy of flow regime identification, thereby presenting a compelling alternative to traditional methodologies. The use of flow physics-based indicator functions brings the additional advantage of exploiting existing structural and functional knowledge in a dynamic and adaptive framework. We conclude this section by referring to recent survey articles that provide further insights into machine learning for turbulence modeling [26,104].

3. Applications of LES for FOM

In this section, we present some of the achievements of the LES concepts and models presented in Section 2, specifically spatial filtering, approximate deconvolution, and the evolve–filter–relax strategy. We present these achievements for both incompressible flows (Section 3.1) and compressible flows (Section 3.2). We believe that presenting applications for both incompressible and compressible flows is important in order to demonstrate that these LES concepts and models are truly universal and are not restricted to particular classes of flows or computational settings. Furthermore, by showcasing the achievements of spatial filtering, approximate deconvolution, and the evolve–filter–relax strategy in challenging applications in biomedicine and weather modeling, rather than in academic benchmarks, we also illustrate the potential that these concepts and models have in reduced-order modeling.

3.1. Incompressible Flows

The use of LES models for incompressible flows covers a wide range of applications. Among these, we want to mention here biomedical applications, with particular reference to computational hemodynamics. This is a field of CFD that has progressively received more attention with the availability of patient-specific data and reliable methods of image processing-enabled numerical methods from the benchmark to the bedside/operating room stage [105]. Using CFD in the clinical routine for the assessment of cardiovascular diseases—and sometimes also for the planning of therapeutic options—is nowadays a consolidated practice, with some outstanding examples involving entrepreneurial activities [106,107]. This circumstance clearly raises some specific challenges that go beyond the mathematical modeling stage, involving the efficiency required by clinical timelines and quality certification/uncertainty quantification.

In general, human blood flow does not feature a turbulent regime in most of the vascular districts of interest. However, in some cases, highly disturbed flow is possible either for physiological reasons (like in the ascending aorta during the systolic phase, when the aortic valve is open and the heart pumps blood into the system) or for pathological conditions, like in the presence of significant stenoses. In some cases, turbulence may be induced by devices required by therapeutical options. In this section, we focus on an application involving a pathological condition called type B aortic dissection (AoD) because it highlights the practical usefulness of the EFR method in a complex real-life setting and the associated challenges.

3.1.1. Computational Hemodynamics in Type B Aortic Dissections

AoD is a pathology consisting of a splitting of the natural lumen of the descending aorta into two separate channels: true lumen and false lumen—see Figure 4. This splitting may have different reasons, including traumatic ones, and represents a danger for the patient: the tissue of the two lumens is weaker than the physiological one, so it may break, causing life-threatening hemorrhages. Clinically, there is a strong interest in predicting the evolution of the false lumen in time, in particular whether it remains stable or grows, the latter condition being the most dangerous for the patient. While drugs are an option for a stable false lumen, surgery is the usual choice for cases in which growth is expected. Since surgery is invasive, it should be considered only when necessary.

Currently, decisions are made only on the geometry of the dissection. The addition of hemodynamics information, in particular of quantities that are difficult or impossible to measure in vivo, is anticipated to improve significantly the reliability of the predictions [8]. In particular, the wall shear stress (WSS), i.e., the tangential component of the normal stress at the wall, may play a significant role for prognostic purposes. It is defined as

$$\boldsymbol{\tau} = 2\mu\boldsymbol{\epsilon}(\mathbf{u})\cdot\mathbf{n} - 2\mu(\mathbf{n}\cdot\boldsymbol{\epsilon}(\mathbf{u})\cdot\mathbf{n})\mathbf{n} \qquad \text{on} \quad \Gamma, \tag{54}$$

where Γ is a portion of $\partial\Omega$ of interest and $\mathbf{n}$ is its outward normal unit vector. Hence, CFD of aortic dissections may become a critical component in the clinical routine. However, the complexity of patient-specific geometries and the high convective fields induced by the small entry tears between the true and the false lumen (see Figure 5) pose significant challenges. It is important to stress that in clinical applications, there is the need to run many patient-specific cases, as required by the concept of *computer-aided clinical trials* (CACTs), to obtain statistically significant data. This means that beyond accuracy, computational efficiency is a critical aspect. In the case of AoD, a one-shot simulation might be possible with DNS, but the computational cost is not compatible with the large number of patients required by CACTs. The trade-off between efficiency and accuracy—where accuracy has to be intended in the clinical sense, i.e., the quantification of indexes to enable correct clinical decisions—is of primary importance, and the EFR scheme, in this context, seems to be the ideal approach.

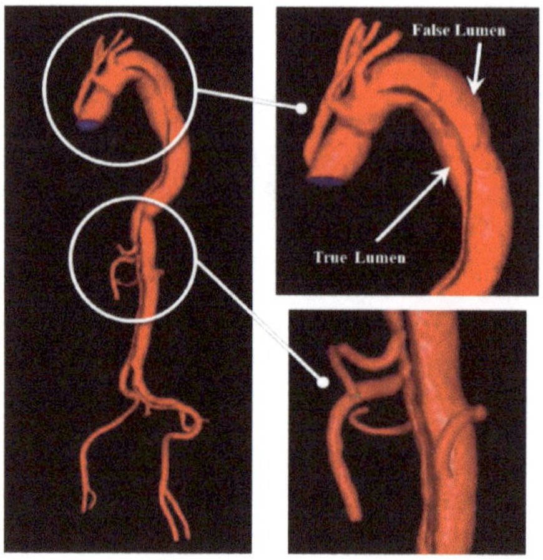

Figure 4. Images of a patient-specific AoD showing the true lumen and the false lumen.

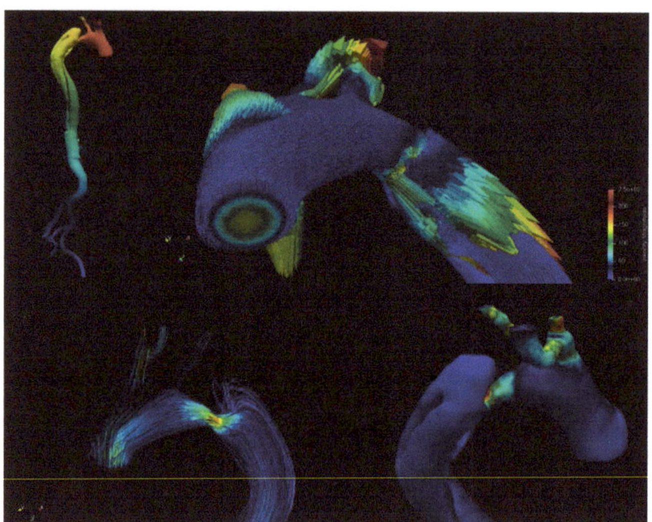

Figure 5. Simulation in a patient-specific AoD. Top left: pressure. Top right: velocity (in cm/s) in the descending aorta and at the entrance of the false lumen. The two bottom panels outline the complexity of the flow induced by the entry tear for the velocity (**left**) and the wall shear stress (**right**).

Geometry

The geometries of AoD may be significantly diverse—see Figure 6. The location of the first tear connecting the true and the false lumen (also known as primal entry tear—PET) has a critical importance. In fact, clinical guidelines suggest that the evolution of the dissection depends on the distance between the PET and the left subclavian artery. A short distance generally indicates the evolution into complicated clinical cases requiring surgery [108]. However, there are cases with a distal PET evolving in cases to be surgically operated [8]. For certain sites of interest in the cardiovascular system (e.g., the coronary arteries), the geometry can be generally templatized, i.e., the patient-specific anatomy can be regarded as a map or a minor modification of a template geometry. In AoDs, it is very hard to find a realistic template. While AI-based methodologies can be used for the segmentation and reconstruction of coronaries, at this time, the classical level-set method is the method of choice for complex geometries like AoD. In fact, the lack of sufficiently large datasets prevents reliable training of data-based methodologies, like neural networks. The geometries presented in this paper were reconstructed with the Vascular Modeling ToolKit [109], a free, open-source tool based on the level-set method, using images from computer tomography. The segmentation was carried out by an expert, trained for the identification of the tears. The mesh was successively obtained by the NetGen mesh generator, with an in-home Python script to guide the refinement of the mesh close to the boundaries (the so-called *mesh boundary layers*). The introduction of a mesh more refined close to the boundary is a common practice in biomedical applications whenever one of the main quantities of interest is computed at the boundary, like the WSS. The meshes we deemed to be appropriate for simulations of AoD with EFR turbulence modeling have a number of elements in the range 6×10^5–6×10^6.

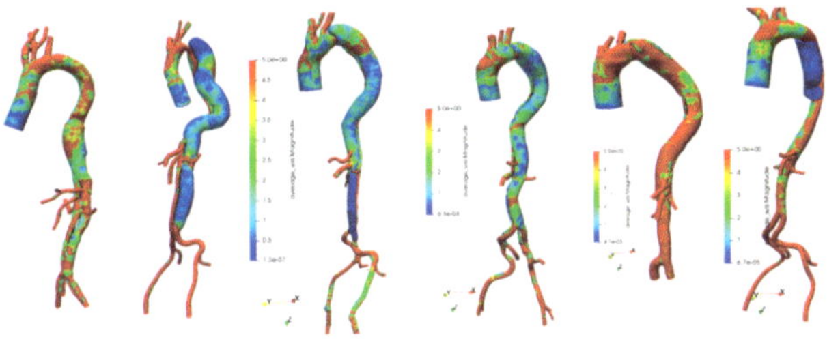

Figure 6. Anatomies of several AoDs, pinpointing the diversity of the possible morphologies. Geometries reconstructed at Emory University with Vascular Modeling ToolKit [110].

Boundary Conditions and Backflows

The patient-specific domain Ω generally features physical boundaries Γ_w represented by the arterial walls and artificial boundaries represented by the regions where the blood enters (Γ_{in}) and leaves (Γ_{out}) the region of interest Ω. The arterial walls include also the flow divider separating the true and the false lumens. On the wall, we postulate the nonslip condition prescribing the continuity of the velocity field of the wall and the blood. Since we assume a rigid structure, this leads to the homogeneous Dirichlet condition

$$\mathbf{u} = \mathbf{0}, \qquad \text{on} \quad \Gamma_w.$$

The role of fluid–structure interaction, in particular at the flap separating the two lumens, has recently attracted the attention of the biomedical engineering community [111,112]. We considered rigid structures for simplicity. However, nothing prevents the extension of the framework presented here to the case of compliant structures (see, e.g., [113]).

When considering the artificial inflow/outflow portions of the boundaries, computational hemodynamics applied to clinical problems generally suffers from a long-term condition, sometimes called the "defective boundary conditions problem" [114] or, in the specific case of medical applications, the "image-legacy problem". Medical doctors used to keep track of images but not of possible hemodynamic conditions to be used as boundary data. This means that we generally have patient-specific geometrical data but with a substantial lack, if not an absence, of patient-specific boundary conditions. To leverage the abundance of geometrical data, many different approaches have been considered in the literature to fill this gap. Standard procedures consider (a) literature-based data for flow rates at the inflow, with an empirical selection of velocity profiles fitting these data; and (b) special boundary conditions at the outflow, based on a lumped parameter modeling of the peripheral circulation. We briefly detail these conditions hereafter.

1. *Inflow conditions.* The available data may generally refer to the flow rate or to the velocity (as a function of time) at one point of the inflow. While this is clearly not enough as a boundary condition, a popular and practical approach consists of assuming a velocity profile constructed around the available data. For instance, one can assume a parabolic profile (corresponding to the well-known Poiseuille solution) or a Womersley profile (corresponding to the time-dependent Womersley solution) [115] that fit the available flow rate or pointwise velocity. This clearly introduces a bias in the solution since the choice of the profile—even if educated—is arbitrary. To mitigate this aspect, an artificial extension of the inflow tract is added to the computational domain, called *flow extension*. An analysis of the error introduced by this approach can be found in [116]. More sophisticated and mathematically sound approaches were introduced with the idea of identifying the best profile through an optimization approach (see [116–120]), yet they require an additional computational cost (and nonstandard solvers)). In the simulation considered in our work on AoD [8,121], we assumed flow rate data provided by the literature, with flow extensions at the inlet (the ascending aorta) and the selection of a constant velocity profile. An analysis of the impact of idealized velocity profiles in AoD simulations can be found in [122].
2. *Outflow Conditions.* For outflow conditions on the portion of the domain denoted by Γ_{out}, a common approach is to combine the 3D model based on the incompressible Navier–Stokes Equations (4) and (5) with surrogate—dimensionally reduced—models representing the downstream circulations, in what has been called the *geometrically multiscale approach* [123–126]. It is important to stress that in this particular case of AoD, we may have some outflow sections referring to collateral vessels like the renal or the mesenteric arteries where not only do we lack patient-specific data but also it is hard to find data in the literature. Yet, the inclusion of the renal flow is critical for a reliable assessment of the hemodynamics in an AoD. A popular approach, in this case, is to resort to the introduction of a lumped parameter model called three-element Windkessel (Windkessel was a device used by firefighters to pump water from a reservoir, converting a periodic action into a quasi-steady flow, which is exactly what happens in the peripheral circulation, where the pulsatile aortic flow is eventually converted into a steady flow in the capillaries [127]), representing the downstream circulation at each outflow boundary (see, e.g., [127]). This approach leads to the prescription of a traction condition of the form

$$p\mathbf{n} - \mu(\nabla\mathbf{u} + \nabla^T\mathbf{u}) \cdot \mathbf{n} = P_{wk}\mathbf{n}, \qquad \text{on} \quad \Gamma_{out}, \tag{55}$$

where P_{wk} is a function of time defined as

$$P_{wk}(t) = R_1 Q(t) + e^{-\frac{t}{R_2 C}}(P_{wk}(0) - R_1 Q(0)) + \int_0^t \frac{e^{\frac{\tau - t}{R_2 C}}}{C} Q(\tau) d\tau.$$

Here, $Q(t) = \int_{\Gamma_{out}} \mathbf{u} \cdot \mathbf{n}\, d\gamma(t)$, and R_1, R_2, and C are parameters representing the proximal and distal flow resistances and the compliance, respectively, of the downstream vascular district. The condition (55) turns out to be a sort of "Robin" condition (see Remark 1) for the NSEs, mixing values of velocity and pressure, as the result of the multiscale coupling between the NSEs and the Windkessel model. Each outflow section features its own set of parameters R_1, R_2, and C. The calibration of these parameters is usually performed with a combination of patient-specific/literature data with some empirical criteria (see, e.g., [128,129]). Although critical, this step is not central to the topics of this paper and was mentioned only for completeness.

It is worth pointing out that in aortic simulations, a numerical instability may occur on the boundary Γ_{out} where traction conditions, like (55), are prescribed. This circumstance goes under the name of "backflow instability," and it is related to a lack of energy control in the solution when a significant amount of fluid enters through Γ_{out}. In detail, when computing the energy of the fluid, the boundedness of the Navier–Stokes solution can be proved under the condition that

$$\int_{\Gamma_{out}} \rho |\mathbf{u}|^2 \mathbf{u} \cdot \mathbf{n} d\gamma \geq 0. \quad (56)$$

Unfortunately, the occurrence of incoming flow (that at outflow sections we can call *backflows*) such that $\mathbf{u} \cdot \mathbf{n} < 0$ may impair the control on the boundedness of the data, eventually leading to the so-called *backflow instabilities* [130]. On the other hand, backflows are physiological in the aortic circulation. The numerical treatment of this problem has been discussed in several contributions, see, e.g., [130–135]. A natural, simple approach in the specific case of Windkessel conditions (55) consists of artificially increasing the proximal resistance R_1, as this results in damping the impact of the backflow on the energy balance. As we will see below, when using LES-EFR models, this turned out to be unnecessary.

EFR in Action

The occurrence of highly disturbed flow in the human ascending aorta is the consequence of the high inlet velocity at the aortic valve [136]. However, the alternation of fast and slow transients associated with the different phases of the heartbeat ("systole" when the aortic valve is open, "diastole" when it is closed) generally prevents the transition to turbulence in physiological cases. Nonetheless, in the presence of small tears where the blood is strongly accelerated, we may have the transition to turbulence. Some studies on AoD report peak Reynolds numbers exceeding 10,000 [137] and consider $k - \omega$ modeling for the turbulence. Other studies [138] find peak Reynolds numbers below 5000 and consider laminar flow appropriate. This is a consequence of the fact that in computational hemodynamics, the patient-specific diversity and the high level of uncertainty may lead to different findings. As we mentioned earlier, CACTs require simulations for a reasonably large number of patients. Hence, computational efficiency is a criterion of paramount relevance, as is the capability to capture transitional and turbulent regimes. The use of LES modeling, and in particular EFR, not only is appropriate for the expected Reynolds numbers but also allows for a significant reduction of the computational costs because it enables the use of coarse meshes that do not necessarily solve all the significant scales.

In [8], we compared a DNS for a patient-specific case with a mesh with 1,063,000 elements vs. an EFR simulation with a mesh with 654,000 elements. Space discretization was performed with classical Taylor–Hood inf–sup compatible finite elements within the C++ library LifeV [139]. The results obtained were similar and equivalent from the clinical point of view. The reduction of the computational time, however, was significant (around 80%).

For the Dirichlet conditions on the wall and at the inflow, we used the conditions for $\mathbf{v}$ and $\overline{\mathbf{v}}$, as specified in Section 2.1.1. For the Windkessel conditions (55), after the time discretization, we have two options:

1. *Implicit*, i.e., we actually evaluate $Q^{n+1} = \int_{\Gamma_{out}} \mathbf{u}^{n+1} \cdot \mathbf{n} d\gamma$ so that we resort to Robin boundary conditions similar to the ones discussed in Remark 1;
2. *Explicit*, i.e., we use a time extrapolation of the velocity consistent with the time-discretization accuracy. For a first-order time advancing, for instance, one has $Q^{n+1} \approx \int_{\Gamma_{out}} \mathbf{u}^{n} \cdot \mathbf{n} d\gamma$, leading to classical traction (Neumann) conditions.

In the former case, for the filtered velocity, we follow the strategy indicated in Remark 1, and in the latter one, we adopt the homogeneous Neumann conditions specified in (26).

Filter Radius and Relaxation Parameter

The selection of the parameters α (radius) and χ (relaxation) is obviously critical and still an open problem. Ideally, one would like to have an automatic selection of these parameters. This calls for the identification of general criteria to be applied to the specific case of interest. In the case of AoD, we identified many different geometrical and fluid dynamic conditions that make the application of general criteria quite problematic. With extensive experience in different cases, we identified some basic principles and heuristic criteria (see [59,110,121]).

The radius α is calibrated to be related to the mesh size h. However, we found that setting $\alpha = h_{max}$, i.e., the maximum size of the elements of the finite element mesh, may generate unnecessary dissipation. We decided to set $\alpha = h_{min}$ (smallest edges of elements in the mesh) to confine the effects of the diffusion introduced by the EFR method.

When selecting α proportional to the mesh size, a consistency argument for the numerical EFR solution with the original equations suggests we take $\chi \propto \Delta t$ [59]. In particular, it has been observed that if we set

$$\chi = \frac{\mu \Delta t}{c_0 \rho \eta \alpha^2} \max(h - \eta, 0), \tag{57}$$

we obtain that the viscous term of the EFR model is comparable to the viscous term in a standard NS solver when the mesh size is the Kolmogorov scale η.

While this approach to set χ is theoretically intriguing, in the case of AoD we need to consider the additional problem of backflow instabilities. The following result was proven [121]:

Theorem 1. *If the data of the problem are regular enough, assuming homogeneous boundary data, for $\chi \to 1^-$ and a suitable choice of the radius α, then*

$$\|\mathbf{u}^n\|_{L^2} \leq \|\mathbf{u}_0\|_{L^2}.$$

This theorem has an important consequence: the energy of the system, even in the presence of backflows, decreases since the diffusive term introduced by the EFR scheme controls the term (56) responsible for the instability. In order to prevent backflow instabilities, the calibration of χ based on (57) needs to be adjusted. This is illustrated in Table 1, referring to a geometry designed to trigger backflows (a domain represented by a portion of a torus with a decreasing radius).

Table 1. Minimal value of the relaxation parameter that prevents backflows for different values of the radius and the time step for two meshes with two different refinements, coarse (C) and fine (F). See [121] for more details.

$\alpha = h_{min}$			$\alpha = 0.9$		
Mesh	**C**	**F**	**Mesh**	**C**	**F**
$\Delta t = 0.01$	0.07	0.08	$\Delta t = 0.01$	0.06	0.06
$\Delta t = 0.005$	0.04	0.06	$\Delta t = 0.005$	0.04	0.05

In general, a larger α allows for a smaller χ and vice versa. However, the stabilization of backflows is obtained for values of the relaxation parameter independent of α and with a sublinear dependence on the time step.

A specific sensitivity analysis to elucidate the impact of the value of α on the simulation is reported below.

Results

With the method described above, we simulated a number of patients in search of possible hemodynamics predictors that could improve the clinical guidelines to diagnose the evolution of the dissection. This work, still in progress [140], is in collaboration with the Department of Surgery at Emory University (Dr. B. Leshnower). While the biomedical implications of our results are not the focus of this paper, we summarize here some major findings as evidence of the practical success of the EFR methodology. In [8], we considered a longitudinal study of a patient presented at Emory University in 2006 and followed for several years. Imaging in 2010 evidenced a significant growth of the false lumen, even though the PET was distal. We co-registered the images in 2006 and 2010 to quantify the false lumen growth and correlate this growth with the hemodynamics computed with our method in the 2006 geometry. The results pointed out a potential role of the time-averaged WSS (TAWSS) defined as

$$\text{TAWSS} \equiv \frac{1}{T}\left\| \int_0^T \boldsymbol{\tau}(t)dt \right\|,$$

where T is the heartbeat duration. The growth is proportional to the TAWSS, but only below a certain threshold. Other hemodynamic indexes were found to be correlated with the possible thrombotic activity leading to a local shrinking of the false lumen (see [110] for more details). After analyzing several more patients with both growth and no growth, we are now focusing on the difference between the TAWSS in the false and true lumens as a potential predictor of pathology evolution. See Figure 7.

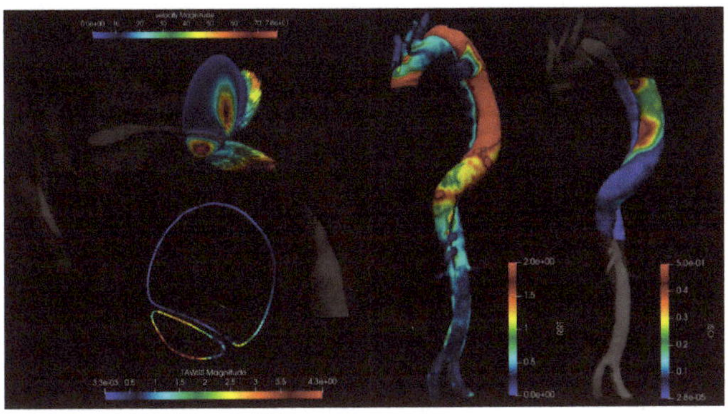

Figure 7. EFR simulation of the hemodynamics in a patient-specific AoD: TAWSS in different regions of the false lumen.

Sensitivity Analysis for the Filtering Radius

Sensitivity analysis can be carried out in different ways. A possible approach is an experimental, yet educated, sampling of the solution for different values of the parameters. In general, this requires several samplings, which, in our case, means running the simulation for several values of α.

Local sensitivity analysis consists of quantifying the dependence of the solution on a parameter by computing the corresponding (Gateaux) derivatives. These are the so-called *sensitivities*. The numerical solution of these equations for a particular value of the parameter allows us to determine in what parts of the computational domain the solution is more sensitive to such a parameter. This is a local analysis since it applies to a particular value of the parameter. A complete derivation of the sensitivity equations to the radius α for Leray linear and nonlinear spatial filtering problems with AD indicator functions (48) and $a_{\mathcal{D}_0}$ can be found in [141].

The results in [141] show that the velocity sensitivity can be a reliable predictor of the regions in the domain with high sensitivity to a specific parameter value. In simulating the flow past an obstacle, for instance, the overdiffusive effects immediately past the obstacle due to an excessively large value of α are correctly identified by the velocity sensitivity. This leads to a possible "adaptive" algorithm for the calibration of the radius. Denoting by $\mathbf{s}_u \equiv \dfrac{\partial \mathbf{u}}{\partial \alpha}$ and by $s_p \equiv \dfrac{\partial p}{\partial \alpha}$ the sensitivities of velocity and pressure, one could solve the LES model and the sensitivity equations with a "large" value α_0 of the radius to attain stability and then refine the solution with a Taylor expansion

$$\mathbf{u}(\alpha) \approx \mathbf{u}(\alpha_0) + \mathbf{s}_u(\alpha_0)(\alpha - \alpha_0), \quad p(\alpha) \approx p(\alpha_0) + s_p(\alpha_0)(\alpha - \alpha_0), \tag{58}$$

with $\alpha < \alpha_0$. In the same paper, the sensitivity of the solution to a scheme with (48) is demonstrated to be significantly higher than with the nonlinear AD filter, pointing out the superiority of the latter.

For specific cardiovascular applications, a *global* sensitivity analysis was performed too [142]. The purpose of this analysis is to understand what inputs affect the solution the most. The inputs are regarded as stochastic processes. Quantities of interest obtained from the numerical solution of the hemodynamics through our EFR method are consequently regarded as stochastic processes too. The sensitivity to the different inputs is quantified in terms of the variance of the outputs as a function of the input variances. In [8], three inputs were considered: the geometry of the domain, the inflow rate, and the radius α. For each of these variables, a probabilistic distribution was postulated. For the radius, a Gaussian distribution with average h_{min} and variance $0.25h_{min}^2$ was deemed to be a correct description of the range of interest for α. The uncertainty in the geometry comes from the noise in the images and the consequent errors in the patient-specific reconstruction. Two cases were considered: (i) an idealized aortic arch, where the uncertainty relies on the geometrical parameters of the ideal geometry; (ii) a patient-specific case, where the geometry was reconstructed in a time range of 6 years. In case (ii), after an image registration procedure, the geometry was parametrized by one parameter specifying the morphing instant from the initial to the final geometry. For this parameter, we postulated a uniform probabilistic distribution. To quantify the uncertainty, we used *polynomial chaos expansion* (PCE). In PCE, a variable-of-interest function of random variables is regarded as a polynomial expansion on a basis incorporating the stochastic features of the inputs. This basis is orthogonal with respect to a suitable probability measure. The coefficients of this (truncated) expansion can be computed with a pseudo-spectral approach through appropriate quadrature rules. This allows us to compute the so-called Sobol' indexes, which are variance-based indicators to characterize the dependence of the output variance on the stochastic inputs [143,144] and rank the influence of the inputs, especially for nonlinear models [145,146]. In short, the Sobol' index for a specific input i on a quantity of interest f is defined as

$$S_i^f \equiv \frac{\mathrm{Var}(f_i^{PC})}{\mathrm{Var}(f^{PC})},$$

where $\mathrm{Var}(f^{PC})$ denotes the variance of the stochastic process f represented by the polynomial chaos expansion and due to all the stochastic inputs, while $\mathrm{Var}(f_i^{PC})$ refers only to the variance induced by the input i.

In [142], the outputs of interest are the turbulent kinetic energy (TKE), the TAWSS, and the oscillatory shear index (OSI). The OSI is a hemodynamic index ranging between 0 and $1/2$, which quantifies how long during the heartbeat the WSS is opposite to the direction of the flow. It is defined as

$$\mathrm{OSI} \equiv \frac{1}{2}\left(1 - \frac{\|\int_0^T \boldsymbol{\tau}(t)dt\|}{\int_0^T \|\boldsymbol{\tau}(t)\|dt}\right).$$

Through the application of PCE, we learned a couple of important points:

1. The geometry is by far the most important factor in the results: an accurate geometrical reconstruction of the region of interest is critical for any biomedical analysis. See Figure 8, rightmost panels.
2. The impact of the radius on the TAWSS and the OSI is minimal. See Figure 8, leftmost and center panels. This means that the selection of the radius with the empirical rules used in our simulations is not expected to have a major impact on the clinical conclusions of the computational analysis.

This corroborates the confidence in our EFR modeling for AoD.

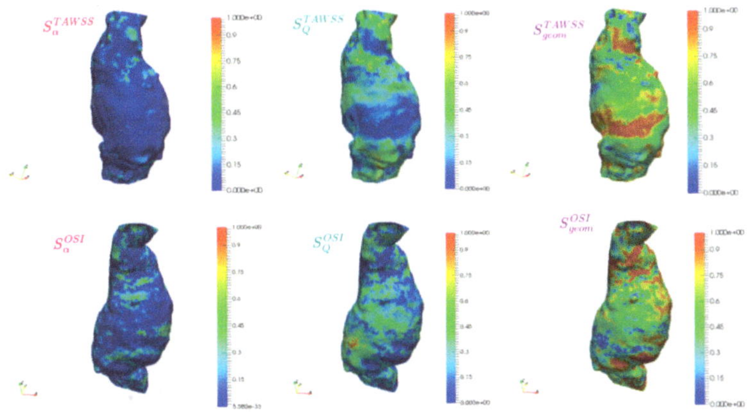

Figure 8. Sobol' indexes in a patient-specific geometry for the sensitivity of the TAWSS and the OSI to the radius α (**left**), the inflow rate Q (**center**), and the geometry (**right**). Blue regions identify parts of the domain weakly affected by variations in the input in comparison with the other uncertainties.

3.1.2. Open Problems

The EFR method gave excellent results in the simulation of AoD, providing the trade-off between the computational efficiency and the accuracy needed to process a significant number of patients. Yet, there are some aspects of the method that still need to be elucidated.

A first important point is the optimal selection of the parameters α and χ. As mentioned earlier, the selection is largely based on empirical and physical criteria. An automatic, somehow optimal, selection would be certainly desirable. As noted, a possible approach for the radius comes from the local sensitivity analysis and the approximate expansion (58). An initial conservative choice of the radius can be refined at each time step so long as some indicators do not suggest the presence of instabilities. This approach has the clear drawback of requiring the solution of the sensitivity equations, which may be quite expensive in some applications, like AoD hemodynamics. Yet, it can provide the backbone for the definition of an adaptive scheme once the sensitivity computations can be properly surrogated. Likewise, the initial setting of χ can be based on linear dependence on Δt and physical arguments, but the occurrence of backflows requires successive refinement. All these aspects need to be further investigated.

Another important problem in hemodynamics applications is related to "data assimilation" procedures that may improve the overall reliability of numerical simulations. With more accurate measurement tools, like 4D MRI, and more storage space available, more hemodynamics data are progressively more available, in particular velocity data in some points of the region of interest. In general, these data are not on the boundary, so they do not provide boundary conditions, yet they give important information about the blood dynamics. However, these data can barely replace the wealth of information provided by a numerical simulation, particularly for hemodynamic indexes like the TAWSS or the OSI. In a cooperative view, where data and models are not alternative but complementary, we need to find a solution both fitting the data (possibly loosely in the presence of high noise) and fulfilling the equations. The problem has been well-known for years (see, e.g., [147–149]), yet a "standard" solution may require even higher—unfeasible—computational costs. In this respect, the combination of physically informed neural networks (PINNs) [150] and our EFR scheme may provide a potential breakthrough that will be investigated in the years to come.

3.2. Compressible Flows

As mentioned in Section 2.1.2, despite many encouraging results for incompressible flows, EFR algorithms have been applied to simulate compressible flows only very recently in [51]. Therein, algorithm (35)–(46) was introduced and tested on classical benchmarks for atmospheric flow. Here, we report the main results to support the following findings: (i) for a given mesh, the EFR algorithm with the AD indicator function is less dissipative than a Smagorinsky-like method (EFR with indicator function $a_{\mathcal{S}}$ in (49)), (ii) the reason for point (i) is the fact that $a_{\mathcal{D}}$ is more selective in identifying the regions of the domain where artificial diffusion is needed, and (iii) the additional cost for using $a_{\mathcal{D}}$ instead of $a_{\mathcal{S}}$ is marginal. For all the results reported here, we set $\chi = \xi = 1$ so that the artificial diffusion introduced by the EFR algorithm can easily be calculated (see [51]). See Section Open Problems below for more on this.

To illustrate the above point (i), we first use results obtained for the well-known rising thermal bubble benchmark. This is a 2D benchmark that entails perturbing with a bubble of warm air a neutrally stratified atmosphere with uniform background potential temperature. In the time interval $(0, 1020]$ s, the air warmer than the ambient rises due to buoyancy and deforms due to shearing motion into a mushroom shape. For the set-up, we refer to, e.g., [51,151,152]. The computational domain, which is 5×10 Km^2, is discretized with two meshes with uniform resolution: a coarser one with $h = 62.5$ m and a finer one with $h = 31.25$ m. The time step is set to $\Delta t = 0.1$ s for all the simulations. Figure 9 shows the spatial distribution of θ' at $t = 1020$ s computed with the EFR algorithm, indicator functions $a_{\mathcal{S}}$ or $a_{\mathcal{D}}$, and the two meshes. On a given mesh, the Rayleigh–Taylor instability at the edge of the bubble is more developed when using $a_{\mathcal{D}}$ instead of $a_{\mathcal{S}}$, which indicates that $a_{\mathcal{D}}$ introduces less artificial viscosity than $a_{\mathcal{S}}$. This is corroborated by the results for a more complex benchmark called density current [153,154] (see also [155–157] for LES of density currents in incompressible flows), in which a neutrally stratified atmosphere with uniform background θ is perturbed with a bubble of cold air. In the time interval $(0, 900]$ s, the cold air descends due to negative buoyancy and when the cold air reaches the ground, it rolls up and forms a front. By time $t = 900$ s, a three-rotor structure emerges as a result of the shear generated at the top boundary of the cold front. The results obtained with the EFR method and indicator functions $a_{\mathcal{S}}$ and $a_{\mathcal{D}}$ are shown in Figure 10. Note that θ' computed with $a_{\mathcal{D}}$ is sharper and less smeared than θ' computed with $a_{\mathcal{S}}$. Additionally, the plots of the indicator functions show that $a_{\mathcal{S}}$ at $t = 900$ s takes larger values over wider regions than $a_{\mathcal{D}}$, illustrating the above point (ii).

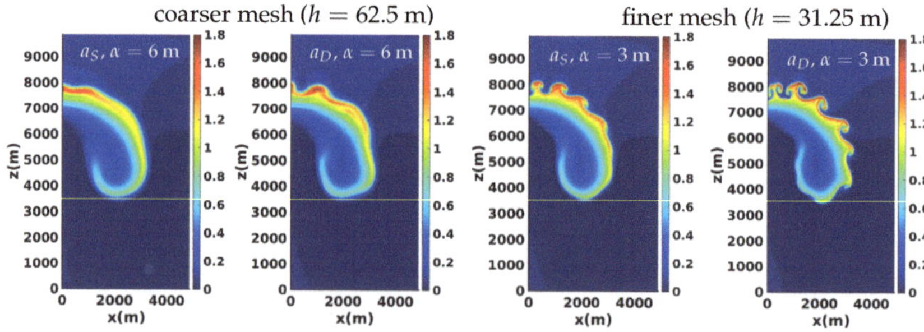

Figure 9. Rising thermal bubble: perturbation of potential temperature θ' at $t = 1020$ s computed with the EFR and $a_{\mathcal{S}}$ and $a_{\mathcal{D}}$ with the coarser mesh (**first two panels**) and the finer mesh (**last two panels**).

The computational time to run the simulations in Figure 10 on a common laptop is reported in Table 2. Obviously, the total computational cost increases when switching from $a_{\mathcal{S}}$ to $a_{\mathcal{D}}$ because while $a_{\mathcal{S}}$ requires a simple post-processing of the velocity field, $a_{\mathcal{D}_0}$ in (52) requires one application of the linear Helmholtz filter. However, note that that increase amounts to only 0.006 s per time step and the filter step with $a_{\mathcal{D}}$ takes about half of the time needed for the evolve step. In total, the simulation takes only about 13 min.

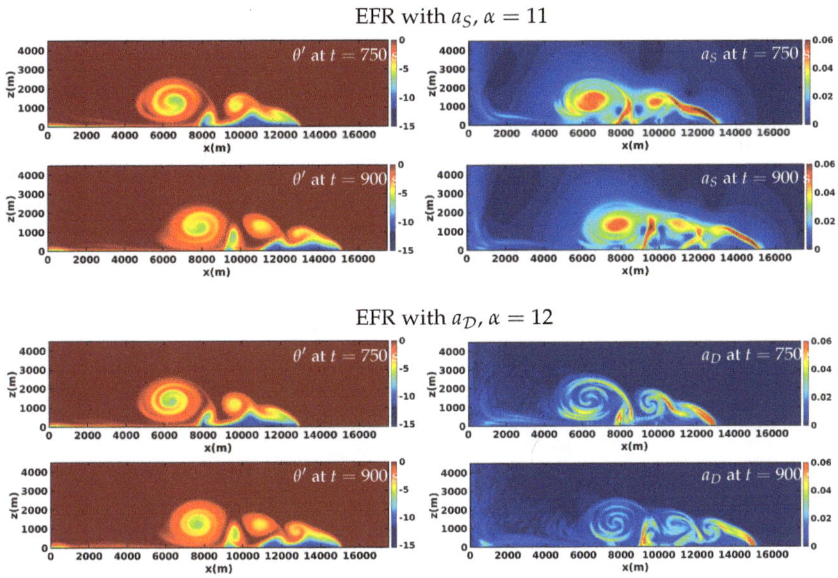

Figure 10. Density potential temperature fluctuation θ' (**left**) and indicator function (**right**) for the EFR method with a_S (**top**) and a_D (**bottom**). The mesh size is $h = 50$ m.

Table 2. Density current: computational time taken by the evolve step and filter step per time step and total simulation time for the EFR algorithm with indicator functions a_S and a_D for mesh $h = 50$ m.

Model	Evolve (s)	Filter (s)	Total (s)
EFR, a_S	0.06	0.023	772
EFR, a_D	0.06	0.029	790

We remark that a DNS for this benchmark would require over 5000 billion DOFs. The results presented here show that the EFR method with a deconvolution-based indicator function is much less dissipative than Smagorinsky-like LES methods while featuring comparable computational efficiency. These findings suggest that EFR is a promising approach for compressible flows, as the studies on incompressible flows suggested.

Open Problems

To obtain the results shown in Figures 9 and 10, we set $\chi = \xi = 1$ (i.e., no relaxation), $N = 0$, and performed a costly tuning of α to obtain results that compare well with the literature [43,81,151,158,159]. A clear understanding of the complex interplay of these parameters and their effect on the solution accuracy is missing for compressible flows. In Section 3.1.1, we saw that even for incompressible flows, for which much more work has been undertaken, this understanding is still partial. A full understanding is obviously central to the success of the EFR algorithm with the AD indicator function and the definition of automatic tuning procedures. Another open question is the identification of the most physically accurate and computationally efficient filter $\mathcal{F}$ for each filtered variable. For this purpose, one could use various filters available in the literature (see, e.g, [5]).

Just like in the case of complex hemodyanmics simulations, care must be taken in setting the boundary conditions for compressible flow problems. Indeed, for compressible flows, the presence of artificial boundaries may trigger unphysical reflections that need to be suppressed by "nonreflecting" conditions. These conditions mimic the evolution of the solution outside the domain along the characteristics in order to suppress artificial inflow components. When nonreflecting boundary conditions are imposed, we need to devise conditions that minimize any artifact in the physical propagation. As the filter adds diffusivity, an initial idea could be to explore noninvasive conditions to minimize the impact of the filter. If this is not enough, one could explore different filter formulations.

Finally, we would like to stress the importance of data assimilation for weather/climate prediction too. In the atmospheric sciences, there is an abundance of data, which are collected either in situ or through remote sensing. Such data (or observations) are routinely incorporated in numerical weather prediction through data assimilation techniques, which seek to optimally combine the weather model with the observations. As mentioned in Section 3.1.2, it would be very interesting to study the computational efficiency and accuracy allowed by a combination of PINN and EFR.

4. LES for Reduced-Order Models

In this section, we build on the LES concepts, ideas, and models (i.e., spatial filtering, approximate deconvolution, and evolve–filter–relax strategy) introduced in Section 2 to construct LES-inspired ROMs, which we call LES-ROMs.

Over the past couple of decades, ROMs (see, e.g., [160–165] for reviews) have emerged as the methodology of choice to reduce the computational burden when traditional flow simulations (also called full-order models—FOMs) have to be carried out for several parameter values, as in the case of uncertainty quantification, optimal control, and inverse problems. ROMs replace the FOM with a lower-dimensional approximation that captures the essential behavior of the flow. The reduction in computational time is achieved through a two-step procedure. In the first step, called the *offline phase*, one constructs a database of several FOM solutions associated with given times and/or physical parameter values. The FOM database is used to generate a reduced basis, which is (hopefully much) smaller than the high-dimensional FOM basis but still preserves the essential features of the flow. One of the most popular reduced bases is proper orthogonal decomposition (POD), which extracts the dominant modes from a FOM database. We remark, however, that the strategies we discuss are not limited to POD and work equally well with other methods (e.g., the reduced basis method [163,166]). In the second step, called the *online phase*, one uses this reduced basis to quickly compute the solution for newly specified times and/or parameter values. Note that while the offline phase is performed only once, the online phase is performed as many times as needed.

ROMs are very efficient surrogate models when the number of reduced basis functions is small, as is typically the case for diffusion-dominated flows. Unfortunately, though, a large number of basis functions is needed to capture the essential features of convection-dominated flows. If one retains such a large number of modes in order for the ROM to be accurate, then the computational efficiency suffers. If the number of modes is otherwise kept low, a severe loss of information hinders the accurate reconstruction of the solution. In fact, Galerkin projection-based ROMs of turbulent flows are affected by energy stability problems related to the fact that, e.g., POD retains the modes biased toward large, high-energy scales, while the TKE is dissipated at the level of the small scales. A possible way to tackle this challenging problem is to introduce dissipation via a closure model [167–169] or stabilization [170–174].

We note, however, that the current ROM closures and stabilizations are typically heuristic and generally lack mathematical support. Traditionally, researchers propose ROM closures and stabilizations that make sense intuitively, test them in limited computational settings, and declare success once they yield more accurate solutions than the corresponding nonstabilized ROM. The main drawback of these ad hoc ROM closures and stabilizations is that while they work well for the specific flow configuration on which they were trained, they can fail in different flow configurations and need to be retrained (which is a costly, tedious process) in order to perform adequately.

Recently, there have been several new research developments aimed at constructing novel ROMs for realistic turbulent flows. The following are three significant research avenues in this direction: LES-ROMs (Section 4.1) aim at "bridging" two distinct research fields, LES and ROMs, to construct efficient and accurate ROMs for turbulent flows. FOM-ROM coupling (Section 4.2) aims at "bridging" the numerical methods used in the FOM and ROM in order to improve the stability and accuracy of the latter. Finally, the numerical analysis of the FOM-ROM framework (Section 4.3) leverages rigorous numerical analysis results to improve the ROM performance.

4.1. LES-ROMs

For the past several years, our groups and collaborators have proposed, analyzed, and investigated a new strategy for constructing ROMs for turbulent flows. This new strategy is based on *spatial filtering*, which is a mathematically and physically sound principle that is central in LES (see Section 2). This new strategy aims squarely at bridging two research fields, LES and ROMs, that have been treated separately until now. We also emphasize that this new strategy is fundamentally different from current ROM closures and stabilizations, which are often disconnected from the numerical methods used at the FOM level.

These new LES-inspired ROMs, which we call *LES-ROMs*, are capable of operating over a range of flow problems with minimal user intervention. Again, this is in stark contrast with current ROM closures and stabilizations, which need to be retrained for each new computational setting.

Just as in the FOM setting, the new LES-ROMs are constructed by leveraging a spatial filter $\mathcal{F}$ to smooth out different terms in the governing equations and eliminate spurious numerical oscillations that generally occur in the convection-dominated, under-resolved regime. In Section 4.1.1, we outline the main types of ROM filters and approximate deconvolution used to construct LES-ROMs. Then, we outline several LES-ROMs: the EFR-ROM (Section 4.1.2), which is arguably the most popular LES-ROM, Leray ROM (Section 4.1.3), approximate deconvolution Leray ROM (Section 4.1.3), and time-relaxation ROM (Section 4.1.3). Other types of LES-ROMs are mentioned in Section 4.1.4.

4.1.1. ROM Filters and Approximate Deconvolution

This section defines the *ROM spatial filters* and *ROM approximate deconvolution* used to construct the LES-ROMs presented in Sections 4.1.2–4.1.4. As discussed in Section 2, the *spatial filter* $\mathcal{F}$ is crucial in the development of LES for FOMs [4,5]. The spatial filter is also increasingly used to construct LES-ROMs for turbulent flows [175,176] because the idea is straightforward: $\mathcal{F}$ is used to eliminate the spurious numerical oscillations in stabilization strategies. Next, we outline the three spatial filters and the approximate deconvolution operators that are currently used for ROMs. In what follows, let $\mathbf{u}$ be a generic solution and $\mathbf{u}_r = \sum_{j=1}^{r} u_{r,j}\boldsymbol{\varphi}_j(\mathbf{x})$ its ROM approximation, with $\boldsymbol{\varphi}_j$ for $j = 1, 2, \ldots, r$ the ROM basis functions. We denote by $\underline{u}_r = (u_{r,1}, \ldots, u_{r,r})^\top$ the vector of ROM coefficients of $\mathbf{u}_r$.

ROM Differential Filter

The *ROM differential filter (DF)* [175,177,178] is probably the most popular ROM spatial filter used to build LES-ROMs. The ROM differential filter is a modular, simple, linear filter, which is a natural extension of the differential filter used to construct the EFR-FOM in Section 2.1. It is defined as follows: find $\overline{u}_r(\mathbf{x}) = \sum_{j=1}^{r} \overline{u}_{r,j}\boldsymbol{\varphi}_j(\mathbf{x})$ such that

$$\left(\overline{u}_r - \alpha^2 \Delta \overline{u}_r, \boldsymbol{\varphi}_i\right) = \left(\mathbf{u}_r, \boldsymbol{\varphi}_i\right) \qquad \forall\, i = 1, \ldots r, \tag{59}$$

where α is the filter radius. The DF weak form (59) yields the following linear system:

$$\left(\mathbb{I} + \alpha^2 A\right)\underline{\overline{u}}_r = \underline{u}_r, \tag{60}$$

where $\underline{\overline{u}}_r$ is the vector of ROM coefficients of $\overline{u}_r$ and $\mathbb{I}$ and A are the identity and ROM stiffness matrices, respectively. We note that $\mathbb{I}$ in (60) is a consequence of the POD basis functions being orthonormal in the L^2 norm. Thus, the ROM mass matrix is the identity matrix.

We emphasize that (60) is a low-dimensional, $r \times r$ linear system, whose computational overhead is negligible (since r is relatively small). Thus, the ROM differential filter will be used to construct LES-ROMs that increase the ROM accuracy without significantly increasing the computational cost.

The differential filter was used in LES of turbulent flows with classical numerical discretizations [5,6,179]. In reduced-order modeling, the ROM DF was used to develop LES-ROMs for the Kuramoto–Sivashinsky equation [177], the NSEs [175,180], and the quasi-geostrophic equations [181].

ROM Higher-Order Algebraic Filter

The second filter we review is the *ROM higher-order algebraic filter (HOAF)*: find $\overline{u}_r(\mathbf{x}) = \sum_{j=1}^{r} \overline{u}_{r,j}\boldsymbol{\varphi}_j(\mathbf{x})$ such that

$$\left(\mathbb{I} + \alpha^{2m} A^m\right)\underline{\overline{u}_r} = \underline{u_r}, \tag{61}$$

where m is a positive integer. As explained in [54] in the Fourier setting, the exponent m controls the percentage of filtering at different wavenumbers: as m increases, the amount of filtering increases for the high wavenumber components and decreases for the low wavenumber components. Just like the DF (60), the HOAF (61) is a relatively low-dimensional, $r \times r$ linear system that can be efficiently solved. Thus, the HOAF also leads to the development of accurate and efficient LES-ROMs. We also note that for $m = 1$, the linear systems (61) and (60) are identical. Thus, DF can be considered a particular case of HOAF with $m = 1$.

The HOAF (61) was proposed in [176] and was based on the HOAF introduced by Fischer and Mullen [54] in a spectral element method (SEM) setting. In [176], the HOAF (61) was called the higher-order differential filter, to be consistent with the SEM nomenclature. In [182], we showed that the HOAF is related to, but slightly different from, the spatial discretization of a higher-order differential operator (they are the same in the periodic case). For clarity, in this paper, we call the operator in (61) the high-order algebraic filter. In [182], a theoretical and numerical investigation of HOAF is presented.

ROM Projection

The third ROM spatial filter used to construct LES-ROMs is the *ROM projection filter*, which is defined as follows: for a fixed $r_1 < r$ and a given $\mathbf{u}_r \in \mathbf{X}^r = \text{span}\{\boldsymbol{\varphi}_1, \ldots, \boldsymbol{\varphi}_r\}$, the ROM projection (Proj) seeks $\overline{\mathbf{u}}_r \in \mathbf{X}^{r_1} = \text{span}\{\boldsymbol{\varphi}_1, \ldots, \boldsymbol{\varphi}_{r_1}\}$ such that

$$(\overline{\mathbf{u}}_r, \boldsymbol{\varphi}_j) = (\mathbf{u}_r, \boldsymbol{\varphi}_j) \qquad \forall j = 1, \ldots, r_1. \tag{62}$$

To our knowledge, the first time the Proj was used as an explicit ROM spatial filter was in [167], where it was leveraged to construct the ROM dynamic subgrid scale (SGS) model. We note that the dynamic SGS is generally considered to be the state-of-the-art closure model in LES [4,5]. Next, the ROM projection was used in [175,178] to construct the Leray ROM (see Section 4.1.3) and EFR-ROM (see Section 4.1.2).

ROM Filter Radius

The filter radius α is fundamental in the definitions of both the ROM differential filter (59) and the ROM higher-order algebraic filter (61): the radius of the ROM spatial filters represents the size of the scales that are captured by the filtered variables. Furthermore, the filter radius is also essential in the approximate deconvolution process, which depends on the particular filter used. Thus, a natural question is "How do we choose the ROM filter radius?" For example, the following practical question could be asked:

$$\boxed{\text{Given 10 ROM basis functions, what is the resolved lengthscale?}} \tag{63}$$

The most straightforward answer to question (63) is to choose the ROM filter radius the same way we choose the FOM filter radius. As mentioned in Section 2, the first choice is a *computational scale*, i.e., a scale based on the underlying numerical discretization. We also mention that other $\alpha(h)$ scalings are possible [5,6]. Another choice for the FOM differential filter radius is a *physical scale*, i.e., based on the physics of the underlying flow. For example, a popular choice in this class is $\alpha \sim \eta$, where we recall η is the Kolmogorov scale.

One limitation of using the computational scale or the physical scale to define the ROM filter radius is that these scales lack adaptivity, i.e., do not change when we vary the ROM dimension, r. To address this limitation, in [183] we put forth a new *energy-based ROM lengthscale*, which was defined by using a judicious partition of the energy. We showed that the new energy-based ROM lengthscale displays the correct asymptotic behavior: it converges to the smallest lengthscale in the system when the ROM dimension increases, and it converges to the largest lengthscale in the system when the ROM dimension decreases. Furthermore, in the numerical simulation of the turbulent channel flow at Reynolds number $Re_{\tau} = 395$, we compared the EFR-ROM equipped with the new energy-based ROM lengthscale and a classical ROM lengthscale based on dimensional arguments [184]. The numerical results showed that the former is more robust with respect to parameter changes than the latter. Finally, we note that the new energy-based ROM lengthscale could also be useful in the preprocessing strategy advocated in [185,186] as a means to filter out the noise in the input data.

Approximate Deconvolution

Although the AD strategy is central in image processing and inverse problems [187,188] and it has been used in LES development (see Section 2.1.3 and the research monographs [4–6,29,30]), it is a relatively new concept in reduced-order modeling. The AD strategy can be formulated as follows: given an approximation of the filtered input variable, $\overline{u} \doteq \mathcal{F}\mathbf{u}$, where $\mathcal{F}$ is an invertible spatial filter, find an approximation of the unfiltered input signal, $\mathbf{u}$. Of course, the exact deconvolution, $\mathcal{F}^{-1}\overline{u}$, would seem a natural choice. Computationally, however, the exact deconvolution is a very bad idea. Indeed, as shown in [187,188], the noise in the high wavenumber components of $\overline{u}$ is amplified by the inverse filter, $\mathcal{F}^{-1}$, in the exact deconvolution. Thus, in practice, AD strategies are used instead of the exact deconvolution [187,188]. In LES of turbulent flows, the AD models were pioneered by Adams and Stolz for classical numerical discretizations [87]. To our knowledge, in reduced-order modeling, the first AD model was proposed in [189], where AD was used to develop a ROM closure model. Recently, the AD strategy was also used to develop the AD Leray ROM [190], where three different AD strategies were investigated: the van Cittert AD approach (Section 3.2 in [190]), the Tikhonov AD approach (Section 3.3 in [190]), and the Lavrentiev AD approach (Section 3.4 in [190]). We also note that the Lavrentiev AD strategy was used in [189] to develop a ROM closure model, and the Tikhonov approach was used in [191–193] for ROM regularizations. These three AD methods were investigated numerically in Section 5.2 in [190].

4.1.2. EFR-ROM

The most popular LES-ROM exports the EFR strategy (11)–(13) to the reduced-order level, in which the FOM variables $\mathbf{U}$ and $\mathbf{V}$ are replaced by the ROM variables $\mathbf{u}_r$ and $\mathbf{v}_r$, respectively, and the filter and deconvolution operators are replaced by their ROM counterparts. This yields the *evolve–filter–relax ROM (EFR-ROM)*:

- *Evolve*: find $\mathbf{v}^{n+1}$ such that

$$\frac{\mathbf{v}_r^{n+1} - \mathbf{u}_r^n}{\Delta t} + \mathcal{M}(\mathbf{v}_r^{n+1}) = \mathbf{0}. \tag{64}$$

- *Filter*: find filtered variable $\overline{\mathbf{v}}_r^{n+1}$ such that

$$\overline{\mathbf{v}_r}^{n+1} = \mathcal{F}\mathbf{v}_r^{n+1}. \tag{65}$$

- *Relax*: set

$$\mathbf{u}_r^{n+1} = (1-\chi)\mathbf{v}_r{}^{n+1} + \chi\overline{\mathbf{v}_r}^{n+1}, \tag{66}$$

with relaxation parameter $\chi \in [0,1]$.

In the evolve step of the EFR-ROM algorithm, one time step of the standard Galerkin projection ROM (G-ROM) time discretization is leveraged to advance the EFR-ROM approximation at the current time step, $\mathbf{u}_r^n$, to an intermediate EFR-ROM approximation, $\mathbf{v}_r^{n+1}$. In the filter step of the EFR-ROM algorithm, one of the ROM spatial filters presented in Section 4.1.1 is used to filter the intermediate EFR-ROM approximation from the evolve step and obtain a smoother ROM approximation without spurious numerical oscillations. Finally, in the relax step of the EFR-ROM algorithm, the EFR-ROM approximation at the next time step is defined as a convex combination of the unfiltered intermediate EFR-ROM approximation obtained in the evolve step, $\mathbf{v}_r^{n+1}$, and its filtered counterpart, $\overline{\mathbf{v}}_r^{n+1}$. Just like in the case of EFR used as FOM (see Section 2.1), the goal of the relax step is to adjust the amount of dissipation introduced in the filter step by using a relaxation parameter, $0 \leq \chi \leq 1$. By varying the relaxation parameter χ, one can produce a range of filter strengths, from no filtering at all ($\chi = 0$) to maximum filtering ($\chi = 1$). As pointed out in Section 3.1.1 for the EFR used at the FOM level, the choice of χ may be critical in the presence of backflows. At the ROM level, the numerical investigation in [180] has shown that EFR-ROM is sensitive with respect to χ.

The evolve–filter ROM was introduced in [175] and EFR-ROM in [176]. Since then, EFR-ROM has been developed in several directions, for example, numerical simulation of turbulent flows (e.g., the turbulent channel flow) [183] and ocean dynamics [181], the FOM-ROM consistency [180], and feedback control [194].

4.1.3. Leray ROM, Approximate Deconvolution Leray ROM, and Time-Relaxation ROM

In this section, we present several LES-ROMs constructed by using ROM spatial filtering and approximate deconvolution.

Leray ROM

The *Leray ROM (L-ROM)* [175,195,196] is an LES-ROM in which the spatial filter, $\mathcal{F}$, is used to smooth out only the convective velocity of the nonlinear term in the NSE, i.e.,

$$\left(\frac{\partial \mathbf{u}_r}{\partial t}, \boldsymbol{\varphi}_i\right) + Re^{-1}\left(\nabla \mathbf{u}_r, \nabla \boldsymbol{\varphi}_i\right) + \left((\overline{\boldsymbol{u}}_r \cdot \nabla)\mathbf{u}_r, \boldsymbol{\varphi}_i\right) = 0, \quad \forall i = 1, \ldots, r, \tag{67}$$

where $\overline{\boldsymbol{u}}_r$ is the ROM velocity filtered with one of the ROM spatial filters, i.e., DF (60), HOAF (61), or Proj (62). By introducing the spatial filtering of the convective term, the Leray ROM smooths out the spurious numerical oscillations that affect the standard G-ROM in the convection-dominated, under-resolved regime.

Leray regularization (see Section 2.1) was first used in the context of reduced-order models in [177] for the Kuramoto–Sivashinsky equations. For fluid flows, L-ROM was first used in [175] for the 3D flow past a circular cylinder at $Re = 1000$. Since then, L-ROM has been successfully used as a stabilization technique for various under-resolved flows: the NSEs [71,197], the stochastic NSEs [176,196], the Boussinesq equations [198,199], the quasi-geostrophic equations [66,181], and the turbulent channel flow [182]. We also note that the ROM projection was leveraged in [198–200] to build the Leray ROM for challenging applications.

Approximate Deconvolution Leray ROM

The *approximate deconvolution Leray ROM (ADL-ROM)* [190] is an extension of the Leray ROM that leverages the AD operator $\mathcal{D}$. The idea of the ADL-ROM is to use AD to limit the amount of dissipation introduced in the Leray ROM when the filter radius is too large: Given u_r^n, find u_r^{n+1} such that

$$\left(\frac{u_r^{n+1} - u_r^n}{\Delta t}, \varphi_i\right) + Re^{-1}\left(\nabla u_r^{n+1}, \nabla \varphi_i\right) + \left(\left(\mathcal{D}(\overline{u}_r^{n+1}) \cdot \nabla\right) u_r^{n+1}, \varphi_i\right) = 0, \quad \forall i = 1, \ldots, r. \tag{68}$$

The numerical investigation in [190] for convection-dominated systems shows that when the filter radius is relatively large, the ADL-ROM is more accurate than the standard Leray ROM and less sensitive to model parameters. In addition, it shows that the increased accuracy allowed by the ADL-ROM (68) is not at the expense of numerical stability.

We emphasize that this is yet another instance in which ideas from completely different fields are synthesized. Indeed, AD is central in the image processing and inverse problems communities and has been successfully used to develop LES models [87,88], as mentioned also in Section 2.1.3.

Time-Relaxation ROM

The *time-relaxation ROM* [182] is an LES-ROM recently proposed with the aim to increase the ROM numerical stability by adding an extra term to the incompressible NSE. Specifically, the time-relaxation ROM reads as follows: Find $\mathbf{u}_r$ such that

$$\left(\frac{\partial \mathbf{u}_r}{\partial t}, \varphi_i\right) + Re^{-1}(\nabla \mathbf{u}_r, \nabla \varphi_i) + \left((\mathbf{u}_r \cdot \nabla)\mathbf{u}_r, \varphi_i\right) + \left(\chi_t(\mathbf{u}_r - \overline{\mathbf{u}}_r), \varphi_i\right) = 0, \quad \forall i = 1, \ldots r, \tag{69}$$

where χ_t is the time-relaxation parameter and $\overline{\mathbf{u}}_r$ is the ROM velocity filtered with one of the ROM spatial filters introduced in Section 4.1.1. The role of this extra term is to add numerical stabilization only at the marginally resolved scales without unnecessarily filtering the resolved scales.

In [182], it was shown that the numerical simulation of the turbulent channel flow at Reynolds numbers $Re_\tau = 180$ and $Re_\tau = 395$ performed with the new time-relaxation ROM equipped with the ROM differential filter and the ROM higher-order algebraic filter yields accurate results.

4.1.4. Other LES-ROMs

There are numerous LES-inspired ROMs, and the list is continuously growing. For comprehensive reviews on ROM closures, the reader is referred to [169,201]. We note, however, that many of the ROM closures presented in [169,201] do not use ROM filtering or approximate deconvolution. Examples include the eddy viscosity ROMs and machine learning ROM closures [169,201]. Since the main goal of this paper is to discuss a specific class of LES-inspired ROMs, which are constructed by using ROM spatial filtering and approximate deconvolution, we do not discuss these alternative ROM closures and instead refer the reader to [169,201]. In this section, however, we outline two types of LES-ROMs that are different from those we presented in Sections 4.1.2–4.1.3.

The *approximate deconvolution ROM (AD-ROM)* was introduced in [189]. To our knowledge, this is the first AD model used in reduced-order modeling. The AD-ROM is constructed by using AD (i.e., a Lavrentiev regularization) to construct an accurate ROM closure model. In [189], the AD-ROM was successfully tested in the numerical simulation of the three-dimensional flow past a circular cylinder at a Reynolds number $Re = 1000$.

The *variational multiscale (VMS)* methods [202,203] are constructed by using the principle of the locality of energy transfer, which states that energy is transferred mainly between neighboring scales or modes. Since ROMs use hierarchical bases in which the large and small structures are clearly displayed, the VMS framework naturally lends itself to be extended to the ROM setting. Although VMS-ROMs do not explicitly use a ROM spatial filter with an explicit filter radius, they display a connection with the LES-ROMs presented in this paper: In the VMS-ROMs, the decomposition of the flow variables into large scales and small scales can be regarded as the result of using the ROM projection operator [173,204,205]. For more discussion on VMS-ROMs, we refer the reader to [169] (Section IV.A.5).

4.2. LES-ROM Consistency

As explained in [180,206], ROMs are of two types:

- *FOM-ROM consistent*, i.e., the ROM uses the same computational model and the same numerical discretization as the FOM (see [206] Definition 1.1).
- *FOM-ROM inconsistent*, i.e., the ROM uses a computational and/or numerical discretization that are different from those used by the FOM.

We note that most ROMs (and data-driven models in general) are FOM-ROM inconsistent. The reason is simple: the FOM-ROM inconsistent ROMs (which are often called "nonintrusive") are much easier to use in practice, e.g., because they allow for the use of legacy codes at the FOM level. We emphasize, however, that recent numerical investigations [180,206] have shown that the FOM-ROM consistent ROMs yield more accurate results than the FOM-ROM inconsistent ROMs. For example, in [180] we generated the FOM data by using the EFR-FOM and we considered two types of ROMs: a FOM-ROM consistent ROM that uses the EFR strategy at the ROM level and a FOM-ROM inconsistent ROM that does not use the EFR strategy at the ROM level. Our numerical investigation showed that the FOM-ROM consistent ROM was more accurate than the FOM-ROM inconsistent ROM. Other examples of FOM-ROM consistent ROMs based on the EFR strategy are presented in [71,197] (see [207–209] for FOM-ROM consistent ROMs based on different stabilizations).

The LES-ROMs that we present in this section are FOM-ROM consistent since they use an LES strategy both at the FOM and the ROM levels. Thus, based on the preliminary studies in [180] for the EFR method, we believe that LES-ROMs will yield more accurate results than their inconsistent counterparts. We also note that by ensuring the modeling consistency between the FOM and the ROM, we increase the ROM robustness with respect to changes in the FOM parameters. For example, when the FOM resolution needs to be adapted by changing the FOM filter radius, the ROM filter radius is adapted automatically and the ROM does not have to be retrained. We emphasize that this is a departure from most of the current ROM closures and stabilizations, which require retraining (a costly process) every time an LES model parameter is changed at the FOM level.

Of course, as pointed out in Section 5 of [180], there are still many open questions. For example, one could investigate the FOM-ROM consistency when different LES models are used at the FOM and ROM levels, e.g., the Leray model is used at the FOM level and the EFR model is used at the ROM level. One could also investigate the *parameter FOM-ROM consistency*, which is complementary to the model FOM-ROM consistency. To investigate the parameter FOM-ROM consistency, one could consider the same LES model at the FOM and ROM levels but use different parameters (e.g., different α values) in these LES models. For example, one could investigate whether using different α or χ values at the FOM and ROM levels (i.e., $\alpha^{FOM} \neq \alpha^{ROM}$ or $\chi^{FOM} \neq \chi^{ROM}$) could yield more accurate ROM solutions.

We also emphasize that the FOM-ROM consistency has been investigated mainly numerically, without providing theoretical support. For notable exceptions, see the numerical analysis performed in [207,210] for FOM-ROM consistency of the streamline upwind Petrov–Galerkin (SUPG) stabilization and [206] for FOM-ROM consistency with respect to the discretization of the nonlinearity of the Navier–Stokes equations. We note, however, that the numerical analysis of the FOM-ROM consistency for LES-ROMs is still an open question.

4.3. LES-ROM Numerical Analysis

At the FOM level, extensive mathematical support exists for LES closures and stabilizations with classical numerical discretizations (e.g., the finite element method). For example, the monographs [5,29,30] present the mathematical analysis for many LES models, as well as the numerical analysis of their discretization. Similarly, the monograph [211] presents the mathematical and numerical analysis of classical stabilization strategies. This extensive literature provides answers not only to fundamental numerical analysis questions (e.g., stability and convergence) but also to *critical practical challenges* (e.g., parameter scalings for critical parameters like the filtering radius). In stark contrast, fundamental questions in the numerical analysis are still wide open for most of the ROM closure models: Is the proposed ROM stable? Does the ROM converge? If so, what does it converge to? What is the ROM rate of convergence?

Only the first steps in the numerical analysis of ROM closures and stabilizations have been taken: numerical analysis (e.g., stability and convergence) of classic ROM closures was performed in [212–216]. Numerical analysis was carried out in [213,214] for the eddy viscosity variational multiscale ROMs and in [215,216] for the Smagorinsky model in a reduced basis method setting. The first numerical analysis of LES-ROMs was performed in [195,196] for the Leray ROM (67) (see also [217] for related work) and in [194] for the EFR-ROM. We have also taken the first steps in proving FOM-ROM *parameter scalings* for ROM stabilizations and closures [207,213,214,218]. For example, in [207], for the SUPG-ROM, we proved scalings between the FOM mesh size and the SUPG-ROM stabilization parameter.

Finally, we note that the most active research area in ROM closure modeling is in the development of data-driven ROM closures in which available data are utilized to build the ROM closure model. An example of data-driven ROM closure is the data-driven variational multiscale ROM (d2-VMS-ROM) proposed in [205,219,220]. The d2-VMS-ROM and its developments have been investigated numerically in [205,219–226]. However, providing mathematical support for the d2-VMS-ROM, and data-driven ROM closures in general, is an open problem. Recently, in [227], we laid the mathematical foundations of data-driven ROM closures. Specifically, in Theorem 2 of [227], we proved that the d2-VMS-ROM is *verifiable*, i.e., the d2-VMS-ROM solution is accurate since the d2-VMS-ROM closure model is accurate.

Despite these first steps that our groups and our collaborators have taken, the numerical analysis of ROM closures and stabilizations is nowhere close to the numerical analysis for FOM closures and stabilizations. Laying the mathematical foundations for LES-ROMs and, more importantly, providing numerical analysis guidance to practical LES-ROM choices (e.g., parameter scalings) are still open questions.

5. Applications of LES for ROM

In this section, we present some results of LES with ROMs for test cases with an academic flavor and some real applications. We also draw a road map of other potential applications for more real-world problems that motivated us in the development of the methodologies described above. We anticipate these applications will provide an excellent benchmark for the assessment and improvement of such methodologies. To illustrate the versatility of the LES-ROMs, we show results for both incompressible flows (Section 5.1) and compressible flows (Section 5.2), as we did in Section 3.

5.1. Incompressible Flows

5.1.1. Flow Past a Cylinder

Although simple (or maybe precisely because of that), the 2D [228,229] and 3D [228,230,231] flow past a cylinder benchmarks have been widely used to assess numerical strategies, even those designed for higher Reynolds number flows. This is due to the fact that to obtain accurate results in the case of the time-dependent Reynolds number $0 \leq Re(t) \leq 100$, one needs a mesh with about 200 K elements for the 2D benchmark and roughly 3 million elements for the 3D benchmark. Thus, one can use these benchmarks to demonstrate that accurate results can be obtained also with much coarser meshes if suitable LES models are used. Indeed, for example, it has been shown that with the EFR algorithm, one can obtain accurate results with meshes as coarse as roughly 15 K elements in both 2D and 3D [58,59,61,62,71,180,232]. Below, we present selected 2D results from [197], which touch upon several topics discussed in Section 4.1.

The work in [71,197] stemmed from the need for FOM-ROM consistency described in Section 4.2. We proposed to use the EFR algorithm at both the full- and reduced-order level. In particular, we generated the snapshots with under-refined meshes. This is unlike other works, e.g., [175,176,233], where the snapshots are obtained by DNS. By using the EFR method as FOM and ROM, we adopted the same mathematical framework during both the *offline* and *online* stages and thus had a ROM that is fully consistent with the FOM. In [197], we proposed to use the POD basis related to the evolve velocity to approximate the filtered velocity and compute the reduced pressure field with a Poisson pressure equation method [234,235]. The main difference between [197] and [71,180,232] lies in the indicator function: it is linear in [71,180,232] and AD-based (52) in [197]. The ROM in [197] is called hybrid projection/data-driven because it exploits a traditional projection method (G-ROM) for the computation of the intermediate reduced velocity and pressure fields and uses a data-driven technique to compute the reduced coefficients of the indicator function field. The data-driven approach leverages interpolation with radial basis functions [236]. Alternatives to the data-driven approach could be using the same set of reduced coefficients for velocity, pressure, and indicator function or an EIM/DEIM technique [237,238]. The former was considered for RANS in [239] and was shown in [240] to provide less accurate results than a hybrid procedure.

The 2D flow past a cylinder for $0 \leq Re(t) \leq 100$ is challenging because the flow is laminar for the first 4 s and then the well-known vortex shedding appears. In [197], the focus was on the time interval $[4, 8]$ s, for which 200 high-fidelity snapshots were collected. The first 50 most energetic POD modes were considered, and a convergence test was performed based on three different energy thresholds: 99% (11 modes for the velocity, 5 modes for the pressure, and 20 modes for the indicator function), 99.9% (26 modes for the velocity, 12 modes for the pressure, and 44 modes for the indicator function), and 99.99% (42 modes for the velocity, 24 modes for the pressure, and 50 modes for the indicator function). Figure 11 compares the lift coefficient given by the FOM with the one given by the ROM. We observe that the evolution of lift coefficient computed by the ROM is very accurate when 99.9% or 99.99% of the eigenvalue energy is retained. Similar accuracy was found for the 3D cylinder test. See [197] for more details.

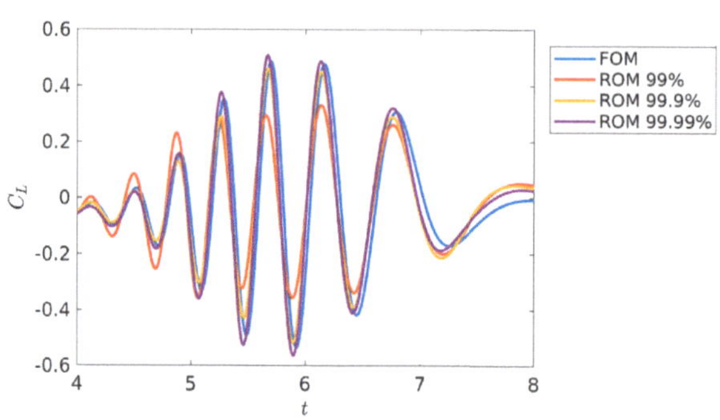

Figure 11. Lift coefficient C_L computed by the FOM and the projection/data-driven ROM from [197] for different thresholds of cumulative energy.

To complete the picture, we need to connect the accuracy with the required computational time. Figure 12 (left) shows the time-averaged relative L^2 error for velocity and pressure versus the corresponding wall time when the number of basis functions for the velocity is varied. The number of basis functions for the pressure is fixed at 24, while the number of basis functions for the indicator function is 50. The relative wall time is given by the ratio between the CPU time needed by the ROM and the CPU time required by a FOM simulation (i.e., 398 s). As one would expect, the errors initially decrease and the relative wall time increases as the number of basis functions increases. The errors reach their minimum value for 46 basis functions for the velocity. When the number of basis functions is increased to 50, two undesirable things happen: the relative wall time is larger than 1, meaning the ROM is more costly than the FOM (an absurdity), and the errors increase. Since 42 velocity modes are enough to retain 99.99% of the eigenvalue energy, we suspect the errors increase because of the noise introduced by the highest modes. We note that the ROM is more computationally efficient in the 3D test. See Figure 12 (right) and [197] for details.

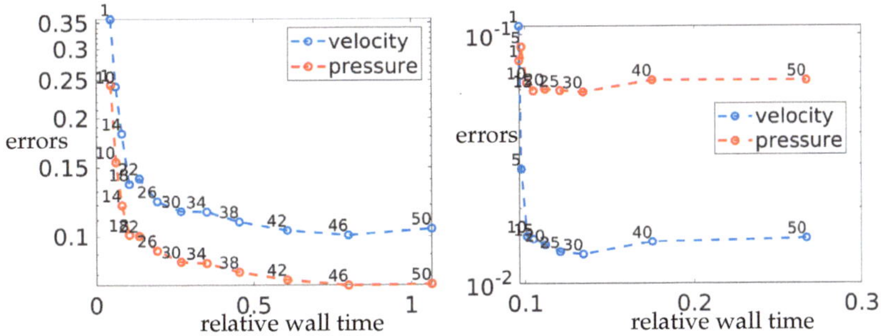

Figure 12. Pareto plots for the velocity and pressure: time-averaged relative L^2 error versus relative wall time when the number of basis functions for the velocity is varied for the 2D (**left**) and 3D (**right**) cylinder tests.

5.1.2. T-Junction

Here, we summarize the results of the numerical investigation performed in [200]. We investigate three of the LES-ROMs discussed in Section 4.1 (i.e., the EFR-ROM, Leray ROM, and time-relaxation ROM) in the numerical simulation of a T-junction problem. When streams of rapidly moving flow merge in a T-junction, large oscillations can occur at the scale of the diameter, D, with a period of $\mathcal{O}(D/U)$, where U is the characteristic flow speed. If the streams are at different temperatures, the oscillations result in fluctuations (see Figure 13). This phenomenon, known as thermal striping, can accelerate thermal–mechanical fatigue at the pipe wall in the outlet branch and ultimately cause pipe failure. Since thermal striping is of critical importance in nuclear engineering, the nuclear energy modeling and simulation community established a T-junction benchmark [241] to test the ability of CFD codes to predict thermal striping.

Figure 13. T-junction test case: instantaneous temperature field at Re = 10,000 for an investigation on thermal striping, which is critical in nuclear engineering [200].

We consider the T-junction problem at the Reynolds number Re = 10,000. The FOM is based on a spectral element discretization with about 21 million mesh points. Further details about the computational setting and FOM are given in [200]. We test the standard G-ROM and the three LES-ROMs in the predictive regime, that is, we test these models on a time interval that is larger than the interval where snapshots are collected. We consider $r = 250$ for the G-ROM and $r = 100$ for the LES-ROMs. To assess the ROMs' performance, we compare the near-wall temperature at several locations ($(x, \pm 0.5, 0)$, $(x, \pm 0.45, 0)$, $(x, 0, \pm 0.5)$, and $(x, 0, \pm 0.45)$ for $x = 2, 3, ..., 9$) with the FOM data (see Figure 14).

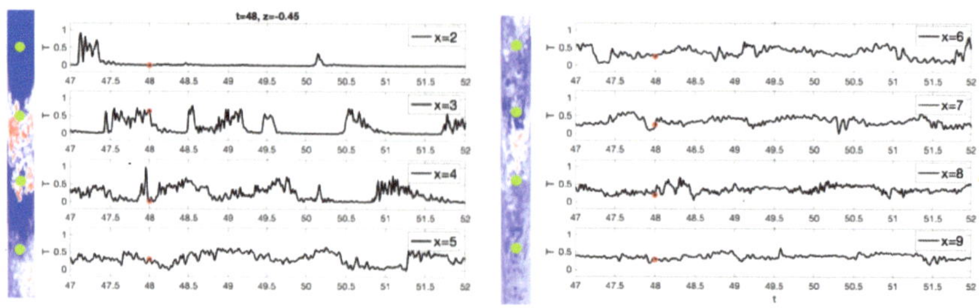

Figure 14. Near-wall temperature history at $y = 0, z = 0.45$ for several x locations in the outlet branch.

Figures 15 and 16 show the near-wall (i.e., $z = 0.45$) temperature of the downstream pipe at several x locations for the G-ROM, the Leray ROM, the EFR-ROM, and the time-relaxation ROM. In these plots, the two vertical red lines denote the time interval in which the snapshots were collected. The results indicate that the near-wall temperature of the G-ROM initially agrees with the FOM but becomes unstable in a relatively short time ($\approx$6 convective time units after the initial condition) even with $r = 250$. This indicates that the G-ROM is not accurate even in the reconstructive regime. Furthermore, larger fluctuations are observed for small x values. In contrast, the LES-ROM temperature stays on the overall FOM trajectory, and in the predictive regime (i.e., outside the region between the two vertical red lines) all three LES-ROMs are significantly more accurate than the G-ROM.

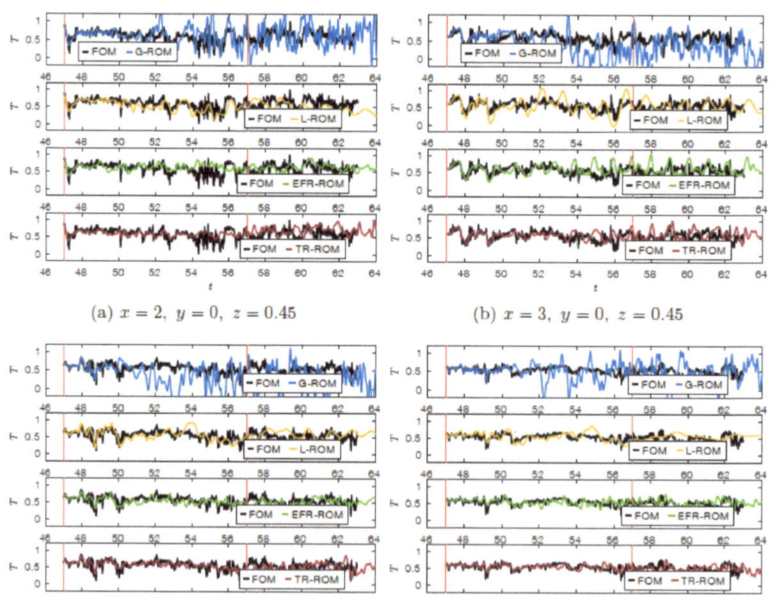

(a) $x = 2,\ y = 0,\ z = 0.45$ (b) $x = 3,\ y = 0,\ z = 0.45$

(c) $x = 4,\ y = 0,\ z = 0.45$ (d) $x = 5,\ y = 0,\ z = 0.45$

Figure 15. T-junction at $Re = 10,000$: comparison of the near-wall temperature history at $z = 0.45$ between the FOM, the G-ROM, and the LES-ROMs for $x = 2,3,4,5$ (outlet branch).

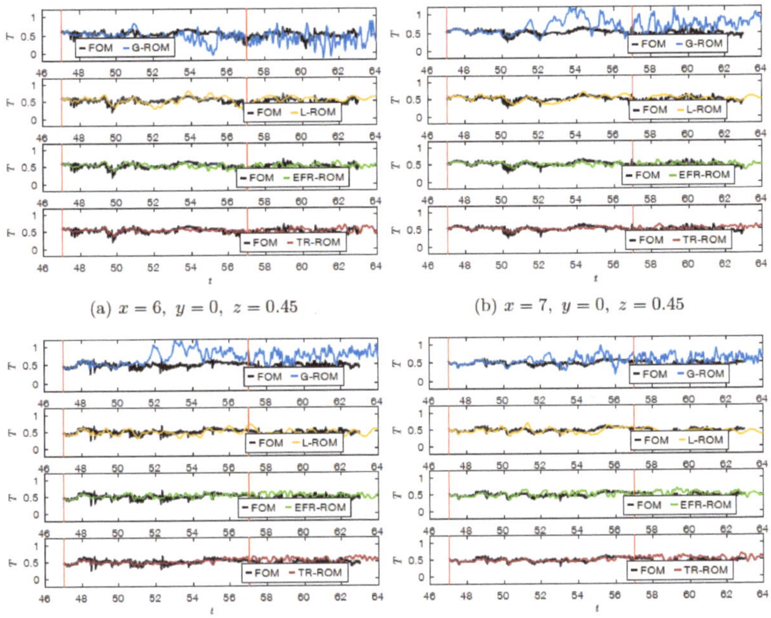

(a) $x = 6,\ y = 0,\ z = 0.45$ (b) $x = 7,\ y = 0,\ z = 0.45$

(c) $x = 8,\ y = 0,\ z = 0.45$ (d) $x = 9,\ y = 0,\ z = 0.45$

Figure 16. T-junction at $Re = 10,000$: comparison of the near-wall temperature history at $z = 0.45$ between the FOM, the G-ROM, and the LES-ROMs for $x = 6,7,8,9$ (outlet branch).

5.1.3. Hemodynamics Applications

As mentioned in Section 3.1, blood flow is generally in a laminar regime, and it does not require specific modeling for turbulence. AoDs are an exception where the combination of the specific flow features induced by the tiny entry tears and the need for an efficient solver to retrieve information from many patients of a (computer-assisted) clinical trial call for specific modeling solutions like EFR.

There are other problems in the field of computational hemodynamics where the need for extremely efficient solvers is critical. Typically, these are problems where the use of specific medical devices requires an optimization procedure. While "optimization" can be performed by trial-and-error approaches, translational mathematicians (i.e., mathematicians engaged in bringing state-of-the-art mathematical and computational procedures to the forefront of medical practice) feel the need for introducing automatic or, at least, semi-automatic rigorous, mathematically sound procedures. The cost of several attempts to find the optimal solution, either as the result of empirically educated guesses or iterations of a mathematical optimization procedure, may be prohibitive. In general, in this kind of application, efficiency is privileged with detrimental consequences for accuracy. For this reason, RANS models are often preferred.

An example of great interest in pediatric surgery is related to the so-called *total cavopulmonary connection (TCPC)*. This is a palliative surgery for newborn babies with a severe heart malformation called left ventricle hypoplasia syndrome. This condition (called *uni-ventricular circulation*) is not compatible with life and requires specific treatments. Years ago, the TCPC was introduced to buy time before a transplant [242]. It consists of an artificial connection between the superior and inferior venae cavae (SVC and IVC, respectively) with the pulmonary artery (PA) to route systemic circulation to the pulmonary circulation. An oversimplified representation of this shape is reported in Figure 17. The functioning ventricle pumps blood into the large circulation. The connection between the venae cavae and the PA has a peculiar cruciform shape, shown in Figure 17, with many interesting implications in terms of fluid dynamics. The biomedical engineering community has investigated these implications in a huge number of excellent contributions, see, e.g., [243–252]. From the computational point of view, the presence of the colliding fronts injected at the SVC and IVC raises some challenges since even moderate Reynolds numbers may induce flow disturbances. In general, the shape of the cruciform connection was speculated to have a major role in the long-term healthy conditions of the patient. For instance, an even splitting of the flow coming from the IVC (called *hepatic flow*) is critical and is determined by the offset between the SVC and IVC left after the surgery (see Figure 17). However, a rigorous shape optimization based on mathematical procedures in the operating room is still a dream for many reasons. One is the identification of all the relevant criteria that define a reliable and robust "optimal" solution. Another one is the huge computational cost of numerical optimization. On top of this, a more recent variant of the surgery requires the introduction of a pump in the cross-shaped junction to establish hemodynamic conditions closer to the physiology of biventricular circulation [253]. The regime induced by the pump requires specific turbulence modeling [7]. High computational costs currently lead to the use of RANS models, even though the superiority of LES modeling for the flow regimes characteristic of TCPC is commonly recognized [254]. Our intention is to bring LES-ROM modeling into this field as a potential breakthrough that enables not only accurate LES modeling as the routine approach but also rigorous shape optimization procedures as part of regular surgical planning.

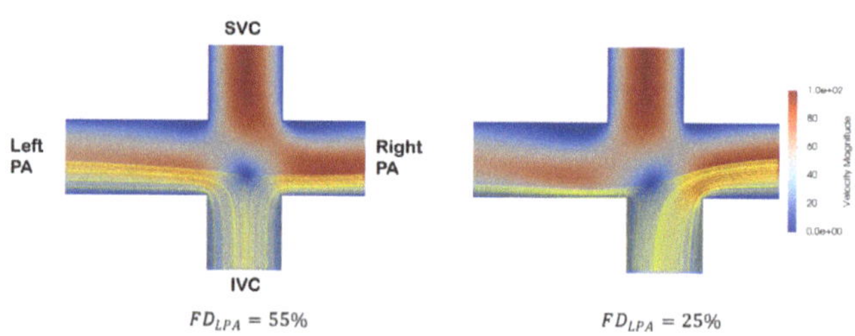

Figure 17. Simplified geometry of a TCPC. The vertical vessel is the vena cava (VC: superior at the top—SVC, inferior at the bottom—IVC). The pulmonary artery (PA) is the horizontal vessel. The inflow sections are at the SVC and at the IVC. This generates colliding fronts. The picture reports the results corresponding to two different surgical options. The difference is in the flow distribution from the IVC (the so-called hepatic flow distribution): FD_{LPA} is the fraction of hepatic flow directed to the left PA. An even flow distribution (i.e., $FD_{LPA} \approx 50\%$) is desirable. Notice that the different FD_{LPA} are created by different offsets between the SVC and IVC.

5.1.4. Wind Energy Applications

LES-based wind modeling presents significant challenges, with key questions posed on turbulence, mostly focused on mean-flow kinetic energy (MKE) entrainment and dissipation in large wind farms [255]. Understanding the upper limit of power production is linked to MKE entrainment. Recent studies, such as [256], identified spatial constraints and the influence of the Coriolis parameter on power output through idealized atmospheric simulations run with the weather research and forecasting model. An entrainment-based model for wind farm flow and power prediction was recently proposed in [257]. Wind energy, projected to meet 35% of global electricity needs by 2050, faces significant challenges, including optimizing wind farm layouts and minimizing wake interactions that reduce power output and affect turbine longevity. With this in mind, it is worth noting that enhancing energy production by just 1% could yield 30 TWh annually, equivalent to adding 3600 turbines and USD 1 billion in revenue [258].

The filtering-based approaches reviewed in this paper can provide accurate modeling tools to predict wind farm wakes, addressing spatio-temporal variability, unsteady wake interactions, and atmospheric turbulence. Moreover, improved turbine wake prediction and cost-effective surrogate models can optimize turbine layout and support wind farm deployment [259]. For further insights on LES of wind applications, we refer the reader to references [256,257,260–264]. Moreover, a recent study [265] offers a comprehensive overview of digital twin technology, with a focus on wind energy applications. The study consolidates digital twin definitions, identifies the current state of the art in modeling and simulation techniques, and outlines research needs in the wind energy sector from an industrial perspective. It proposes solutions to these challenges from research institutes' viewpoints and provides recommendations for stakeholders to facilitate technology adoption, with the methodologies highlighted in our article seen as key enablers for this digital technology adaptation.

5.2. Compressible Flows

ROMs combined with LES modeling has the potential to be a major player in reducing the computational cost associated with weather forecasting. This section is meant to convey an ideal of such potential and show some limitations of current methodologies as an incentive to undertake more work in this research area. Mainly, we report results from [266] obtained with selected data-driven ROMs for the same benchmarks for atmospheric flow described in Section 3.2. However, let us start with a brief literature review.

While POD, which is also referred to as empirical orthogonal function (EOF) analysis in the geophysical fluid dynamics community, has been used for a long time, it is only recently that ROMs have been applied for the simulation of atmospheric flows. EOF has been applied to identify spatio-temporal coherent meteorological patterns, e.g., the Madden–Julian oscillation, the quasi-biennial oscillation, and the El Niño–southern oscillation. See, e.g., [267–271]. The EOF analysis for these phenomena uses data on the global scale and considers time (ranging from several months to many years) as the only parameter. In addition, it is mostly limited to data analysis as a mean to understand the weather system, i.e., EOF is not used for forecasts. Only very recently, a data-driven ROM based on EOF analysis has been used for pattern prediction, specifically to forecast the weekly average sea surface temperature [268]. Other data-driven methods borrowed from machine learning have been applied to global weather forecasting. See, e.g., [272–278].

The work in [266] is a first attempt to apply ROMs that generate a reduced-order basis (unlike machine learning methods) to both reconstruct and forecast regional atmospheric flows (unlike EOF, which is not used to forecast). Additionally, since the spatial scale is a few kilometers and the time scale is a few hours, there is an obvious difference in resolution from EOF analysis. The considered data-driven ROMs are dynamic mode decomposition (DMD), Hankel dynamic mode decomposition (HDMD), and proper orthogonal decomposition with interpolation (PODI). DMD was specifically designed to predict the future behavior of a system [279–282] and has been successfully applied to several incompressible fluid dynamics problems [283–285]. HDMD enhances the DMD algorithm with time-delay embedding [286–291] and is able to predict more accurately and for longer periods of time systems exhibiting strong nonlinear dynamics [290,292]. Successful applications include periodic cavity flow [286,293], electromechanical systems [291], and biological systems [288]. Unlike DMD and HDMD, PODI was not designed to forecast system evolution but rather to interpolate solutions in a parameter space, where time is one of possibly many parameters of interest. To date, PODI has been applied to perform parametric studies for problems in hemodynamics [294], chemical [295] and naval [296,297] engineering, and aeronautics [298].

For the rising thermal bubble, we collect a database of potential temperature perturbations consisting of 204 snapshots, i.e., the computed θ' every 5 s. For simplicity, we used a constant eddy viscosity $\nu_a = 15$ m^2/s, which is frequently used in the literature (see, e.g., [43,151,159]). Then, 90% of the database (i.e., 184 snapshots) is used for training. In the case of DMD and HDMD, these 184 solutions are the first 184 in the database (associated with the time interval $(0, 920]$ s). In the case of PODI, these 184 solutions are selected randomly over the entire time interval $[0, 1020]$ s. The remaining 20 solutions form the validation set. This difference in the training and validation sets reflects the different nature of PODI and DMD/HDMD algorithms. To retain 99% of the eigenvalue energy in the database, one needs 17 POD modes for DMD and PODI and 46 modes for HDMD. Figure 18 shows a qualitative comparison of the ROM solutions with the FOM solution. Among the four time instants chosen for the visualization, two of them ($t = 255$ and $t = 505$ s) correspond to solutions belonging to the training set and allow us to asses the ability of each ROM technique to identify the system dynamics. We see that all three ROMs perform well in system identification. The remaining two times ($t = 980$ and $t = 1020$ s) are not associated with the training set and thus are used to check the accuracy of the ROM in predicting (for DMD and HDMD) or interpolating (for PODI) the system dynamics. For these times, DMD provides a poor approximation of θ', while HDMD and PODI reconstruct the solution well. The problem with the reconstruction provided by DMD appears to be related to assigning large weights to basis functions associated with previous times. Indeed, in the DMD solution for $t = 1020$ s, we can observe the time history of the system dynamics, i.e., the rising of the bubble.

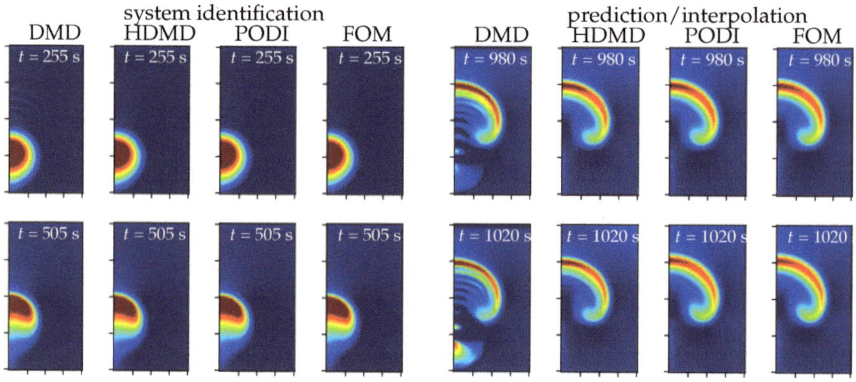

Figure 18. Rising thermal bubble: θ' given by the ROMs and the FOM at time values within (**left**) and outside (**right**) the training dataset.

All the simulations whose results are reported in Figure 18 were run on a common laptop. Table 3 reports the computational time needed to construct the reduced basis offline and to perform a simulation online for each of the ROMs we considered. The computational times for DMD and PODI are comparable, which for the online phase is explained by the fact that the DMD and PODI reduced basis have the same size, while HDMD is computationally more expensive.

Table 3. Rising thermal bubble: computational time needed to construct the reduced basis offline and to perform a simulation online for DMD, HDMD, and PODI.

Method	Basis Construction	Online Run
DMD	0.085 s	0.02 s
HDMD	2.865 s	2.95 s
PODI	0.1 s	0.018 s

The density current benchmark proved to be more challenging for the fact that it features mainly vertical dynamics for part of the time interval (bubble descending) and mainly horizontal dynamics for the rest of the time (after the bubble hits the ground and a cold front propagates). To separate the two dynamics [180,197], we collect snapshots associated only with the cold front propagation, i.e., the computed θ' every 4 s for $t \geq 280$. Furthermore, in this case, we set the eddy viscosity to a constant value ($\mu_a = 65$ m^2/s) taken from the literature [151,154]. Out of the snapshot database, 85% is used for training and $t = [808, 900]$ s is used as the prediction window. We set again the eigenvalue energy threshold to 99%, which leads to retaining 49 POD modes for DMD, 96 modes for HDMD, and 52 modes for PODI. Figure 19 compares the evolution of the potential temperature perturbation given by DMD, HDMD, and PODI with the evolution computed by the FOM. The top two rows in Figure 19 correspond to solutions belonging to the training set. Once again, we see that all three ROMs perform well in system identification. The bottom two rows in Figure 19 correspond to solutions not within the training set. The DMD solutions are affected by the same problem observed in Figure 18: we can see the evolution of θ' in each panel because large weights are assigned to basis functions associated with previous times. The HDMD solutions are a clear improvement, but the PODI solutions compare more favorably with the FOM solutions.

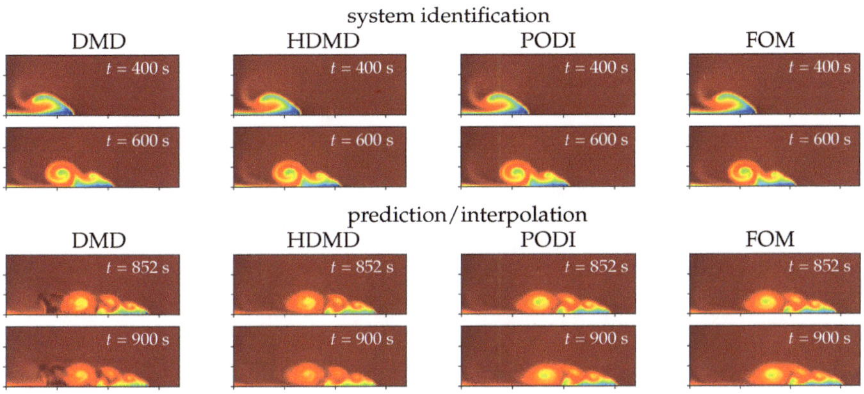

Figure 19. Density current: θ' given by the ROMs and the FOM at time values within (**top**) and outside (**bottom**) the training dataset

In conclusion, although DMD and HDMD are intended for forecasts, the accuracy in the prediction of the system dynamics is low even when 99% of the eigenvalue energy is retained and the snapshots in the database are tailored to the problem at hand. Thanks to the interpolatory approach, PODI maintains a good level of accuracy during the entire time interval of interest. In [266], it is shown that this is true also when a physical parameter (not just time) is varied within a parametric study and for 3D results. While still preliminary, these results show that a lot more work is needed if we want to use ROMs to cut the computational time required to forecast the weather for long prediction windows and in high-dimensional parameter space. The results for the T-junction benchmark test summarized in Section 5.1 suggest that L-ROM, EFR-ROM, and TR-ROM could likely improve the results presented in this section in terms of accuracy. Additional improvements in terms of efficiency could come from machine learning-based techniques that can better detect and reproduce the nonlinear behavior exhibited by the FOM, e.g., convolutional autoencoders and long short-term memory [299–302].

6. Concluding Remarks and Outlook

In this paper, we connect two important research fields that have been treated separately until now, LES and ROMs, by reviewing ROMs that are inspired from LES strategies, which we call LES-ROMs. In our review, we tried to emphasize two essential features of LES-ROMs: (i) The LES-ROM achievements in reduced-order modeling of realistic, convection-dominated, under-resolved flows; (ii) Their rigorous, mathematical foundations, which enable a clear physical interpretation of the results.

Convection-dominated flows in the under-resolved regime are central in important engineering, scientific, and medical applications. Standard ROMs generally fail to produce acceptable results in these challenging settings. Thus, alternatives are urgently needed.

In our review, we emphasized the significant achievements of LES-ROMs in the efficient and accurate numerical simulation of convection-dominated, under-resolved flows in important applications in engineering (e.g., turbulent flows in aerospace industry), science (e.g., weather/climate modeling), and medicine (e.g., cardiovascular flows). In these critical applications, LES-ROMs have yielded accurate results, while their cost was orders of magnitude lower than the cost of traditional numerical methods (e.g., the finite element method), also called FOMs. We also emphasized that an important advantage of LES-ROMs is that they often are extremely easy to implement. Indeed, one can start with a "legacy" ROM code (e.g., one of the current ROM software libraries [303,304]) and, in a matter of *minutes*, implement an efficient and accurate LES-ROM. We illustrated the significant achievements of LES-ROMs for both incompressible flows (e.g., the incompressible Navier–Stokes equations used in computational hemodynamics) and compressible flows (e.g., the weakly compressible Euler equations used in weather prediction).

Our review revolves around the mathematical tools that are used to construct LES-ROMs, which not only provide sound support for LES-ROMs but also connect ("bridge") two important research fields, LES and ROMs, that have been investigated separately. Our presentation focused on *spatial filters*, both at the FOM level and at the ROM level, and *approximate deconvolution (AD)* operators, which provide efficient and stable approximations to the inverse of the spatial filter. The FOM spatial filters we discussed are the differential filter and the nonlinear filter. We used these filters to build LES models. The ROM spatial filters we presented are the ROM differential filter, the ROM higher-order algebraic filter, and the ROM projection. We then illustrated how these ROM spatial filters are leveraged to construct LES-ROMs. As for the AD, we presented several strategies (e.g., Lavrentiev, Tikhonov, and Van Cittert) and showed how they can be used to increase the accuracy of both LES models and LES-ROMs.

Equipped with these mathematical tools (i.e., spatial filtering and AD), we then showed how they can be used to construct LES models and LES-ROMs. For clarity of presentation, we focused on one of the most popular and successful models, the *evolve–filter–relax (EFR)* strategy, and discussed it at both the FOM and the ROM levels. At the FOM level, we showed how the EFR model is constructed for both the incompressible Navier–Stokes equations and the weakly compressible Euler equations. Then, we illustrated some of the EFR's achievements in incompressible flows (computational hemodynamics) and compressible flows (the rising thermal bubble and the density current benchmarks). At the ROM level, we constructed the EFR-ROM and several other LES-ROMs (the Leray ROM, AD Leray ROM, and time-relaxation ROM). Then, we illustrated some LES-ROMs' achievements for both incompressible flows (the classical flow past a cylinder and the T-junction test case with thermal striping) and compressible flows (again, the rising thermal bubble benchmark and density currents).

We are aware that this is not an exhaustive review of LES-ROMs. However, in writing this paper, our hope was to provide a brief and friendly introduction to this exciting research area, which we believe has a lot of potential in practical numerical simulation of convection-dominated flows in engineering, science, and medicine. We hope we conveyed to the reader the reason for our excitement: LES-ROMs are extremely easy to implement and use, highly efficient, and accurate in reproducing average flow quantities of interest, even in the case of challenging, critical applications of convection-dominated flows.

Despite the LES-ROMs' success, there are still many research directions that need to be further explored. This is natural since the LES-ROMs are much more recent than classical LES models. Among the promising research directions in LES-ROM development, we mention the construction, analysis, and investigation of other LES-ROMs, spatial filters, and approximate deconvolution operators. Indeed, at the FOM level, there is a plethora of LES models that are constructed with various types of spatial filters and closure modeling strategies. A few of those have been already used to build LES-ROMs, but there are many more that have not been investigated. Another interesting research direction is the construction of new, data-driven spatial scales (lengthscales) at the ROM level. Indeed, at the FOM level, the spatial filter radius can be defined based on the mesh used in the numerical discretization. At the ROM level, however, how should we define the filter radius? In this paper, we outlined a few possibilities, but much more remains to be undertaken. Probably one of the most important research directions is the investigation of the new LES-ROMs in classical test problems for numerical methods for turbulent flows, e.g., the turbulent channel flow [305]. These test problems will yield new insight into the practical application of the proposed LES-ROMs in the numerical simulation of turbulent flows. We note, however, that the investigations of ROMs for turbulent flows is relatively scarce (see, e.g., [182] for steps in this direction). Finally, an important research direction is continuing to lay the mathematical foundations of LES-ROMs. Only the first steps in this direction have been made, as we briefly discussed in this paper. There is, however, much more that needs to be done to bring the mathematical support of LES-ROMs to the same level as the mathematical support of LES (see, e.g., the research monographs [5,6,29,30] on the mathematical theory of LES).

Author Contributions: Conceptualization, A.Q., O.S., A.V. and T.I.; methodology, A.Q., O.S., A.V. and T.I.; software, A.Q., O.S., A.V. and T.I.; validation, A.Q., O.S., A.V. and T.I.; formal analysis, A.Q., O.S., A.V. and T.I.; investigation, A.Q., O.S., A.V. and T.I.; resources, A.Q., O.S., A.V. and T.I.; data curation, A.Q., O.S., A.V. and T.I.; writing—original draft preparation, A.Q., O.S., A.V. and T.I.; writing—review and editing, A.Q., O.S., A.V. and T.I.; visualization, A.Q., O.S., A.V. and T.I.; supervision, A.Q., O.S., A.V. and T.I.; project administration, A.Q., O.S., A.V. and T.I.; funding acquisition, A.Q., O.S., A.V. and T.I. All authors have read and agreed to the published version of the manuscript.

Funding: A.Q. was supported by the U.S. National Science Foundation through grant number DMS-1953535. O.S. was supported by the U.S. National Science Foundation through grant number DMS-2345048. A.V. was supported by the U.S. National Science Foundation through grant number DMS-2012286 and grant number DMS-2038118, the U.S. National Institutes of Health through grant number R01EB031101, and Emory University Research Committee. T.I. was supported by the U.S. National Science Foundation through grant number DMS-2012253 and grant number CDS&E-MSS-1953113.

Data Availability Statement: No new data were created or analyzed in this study. Data sharing is not applicable to this article.

Acknowledgments: We thank Jiaqi Yang for the simulations of aortic dissections, and Marina Piccinelli and Bradley Leshnower for providing the data and elaborating the images. We also thank Gianluigi Rozza, Michele Girfoglio, Francesco Ballarin, and Maria Strazzullo for carefully reading the manuscript and providing feedback that has improved the quality and clarity of the paper.

Conflicts of Interest: The authors declare no conflicts of interest.

References

1. Kolmogorov, A.N. The local structure of turbulence in incompressible viscous fluids at very large Reynolds numbers. *Dokl. Akad. Nauk. SSSR* **1941**, *30*, 299–303.
2. Kolmogorov, A.N. Dissipation of energy in isotropic turbulence. *Dokl. Akad. Nauk. SSSR* **1941**, *32*, 19–21.
3. Duraisamy, K.; Iaccarino, G.; Xiao, H. Turbulence modeling in the age of data. *Annu. Rev. Fluid Mech.* **2019**, *51*, 357–377. [CrossRef]
4. Sagaut, P. *Large Eddy Simulation for Incompressible Flows*, 3rd ed.; Scientific Computation; Springer: Berlin/Heidelberg, Germany, 2006; p. xxx+556.
5. Berselli, L.C.; Iliescu, T.; Layton, W.J. *Mathematics of Large Eddy Simulation of Turbulent Flows*; Scientific Computation; Springer: Berlin/Heidelberg, Germany, 2006; p. xviii+348.
6. Layton, W.J.; Rebholz, L.G. *Approximate Deconvolution Models of Turbulence: Analysis, Phenomenology and Numerical Analysis*; Springer Science & Business Media: Berlin/Heidelberg, Germany, 2012; Volume 2042.
7. Delorme, Y.; Anupindi, K.; Kerlo, A.; Shetty, D.; Rodefeld, M.; Chen, J.; Frankel, S. Large eddy simulation of powered Fontan hemodynamics. *J. Biomech.* **2013**, *46*, 408–422. [CrossRef] [PubMed]
8. Xu, H.; Piccinelli, M.; Leshnower, B.G.; Lefieux, A.; Taylor, W.R.; Veneziani, A. Coupled morphological–hemodynamic computational analysis of type B aortic dissection: A longitudinal study. *Ann. Biomed. Eng.* **2018**, *46*, 927–939. [CrossRef]
9. Manchester, E.L.; Pirola, S.; Salmasi, M.Y.; O'Regan, D.P.; Athanasiou, T.; Xu, X.Y. Analysis of turbulence effects in a patient-specific aorta with aortic valve stenosis. *Cardiovasc. Eng. Technol.* **2021**, *12*, 438–453. [CrossRef] [PubMed]

10. Caldwell, P.M.; Mametjanov, A.; Tang, Q.; Van Roekel, L.P.; Golaz, J.C.; Lin, W.; Bader, D.C.; Keen, N.D.; Feng, Y.; Jacob, R.; et al. The DOE E3SM coupled model version 1: Description and results at high resolution. *J. Adv. Model. Earth Syst.* **2019**, *11*, 4095–4146. [CrossRef]
11. Terai, C.; Caldwell, P.; Klein, S.; Tang, Q.; Branstetter, M. The atmospheric hydrologic cycle in the ACME v0.3 model. *Clim. Dyn.* **2018**, *50*, 3251–3279. [CrossRef]
12. Wehner, M.; Reed, K.; Li, F.; Prabhat; Bacmeister, J.; Chen, C.T.; Paciorek, C.; Gleckler, P.J.; Sperber, K.R.; Collins, W.D.; et al. The effect of horizontal resolution on simulation quality in the Community Atmospheric Model, CAM5.1. *J. Adv. Model. Earth Syst.* **2014**, *6*, 980–997. [CrossRef]
13. Bacmeister, J.; Wehner, M.; Neale, R.; Gettelman, A.; Hannay, C.; Lauritzen, P.; Caron, J.M.; Truesdale, J.E. Exploratory high-resolution climate simulations using the Community Atmosphere Model (CAM). *J. Clim.* **2014**, *27*, 3073–3099. [CrossRef]
14. Delworth, T.; Rosati, A.; Anderson, W.; Adcroft, A.J.; Balaji, V.; Benson, R.; Dixon, K.; Griffies, S.M.; Lee, H.C.; Pacanowski, R.C.; et al. Simulated climate and climate change in the GFDL CM2.5 high-resolution coupled climate model. *J. Clim.* **2012**, *25*, 2755–2781. [CrossRef]
15. Love, B.; Matthews, A.; Lister, G. The diurnal cycle of precipitation over the maritime continent in a high-resolution atmospheric model. *Quaterly J. R. Meteorol. Soc.* **2011**, *137*, 934–947. [CrossRef]
16. Atlas, R.; Reale, O.; Shen, B.W.; Lin, S.J.; Chern, J.D.; Putman, W.; Lee, T.; Yeh, K.S.; Bosilovich, M.; Radakovich, J. Hurricane forecasting with the high-resolution NASA finite volume general circulation model. *Geophys. Res. Lett.* **2005**, *32*. [CrossRef]
17. Iorio, J.; Duffy, P.; Govindasamy, B.; Thompson, S.; Khairoutdinov, M.; Randall, D. Effects of model resolution and subgrid-scale physics on the simulation of precipitation in the continental United States. *Clim. Dyn.* **2004**, *23*, 243–258. [CrossRef]
18. Duffy, P.; Govindasamy, B.; Iorio, J.; Milanovich, J.; Sperber, K.; Taylor, K.; Wehner, M.F.; Thompson, S.L. High-resolution simulations of global climate, Part 1: Present climate. *Clim. Dyn.* **2003**, *21*, 371–390. [CrossRef]
19. Pope, V.; Stratton, R. The processes governing horizontal resolution sensitivity in a climate model. *Clim. Dyn.* **2002**, *19*, 211–236.
20. Leonard, A. Energy Cascade in Large-Eddy Simulations of Turbulent Fluid Flows. In *Turbulent Diffusion in Environmental Pollution*; Frenkiel, F., Munn, R., Eds.; Advances in Geophysics; Elsevier: Amsterdam, The Netherlands, 1975; Volume 18, pp. 237–248. [CrossRef]
21. Germano, M. Turbulence: The filtering approach. *J. Fluid Mech.* **1992**, *238*, 325–336. [CrossRef]
22. Moser, R.D.; Haering, S.W.; Yalla, G.R. Statistical Properties of Subgrid-Scale Turbulence Models. *Annu. Rev. Fluid Mech.* **2021**, *53*, 255–286. [CrossRef]
23. Beck, A.; Flad, D.; Munz, C.D. Deep neural networks for data-driven LES closure models. *J. Comput. Phys.* **2019**, *398*, 108910. [CrossRef]
24. Sirignano, J.; MacArt, J.F.; Freund, J.B. DPM: A deep learning PDE augmentation method with application to large-eddy simulation. *J. Comput. Phys.* **2020**, *423*, 109811. [CrossRef]
25. Xie, C.; Wang, J.; E, W. Modeling subgrid-scale forces by spatial artificial neural networks in large eddy simulation of turbulence. *Phys. Rev. Fluids* **2020**, *5*, 054606. [CrossRef]
26. Duraisamy, K. Perspectives on machine learning-augmented Reynolds-averaged and large eddy simulation models of turbulence. *Phys. Rev. Fluids* **2021**, *6*, 050504. [CrossRef]
27. Raissi, M.; Yazdani, A.; Karniadakis, G.E. Hidden fluid mechanics: Learning velocity and pressure fields from flow visualizations. *Science* **2020**, *367*, 1026–1030. [CrossRef] [PubMed]
28. Di Leoni, P.C.; Zaki, T.A.; Karniadakis, G.; Meneveau, C. Two-point stress–strain-rate correlation structure and non-local eddy viscosity in turbulent flows. *J. Fluid Mech.* **2021**, *914*, A6. [CrossRef]
29. John, V. Large Eddy Simulation of Turbulent Incompressible Flows. In *Lecture Notes in Computational Science and Engineering*; Springer: Berlin/Heidelberg, Germany, 2004; Volume 34, p. xii+261.
30. Rebollo, T.C.; Lewandowski, R. *Mathematical and Numerical Foundations of Turbulence Models and Applications*; Springer: Berlin/Heidelberg, Germany, 2014.
31. Frisch, U. *Turbulence: The Legacy of A.N. Kolmogorov*; Cambridge University Press: Cambridge, UK, 1995.
32. Tennekes, H.; Lumley, J. *A First Course in Turbulence*; MIT Press: Cambridge, MA, USA, 1972.
33. Smagorinsky, J. General Circulation Experiments with the Primitive Equations: I. The basic experiement. *Mon. Wea. Rev.* **1963**, *91*, 99–164. [CrossRef]
34. Abgrall, R. Toward the Ultimate Conservative Scheme: Following the Quest. *J. Comput. Phys.* **2001**, *167*, 277–315. [CrossRef]
35. Prandtl, L. Turbulent Flow. In *Lecture Delivered before the International Congress for Applied Mechanics*; US Department of Commerce: Springfield, VA, USA, 1926. Available online: https://ntrs.nasa.gov/api/citations/19930090799/downloads/19930090799.pdf (accessed on 28 June 2024).
36. Kloeckner, A.; Warburton, T.; Hesthaven, J.S. Viscous Shock Capturing in a Time-Explicit Discontinuous Galerkin Method. *Math. Model. Nat. Phenom.* **2011**, *6*, 57–83. [CrossRef]
37. Rispoli, F.; Saavedra, R. A stabilized finite element method based on SGS models for compressible flows. *Comp. Meth. Appl. Mech. Eng.* **2006**, *196*, 652–664. [CrossRef]
38. Persson, P.O.; Peraire, J. Sub-cell shock capturing for discontinuous Galerkin methods. In Proceedings of the 44th AIAA Aerospace Sciences Meeting and Exhibit, Reno, NV, USA, 9–12 January 2006; p. 112.

39. Guermond, J.L.; Pasquetti, R.; Popov, B. Entropy viscosity method for nonlinear conservation laws. *J. Comput. Phys.* **2011**, *230*, 4248–4267. [CrossRef]
40. Guermond, J.L.; Pasquetti, R. Entropy-based nonlinear viscosity for Fourier approximations of conservation laws. *C. R. Acad. Sci. Ser. I* **2008**, *346*, 801–806. [CrossRef]
41. Guermond, J.L.; Popov, B. Viscous regularization of the Euler equations and entropy principles. *SIAM J. Appl. Math.* **2014**, *74*, 284–305. [CrossRef]
42. Kurganov, A.; Liu, Y. New adaptive artificial viscosity method for hyperbolic systems of conservation laws. *J. Comput. Phys.* **2012**, *231*, 8114–8132. [CrossRef]
43. Marras, S.; Nazarov, M.; Giraldo, F.X. Stabilized high-order Galerkin methods based on a parameter-free dynamic SGS model for LES. *J. Comput. Phys.* **2015**, *301*, 77–101. [CrossRef]
44. Wang, Z.; Triantafyllou, M.S.; Constantinides, Y.; Karniadakis, G.E. An entropy-viscosity large eddy simulation study of turbulent flow in a flexible pipe. *J. Fluid Mech.* **2019**, *859*, 691–730. [CrossRef]
45. Bazilevs, Y.; Calo, V.; Cottrell, J.A.; Hughes, T.J.R.; Reali, A.; Scovazzi, G. Variational Multiscale Residual-based Turbulence Modeling for Large Eddy Simulation of Incompressible Flows. *Comput. Methods Appl. Mech. Eng.* **2007**, *197*, 173–201. [CrossRef]
46. Codina, R. Stabilized finite element approximation of transient incompressible flows using orthogonal subscales. *Comput. Methods Appl. Mech. Eng.* **2002**, *191*, 4295–4321. [CrossRef]
47. Codina, R.; Badia, S.; Baiges, J.; Principe, J. Variational Multiscale Methods in Computational Fluid Dynamics. In *Encyclopedia of Computational Mechanics*, 2nd ed.; John Wiley & Sons, Ltd: Hoboken, NJ, USA, 2017; pp. 1–28.
48. Hughes, T.J.R.; Feijóo, G.; Mazzei, L.; Quincy, J. The variational multiscale method—A paradigm for computational mechanics. *Comput. Methods Appl. Mech. Eng.* **1998**, *166*, 3–24. [CrossRef]
49. Guermond, J.L.; Pasquetti, R.; Popov, B. From suitable weak solutions to entropy viscosity. *J. Sci. Comput.* **2011**, *49*, 35–50. [CrossRef]
50. Olshanskii, M.; Xiong, X. A connection between filter stabilization and eddy viscosity models. *Numer. Methods Partial. Differ. Equations* **2013**, *29*, 2061–2080. [CrossRef]
51. Clinco, N.; Girfoglio, M.; Quaini, A.; Rozza, G. Filter stabilization for the mildly compressible Euler equations with application to atmosphere dynamics simulations. *Comput. Fluids* **2023**, *266*, 106057. [CrossRef]
52. Boyd, J.P. Two Comments on Filtering (Artificial Viscosity) for Chebyshev and Legendre Spectral and Spectral Element Methods: Preserving Boundary Conditions and Interpretation of the Filter as a Diffusion. *J. Comput. Phys.* **1998**, *143*, 283–288. [CrossRef]
53. Fischer, P.; Mullen, J. Filter-based stabilization of spectral element methods. *Comptes Rendus Acad. Sci.-Ser.-Math.* **2001**, *332*, 265–270. [CrossRef]
54. Mullen, J.S.; Fischer, P.F. Filtering techniques for complex geometry fluid flows. *Commun. Numer. Meth. Eng.* **1999**, *15*, 9–18. [CrossRef]
55. Mathew, J.; Lechner, R.; Foysi, H.; Sesterhenn, J.; Friedrich, R. An explicit filtering method for large eddy simulation of compressible flows. *Phys. Fluids* **2003**, *15*, 2279–2289. [CrossRef]
56. Visbal, M.; Rizzetta, D. Large eddy simulation on curvilinear grids using compact differencing and filtering schemes. *J. Fluids Eng.* **2002**, *124*, 836–847. [CrossRef]
57. Garnier, E.; Adams, N.; Sagaut, P. *Large Eddy Simulation for Compressible Flows*; Springer: Berlin/Heidelberg, Germany, 2009.
58. Girfoglio, M.; Quaini, A.; Rozza, G. A Finite Volume approximation of the Navier-Stokes equations with nonlinear filtering stabilization. *Comput. Fluids* **2019**, *187*, 27–45. [CrossRef]
59. Bertagna, L.; Quaini, A.; Veneziani, A. Deconvolution-based nonlinear filtering for incompressible flows at moderately large Reynolds numbers. *Int. J. Num. Meth. Fluids* **2016**, *81*, 463–488. [CrossRef]
60. Bowers, A.L.; Rebholz, L.G.; Takhirov, A.; Trenchea, C. Improved accuracy in regularization models of incompressible flow via adaptive nonlinear filtering. *Int. J. Numer. Methods Fluids* **2012**, *70*, 805–828. [CrossRef]
61. Bowers, A.; Rebholz, L. Numerical study of a regularization model for incompressible flow with deconvolution-based adaptive nonlinear filtering. *Comput. Methods Appl. Mech. Eng.* **2013**, *258*, 1–12. [CrossRef]
62. Layton, W.; Rebholz, L.G.; Trenchea, C. Modular nonlinear filter stabilization of methods for higher Reynolds numbers flow. *J. Math. Fluid Mech.* **2012**, *14*, 325–354. [CrossRef]
63. Layton, W.; Röhe, L.; Tran, H. Explicitly uncoupled VMS stabilization of fluid flow. *Comput. Methods Appl. Mech. Eng.* **2011**, *200*, 3183–3199. [CrossRef]
64. Viguerie, A.; Veneziani, A. Deconvolution-based stabilization of the incompressible Navier–Stokes equations. *J. Comput. Phys.* **2019**, *391*, 226–242. [CrossRef]
65. Brezis, H. *Functional Analysis, Sobolev Spaces and Partial Differential Equations*; Springer: New York, NY, USA, 2011.
66. Girfoglio, M.; Quaini, A.; Rozza, G. A novel Large Eddy Simulation model for the Quasi-Geostrophic equations in a Finite Volume setting. *J. Comput. Appl. Math.* **2023**, *418*, 114656. [CrossRef]
67. Besabe, L.; Girfoglio, M.; Quaini, A.; Rozza, G. Linear and nonlinear filtering for a two-layer quasi-geostrophic ocean model. *arXiv* **2024**, arXiv:2404.11718.
68. Quarteroni, A.; Valli, A. *Domain Decomposition Methods for Partial Differential Equations*; Oxford Science Publications: Oxford, UK, 1999.

69. Deparis, S.; Fernandez, M.A.; Formaggia, L. Acceleration of a fixed point algorithm for fluid-structure interaction using transpiration conditions. *ESAIM Math. Model. Numer. Anal.* **2003**, *37*, 601–616. [CrossRef]
70. Ervin, V.; Layton, W.; Neda, M. Numerical Analysis of Filter-Based Stabilization for Evolution Equations. *SIAM J. Numer. Anal.* **2012**, *50*, 2307–2335. [CrossRef]
71. Girfoglio, M.; Quaini, A.; Rozza, G. A POD-Galerkin reduced order model for a LES filtering approach. *J. Comput. Phys.* **2021**, *436*, 110260. [CrossRef]
72. Leray, J. Sur le mouvement d'un fluide visqueux emplissant l'espace. *Acta Math.* **1934**, *63*, 193–248. [CrossRef]
73. Geurts, B.J.; Holm, D.D. Regularization modeling for large-eddy simulation. *Phys. Fluids* **2003**, *15*, L13–L16. [CrossRef]
74. Guermond, J.L.; Oden, J.T.; Prudhomme, S. Mathematical perspectives on large eddy simulation models for turbulent flows. *J. Math. Fluid Mech.* **2004**, *6*, 194–248. [CrossRef]
75. Foiaş, C.; Holm, D.; Titi, E. The Navier-Stokes-alpha model of fluid turbulence. *Phys. D* **2001**, *152/153*, 505–519. [CrossRef]
76. Chehab, J.P. Damping, Stabilization and Numerical Filtering for the Modeling and the Simulation of time dependent PDEs. *Discret. Contin. Dyn. Syst. Ser. S* **2021**, *14*, 2693–2728. [CrossRef]
77. Hesthaven, J.S.; Warburton, T. *Nodal Discontinuous Galerkin Methods: Algorithms, Analysis, and Applications*; Springer Publishing Company, Incorporated: Berlin/Heidelberg, Germany, 2007.
78. Holm, D.D. Averaged Lagrangians and the mean effects of fluctuations in ideal fluid dynamics. *Phys. D Nonlinear Phenom.* **2002**, *170*, 253–286. [CrossRef]
79. Secchi, P. An alpha model for compressible fluids. *Discrete Contin. Dyn. Syst. S.* **2008**, *3*, 351–359. [CrossRef]
80. Giraldo, F.X.; Restelli, M. A study of spectral element and discontinuous Galerkin methods for the Navier-Stokes equations in nonhydrostatic mesoscale atmospheric modeling: Equation sets and test cases. *J. Comput. Phys.* **2008**, *227*, 3849–3877. [CrossRef]
81. Marras, S.; Moragues, M.; Vázquez, M.; Jorba, O.; Houzeaux, G. A Variational Multiscale Stabilized Finite Element Method for the Solution of the Euler Equations of Nonhydrostatic Stratified Flows. *J. Comput. Phys.* **2013**, *236*, 380–407. [CrossRef]
82. Marras, S.; Kelly, J.; Moragues, M.; Müller, A.; Kopera, M.; Vázquez, M.; Giraldo, F.; Houzeaux, G.; Jorba, O. A Review of Element-Based Galerkin Methods for Numerical Weather Prediction: Finite Elements, Spectral Elements, and Discontinuous Galerkin. *Arch. Comput. Methods Eng.* **2016**, *23*, 673–722. [CrossRef]
83. Marchuk, G.I. *Numerical Methods in Weather Prediction*; Academic Press: Cambridge, MA, USA, 1974; pp. 1–277.
84. Borggaard, J.; Iliescu, T.; Roop, J.P. A bounded artificial viscosity large eddy simulation model. *SIAM J. Num. Anal.* **2009**, *47*, 622–645. [CrossRef]
85. Hunt, J.; Wray, A.; Moin, P. *Eddies Stream and Convergence Zones in Turbulent Flows*; Technical Report CTR-S88, CTR Report; NASA: Washington, DC, USA, 1988.
86. Vreman, A. An eddy-viscosity subgrid-scale model for turbulent shear flow: Algebraic theory and applications. *Phys. Fluids* **2004**, *16*, 3670–3681. [CrossRef]
87. Stolz, S.; Adams, N. An approximate deconvolution procedure for large-eddy simulation. *Phys. Fluids* **1999**, *11*, 1699–1701. [CrossRef]
88. Stolz, S.; Adams, N.; Kleiser, L. An approximate deconvolution model for large-eddy simulation with application to incompressible wall-bounded flows. *Phys. Fluids* **2001**, *13*, 997–1015. [CrossRef]
89. Dunca, A.; Epshteyn, Y. On the Stolz-Adams deconvolution model for the large-eddy simulation of turbulent flows. *SIAM J. Math. Anal.* **2005**, *37*, 1890–1902. [CrossRef]
90. Brunton, S.L.; Noack, B.R.; Koumoutsakos, P. Machine learning for fluid mechanics. *Annu. Rev. Fluid Mech.* **2020**, *52*, 477–508. [CrossRef]
91. Ling, J.; Kurzawski, A.; Templeton, J. Reynolds averaged turbulence modelling using deep neural networks with embedded invariance. *J. Fluid Mech.* **2016**, *807*, 155–166. [CrossRef]
92. Beck, A.; Kurz, M. A perspective on machine learning methods in turbulence modeling. *GAMM-Mitteilungen* **2021**, *44*, e202100002. [CrossRef]
93. Sirignano, J.; MacArt, J.F. Deep learning closure models for large-eddy simulation of flows around bluff bodies. *J. Fluid Mech.* **2023**, *966*, A26. [CrossRef]
94. Schmelzer, M.; Dwight, R.P.; Cinnella, P. Discovery of algebraic Reynolds-stress models using sparse symbolic regression. *Flow Turbul. Combust.* **2020**, *104*, 579–603. [CrossRef]
95. Reissmann, M.; Hasslberger, J.; Sandberg, R.D.; Klein, M. Application of gene expression programming to a-posteriori LES modeling of a Taylor Green vortex. *J. Comput. Phys.* **2021**, *424*, 109859. [CrossRef]
96. Wu, C.; Zhang, Y. Enhancing the shear-stress-transport turbulence model with symbolic regression: A generalizable and interpretable data-driven approach. *Phys. Rev. Fluids* **2023**, *8*, 084604. [CrossRef]
97. Ren, P.; Erichson, N.B.; Subramanian, S.; San, O.; Lukic, Z.; Mahoney, M.W. SuperBench: A Super-Resolution Benchmark Dataset for Scientific Machine Learning. *arXiv* **2023**, arXiv:2306.14070.
98. Maulik, R.; San, O.; Jacob, J.D.; Crick, C. Sub-grid scale model classification and blending through deep learning. *J. Fluid Mech.* **2019**, *870*, 784–812. [CrossRef]
99. Maulik, R.; San, O.; Jacob, J.D. Spatiotemporally dynamic implicit large eddy simulation using machine learning classifiers. *Phys. Nonlinear Phenom.* **2020**, *406*, 132409. [CrossRef]

100. Kim, B.; Azevedo, V.C.; Thuerey, N.; Kim, T.; Gross, M.; Solenthaler, B. Deep fluids: A generative network for parameterized fluid simulations. In *Computer Graphics Forum*; Wiley Online Library: Hoboken, NJ, USA, 2019; Volume 38, pp. 59–70.
101. Kontolati, K.; Goswami, S.; Karniadakis, G.E.; Shields, M.D. Learning in latent spaces improves the predictive accuracy of deep neural operators. *arXiv* **2023**, arXiv:2304.07599.
102. Du, P.; Parikh, M.H.; Fan, X.; Liu, X.Y.; Wang, J.X. CoNFiLD: Conditional Neural Field Latent Diffusion Model Generating Spatiotemporal Turbulence. *arXiv* **2024**, arXiv:2403.05940.
103. Lozano-Durán, A.; Bae, H.J. Machine learning building-block-flow wall model for large-eddy simulation. *J. Fluid Mech.* **2023**, *963*, A35. [CrossRef]
104. Vadrot, A.; Yang, X.I.; Abkar, M. Survey of machine-learning wall models for large-eddy simulation. *Phys. Rev. Fluids* **2023**, *8*, 064603. [CrossRef]
105. Veneziani, A.; Vergara, C. Inverse problems in cardiovascular mathematics: Toward patient-specific data assimilation and optimization. *Int. J. Numer. Methods Biomed. Eng.* **2013**, *29*, 723–725. [CrossRef] [PubMed]
106. Taylor, C.A.; Fonte, T.A.; Min, J.K. Computational fluid dynamics applied to cardiac computed tomography for noninvasive quantification of fractional flow reserve: Scientific basis. *J. Am. Coll. Cardiol.* **2013**, *61*, 2233–2241. [CrossRef]
107. Taylor, C.A.; Petersen, K.; Xiao, N.; Sinclair, M.; Bai, Y.; Lynch, S.R.; UpdePac, A.; Schaap, M. Patient-specific modeling of blood flow in the coronary arteries. *Comput. Methods Appl. Mech. Eng.* **2023**, *417*, 116414. [CrossRef]
108. Weiss, G.; Wolner, I.; Folkmann, S.; Sodeck, G.; Schmidli, J.; Grabenwöger, M.; Carrel, T.; Czerny, M. The location of the primary entry tear in acute type B aortic dissection affects early outcome. *Eur. J.-Cardio-Thorac. Surg.* **2012**, *42*, 571–576. [CrossRef]
109. Izzo, R.; Steinman, D.; Manini, S.; Antiga, L. The vascular modeling toolkit: A python library for the analysis of tubular structures in medical images. *J. Open Source Softw.* **2018**, *3*, 745. [CrossRef]
110. Xu, H. Efficient Modeling of the Incompressible Flow with Moderate Large Reynolds Numbers Using a Deconvolution-Based Leray Model: Analysis, Uncertainty Quantification and Application in Aortic Dissections. Ph.D. Thesis, Georgia Institute of Technology, Atlanta, GA, USA, 2020.
111. Bäumler, K.; Vedula, V.; Sailer, A.M.; Seo, J.; Chiu, P.; Mistelbauer, G.; Chan, F.P.; Fischbein, M.P.; Marsden, A.L.; Fleischmann, D. Fluid–structure interaction simulations of patient-specific aortic dissection. *Biomech. Model. Mechanobiol.* **2020**, *19*, 1607–1628. [CrossRef] [PubMed]
112. Zimmermann, J.; Bäumler, K.; Loecher, M.; Cork, T.E.; Marsden, A.L.; Ennis, D.B.; Fleischmann, D. Hemodynamic effects of entry and exit tear size in aortic dissection evaluated with in vitro magnetic resonance imaging and fluid–structure interaction simulation. *Sci. Rep.* **2023**, *13*, 22557. [CrossRef]
113. Girfoglio, M.; Quaini, A.; Rozza, G. Fluid-structure interaction simulations with a LES filtering approach in solids4Foam. *Commun. Appl. Ind. Math.* **2022**, *13*, 13–28. [CrossRef]
114. Heywood, J.G.; Rannacher, R.; Turek, S. Artificial boundaries and flux and pressure conditions for the incompressible Navier–Stokes equations. *Int. J. Numer. Methods Fluids* **1996**, *22*, 325–352. [CrossRef]
115. Womersley, J.R. Method for the calculation of velocity, rate of flow and viscous drag in arteries when the pressure gradient is known. *J. Physiol.* **1955**, *127*, 553. [CrossRef] [PubMed]
116. Veneziani, A.; Vergara, C. An approximate method for solving incompressible Navier–Stokes problems with flow rate conditions. *Comput. Methods Appl. Mech. Eng.* **2007**, *196*, 1685–1700. [CrossRef]
117. Formaggia, L.; Gerbeau, J.F.; Nobile, F.; Quarteroni, A. Numerical treatment of defective boundary conditions for the Navier–Stokes equations. *SIAM J. Numer. Anal.* **2002**, *40*, 376–401. [CrossRef]
118. Veneziani, A.; Vergara, C. Flow rate defective boundary conditions in haemodynamics simulations. *Int. J. Numer. Methods Fluids* **2005**, *47*, 803–816. [CrossRef]
119. Formaggia, L.; Veneziani, A.; Vergara, C. A new approach to numerical solution of defective boundary value problems in incompressible fluid dynamics. *SIAM J. Numer. Anal.* **2008**, *46*, 2769–2794. [CrossRef]
120. Formaggia, L.; Veneziani, A.; Vergara, C. Flow rate boundary problems for an incompressible fluid in deformable domains: Formulations and solution methods. *Comput. Methods Appl. Mech. Eng.* **2010**, *199*, 677–688. [CrossRef]
121. Xu, H.; Baroli, D.; Di Massimo, F.; Quaini, A.; Veneziani, A. Backflow stabilization by deconvolution-based large eddy simulation modeling. *J. Comput. Phys.* **2020**, *404*, 109103. [CrossRef]
122. Armour, C.H.; Guo, B.; Pirola, S.; Saitta, S.; Liu, Y.; Dong, Z.; Xu, X.Y. The influence of inlet velocity profile on predicted flow in type B aortic dissection. *Biomech. Model. Mechanobiol.* **2021**, *20*, 481–490. [CrossRef] [PubMed]
123. Formaggia, L.; Nobile, F.; Quarteroni, A.; Veneziani, A. Multiscale modelling of the circulatory system: A preliminary analysis. *Comput. Vis. Sci.* **1999**, *2*, 75–83. [CrossRef]
124. Vignon-Clementel, I.E.; Figueroa, C.A.; Jansen, K.E.; Taylor, C.A. Outflow boundary conditions for three-dimensional finite element modeling of blood flow and pressure in arteries. *Comput. Methods Appl. Mech. Eng.* **2006**, *195*, 3776–3796. [CrossRef]
125. Formaggia, L.; Quarteroni, A.; Veneziani, A. Multiscale models of the vascular system. *Cardiovasc. Math. Model. Simul. Circ. Syst.* **2009**, 395–446.
126. Quarteroni, A.; Veneziani, A.; Vergara, C. Geometric multiscale modeling of the cardiovascular system, between theory and practice. *Comput. Methods Appl. Mech. Eng.* **2016**, *302*, 193–252. [CrossRef]
127. Peiró, J.; Veneziani, A. Reduced models of the cardiovascular system. In *Cardiovascular Mathematics: Modeling and sImulation of the Circulatory System*; Springer: Berlin/Heidelberg, Germany, 2009; pp. 347–394.

128. Romarowski, R.M.; Lefieux, A.; Morganti, S.; Veneziani, A.; Auricchio, F. Patient-specific CFD modelling in the thoracic aorta with PC-MRI–based boundary conditions: A least-square three-element Windkessel approach. *Int. J. Numer. Methods Biomed. Eng.* **2018**, *34*, e3134. [CrossRef] [PubMed]
129. Pirola, S.; Guo, B.; Menichini, C.; Saitta, S.; Fu, W.; Dong, Z.; Xu, X.Y. 4-D flow MRI-based computational analysis of blood flow in patient-specific aortic dissection. *IEEE Trans. Biomed. Eng.* **2019**, *66*, 3411–3419. [CrossRef]
130. Bertoglio, C.; Caiazzo, A.; Bazilevs, Y.; Braack, M.; Esmaily, M.; Gravemeier, V.; L. Marsden, A.; Pironneau, O.; Vignon-Clementel, I.; Wall, W. Benchmark problems for numerical treatment of backflow at open boundaries. *Int. J. Numer. Methods Biomed. Eng.* **2018**, *34*, e2918. [CrossRef]
131. Bruneau, C.H.; Fabrie, P. Effective downstream boundary conditions for incompressible Navier–Stokes equations. *Int. J. Numer. Methods Fluids* **1994**, *19*, 693–705. [CrossRef]
132. Bruneau, C.H.; Fabrie, P. New efficient boundary conditions for incompressible Navier-Stokes equations: A well-posedness result. *ESAIM: Math. Model. Numer. Anal.* **1996**, *30*, 815–840. [CrossRef]
133. Arbia, G.; Vignon-Clementel, I.; Hsia, T.Y.; Gerbeau, J.F.; Modeling of Congenital Hearts Alliance (MOCHA) Investigators; et al. Modified Navier–Stokes equations for the outflow boundary conditions in hemodynamics. *Eur. J.-Mech.-B/Fluids* **2016**, *60*, 175–188. [CrossRef]
134. Bertoglio, C.; Caiazzo, A. A tangential regularization method for backflow stabilization in hemodynamics. *J. Comput. Phys.* **2014**, *261*, 162–171. [CrossRef]
135. Bertoglio, C.; Caiazzo, A. A Stokes-residual backflow stabilization method applied to physiological flows. *J. Comput. Phys.* **2016**, *313*, 260–278. [CrossRef]
136. Nichols, W.W.; O'Rourke, M.; Edelman, E.R.; Vlachopoulos, C. *McDonald's Blood Flow in Arteries: Theoretical, Experimental and Clinical Principles*; CRC Press: Boca Raton, FL, USA, 2022.
137. Chen, D.; Müller-Eschner, M.; von Tengg-Kobligk, H.; Barber, D.; Böckler, D.; Hose, R.; Ventikos, Y. A patient-specific study of type-B aortic dissection: Evaluation of true-false lumen blood exchange. *Biomed. Eng. Online* **2013**, *12*, 65. [CrossRef] [PubMed]
138. Fatma, K.; Carine, G.C.; Marine, G.; Philippe, P.; Valérie, D. Numerical modeling of residual type B aortic dissection: Longitudinal analysis of favorable and unfavorable evolution. *Med. Biol. Eng. Comput.* **2022**, *60*, 769–783. [CrossRef] [PubMed]
139. Bertagna, L.; Deparis, S.; Formaggia, L.; Forti, D.; Veneziani, A. The LifeV library: Engineering mathematics beyond the proof of concept. *arXiv* **2017**, arXiv:1710.06596.
140. Yang, J.; Piccinelli, M.; Leshnower, B.; Veneziani, A. Predicting The Evolution Of Type-b Aortic Dissection: Combining Computational Hemodynamics And Data Analysis. In Proceedings of the 8th International Conference on Computational and Mathematical Biomedical Engineering—CMBE2024, Arlington, VA, USA, 24–26 June 2024.
141. Bertagna, L.; Quaini, A.; Rebholz, L.G.; Veneziani, A. On the sensitivity to the filtering radius in Leray models of incompressible flow. In *Contributions to Partial Differential Equations and Applications*; Springer: Berlin/Heidelberg, Germany, 2018; pp. 111–130.
142. Xu, H.; Baroli, D.; Veneziani, A. Global sensitivity analysis for patient-specific aortic simulations: The role of geometry, boundary condition and large Eddy simulation modeling parameters. *J. Biomech. Eng.* **2021**, *143*, 021012. [CrossRef]
143. Sobol, I.M. Sensitivity estimates for nonlinear mathematical models. *Math. Model. Comput. Exp.* **1993**, *1*, 407–414.
144. Sobol, I.M. Global sensitivity indices for nonlinear mathematical models and their Monte Carlo estimates. *Math. Comput. Simul.* **2001**, *55*, 271–280. [CrossRef]
145. Sudret, B. Global sensitivity analysis using polynomial chaos expansions. *Reliab. Eng. Syst. Saf.* **2008**, *93*, 964–979. [CrossRef]
146. Crestaux, T.; Le Maıtre, O.; Martinez, J.M. Polynomial chaos expansion for sensitivity analysis. *Reliab. Eng. Syst. Saf.* **2009**, *94*, 1161–1172. [CrossRef]
147. D'Elia, M.; Mirabella, L.; Passerini, T.; Perego, M.; Piccinelli, M.; Vergara, C.; Veneziani, A. Applications of variational data assimilation in computational hemodynamics. *Model. Physiol. Flows* **2012**, 363–394.
148. D'Elia, M.; Perego, M.; Veneziani, A. A variational data assimilation procedure for the incompressible Navier-Stokes equations in hemodynamics. *J. Sci. Comput.* **2012**, *52*, 340–359. [CrossRef]
149. D'Elia, M.; Veneziani, A. Uncertainty quantification for data assimilation in a steady incompressible Navier-Stokes problem. *ESAIM Math. Model. Numer. Anal.* **2013**, *47*, 1037–1057. [CrossRef]
150. Cai, S.; Mao, Z.; Wang, Z.; Yin, M.; Karniadakis, G.E. Physics-informed neural networks (PINNs) for fluid mechanics: A review. *Acta Mech. Sin.* **2021**, *37*, 1727–1738. [CrossRef]
151. Ahmad, N.N.; Lindeman, J. Euler solutions using flux-based wave decomposition. *Int. J. Numer. Meth. Fluids* **2007**, *54*, 47–72. [CrossRef]
152. Feng, Y.; Miranda-Fuentes, J.; Jacob, J.; Sagaut, P. Hybrid lattice Boltzmann model for atmospheric flows under anelastic approximation. *Phys. Fluids* **2021**, *33*, 036607. [CrossRef]
153. Carpenter, R.; Droegemeier, K.; Woodward, P.; Hane, C. Application of the piecewise parabolic method (PPM) to meteorological modeling. *Mon. Wea. Rev.* **1990**, *118*, 586–612. [CrossRef]
154. Straka, J.; Wilhelmson, R.; Wicker, L.; Anderson, J.; Droegemeier, K. Numerical solution of a nonlinear density current: A benchmark solution and comparisons. *Int. J. Num. Meth. Fluids* **1993**, *17*, 1–22. [CrossRef]
155. Özgökmen, T.; Iliescu, T.; Fischer, P.; Srinivasan, A.; Duan, J. Large Eddy Simulation of Stratified Mixing in Two-Dimensional Dam-Break Problem in a Rectangular Enclosed Domain. *Ocean. Model.* **2007**, *16*, 106–140. [CrossRef]

156. Özgökmen, T.; Iliescu, T.; Fischer, P. Large Eddy Simulation of Stratified Mixing in a Three-Dimensional Lock-Exchange System. *Ocean. Model.* **2009**, *26*, 134–155. [CrossRef]
157. Özgökmen, T.; Iliescu, T.; Fischer, P. Reynolds number dependence of mixing in a lock-exchange system from direct numerical and large eddy simulations. *Ocean. Model.* **2009**, *30*, 190–206. [CrossRef]
158. Ahmad, N.N. High-Resolution Wave Propagation Method for Stratified Flows. In Proceedings of the AIAA Aviation Forum, Atlanta, GA, USA, 25–29 June 2018. [CrossRef]
159. Girfoglio, M.; Quaini, A.; Rozza, G. Validation of an OpenFOAM®-based solver for the Euler equations with benchmarks for mesoscale atmospheric modeling. *AIP Adv.* **2023**, *13*, 055024. [CrossRef]
160. Benner, P.; Schilders, W.; Grivet-Talocia, S.; Quarteroni, A.; Rozza, G.; Miguel Silveira, L. *Model Order Reduction: Volume 1: System- and Data-Driven Methods and Algorithms*; De Gruyter: Berlin, Germany, 2021.
161. Benner, P.; Schilders, W.; Grivet-Talocia, S.; Quarteroni, A.; Rozza, G.; Miguel Silveira, L. *Model Order Reduction: Volume 2: Snapshot-Based Methods and Algorithms*; De Gruyter: Berlin, Germany, 2021.
162. Benner, P.; Schilders, W.; Grivet-Talocia, S.; Quarteroni, A.; Rozza, G.; Miguel Silveira, L. *Model Order Reduction: Volume 3: Applications*; De Gruyter: Berlin, Germany, 2021.
163. Hesthaven, J.S.; Rozza, G.; Stamm, B. *Certified Reduced Basis Methods for Parametrized Partial Differential Equations*; Springer: Berlin/Heidelberg, Germany, 2016; Volume 590.
164. Malik, M.H. Reduced Order Modeling for Smart Grids' Simulation and Optimization. Ph.D. Thesis, École Centrale de Nantes; Universitat Politécnica de Catalunya, Barcelona, Spain, 2017.
165. Rozza, G.; Huynh, D.B.P.; Patera, A.T. Reduced basis approximation and a posteriori error estimation for affinely parametrized elliptic coercive partial differential equations: Application to transport and continuum mechanics. *Arch. Comput. Methods Eng.* **2008**, *15*, 229–275. [CrossRef]
166. Quarteroni, A.; Manzoni, A.; Negri, F. *Reduced Basis Methods for Partial Differential Equations: An Introduction*; Springer: Berlin/Heidelberg, Germany, 2015; Volume 92.
167. Wang, Z.; Akhtar, I.; Borggaard, J.; Iliescu, T. Proper orthogonal decomposition closure models for turbulent flows: A numerical comparison. *Comput. Meth. Appl. Mech. Eng.* **2012**, *237–240*, 10–26. [CrossRef]
168. Aubry, N.; Holmes, P.; Lumley, J.L.; Stone, E. The dynamics of coherent structures in the wall region of a turbulent boundary layer. *J. Fluid Mech.* **1988**, *192*, 115–173. [CrossRef]
169. Ahmed, S.E.; Pawar, S.; San, O.; Rasheed, A.; Iliescu, T.; Noack, B.R. On closures for reduced order models—A spectrum of first-principle to machine-learned avenues. *Phys. Fluids* **2021**, *33*, 091301. [CrossRef]
170. Carlberg, K.; Bou-Mosleh, C.; Farhat, C. Efficient non-linear model reduction via a least-squares Petrov–Galerkin projection and compressive tensor approximations. *Int. J. Num. Meth. Eng.* **2011**, *86*, 155–181. [CrossRef]
171. Grimberg, S.; Farhat, C.; Youkilis, N. On the stability of projection-based model order reduction for convection-dominated laminar and turbulent flows. *J. Comput. Phys.* **2020**, *419*, 109681. [CrossRef]
172. Baiges, J.; Codina, R.; Idelsohn, S. Explicit Reduced Order Models for the stabilized finite element approximation of the incompressible Navier-Stokes equations. *Int. J. Num. Meth. Fluids* **2013**, *72*, 1219–1243. [CrossRef]
173. Reyes, R.; Codina, R. Projection-based reduced order models for flow problems: A variational multiscale approach. *Comput. Methods Appl. Mech. Eng.* **2020**, *363*, 112844. [CrossRef]
174. Parish, E.J.; Wentland, C.; Duraisamy, K. The adjoint Petrov–Galerkin method for non-linear model reduction. *Comput. Meth. Appl. Mech. Eng.* **2020**, *365*, 112991. [CrossRef]
175. Wells, D.; Wang, Z.; Xie, X.; Iliescu, T. An evolve-then-filter regularized reduced order model for convection-dominated flows. *Int. J. Num. Meth. Fluids* **2017**, *84*, 598–615. [CrossRef]
176. Gunzburger, M.; Iliescu, T.; Mohebujjaman, M.; Schneier, M. An evolve-filter-relax stabilized reduced order stochastic collocation method for the time-dependent Navier-Stokes equations. *SIAM-ASA J. Uncertain.* **2019**, *7*, 1162–1184. [CrossRef]
177. Sabetghadam, F.; Jafarpour, A. α regularization of the POD-Galerkin dynamical systems of the Kuramoto–Sivashinsky equation. *Appl. Math. Comput.* **2012**, *218*, 6012–6026. [CrossRef]
178. Wells, D. Stabilization of POD-ROMs. Ph.D. Thesis, Virginia Tech, Blacksburg, VA, USA, 2015. Available online: http://vtechworks.lib.vt.edu/bitstream/handle/10919/52960/Wells_DR_D_2015.pdf?sequence=1&isAllowed=y (accessed on 28 June 2024).
179. Germano, M. Differential filters of elliptic type. *Phys. Fluids* **1986**, *29*, 1757–1758. [CrossRef]
180. Strazzullo, M.; Girfoglio, M.; Ballarin, F.; Iliescu, T.; Rozza, G. Consistency of the full and reduced order models for evolve-filter-relax regularization of convection-dominated, marginally-resolved flows. *Int. J. Numer. Methods Eng.* **2022**, *123*, 3148–3178. [CrossRef]
181. Girfoglio, M.; Quaini, A.; Rozza, G. A linear filter regularization for POD-based reduced-order models of the quasi-geostrophic equations. *C. R. Mech.* **2023**, *351*, 1–21. [CrossRef]
182. Tsai, P.H.; Fischer, P.; Iliescu, T. A Time-Relaxation Reduced Order Model for the Turbulent Channel Flow. *arXiv* **2023**, arXiv:2312.13272.
183. Mou, C.; Merzari, E.; San, O.; Iliescu, T. An energy-based lengthscale for reduced order models of turbulent flows. *Nucl. Eng. Des.* **2023**, *412*, 112454. [CrossRef]

184. Holmes, P.; Lumley, J.L.; Berkooz, G. *Turbulence, Coherent Structures, Dynamical Systems and Symmetry*; Cambridge University Press: Cambridge, UK, 1996.
185. Aradag, S.; Siegel, S.; Seidel, J.; Cohen, K.; McLaughlin, T. Filtered POD-based low-dimensional modeling of the 3D turbulent flow behind a circular cylinder. *Int. J. Num. Meth. Fluids* **2011**, *66*, 1–16. [CrossRef]
186. Farcas, I.; Munipalli, R.; Willcox, K.E. On filtering in non-intrusive data-driven reduced-order modeling. In Proceedings of the AIAA AVIATION 2022 Forum, Chicago, IL, USA, 27 June–1 July 2022; p. 3487.
187. Bertero, M.; Boccacci, P. *Introduction to Inverse Problems in Imaging*; Institute of Physics Publishing: Bristol, UK, 1998; p. xii+351.
188. Hansen, P.C. *Discrete Inverse Problems: Insight and Algorithms*; Society for Industrial and Applied Mathematics: Philadelphia, PA, USA, 2010; Volume 7.
189. Xie, X.; Wells, D.; Wang, Z.; Iliescu, T. Approximate Deconvolution Reduced Order Modeling. *Comput. Methods Appl. Mech. Eng.* **2017**, *313*, 512–534. [CrossRef]
190. Sanfilippo, A.; Moore, I.R.; Ballarin, F.; Iliescu, T. Approximate deconvolution Leray reduced order model. *Finite Elem. Anal. Des.* **2023**, *226*, 104021. [CrossRef]
191. Cordier, L.; Abou El Majd, B.; Favier, J. Calibration of POD reduced-order models using Tikhonov regularization. *Int. J. Num. Meth. Fluids* **2010**, *63*, 269–296. [CrossRef]
192. Wang, Y.; Navon, I.M.; Wang, X.; Cheng, Y. 2D Burgers equation with large Reynolds number using POD/DEIM and calibration. *Int. J. Num. Meth. Fluids* **2016**, *82*, 909–931. [CrossRef]
193. Weller, J.; Lombardi, E.; Iollo, A. Robust model identification of actuated vortex wakes. *Phys. D* **2009**, *238*, 416–427. [CrossRef]
194. Strazzullo, M.; Ballarin, F.; Iliescu, T.; Canuto, C. New Feedback Control and Adaptive Evolve-Filter-Relax Regularization for the Navier-Stokes Equations in the Convection-Dominated Regime. *arXiv* **2023**, arXiv:2307.00675.
195. Xie, X.; Wells, D.; Wang, Z.; Iliescu, T. Numerical Analysis of the Leray Reduced Order Model. *J. Comput. Appl. Math.* **2018**, *328*, 12–29. [CrossRef]
196. Gunzburger, M.; Iliescu, T.; Schneier, M. A Leray regularized ensemble-proper orthogonal decomposition method for parameterized convection-dominated flows. *IMA J. Numer. Anal.* **2020**, *40*, 886–913. [CrossRef]
197. Girfoglio, M.; Quaini, A.; Rozza, G. A hybrid projection/data-driven reduced order model for the Navier-Stokes equations with nonlinear filtering stabilization. *J. Comput. Phys.* **2023**, *486*, 112127. [CrossRef]
198. Kaneko, K.; Tsai, P.H.; Fischer, P. Towards model order reduction for fluid-thermal analysis. *Nucl. Eng. Des.* **2020**, *370*, 110866. [CrossRef]
199. Tsai, P.H.; Fischer, P. Parametric model-order-reduction development for unsteady convection. *Front. Phys.* **2022**, *10*, 903169. [CrossRef]
200. Tsai, P.H. Parametric Model Order Reduction Development for Navier-Stokes Equations from 2D Chaotic to 3D Turbulent Flow Problems. Ph.D. Thesis, University of Illinois at Urbana-Champaign, Champaign, IL, USA, 2023.
201. Sanderse, B.; Stinis, P.; Maulik, R.; Ahmed, S.E. Scientific machine learning for closure models in multiscale problems: A review. *arXiv* **2024**, arXiv:2403.02913.
202. Hughes, T.J.R. Multiscale phenomena: Green's functions, the Dirichlet-to-Neumann formulation, subgrid scale models, bubbles and the origins of stabilized methods. *Comput. Methods Appl. Mech. Engrg.* **1995**, *127*, 387–401. [CrossRef]
203. Hughes, T.J.R.; Mazzei, L.; Oberai, A.; Wray, A. The multiscale formulation of large eddy simulation: Decay of homogeneous isotropic turbulence. *Phys. Fluids* **2001**, *13*, 505–512. [CrossRef]
204. Codina, R.; Reyes, R.; Baiges, J. A posteriori error estimates in a finite element VMS-based reduced order model for the incompressible Navier-Stokes equations. *Mech. Res. Commun.* **2021**, *112*, 103599. [CrossRef]
205. Mou, C.; Koc, B.; San, O.; Rebholz, L.G.; Iliescu, T. Data-Driven Variational Multiscale Reduced Order Models. *Comput. Methods Appl. Mech. Eng.* **2021**, *373*, 113470. [CrossRef]
206. Ingimarson, S.; Rebholz, L.G.; Iliescu, T. Full and Reduced Order Model Consistency of the Nonlinearity Discretization in Incompressible Flows. *Comput. Meth. Appl. Mech. Eng.* **2022**, *401*, 115620. [CrossRef]
207. Giere, S.; Iliescu, T.; John, V.; Wells, D. SUPG Reduced Order Models for Convection-Dominated Convection-Diffusion-Reaction Equations. *Comput. Methods Appl. Mech. Eng.* **2015**, *289*, 454–474. [CrossRef]
208. Zoccolan, F.; Strazzullo, M.; Rozza, G. Stabilized weighted reduced order methods for parametrized advection-dominated optimal control problems governed by partial differential equations with random inputs. *arXiv* **2023**, arXiv:2301.01975.
209. Zoccolan, F.; Strazzullo, M.; Rozza, G. A streamline upwind Petrov-Galerkin reduced order method for advection-dominated partial differential equations under optimal control. *Comput. Methods Appl. Math.* **2024**. [CrossRef]
210. Pacciarini, P.; Rozza, G. Stabilized reduced basis method for parametrized advection–diffusion PDEs. *Comput. Meth. Appl. Mech. Eng.* **2014**, *274*, 1–18. [CrossRef]
211. Roos, H.G.; Stynes, M.; Tobiska, L. *Robust Numerical Methods for Singularly Perturbed Differential Equations: Convection-Diffusion-Reaction and Flow Problems*, 2nd ed.; Springer Series in Computational Mathematics; Springer: Berlin/Heidelberg, Germany, 2008; Volume 24.
212. Borggaard, J.; Iliescu, T.; Wang, Z. Artificial Viscosity Proper Orthogonal Decomposition. *Math. Comput. Modelling* **2011**, *53*, 269–279. [CrossRef]
213. Iliescu, T.; Wang, Z. Variational multiscale proper orthogonal decomposition: Convection-dominated convection-diffusion-reaction equations. *Math. Comput.* **2013**, *82*, 1357–1378. [CrossRef]

214. Iliescu, T.; Wang, Z. Variational Multiscale Proper Orthogonal Decomposition: Navier-Stokes Equations. *Num. Meth. P.D.E.s* **2014**, *30*, 641–663. [CrossRef]
215. Ballarin, F.; Rebollo, T.C.; Ávila, E.D.; Mármol, M.G.; Rozza, G. Certified reduced basis VMS-Smagorinsky model for natural convection flow in a cavity with variable height. *Comput. Math. Appl.* **2020**, *80*, 973–989. [CrossRef]
216. Rebollo, T.C.; Ávila, E.D.; Mármol, M.G.; Ballarin, F.; Rozza, G. On a certified Smagorinsky reduced basis turbulence model. *SIAM J. Numer. Anal.* **2017**, *55*, 3047–3067. [CrossRef]
217. Azaïez, M.; Rebollo, T.C.; Rubino, S. A cure for instabilities due to advection-dominance in POD solution to advection-diffusion-reaction equations. *J. Comput. Phys.* **2021**, *425*, 109916. [CrossRef]
218. John, V.; Moreau, B.; Novo, J. Error analysis of a SUPG-stabilized POD-ROM method for convection-diffusion-reaction equations. *Comput. Math. Appl.* **2022**, *122*, 48–60. [CrossRef]
219. Xie, X.; Mohebujjaman, M.; Rebholz, L.G.; Iliescu, T. Data-Driven Filtered Reduced Order Modeling of Fluid Flows. *SIAM J. Sci. Comput.* **2018**, *40*, B834–B857. [CrossRef]
220. Ivagnes, A.; Stabile, G.; Mola, A.; Iliescu, T.; Rozza, G. Pressure data-driven variational multiscale reduced order models. *J. Comput. Phys.* **2023**, *476*, 111904. [CrossRef]
221. Ivagnes, A.; Stabile, G.; Mola, A.; Iliescu, T.; Rozza, G. Hybrid data-driven closure strategies for reduced order modeling. *Appl. Math. Comput.* **2023**, *448*, 127920. [CrossRef]
222. Koc, B.; Mohebujjaman, M.; Mou, C.; Iliescu, T. Commutation error in reduced order modeling of fluid flows. *Adv. Comput. Math.* **2019**, *45*, 2587–2621. [CrossRef]
223. Mohebujjaman, M.; Rebholz, L.G.; Iliescu, T. Physically-constrained data-driven correction for reduced order modeling of fluid flows. *Int. J. Num. Meth. Fluids* **2019**, *89*, 103–122. [CrossRef]
224. Mou, C.; Liu, H.; Wells, D.R.; Iliescu, T. Data-Driven Correction Reduced Order Models for the Quasi-Geostrophic Equations: A Numerical Investigation. *Int. J. Comput. Fluid Dyn.* **2020**, *34*, 147–159. [CrossRef]
225. Xie, X.; Webster, C.; Iliescu, T. Closure Learning for Nonlinear Model Reduction Using Deep Residual Neural Network. *Fluids* **2020**, *5*, 39. [CrossRef]
226. Ahmed, S.E.; San, O.; Rasheed, A.; Iliescu, T.; Veneziani, A. Physics guided machine learning for variational multiscale reduced order modeling. *SIAM J. Sci. Comput.* **2023**, *45*, B283–B313. [CrossRef]
227. Koc, B.; Mou, C.; Liu, H.; Wang, Z.; Rozza, G.; Iliescu, T. Verifiability of the Data-Driven Variational Multiscale Reduced Order Model. *J. Sci. Comput.* **2022**, *93*, 1–26. [CrossRef]
228. Turek, S.; Schäfer, M. Benchmark computations of laminar flow around cylinder. In *Flow Simulation with High-Performance Computers II*; Hirschel, E., Ed.; Notes on Numerical Fluid Mechanics; Vieweg: Decatur, IL, USA, 1996; Volume 52.
229. John, V. Reference values for drag and lift of a two dimensional time-dependent flow around a cylinder. *Int. J. Numer. Methods Fluids* **2004**, *44*, 777–788. [CrossRef]
230. John, V. On the efficiency of linearization schemes and coupled multigrid methods in the simulation of a 3D flow around a cylinder. *Int. J. Numer. Methods Fluids* **2006**, *50*, 845–862. [CrossRef]
231. Bayraktar, E.; Mierka, O.; Turek, S. Benchmark computations of 3D laminar flow around a cylinder with CFX, OpenFOAM and FeatFlow. *Int. J. Comput. Sci. Eng.* **2012**, *7*, 253–266. [CrossRef]
232. Girfoglio, M.; Quaini, A.; Rozza, G. Pressure stabilization strategies for a LES filtering Reduced Order Model. *Fluids* **2021**, *6*, 302. [CrossRef]
233. Xie, X.; Bao, F.; Webster, C. Evolve Filter Stabilization Reduced-Order Model for Stochastic Burgers Equation. *Fluids* **2018**, *3*, 84. [CrossRef]
234. Stabile, G.; Rozza, G. Finite volume POD-Galerkin stabilised reduced order methods for the parametrised incompressible Navier‚ÄìStokes equations. *Comput. Fluids* **2018**, *173*, 273–284. [CrossRef]
235. Akhtar, I.; Nayfeh, A.H.; Ribbens, C.J. On the stability and extension of reduced-order Galerkin models in incompressible flows. *Theor. Comp. Fluid Dyn.* **2009**, *23*, 213–237. [CrossRef]
236. Lazzaro, D.; Montefusco, L. Radial Basis Functions for the Multivariate Interpolation of Large Scattered Data Sets. *J. Comput. Appl. Math.* **2002**, *140*, 521–536. [CrossRef]
237. Barrault, M.; Nguyen, N.C.; Maday, Y.; Patera, A.T. An "Empirical Interpolation" Method: Application to Efficient Reduced-Basis Discretization of Partial Differential Equations. *Comptes Rendus Math.* **2004**, *339*, 667–672. [CrossRef]
238. Chaturantabut, S.; Sorensen, D. Nonlinear Model Reduction via Discrete Empirical Interpolation. *SIAM J. Sci. Comput.* **2010**, *32*, 2737–2764. [CrossRef]
239. Lorenzi, S.; Cammi, A.; Luzzi, L.; Rozza, G. POD-Galerkin method for finite volume approximation of Navier-Stokes and RANS equations. *Comput. Methods Appl. Mech. Eng.* **2016**, *311*, 151–179. [CrossRef]
240. Hijazi, S.; Stabile, G.; Mola, A.; Rozza, G. Data-Driven POD-Galerkin reduced order model for turbulent flows. *J. Comput. Phys.* **2020**, *416*, 109513. [CrossRef]
241. Obabko, A.V.; Fischer, P.F.; Tautges, T.J.; Karabasov, S.; Goloviznin, V.M.; Zaytsev, M.A.; Chudanov, V.V.; Pervichko, V.A.; Aksenova, A.E. *CFD Validation in OECD/NEA T-Junction Benchmark*; Technical Report; Argonne National Lab. (ANL): Argonne, IL, USA, 2011.

242. de Leval, M.R.; Kilner, P.; Gewillig, M.; Bull, C.; McGoon, D.C. Total cavopulmonary connection: A logical alternative to atriopulmonary connection for complex Fontan operations: Experimental studies and early clinical experience. *J. Thorac. Cardiovasc. Surg.* **1988**, *96*, 682–695. [CrossRef]
243. Sharma, S.; Goudy, S.; Walker, P.; Panchal, S.; Ensley, A.; Kanter, K.; Tam, V.; Fyfe, D.; Yoganathan, A. In vitro flow experiments for determination of optimal geometry of total cavopulmonary connection for surgical repair of children with functional single ventricle. *J. Am. Coll. Cardiol.* **1996**, *27*, 1264–1269. [CrossRef]
244. Migliavacca, F.; Kilner, P.J.; Pennati, G.; Dubini, G.; Pietrabissa, R.; Fumero, R.; de Leval, M.R. Computational fluid dynamic and magnetic resonance analyses of flow distribution between the lungs after total cavopulmonary connection. *IEEE Trans. Biomed. Eng.* **1999**, *46*, 393–399. [CrossRef] [PubMed]
245. Ensley, A.E.; Lynch, P.; Chatzimavroudis, G.P.; Lucas, C.; Sharma, S.; Yoganathan, A.P. Toward designing the optimal total cavopulmonary connection: An in vitro study. *Ann. Thorac. Surg.* **1999**, *68*, 1384–1390. [CrossRef] [PubMed]
246. Dubini, G.; Migliavacca, F.; Pennati, G.; De Leval, M.R.; Bove, E.L. Ten years of modelling to achieve haemodynamic optimisation of the total cavopulmonary connection. *Cardiol Young* **2004**, *14*, 48–52. [CrossRef] [PubMed]
247. Hsia, T.Y.; Figliola, R.; Multiscale of Congenital Hearts Alliance (MOCHA) Investigators; Bove, E.; Dorfman, A.; Taylor, A.; Giardini, A.; Khambadkone, S.; Schievano, S.; Hsia, T.Y.; et al. Multiscale modelling of single-ventricle hearts for clinical decision support: A Leducq Transatlantic Network of Excellence. *Eur. J.-Cardio-Thorac. Surg.* **2016**, *49*, 365–368. [CrossRef]
248. Schiavazzi, D.E.; Kung, E.O.; Marsden, A.L.; Baker, C.; Pennati, G.; Hsia, T.Y.; Hlavacek, A.; Dorfman, A.L.; Modeling of Congenital Hearts Alliance (MOCHA) Investigators. Hemodynamic effects of left pulmonary artery stenosis after superior cavopulmonary connection: A patient-specific multiscale modeling study. *J. Thorac. Cardiovasc. Surg.* **2015**, *149*, 689–696. [CrossRef]
249. Pekkan, K.; Kitajima, H.D.; De Zelicourt, D.; Forbess, J.M.; Parks, W.J.; Fogel, M.A.; Sharma, S.; Kanter, K.R.; Frakes, D.; Yoganathan, A.P. Total cavopulmonary connection flow with functional left pulmonary artery stenosis: Angioplasty and fenestration in vitro. *Circulation* **2005**, *112*, 3264–3271. [CrossRef]
250. Tang, E.; Restrepo, M.; Haggerty, C.M.; Mirabella, L.; Bethel, J.; Whitehead, K.K.; Fogel, M.A.; Yoganathan, A.P. Geometric characterization of patient-specific total cavopulmonary connections and its relationship to hemodynamics. *JACC Cardiovasc. Imaging* **2014**, *7*, 215–224. [CrossRef] [PubMed]
251. Khiabani, R.H.; Restrepo, M.; Tang, E.; De Zélicourt, D.; Sotiropoulos, F.; Fogel, M.; Yoganathan, A.P. Effect of flow pulsatility on modeling the hemodynamics in the total cavopulmonary connection. *J. Biomech.* **2012**, *45*, 2376–2381. [CrossRef] [PubMed]
252. Dasi, L.P.; KrishnankuttyRema, R.; Kitajima, H.D.; Pekkan, K.; Sundareswaran, K.S.; Fogel, M.; Sharma, S.; Whitehead, K.; Kanter, K.; Yoganathan, A.P. Fontan hemodynamics: Importance of pulmonary artery diameter. *J. Thorac. Cardiovasc. Surg.* **2009**, *137*, 560–564. [CrossRef] [PubMed]
253. Rodefeld, M.D.; Boyd, J.H.; Myers, C.D.; LaLone, B.J.; Bezruczko, A.J.; Potter, A.W.; Brown, J.W. Cavopulmonary assist: Circulatory support for the univentricular Fontan circulation. *Ann. Thorac. Surg.* **2003**, *76*, 1911–1916. [CrossRef] [PubMed]
254. Sarfare, S.; Ali, M.S.; Palazzolo, A.; Rodefeld, M.; Conover, T.; Figliola, R.; Giridharan, G.; Wampler, R.; Bennett, E.; Ivashchenko, A. Computational Fluid Dynamics Turbulence Model and Experimental Study for a Fontan Cavopulmonary Assist Device. *J. Biomech. Eng.* **2023**, *145*, 111008. [CrossRef] [PubMed]
255. Meneveau, C. Big wind power: Seven questions for turbulence research. *J. Turbul.* **2019**, *20*, 2–20. [CrossRef]
256. Antonini, E.G.; Caldeira, K. Spatial constraints in large-scale expansion of wind power plants. *Proc. Natl. Acad. Sci. USA* **2021**, *118*, e2103875118. [CrossRef] [PubMed]
257. Bempedelis, N.; Laizet, S.; Deskos, G. Turbulent entrainment in finite-length wind farms. *J. Fluid Mech.* **2023**, *955*, A12. [CrossRef]
258. Howland, M.F.; Quesada, J.B.; Martinez, J.J.P.; Larrañaga, F.P.; Yadav, N.; Chawla, J.S.; Sivaram, V.; Dabiri, J.O. Collective wind farm operation based on a predictive model increases utility-scale energy production. *Nature Energy* **2022**, *7*, 818–827. [CrossRef]
259. Pawar, S.; Sharma, A.; Vijayakumar, G.; Bay, C.J.; Yellapantula, S.; San, O. Towards multi-fidelity deep learning of wind turbine wakes. *Renew. Energy* **2022**, *200*, 867–879. [CrossRef]
260. Luzzatto-Fegiz, P.; Caulfield, C.c.P. Entrainment model for fully-developed wind farms: Effects of atmospheric stability and an ideal limit for wind farm performance. *Phys. Rev. Fluids* **2018**, *3*, 093802. [CrossRef]
261. Sedaghatizadeh, N.; Arjomandi, M.; Kelso, R.; Cazzolato, B.; Ghayesh, M.H. Modelling of wind turbine wake using large eddy simulation. *Renew. Energy* **2018**, *115*, 1166–1176. [CrossRef]
262. Mehta, D.; Van Zuijlen, A.; Koren, B.; Holierhoek, J.; Bijl, H. Large Eddy Simulation of wind farm aerodynamics: A review. *J. Wind. Eng. Ind. Aerodyn.* **2014**, *133*, 1–17. [CrossRef]
263. Meneveau, C. The top-down model of wind farm boundary layers and its applications. *J. Turbul.* **2012**, N7. [CrossRef]
264. Porté-Agel, F.; Wu, Y.T.; Lu, H.; Conzemius, R.J. Large-eddy simulation of atmospheric boundary layer flow through wind turbines and wind farms. *J. Wind Eng. Ind. Aerod.* **2011**, *99*, 154–168. [CrossRef]
265. Stadtmann, F.; Rasheed, A.; Kvamsdal, T.; Johannessen, K.A.; San, O.; Kölle, K.; Tande, J.O.; Barstad, I.; Benhamou, A.; Brathaug, T.; et al. Digital twins in wind energy: Emerging technologies and industry-informed future directions. *IEEE Access* **2023**, *11*, 110762–110795. [CrossRef]
266. Hajisharifi, A.; Girfoglio, M.; Quaini, A.; Rozza, G. A comparison of data-driven reduced order models for the simulation of mesoscale atmospheric flow. *Finite Elem. Anal. Des.* **2024**, *228*, 104050. [CrossRef]

267. Lario, A.; Maulik, R.; Schmidt, O.T.; Rozza, G.; Mengaldo, G. Neural-network learning of SPOD latent dynamics. *J. Comput. Phys.* **2022**, *468*, 111475. [CrossRef]
268. Pawar, S.; San, O. Equation-Free Surrogate Modeling of Geophysical Flows at the Intersection of Machine Learning and Data Assimilation. *J. Adv. Model. Earth Syst.* **2022**, *14*, e2022MS003170. [CrossRef]
269. Schmidt, O.T.; Mengaldo, G.; Balsamo, G.; Wedi, N.P. Spectral empirical orthogonal function analysis of weather and climate data. *Mon. Weather. Rev.* **2019**, *147*, 2979–2995. [CrossRef]
270. Chen, N.; Majda, A.J.; Giannakis, D. Predicting the cloud patterns of the Madden-Julian Oscillation through a low-order nonlinear stochastic model. *Geophys. Res. Lett.* **2014**, *41*, 5612–5619. [CrossRef]
271. Chen, N.; Majda, A.J.; Sabeerali, C.T.; Ajayamohan, R.S. Predicting monsoon intraseasonal precipitation using a low-order nonlinear stochastic model. *J. Clim.* **2018**, *31*, 4403–4427. [CrossRef]
272. Pathak, J.; Hunt, B.; Girvan, M.; Lu, Z.; Ott, E. Model-free prediction of large spatiotemporally chaotic systems from data: A reservoir computing approach. *Phys. Rev. Lett.* **2018**, *120*, 024102. [CrossRef] [PubMed]
273. Rasp, S.; Thuerey, N. Data-driven medium-range weather prediction with a resnet pretrained on climate simulations: A new model for weatherbench. *J. Adv. Model. Earth Syst.* **2021**, *13*, e2020MS002405. [CrossRef]
274. Schultz, M.G.; Betancourt, C.; Gong, B.; Kleinert, F.; Langguth, M.; Leufen, L.H.; Mozaffari, A.; Stadtler, S. Can deep learning beat numerical weather prediction? *Philos. Trans. R. Soc. A* **2021**, *379*, 20200097. [CrossRef] [PubMed]
275. Weyn, J.A.; Durran, D.R.; Caruana, R. Can machines learn to predict weather? Using deep learning to predict gridded 500-hPa geopotential height from historical weather data. *J. Adv. Model. Earth Syst.* **2019**, *11*, 2680–2693. [CrossRef]
276. Lam, R.; Sanchez-Gonzalez, A.; Willson, M.; Wirnsberger, P.; Fortunato, M.; Pritzel, A.; Ravuri, S.; Ewalds, T.; Alet, F.; Eaton-Rosen, Z.; et al. GraphCast: Learning skillful medium-range global weather forecasting. *arXiv* **2022**, arXiv:2212.12794.
277. Pathak, J.; Subramanian, S.; Harrington, P.; Raja, S.; Chattopadhyay, A.; Mardani, M.; Kurth, T.; Hall, D.; Li, Z.; Azizzadenesheli, K.; et al. FourCastNet: A global data-driven high-resolution weather model using adaptive fourier neural operators. *arXiv* **2022**, arXiv:2202.11214.
278. Bi, K.; Xie, L.; Zhang, H.; Chen, X.; Gu, X.; Tian, Q. Accurate medium-range global weather forecasting with 3D neural network. *Nature* **2023**, *619*, 533–538. [CrossRef]
279. Kutz, J.N.; Brunton, S.L.; Brunton, B.W.; Proctor, J.L. *Dynamic Mode Decomposition: Data-Driven Modeling of Complex Systems*; Society for Industrial and Applied Mathematics (SIAM): Philadelphia, PA, USA, 2016.
280. Schmid, P.J. Dynamic mode decomposition of numerical and experimental data. *J. Fluid Mech.* **2010**, *656*, 5–28. [CrossRef]
281. Schmid, P.J.; Li, L.; Juniper, M.P.; Pust, O. Applications of the dynamic mode decomposition. *Theor. Comput. Fluid Dyn.* **2011**, *25*, 249–259. [CrossRef]
282. Tu, J.H.; Rowley, C.W.; Luchtenburg, D.M.; Brunton, S.L.; Kutz, J.N. On Dynamic Mode Decomposition: Theory and Applications. *J. Comput. Dyn.* **2014**, *1*, 391–421. [CrossRef]
283. Schmid, P.J. Application of the dynamic mode decomposition to experimental data. *Exp. Fluids* **2011**, *50*, 1123–1130. [CrossRef]
284. Duke, D.; Honnery, D.; Soria, J. Experimental investigation of nonlinear instabilities in annular liquid sheets. *J. Fluid Mech.* **2012**, *691*, 594–604. [CrossRef]
285. Seena, A.; Sung, H.J. Dynamic mode decomposition of turbulent cavity flows for self-sustained oscillations. *Int. J. Heat Fluid Flow* **2011**, *32*, 1098–1110. [CrossRef]
286. Arbabi, H.; Mezic, I. Ergodic theory, dynamic mode decomposition, and computation of spectral properties of the Koopman operator. *SIAM J. Appl. Dyn. Syst.* **2017**, *16*, 2096–2126. [CrossRef]
287. Curtis, C.W.; Alford-Lago, D.J.; Bollt, E.; Tuma, A. Machine Learning Enhanced Hankel Dynamic-Mode Decomposition. *arXiv* **2023**, arXiv:2303.06289.
288. Fujii, K.; Takeishi, N.; Kibushi, B.; Kouzaki, M.; Kawahara, Y. Data-driven spectral analysis for coordinative structures in periodic human locomotion. *Sci. Rep.* **2019**, *9*, 16755. [CrossRef] [PubMed]
289. Jiang, H.; Chen, J.; Dong, G.; Liu, T.; Chen, G. Study on Hankel matrix-based SVD and its application in rolling element bearing fault diagnosis. *Mech. Syst. Signal Process.* **2015**, *52*, 338–359. [CrossRef]
290. Vasconcelos Filho, E.; dos Santos, P.L. A dynamic mode decomposition approach with Hankel blocks to forecast multi-channel temporal series. *IEEE Control. Syst. Lett.* **2019**, *3*, 739–744. [CrossRef]
291. Yang, D.; Gao, H.; Cai, G.; Chen, Z.; Wang, L.; Ma, J.; Li, D. Synchronized ambient data-based extraction of interarea modes using Hankel block-enhanced DMD. *Int. J. Electr. Power Energy Syst.* **2021**, *128*, 106687. [CrossRef]
292. Frame, P.; Towne, A. Space-time POD and the Hankel matrix. *arXiv* **2022**, arXiv:2206.08995.
293. Hess, M.W.; Quaini, A.; Rozza, G. A data-driven surrogate modeling approach for time-dependent incompressible Navier-Stokes equations with dynamic mode decomposition and manifold interpolation. *Adv. Comput. Math.* **2023**, *49*, 22. [CrossRef]
294. Girfoglio, M.; Ballarin, F.; Infantino, G.; Nicoló, F.; Montalto, A.; Rozza, G.; Scrofani, R.; Comisso, M.; Musumeci, F. Non-intrusive PODI-ROM for patient-specific aortic blood flow in presence of a LVAD device. *Med. Eng. Phys.* **2022**, *107*, 103849. [CrossRef] [PubMed]
295. Hajisharifi, A.; Romano', F.; Girfoglio, M.; Beccari, A.; Bonanni, D.; Rozza, G. A non-intrusive data-driven reduced order model for parametrized CFD-DEM numerical simulations. *J. Comput. Phys.* **2023**, *491*, 112355. [CrossRef]

296. Demo, N.; Tezzele, M.; Mola, A.; Rozza, G. An efficient shape parametrisation by free-form deformation enhanced by active subspace for hull hydrodynamic ship design problems in open source environment. In Proceedings of the ISOPE International Ocean and Polar Engineering Conference. ISOPE, Sapporo, Japan, 10–15 June 2018; pp. 565–572
297. Demo, N.; Tezzele, M.; Gustin, G.; Lavini, G.; Rozza, G. Shape optimization by means of proper orthogonal decomposition and dynamic mode decomposition. In Proceedings of the Technology and Science for the Ships of the Future: NAV 2018: 19th International Conference on Ship & Maritime Research, Trieste, Italy, 20–22 June 2018; pp. 212–219.
298. Ripepi, M.; Verveld, M.J.; Karcher, N.; Franz, T.; Abu-Zurayk, M.; Görtz, S.; Kier, T. Reduced-order models for aerodynamic applications, loads and MDO. *CEAS Aeronaut. J.* **2018**, *9*, 171–193. [CrossRef]
299. Gonzalez, F.J.; Balajewicz, M. Deep convolutional recurrent autoencoders for learning low-dimensional feature dynamics of fluid systems. *arXiv* **2018**, arXiv:1808.01346.
300. Maulik, R.; Lusch, B.; Balaprakash, P. Reduced-order modeling of advection-dominated systems with recurrent neural networks and convolutional autoencoders. *Phys. Fluids* **2021**, *33*, 037106. [CrossRef]
301. Mohan, A.; Daniel, D.; Chertkov, M.; Livescu, D. Compressed convolutional LSTM: An efficient deep learning framework to model high fidelity 3D turbulence. *arXiv* **2019**, arXiv:1903.00033.
302. Shi, X.; Chen, Z.; Wang, H.; Yeung, D.Y.; Wong, W.K.; Woo, W.C. Convolutional LSTM network: A machine learning approach for precipitation nowcasting. *Adv. Neural Inf. Process. Syst.* **2015**, *28*.
303. RBniCSx—Reduced Order Modelling in FEniCSx. Available online: https://github.com/rbnics/rbnicsx (accessed on 28 June 2024).
304. libROM—Library for Reduced Order Models. Available online: https://www.librom.net/ (accessed on 28 June 2024).
305. Moser, D.R.; Kim, J.; Mansour, N.N. Direct numerical simulation of turbulent channel flow up to $Re_\tau = 590$. *Phys. Fluids* **1999**, *11*, 943–945. [CrossRef]

Disclaimer/Publisher's Note: The statements, opinions and data contained in all publications are solely those of the individual author(s) and contributor(s) and not of MDPI and/or the editor(s). MDPI and/or the editor(s) disclaim responsibility for any injury to people or property resulting from any ideas, methods, instructions or products referred to in the content.

fluids

MDPI

Review

Numerical Simulations of Scalar Transport on Rough Surfaces

Zvi Hantsis * and Ugo Piomelli

Department of Mechanical and Materials Engineering, Queen's University, Smith Engineering, Kingston, ON K7L 3N6, Canada
* Correspondence: zvi.hantsis@queensu.ca

Abstract: Numerical simulations provide unfettered access to details of the flow where experimental measurements are difficult to obtain. This paper summarises the progress achieved in the study of passive scalars in flows over rough surfaces thanks to recent numerical simulations. Townsend's similarity applies to various scalar statistics, implying the differences due to roughness are limited to the roughness sublayer (RSL). The scalar field exhibits a diffusive sublayer that increasingly conforms to the roughness surface as k_s^+ or Pr increase. The scalar wall flux is enhanced on the windward slopes of the roughness, where the analogy between momentum and scalar holds well; the momentum and scalar fields, however, have very different behaviours downwind of the roughness elements, due to recirculation, which reduces the scalar wall flux. Roughness causes breakdown of the Reynolds analogy: any increase in St is accompanied by a larger increase in c_f. A flattening trend for the scalar roughness function, $\Delta\Theta^+$, is observed as k_s^+ increases, suggesting the possibility of a scalar fully rough regime, different from the velocity one. The form-induced (FI) production of scalar fluctuations becomes dominant inside the RSL and is significantly different from the FI production of turbulent kinetic energy, resulting in notable differences between the scalar and velocity fluctuations. Several key questions remain open, in particular regarding the existence of a fully rough scalar regime and its characteristics. With the increase in Re and Pr, various quantities such as scalar roughness function, the dispersive fluxes, FI wall flux, etc., appear to trend towards saturation. However, the limited range of Re and Pr achieved by numerical simulations only allows us to speculate regarding such asymptotic behaviour. Beyond extending the range of Re and Pr, systematic coverage of different roughness types and topologies is needed, as the scalar appears to remain sensitive to the geometrical details.

Keywords: roughness; scalar transport; numerical simulations

1. Introduction

1.1. Scalar Transport

The transport of scalar quantities in turbulent flows is of great importance in many engineering applications, as well as a central phenomenon in nature. The transfer of internal energy or enthalpy affects applications such as heat exchangers, gas turbines, and nuclear reactors, while mass transport governs the dispersion of particles in the atmosphere, or hazardous material release. Scalar transport adds additional transport equations with a new parameter, the Prandtl number $Pr = \nu/\alpha$ or Schmidt number $Sc = \nu/D$, where ν is the fluid kinematic viscosity, and α or D are the conductivity or diffusivity of the scalar, respectively. Scalars are defined as "passive" when their effect on the fluid or flow is negligible, that is, they do not change significantly the fluid viscosity, density, etc.

1.2. Stanton Number and the Reynolds Analogy

Scalar transport (especially for heat transfer applications) has been studied for over a century. Reynolds [1] suggested that momentum and heat in a fluid are transferred in the same way and, therefore, proportionality must exist in geometrically similar systems; this relation is commonly referred to as the "Reynolds analogy". Generally, it applies to gases where the fluid viscosity and scalar diffusivity are similar, and is based on the notion that the same turbulent eddies transport both scalar quantities and momentum [2–4]. The Reynolds analogy is based on the formal similarity between the transport equations for momentum and scalar quantities, as long as the pressure gradient is insignificant and no additional forces are present [1,2]. The skin-friction coefficient and the scalar-flux coefficient, St, are defined as

$$C_f = \frac{2\tau_w}{\rho U_b^2}; \qquad St = \frac{q_w}{\rho c_p \Theta_m U_b} = \frac{1}{\Theta_m^+ U_b^+}, \tag{1}$$

where τ_w is the wall stress, q_w the wall flux, ρ the fluid density, c_p the specific heat capacity, Θ_m is the mixed-mean scalar, and U_b is the average velocity. The Reynolds analogy implies that the ratio $2C_f/St$, known as the *Reynolds analogy factor*, $RA = U_b^+/\Theta_m^+$, is constant and equal to one for Pr near unity [2–5]. While this analogy also applies to mass transfer (where the transfer coefficient is the Sherwood number, Sh) or any other type of scalar transport, the rest of the paper will refer to heat transfer and temperature. It is to be understood that other scalars can be treated in the same way.

The analogy may also hold when $Pr \neq 1$; in this case, $RA \neq 1$ (but remains constant) [4,6]. Classic approximations of RA for $Pr \leq 5$ are power laws of the form $RA \sim Pr^m$, and various values of m have been suggested [3,7,8]. For larger Pr, a more complicated formulation is necessary [4,8,9].

1.3. Effects of Roughness on Scalar Transport

The Reynolds analogy has been extensively verified for smooth surfaces [10–14]. However, hydraulically smooth behaviour can only be achieved at sufficiently low Re [15,16]. At higher Re, when the roughness scale becomes comparable to the viscous sublayer, any surface becomes rough. Roughness has many effects on momentum transfer; the main ones are (1) an increase in the drag [17,18]; (2) the amplification of the wall-normal and spanwise fluctuations, at the expense of the streamwise ones, resulting in decreased flow anisotropy [19–22]; and (3) the breakup of the near-wall structures and modifications of the near-wall turbulence generation cycle [23,24]. The increased wall stress results in a downward shift of the logarithmic layer, the "roughness function", ΔU^+ (here, $+$ denotes wall units, i.e., quantities normalised using u_τ and ν, where $u_\tau = \sqrt{\tau_w/\rho}$ is the friction velocity). Townsend [25] proposed a similarity hypothesis stating that, at sufficiently high Reynolds numbers, the only role of the region below the roughness crest is to set the length and velocity scales for the outer flow. Approximately $3-5$ roughness heights above the crest, the turbulence statistics thus normalised collapse with the corresponding smooth-wall case [15,16,26,27]. This region is known as the *roughness sublayer*, RSL. Reviews of this topic can be found in Flack and Schultz [15], Chung et al. [16], Raupach et al. [23], Jiménez [28].

There are many similarities between the effects of roughness on momentum and scalar transport. St increases, and the mean scalar profile, which for smooth wall presents a log-law behaviour, is also shifted down by a *scalar roughness function*, $\Delta\Theta^+$. Townsend's similarity hypothesis was found to apply to the scalar variance and turbulent fluxes [29–33], to the terms in their budget [29,31,34], to skewness and flatness [29], and to some of the higher-order moments [29]. However, experimental studies [9,35] showed that when roughness is introduced St remains sensitive to scalar diffusivity and decreases with Re, while c_f approaches a constant value in the fully rough regime [9,15,17,35]. This implies the breakdown of the Reynolds analogy when the surface is sufficiently rough. Since the roughness elements create stagnation points and recirculation regions, significant pressure gradients occur that break the formal similarity between momentum and scalar transport equations. Away from the wall, on the other hand, the scalar transport is dominated by the turbulent transport, so that Townsend's similarity hypothesis [25] also applies to the scalar.

1.4. Numerical Simulations of Rough-Wall Scalar Transport

Numerical simulations have recently begun to contribute to our understanding of scalar-transport physics over rough surfaces. Direct numerical simulations (DNSs) (which requires that all the scales of motion are resolved) and large-eddy simulations (LESs) (in which only the large, energy-carrying eddies are resolved) have been used to highlight the flow physics and validate turbulence models for the Reynolds-averaged Navier–Stokes (RANS) equations. Early simulations were limited to simple roughness shapes, such as two-dimensional (2D) square bars, ribs with various orientations, circular rods, wavy walls, or ridges [36–42]. Direct and large-eddy simulations have also been used to study three-dimensional roughness [5,30–32,34,41,43–54]. These studies considered Pr = 0.2–6 and k_s^+ up to 700, where k_s^+ is the equivalent sand-grain size (that is, the sand-grain height from Nikuradse's experiments [17] for which the same roughness function, ΔU^+, value as the present surface is achieved); however, most of them were focused on Pr near unity (in particular 0.7 and 1.0), and $k_s^+ \leq 400$. Table 1 contains a non-exhaustive list of key studies.

Table 1. Summary of recent numerical simulations of wall-resolved passive scalar transport with roughness.

Study	Max k_s^+ I	Pr or Sc	Roughness Type
Doosttalab et al. [29]	11	0.7	Grit blasted
Stroh et al. [42]	12 I	0.7	Streamwise-aligned triangular ridges
Peeters [48]	52 II	2.0–6.0	Grit blasted
Peeters and Sandham [46]	102	1.0	Grit blasted
Hantsis and Piomelli [31]	110	0.7–1.4	Sand grain
Yang et al. [55]	120	0.7	Ice accretion, sandpaper, combustion chamber deposits
Miyake et al. [36]	130 I	0.7	Sand grain III, 2D ribs
Jelly et al. [5]	130	0.7	X-ray computed tomography
Orlandi et al. [43]	200 I	1.0	Regularly arranged protrusions
Kuwata [52]	200	1.0	Distributed hyperbolic elements
Nagura et al. [56]	230	0.7	Sinusoidal wall
Forooghi et al. [51]	300	0.7	Real (scanned) surfaces, various distributed protrusions
Zhong et al. [45]	300	0.5–2.0	3D sinusoidal
Rowin et al. [57]	340	0.7	3D sinusoidal
MacDonald et al. [30,44]	380 IV	0.7	3D sinusoidal
Hantsis [32]	485	0.2–2.0	Sand grain
Leonardi et al. [39]	600 I	1.0	2D circular and square bars V
Zhang et al. [58,59]	600 I	0.7	2D sinusoidal wall
Garg et al. [54]	633	0.7	Additive manufacturing scanned surfaces
Kuwata [60]	700	1.0	Distributed hyperbolic elements
Zhang et al. [58]	700	0.7	Distributed hyperbolic elements

I If not explicitly reported, value is derived from the given values of ΔU^+. II The DNS conducted in the study was limited; however, a wider range of k_s^+ was considered by using results from other studies. III Not explicitly solved but represented by a line force emanating from the wall. IV This value was later revised to 233 by [45]. V Only on one side; other wall is smooth.

1.5. Scope of This Paper

The focus of this paper is on studies that resolve the near-wall region and provide physical understanding of the flow behaviour. First, we will present the formulation of the models discussed here, and then we will discuss some of the novel findings obtained from numerical simulations. A list of open questions and some concluding remarks will follow.

2. Problem Description

The problem at hand can be summarised as follows: for a given roughness topology, knowing the velocity and pressure fields, we would like to predict the behaviour of the scalar field and its statistics. A particular question is whether a regime equivalent to the "fully rough" regime exists for the scalar quantity and, if so, how it manifests itself. Since Townsend's similarity hypothesis extends to scalar transport [29,31,44,46,52], the focus of this review will be the roughness sublayer (RSL), which extends a few roughness heights, k (typically 3–5) above the roughness crest. Most of the studies considered focus on canonical flows: plane channels or flat-plate boundary layers. A passive scalar is then transported by the flow, with the rough boundaries either kept at a constant scalar value, or providing a constant and uniform flux.

2.1. Governing Equations

In incompressible flows, conservation of mass, momentum, and passive scalar can be written as

$$\frac{\partial u_k}{\partial x_k} = 0 \tag{2}$$

$$\frac{\partial u_i}{\partial t} + \frac{\partial (u_i u_k)}{\partial x_k} = -\frac{\partial p}{\partial x_i} + \frac{\partial}{\partial x_k}(2\nu S_{ik}) \tag{3}$$

$$\frac{\partial \theta}{\partial t} + \frac{\partial (\theta u_k)}{\partial x_k} = \frac{\partial}{\partial x_k}\left(\alpha \frac{\partial \theta}{\partial x_k}\right) + Q \tag{4}$$

where x_i, for $i = 1-3$ (or x, y, and z), are the Cartesian coordinates in the streamwise, wall-normal and spanwise directions, respectively. u_i (or u, v, and w) are the velocity components in the Cartesian directions. $p = P/\rho$ is the hydrodynamic pressure divided by the density, ν is the kinematic molecular viscosity (assumed constant), and S_{ij} is the strain-rate tensor:

$$S_{ij} = \frac{1}{2}\left(\frac{\partial u_i}{\partial x_j} + \frac{\partial u_j}{\partial x_i}\right). \tag{5}$$

In Equation (4), α is the molecular diffusivity and $\theta = T - T_w$, where T is the transported scalar and T_w is its value at the wall. Q is a source term used to facilitate the use of periodic boundary conditions. For a constant-flux condition,

$$Q = u\frac{dT_w}{dx} = \frac{q_w}{\rho c_p \delta}\frac{u}{U_b}, \tag{6}$$

where c_p is the specific heat, δ is the boundary layer thickness (or channel half-width), and q_w is the prescribed scalar wall flux. For the constant temperature (Dirichlet) boundary condition, $Q = 0$. A discussion of this approach and its limitations can be found in [61].

2.2. Averaging Process and Mean Equations

Roughness causes the time-averaged fields to be spatially inhomogeneous. Consequently, three types of averages can be defined: the usual time average, and two types of spatial averages, the *superficial average*, $\langle f \rangle_s$, and the *intrinsic* one, $\langle f \rangle$. The intrinsic average is only performed over the fluid portion of the averaging region, while the superficial one is performed over the averaging region, including both fluid and solid [23,62,63]. If the averaging region does not contain any solid portion, both averages are identical and correspond to the spatial mean of the averaged quantity; this framework allows for consistent comparison between averages of smooth- and rough-wall cases. Using the intrinsic and time averages, a triple decomposition can be defined. Any variable can be written as

$$f(x_i,t) = \overline{f}(x_i) + f'(x_i,t) = \left\langle \overline{f} \right\rangle + \widetilde{f}(x_i) + f'(x_i,t). \tag{7}$$

Here, $\left\langle \overline{f} \right\rangle$ is the *double-averaged* value of f, $\widetilde{f}$ represents the spatial inhomogeneity of the time-averaged f, and is known as the *dispersive* or *wake* component, and f' is the stochastic part. The differentiation of the averaged quantities introduces additional terms in the governing equations. The double-averaged momentum and passive-scalar equations are [23,63]

$$-\langle\overline{\Pi}\rangle_s = -\langle\overline{u_j}\rangle_s\frac{\partial\langle\overline{u_i}\rangle}{\partial x_j} - \frac{\partial}{\partial x_j}\left\langle\overline{u_i'u_j'}\right\rangle_s - \frac{\partial}{\partial x_j}\langle\widetilde{u}_i\,\widetilde{u}_j\rangle_s + \frac{\partial}{\partial x_j}\left(\nu\frac{\partial\langle\overline{u}_i\rangle_s}{\partial x_j}\right) + \left\langle\frac{\partial}{\partial x_j}\left(\nu\frac{\partial\widetilde{u}_i}{\partial x_j}\right)\right\rangle_s - \left\langle\frac{\partial\widetilde{p}}{\partial x_i}\right\rangle_s \tag{8}$$

$$-\langle\overline{Q}\rangle_s = -\langle\overline{u_j}\rangle_s\frac{\partial\langle\overline{\theta}\rangle}{\partial x_j} - \frac{\partial}{\partial x_j}\left\langle\overline{\theta'u_j'}\right\rangle_s - \frac{\partial}{\partial x_j}\left\langle\widetilde{\theta}\,\widetilde{u}_j\right\rangle_s + \frac{\partial}{\partial x_j}\left(\alpha\frac{\partial\langle\overline{\theta}\rangle_s}{\partial x_j}\right) + \left\langle\frac{\partial}{\partial x_j}\left(\alpha\frac{\partial\widetilde{\theta}}{\partial x_j}\right)\right\rangle_s. \tag{9}$$

The third term on the RHS of Equation (8) is the *dispersive stress*, while the last two terms are the viscous and pressure fluxes; integrating them over the RSL yields the viscous and form drag, respectively. In Equation (9), conversely, the third and fifth terms on the RHS are the dispersive flux and the form-induced (FI) flux, sometimes referred to as "wall-interaction term" [33,60].

3. Achievements

3.1. Instantaneous Scalar Field

Numerical simulations provide unfettered access to details of the flow field, both for momentum and scalar. Examining the instantaneous and time-averaged scalar field is the most direct way to understand the effects of roughness on scalar quantities.

The key differences between the scalar and momentum fields occur near and around the roughness elements, which frequently have stagnation points in front and recirculation regions behind them, leading to significant pressure gradients. As is well known, in the fully rough regime closely packed roughness elements "shelter" each other [62], and the pressure drag (which is weakly dependent on *Re*) is much larger than the *Re*-dependent viscous one. This behaviour has been observed in many investigations [30–32,39,41,44,46,52].

Figure 1 shows the instantaneous velocity (a, e) and scalar contours (b–d, f–h) for a range of *Pr* and k_s^+. It is apparent that, as k_s^+ increases, the scalar gradients are confined to a thinner and thinner layer that closely follows the geometry [30,44]. This is demonstrated by the thickness of the regions covered by $u^+ \leq 5$ or $\theta^+ \leq 5$, respectively, which indicate the extent of viscous and diffusive sublayers.

The same effect is observed when *Pr* is increased, and the low diffusivity also acts to confine the gradients to a thin layer. For the lower values of *Pr* and k_s^+, however, we still observe some effect of the sheltering and of the recirculation regions. When the convection behind the roughness elements is less vigorous, the scalar forms regions where it is well mixed, where the small gradient limits the local transfer rate of the scalar quantity [30,46,51]. The much thicker recirculation regions behind the roughness protrusions act as "resistance" to the scalar wall flux [30,32,45,51].

In a smooth-wall case, an increase in *Re* affects the thickness of both viscous and diffusive sublayers (and the boundary layer thickness itself). For a fully rough case, on the other hand, the increase in *Re* does not affect the viscous sublayer while still thinning the diffusive sublayer; therefore, the mechanisms that govern the viscous sublayer in the smooth-wall case do not apply directly to the rough-wall case. It is then clear that the effects of *Re* and *Pr*, while similar, work separately from each other—a Péclet-number scaling $Pe_\tau = Re_\tau Pr$, commonly used in the smooth-wall case, is not directly applicable in the presence of roughness [34,45,57].

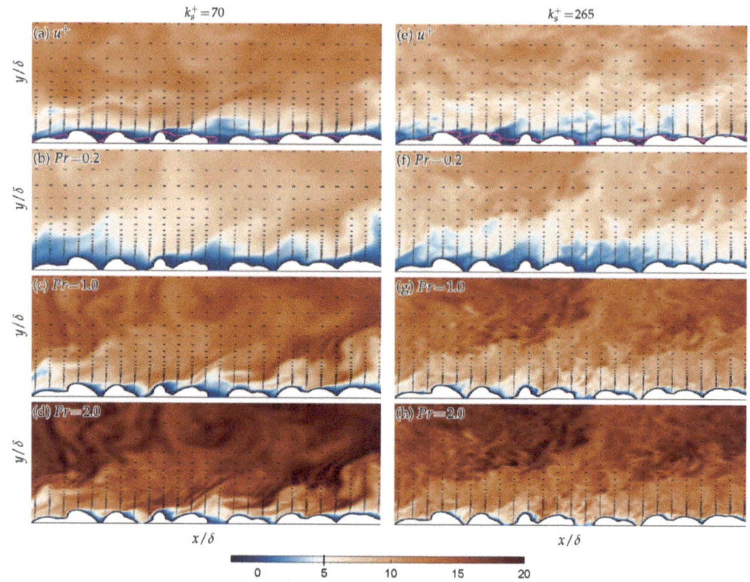

Figure 1. Instantaneous velocity u^+ (**a**,**e**) and scalar (**b**–**h**) at various Pr ranging from 0.2 to 2.0 shown across a vertical plane $(x - y)$ of dimensions $\delta \times 0.2\delta$. k_s^+ is 70 in the first column (**a**–**d**) and 265 in the second (**e**–**h**). Velocity quivers show the direction of the instantaneous velocity. The white region corresponds to ≈ 5.0; purple lines indicate $u^+ = 0$. All quantities are scaled in inner units. Figure reprinted with permission from Hantsis [32].

Repeated impingement of bulk fluid on the exposed (i.e., not sheltered by upstream elements) windward side of the roughness produces an attached thin diffusive sublayer over the windward surface [45,57]. Zhong et al. [45] found that the ratio between the local diffusive and viscous sublayer thicknesses $\delta_\alpha/\delta_\nu \propto Pr^{-1/3}$ at the crests, supporting an argument originally put forward by Owen and Thomson [35] and Yaglom and Kader [64] that "smooth-wall-like" boundary-layer behaviour is seen at the local roughness crests. The diffusive sublayer is observed to be thinnest on the exposed windward side and around the crest, resulting in sharp scalar gradients that enhance the scalar wall flux [45,46]. Both viscous and diffusive sublayers at the roughness crests were shown to be thinner compared with an equivalent smooth-wall case (under the same Re_τ and Pr) [45,57]; however, as k_s^+ increased, their thickness approached that of the smooth-wall case (this was shown for Pr = 0.5–2.0; it is assumed that this behaviour is maintained at higher Pr) [30,32,45].

Conversely, on the leeward side, the diffusive sublayer is much thicker than the equivalent smooth-wall one [45,57]. This can be seen best on the leeward side right behind the crest in Figure 1, where the diffusive sublayer can sometimes be the same thickness as the roughness elements themselves, and this is further supported by the effective reduction in the scalar wall flux in the recirculation regions. On the windward side, the rough-wall diffusive sublayer is analogous to a contorted, smooth-wall diffusive sublayer (at sufficiently high k_s^+) [44,45]; a similar analogy cannot be made on the leeward side. The variation of the diffusive-sublayer thickness (compared with an equivalent smooth-wall case) results in enhanced scalar transport in some regions but attenuation in others; the integral of all of these results in a net increase or decrease in scalar transport relative to the smooth-wall case.

The recirculation affects the fluid that impinges on the windward slope, drawing it down and backwards (opposite to the direction of the bulk flow) towards the base of the roughness [44,46]. The lack of a pressure-equivalent mechanism for the scalar results in reduced advection of the scalar away from the troughs and, therefore, results in some accumulation of the scalar in "pockets" [45,46]. These "pockets" extend into sheltered regions where the velocity is much smaller, or even negative [44,46]; their extent seems to be well correlated with the location where $\overline{u}$ becomes zero [45]. This mechanism generates larger scalar gradients locally, which partially compensates the reduced scalar transfer in the recirculation regions (more details in Section 3.3).

3.2. Mean Profiles and the Scalar Roughness Function

Mean velocity and scalar profiles have been extensively measured experimentally ([11–13,17,23,26–28,35,62,65–69] and others). These studies produced empirical correlations and phenomenological models that, in turn, led to predictions of the scalar roughness function, $\Delta\Theta^+$, in the fully rough regime.

Disagreement amongst these models resulted in very different predicted behaviours of $\Delta\Theta^+$ when $k_s^+ \to \infty$. The underlying physical assumptions that underpin these models were difficult to evaluate, due to a lack of high-fidelity data, in particular for dense roughness. Numerical simulations provide a tool to test the underlying physical assumptions, and to sweep systematically across a range of parameters (geometry, Prandtl number, Reynolds number, etc.), which may be difficult to achieve experimentally. In particular, the minimal-channel approach facilitates a relatively low-cost way to achieve such parameter sweeps.

Figure 2 show the profile of the mean scalar as Re and Pr are varied. Similar to the mean velocity, the mean scalar also shows a logarithmic region, which can be expressed as:

$$\Theta^+ = \langle \overline{\theta} \rangle^+ = \frac{1}{\kappa_\theta} \log\left[(y-d)^+\right] + A_\theta - \Delta\Theta^+ \tag{10}$$

where $\kappa_\theta \approx 0.46$ [31,70] and $A_\theta(Pr)$ is the smooth-wall intercept. Zhong et al. [45] report other equivalent forms of Θ^+. A_θ increases with Pr [3,70], resulting in an upwards shift of the logarithmic region. In contrast, the increase in Re (and therefore k_s^+) results in a downward shift of the logarithmic region.

The downwards shift of the logarithmic region between the smooth- and rough-wall cases is quantified by the scalar roughness function, $\Delta\Theta^+$, which is equivalent to the Clauser–Hama roughness function, ΔU^+. $\Delta\Theta^+$ depends on Pr and various geometrical roughness parameters, as illustrated in Figure 2. In Figure 2a, $Pr = 1.0$ is fixed and k^+ is increased (by varying Re); $\Delta\Theta^+$ increases with k^+, but seems to approach an asymptotic state where all curves collapse at high k^+. This trend was also noticed in other types of roughness [5,30,34,45,57]. In Figure 2b, the roughness height is fixed and Pr is varied. Again, there is an upwards shift of the logarithmic layer with Pr, which is much larger in the smooth-wall case [34,64]. As a consequence, the net shift in $\Delta\Theta^+$ between the smooth- and rough-wall cases (for each Pr) increases with Pr [34,48,57].

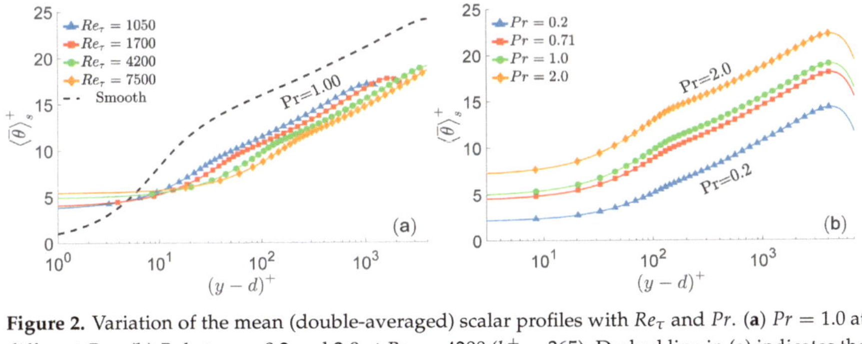

Figure 2. Variation of the mean (double-averaged) scalar profiles with Re_τ and Pr. (**a**) $Pr = 1.0$ at different Re_τ; (**b**) Pr between 0.2 and 2.0 at $Re_\tau = 4200$ ($k_s^+ = 265$). Dashed line in (**a**) indicates the smooth-wall curve for $Pr = 1.0$. Reprinted with permission from Hantsis [32] .

Several empirical models have been proposed to relate $\Delta\Theta^+$ to the equivalent sand-grain roughness, Prandtl number, and scalar von Kármán constant [3,9,35,64,69]. These models result in disparate predictions of $\Delta\Theta^+$, generally suggesting a monotonic decrease at higher k_s^+. These studies also cannot agree on the value of k_s^+ at which this decrease starts or on the decrease rate.

A different trend was observed by MacDonald et al. [30,44], who suggested a flattening of $\Delta\Theta^+$ and the possibility of an asymptotic value $\Delta\Theta^+_{\mathrm{FR}}$ for $k_s^+ \to \infty$. This behaviour is consistent with the observation of a thin diffusive sublayer that conforms to the topography in the fully rough regime. Other studies [5,31,41,46,47,52,53,57,71], covering various types of roughness and a wider range of Pr, support this trend (Figure 3).

Zhong et al. [45], however, found that the scalar roughness function is largest at $k^+ \approx 40$, and then gradually decreases. They suggest that this trend would lead to a decrease of the wall flux, which contradicts the current understanding of rough-wall scalar transfer [72,73]. This discrepancy may result from the use of the minimal-channel approach, in which the mean profiles away from the wall may be artificially altered.

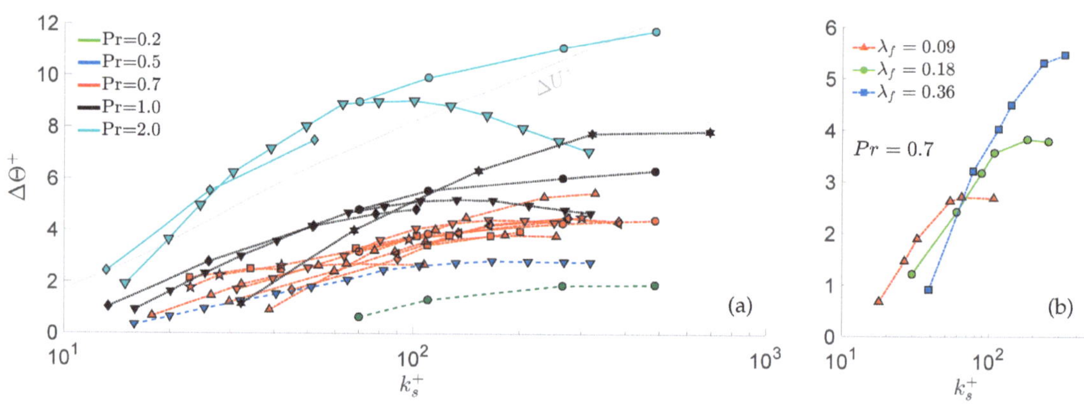

Figure 3. Scalar roughness function, $\Delta\Theta^+$, vs k_s^+ from selected studies; (**a**) shows the effect of Pr and (**b**) the effect of the frontal solidity, λ_f. For (**a**): Colours and lines indicate the Prandtl number: - - $Pr=0.2$; - - $Pr=0.5$; -·- $Pr=0.7$; ···· $Pr=1.0$; — $Pr=2.0$. Markers indicate: ● Hantsis [32]; ▼ Zhong et al. [45]; ▲ Rowin et al. [57]; ◀ MacDonald et al. [44]; ■ Jelly et al. [5]; ◆ Peeters and Sandham [46], Peeters [48]; ★ De Maio [74]; ✶ Kuwata [60]. Grey dash–dot line: The velocity roughness function, ΔU^+, from Nikuradse's sandpaper experiment [17]. For (**b**): $Pr = 0.7$, data from Rowin et al. [57] for -▲- $\Lambda=0.09$; —●— $\Lambda=0.18$; -·■·- $\Lambda=0.36$. Reprinted with permission from [57].

Figure 3a shows the variation of $\Delta\Theta^+$ with k_s^+ at different Pr. One commonly observed feature is $\Delta\Theta^+ < \Delta U^+$ at sufficiently high k_s^+. For $Pr > 1.0$, there exists a finite range of k_s^+ for which the opposite is true; the value of k_s^+ for which $\Delta\Theta^+$ intersects the velocity asymptote increases with Pr. Since higher values of ΔU^+ and $\Delta\Theta^+$ indicate enhancement of momentum and scalar transport, respectively, the range in which $\Delta U^+ < \Delta\Theta^+$ is of significant interest.

The implication of an asymptotic value $\Delta\Theta^+_{\mathrm{FR}}$ for all Pr is the possibility that an "ultimate regime", a scalar equivalent of the fully rough regime, exists, characterised by $\Delta\Theta^+ \to \Delta\Theta^+_{\mathrm{FR}}$. Peeters [47,48] suggested a model for $\Delta\Theta^+_{\mathrm{FR}}$, noting that the flattening of $\Delta\Theta^+$ at sufficiently high k_s^+ was reproduced when only the viscous–convective and diffusive ranges were considered; otherwise an increase in $\Delta\Theta^+$ would be observed. Peeters [48] concluded that the flattening must be the result of the interaction between the viscous–convective and diffusive scales with the rough wall.

$\Delta\Theta^+_{\mathrm{FR}}$ increases with Pr for a fixed roughness geometry [31,32] (Figure 3a). For a fixed Pr, it is sensitive to topological details such as the frontal solidity, λ_f (the ratio between the frontal area of the roughness and the total plan area), as shown in Figure 3b. $\Delta\Theta^+_{FR}$ also increases with λ_f, unlike ΔU^+, which saturates. Figure 3a also raises the possibility of $\Delta\Theta^+_{FR}$ clustering with Pr, even for different roughness types and topologies, suggesting stronger dependence on Pr and weaker dependence on the roughness details.

Because of the limited range of Pr and k_s^+ achieved by numerical studies, we can only speculate on the behaviour of $\Delta\Theta^+$ at higher k_s^+, and further investigations are required to clarify the issue. However, it is expected that at the ultimate regime the enhancement of the scalar transport will saturate and will be accompanied by increasing drag.

MacDonald et al. [30] and Rowin et al. [57] analytically derived a relation between $\Delta\Theta^+$ and the Stanton number, St:

$$\begin{aligned}\Delta\Theta^+ &= \sqrt{\frac{2}{c_{f0}}}\left(\frac{1}{St_0} - \frac{1}{\kappa_\theta\kappa}\right) - \sqrt{\frac{2}{c_f}}\left(\frac{1}{St} - \frac{1}{\kappa_\theta\kappa}\right) \\ &= \underbrace{A_\theta + \frac{\kappa}{\kappa_\theta}(\Delta U^+ - A)}_{\text{Term 1}} - \underbrace{\sqrt{\frac{2}{c_f}}\left(\frac{c_f}{2St} - \frac{\kappa}{\kappa_\theta}\right)}_{\text{Term 2}},\end{aligned} \tag{11}$$

where the subscript 0 indicates the equivalent smooth-wall case value; and A and $A_\theta(Pr)$ are the previously mentioned velocity and scalar log intercepts for the smooth-wall case. This expression identifies two types of contributions: Term 1 depends only on inner-scale quantities that are invariant to bulk (outer-scale) Re, while Term 2 only involves bulk quantities (such as the skin-friction coefficient and Stanton number), which inherently depend on the bulk Re; note that the fraction in Term 2 is the inverse of the Reynolds analogy factor. The Reynolds number does not appear explicitly, implying that $\Delta\Theta^+$ does not scale with Re. Furthermore, in the fully rough regime, only ΔU^+ and St change for a fixed Prandtl number, with $\Delta U^+ = \log(k_s^+)/\kappa - 3.5$ [15–17]. Thus, we can then rewrite Equation (11) as:

$$\Delta\Theta^+ = \left[A_\theta + \frac{\kappa}{\kappa_\theta}\left(\sqrt{\frac{2}{c_f}} - A\right)\right] + \left[\frac{\kappa}{\kappa_\theta}\left(\frac{1}{\kappa}\log(k_s^+) - 3.5\right) - \sqrt{\frac{c_f}{2}\frac{1}{St}}\right] \tag{12}$$

where the expression in the first bracket is constant for fixed Pr and given roughness geometry. An asymptotic value $\Delta\Theta^+_{FR}$ implies that the terms in the second bracket must also be constant and that $St \sim \log^{-1}(k_s^+)$. This provides an alternative indicator to verify the behaviour of $\Delta\Theta^+$ at $k_s^+ \to \infty$ using a bulk quantity (St), which is easier to measure experimentally.

3.3. Scalar-Flux Coefficient

The scalar-flux coefficient, known, in normalised form, as the Stanton number, St, is the scalar equivalent of the skin-friction coefficient, c_f. The Reynolds analogy factor, RA, is the ratio between St and c_f. Both St and RA are commonly used to quantifying the effects of roughness, by comparing them to the equivalent smooth-wall case. Hereinafter, we use the subscript 0 to denote the values from the equivalent smooth-wall case (same Re_τ and Pr), i.e., St_0, c_{f0}, and RA_0.

Many experimental studies (such as [3,9,12,35,64,65,75] and others) focus on the bulk values of c_f and RA.

A common objective is to find a predictive empirical correlation for St in various roughness conditions which, in conjunction with known c_f correlations, can be used to calculate RA. Numerical simulations, on the other hand, have allowed us to examine the local distributions of their quantities, and to access the flow properties below the roughness crest.

Local values of both St and c_f (and therefore RA) can be defined using either the instantaneous or time-averaged scalar flux, q_w, shear velocity, u_τ, and shear scalar, θ_τ [30,44–46].

MacDonald et al. [44] separated the friction- and pressure-drag components of c_f (which is difficult or impossible in experimental studies). The viscous component, c_{fv}, was then compared with St locally. They are shown over a section of the roughness surface (which was a 3D sinusoid) in Figure 4. For $Pr = 0.7$, the behaviours of St and c_{fv} are overall similar; the most distinct differences occur in the troughs between roughness elements. Unlike St, which is a non-negative quantity, the viscous component, c_{fv}, is negative in the recirculation regions on the leeward side of the roughness elements. For this sinusoidal roughness, the negative region covers approximately 40% of the plan area, both in the transitionally and fully rough regimes [44]. Thus, RA is more likely to be close to one on the windward side of the roughness elements, or near the crests [44].

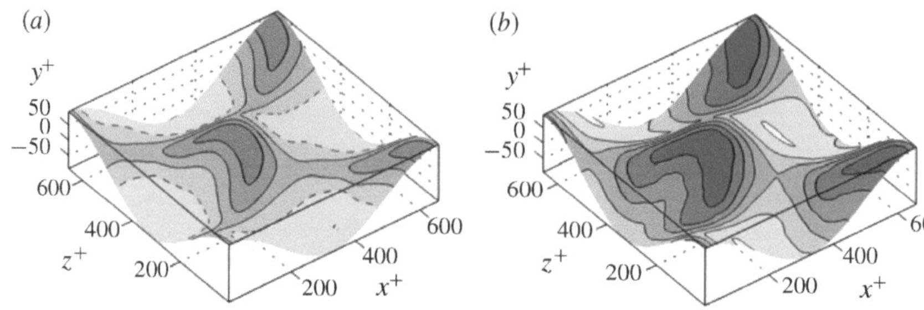

Figure 4. Spatial distribution of the time-averaged viscous skin-friction component c_{fv} (**a**) and St (**b**). A dashed contour in (**a**) show negative stress. Reprinted with permission from [44].

A different approach was employed by Zhong et al. [45], who used a compensated viscous-to-diffusive sublayer ratio, $\mathcal{R}_{v\alpha} = (\delta_v/\delta_\alpha)Pr^{1/3}$, to identify where a Reynolds analogy-type behaviour exists. The Reynolds analogy can be assumed to be valid when this ratio is approximately constant (with a value depending on Pr). Departures from a constant value indicate weakening of the analogy. It was found that, at the shear-driven windward side of the roughness elements, the Reynolds analogy usually holds locally, as the ratio is almost constant, in line with the findings of [30]. The thicknesses of the viscous and diffusive sublayers on the windward side are comparable to those in the equivalent smooth-wall case. At the crests, a mild departure from a constant is reported, suggesting a partial analogy [45,58]. It is in the detached regions behind the roughness elements where the analogy clearly breaks down, due to the prominence of reversed flow and weak shear. Zhong et al. [45] noted that, while a Reynolds analogy-type mechanism is dominant on the windward side, behind the roughness elements an ensemble of different local behaviours determine the attenuation of the scalar wall flux. The degree of dissimilarity also depends on the relaxation of the detached region, and whether reattachment occurs before the next roughness element (i.e., on the level of sheltering).

This is also confirmed by examining the probability density functions (PDFs) of the instantaneous c_f and St, as well as their joint PDF (jPDF), conditioned on flow reversal. Indeed, the conditioning on $u > 0$ (attached flow) resulted in similar PDFs for c_f and St. In contrast, the condition $u < 0$ (reversed-flow region) produced significantly different PDFs for c_f and St [45,46]. Similarly, both the shape and trend of the jPDFs between c_f and $StPr^{2/3}$ (the components of $\mathcal{R}_{v\alpha}$) on the windward side were very similar to a smooth-wall case; in contrast, the trough regions resulted in very different-looking jPDFs for all the cases examined [45]. These results were consistent over a wide range of roughness heights.

Another qualitative observation of [44] was quantitatively confirmed by the PDFs: higher values of St are associated with the larger positive slopes on the windward side, while lower St values are associated with the negative slopes where the flow is reversed in the troughs, which act as local resistance [40,46]. That means that taller roughness elements are likely to have higher-than-average St on the windward side, and lower-than-average St on the leeward side [30,46,56]. This indicates that describing the rough wall as a "contorted smooth wall" with increased effective wall area (compared with the plan area) is only reasonable in regions where Reynolds analogy-type behaviour exists.

Comparing δ_ν and δ_α on the rough surface to the equivalent smooth-wall value is a measure of the local augmentation or attenuation of the scalar wall flux [45]. The deviation of the rough-wall values from the smooth-wall ones can be integrated over the entire wall to estimate the bulk effect of the roughness.

On the windward side of the roughness elements, where the flow is attached and exposed to high shear, a smooth-like boundary layer is observed, thickest at the crest [35,45,64]. Here, both δ_ν and δ_α are smaller than the equivalent smooth-wall values, and δ_ν is attenuated more significantly than δ_α at $Pr = 1.0$ [45]. As k_s^+ increases, however, both δ_ν and δ_α approach their respective smooth-wall values, albeit at a decreasing rate. This trend appears to become insensitive to Pr as k_s^+ increases; the slowdown of the growth rate is much more drastic for δ_α than for δ_ν. This implies a local Reynolds analogy-type behaviour at higher k_s^+, and an increase in St relative to the smooth-wall case on the windward side. Furthermore, for a wavy rough wall, [45] suggested the following correlations for c_f and St at the crests:

$$c_{f\,\text{crest}} = 2\frac{u_{\delta_\nu}}{U_k}\frac{1}{U_k^+\delta_\nu^+} \approx 0.46\left(k^+\right)^{-0.5} \tag{13}$$

$$St_{\text{crest}} = \frac{\theta_{\delta_\alpha}}{\Theta_k}\frac{1}{U_k^+\delta_\alpha^+} \approx 0.16\left(k^+\right)^{-0.42}Pr^{-0.72} \tag{14}$$

where u_{δ_ν} and θ_{δ_α} are the velocity and scalar at the edge of the viscous or conductive sublayers, and $U_k = \langle\overline{u}\rangle_{y=y_{\text{crest}}}$ and $\Theta_k = \langle\overline{\theta}\rangle_{y=y_{\text{crest}}}$ are the averaged crest velocity and scalar, respectively. In this context, the roughness height, k, is the wavy-wall amplitude, related to the equivalent sand-grain size by $k_s^+ = 2.7k^+$.

Numerical simulations also help understand the effect of Pr on St in the different regions. In general, $St \sim Pr^{-m}$, where for a smooth wall m = 0.48–0.5 [3,7,67]. On the windward side and crests, $m \geq 0.5$, while on the leeward side and in recirculation bubbles, $m \leq 0.5$. In the wavy-wall example mentioned earlier, $St \sim Pr^{-0.38}$ in the troughs, while $St \sim Pr^{-0.72}$ at the crests [45]. Rowin et al. [57] further refined this observation, noting that, when the windward slope is partially sheltered, $St \sim Pr^{-2/3}$ (corresponding to $RA = 1$) on the exposed side of the slope, while $St \sim Pr^{-0.45}$ on the sheltered side. They used correlations similar to Equations (13) and (14) to develop a model predicting the scalar transport for rough walls, accounting for the "sheltered" and "exposed" regions of the roughness. The model gives good results for the cases tested (3D sinusoidal roughness with high frontal solidity, and Pr = 0.5–2.0). Further improvements of the model for lower solidity are, however, desirable.

The local contributions can be integrated over the roughness surface to determine the net effect of roughness on the bulk scalar, which is a more relevant measure for practical engineering applications. The increased contribution of the windward side overcomes the resistance due to the recirculating regions, resulting in a net increase in the scalar transport compared with the smooth-wall case, that is, $St > St_0$ [9,35,44,46].

The smooth-wall Stanton number and skin-friction coefficients can be obtained from correlations of the experimental data [3,76]:

$$St_0 = 0.021Re^{-1/5}Pr^{-1/2}; \quad c_{f0} = 0.073Re^{-1/4}. \tag{15}$$

Both quantities decrease with Re at different rates. In the rough-wall case, St increases with equivalent sand-grain size k_s^+, reaching a peak in the transitionally rough regime, followed by a monotonic decrease in the fully rough regime [35,39,44,46,48,51]. In contrast, the bulk c_f approaches an asymptotic value in the fully rough regime [17].

Several numerical studies [40,44,45,51,56,57,60,77] examined various types of roughness, systematically changing some topological parameters while maintaining the others fixed. They included variation of the frontal solidity, spacing, density, orientation, slope, etc. They showed that St is both sensitive to the type of roughness and to its topology. The rate of decrease for St seems to be specific to the type of roughness and details of the topology [5,51,52] and, perhaps, also to the value of Pr.

This corresponds to the scalar roughness function behaviour discussed in Section 3.2. Some authors (e.g., [5,30]) reported that St decreases at a slower rate than St_0, indicating that the St increase by roughness becomes larger with Re (although at a decreasing rate), while others [45,58] reported the opposite. This suggests that, at high enough k_s^+, St may be smaller than in the corresponding smooth-wall case at the same values of Re and Pr. It would be worthwhile to explore this issue in a systematic manner.

The frontal solidity, λ_f, is a key parameter associated with impingement and recirculation. The effective slope, ES, plays the same role as λ_f, as the two are directly related by $ES = 2\lambda_f$ [78]. Therefore, low values of λ_f correspond to sparsely packed shallow roughness, while high values of λ_f correspond to densely packed steep roughness [16]. Depending on the roughness geometry, a peak of St can be seen in $0.1 < \lambda_f < 0.3$, usually corresponding to maximum c_f [39,45,51].

Forooghi et al. [51] used distribution of truncated cones, and varied independently the height, effective diameter (i.e., the diameter of a cylinder with the same frontal area), and the spacing between the roughness elements. They found that as the spacing decreases (while maintaining the other parameters, i.e., the roughness becomes more "dense") St reaches a maximum and then decreases. It was suggested that this behaviour occurs because more of the upwind portions becomes sheltered (partially or fully), which limits the amount of unmixed fluid that could impinge on the surface [45,51]. In the limit of zero spacing (or infinite density), a smooth-wall equivalent is expected with $St \rightarrow St_0$, consistent with this trend. In contrast, increasing in the slope by scaling the geometry while maintaining the elevation resulted in a monotonic increase in St.

As discussed, c_f and St behave differently at increasing Re and for different roughness geometries, and, therefore, RA depends on the morphological parameters for both regular and irregular roughness [5,44,51]; this is a result of the large differences between momentum and scalar in the recirculation region downstream of roughness elements.

An expression for RA can be directly derived from a semi-empirical expression for the scalar roughness function proposed by Kays and Crawford [3]. Kuwata [60] modified this expression based on the phenomenological behaviour of $\Delta\Theta^+$ discussed in Section 3.2 and on their DNS data; their expression predicted all the results within a 4.2% error. Its universality, however, was not established and the authors suggest the possibility of adding a coefficient dependent on roughness type.

Figure 5 compares c_f/c_{f0} and St/St_0. In most cases, the skin friction increases faster than the Stanton number, implying $RA < 1$; in Ref. [41], however, the opposite behaviour was observed. It should be pointed out that in this paper high-aspect-ratio streamwise ribs were used, whose behaviour differs significantly from that of distributed roughness because of the rollers generated by the Kelvin–Helmoltz instability, which suppresses sweeps and ejections, resulting in lower Reynolds stress [41,77]. A similar effect can be achieved using blowing or suction at the wall [77]. It is important to note that $RA > 1$ can only be attained for a limited range of Re; once the rib spacing, s^+, becomes too large the increased drag overcomes the augmentation of scalar transfer, leading to $RA < 1$; this is true for various types of longitudinal ribs and riblets [41,77] where the range of s^+ at which $RA > 1$ depends on the geometry.

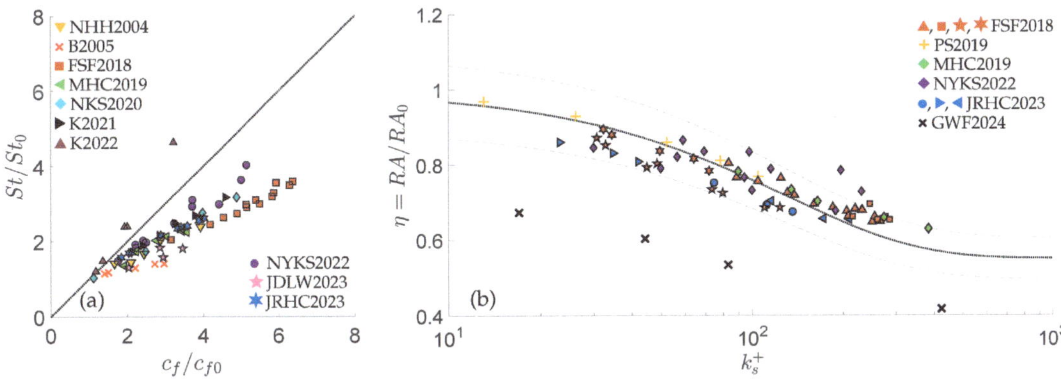

Figure 5. (**a**) St/St_0 vs. c_f/c_{f0} for various types of roughness. NHH2004: Nagano et al. [37] ▼; B2005: Bons [72] ×; FSF2018: Forooghi et al. [51] ■; MHC2019: MacDonald et al. [44] ◀; NKS2020: Nishiyama et al. [79] ◆; K2021: Kuwata [52] ▶; NYKS2022: Nagura et al. [56] ●; JDLW2023: Ji et al. [40] ★; JRHC2023: Jelly et al. [5] ✱; Black dotted line indicates $St/St_0 = c_f/c_{f0}$. K2022: Kuwata [41] ▲—reference case with longitudinal ribs. (**b**) Variation of $\eta = RA/RA_0$ with the equivalent sand-grain roughness k_s^+. FSF2018: Forooghi et al. [51] ▲ truncated cones, ■ cylinders, ★ IC piston surface, ✱ turbine blades; PS2019: Peeters and Sandham [46] +; MHC2019: MacDonald et al. [44] ◆; NYKS2022: Nagura et al. [56] ◆; JRHC2023: Jelly et al. [5] ● Isotropic, ▶ streamwise bars, ◀ spanwise bars; GWF2024: Garg et al. [54] ×; Black dotted line is Equation (18) with the grey dashed lines indicating a 10% error interval.

It is also possible to examine how RA compares to the equivalent smooth-wall value RA_0 as function of k_s^+. The rough-to-smooth ratio of the Reynolds analogy factor, η,

$$\eta = \frac{RA}{RA_0} = \frac{St}{St_0}\frac{c_{f0}}{c_f}. \tag{16}$$

is an efficiency parameter commonly known as the "aero-thermal efficiency factor". We will use the same terminology here, understanding it is applicable to the various scalar quantities. In most cases, $\eta < 1$ (i.e., $\Delta U^+ > \Delta\Theta^+$); riblets or other high-aspect-ratio streamwise longitudinal ribs have a different behaviour [41,77]. Using the discrete element approach (in which the flow is spatially averaged over the roughness elements, modifying the mean flow equations to account for blockage effects, as well as the drag and scalar flux), Aupoix [80] suggested:

$$\eta = \frac{1 - \Delta U^+ \sqrt{c_{f0}/2}}{1 - RA_0 \Delta\Theta^+ \sqrt{c_{f0}/2}} \tag{17}$$

which was found to give good predictions for various types of roughness [46,51].

Forooghi et al. [51] observed that η is a convex function of the roughness density and the frontal solidity, λ_f; predicting the λ_f value of the minimum for different types of roughness is still an open question. It was also shown [5,51,56] that η decreases both with an increase in k_s^+ or in the effective slope, ES. This decrease eventually saturates, Figure 5b. Forooghi et al. [51] derived an expression for η based on a systematic DNS study, using various types of roughness for $Pr = 0.7$:

$$\eta = 0.55 + 0.45e^{-k_s^+/130} \tag{18}$$

similar to the proposal by Bons [72], but using k_s^+ instead of the outer-scaled k_s. This is meant to be a simplified single-parameter correlation rather than a general expression that accounts for all possible effects; nevertheless, although only requiring the knowledge of k_s^+, Equation (18) predicts all the roughness morphologies considered by Forooghi et al. [51] (with a single exception) within a 10% error, as well as the results of Jelly et al. [5], Nagura et al. [56], Aupoix [80]. Interestingly enough, Equation (18) also predicts well the data of Peeters and Sandham [46] (grit-blasted surfaces) and Kuwata and Kawaguchi [33] (irregular roughness) for $Pr = 1.0$. One exception is the additive-manufacturing (AM) roughness of [54], which follows the same trend but falls below all the other results. This may indicate some significant difference in the properties of AM compared with other types of roughness.

3.4. Dispersive Flux

The roughness geometry generates a stationary inhomogeneity in the velocity and scalar fields, referred to as a *wake* or *dispersive field*. The wake velocity, $\widetilde{u}$, and scalar, $\widetilde{\theta}$, fields are compared in Figure 6 for $Pr = 1$, showing overall similarity throughout the RSL. A difference can be observed in the troughs between roughness elements, where $\widetilde{u}$ and $\widetilde{\theta}$ tend to have opposite signs [31,58], the result of the accumulation of scalar near the wall due to the lack of a pressure-like mechanism, discussed in Section 3.1. For $Pr \geq 1.0$, it is expected that the $\widetilde{\theta} > 0$ regions would be more extensive than the $\widetilde{u} > 0$ regions.

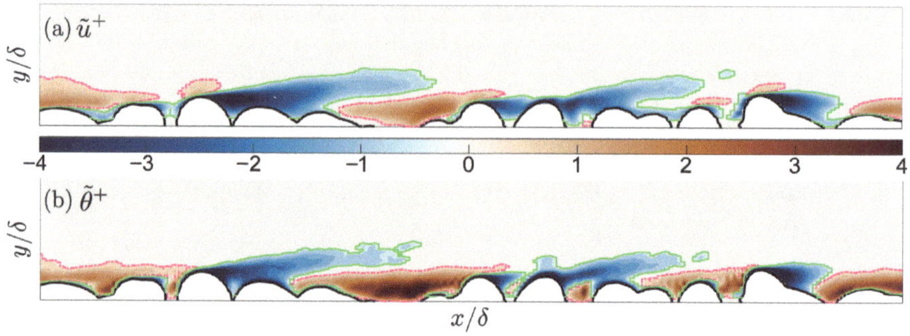

Figure 6. Dispersive component of the streamwise velocity, $\widetilde{u}$ (**a**) and scalar, $\widetilde{\theta}$, with $Pr = 1.0$ (**b**), normalised in wall units. The vertical slice has dimensions $1.4\delta \times 0.2\delta$. Colour denotes the magnitude of each component. Isolines denote the $\widetilde{u} = 0$ or $\widetilde{\theta} = 0$ contours.

The normal dispersive stresses, $\widetilde{u}_k\widetilde{u}_k$, tend to be largest at the crest of the elements. Their streamwise component, $\widetilde{u}\widetilde{u}$, is the largest in magnitude, and has been found to be insensitive to k_s^+ [58,60]. A secondary peak of high $\widetilde{u}_k\widetilde{u}_k$ can be found around the troughs, usually at the edge of the recirculation bubble. These locations generally correspond to the points where the stochastic normal Reynolds stresses, $\langle \overline{u_k' u'k} \rangle$, are smaller [32,58].

Contours of the shear components that contribute to the momentum and scalar transport, $\langle \widetilde{u}\widetilde{v} \rangle$ and $\left\langle \widetilde{\theta}\widetilde{v} \right\rangle$, from Zhang et al. [58], are shown in Figure 7. The wall-normal wake velocity, $\widetilde{v}$, is particularly large in magnitude around the crests and close to the trough, and is directed towards the boundary [58,60,81]. The difference between the dispersive stresses and fluxes is due to the differences in size and magnitude between $\widetilde{\theta}$ and $\widetilde{u}$, as well as their correlation with locations of intense $\widetilde{v}$ (note that in a channel $\langle \overline{v} \rangle = \langle \overline{w} \rangle = 0$, so that $\widetilde{v} = \overline{v}$ and $\widetilde{w} = \overline{w}$). Both $\widetilde{u}\widetilde{v}$ and $\widetilde{\theta}\widetilde{v}$ are large and negative around the crest but positive on the leeward slope, where $\widetilde{\theta}\widetilde{v}$ is more dominant. In the secondary peak near the trough, $\widetilde{u}\widetilde{v}$ is positive, while $\widetilde{\theta}\widetilde{v}$ is negative, and peaks slightly further downstream, close to the reattachment point [58]. This is due to the positive region of $\widetilde{\theta}$ extending towards the recirculation region and into regions of negative $\widetilde{v}$, as discussed before.

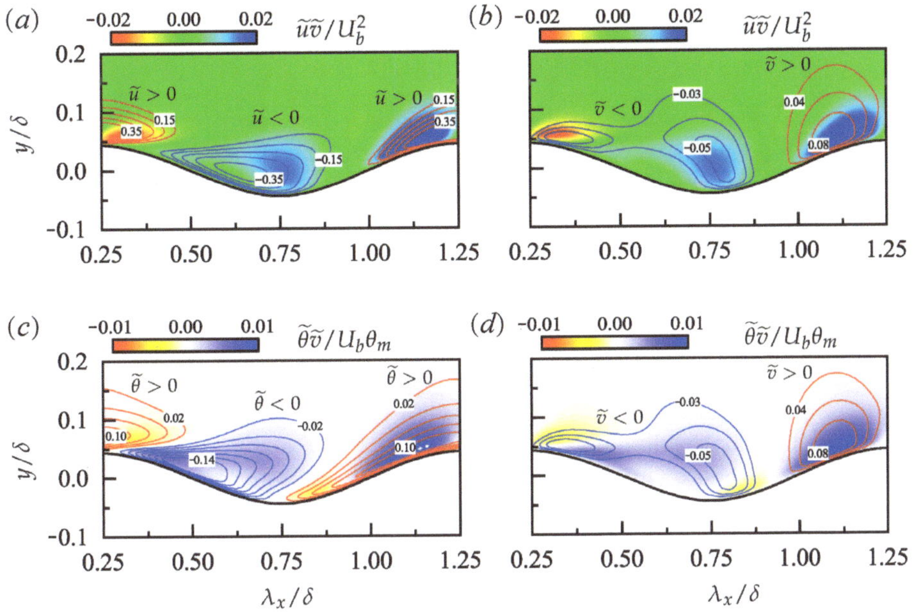

Figure 7. Dispersive stresses (**a**,**b**) and fluxes for $Pr = 0.7$ (**c**,**d**) shown on vertical XY slices over a sinusoidal surface. Isolines denote values of $\widetilde{u}$ (**a**), $\widetilde{\theta}$ (**c**), or $\widetilde{v}$ (**b**,**d**). λ_x is the sinusoidal wavelength. Reprinted with permission from Zhang et al. [58].

Zhang et al. [58] noted that the maximum value of the various dispersive quantities, in outer units, increased linearly as the slope became steeper. The dispersive normal stresses, shear stress, and vertical scalar dispersive flux all showed similar trends, while the streamwise dispersive scalar flux showed a more moderate increase.

From an average perspective, the streamwise component is the the dominant one for both dispersive fluxes and stresses [60]. The flux component, $\left\langle \widetilde{\theta}\widetilde{v} \right\rangle_s$, was reported to vary between 25% [32] and 50% [82] of the total scalar flux inside the RSL. At sufficiently large k_s^+, the mean dispersive stresses and fluxes become independent of k_s^+ and Re [32,40,52,60]; a similar trend was found for $\left\langle \widetilde{\theta}\widetilde{u}_j \right\rangle_s$ with increasing Pr [32]. Both the dispersive stresses and fluxes remain dependent on the roughness topological details.

Overall, the dispersive fluxes are not well studied and may be a very useful avenue of numerical exploration, since they are very difficult to measure experimentally. They seem to be less sensitive to k_s^+ [60] and even reach saturation [32,58,60]. They are also important for the production of turbulent fluxes [31,32] and in the context of modelling, especially if the double-averaged NS equations (DANS) are used [80,82,83].

3.5. Form-Induced Scalar Flux

A key difference between the transport of momentum and scalar, in the context of roughness, is the mechanism by which the flux across the solid boundary takes place. The transfer of momentum becomes progressively dominated by the pressure drag, F_p, as k_s^+ increases, while the viscous drag, F_v, decreases. The scalar wall transfer, however, is only due to diffusive mechanisms [3,9,35], and is denoted here as"form-induced (FI) transfer", F_α (sometimes referred to as "wall-interaction transfer term" [52,56,60]). The spatial distributions of F_v, F_p, and F_α are difficult to obtain experimentally but can be easily calculated numerically [23,62,84].

The FI scalar transfer, F_α, is the spatial integral of the local FI flux f_α, found on the RHS of Equation (9):

$$f_\alpha = \left\langle \frac{\partial}{\partial x_j}\left(\alpha \frac{\partial \widetilde{\theta}}{\partial x_j}\right)\right\rangle_s = -\frac{1}{A_0} \oint_{\partial S_f} \alpha \frac{\partial \widetilde{\theta}}{\partial x_k}(n_k dl)$$

and the viscous drag, F_ν, and pressure drag, F_p, are the integrals of

$$f_\nu = \left\langle \frac{\partial}{\partial x_j}\left(\nu \frac{\partial \widetilde{u}_i}{\partial x_j}\right)\right\rangle_s = -\frac{1}{A_0} \oint_{\partial S_f} \nu \frac{\partial \widetilde{u}_i}{\partial x_k}(n_k dl)$$

$$f_p = -\left\langle \frac{\partial \widetilde{p}}{\partial x_i}\right\rangle_s = -\frac{1}{A_0} \oint_{\partial S_f} \widetilde{p}(n_k dl)$$

respectively, found on the RHS of Equation (8). The integration domain, ∂S_f, is the part of the roughness surface that intersects the averaging domain of size A_0 (i.e., the perimeter of the roughness surface intersecting a horizontal plane of size A_0 for a channel) and n_k is the k-th component of the outward-pointing unit vector normal to the roughness surface.

As the bottom of the RSL is approached, F_p and F_ν dominate the mean momentum transfer (with F_p being more significant at increasing k_s^+), while F_α dominates the mean scalar transfer [31–33].

The total FI momentum flux, $f_\nu^+ + f_p^+$, is generally larger than the scalar FI one, f_α^+, [32,33], which results in the total drag ($F_\nu^+ + F_p^+$) being larger than the scalar wall transfer (F_α^+). Kuwata and Kawaguchi [33] noted that, in the region closest to the bottom of the roughness, the opposite ($f_\nu^+ + f_p^+ < f_\alpha^+$) can occur locally due to the negative local shear stress.

Kuwata [52] showed that the total momentum FI, $f_\nu^+ + f_p^+$, and scalar FI flux, f_α^+, have similar behaviours when scaled in inner units and plotted against non-dimensionalised wall distance. Furthermore, the increase in the effective slope, ES, shifts the main contribution of f_α^+ away from the base of the roughness. A possible explanation for this is the increase in surface area away from the base, as well as the presence of the well-mixed fluid between the roughness elements (see Section 3.1), referred to as an "insulation layer", near the base of the roughness. Consistent with the discussion of the thinning of the diffusive layer, which tends to conform to the roughness shape when either Re or Pr increase, F_α^+ becomes proportional to the solid fraction [31,32], as shown in Figure 8.

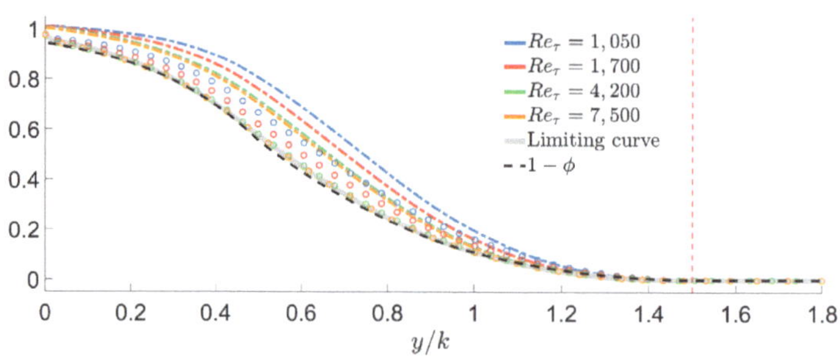

Figure 8. Comparison of the FI momentum and scalar fluxes for a range of Re_τ with roughness height $k/\delta = 0.04$. Dash–dot lines (-·-·) indicate the total drag ($F_p^+ + F_\nu^+$); circle markers (○) indicate the scalar wall transfer F_α^+ for $Pr = 1.0$. Figure reprinted with permission from Hantsis [32].

3.6. Scalar Fluctuations and Turbulent Structures

Roughness has a profound impact on both velocity and scalar fluctuations, and numerical simulations are able to access them even below the roughness crest. Several studies compared stochastic stresses and scalar correlations inside the RSL, contributing to a better understanding of the dynamics of momentum and scalar transport.

Roughness breaks up both velocity and scalar streaks, resulting in more isotropic turbulence, evidenced by both stochastic stresses, $\left\langle \overline{u_i' u_j'} \right\rangle_s$, and scalar correlations, $\left\langle \overline{u_i' \theta'} \right\rangle_s$. Numerical studies show that both the magnitude and peak location of $\left\langle \overline{\theta' \theta'} \right\rangle_s$ are affected: for $Pr \geq 1$, the peak is lower in the rough-wall case [32,46,52], while for lower values of Pr it increases [32]. Roughness attenuates the mean streamwise velocity fluctuations (normalised in wall units), $\left\langle \overline{u' u'} \right\rangle_s$, more than the scalar ones, $\left\langle \overline{\theta' \theta'} \right\rangle_s$, for Pr near unity [29,31,46,52]. It could be expected that the mean wall-normal stochastic flux, $\left\langle \overline{\theta' v'} \right\rangle_s$, in wall units, is less affected by roughness than $\left\langle \overline{u' v'} \right\rangle_s$. For $Pr = 1.0$, however, they are similar at low-to-moderate k_s^+. However, the differences between the two inside the RSL increase with roughness size [34,46,60]; outside the RSL, $\left\langle \overline{\theta' v'} \right\rangle_s$ obey Townsend's outer-layer similarity [25,32].

Various studies [29,31,32,39,46,52,60] examined the time-averaged field, $\overline{\theta' v'}$. As mentioned in Section 3.1, the main differences between the scalar and velocity fields occur in recirculation regions behind roughness elements and in the sheltered portion of the windward slopes. This is also evident for the turbulent scalar fluxes, Figure 9, where the time-averaged wall-normal Reynolds stress, $-\overline{u'v'}^+$, and scalar flux, $-\overline{\theta'v'}^+$, are shown in a vertical plane, for $k_s^+ \approx 200$ and $Pr = 1.0$. For the most part, the two fields are visually similar, which is expected due to the fact that both are driven by the same turbulent eddies and associated with v'. There are noticeable differences in the recirculation regions downstream of larger roughness elements, where the scalar flux is more intense than the stress (both in positive and negative values). Quasi-streamwise elongated vortices develop in narrow gaps in the roughness, which strengthen both the turbulent stress and flux close to the rough surface [60]. For $Pr > 1$, this enhances the scalar turbulent flux more than the stress.

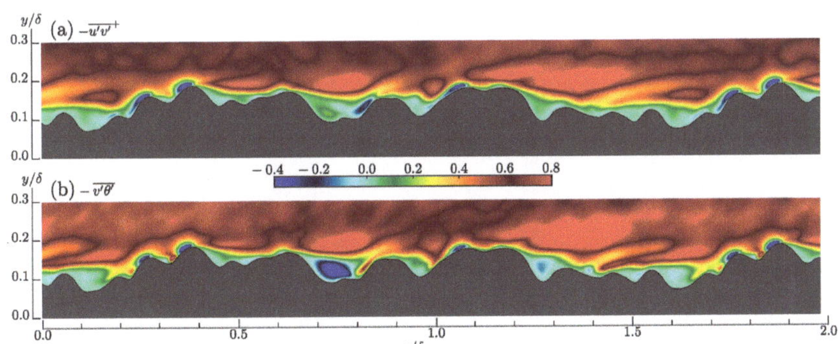

Figure 9. Contours of (**a**) time-averaged Reynolds stress, $-\overline{u'v'}^+$, and (**b**) scalar flux, $-\overline{\theta'v'}^+$, with $Pr = 1.0$ in an xy-plane. The dimensions of the region considered are $2\delta \times 0.3\delta$ and $k_s^+ \approx 200$. All quantities are scaled in inner units. Reprinted with permission from Kuwata [52].

Quadrant analysis (QA) [85] can be used to examine the instantaneous behaviour through the probabilities of turbulent events, as well as their relative contribution to the Reynolds stresses and scalar fluxes. The quadrants are defined as Q1: $u' > 0$, $v' > 0$; Q2: $u' < 0$, $v' > 0$; Q3: $u' < 0$ $v' < 0$; and Q4: $u' > 0$, $v' < 0$. Q2 and Q4 events, in flat-plate boundary layers and plane channels, correspond to ejections (slow fluid moving away from the wall) and sweeps (fast fluid moving towards the wall), respectively. They are the main contributors to the Reynolds shear stress, $\overline{u'v'}$. Replacing u' with θ' provides an additional way to examine the relationship between the flow topology and the scalar fluxes. Note that in the classical quadrant analysis the fluctuations normal to the wall are used; here, instead, u' and v' are fluctuations parallel and normal to the surface at the bottom of the roughness, and not to the roughness surface. Thus, Q2 events represent fluid going away from the roughness layer towards the outer flow, and, conversely, Q4 events represent outer-layer fluid advected into the troughs.

Figure 10 shows the distribution of events, sorted by quadrants, in the near-wall region. Above the roughness crest, the smooth-wall distribution is found both in the $u' - v'$ and $\theta' - v'$ planes, with Q2 and Q4 events dominating. Inside the RSL, on the other hand, the probability of the four quadrants becomes roughly equal for the stress, tending to approximately 25% as the base of the roughness is approached. For the scalar flux (at $Pr = 1.0$), however, the probability of Q4 motions, where $\theta' > 0$ & $v' < 0$, is meaningfully diminished, while the probability of Q3 events, where $\theta' < 0$ & $v' > 0$, is significantly enhanced (Figure 10b) [39,46]. The difference in event probability, however, does not translate into a contribution to the mean flux and stress [46,82]. Examining the relative contributions to the mean (double-averaged) scalar flux, $\left\langle\overline{\theta' v'}\right\rangle_i / \left\langle\overline{\theta' v'}\right\rangle$ shows that the relative contribution of Q4 events remains larger than that of Q2 events (Figure 10d, similar to the counterpart quadrants $\left\langle\overline{u'v'}\right\rangle_i$. This implies that fewer Q4 motions occur (e.g., motions of hot fluid towards the troughs) but they are generally higher in intensity compared with the smooth-wall case [29,39,46]. These results are consistent with the observations of Doosttalab et al. [29] that roughness enhances the transport of both $\overline{v'u'v'}^+$ and $\overline{v'\theta' v'}^+$ towards the troughs, as well as the increasing the skewness of v' and θ', mentioned earlier.

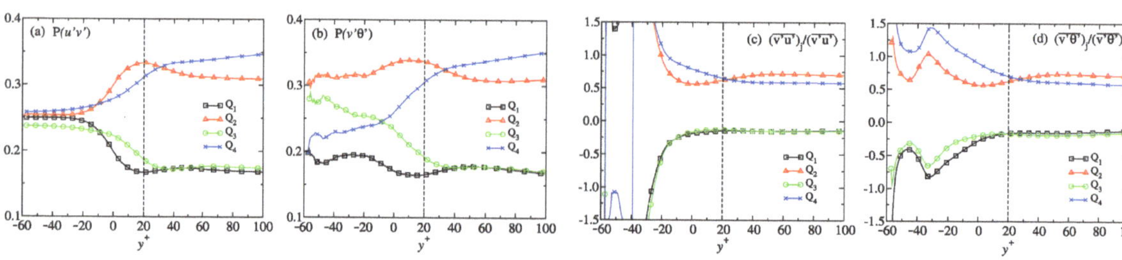

Figure 10. Quadrant analysis for the Reynolds stress, $u'v'^+$ (**a**,**c**) and scalar flux, $\theta' v'^+$ (**b**,**d**) at $k_s^+ \approx 52$. (**a**,**b**) Probabilities of the various events; (**c**,**d**) relative contributions to the Reynolds stress and scalar flux, respectively. Vertical dashed line indicates $k/\delta \approx 1/3$. Reprinted with permission from Peeters and Sandham [46].

Despite the conclusion that the relative contributions of Q1–Q4 to the mean are similar between momentum and scalar, there is meaningful dissimilarity in the type of motions experienced by the momentum and scalar, implying a breakdown of analogy locally between the two in the near-wall region [46], which corresponds to $y/k < 1/3$ in the case presented in Figure 10.

3.7. Scalar-Fluctuation Budget

Roughness was shown to affect the scalar fluctuations and the TKE differently [31,32,46,60]. These differences are reflected in the budgets for the double-averaged scalar fluctuation variance $\mathcal{K}_\theta \equiv \left\langle\overline{\theta'\theta'}\right\rangle_s /2$ and TKE. Figure 11 shows the various budget terms in the fully rough regime for $Pr = 1$. A key reason for the difference between velocity and scalar fluctuations lies in the production of the TKE and scalar variance near the roughness crests. In addition to the shear production, $\mathcal{P}^s$, which also exists in a smooth-wall case, a second "bypass" production mechanism exists, which generates velocity and scalar fluctuations at a scale comparable to the roughness height [23,31,62,63]. This form-induced (FI) production, $\mathcal{P}^{fi}$, stems from the gradients of the wake fields of the velocity and scalar:

$$\mathcal{P}_{ij}^{fi} = -\left\langle \overline{u_i' u_m'} \frac{\partial \widetilde{u}_j}{\partial x_m} + \overline{u_j' u_m'} \frac{\partial \widetilde{u}_i}{\partial x_m} \right\rangle_s \tag{19}$$

$$\mathcal{P}_{\theta}^{fi} = -2\left\langle \overline{\theta u_m'} \frac{\partial \widetilde{\theta}}{\partial x_m} \right\rangle_s . \tag{20}$$

The total production is $\mathcal{P} = \mathcal{P}^s + \mathcal{P}^{fi}$.

If the wall is rough, the total production (in wall units) of both velocity and scalar fluctuations is reduced, compared with the smooth-wall case; this reduction depends on k_s^+ [29,31,32]. The reduction in the velocity fluctuation production is more significant than that of the scalar ones: for transitional roughness, Doosttalab et al. [29] reported a reduction of ∼5% in the production of $\mathcal{K}_\theta$ compared with ∼17% for TKE production, while for a fully rough regime Hantsis and Piomelli [31] observed a reduction closer to 50% and 70% for $\mathcal{K}_\theta$ and TKE, respectively.

While the scalar shear production, $\mathcal{P}_\theta^s$, increases with Pr, it quickly reaches saturation near $Pr \simeq 1$, where the curves for increasing Pr collapse [32]; this is consistent with the observation that the bulk contribution of the shear production is in the upper portion of the roughness due to strong shear layers emanating from the crests, where the motions are generally insensitive to Pr. The shear production term for both TKE and $\mathcal{K}_\theta$ decreases with k_s^+; the two remain similar for $Pr \geq 1$ across the range of k_s^+ [31,32]. It is not known if the shear production continues to decrease with k_s^+ or eventually reaches saturation.

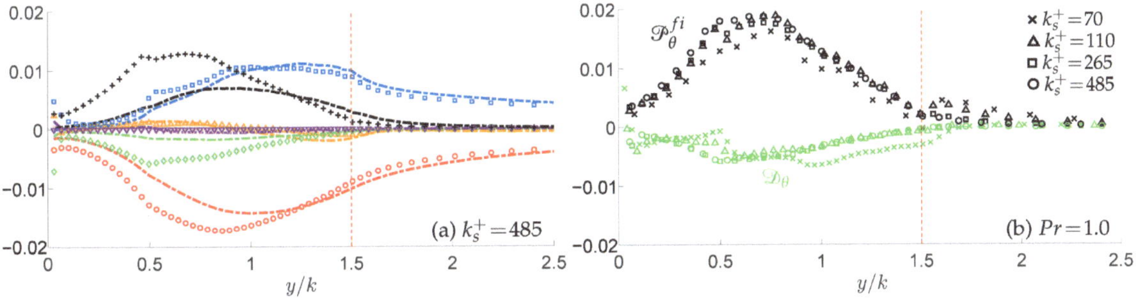

Figure 11. (**a**) Comparison of TKE (lines) and scalar-variance (symbols) budgets at $k_s^+ = 485$; (**b**) scalar-variance FI production and diffusion term for k_s^+ ranging from 70 to 485; $Pr = 1.0$ in all cases. Lines: budget terms of $\mathcal{K} = \frac{1}{2}\left\langle \overline{u_k' u_k'} \right\rangle^+$; symbols: budget terms for $\mathcal{K}_\theta = \frac{1}{2}\left\langle \overline{\theta'\theta'} \right\rangle^+$. In (**a**), the various budget terms are indicated by colour: --□-- : shear production, $\mathcal{P}^s$; --+-- : form-induced production, $\mathcal{P}^{fi}$; --∗-- : convective transport, $\mathcal{A}$; --Δ-- : turbulent transport, $\mathcal{T}$; --◇-- : diffusion, $\mathcal{D}$; --○-- : dissipation, $-\varepsilon$; --▽-- : pressure work, Π_{kk}, for TKE or source term, Q_θ, for scalar. In (**b**), different symbols indicate the equivalent sand-grain height: × $k_s^+=70$; Δ $k_s^+=110$; □ $k_s^+=265$; ○ $k_s^+=485$. All terms are in wall units. The vertical line identifies the top of the roughness. Figure adapted from Hantsis [32].

In contrast, the scalar FI production, $\mathcal{P}_\theta^{fi}$, becomes increasingly dominant inside the RSL as k_s^+ increases, compared with the TKE counterpart, $\mathcal{P}_{kk}^{fi}$; the magnitude of $\mathcal{P}_\theta^{fi}$ increased with Pr and does not seem to saturate [31,32]. Furthermore, for a given Pr, the $\mathcal{P}_\theta^{fi}$ curves collapse as k_s^+ increases [32], as seen in Figure 11b, which corresponds to the geometry-conforming behaviour of the scalar field discussed in Section 3.1. Thus, it is the FI portion that is responsible for the increased scalar fluctuations' production. The bulk contribution of $\mathcal{P}_\theta^{fi}$ occurs in the lower portion of the roughness, and is small around the crest. As k_s^+ increases, the scalar variance budget becomes dominated by the FI production,

$\mathcal{P}_{\theta}^{fi'}$, diffusion, and dissipation; both diffusion and FI production tend to saturate with increasing k_s^+ [32]. A saturation in the shear production then implies saturation in the dissipation as well, resulting in a budget that is insensitive to k_s^+. A more systematic study is required to ascertain these trends at higher k_s^+, for a wider range of Pr and various types of roughness. Asymptotic trends such as these are useful, in particular in the context of modelling.

3.8. Turbulent Prandtl Number

The turbulent Prandtl number is the ratio between the eddy viscosity and the scalar eddy diffusivity, $Pr_T = \nu_T/\alpha_T$. It is used to model scalar transport [3] in analogy to momentum transport. A direct consequence of the Reynolds analogy is that the turbulent Prandtl number, Pr_T, needs to be constant and close to unity for $Pr \approx 1.0$. Although many experimental studies examined Pr_T, to properly calculate it, an accurate and simultaneous measurement of at least four quantities (stochastic stress and flux, mean velocity, and scalar gradients) is required, at multiple locations—which is extremely difficult experimentally [3,86]. This difficulty resulted in a relatively large scatter between experimental results [3,29]; numerical simulations can compute these quantities more accurately.

While Pr_t is generally obtained, in thin shear-layer flows, from the only non-zero Reynolds shear stress (or scalar flux) and velocity (or scalar) gradient, a more general form is obtained through a contraction with the strain rate or scalar gradient [87]:

$$Pr_T = \frac{\nu_T}{\alpha_T} = \frac{\left\langle \overline{u_i'u_j'} \right\rangle \langle \bar{S}_{ij} \rangle}{2\langle \bar{S}_{lm} \rangle \langle \bar{S}_{lm} \rangle} \times \frac{\left\langle \frac{\partial \bar{\theta}}{\partial x_k} \right\rangle \left\langle \frac{\partial \bar{\theta}}{\partial x_k} \right\rangle}{\left\langle \overline{u_n'\theta'} \right\rangle \left\langle \frac{\partial \bar{\theta}}{\partial x_n} \right\rangle}. \tag{21}$$

Pr_T decreases with wall-normal distance, starting from a value near one at the wall (e.g., $Pr_T = 1.1$ for $Pr = 0.7$ [88]) to $0.6 - 0.8$ away from the wall [89]. It tends to a roughly constant value in the logarithmic layer, with reported values falling within the range of 0.8–1.0, where the most common values are around 0.85–0.9 (see [3] for comprehensive discussion). In the context of modelling, however, it is often taken as a constant, usually the logarithmic layer value.

All the terms comprising Pr_T are affected by roughness within the RSL, while away from the wall, corresponding with Townsend's outer-layer similarity hypothesis [25], the various quantities collapse on the smooth-wall curves; the differences in Pr_T between smooth- and rough-wall cases are limited to the RSL, where numerical simulations can provide more information than experimental measurements.

In the presence of roughness, Pr_T is ill-behaved close to the bottom of the roughness, as the denominator becomes small [29,34,46,52]; the magnitude of these changes increases with k_s^+ and seems to depend on the roughness geometry [52,60]. For $Pr = 1$, the rough-wall Pr_T can be much greater than one; for example, [46] reported peak values of 5. A higher peak indicates a decrease in the effective scalar diffusivity, resulting in a decreased scalar flux/transfer (i.e., in thermal resistance). The erratic behaviour near the bottom can be related to the difference in the behaviour of the stochastic stresses and fluxes in the recirculation regions [33,34,46], as is demonstrated in Figure 9. In fact, it may be argued that in the vicinity of the wall, where the Reynolds stress and turbulent scalar flux are nearly damped out, Pr_T does not have physically meaningful values [33,46].

As the roughness crest is approached, a more predictable behaviour is observed, with Pr_T peaking and then decreasing towards the smooth-wall value [29,34,46]. The peak magnitude increases with k_s^+ and seems to depend on the details of the geometry as well [46,52,60]. The location of the peak may depend on the roughness geometry [29,34,46,52].

While Pr_T is a key element for modelling, it may be too sensitive to the details of the roughness to be used effectively. The existence of additional dispersive stresses and fluxes (discussed in Section 3.4) in the mean transport equations, which are neglected in the classical definition Equation (21), motivated the definition of an "effective" Prandtl number [34,52]:

$$Pr_{T,\text{eff}} = \frac{\nu_T + \widetilde{\nu}_T}{\alpha_T + \widetilde{\alpha}_T} = \frac{\left(\left\langle\overline{u_i'u_j'}\right\rangle + \langle\widetilde{u}_i\widetilde{u}_j\rangle\right)\langle\overline{S}_{ij}\rangle}{2\langle\overline{S}_{lm}\rangle\langle\overline{S}_{lm}\rangle} \times \frac{\left\langle\frac{\partial\overline{\theta}}{\partial x_k}\right\rangle\left\langle\frac{\partial\overline{\theta}}{\partial x_k}\right\rangle}{\left(\left\langle\overline{u_i'\theta'}\right\rangle + \left\langle\widetilde{u}_i\widetilde{\theta}\right\rangle\right)\left\langle\frac{\partial\overline{\theta}}{\partial x_i}\right\rangle}. \tag{22}$$

The expectation is that the inclusion of the dispersive stresses and fluxes will make $Pr_{T,\text{eff}}$ better behaved, even where the stochastic stresses and fluxes go to zero near the wall. Furthermore, in the context of modelling, the dispersive stresses and fluxes are required for the closure relations [83].

Hantsis and Piomelli [34] compared $Pr_{T,\text{eff}}$ and Pr_T, and found them to be generally similar in the RSL, collapsing above the roughness crest, where the dispersive stress and fluxes vanish. $Pr_{T,\text{eff}}$ is also ill-behaved near the bottom of the rough surface. For $Pr \simeq 1$, ref. [34] found that $Pr_{T,\text{eff}}$ is larger in magnitude than Pr_T. A peak, larger than that observed in Pr_T, was also observed for $Pr_{T,\text{eff}}$, but occurred closer to the roughness crest. Due to the scarcity of studies considering $Pr_{T,\text{eff}}$, and comparing $Pr_{T,\text{eff}}$ to Pr_T under comparable conditions, it is unclear if this behaviour is particular to this setup or has more general validity.

The dependence on the details of the roughness is illustrated in Figure 12, showing $Pr_{T,\text{eff}}$ for different types of roughness and topologies, for Pr of unity. Four representative roughness surfaces from [52] show the behaviour at different skewness (S_k) and effective-slope (ES) values for an irregular roughness, while the sand-grain roughness of [34] shows the change of roughness type. A common pattern of positive and negative peaks can be noticed as the wall is approached. The magnitudes and locations of the peak values (where the jumps occur), however, differ significantly between the various cases.

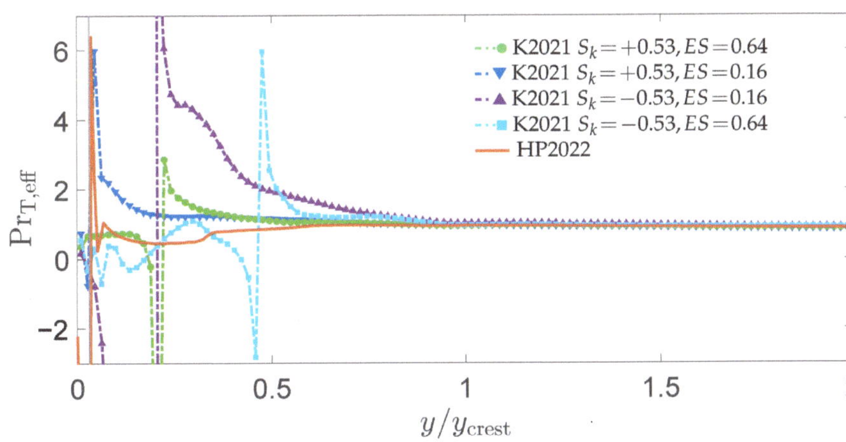

Figure 12. Comparison of the effective turbulent Prandtl number, $Pr_{T,\text{eff}}$, between various roughness types. Data collected from Kuwata [52] (KW2021, irregular roughness); Hantsis and Piomelli [34] (HP2022, sandpaper). Wall distance is normalised by the crest height, y_{crest}; $Pr = 1.0$ for all cases.

Neither Pr_T nor $Pr_{T,\text{eff}}$ seem to be effective modelling tools in generalised rough-wall flows, and an alternative might be required for modelling. In the context of smooth-wall flows, it was found that quantities such as the time-scale ratio, $\mathscr{R}$ (between the scalar variance and the turbulent kinetic energy) are more useful in modelling, and can replace the need for Pr_T Abe and Antonia [90]. While a similar observation was made for a rough-wall case [34], only one case was considered, and further study would help in examining the viability of this alternative.

4. Open Issues

4.1. Boundary Conditions

The mixed-boundary condition proposed by Kasagi et al. [61] is commonly applied for the uniform scalar isoflux condition, when periodic domains are used. Kasagi et al. [61] derived this condition for a smooth-wall case, where, for a limited range of Pr around unity, the fluctuations at the wall were negligible, and the wall could be assumed to be locally isothermal [91]. As was discussed in Section 3.6, roughness tends to reduce scalar fluctuations near the wall, which suggests the assumption is still valid for a rough-wall case; indeed, many studies implemented this approach [5,30–32,34,44–46,51,54,55,57,59,60,92].

Two key issues arise from this setup: (1) the use of this condition is limited to a narrow range of Pr, and (2) a proper validation of the locally isothermal assumption was not conducted in the context of rough-wall flows (see [93] for a smooth-wall case comparison). Thus, a boundary condition that allows for fluctuations at the wall is desirable.

Another issue arises from the use of the mean wall-scalar, T_w, in the case of a wall flux condition: for a smooth wall, T_w has a constant streamwise gradient, equal to that of the bulk scalar, T_b. For the rough-wall case, the spatial inhomogeneity of the roughness implies that the streamwise gradient of T_w depends on the surface area of the roughness [52], and a local spanwise gradient may also exist. Kuwata [52] suggested a correction by considering the local time-averaged scalar budget at the roughness surface and relating T_w to T_b; this yields

$$\frac{dT_b}{dx} = \frac{q_w}{\rho c_p \dot{V}} \ell_{yz} \tag{23}$$

where $\dot{V}$ is the flow rate and ℓ_{yz} is the local roughness length scale. The source term Q then becomes

$$Q = \frac{q_w}{\rho c_p \dot{V}} \left(u\ell_{yz} - \alpha \frac{d\ell_{yz}}{dx} \right) \tag{24}$$

For a sufficiently large roughness surface of a homogeneous and isotropic nature, ℓ_{yz} can be approximated as $\ell_{yz} \approx A_r / L_x$, where A_r is the total roughness wetted area and L_x is the streamwise domain length (assuming roughness covers the entire boundary). The source term Q then can be approximated as

$$Q \approx \frac{A_r}{A} \frac{q_w}{\rho c_p \delta_e} \frac{u}{U_b} \tag{25}$$

where δ_e is the effective half-channel height ($\delta_e = \delta - d$). While this formulation better accounts for the local wall flux, there has not been a comparative study examining the significance of this correction.

4.2. Limited Coverage of the Parameter Space

The study of flow over rough surfaces is a complex problem with a large parameter space [16]. As discussed in earlier sections, the parameter space associated with scalar transport is larger than that of momentum transport, at the very least by the addition of a Prandtl number (or equivalent). The need to resolve all (for DNS) or most (for LES) of the scales of motion becomes prohibitive as the Reynolds number increases. For $Pr > 1$, the resolution requirements for the scalar are stricter than those for momentum, making the calculations even more expensive.

The presence of roughness increases the parameter space further. Jiménez [28] recommends $\delta/k > 40$ and $k^+ > 100$ to achieve the fully rough regime with negligible blockage effects. Most present studies relax one or both of these requirements because of the costs associated with them.

Some studies [32,45] indicated the possibility that, with proper scaling at sufficiently high k_s^+, various scalar statistics may become independent of Pr. A much wider range of Pr is required to validate this conjecture, which could have repercussions on modelling of these problems.

The roughness geometry is arguably the parameter space that is most difficult to cover. Beyond covering an immense spectrum of roughness types (i.e., regular, irregular, k-type, d-type, etc.), it also includes different morphological characteristics of those roughness types. To examine if any of the findings and trends have some level of universality, a wider portion of this parameter space needs to be covered systematically. In most cases, a single roughness type is considered, with varying the roughness height, usually by increasing *Re*, and usually for a single value of *Pr*.

Since only near-wall phenomena are of interest, the minimal channel approach, proposed by Jiménez and Moin [94], can be used to decrease the computational expense of channel-flow calculations and facilitate sweeping across the parameter space. Chung et al. [95] demonstrated the applicability of the approach to rough-wall flows; MacDonald et al. [44] extended it to also include scalar transport. Limiting the width or length of the domain (see [95,96] for more details) removes the larger motions while retaining many of the dynamically relevant scales within the RSL [97], in particular at higher *Re*. This approach was used successfully to sweep across the topological space of 3D sinusoidal roughness and k_s^+ [45,57,84,95–98].

Relatively few studies ([51,55,80]) systematically cover a range of roughness types. Thus, it is difficult to form conclusions on specific quantities across the parameter space by mapping out results from various studies, and a more systematic approach is desirable. Forooghi et al. [51] performed a systematic study of several roughness types, and several investigations consider variations of a single roughness type [30,44,45,57,92]. The importance of the geometrical detail, however, remains an open question.

Because of the wide parameter space and its associated computational cost, only a limited number of the studies mentioned here reached k_s^+ above 500 (see Table 1 for a non-exhaustive list), while some remained in the transitional regime. Most studies tend to be in the range $k_s^+ = 200 - 400$. While the fully rough regime starts at $k_s^+ \approx 100$ [28], the onset of the scalar "'ultimate regime" seems to depend on both the roughness geometry and Pr. For $Pr = 0.7$, MacDonald et al. [44] suggested the onset occurring beyond $k_s^+ = 250$. Hantsis [32] observed that as Pr the onset of the ultimate regime occurs at higher k_s^+ (see Figure 3a).

As such, much of the discussion around the ultimate regime (e.g., Section 3.2) and the behaviour of various quantities such as $\Delta\Theta^+$ requires a wider range of k_s^+ to determine if the trends and findings extend in their applicability beyond the limited scope of k_s^+ examined by these studies, leaving much room for speculation. This is particularly true for higher Prandtl numbers.

The increase in Prandtl numbers also results in improved scalar transfer for both smooth- and rough-wall cases, which is important for various applications. For example, in the context of nuclear reactor cooling, Gowen and Smith [65] considered a Prandtl number range of $0.7 - 14.3$. Most wall-resolved numerical studies remained in the vicinity of $Pr = 0.7 - 1.0$, with $Pr = 0.7$ being of particular interest. In some cases it was due to the limitation computational cost, while in other cases it was either due to the relevance to practical applications (e.g., $Pr \approx 0.7$ for air) or existing experimental data to compare against, or making the most direct comparison between the velocity and scalar (with $Pr = 1.0$ providing this case). Furthermore, most studies only considered a single Pr, while the few studies that examined a range of Pr were usually limited to a small range around unity, say $Pr = 0.5 - 2.0$ [31,32,45,48]. While the latter can be due to the increased computational costs associated with higher Pr values, it is also (at least in part) due to the limitation on the boundary conditions, as discussed in Section 4.1. Some studies [32,45] indicated the possibility that at sufficiently high k_s^+ various scalar statistics may become independent of Pr when scaled accordingly. A much wider range of Pr is required, in particular $Pr \gg 1.0$, to see if the various findings and trends, identified for the much more limited range of Pr, are valid and applicable at much higher Pr.

4.3. Effect of Blockage

Roughness generates drag and scalar flux, along with blockage of the flow [2]. As mentioned above, Jiménez [28] recommends roughness no higher than $1/40$ of the channel half-height or boundary-layer thickness.

This limitation seems to be less strict in the transitionally rough regime; Thakkar et al. [78] demonstrated that $k/\delta \approx 1/6$ hardly altered the flow in the outer layer by using 17 different irregular surfaces. Even if the outer flow remains mostly unaffected, a large blockage ratio will modify the effective boundary-layer height ($\delta_{\text{eff}} < \delta$), impact k_s^+/k^+, and/or modify u_τ, etc. [28,57].

For a given roughness topology, increasing k_s^+ can be achieved by either increasing Re or by increasing the physical roughness height, k. The former is associated with higher computational costs, leading authors to opt for the latter option. The latter option, however, increases the blockage; if the flow is significantly constricted, it should be considered flow over obstacles rather than roughness.

The limits on the blockage ratio are acknowledged by the various studies but it is not uncommon for numerical simulations to reach a blockage factor, k/δ, of 0.2 or more, which is significant. In contrast, the work performed on minimal channels by MacDonald et al. [30], Zhong et al. [45], Rowin et al. [57] achieves better blockage ratios, with k/δ between 0.015 and 0.05. The sand-grain roughness of Hantsis and Piomelli [31], Hantsis [32], Hantsis and Piomelli [34] also maintained $k/\delta = 0.04$.

4.4. Virtual Origin

In a smooth-wall case, the outer turbulent flow "perceives" its origin to be at $y = 0$, where the no-slip condition is applied. This is not the case when the wall is rough, and the "virtual origin" of the flow is located somewhere between the base and the crest of the roughness [23,99]. The virtual origin is sometimes referred to as "wall offset" or "zero-plane displacement", and is commonly denoted by d or h_m. This is equivalent to the "canopy penetration depth" concept used in atmospheric flows and dense vegetation flows.

To obtain robust measures of the logarithmic intercepts, the wall distance need to be measured relative to this virtual origin [23,28,99]. The wall distance thus modified, y_e, yields the required collapse of various rough-wall statistics on the corresponding smooth-wall ones in the outer region, satisfying Townsend's outer-layer similarity [45].

While the concept is widely recognised, there is debate over the proper choice of the virtual origin, which manifests itself in various definitions of d. Two of the most common ones are based either on the roughness geometry or on the center of drag. The former commonly defines d as the geometrical mean of the roughness, and is insensitive to the conditions of the flow. The latter usually defines the virtual origin as the displacement that situates the centre of drag acting on the rough surface [99]. Both definitions are common throughout the literature and generally result in similar values. However, this creates uncertainty, which propagates into the results and statistics, in particular when comparing studies that use different definitions for d.

One limitation of the centroid method suggested by Jackson [99] is that it assumes a zero pressure gradient and fully developed flow. This was partially amended by [100], who suggested an extension to include pressure gradients, which seems promising.

Zhong et al. [45] studied the variation of d/k, and concluded that the influence of the uncertainty in d is largely inconsequential when measuring the logarithmic intercept, which translates into small variations in the roughness functions for velocity, ΔU^+, and scalar, $\Delta\Theta^+$. Thus, while there is a level of unreliability when comparing between studies that use different definitions of d, it is not significant enough to affect the conclusions.

Another geometry-based approach to define the centroid-shifted coordinate was suggested by [33] and used by [52,60,101]. The effective wall-normal distance from the rough surface is introduced as

$$y_e = \int_0^y \phi dy \tag{26}$$

where the plane porosity $\phi = A_f/A_0$, and A_f is the plane area occupied by the fluid. While this formulation produces the same shifted coordinate, $y - d$, outside the rough wall, its origin remains at zero. This is in contrast to the previously mentioned simple shifting, whose origin becomes negative (i.e., $y_e = y - d < 0$).

The uncertainty of d also influences the ratio of ks/δ [57], which may cause difficulties properly determining k_s^+. Not only does the choice of virtual origin tend to have more significant impact especially for larger k_s^+ [16,45], but also there is no reason for the scalar virtual origin, d_θ, to correspond with the momentum one [57] (despite it being common practice).

While these differences may not change trends and conclusions, they can affect the resulting values. For example, both MacDonald et al. [30] and Rowin et al. [57] used the same 3D sinusoidal roughness at $Pr = 0.71$ and identified a flattening of $\Delta\Theta^+$ at the ultimate regime. Using different approaches to determine the virtual origin (assuming $d_\theta = d$ in both studies), [30] reported $\Delta\Theta^+_{\mathrm{FR}} \approx 4.4$, while [57] reported $\Delta\Theta^+_{\mathrm{FR}} \approx 3.8$—a difference of approximately 13%, which is not insignificant. This may cause discrepancies between studies that are not physical, but rather due to a methodological choice.

A similar problem also exists with respect to experimental studies. Therefore, a more universal and consistent method of determining d and d_θ is desirable.

5. Conclusions

Roughness is known to have significant effects on the transport of momentum and scalar quantities. Despite significant research, our understanding of the latter lags behind that of the former. Wall-resolved numerical simulations provide unfettered access to details of the flow, both for momentum and scalar; this is particularly important when experimental measurements are difficult to obtain (e.g., dense roughness). This provides a tool to study the physics of the flow inside the roughness sublayer (RSL) and examine the physical assumptions taken by various phenomenological modes. Moreover, numerical simulations provide an effective technique for systematic studies and sweeping across the parameter spectrum. Here we summarise the progress in the study of passive scalars in flows over rough surfaces, thanks to recent numerical simulations.

Wall-resolved numerical simulations provided a high-fidelity description of the instantaneous scalar field, showing that the scalar field exhibits a diffusive sublayer that increasingly conforms to the roughness surface as k_s^+ or Pr increase. The diffusive sublayer is thinnest on the windward roughness surface and around the crest, resulting in sharp gradients that enhance the scalar wall flux. The Reynolds analogy was found to hold locally on the exposed portion of the windward slope; it breaks on the sheltered portion and in the recirculation regions behind the roughness elements, where the scalar wall flux is attenuated. The differences between the velocity and scalar fields are focused on the leeward side of the roughness elements and in the recirculation region behind them. On the windward side, the rough-wall diffusive sublayer is analogous to a contorted, smooth-wall diffusive sublayer (at sufficiently high k_s^+); a similar analogy cannot be made on the leeward side.

The downwards shift of the logarithmic region between the smooth- and rough-wall profiles of the mean scalar is quantified by the scalar roughness function, $\Delta\Theta^+$, which is equivalent to the Clauser–Hama roughness function, ΔU^+. It was observed that $\Delta U^+ < \Delta\Theta^+$ for sufficiently large k_s^+, regardless of the Prandtl number, Pr. For $Pr > 1.0$, there exists a finite range of k_s^+ for which the opposite is true; the value of k_s^+ for which $\Delta\Theta^+$ intersects the velocity asymptote increases with Pr. Various numerical studies observed a flattening of $\Delta\Theta^+$ as $k_s^+ \to \infty$, suggesting the possibility that an "ultimate regime", a scalar equivalent of the fully rough regime, exists. This regime is characterised by a finite asymptotic value $\Delta\Theta^+_{\mathrm{FR}}$, which depends strongly on the Prandtl number, Pr, and weakly on the roughness geometry. While this behaviour deviates from experimental findings, it is consistent with the observed topography-conforming behaviour for the scalar field. Further investigations are required to clarify the issue.

The scalar roughness function, $\Delta\Theta^+$, is related to the total scalar wall flux and the Stanton number, St. As such, St also depends on Pr and the roughness parameters such as the frontal solidity, λ_f. A peak in St was found for $0.1 < \lambda_f < 0.3$, usually corresponding to the maximum of the skin-friction coefficient, c_f. This is a result of an interplay between the impingement of bulk fluid and the sheltering by upstream roughness elements; the earlier enhances St, while the latter attenuates it. The rough-wall St is greater than the smooth-wall case, St_0, and approaches it in the limit of infinite roughness density. The existence of an ultimate regime implies that the enhancement of the scalar transport will saturate and will be accompanied by increasing drag, resulting in the Reynolds analogy factor $RA < 1$.

The rough-to-smooth ratio of the Reynolds analogy factor, η, is an efficiency parameter indicating the augmentation to St relative to the augmentation to c_f. Roughness results in $\eta < 1$, which tends to decrease with k_s^+ until reaching saturation; it was shown to be a convex function of the roughness density and λ_f. Predicting the λ_f value of the minimum for different types of roughness is still an open question.

The wake velocity, $\widetilde{u}$, and scalar, $\widetilde{\theta}$, fields are overall similar throughout the RSL, with differences observed in the troughs between roughness elements, where $\widetilde{u}$ and $\widetilde{\theta}$ tend to have opposite signs. The mean vertical scalar dispersive flux, $\left\langle \widetilde{\theta}\widetilde{v} \right\rangle_s$, is larger than $\langle \widetilde{u}\widetilde{v} \rangle_s$ for $Pr \geq 1.0$. It can account for a significant portion of the mean vertical scalar flux inside the RSL and is also important for the production of turbulent fluxes. Overall, the dispersive stresses, $\langle \widetilde{u}_i\widetilde{u}_k j \rangle_s$, fluxes, $\left\langle \widetilde{\theta}\widetilde{u}_j \right\rangle_s$, are not well studied and may be a very useful avenue of numerical exploration, since they are very difficult to measure experimentally. Additionally, they seem to be less sensitive to k_s^+ and even reach saturation with either k_s^+ or Pr. Dispersive fluxes are also needed for closing models that are based on the discrete-element method or the double-averaged NS equations (DANS).

When scaled in inner units, both the velocity and scalar fluctuations are attenuated by roughness; the former is attenuated more significantly than the latter. This may be due to the meaningful difference in the production of the velocity and scalar fluctuations, particularly in the form-induced (FI) production term, which produces fluctuations at a scale comparable to the roughness scale, k. As k_s^+ increases, the scalar FI production, $\mathcal{P}_\theta^{fi}$, becomes dominant in the RSL and is much more significant than the corresponding FI production of TKE, $\mathcal{P}_{kk}^{fi}$. $\mathcal{P}_\theta^{fi}$ increases with Pr and does not seem to saturate; however, for a fixed Pr, the $\mathcal{P}_\theta^{fi}$ curves collapse as k_s^+ increases. The shear production component remains similar between the TKE and scalar variance; it continually decreases with k_s^+ and may reach saturation. A possible result is that for a fixed Pr the scalar variance budget becomes insensitive to k_s^+. Asymptotic trends such as these are useful, in particular in the context of modelling. A more systematic study is required to ascertain these trends at higher k_s^+, and for a wider range of Pr and various types of roughness.

Both the turbulent Prandtl number, Pr_T, and the effective turbulent Prandtl number, $Pr_{T,\text{eff}}$, are ill-behaved near the bottom of the rough surface. Neither seem to be an effective modelling tool in generalised rough-wall flows and an alternative might be required for modelling.

Several key issues, however, remain open. Despite providing an effective technique for systematic studies and sweeping across the parameter spectrum, the computational cost associated with higher Reynolds and Prandtl numbers has limited the parameter range covered by numerical simulations. One way to circumvent this limitation and reach higher k_s^+ is to increase the roughness height, k, at relatively low Re. This approach, however, results in excessive blockage, which can have a detrimental effect on the simulation, turning it effectively to a flow over obstacles with significant acceleration.

The boundary conditions commonly used to implement constant scalar flux at the wall assume negligible fluctuations at the wall surface. These assumptions are only valid for relatively low Prandtl numbers, say $Pr < 5$; these may also be invalid at sufficiently high Reynolds numbers.

Finally, a common problem with numerical and experimental studies is the definition of the virtual origin, d, or the location of the zero-plane displacement, which is where the outer turbulent flow "perceives" its origin to be. Various definitions of d are employed, such as the geometric average or the location where the center of drag is acting on the rough surface; the latter is sensitive to the flow conditions, while the former is not. Both definitions are common throughout the literature and all the statistical quantities are measured with respect to the displaced coordinate. This creates uncertainty, which propagates into the results and statistics, in particular when comparing studies that use different definitions for d.

Author Contributions: Conceptualization, Z.H. and U.P.; methodology, Z.H.; software, Z.H.; formal analysis, Z.H. and U.P.; investigation, Z.H.; resources, U.P.; data curation, Z.H.; writing—original draft preparation, Z.H.; writing—review and editing, Z.H. and U.P.; visualization, Z.H.; supervision, U.P.; project administration, U.P.; funding acquisition, U.P. All authors have read and agreed to the published version of the manuscript.

Funding: This research was funded by the Natural Sciences and Engineering Research Council of Canada (NSERC-CRNSG) under the Discovery Grant program.

Data Availability Statement: No new data were created or analyzed in this study. Data sharing is not applicable to this article.

Acknowledgments: The authors acknowledge the support from the Natural Sciences and Engineering Research Council of Canada (NSERC-CRNSG).

Conflicts of Interest: The authors declare no conflicts of interest.

Abbreviations

The following abbreviations are used in this manuscript:

FI	form-induced
RSL	roughness sublayer
DNS	direct numerical simulation
LES	large-eddy simulation
RANS	Reynolds-averaged Navier–Stokes
PDF	probability density function

References

1. Reynolds, O. On the Extent and Action of the Heating Surface of Steam Boilers. *Int. J. Heat Mass Transf.* **1961**, *3*, 163–166. [CrossRef]
2. Schlichting, H. *Experimental Investigation of the Problem of Surface Roughness*; National Advisory Committee for Aeronautics: Washington, DC, USA, 1937.
3. Kays, W.M.; Crawford, M. *Convective Heat and Mass Transfer*, 3rd ed.; McGraw-Hill: New York, NY, USA, 1993.
4. Belnap, B.J.; van Rij, J.A.; Ligrani, P.M. A Reynolds Analogy for Real Component Surface Roughness. *Int. J. Heat Mass Transf.* **2002**, *45*, 3089–3099. [CrossRef]
5. Jelly, T.O.; Abu Rowin, W.; Hutchins, N.; Chung, D.; Tanimoto, K.; Oda, T.; Sandberg, R.D. High-Fidelity Computational Assessment of Aero-Thermal Performance and the Reynolds' Analogy for Additively Manufactured Anisotropic Surface Roughness. *J. Turbomach.* **2023**, *145*, 111005. [CrossRef]
6. Wenzel, C.; Gibis, T.; Kloker, M.; Rist, U. Reynolds Analogy Factor in Self-Similar Compressible Turbulent Boundary Layers with Pressure Gradients. *J. Fluid Mech.* **2021**, *907*, R4. [CrossRef]
7. Von Kármán, T. The Analogy between Fluid Friction and Heat Transfer. *Trans. Am. Soc. Mech. Eng.* **1939**, *61*, 705–710. [CrossRef]
8. Vigdorovich, I.I.; Leont'ev, A.I. The Restoration Coefficient and Reynolds Analogy in a Boundary Layer with Injection and Suction over the Entire Prandtl Number Range. *Fluid Dyn.* **2011**, *46*, 565–578. [CrossRef]
9. Dipprey, D.F.; Sabersky, R.H. Heat and Momentum Transfer in Smooth and Rough Tubes at Various Prandtl Numbers. *Int. J. Heat Mass Transf.* **1963**, *6*, 329–353. [CrossRef]
10. Chilton, T.H.; Colburn, A.P. Mass Transfer (Absorption) Coefficients Prediction from Data on Heat Transfer and Fluid Friction. *Ind. Eng. Chem.* **1934**, *26*, 1183–1187. [CrossRef]
11. Cope, W.F. The Friction and Heat Transmission Coefficients of Rough Pipes. *Proc. Inst. Mech. Eng.* **1941**, *145*, 99–105. [CrossRef]

12. Nunner, W.; Hudswell, F.; United Kingdom Atomic Energy Authority Research Group; Atomic Energy Research Establishment. *Heat Transfer and Pressure Drop in Rough Tubes*; AERE Lib/Trans 786; Atomic Energy Research Establishment: Berkshire, UK, 1958; Chapter 58, p. 9.
13. Lancet, R.T. The Effect of Surface Roughness on the Convection Heat-Transfer Coefficient for Fully Developed Turbulent Flow in Ducts With Uniform Heat Flux. *J. Heat Transf.* **1959**, *81*, 168–173. [CrossRef]
14. Zhang, Y.S.; Bi, W.T.; Hussain, F.; She, Z.S. A Generalized Reynolds Analogy for Compressible Wall-Bounded Turbulent Flows. *J. Fluid Mech.* **2014**, *739*, 392–420. [CrossRef]
15. Flack, K.A.; Schultz, M.P. Review of Hydraulic Roughness Scales in the Fully Rough Regime. *J. Fluids Eng.* **2010**, *132*. [CrossRef]
16. Chung, D.; Hutchins, N.; Schultz, M.P.; Flack, K.A. Predicting the Drag of Rough Surfaces. *Annu. Rev. Fluid Mech.* **2021**, *53*, 439–471. [CrossRef]
17. Nikuradse, J. *Laws of Flow in Rough Pipes*; Technical Report; National Advisory Committee for Aeronautics: Washington, DC, USA, 1950.
18. Moody, L.F. Friction Factors for Pipe Flow. *Trans. Am. Soc. Mech. Eng.* **1944**, *66*, 671–678. [CrossRef]
19. Krogstad, P.å.; Antonia, R.A.; Browne, L.W.B. Comparison between Rough- and Smooth-Wall Turbulent Boundary Layers. *J. Fluid Mech.* **1992**, *245*, 599–617. [CrossRef]
20. Krogstadt, P.Å.; Antonia, R. Surface Roughness Effects in Turbulent Boundary Layers. *Exp. Fluids* **1999**, *27*, 450–460. [CrossRef]
21. Shafi, H.S.; Antonia, R.A. Anisotropy of the Reynolds Stresses in a Turbulent Boundary Layer on a Rough Wall. *Exp. Fluids* **1995**, *18*, 213–215. [CrossRef]
22. Antonia, R.A.; Krogstad, P.Å. Turbulence Structure in Boundary Layers over Different Types of Surface Roughness. *Fluid Dyn. Res.* **2001**, *28*, 139–157. [CrossRef]
23. Raupach, M.R.; Antonia, R.A.; Rajagopalan, S. Rough-Wall Turbulent Boundary Layers. *Appl. Mech. Rev.* **1991**, *44*, 1–25. [CrossRef]
24. Finnigan, J. Turbulence in Plant Canopies. *Annu. Rev. Fluid Mech.* **2000**, *32*, 519–571. [CrossRef]
25. Townsend, A.A. *The Structure of Turbulent Shear Flow*, 2nd ed.; Cambridge University Press: London, UK, 1976.
26. Flack, K.A.; Schultz, M.P.; Shapiro, T.A. Experimental Support for Townsend's Reynolds Number Similarity Hypothesis on Rough Walls. *Phys. Fluids* **2005**, *17*, 035102. [CrossRef]
27. Flack, K.A.; Schultz, M.P. Roughness Effects on Wall-Bounded Turbulent Flows. *Phys. Fluids* **2014**, *26*, 101305. [CrossRef]
28. Jiménez, J. Turbulent Flows Over Rough Walls. *Annu. Rev. Fluid Mech.* **2004**, *36*, 173–196. [CrossRef]
29. Doosttalab, A.; Araya, G.; Newman, J.; Adrian, R.J.; Jansen, K.; Castillo, L. Effect of Small Roughness Elements on Thermal Statistics of a Turbulent Boundary Layer at Moderate Reynolds Number. *J. Fluid Mech.* **2016**, *787*, 84–115. [CrossRef]
30. MacDonald, M.; Hutchins, N.; Lohse, D.; Chung, D. Heat Transfer in Rough-Wall Turbulent Thermal Convection in the Ultimate Regime. *Phys. Rev. Fluids* **2019**, *4*, 071501. [CrossRef]
31. Hantsis, Z.; Piomelli, U. Roughness Effects on Scalar Transport. *Phys. Rev. Fluids* **2020**, *5*, 114607. [CrossRef]
32. Hantsis, Z. Effects of Roughness on Passive-Scalar Transport. Ph.D. Thesis, Queen's University, Kingston, ON, Canada, 2022.
33. Kuwata, Y.; Kawaguchi, Y. Direct Numerical Simulation of Turbulence over Systematically Varied Irregular Rough Surfaces. *J. Fluid Mech.* **2019**, *862*, 781–815. [CrossRef]
34. Hantsis, Z.; Piomelli, U. Effects of Roughness on the Turbulent Prandtl Number, Timescale Ratio, and Dissipation of a Passive Scalar. *Phys. Rev. Fluids* **2022**, *7*, 124601. [CrossRef]
35. Owen, P.R.; Thomson, W.R. Heat Transfer across Rough Surfaces. *J. Fluid Mech.* **1963**, *15*, 321–334. [CrossRef]
36. Miyake, Y.; Tsujimoto, K.; Nakaji, M. Direct Numerical Simulation of Rough-Wall Heat Transfer in a Turbulent Channel Flow. *Int. J. Heat Fluid Flow* **2001**, *22*, 237–244. [CrossRef]
37. Nagano, Y.; Hattori, H.; Houra, T. DNS of Velocity and Thermal Fields in Turbulent Channel Flow with Transverse-Rib Roughness. *Int. J. Heat Fluid Flow* **2004**, *25*, 393–403. [CrossRef]
38. Choi, H.S.; Suzuki, K. Large Eddy Simulation of Turbulent Flow and Heat Transfer in a Channel with One Wavy Wall. *Int. J. Heat Fluid Flow* **2005**, *26*, 681–694. [CrossRef]
39. Leonardi, S.; Orlandi, P.; Djenidi, L.; Antonia, R.A. Heat Transfer in a Turbulent Channel Flow with Square Bars or Circular Rods on One Wall. *J. Fluid Mech.* **2015**, *776*, 512–530. [CrossRef]
40. Ji, F.; Ding, J.; Lu, J.; Wang, W. Direct Numerical Simulation of Thermal Turbulent Boundary Layer Flow over Multiple V-Shaped Ribs at Different Angles. *Energies* **2023**, *16*, 3831. [CrossRef]
41. Kuwata, Y. Dissimilar Turbulent Heat Transfer Enhancement by Kelvin–Helmholtz Rollers over High-Aspect-Ratio Longitudinal Ribs. *J. Fluid Mech.* **2022**, *952*, A21. [CrossRef]
42. Stroh, A.; Schäfer, K.; Forooghi, P.; Frohnapfel, B. Secondary Flow and Heat Transfer in Turbulent Flow over Streamwise Ridges. *Int. J. Heat Fluid Flow* **2020**, *81*, 108518. [CrossRef]
43. Orlandi, P.; Sassun, D.; Leonardi, S. DNS of Conjugate Heat Transfer in Presence of Rough Surfaces. *Int. J. Heat Mass Transf.* **2016**, *100*, 250–266. [CrossRef]
44. MacDonald, M.; Hutchins, N.; Chung, D. Roughness Effects in Turbulent Forced Convection. *J. Fluid Mech.* **2019**, *861*, 138–162. [CrossRef]
45. Zhong, K.; Hutchins, N.; Chung, D. Heat-Transfer Scaling at Moderate Prandtl Numbers in the Fully Rough Regime. *J. Fluid Mech.* **2023**, *959*, A8. [CrossRef]

46. Peeters, J.; Sandham, N. Turbulent Heat Transfer in Channels with Irregular Roughness. *Int. J. Heat Mass Transf.* **2019**, *138*, 454–467. [CrossRef]
47. Peeters, J.W.R. Modelling Turbulent Heat Transfer in Rough Channels Using Phenomenological Theory. *J. Phys. Conf. Ser.* **2021**, *2116*, 012025. [CrossRef]
48. Peeters, J.W.R. Spectral Theory of Turbulent Heat Transfer in the Presence of a Rough Wall. *Phys. Rev. Lett.* **2023**, *131*, 134001. [CrossRef] [PubMed]
49. Forooghi, P.; Flory, M.; Bertsche, D.; Wetzel, T.; Frohnapfel, B. Heat Transfer Enhancement on the Liquid Side of an Industrially Designed Flat-Tube Heat Exchanger with Passive Inserts—Numerical Investigation. *Appl. Therm. Eng.* **2017**, *123*, 573–583. [CrossRef]
50. Forooghi, P.; Weidenlener, A.; Magagnato, F.; Böhm, B.; Kubach, H.; Koch, T.; Frohnapfel, B. DNS of Momentum and Heat Transfer over Rough Surfaces Based on Realistic Combustion Chamber Deposit Geometries. *Int. J. Heat Fluid Flow* **2018**, *69*, 83–94. [CrossRef]
51. Forooghi, P.; Stripf, M.; Frohnapfel, B. A Systematic Study of Turbulent Heat Transfer over Rough Walls. *Int. J. Heat Mass Transf.* **2018**, *127*, 1157–1168. [CrossRef]
52. Kuwata, Y. Direct Numerical Simulation of Turbulent Heat Transfer on the Reynolds Analogy over Irregular Rough Surfaces. *Int. J. Heat Fluid Flow* **2021**, *92*, 108859. [CrossRef]
53. Kuwata, Y.; Yagasaki, W.; Suga, K. Scaling of Turbulent Heat Transfer over Sinusoidal Rough Surfaces. In Proceedings of the International Heat Transfer Conference 17, Cape Town, South Africa, 14–18 August 2023; Begel House Inc.: Danbury, CT, USA, 2023.
54. Garg, H.; Wang, L.; Fureby, C. Heat Transfer Enhancement with Additively Manufactured Rough Surfaces: Insights from Large-Eddy Simulations. *Phys. Fluids* **2024**, *36*, 025109. [CrossRef]
55. Yang, J.; Velandia, J.; Bansmer, S.; Stroh, A.; Forooghi, P. A Comparison of Hydrodynamic and Thermal Properties of Artificially Generated against Realistic Rough Surfaces. *Int. J. Heat Fluid Flow* **2023**, *99*, 109093. [CrossRef]
56. Nagura, R.; Yagasaki, W.; Kuwata, Y.; Suga, K. Discussion on the Effects of Structural Parameters of Roughness on Heat Transfer Similarity. In Proceedings of the 17th UK Heat Transfer Conference, Manchester, UK, 4–6 April 2022.
57. Rowin, W.A.; Zhong, K.; Saurav, T.; Jelly, T.; Hutchins, N.; Chung, D. Modelling the Effect of Roughness Density on Turbulent Forced Convection. *J. Fluid Mech.* **2024**, *979*, A22. [CrossRef]
58. Zhang, E.; Wang, X.; Liu, Q. Numerical Investigation on the Temporal and Spatial Statistical Characteristics of Turbulent Mass Transfer above a Two-Dimensional Wavy Wall. *Int. J. Heat Mass Transf.* **2022**, *184*, 122260. [CrossRef]
59. Zhang, E.; Wu, W.; Liu, Q.; Wang, X. Effects of Vortex Formation and Interaction on Turbulent Mass Transfer over a Two-Dimensional Wavy Wall. *Phys. Rev. Fluids* **2022**, *7*, 114607. [CrossRef]
60. Kuwata, Y. Reynolds Number Dependence of Turbulent Heat Transfer over Irregular Rough Surfaces. *Phys. Fluids* **2022**, *34*, 045118. [CrossRef]
61. Kasagi, N.; Tomita, Y.; Kuroda, A. Direct Numerical Simulation of Passive Scalar Field in a Turbulent Channel Flow. *J. Heat Transf.* **1992**, *114*, 598–606. [CrossRef]
62. Raupach, M.R. Drag and Drag Partition on Rough Surfaces. *Bound.-Layer Meteorol.* **1992**, *60*, 375–395. [CrossRef]
63. Nikora, V.; Ballio, F.; Coleman, S.; Pokrajac, D. Spatially Averaged Flows over Mobile Rough Beds: Definitions, Averaging Theorems, and Conservation Equations. *J. Hydraul. Eng.* **2013**, *139*, 803–811. [CrossRef]
64. Yaglom, A.M.; Kader, B.A. Heat and Mass Transfer between a Rough Wall and Turbulent Fluid Flow at High Reynolds and Péclet Numbers. *J. Fluid Mech.* **1974**, *62*, 601–623. [CrossRef]
65. Gowen, R.A.; Smith, J.W. Turbulent Heat Transfer from Smooth and Rough Surfaces. *Int. J. Heat Mass Transf.* **1968**, *11*, 1657–1674. [CrossRef]
66. Garratt, J.R.; Hicks, B.B. Momentum, Heat and Water Vapour Transfer to and from Natural and Artificial Surfaces. *Q. J. R. Meteorol. Soc.* **1973**, *99*, 680–687. [CrossRef]
67. Kader, B.A.; Yaglom, A.M. Heat and Mass Transfer Laws for Fully Turbulent Wall Flows. *Int. J. Heat Mass Transf.* **1972**, *15*, 2329–2351. [CrossRef]
68. Kader, B.A.; Yaglom, A.M. Turbulent Heat and Mass Transfer from a Wall with Parallel Roughness Ridges. *Int. J. Heat Mass Transf.* **1977**, *20*, 345–357. [CrossRef]
69. Brutsaert, W. A Theory for Local Evaporation (or Heat Transfer) from Rough and Smooth Surfaces at Ground Level. *Water Resour. Res.* **1975**, *11*, 543–550. [CrossRef]
70. Pirozzoli, S.; Bernardini, M.; Orlandi, P. Passive Scalars in Turbulent Channel Flow at High Reynolds Number. *J. Fluid Mech.* **2016**, *788*, 614–639. [CrossRef]
71. Rowin, W.A.; Saurav, T.M.; Jelly, T.O.; Hutchins, N.; Chung, D. Turbulent Forced Convection over Roughness with Different Heights and Densities. In Proceedings of the 12th Australasian Heat and Mass Transfer Conference, Sydney, Australia, 8–9 July 2021.
72. Bons, J. A Critical Assessment of Reynolds Analogy for Turbine Flows. *J. Heat Transf.* **2005**, *127*, 472–485. [CrossRef]
73. Bons, J.P.; Taylor, R.P.; McClain, S.T.; Rivir, R.B. The Many Faces of Turbine Surface Roughness. In Proceedings of the ASME Turbo Expo 2001: Power for Land, Sea, and Air, American Society of Mechanical Engineers Digital Collection, New Orleans, LA, USA, 4–7 June 2001. [CrossRef]

74. De Maio, M. DNS of Momentum and Heat Transfer inside Rough Pipes. *Mater. Res. Proc.* **2023**, *33*, 41–48. [CrossRef]
75. Kolář, V. Heat Transfer in Turbulent Flow of Fluids through Smooth and Rough Tubes. *Int. J. Heat Mass Transf.* **1965**, *8*, 639–653. [CrossRef]
76. Dean, R.B. Reynolds Number Dependence of Skin Friction and Other Bulk Flow Variables in Two-Dimensional Rectangular Duct Flow. *J. Fluids Eng.* **1978**, *100*, 215–223. [CrossRef]
77. Rouhi, A.; Endrikat, S.; Modesti, D.; Sandberg, R.D.; Oda, T.; Tanimoto, K.; Hutchins, N.; Chung, D. Riblet-Generated Flow Mechanisms That Lead to Local Breaking of Reynolds Analogy. *J. Fluid Mech.* **2022**, *951*, A45. [CrossRef]
78. Thakkar, M.; Busse, A.; Sandham, N. Surface Correlations of Hydrodynamic Drag for Transitionally Rough Engineering Surfaces. *J. Turbul.* **2017**, *18*, 138–169. [CrossRef]
79. Nishiyama, Y.; Kuwata, Y.; Suga, K. Direct Numerical Simulation of Turbulent Heat Transfer over Fully Resolved Anisotropic Porous Structures. *Int. J. Heat Fluid Flow* **2020**, *81*, 108515. [CrossRef]
80. Aupoix, B. Improved Heat Transfer Predictions on Rough Surfaces. *Int. J. Heat Fluid Flow* **2015**, *56*, 160–171. [CrossRef]
81. Kuwata, Y.; Nagura, R. Direct Numerical Simulation on the Effects of Surface Slope and Skewness on Rough-Wall Turbulence. *Phys. Fluids* **2020**, *32*, 105113. [CrossRef]
82. Li, Q.; Bou-Zeid, E. Contrasts between Momentum and Scalar Transport over Very Rough Surfaces. *J. Fluid Mech.* **2019**, *880*, 32–58. [CrossRef]
83. Chedevergne, F. A Double-Averaged Navier-Stokes $k-\omega$ Turbulence Model for Wall Flows over Rough Surfaces with Heat Transfer. *J. Turbul.* **2021**, *22*, 713–734. [CrossRef]
84. Chan, L.; MacDonald, M.; Chung, D.; Hutchins, N.; Ooi, A. A Systematic Investigation of Roughness Height and Wavelength in Turbulent Pipe Flow in the Transitionally Rough Regime. *J. Fluid Mech.* **2015**, *771*, 743–777. [CrossRef]
85. Wallace, J.M.; Eckelmann, H.; Brodkey, R.S. The Wall Region in Turbulent Shear Flow. *J. Fluid Mech.* **1972**, *54*, 39–48. [CrossRef]
86. Crimaldi, J.P.; Koseff, J.R.; Monismith, S.G. A Mixing-Length Formulation for the Turbulent Prandtl Number in Wall-Bounded Flows with Bed Roughness and Elevated Scalar Sources. *Phys. Fluids* **2006**, *18*, 095102. [CrossRef]
87. Spalart, P.R.; Strelets, M.K. Mechanisms of Transition and Heat Transfer in a Separation Bubble. *J. Fluid Mech.* **2000**, *403*, 329–349. [CrossRef]
88. Kawamura, H.; Abe, H.; Matsuo, Y. Very Large-Scale Structures Observed in DNS of Turbulent Channel Flow with Passive Scalar Transport. In Proceedings of the 15th Australasian Fluid Mechanics Conference, The University of Sydney, Sydney, Australia, 13–17 December 2004.
89. Cebeci, T. A Model for Eddy Conductivity and Turbulent Prandtl Number. *J. Heat Transf.* **1973**, *95*, 227–234. [CrossRef]
90. Abe, H.; Antonia, R.A. Scaling of Normalized Mean Energy and Scalar Dissipation Rates in a Turbulent Channel Flow. *Phys. Fluids* **2011**, *23*, 055104. [CrossRef]
91. Kasagi, N.; Kuroda, A.; Hirata, M. Numerical Investigation of Near-Wall Turbulent Heat Transfer Taking Into Account the Unsteady Heat Conduction in the Solid Wall. *J. Heat Transfer* **1989**, *111*, 385–392. [CrossRef]
92. Zhong, K.; Hutchins, N.; Chung, D. Rough-Wall Heat Transfer at Moderate Prandtl Numbers: Towards Reconciling the Diverse Model Predictions. In Proceedings of the 12th International Symposium on Turbulence and Shear Flow Phenomena (TSFP12), Osaka, Japan, 19–22 July 2022.
93. Tiselj, I.; Bergant, R.; Mavko, B.; Bajsic´, I.; Hetsroni, G. DNS of Turbulent Heat Transfer in Channel Flow With Heat Conduction in the Solid Wall. *J. Heat Transf.* **2001**, *123*, 849–857. [CrossRef]
94. Jiménez, J.; Moin, P. The Minimal Flow Unit in Near-Wall Turbulence. *J. Fluid Mech.* **1991**, *225*, 213–240. [CrossRef]
95. Chung, D.; Chan, L.; MacDonald, M.; Hutchins, N.; Ooi, A. A Fast Direct Numerical Simulation Method for Characterising Hydraulic Roughness. *J. Fluid Mech.* **2015**, *773*, 418–431. [CrossRef]
96. MacDonald, M.; Chung, D.; Hutchins, N.; Chan, L.; Ooi, A.; García-Mayoral, R. The Minimal Channel: A Fast and Direct Method for Characterising Roughness. *J. Phys. Conf. Ser.* **2016**, *708*, 012010. [CrossRef]
97. MacDonald, M.; Chung, D.; Hutchins, N.; Chan, L.; Ooi, A.; García-Mayoral, R. The Minimal-Span Channel for Rough-Wall Turbulent Flows. *J. Fluid Mech.* **2017**, *816*, 5–42. [CrossRef]
98. Belcher, S.E.; Coceal, O.; Goulart, E.V.; Rudd, A.C.; Robins, A.G. Processes Controlling Atmospheric Dispersion through City Centres. *J. Fluid Mech.* **2015**, *763*, 51–81. [CrossRef]
99. Jackson, P.S. On the Displacement Height in the Logarithmic Velocity Profile. *J. Fluid Mech.* **1981**, *111*, 15–25. [CrossRef]
100. Breugem, W.P.; Boersma, B.J.; Uittenbogaard, R.E. The Influence of Wall Permeability on Turbulent Channel Flow. *J. Fluid Mech.* **2006**, *562*, 35–72. [CrossRef]
101. Kuwata, Y.; Yamamoto, Y.; Tabata, S.; Suga, K. Scaling of the Roughness Effects in Turbulent Flows over Systematically-Varied Irregular Rough Surfaces. *Int. J. Heat Fluid Flow* **2023**, *101*, 109130. [CrossRef]

Disclaimer/Publisher's Note: The statements, opinions and data contained in all publications are solely those of the individual author(s) and contributor(s) and not of MDPI and/or the editor(s). MDPI and/or the editor(s) disclaim responsibility for any injury to people or property resulting from any ideas, methods, instructions or products referred to in the content.

fluids

MDPI

Review

Deep Reinforcement Learning for Fluid Mechanics: Control, Optimization, and Automation

Innyoung Kim †, Youngmin Jeon †, Jonghyun Chae † and Donghyun You *

Department of Mechanical Engineering, Pohang University of Science and Technology, 77 Cheongam-ro, Nam-gu, Pohang-si 37673, Gyeongsangbuk-do, Republic of Korea; innykim@postech.ac.kr (I.K.); jym1016@postech.ac.kr (Y.J.); jhchae96@postech.ac.kr (J.C.)

* Correspondence: dhyou@postech.ac.kr

† These authors contributed equally to this work.

Abstract: A comprehensive review of recent advancements in applying deep reinforcement learning (DRL) to fluid dynamics problems is presented. Applications in flow control and shape optimization, the primary fields where DRL is currently utilized, are thoroughly examined. Moreover, the review introduces emerging research trends in automation within computational fluid dynamics, a promising field for enhancing the efficiency and reliability of numerical analysis. Emphasis is placed on strategies developed to overcome challenges in applying DRL to complex, real-world engineering problems, such as data efficiency, turbulence, and partial observability. Specifically, the implementations of transfer learning, multi-agent reinforcement learning, and the partially observable Markov decision process are discussed, illustrating how these techniques can provide solutions to such issues. Finally, future research directions that could further advance the integration of DRL in fluid dynamics research are highlighted.

Keywords: deep reinforcement learning; fluid mechanics; flow control; shape optimization; CFD automation

1. Introduction

The mechanisms through which the human brain acquires and incorporates knowledge have long been intriguing research topics. Emulating such abilities through self-developing algorithms has become a primary goal in machine learning. Reinforcement learning (RL) is a framework in which an agent learns to make decisions by interacting with the environment, aiming to maximize cumulative reward through trial and error, akin to human learning [1]. Since its inception, RL has shown considerable potential [2], although its applications have traditionally been confined to low-dimensional problems. However, the advent of deep reinforcement learning (DRL) represents a significant advancement, integrating deep neural networks (DNNs) with RL algorithms. DRL leverages the capacity of DNNs to approximate nonlinear functions in high-dimensional spaces [3], leading to groundbreaking achievements in various fields such as robotics [4], autonomous driving [5,6], natural language processing [7], and gaming [8–10].

The field of fluid mechanics also struggles with challenges due to the nonlinear and high-dimensional nature of fluid motions governed by the Navier–Stokes equations. The ability of DRL to address complex decision-making problems offers new possibilities for fluid dynamics research. Common applications of DRL in fluid mechanics include flow control and shape optimization. Flow control aims to appropriately alter system dynamics to achieve desirable performance. An early and widely studied example is drag reduction for a two-dimensional cylinder, which now serves as a benchmark case for validating DRL-based flow control methods [11–24]. In shape optimization, DRL agents modify geometries to optimize specific objective functions. A classical example is the optimization of airfoil shapes to enhance aerodynamic characteristics, such as reducing drag or increasing lift [25–31].

These foundational applications demonstrate the potential of DRL to resolve fluid dynamics problems that were previously intractable with conventional methods.

However, a variety of challenges have emerged as DRL-based research in fluid dynamics advances to tackle more intricate real-world problems. The most prominent challenge is the computational cost of acquiring data. The computational cost needed to evaluate system dynamics escalates as geometries and flow conditions become more complex. This issue is especially pronounced when dealing with turbulent flow regimes, where the multi-scale nature of turbulent flow requires extensive data to accurately capture down to small-scale motions. Moreover, the inherent stochasticity of turbulence introduces high variance in the training data, negatively impacting both the performance and robustness of DRL algorithms. The partial observability of the state space further complicates the problem, necessitating techniques to effectively utilize incomplete information. Consequently, the recent research trend is to adopt supplementary strategies alongside DRL algorithms to address these challenges effectively.

The aim of the present paper is to provide a comprehensive review of recent literature on the application of DRL to fluid dynamics problems. Earlier reviews of DRL-based research in fluid mechanics were introduced by Rabault et al. [32] and updated by Garnier et al. [33] and Viquerat et al. [34]. Vignon et al. [35] focused on studies utilizing DRL with an emphasis on flow control. Compared to the previous reviews, the present review concentrates on the strategies employed in recent research to address the challenges encountered when applying DRL to more complex fluid dynamics problems. Specifically, transfer learning, multi-agent reinforcement learning (MARL), and the partially observable Markov decision process (POMDP) are discussed, detailing how these techniques can provide solutions to such issues. Additionally, unlike previous reviews that primarily focused on flow control and shape optimization, the current review introduces research trends in the automation of computational fluid dynamics (CFD), a newly emerging field that holds promise for advancing the efficiency and reliability of numerical analysis.

The paper is organized as follows: The theoretical background and the main DRL algorithms applied in fluid dynamics research are described in Section 2. The challenges encountered in DRL-based fluid dynamics research and the strategies used to address these issues are explained in Section 3. Comprehensive reviews of the applications in flow control, shape optimization, and automation of CFD are provided in Sections 4–6, respectively, followed by concluding remarks in Section 7.

2. Deep Reinforcement Learning

RL is a framework for acquiring an optimal strategy in a sequential decision-making process [1]. In the general RL problem, the process is often modeled as a Markov decision process (MDP), where an agent interacts with the environment, as illustrated in Figure 1. At each step t, a state $\mathbf{s}_t \in \mathcal{S}$ is given to the agent by the environment, and the agent executes an action $\mathbf{a}_t \in \mathcal{A}$ according to its policy π, where $\mathcal{S}$ and $\mathcal{A}$ refer to the state and action spaces, respectively. Subsequently, a reward $r_t(\mathbf{s}_t, \mathbf{a}_t) \in \mathbb{R}$ is obtained, and the state transitions to $\mathbf{s}_{t+1}$. The probability of transitioning to the next state $\mathbf{s}_{t+1}$ is determined by the state transition function $T(\mathbf{s}_t, \mathbf{a}_t, \mathbf{s}_{t+1}) : \mathcal{S} \times \mathcal{A} \times \mathcal{S} \to [0,1]$, and the reward is defined by the reward function $R(\mathbf{s}_t, \mathbf{a}_t, \mathbf{s}_{t+1}) : \mathcal{S} \times \mathcal{A} \times \mathcal{S} \to \mathcal{R}$, where $\mathcal{R}$ denotes a set of possible rewards. The procedure continues to the terminal step T, where one episode ends. The goal of the agent is to discover an optimal policy that maximizes the discounted cumulative reward, denoted as the return G_t, which is defined as follows:

$$G_t = \sum_{i=t}^{T} \gamma^{(i-t)} \, r_i(\mathbf{s}_i, \mathbf{a}_i), \tag{1}$$

where $\gamma \in [0,1)$ is a discount factor that defines the weight between long-term and short-term rewards.

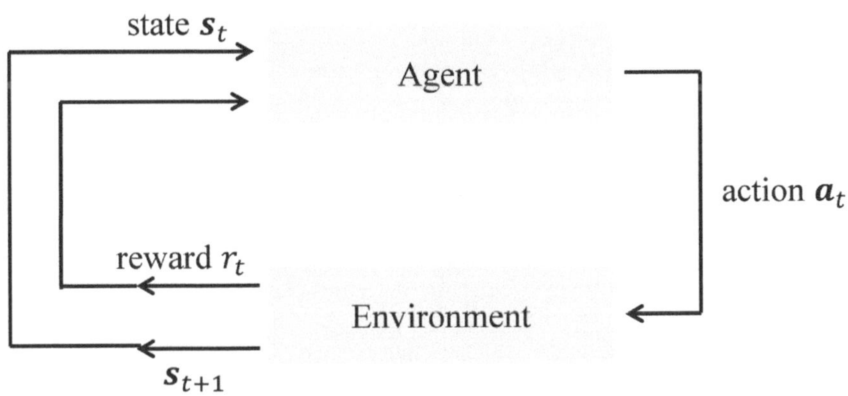

Figure 1. Reinforcement learning framework.

To train the agent, value functions are utilized to evaluate the performance of the policy. The state value function $V^{\pi}(\mathbf{s})$ is defined as the expected return as follows:

$$V^{\pi}(\mathbf{s}) = \mathbb{E}_{\pi}[G_t|\mathbf{s}_t = \mathbf{s}]. \tag{2}$$

Similarly, the action value function $Q^{\pi}(\mathbf{s},\mathbf{a})$ is defined as follows:

$$Q^{\pi}(\mathbf{s},\mathbf{a}) = \mathbb{E}_{\pi}[G_t|\mathbf{s}_t = \mathbf{s}, \mathbf{a}_t = \mathbf{a}]. \tag{3}$$

In Equations (2) and (3), the expectation is taken over the policy π with environment transitions. The optimal policy π^* refers to the policy achieving optimal value functions, as follows:

$$V^*(\mathbf{s}) = \max_{\pi} V^{\pi}(\mathbf{s}), \tag{4}$$

$$Q^*(\mathbf{s},\mathbf{a}) = \max_{\pi} Q^{\pi}(\mathbf{s},\mathbf{a}). \tag{5}$$

In early RL algorithms, tables were utilized to calculate policy measures such as the value functions by storing the values for every state–action pair. However, such an approach can easily become intractable with the increasing dimensions of state and action spaces. DRL incorporates deep learning techniques, leveraging the capacity of artificial neural networks to approximate complex functions and learn from high-dimensional inputs using nonlinear activation functions. This capability is particularly advantageous for fluid mechanics problems, where the dynamics are challenging due to the nonlinear and high-dimensional nature of fluid motions. Based on whether the neural network is used to approximate value functions or directly map policies, DRL algorithms can be categorized as value-based methods and policy-based methods, respectively.

2.1. Value-Based Methods

Value-based methods utilize neural networks to approximate value functions, from which the policy is derived. One of the most popular value-based algorithms is Q-learning [36], which aims to learn the action-value function in Equation (3) to obtain an optimal policy. To achieve the optimal action-value function, the Bellman optimality equation is employed as follows:

$$Q^*(\mathbf{s},\mathbf{a}) = \sum_{\mathbf{s}' \in \mathcal{S}} T(\mathbf{s},\mathbf{a},\mathbf{s}')\left(R(\mathbf{s},\mathbf{a},\mathbf{s}') + \gamma \max_{\mathbf{a}'} Q^*(\mathbf{s}',\mathbf{a}')\right). \tag{6}$$

In the early version of Q-learning, a Q-table is used to store the values of the action-value function for every state–action pair, with values gradually updated until Equation (6) is satisfied. The optimal policy is then achieved as follows:

$$\pi^*(\mathbf{s}) = \underset{\mathbf{a}}{\operatorname{argmax}}\, Q^*(\mathbf{s}, \mathbf{a}). \tag{7}$$

Deep Q-networks. The use of a Q-table becomes intractable as the dimensions of the state and action spaces increase in Q-learning. Deep Q-networks (DQN) [8,37] address this challenge by using neural networks to approximate the action-value function as $Q_{\boldsymbol{\theta}}(\mathbf{s}, \mathbf{a})$, where $\boldsymbol{\theta}$ denotes the network parameters. The network parameters are updated using a gradient descent algorithm to minimize the loss function derived from Equation (6) as follows:

$$L(\boldsymbol{\theta}) = \mathbb{E}_{(\mathbf{s},\mathbf{a},r,\mathbf{s}') \sim \mathcal{D}}\left[\left(r + \gamma \max_{\mathbf{a}'} Q_{\boldsymbol{\theta}^-}(\mathbf{s}', \mathbf{a}') - Q_{\boldsymbol{\theta}}(\mathbf{s}, \mathbf{a})\right)^2\right]. \tag{8}$$

Two key techniques are employed to improve stability and convergence. Firstly, a target network is used to stabilize the training process by periodically updating its parameters $\boldsymbol{\theta}^-$ to match the main network $\boldsymbol{\theta}$. Secondly, the experience replay buffer $\mathcal{D}$ stores transition tuples $(\mathbf{s}, \mathbf{a}, r, \mathbf{s}')$, which are sampled randomly to break the correlation between successive updates. For action selection, the ϵ-greedy strategy is employed to balance exploration and exploitation. The agent chooses a random action with probability ϵ and, otherwise, executes the optimal action of $\operatorname{argmax}_{\mathbf{a}} Q_{\boldsymbol{\theta}}(\mathbf{s}, \mathbf{a})$. Various algorithms and techniques continue to improve the performance of DQN, including double DQN [38], dueling DQN [39], prioritized replay DQN [40], and rainbow DQN [41].

2.2. Policy-Based Methods

When dealing with high-dimensional or continuous action spaces, value-based methods become computationally inefficient due to the need for an extensive number of discretizations of the action space. Policy-based methods address this issue by directly mapping the policy using neural networks with parameters $\boldsymbol{\theta}$ as $\pi_{\boldsymbol{\theta}}$. The objective is to maximize the expected return, defined as $J(\boldsymbol{\theta}) = \mathbb{E}_\pi[G_t]$, by adjusting the parameters using a gradient ascent algorithm as follows:

$$\boldsymbol{\theta} \leftarrow \boldsymbol{\theta} + \alpha \nabla_{\boldsymbol{\theta}} J(\boldsymbol{\theta}), \tag{9}$$

where α represents a learning rate. Policy-based methods can be categorized based on whether the policy is stochastic or deterministic.

2.2.1. Stochastic Policy-Based Methods

A stochastic policy defines a probability distribution over actions given a state as $\pi_{\boldsymbol{\theta}}(\mathbf{s}, \mathbf{a}) : \mathcal{S} \times \mathcal{A} \to [0, 1]$. The gradient required in Equation (9) is then computed by the policy gradient theorem with the log-probability trick [42,43] as follows:

$$\nabla_{\boldsymbol{\theta}} J(\boldsymbol{\theta}) = \mathbb{E}_\pi\left[\sum_{t=0}^{T} \nabla_{\boldsymbol{\theta}} \log \pi_{\boldsymbol{\theta}}(\mathbf{s}_t, \mathbf{a}_t) G_t\right]. \tag{10}$$

A classical method to evaluate the above equation is the REINFORCE algorithm [43], which utilizes the Monte Carlo approach. However, such an approach requires a complete trajectory for sampling to compute the return G_t, leading to high variance and slow learning. Thus, instead of using G_t, several other expressions are employed in practice that do not modify the value of the computed gradient. One of the most widely used approaches is the actor–critic method [44–46], which combines policy-based and value-based approaches. The policy is parameterized by the actor network, while the critic network estimates value functions to provide a lower-variance estimate of the expected return.

Advantage actor–critic. Advantage actor–critic methods [47] replace G_t in Equation (10) with the advantage function $A^{\pi}(\mathbf{s}, \mathbf{a})$ to mitigate high variance. The advantage function is defined as follows:

$$A^{\pi}(\mathbf{s}, \mathbf{a}) = Q^{\pi}(\mathbf{s}, \mathbf{a}) - V^{\pi}(\mathbf{s}). \tag{11}$$

Consequently, the gradient of the loss function is expressed as:

$$\nabla_{\theta} J(\boldsymbol{\theta}) = \mathbb{E}_{\pi}\left[\sum_{t=0}^{T} \nabla_{\theta} \log \pi_{\theta}(\mathbf{s}_t, \mathbf{a}_t) A^{\pi_{\theta}}(\mathbf{s}_t, \mathbf{a}_t)\right], \tag{12}$$

where $\pi_{\theta}(\mathbf{s}_t, \mathbf{a}_t)$ is the actor network, and $A^{\pi_{\theta}}(\mathbf{s}, \mathbf{a})$ is computed from the critic network. To avoid learning both value functions in Equation (11), the advantage function is approximated as follows:

$$A^{\pi}(\mathbf{s}_t, \mathbf{a}_t) \approx r_t(\mathbf{s}_t, \mathbf{a}_t) + \gamma V^{\pi}(\mathbf{s}_{t+1}) - V^{\pi}(\mathbf{s}_t), \tag{13}$$

which is derived using the definition of the action-value function. Thus, the critic network only needs to predict the state value function $V^{\pi}(\mathbf{s})$. Often, multiple agents run in parallel on different environment instances to obtain diverse data and accelerate learning. Two types of approaches exist: synchronous (A2C) and asynchronous (A3C) [47], differing in whether the parameter updates occur in a coordinated manner or not.

Proximal policy optimization. Proximal policy optimization (PPO) [48] not only reduces variance by employing the advantage function but also improves sample efficiency. To this end, PPO leverages data obtained by the old policy $\pi_{\theta_{\text{old}}}$ through the concept of importance sampling, which accounts for the ratio between the old and current policies in the objective function. At the same time, to prevent the algorithm from diverging due to excessively large updates when the old policy significantly differs from the current policy, PPO uses a heuristic clipping function to bound the amount of updates. Thus, the objective function is defined as follows:

$$J(\boldsymbol{\theta}) = \mathbb{E}_{\pi_{\theta_{\text{old}}}}\left[\min\left(\frac{\pi_{\theta}(\mathbf{s}, \mathbf{a})}{\pi_{\theta_{\text{old}}}(\mathbf{s}, \mathbf{a})} A^{\pi_{\theta_{\text{old}}}}(\mathbf{s}, \mathbf{a}), \text{clip}\left(\frac{\pi_{\theta}(\mathbf{s}, \mathbf{a})}{\pi_{\theta_{\text{old}}}(\mathbf{s}, \mathbf{a})}, 1-\epsilon, 1+\epsilon\right) A^{\pi_{\theta_{\text{old}}}}(\mathbf{s}, \mathbf{a})\right)\right], \tag{14}$$

where ϵ defines how far the old policy is allowed to differ from the current policy. Due to its sample efficiency and robustness, PPO is by far the most widely applied DRL algorithm for flow control.

Soft actor–critic. Soft actor–critic (SAC) [49,50] is an algorithm designed to enhance sample efficiency and robustness. A distinct feature of SAC is that it operates as an off-policy algorithm, in contrast to on-policy algorithms such as A2C, A3C, and PPO, which update the policy using data generated by the current policy. The on-policy approach often results in inefficient data usage, as experiences are typically discarded after each update. Although PPO allows for some degree of data reuse when the policy remains relatively unchanged, it fundamentally adheres to the on-policy framework. In contrast, SAC updates the policy using data generated from entirely different policies, enabling the reuse of past experiences and thereby improving sample efficiency.

A central aspect of SAC is the inclusion of an entropy term in its objective function. Entropy, which quantifies the randomness of the policy, is defined as:

$$\mathcal{H}(\pi) = -\mathbb{E}_{\pi}[\log \pi(\mathbf{s}, \mathbf{a})]. \tag{15}$$

By maximizing the discounted cumulative reward, regularized by the entropy term, SAC achieves robustness, effectively balancing the exploration-exploitation trade-off. Higher entropy encourages greater exploration, accelerating learning and preventing the policy from prematurely converging to suboptimal solutions.

2.2.2. Deterministic Policy-Based Methods

A deterministic policy directly defines an optimal action for a given state as $\pi_\theta(\mathbf{s}) : \mathcal{S} \to \mathcal{A}$. The network parameter $\boldsymbol{\theta}$ is updated according to Equation (9), where $\nabla_\theta J(\boldsymbol{\theta})$ is computed using the deterministic policy gradient theorem [51] as follows:

$$\nabla_\theta J(\boldsymbol{\theta}) = \mathbb{E}_\pi \left[\nabla_a Q^\pi(\mathbf{s}, \mathbf{a})|_{a=\pi_\theta(\mathbf{s})} \nabla_\theta \pi_\theta(\mathbf{s}) \right]. \tag{16}$$

Here, the chain rule is employed to first compute the gradient of $Q^\pi(\mathbf{s}, \mathbf{a})$ with respect to the action, followed by the gradient of the action (policy) with respect to $\boldsymbol{\theta}$.

Compared to stochastic policies, where importance sampling is often used to leverage data obtained from different policies, the deterministic policy gradient eliminates the integral over the action probability distribution. This allows direct usage of data obtained from different policies without additional adjustments in Equation (16), thereby increasing sample efficiency.

Deep deterministic policy gradient. Deep deterministic policy gradient (DDPG) [52] is an actor–critic algorithm that leverages the strengths of DQN in a deterministic policy-based approach. Similar to DQN, DDPG employs replay buffer and target network techniques to increase sample efficiency and stabilize training. Advancing beyond DQN, DDPG utilizes a critic network to learn the value function $Q^\pi(s, a)$, and an actor network to learn the optimal policy, where the action is deterministically defined as $\pi(s) : \mathcal{S} \to \mathcal{A}$. Thus, the critic is updated by minimizing the loss between the predicted Q-values and the target Q-values using a gradient descent algorithm. The actor is updated by maximizing the expected return, as given by Equation (16), using a gradient ascent algorithm, where $\nabla_a Q^\pi(s, a)|_{a=\pi_\theta(s)}$ is provided by the critic network. Gaussian noise is typically added to the action selection to efficiently balance exploration and exploitation.

Twin-delayed deep deterministic policy gradient. The twin-delayed deep deterministic policy gradient (TD3) [53] algorithm addresses the overestimation bias and error accumulation issues often encountered with DDPG. To that end, three techniques are employed. Firstly, twin critic networks are trained separately, and the minimum value predicted by the two critics is used to reduce overestimation bias. Secondly, additional noise is incorporated into the target action to smooth out the target policy, preventing the network from exploiting incorrect sharp peaks in the value estimates. Finally, updating the actor network less frequently than the critic network allows value estimates with less error to be used in updating the policy. Along with DDPG, TD3 has recently been widely used in fluid dynamics research utilizing DRL.

2.3. Single-Step Deep Reinforcement Learning

Single-step deep reinforcement learning. Single-step DRL is an algorithm particularly employed in fluid mechanics for optimization and open-loop control problems. In the general DRL framework, an agent interacts with the environment by sequentially choosing actions at each discrete step t to maximize the return, which is the expectation of the discounted cumulative reward. Consequently, DRL is widely applied in closed-loop control problems where the effect of the executed action influences the transitioned state. However, in optimization or open-loop control problems, the task is completed when the optimal solution for a given problem is identified without the need for continuous interactions. Single-step DRL adapts the DRL framework to solve such problems by formalizing each learning episode as a single step. Each episode terminates once the agent executes an action and receives a reward without transitioning to a subsequent state. The aim here is to maximize solely the immediate reward r, as future reward terms in Equation (1) are eliminated. Thus, the optimal solution can be precisely identified by defining the objective function as the reward. This facilitates single-step DRL to be utilized as a black-box optimizer when only a single state variable is considered [25,54–56].

Leveraging the adaptability of DRL to changing state variables, single-step DRL can efficiently solve optimization problems involving multiple and continuous ranges of condition and objective variables, a recently introduced concept known as multi-condition multi-objective optimization [30]. Multiple conditions and objectives can be considered within a single optimization problem, unlike conventional optimization methods, where multiple conditions must be treated independently, requiring extensive discretization for continuous condition ranges. This approach is especially useful for shape optimization in fluid dynamics where operating conditions change significantly. Another key feature of single-step DRL is that the optimal action can be defined in a single attempt, eliminating the repetitive process required for gradual improvement. This provides a significant advantage in automating CFD, which typically requires repeated trial-and-error procedures [57,58]. The studies reviewed in the present paper are categorized in Table 1 according to the DRL algorithms discussed above.

Table 1. Deep reinforcement learning (DRL) algorithms in the studies reviewed in the present paper.

Deep Q-networks (DQN)	[27,59–61]
Advantage actor–critic	[61,62]
Proximal policy optimization (PPO)	[11,13–21,23,24,26,28,29,31,61–73]
Soft actor–critic (SAC)	[74–77]
Deep deterministic policy gradient (DDPG)	[12,22,78–81]
Twin-delayed deep deterministic policy gradient (TD3)	[82–84]
Single-step deep reinforcement learning	[25,30,57,58]

2.4. Computational Fluid Dynamics

Various CFD methods are employed to acquire flow data for training DRL. In the laminar regime at low Reynolds numbers (*Re*), the flow data are obtained by directly solving the Navier–Stokes equations, which govern fluid motion [11–16,25,63].

In the turbulent regime, direct numerical simulation (DNS) is performed by solving the Navier–Stokes equations, resolving all scales of fluid motion in space and time without any turbulence model. Although DNS is a highly accurate method, the computational cost increases exponentially with a scale of Re^3 [85] to resolve the smallest scales of turbulent eddies. Thus, research on DRL using DNS in turbulent regions has been reported only to a limited extent [79].

On the other hand, to reduce computational costs, Reynolds-averaged Navier–Stokes (RANS) equations are solved for the mean flow variables, while the unknown terms (i.e., Reynolds stresses) are modeled with a turbulence model [85]. Since it is an efficient simulation method to obtain flow data for various turbulent flows, many studies have utilized RANS for DRL data acquisition [28,29,58,73,80,81].

Large-eddy simulation (LES), which directly resolves large-scale turbulent motions while modeling the effects of small-scale turbulent motions using a subgrid-scale model, is also used for data acquisition [72,82,84,86]. The computational cost of LES is between that of DNS and RANS [85]. Since LES explicitly solves the large-scale energetic motions, it is known to be more accurate than RANS in simulating flows with large-scale unsteadiness, such as unsteady separation and vortex shedding [85].

Additionally, attempts to train DRL using approximated solutions based on potential flow theory [27,28,30] or empirical correlations [80] have been reported.

3. Addressing Challenges in Deep Reinforcement Learning for Fluid Dynamics

The nonlinear and high-dimensional characteristics of fluid flow pose significant challenges in incorporating DRL into fluid dynamics research [34]. The key challenges include:

- Sample efficiency: The computational expense required to evaluate system dynamics increases with the complexity of geometries and flow conditions.
- Turbulence: Capturing the multi-scale nature of turbulence necessitates extensive data collection, and the stochastic properties of turbulence introduce significant variance in training data.

- Partial observability: The state space is often partially observable, necessitating methods that can effectively utilize incomplete information.

In this section, beyond the DRL algorithms discussed earlier, additional strategies recently adopted in fluid mechanics research to mitigate these challenges are introduced.

3.1. Transfer Learning

Transfer learning leverages knowledge gained from training on one task to improve learning performance on a related but different task. The primary advantage of transfer learning is increased data efficiency. The approach involves initially training on simpler tasks with lower computational costs and then utilizing the trained network to address the target problem. In recent flow control applications, networks trained on two-dimensional problems are often used to solve three-dimensional configurations [63,75,82]. For optimization problems, a multi-fidelity approach is often employed: low-cost, low-fidelity computations are performed first and progressively fine-tuned with high-cost, high-fidelity simulations [28,80].

3.2. Multi-Agent Reinforcement Learning

MARL is a framework in which multiple agents coexist within a shared environment, either operating independently or interacting with each other. A direct consequence is the acceleration of data acquisition, as multiple agents conduct tasks in parallel, improving exploration efficiency and robustness. Furthermore, when dealing with a large number of actions, assigning all actions to a single agent can lead to an excessively high-dimensional action space. MARL addresses this issue by distributing the action set among multiple agents, each responsible for a subset of actions, thereby reducing the dimensionality of the action space for each agent and improving data efficiency. Recently, MARL has been successfully applied to flow control in three-dimensional configurations [64–66] and the automation of turbulence modeling [86,87]. One approach to implementing MARL involves placing multiple agents at different positions within a fixed environment [86]. Another approach replicates multiple environments by translating the system in the invariant direction (i.e., the periodic direction in the domain). In this case, actions in each environment are determined by a shared network [64–66].

3.3. Partially Observable Markov Decision Process

In practice, an agent often receives partially observable or noisy information from the environment, without full access to the true state. A POMDP is a generalized formulation of the MDP, where the agent receives observations from the environment that provide incomplete information about the state. The agent then makes decisions based on a belief state, a probabilistic representation of potential true states derived from the history of observations and actions. Since past information is necessary to develop a reliable belief state, a time history of observations and actions is directly provided to the network [74,76], or memory-based neural networks are utilized. In the latter case, transformer architectures have recently been used more frequently than recurrent neural networks (RNNs) and long short-term memory (LSTM) networks due to the ability to handle long-term dependencies in sequential data effectively [88,89].

When selecting DRL algorithms for POMDP problems, determining the most suitable approach is challenging, as the optimal choice often depends on the specific characteristics of the environment and necessitates comprehensive empirical evaluation. In the study by Wang et al. [89], a comparative analysis of control performance between SAC, an off-policy algorithm, and PPO, an on-policy algorithm, revealed that PPO exhibited superior performance to SAC. While this outcome is not universally established in the literature, it aligns with the theoretical perspective that on-policy methods may offer greater robustness in dynamic environments by mitigating issues associated with learning from outdated or irrelevant experiences [1,48].

4. Flow Control

Various DRL-based flow control strategies have been widely developed, including drag reduction, control of biological systems, and control of heat exchange performance. Specifically, the active flow control of cylinder wakes, a representative canonical flow, is considered a benchmark case to evaluate the developed control methods.

4.1. Active Flow Control of Cylinder Wakes

4.1.1. Two-Dimensional and Low Reynolds Number Flows

The most representative case is the control of the flow over a two-dimensional cylinder for drag reduction [11–24]. The two primary control strategies for drag reduction are rotating the cylinders [14,15] and using zero-mass-flow-rate jets ejected from the surface of the cylinders [11–13,16–24], as illustrated in Figure 2. The DRL algorithms widely used for flow control of a two-dimensional cylinder are PPO [11,13–21,24] and DDPG [12,22], which are explained in Sections 2.2.1 and 2.2.2, respectively. The aforementioned studies have generally demonstrated successful drag reduction performance through mechanisms that suppress flow separation.

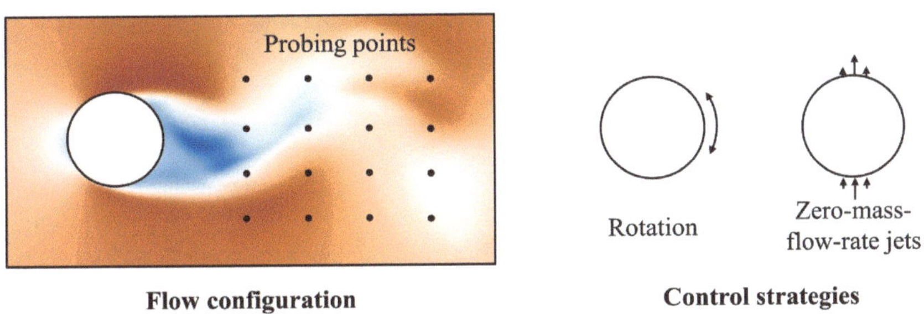

Figure 2. Flow configuration and a probe array for active flow control of cylinder wakes. Two representative control strategies for drag reduction methods using DRL: control with rotation of the cylinder and zero-mass-flow-rate jets.

The flow regimes of many previous studies are limited to the laminar regime, covering Re from 100 to 400 [11–18,20–22,24]. Tang et al. [16] simultaneously trained a single agent over a range of Re = 100 to 400 to expand the applicability of DRL to various flow conditions, reporting that a high level of irregular drag and actuation fluctuations are the main difficulties in the turbulent regime. Ren et al. [19] also reported difficulties in flow control in a weakly turbulent regime at $Re = 1000$ due to the different flow dynamics and the increased number of episodes needed to train a policy.

Thus, common challenges of DRL in flow control at high Reynolds numbers exist due to the complex and nonlinear flow physics in turbulent regimes, where small-scale instabilities and secondary vortices occur within high frequency ranges. Moreover, three-dimensional wake phenomena involving vortex shedding parallel or oblique to the span-wise direction appear at high Reynolds numbers, making three-dimensional flow control essential. In short, the objective of recent studies in DRL-based flow control is to develop efficient methods for highly turbulent conditions and to expand to more complex three-dimensional problems at the same time.

4.1.2. Three-Dimensional or Turbulent Flows

Transfer learning (Section 3.1) has been widely utilized in attempts where DRL agents are initially trained with low-cost two-dimensional flow conditions and later fine-tuned with high-cost three-dimensional flow conditions [63,75].

Yan et al. [75] utilized a transfer learning technique to develop a DRL control method to reduce aerodynamic forces acting on a three-dimensional square cylinder at a high

Reynolds number. The main idea is to pre-train the DRL agent with two-dimensional flow fields at $Re = 1000$ with lower computational cost and then transfer the agent to a three-dimensional flow condition at Re = 22,000 to reduce the total computational cost for the training. Figure 3 illustrates the process of transfer learning from two-dimensional to three-dimensional environments. They controlled flow over a square cylinder by utilizing jets at the corners of the square cylinder and the pressure measured at the surface of the cylinder as actuators and observers, respectively. As a result, they reported that the developed control method achieved a substantial drag reduction of 52.3%. It is also reported that transfer learning accelerated the training process at the high Reynolds number, achieving a 51.1% reduction in training time compared to DRL trained from scratch without transfer learning.

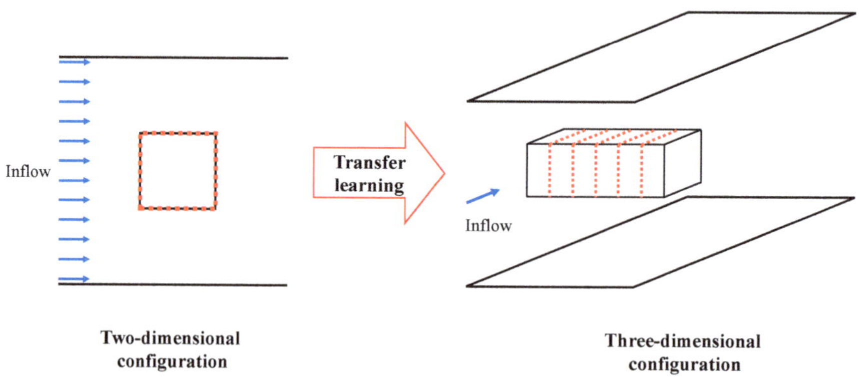

Figure 3. Schematic of a transfer learning technique from a two-dimensional configuration to a three-dimensional configuration for flow control of the wake behind a square cylinder, as adopted in the work of Yan et al. [75].

He et al. [63] employed transfer learning to utilize a DRL agent trained on a two-dimensional flow condition to a three-dimensional flow condition, aiming for drag reduction. The control method was developed with a PPO algorithm, and flow was controlled using zero-mass synthetic jets. The policy trained in the two-dimensional environment generalized well in the three-dimensional environment, confirming that transfer learning is an effective method to expand the DRL-based control method to new three-dimensional configurations.

Wang et al. [82] utilized transfer learning from low Reynolds number flows to high Reynolds number flows to develop a DRL-based control method for three-dimensional cylinder wakes, aiming to reduce drag and maximize system power by controlling the rotation of two small cylinders behind the main cylinder. To overcome the limitations of training a DRL from scratch due to the computational cost, the agent was initially trained at a low Reynolds number of $Re = 500$, which requires less computational time, and then fine-tuned at high Reynolds numbers up to $Re = 1.4 \times 10^5$. This research was the first attempt to analyze control for the wake flow at $Re = 1.4 \times 10^5$, and it was reported that transfer learning is capable of significantly reducing the training episodes.

MARL (Section 3.2) is another approach to expand DRL-based flow control methods to three-dimensional configurations by assigning agents along the spanwise direction [64,65], as illustrated in Figure 4. The main advantage of MARL is its ability to significantly reduce training costs by dividing the total action space required for an entire three-dimensional domain among multiple agents.

Suárez et al. [65] employed MARL for flow control of a three-dimensional cylinder over a range of $Re = 100$ to 400 to achieve drag reduction. They distributed multiple DRL agents to control the zero-net-mass-flux jets located along the spanwise direction of the cylinder and obtained drag reductions of 21% and 16.5% at $Re = 300$ and $Re = 400$, respectively.

More recently, Suárez et al. [64] also reported an attempt to apply MARL to turbulent flow over a cylinder at $Re = 3900$, corresponding to the subcritical flow regime. The main idea of their study is similar to that of Suárez et al. [65], involving the distribution of multiple agents in the spanwise direction to control zero-net-mass-flux jets independently. Their method achieved an 8% drag reduction in a three-dimensional turbulent flow regime.

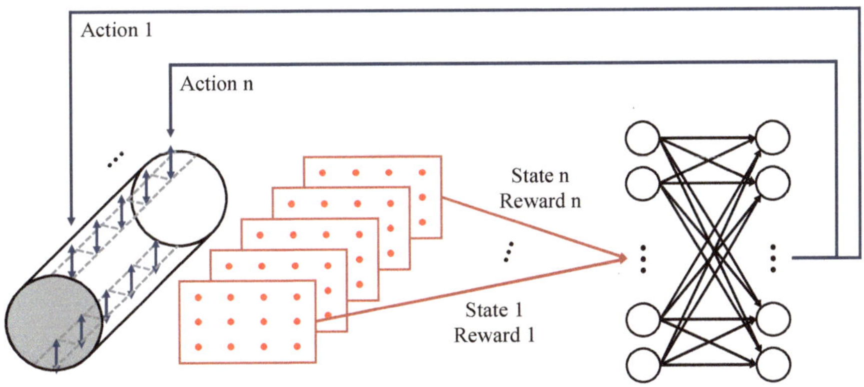

Figure 4. Multi-agent reinforcement learning (MARL) framework for control of flow over a cylinder in a three-dimensional configuration, as adopted in the work of Suárez et al. [65]. The framework involves distributing DRL agents, which control the zero-net-mass-flux jets, along the spanwise direction.

Studies addressing the practicality of control methods have been conducted considering POMDP (Section 3.3) by solely using sensor information on the surface of an object, as it is not practical to measure the entire flow field with sensor arrays in real applications.

Wang et al. [74] implemented a control method for the wake behind a three-dimensional cylinder at $Re = 10{,}000$, using only sparse surface pressure probes instead of a sensor array in the wake. Their main idea is that the agent collects data from the recent 30 time steps to include the history of observations and actions, which are essential for POMDP. In their study, a DRL agent utilizing surface pressure sensors achieved similar performance to the base DRL method that uses a sensor array fully covering the wake region, in terms of reducing drag and lift fluctuations.

Similarly, Xia et al. [76] suggested a control method for the wake behind a two-dimensional square cylinder in laminar regimes using partial surface pressure measurements. They augmented the state by including current and past observations and actions, addressing the POMDP problem, and were able to achieve drag reduction performance similar to that of using near-wake sensors.

4.2. Control of Biological Systems

One of the promising research directions that leverages DRL in more complex and realistic scenarios involves controlling the biological motion of living organisms. The ability of DRL to handle highly nonlinear data facilitates the characterization of the intricate behavior dynamics of living organisms, which were challenging to define analytically.

Verma et al. [59] investigated the energy-saving swimming dynamics of a school of fish. Unlike individual fish, the movement of each fish within a school is adjusted in response to the motions of others to optimize energy use. However, the mechanisms governing such interactions have not been well understood due to the complexity of modeling the intricate swimming patterns analytically. Instead, the authors acquired an energy-efficient swimming model by training the fish school to minimize energy consumption using DRL. The trained policy was then analyzed to examine the formation and harnessing of vortical structures within the school, elucidating how these mechanisms contribute to overall energy efficiency.

Hong et al. [83] controlled a flexible-winged flyer, approximately the size of a fruit fly, using DRL. A fluid-structure interaction simulation was incorporated to solve the complicated dynamics of the flyer accurately (see Figure 5). The position of the flyer and the surrounding flow field were used as state variables, while the frequency and amplitude of the wing movements were controlled as actions. By designing the reward function based on the distance to the destination, stable control was achieved even in complex flow environments with external disturbances such as sudden gusts of wind and raindrops.

Figure 5. Control of a flexible-winged flyer using DRL, as demonstrated by Hong et al. [83]. A fluid–structure interaction simulation is incorporated to model the complex dynamics of the flyer accurately.

4.3. Other Applications

DRL-based flow control methods have also been actively developed for various other applications in fluid mechanics, including drag reduction in turbulent channel flow [78,79] and airfoils [67], Rayleigh–Bénard convection [66], and optimal control of a wind farm [90].

For example, DRL-based control strategies for drag reduction in turbulent channel flow have been developed [78,79]. Sonoda et al. [79] achieved a 37% drag reduction by adapting the blowing or suction of normal velocity at the wall, using velocity fluctuations observed near the wall as the state. Guastoni et al. [78] further developed the control method by arranging DRL agents in a grid covering the bottom wall of the channel.

Additionally, a DRL-based control method for drag reduction of a NACA airfoil using synthetic jets, a strategy similar to those used for controlling cylinder wakes, was developed by Wang et al. [67]. In their study, a drag reduction of 27.0% and a lift enhancement of 27.7% were achieved after applying the flow control.

Vignon et al. [66] employed MARL to control Rayleigh–Bénard convection in a domain bounded by adiabatic walls on the top and the bottom. The control objective was to improve heat exchange performance, represented as a decrease in Nusselt number, by controlling the lower wall temperature. They successfully developed the control method with MARL by distributing the agents to multiple control segments of the lower wall, whereas the single-agent DRL failed to obtain an effective control policy.

5. Shape Optimization

Another area that has been actively studied using DRL is shape optimization. Numerous studies have focused on aerodynamic shape optimization [25–31,68,80,81], while other applications in shape optimization have also been reported [60,69,70].

5.1. Aerodynamic Shape Optimization

Aerodynamic shape optimization has primarily been conducted to enhance aerodynamic performance such as reducing drag or improving lift. DRL-based optimization is frequently integrated with numerical simulations. However, optimization utilizing CFD becomes computationally expensive as the Reynolds number increases due to the complexity and nonlinearity of the fluid motions. As a result, many studies have focused on using low-fidelity models or on flows at relatively low Reynolds numbers.

Viquerat et al. [25] introduced a direct shape optimization method using DRL and CFD to determine the optimal two-dimensional shape of an airfoil that maximizes the lift-to-drag ratio. The shape was represented by a set of Bézier curves defined by multiple control points. The DRL agent controls the position and the local curvature of each control point. Consequently, without prior knowledge, the DRL network generated an airfoil-like shape that enhances the lift-to-drag ratio. A single-step DRL approach was employed to generate the optimal shape from the initial state in a single attempt. However, the study was limited to relatively low Reynolds numbers.

Recently, transfer learning has been widely used to address issues related to computational cost. Bhola et al. [28] conducted a study on airfoil shape optimization to minimize drag at high Reynolds numbers using transfer learning. The method employed a multi-fidelity approach integrating potential flow and RANS simulations. The airfoil shapes were parameterized using Bézier curves, defined by six control points and the leading-edge radius. Initially, the network was trained with low-fidelity data from potential flow simulations. The trained weights were then fine-tuned using high-fidelity data from RANS simulations, as illustrated in Figure 6. By utilizing the multi-fidelity transfer learning approach, computational costs were significantly reduced compared to retraining from scratch. Similarly, Yan et al. [80] applied multi-fidelity transfer learning to enhance the lift-to-drag ratio in the aerodynamic shape optimization of missile control surfaces. The study integrated semi-empirical methods with CFD, resulting in significant computational cost savings.

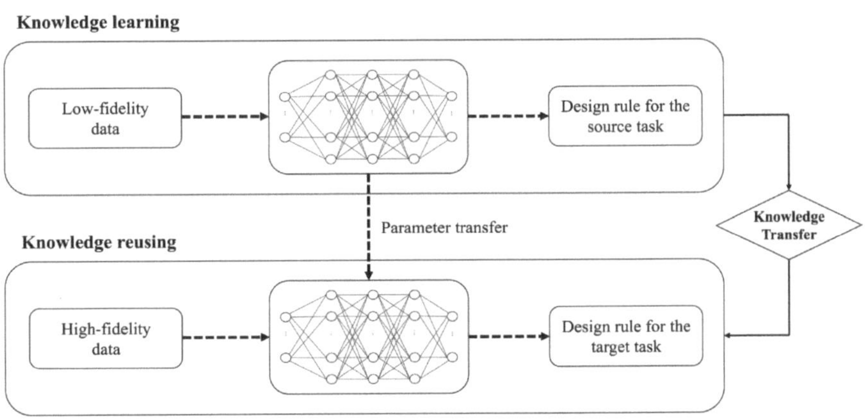

Figure 6. Schematic diagram of the transfer learning framework, as adopted in the work of Bhola et al. [28]. The network is initially trained with low-fidelity data, and then the pre-trained weights are fine-tuned with high-fidelity data.

Another significant issue in DRL-based shape optimization is that the optimization process must be restarted from scratch for different problems, such as different target geometries or flow conditions. Such a requirement adversely affects the generalization performance of shape optimization using DRL. Therefore, methods that can be applied to various problem scenarios have recently been developed.

Li et al. [29] utilized DRL and a DNN-based surrogate model to develop a policy for reducing the aerodynamic drag of supercritical airfoils. The surrogate model, trained with CFD data, was used for performance evaluation to reduce computational costs. The agent was trained to execute actions based on features of the wall Mach number distribution to enhance the generalizability across different airfoils and flow conditions. In another study, Kim et al. [30] developed a multi-condition multi-objective optimization method and applied it to airfoil shape optimization. The method not only accommodated multi-objective problems (i.e., lift and lift-to-drag ratio) but also addressed various conditions (i.e., Reynolds and Mach numbers). The developed method was able to obtain the optimal shape without the need for retraining the optimization process from scratch, even if conditions or objectives change (see Figure 7).

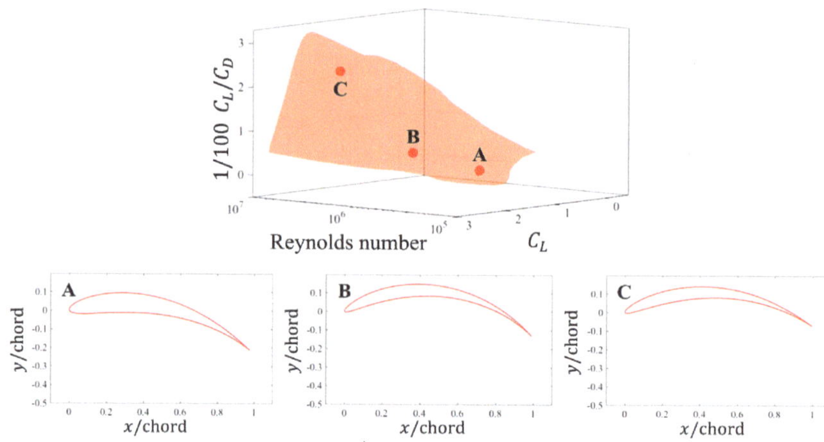

Figure 7. Results of multi-condition multi-objective optimization in the work of Kim et al. [30]. Each point represents the optimal shape under different conditions and objectives.

5.2. Other Applications

In addition to research focusing on aerodynamic shape optimization, other applications of DRL-based shape optimization have also been studied. Keramati et al. [69] conducted a study to optimize the fin shape at solid–fluid interfaces of a heat exchanger to maximize the heat transfer-to-pressure drop ratio. The fin geometry was represented by a set of Bézier curves, and the training of DRL was performed by gradually adjusting the positions of control points. As a result, the overall heat transfer of the optimized shape was improved and the pressure drop was reduced compared to the rectangular reference geometry.

Ma et al. [60] optimized a rocket nozzle to improve the thrust using DRL and a surrogate model parameterized by a convolutional neural network (CNN). The surrogate model was initially developed to predict the flow field based on geometric information and was used for performance evaluation with reduced computational cost. The nozzle shape was parameterized by basis spline (B-spline) curves, where the positions of the control points were optimized during the training. The DRL agent was able to produce an optimized nozzle shape that resembles the theoretically driven solution exhibiting improved thrust compared to the baseline model.

6. Automation of Computational Fluid Dynamics

Developing accurate, reliable, and efficient CFD methods, such as constructing high-quality meshes, designing precise turbulence models, and formulating effective numerical schemes, is a highly expert-intensive task. Given that the Navier–Stokes equations do not have a priori solutions, even experts must rely on iterative and experience-dependent approaches to develop effective techniques based on physical intuition. Recently, DRL has been explored as an automatic development tool for CFD methods, utilizing its trial-and-error learning approach to replace the heuristic methods traditionally dependent on expert intervention. This section reviews recent research on the automation of CFD using DRL, focusing on mesh generation, turbulence modeling, and numerical schemes.

6.1. Mesh Generation

Mesh generation is a foundational step in conducting CFD, but it remains one of the most time-consuming and labor-intensive tasks. Despite advancements in automation methods, developing techniques that achieve both the diversity of applications and the high quality required for accuracy solely through human intuition is challenging. Manual interventions are often required to obtain optimal meshes iteratively for each new configuration. Recently, DRL, with its ability to dynamically adapt to changing state variables for maximum cumulative reward, has emerged as a promising solution to the generality and optimality challenges inherent in traditional automation techniques.

Pan et al. [77] employed DRL to automatically generate high-quality quadrilateral meshes within two-dimensional domains. In their approach, the current boundary configuration was defined as the state, and creating a quadrilateral element constituted the action. Upon executing the action, the subsequent state was updated to reflect the new boundary formed after the element was added. By defining partially observable boundaries based on specified vertices, rather than the entire geometry, the method was able to achieve generalizability across various geometries.

While the previous study focused on automating mesh generation from a geometrical perspective, Foucart et al. [61] and Dzanic et al. [71] optimized mesh resolution to balance simulation accuracy and computational cost. The method employed DRL to explore adaptive mesh refinement strategies, dynamically adjusting the resolution in critical regions. The trained DRL agent evaluated each cell to determine whether further refinement or coarsening is necessary, defining such decisions as actions. By utilizing a POMDP formulation, the agent effectively learned from the partial information surrounding the cell where the action was performed.

Kim et al. [58] developed a method to generate optimal meshes in terms of both geometric quality and resolution. The study targeted RANS simulations of flow through a blade passage. Given various blade profiles and flow conditions, the DRL agents generated meshes and conducted CFD simulations, optimizing the accuracy and cost of the solutions within a multi-objective framework. The novelty of the study lies in the utilization of a single-step DRL approach, enabling the acquisition of the optimal mesh and solution in a single attempt based on the given configuration. Consequently, a grid-converged solution was obtained in a non-iterative manner, eliminating the need for repetitive CFD processes required for grid convergence tests, thereby automating the CFD workflow with great efficiency.

The study strategically separated the training phase into two steps to alleviate the computational burden of data acquisition with CFD. The approach allowed for the segmentation of the overall action space dimension. The initial phase, as outlined in the work of Kim et al. [57], focused on optimizing the geometric quality of the mesh independently of CFD simulations. The agents were trained to improve quality metrics, such as the min-max ratio of determinants of the Jacobian matrices and cell skewness, for randomly assigned blade profiles in each training episode. Figure 8 illustrates an example of the training process for a specific blade profile, clearly demonstrating the capability of DRL to generate higher-quality meshes. It is important to note that the actual training was conducted with randomly assigned blade profiles to ensure generalizability.

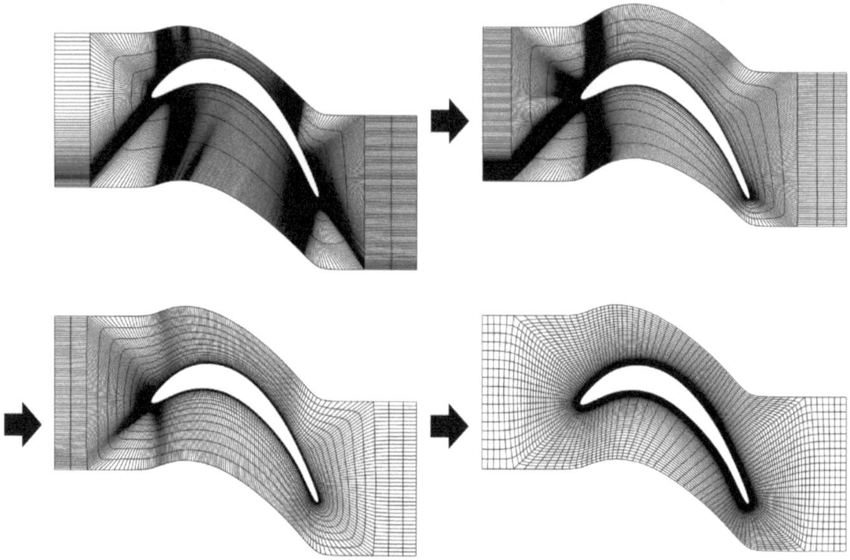

Figure 8. Training process for automatic mesh generation for a blade passage aimed at maximizing geometric quality, as described in the work of Kim et al. [57].

In the subsequent phase, the pre-trained network for geometric quality was refined to define the optimal resolution by incorporating CFD simulations. To further accelerate data acquisition, the study employed the MARL framework. As illustrated in Figure 9, multiple agents simultaneously performed tasks of mesh generation and CFD analysis in parallel, continuously providing data to the main network to optimize the accuracy and cost of the solutions. The trained network was able to non-iteratively produce grid-converged solutions at a desired computational expense for various blade configurations, effectively capturing complex physical phenomena such as shock waves and flow separation.

6.2. Turbulence Modeling

In the field of data-driven turbulence modeling, there have been many attempts to develop turbulence models using supervised learning methods. Regarding turbulence models for LES, for example, the models with supervised learning have shown great possibilities showing partial improvement over algebraic turbulence models for canonical turbulent flows, such as homogeneous isotropic turbulence [91–96], turbulent channel flow [97–99], and flow over a circular cylinder [100]. However, since labeled data are required for supervised learning, a higher fidelity simulation (i.e., DNS) needs to be conducted a priori to obtain the unresolved flow quantities as labeled target data, which limits the practicality.

Therefore, DRL frameworks for turbulence closure modeling have been newly proposed by controlling the unresolved closure term as an action and using the errors in the statistical properties of the reference simulation as a reward. Novati et al. [86] integrated MARL with an LES solver by deploying DRL agents at sampling points distributed throughout a computational domain of homogeneous isotropic turbulence, and achieved good generalization performance across Reynolds numbers and grid sizes. In the study of Kurz et al. [72], a similar DRL-based closure model for LES of homogeneous isotropic turbulence, in which a CNN is used for the policy network instead of fully connected layers, was developed.

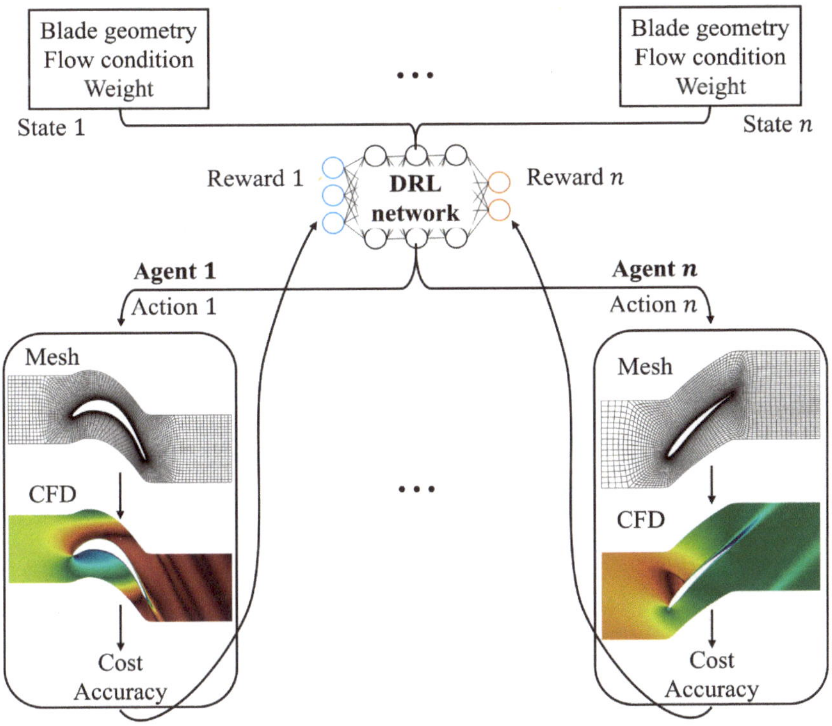

Figure 9. Schematic of the MARL framework for automatic mesh generation to optimize the accuracy and cost of computational fluid dynamics solutions for a blade passage, as adopted in the work of Kim et al. [58].

Kim et al. [84] applied a similar DRL approach to turbulence modeling for LES in turbulent channel flow. In their method, the DRL produces the subgrid-scale stress for wall-bounded turbulent flow during the simulation, based on a reward designed to reduce errors in the statistical properties (see Figure 10). They reported better performance of the developed DRL model compared to algebraic subgrid-scale models in various turbulence statistics.

Bae and Koumoutsakos [87] further developed the MARL approach for wall modeling for LES. Agents are distributed at the walls of the channel, and their actions are given as the magnitude of the modeled wall shear stress, while the reward is defined to minimize the error between the modeled and the true mean wall shear stress. The developed wall model successfully generalized to extremely high Reynolds numbers, exhibiting similar off-wall velocity profiles to those from fully resolved simulations.

In the field of RANS closure modeling, Fuchs et al. [73] introduced a DRL-based model by adding a source term into the transport equation of the Spalart–Allmaras RANS model, controlled by a DRL algorithm. A reward is defined to reduce the error between the velocity field using the developed RANS model and the time-averaged LES field. Their model is trained for turbulent jet flow at Re = 10,000 and successfully generalized to Re = 15,000.

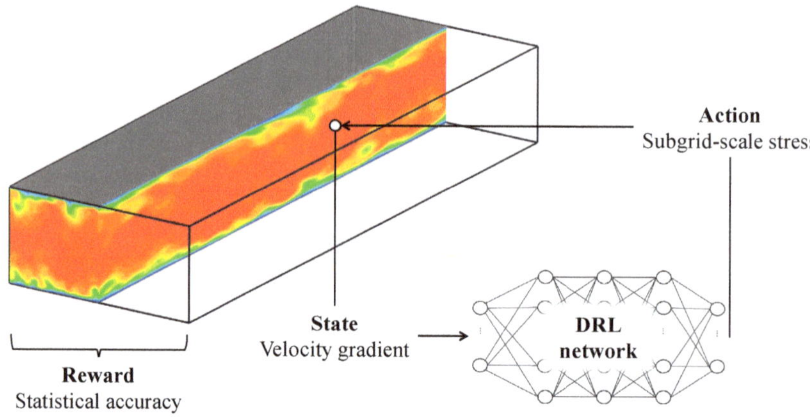

Figure 10. Schematic of a DRL-based subgrid-scale model for large-eddy simulation (LES) of turbulent channel flow, as adopted in the work of Kim et al. [84]. The DRL produces the subgrid-scale stress as an action during the simulation, based on a state defined as local velocity gradient. Statistical errors in the mean viscous stress and mean Reynolds shear stress are used to calculate a reward.

6.3. Numerical Scheme

Feng et al. [62] optimized the numerical scheme for compressible flow simulations to dynamically balance dispersion and dissipation according to flow evolution. To analyze compressible flow, significant gradients, such as shock waves, necessitate numerical dissipation to stabilize the simulation. However, most existing methods apply dissipation non-selectively across scales, leading to physical inconsistencies. The study aimed to develop a physically consistent numerical scheme by dynamically adjusting the scheme to different subgrid-scale types: genuine (discontinuities and interfaces) and non-genuine (fluctuations and turbulence). Specifically, tunable parameters of the fifth-order targeted essentially non-oscillatory scheme (TENO5), which determine the dissipation and dispersion terms, were optimized using DRL. The current flow variables are defined as the state, and the parameters of the TENO5 scheme are optimized as the action. As a result, the trained network was able to capture small-scale dynamics while maintaining the stability of the simulation. Although there is currently limited literature in this area, DRL has the potential to develop a self-adaptive scheme across different flow configurations, automating the definition of appropriate numerical schemes in the CFD process.

7. Concluding Remarks

Recent advancements in the application of deep reinforcement learning (DRL) to fluid dynamics problems have been comprehensively reviewed. The present review not only synthesizes current research but also highlights strategies employed to overcome challenges in applying DRL to complex, real-world engineering problems, including data efficiency, turbulence, and partial observability. Specifically, the use of transfer learning, multi-agent reinforcement learning, and the partially observable Markov decision process in DRL-based fluid dynamics research is explained. The reviewed literature is categorized by application: flow control and shape optimization, which are the main fields where DRL is currently applied, as well as automation of computational fluid dynamics, an emerging and promising field for advancing the efficiency and reliability of numerical analysis. Despite such progress, ongoing research is essential toward the ultimate objective of training control, optimization, and automation policies capable of addressing intricate geometries and highly turbulent regimes and generalizing such capabilities across diverse configurations. Continued innovation in algorithm development, data handling methods, and computational techniques will be crucial. Exploring advancements to address related problems in different fields, such as computer vision, robotics, and natural language processing, can offer valuable insights.

Funding: The work was supported by the National Research Foundation of Korea (NRF) under the Grant Number NRF-2021R1A2C2092146 and the Korea Institute of Energy Technology Evaluation and Planning (KETEP) under the Grant Number RS-2023-00242282.

Data Availability Statement: Data sharing is not applicable since no new data were created or analyzed in this study.

Conflicts of Interest: The authors declare no conflicts of interest.

References

1. Sutton, R.S.; Barto, A.G. *Reinforcement Learning: An Introduction;* MIT Press: Cambridge, MA, USA, 2018.
2. Tesauro, G. Temporal difference learning and TD-Gammon. *Commun. ACM* **1995**, *38*, 58–68. [CrossRef]
3. Hornik, K.; Stinchcombe, M.; White, H. Multilayer feedforward networks are universal approximators. *Neural Netw.* **1989**, *2*, 359–366. [CrossRef]
4. Pinto, L.; Andrychowicz, M.; Welinder, P.; Zaremba, W.; Abbeel, P. Asymmetric actor critic for image-based robot learning. *arXiv* **2017**, arXiv:1710.06542.
5. Kendall, A.; Hawke, J.; Janz, D.; Mazur, P.; Reda, D.; Allen, J.M.; Lam, V.D.; Bewley, A.; Shah, A. Learning to drive in a day. In Proceedings of the 2019 International Conference on Robotics and Automation (ICRA), Montreal, QC, Canada, 20–24 May 2019; pp. 8248–8254.
6. Bewley, A.; Rigley, J.; Liu, Y.; Hawke, J.; Shen, R.; Lam, V.D.; Kendall, A. Learning to drive from simulation without real world labels. In Proceedings of the 2019 International Conference on Robotics and Automation (ICRA), Montreal, QC, Canada, 20–24 May 2019; pp. 4818–4824.
7. Bahdanau, D.; Brakel, P.; Xu, K.; Goyal, A.; Lowe, R.; Pineau, J.; Courville, A.; Bengio, Y. An actor-critic algorithm for sequence prediction. *arXiv* **2016**, arXiv:1607.07086.
8. Mnih, V.; Kavukcuoglu, K.; Silver, D.; Graves, A.; Antonoglou, I.; Wierstra, D.; Riedmiller, M. Playing atari with deep reinforcement learning. *arXiv* **2013**, arXiv:1312.5602.
9. Silver, D.; Schrittwieser, J.; Simonyan, K.; Antonoglou, I.; Huang, A.; Guez, A.; Hubert, T.; Baker, L.; Lai, M.; Bolton, A.; et al. Mastering the game of go without human knowledge. *Nature* **2017**, *550*, 354–359. [CrossRef] [PubMed]
10. Brown, N.; Sandholm, T. Superhuman AI for multiplayer poker. *Science* **2019**, *365*, 885–890. [CrossRef]
11. Rabault, J.; Kuchta, M.; Jensen, A.; Réglade, U.; Cerardi, N. Artificial neural networks trained through deep reinforcement learning discover control strategies for active flow control. *J. Fluid Mech.* **2019**, *865*, 281–302. [CrossRef]
12. Koizumi, H.; Tsutsumi, S.; Shima, E. Feedback control of Karman vortex shedding from a cylinder using deep reinforcement learning. In Proceedings of the 2018 Flow Control Conference, Atlanta, GA, USA, 25–29 June 2018; p. 3691.
13. Rabault, J.; Kuhnle, A. Accelerating deep reinforcement learning strategies of flow control through a multi-environment approach. *Phys. Fluids* **2019**, *31*, 094105. [CrossRef]
14. Tokarev, M.; Palkin, E.; Mullyadzhanov, R. Deep reinforcement learning control of cylinder flow using rotary oscillations at low Reynolds number. *Energies* **2020**, *13*, 5920. [CrossRef]
15. Xu, H.; Zhang, W.; Deng, J.; Rabault, J. Active flow control with rotating cylinders by an artificial neural network trained by deep reinforcement learning. *J. Hydrodyn.* **2020**, *32*, 254–258. [CrossRef]
16. Tang, H.; Rabault, J.; Kuhnle, A.; Wang, Y.; Wang, T. Robust active flow control over a range of Reynolds numbers using an artificial neural network trained through deep reinforcement learning. *Phys. Fluids* **2020**, *32*, 053605. [CrossRef]
17. Paris, R.; Beneddine, S.; Dandois, J. Robust flow control and optimal sensor placement using deep reinforcement learning. *J. Fluid Mech.* **2021**, *913*, A25. [CrossRef]
18. Qin, S.; Wang, S.; Rabault, J.; Sun, G. An application of data driven reward of deep reinforcement learning by dynamic mode decomposition in active flow control. *arXiv* **2021**, arXiv:2106.06176.
19. Ren, F.; Rabault, J.; Tang, H. Applying deep reinforcement learning to active flow control in weakly turbulent conditions. *Phys. Fluids* **2021**, *33*, 037121. [CrossRef]
20. Li, J.; Zhang, M. Reinforcement-learning-based control of confined cylinder wakes with stability analyses. *J. Fluid Mech.* **2022**, *932*, A44. [CrossRef]
21. Castellanos, R.; Cornejo Maceda, G.; De La Fuente, I.; Noack, B.; Ianiro, A.; Discetti, S. Machine-learning flow control with few sensor feedback and measurement noise. *Phys. Fluids* **2022**, *34*, 047118. [CrossRef]
22. Pino, F.; Schena, L.; Rabault, J.; Mendez, M.A. Comparative analysis of Mach. Learn. methods for active flow control. *J. Fluid Mech.* **2023**, *958*, A39. [CrossRef]
23. Mei, Y.F.; Zheng, C.; Hua, Y.; Zhao, Q.; Wu, P.; Wu, W.T. Active control for the flow around various geometries through deep reinforcement learning. *Fluid Dyn. Res.* **2022**, *54*, 015510. [CrossRef]
24. Mao, Y.; Zhong, S.; Yin, H. Active flow control using deep reinforcement learning with time delays in Markov decision process and autoregressive policy. *Phys. Fluids* **2022**, *34*, 053602. [CrossRef]

25. Viquerat, J.; Rabault, J.; Kuhnle, A.; Ghraieb, H.; Larcher, A.; Hachem, E. Direct shape optimization through deep reinforcement learning. *J. Comput. Phys.* **2021**, *428*, 110080. [CrossRef]
26. Dussauge, T.P.; Sung, W.J.; Pinon Fischer, O.J.; Mavris, D.N. A reinforcement learning approach to airfoil shape optimization. *Sci. Rep.* **2023**, *13*, 9753. [CrossRef] [PubMed]
27. Lou, J.; Chen, R.; Liu, J.; Bao, Y.; You, Y.; Chen, Z. Aerodynamic optimization of airfoil based on deep reinforcement learning. *Phys. Fluids* **2023**, *35*, 037128. [CrossRef]
28. Bhola, S.; Pawar, S.; Balaprakash, P.; Maulik, R. Multi-fidelity reinforcement learning framework for shape optimization. *J. Comput. Phys.* **2023**, *482*, 112018. [CrossRef]
29. Li, R.; Zhang, Y.; Chen, H. Learning the aerodynamic design of supercritical airfoils through deep reinforcement learning. *AIAA J.* **2021**, *59*, 3988–4001. [CrossRef]
30. Kim, S.; Kim, I.; You, D. Multi-condition multi-objective optimization using deep reinforcement learning. *J. Comput. Phys.* **2022**, *462*, 111263. [CrossRef]
31. Hui, X.; Wang, H.; Li, W.; Bai, J.; Qin, F.; He, G. Multi-object aerodynamic design optimization using deep reinforcement learning. *AIP Adv.* **2021**, *11*, 085311. [CrossRef]
32. Rabault, J.; Ren, F.; Zhang, W.; Tang, H.; Xu, H. Deep reinforcement learning in fluid mechanics: A promising method for both active flow control and shape optimization. *J. Hydrodyn.* **2020**, *32*, 234–246. [CrossRef]
33. Garnier, P.; Viquerat, J.; Rabault, J.; Larcher, A.; Kuhnle, A.; Hachem, E. A review on deep reinforcement learning for fluid mechanics. *Comput. Fluids* **2021**, *225*, 104973. [CrossRef]
34. Viquerat, J.; Meliga, P.; Larcher, A.; Hachem, E. A review on deep reinforcement learning for fluid mechanics: An update. *Phys. Fluids* **2022**, *34*, 111301. [CrossRef]
35. Vignon, C.; Rabault, J.; Vinuesa, R. Recent advances in applying deep reinforcement learning for flow control: Perspectives and future directions. *Phys. Fluids* **2023**, *35*, 031301. [CrossRef]
36. Watkins, C.J.C.H. Learning from Delayed Rewards. Ph.D. Thesis, King's College, Cambridge, UK, 1989.
37. Mnih, V.; Kavukcuoglu, K.; Silver, D.; Rusu, A.A.; Veness, J.; Bellemare, M.G.; Graves, A.; Riedmiller, M.; Fidjeland, A.K.; Ostrovski, G.; et al. Human-level control through deep reinforcement learning. *Nature* **2015**, *518*, 529–533. [CrossRef] [PubMed]
38. Van Hasselt, H.; Guez, A.; Silver, D. Deep reinforcement learning with double Q-Learning. In Proceedings of the AAAI Conference on Artificial Intelligence, Phoenix, AZ, USA, 12–17 February 2016; Volume 30.
39. Wang, Z.; Schaul, T.; Hessel, M.; Hasselt, H.; Lanctot, M.; Freitas, N. Dueling network architectures for deep reinforcement learning. In Proceedings of the 33rd International Conference on Machine Learning, New York, NY, USA, 19–24 June 2016; pp. 1995–2003.
40. Schaul, T.; Quan, J.; Antonoglou, I.; Silver, D. Prioritized experience replay. *arXiv* **2015**, arXiv:1511.05952.
41. Hessel, M.; Modayil, J.; Van Hasselt, H.; Schaul, T.; Ostrovski, G.; Dabney, W.; Horgan, D.; Piot, B.; Azar, M.; Silver, D. Rainbow: Combining improvements in deep reinforcement learning. In Proceedings of the AAAI Conference on Artificial Intelligence, New Orleans, LA, USA, 2–7 February 2018; Volume 32.
42. Sutton, R.S.; McAllester, D.; Singh, S.; Mansour, Y. Policy gradient methods for reinforcement learning with function approximation. *Adv. Neural Inf. Process. Syst.* **1999**, *12*.
43. Williams, R.J. Simple statistical gradient-following algorithms for connectionist reinforcement learning. *Mach. Learn.* **1992**, *8*, 229–256. [CrossRef]
44. Konda, V.; Tsitsiklis, J. Actor-critic algorithms. *Adv. Neural Inf. Process. Syst.* **1999**, *12*.
45. Peters, J.; Schaal, S. Natural actor-critic. *Neurocomputing* **2008**, *71*, 1180–1190. [CrossRef]
46. Bhatnagar, S.; Ghavamzadeh, M.; Lee, M.; Sutton, R.S. Incremental natural actor-critic algorithms. *Adv. Neural Inf. Process. Syst.* **2007**, *20*.
47. Mnih, V.; Badia, A.P.; Mirza, M.; Graves, A.; Lillicrap, T.; Harley, T.; Silver, D.; Kavukcuoglu, K. Asynchronous methods for deep reinforcement learning. In Proceedings of the International conference on Machine Learning, New York City, NY, USA, 19–24 June 2016; pp. 1928–1937.
48. Schulman, J.; Wolski, F.; Dhariwal, P.; Radford, A.; Klimov, O. Proximal policy optimization algorithms. *arXiv* **2017**, arXiv:1707.06347.
49. Haarnoja, T.; Zhou, A.; Abbeel, P.; Levine, S. Soft actor-critic: Off-policy maximum entropy deep reinforcement learning with a stochastic actor. In Proceedings of the International Conference on Machine Learning, PMLR, Stockholm, Sweden, 10–15 July 2018; pp. 1861–1870.
50. Haarnoja, T.; Zhou, A.; Hartikainen, K.; Tucker, G.; Ha, S.; Tan, J.; Kumar, V.; Zhu, H.; Gupta, A.; Abbeel, P.; et al. Soft actor-critic algorithms and applications. *arXiv* **2018**, arXiv:1812.05905.
51. Silver, D.; Lever, G.; Heess, N.; Degris, T.; Wierstra, D.; Riedmiller, M. Deterministic policy gradient algorithms. In Proceedings of the International Conference on Machine Learning, PMLR, Beijing, China, 22–24 June 2014; pp. 387–395.
52. Lillicrap, T.P.; Hunt, J.J.; Pritzel, A.; Heess, N.; Erez, T.; Tassa, Y.; Silver, D.; Wierstra, D. Continuous control with deep reinforcement learning. *arXiv* **2015**, arXiv:1509.02971.
53. Fujimoto, S.; Hoof, H.; Meger, D. Addressing function approximation error in actor-critic methods. In Proceedings of the International conference on Machine Learning, PMLR, Stockholm, Sweden, 10–15 July 2018; pp. 1587–1596.

54. Ghraieb, H.; Viquerat, J.; Larcher, A.; Meliga, P.; Hachem, E. Single-step deep reinforcement learning for open-loop control of laminar and turbulent flows. *Phys. Rev. Fluids* **2021**, *6*, 053902. [CrossRef]
55. Hachem, E.; Ghraieb, H.; Viquerat, J.; Larcher, A.; Meliga, P. Deep reinforcement learning for the control of conjugate heat transfer. *J. Comput. Phys.* **2021**, *436*, 110317. [CrossRef]
56. Viquerat, J.; Duvigneau, R.; Meliga, P.; Kuhnle, A.; Hachem, E. Policy-based optimization: Single-step policy gradient method seen as an evolution strategy. *Neural Comput. Appl.* **2023**, *35*, 449–467. [CrossRef]
57. Kim, I.; Kim, S.; You, D. Non-iterative generation of an optimal mesh for a blade passage using deep reinforcement learning. *Comput. Phys. Commun.* **2024**, *294*, 108962. [CrossRef]
58. Kim, I.; Chae, J.; You, D. Optimal mesh generation for a non-iterative grid-converged solution of flow through a blade passage using deep reinforcement learning. *arXiv* **2024**, arXiv:2402.15079.
59. Verma, S.; Novati, G.; Koumoutsakos, P. Efficient collective swimming by harnessing vortices through deep reinforcement learning. *Proc. Natl. Acad. Sci. USA* **2018**, *115*, 5849–5854. [CrossRef]
60. Ma, H.; Liu, J.; Ye, M.; Haidn, O.J. A comprehensive deep learning geometric shape optimization framework with field prediction surrogate and reinforcement learning. *Phys. Fluids* **2024**, *36*, 045154. [CrossRef]
61. Foucart, C.; Charous, A.; Lermusiaux, P.F. Deep reinforcement learning for adaptive mesh refinement. *J. Comput. Phys.* **2023**, *491*, 112381. [CrossRef]
62. Feng, Y.; Schranner, F.S.; Winter, J.; Adams, N.A. A deep reinforcement learning framework for dynamic optimization of numerical schemes for compressible flow simulations. *J. Comput. Phys.* **2023**, *493*, 112436. [CrossRef]
63. He, X.J.; Wang, Y.Z.; Hua, Y.; Chen, Z.H.; Li, Y.B.; Wu, W.T. Policy transfer of reinforcement learning-based flow control: From two-to three-dimensional environment. *Phys. Fluids* **2023**, *35*, 055116.
64. Suárez, P.; Alcantara-Avila, F.; Miró, A.; Rabault, J.; Font, B.; Lehmkuhl, O.; Vinuesa, R. Active flow control for drag reduction through multi-agent reinforcement learning on a turbulent cylinder at $Re_D = 3900$. *arXiv* **2024**, arXiv:2405.17655.
65. Suárez, P.; Álcantara-Ávila, F.; Rabault, J.; Miró, A.; Font, B.; Lehmkuhl, O.; Vinuesa, R. Flow control of three-dimensional cylinders transitioning to turbulence via multi-agent reinforcement learning. *arXiv* **2024**, arXiv:2405.17210.
66. Vignon, C.; Rabault, J.; Vasanth, J.; Alcántara-Ávila, F.; Mortensen, M.; Vinuesa, R. Effective control of two-dimensional Rayleigh–Bénard convection: Invariant multi-agent reinforcement learning is all you need. *Phys. Fluids* **2023**, *35*, 065146. [CrossRef]
67. Wang, Y.Z.; Mei, Y.F.; Aubry, N.; Chen, Z.; Wu, P.; Wu, W.T. Deep reinforcement learning based synthetic jet control on disturbed flow over airfoil. *Phys. Fluids* **2022**, *34*, 033606. [CrossRef]
68. Patel, J.; Spyridis, Y.; Argyriou, V. Enhancing vehicle aerodynamics with deep reinforcement learning in voxelised models. *arXiv* **2024**, arXiv:2405.11492.
69. Keramati, H.; Hamdullahpur, F.; Barzegari, M. Deep reinforcement learning for heat exchanger shape optimization. *Int. J. Heat Mass Transfer* **2022**, *194*, 123112. [CrossRef]
70. Wolff, D.; Fricke, C.; Kemmerling, M.; Elgeti, S. Towards shape optimization of flow channels in profile extrusion dies using reinforcement learning. *Proc. Appl. Math. Mech.* **2023**, *22*, e202200009. [CrossRef]
71. Dzanic, T.; Mittal, K.; Kim, D.; Yang, J.; Petrides, S.; Keith, B.; Anderson, R. DynAMO: Multi-agent reinforcement learning for dynamic anticipatory mesh optimization with applications to hyperbolic conservation laws. *J. Comput. Phys.* **2024**, *506*, 112924. [CrossRef]
72. Kurz, M.; Offenhäuser, P.; Beck, A. Deep reinforcement learning for turbulence modeling in large eddy simulations. *Int. J. Heat Fluid Flow* **2023**, *99*, 109094. [CrossRef]
73. Fuchs, L.M.; von Saldern, J.G.; Kaiser, T.L.; Oberleithner, K. Deep reinforcement learning-augmented Spalart–Allmaras turbulence model: Application to a turbulent round jet flow. *Fluids* **2024**, *9*, 88. [CrossRef]
74. Wang, Q.; Yan, L.; Hu, G.; Chen, W.; Rabault, J.; Noack, B.R. Dynamic feature-based deep reinforcement learning for flow control of circular cylinder with sparse surface pressure sensing. *J. Fluid Mech.* **2024**, *988*, A4. [CrossRef]
75. Yan, L.; Hu, G.; Chen, W.; Noack, B.R. Deep reinforcement transfer learning for active flow control of a 3D square cylinder under state dimension mismatch. *arXiv* **2024**, arXiv:2401.12543.
76. Xia, C.; Zhang, J.; Kerrigan, E.C.; Rigas, G. Active flow control for bluff body drag reduction using reinforcement learning with partial measurements. *J. Fluid Mech.* **2024**, *981*, A17. [CrossRef]
77. Pan, J.; Huang, J.; Cheng, G.; Zeng, Y. Reinforcement learning for automatic quadrilateral mesh generation: A soft actor–critic approach. *Neural Netw.* **2023**, *157*, 288–304. [CrossRef] [PubMed]
78. Guastoni, L.; Rabault, J.; Schlatter, P.; Azizpour, H.; Vinuesa, R. Deep reinforcement learning for turbulent drag reduction in channel flows. *Eur. Phys. J. E* **2023**, *46*, 27. [CrossRef]
79. Sonoda, T.; Liu, Z.; Itoh, T.; Hasegawa, Y. Reinforcement learning of control strategies for reducing skin friction drag in a fully developed turbulent channel flow. *J. Fluid Mech.* **2023**, *960*, A30. [CrossRef]
80. Yan, X.; Zhu, J.; Kuang, M.; Wang, X. Aerodynamic shape optimization using a novel optimizer based on Mach. Learn. techniques. *Aerosp. Sci. Technol.* **2019**, *86*, 826–835. [CrossRef]
81. Qin, S.; Wang, S.; Wang, L.; Wang, C.; Sun, G.; Zhong, Y. Multi-objective optimization of cascade blade profile based on reinforcement learning. *Appl. Sci.* **2020**, *11*, 106. [CrossRef]

82. Wang, Z.; Fan, D.; Jiang, X.; Triantafyllou, M.S.; Karniadakis, G.E. Deep reinforcement transfer learning of active control for bluff body flows at high Reynolds number. *J. Fluid Mech.* **2023**, *973*, A32. [CrossRef]
83. Hong, S.; Kim, S.; You, D. Control of a fly-mimicking flyer in complex flow using deep reinforcement learning. *arXiv* **2021**, arXiv:2111.03454.
84. Kim, J.; Kim, H.; Kim, J.; Lee, C. Deep reinforcement learning for large-eddy simulation modeling in wall-bounded turbulence. *Phys. Fluids* **2022**, *34*, 105132. [CrossRef]
85. Pope, S.B. *Turbulent Flows*; Cambridge University Press: Cambridge, UK, 2001.
86. Novati, G.; de Laroussilhe, H.L.; Koumoutsakos, P. Automating turbulence modelling by multi-agent reinforcement learning. *Nat. Mach. Intell.* **2021**, *3*, 87–96. [CrossRef]
87. Bae, H.J.; Koumoutsakos, P. Scientific multi-agent reinforcement learning for wall-models of turbulent flows. *Nat. Commun.* **2022**, *13*, 1443. [CrossRef] [PubMed]
88. Nasir, Y.; Durlofsky, L.J. Deep reinforcement learning for optimal well control in subsurface systems with uncertain geology. *J. Comput. Phys.* **2023**, *477*, 111945. [CrossRef]
89. Wang, Z.; Lin, R.; Zhao, Z.; Chen, X.; Guo, P.; Yang, N.; Wang, Z.; Fan, D. Learn to flap: Foil non-parametric path planning via deep reinforcement learning. *J. Fluid Mech.* **2024**, *984*, A9. [CrossRef]
90. Kim, T.; Kim, C.; Song, J.; You, D. Optimal control of a wind farm in time-varying wind using deep reinforcement learning. *Energy* **2024**, 303, 131950. [CrossRef]
91. Xie, C.; Yuan, Z.; Wang, J. Artificial neural network-based nonlinear algebraic models for large eddy simulation of turbulence. *Phys. Fluids* **2020**, *32*, 115101. [CrossRef]
92. Xie, C.; Wang, J.; Weinan, E. Modeling subgrid-scale forces by spatial artificial neural networks in large eddy simulation of turbulence. *Phys. Rev. Fluids* **2020**, *5*, 054606. [CrossRef]
93. Yuan, Z.; Xie, C.; Wang, J. Deconvolutional artificial neural network models for large eddy simulation of turbulence. *Phys. Fluids* **2020**, *32*, 115106. [CrossRef]
94. Xie, C.; Wang, J. Artificial neural network approach to large-eddy simulation of compressible isotropic turbulence. *Phys. Rev. E* **2019**, *99*, 053113. [CrossRef] [PubMed]
95. Xie, C.; Li, K.; Ma, C.; Wang, J. Modeling subgrid-scale force and divergence of heat flux of compressible isotropic turbulence by artificial neural network. *Phys. Rev. Fluids* **2019**, *4*, 104605. [CrossRef]
96. Wang, Z.; Luo, K.; Li, D.; Tan, J.; Fan, J. Investigations of data-driven closure for subgrid-scale stress in large-eddy simulation. *Phys. Fluids* **2018**, *30*, 125101. [CrossRef]
97. Park, J.; Choi, H. Toward neural-network-based large eddy simulation: Application to turbulent channel flow. *J. Fluid Mech.* **2021**, *914*, A16. [CrossRef]
98. Gamahara, M.; Hattori, Y. Searching for turbulence models by artificial neural network. *Phys. Rev. Fluids* **2017**, *2*, 054604. [CrossRef]
99. Kang, M.; Jeon, Y.; You, D. Neural-network-based mixed subgrid-scale model for turbulent flow. *J. Fluid Mech.* **2023**, *962*, A38. [CrossRef]
100. Kim, M.; Park, J.; Choi, H. Large eddy simulation of flow over a circular cylinder with a neural-network-based subgrid-scale model. *J. Fluid Mech.* **2024**, *984*, A6. [CrossRef]

Disclaimer/Publisher's Note: The statements, opinions and data contained in all publications are solely those of the individual author(s) and contributor(s) and not of MDPI and/or the editor(s). MDPI and/or the editor(s) disclaim responsibility for any injury to people or property resulting from any ideas, methods, instructions or products referred to in the content.

fluids

MDPI

Review

Indoor Air Quality Control for Airborne Diseases: A Review on Portable UV Air Purifiers

Shriram Sankurantripati *,† and Florent Duchaine †

CERFACS, 42 Av. Gaspard Coriolis, 31100 Toulouse, France; duchaine@cerfacs.fr
* Correspondence: sankurantripati@cerfacs.fr
† These authors contributed equally to this work.

Abstract: The spread of airborne diseases such as COVID-19 underscores the need for effective indoor air quality control. This review focuses on ventilation strategies and portable air purifiers as key mitigation solutions. Ventilation systems, including natural and mechanical approaches, can reduce pathogen concentrations by improving airflow. However, combining ventilation with portable air purifiers, particularly those using HEPA filters, ESP filters, and UV-C radiation, can enhance Indoor air quality. While HEPA and ESP filters focus on trapping airborne particles, UV-C radiation can inactivate pathogens by disrupting their RNA. A review of UV air purifiers reveals a lack of studies on their efficacy and effectiveness in real-world settings. A thorough investigation into the performance of this mitigation solution is necessary, focusing on varying key factors, such as purifier placement, airflow dynamics, and UV dosage, to ensure optimal effectiveness. High-fidelity computational methods are essential in accurately assessing these factors, as informed by the physics of airborne transmission. Such advanced computations are necessary to determine the viability of portable UV air purifiers in mitigating airborne transmission in enclosed environments such as hospitals and public spaces. Integrating advanced air purification technologies with proper ventilation can improve safety in indoor environments and prevent future disease-related outbreaks.

Keywords: airborne viruses; UV air purifiers; mitigation solutions

1. Introduction

Disease transmission, a multifaceted process involving diverse routes and mechanisms, can be caused by various pathogens, including bacteria, viruses, fungi, parasites, and prions. Transmission modes vary across pathogen types, with some capable of utilizing multiple routes. The Healthcare Infection Control Practices Advisory Committee (HICPAC) classifies these routes into three categories: airborne, droplet, and contact [1]. Contact transmission occurs when an individual interacts with the pathogen either directly through exposure to blood or bodily fluids from an infected individual or indirectly through intermediary sources such as shared surfaces or equipment. Both droplet and airborne transmission occur when pathogens originating from the respiratory tract of an infected individual are disseminated to a susceptible individual. Droplet transmission is characterized by the transfer of pathogens via droplets larger than 5 μm in diameter, which typically traverse short distances before reaching the recipient's mucosal surfaces. Conversely, airborne transmission involves smaller droplets of less than 5 μm that are capable of traveling greater distances from the infection source [2,3]. However, these definitions, established in 1934, are based on research methods limited in their ability to accurately measure airborne particles near the source of infection. To address this, the concept of aerosol transmission [4] was introduced, defining an aerosol as a suspension of solid or liquid particles in air [5]. Despite the emergence of aerosol transmission, the terms "airborne" and "droplet" transmission remain widely used within the scientific community are therefore employed throughout this thesis.

Infectious diseases transmitted via the droplet route are primarily caused by three major pathogen classes: viruses, bacteria, and fungi. The earliest documented airborne contagious disease, measles, caused by the measles virus, was identified in the 9th century [6]. Table 1 provides a chronological overview of airborne diseases and their causative pathogens, spanning from the 9th century to the present day, highlighting the diverse array of airborne pathogens. Airborne diseases have inflicted devastating consequences on global populations. For instance, while World War I resulted in 16 million deaths, the subsequent influenza pandemic caused an estimated 50 million fatalities worldwide [7]. The COVID-19 pandemic, as of 1 June 2024, has affected over 700 million individuals and claimed more than 7 million lives [8]. Despite the World Health Organization (WHO) acknowledging the presence of COVID-19 in December 2019, their official recognition of its airborne transmission was delayed for two years [9]. Figure 1 illustrates the timeline of the pandemic alongside corresponding WHO declarations. Initially, the WHO asserted that COVID-19 transmission occurred solely through direct or indirect contact with infected individuals. A year into the pandemic, with over 1 million deaths globally, the WHO revised its stance, suggesting the possibility of airborne transmission in crowded, poorly ventilated spaces. By the time they acknowledged the potential for droplets to remain airborne for extended periods and travel beyond 1 m, the death toll had surpassed 3 million. On 23 December 2021, after a comprehensive review of case studies and transmission patterns, the WHO officially declared COVID-19 as airborne. By this time, the global death toll had already reached 5.5 million.

Table 1. Airborne diseases and their pathogens (sorted by discovery year).

Airborne Disease	Pathogen (Type)	Year Discovered
Measles	Measles virus (virus)	9th century [6]
Small pox	Variola virus (virus)	1796 [10]
Aspergillosis	Aspergillus fungus (fungus)	1842 [11]
Tuberculosis (TB)	Mycobacterium tuberculosis (bacterium)	1882 [12]
Diphtheria	Corynebacterium diphtheriae (bacterium)	1883 [13]
Meningococcal meningitis	Neisseria meningitidis (bacterium)	1887 [14]
Coccidioidomycosis	Coccidioides fungus (fungus)	1892 [15]
Common cold	Rhinoviruses (virus)	Early 1900s [16]
Whooping cough (pertussis)	Bordetella pertussis (bacterium)	1906 [17]
Histoplasmosis	Histoplasma capsulatum fungus (fungus)	1906 [18]
Influenza (flu)	Influenza viruses (Virus)	1918 [19]
Adenovirus	Adenovirus (virus)	1953 [20]
Chickenpox	Varicella-zoster virus (VZV) (virus)	1953 [21]
Mumps	Mumps virus (virus)	1967 [22]
COVID-19	SARS-CoV-2 (virus)	2019 [23]

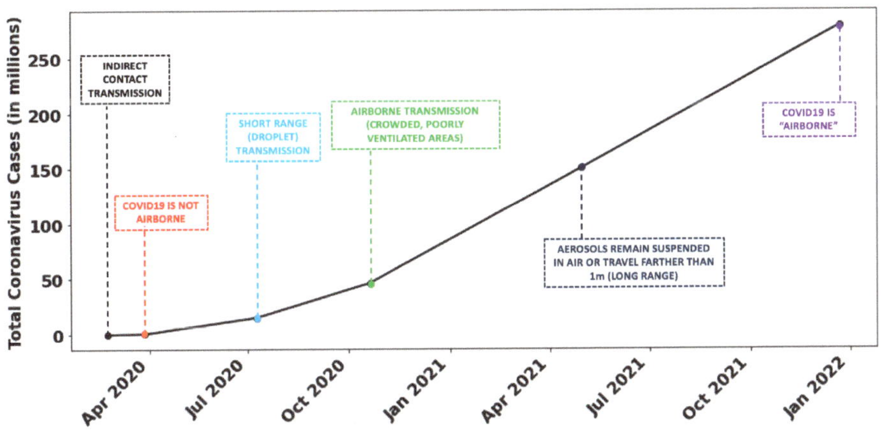

Figure 1. Evolution of global infected cases and the World Health Organization's corresponding declarations during the COVID-19 pandemic.

This historical context underscores the critical importance of comprehensively understanding disease transmission mechanisms, integrating insights from existing literature, as well as from experimental and computational studies and case studies, to facilitate the timely implementation of effective mitigation strategies and prevent future large-scale outbreaks. Given the wealth of research generated during the COVID-19 pandemic, this thesis focuses on airborne viruses, specifically SARS-CoV-2. The remainder of this section delves into case studies that shed light on the transmission routes of SARS-CoV-2 and the efficacy of various mitigation strategies.

2. Case Studies

Case studies of COVID-19 transmission are vital in understanding airborne transmission mechanisms, pandemic trends, and the effectiveness of mitigation strategies. Early research suggested airborne transmission as a possibility [24–26], and geographical analyses of infection spread, in conjunction with mitigation measures, support this as the dominant route [27,28]. A key finding is the significantly higher risk of transmission indoors [29] compared to outdoors [30]. Studies show an 18.7-fold increased odds of COVID-19 transmission in closed environments [31]. While outdoor transmission is less common, it often occurs during large gatherings, such as at parks, construction sites, camps, etc. [32–34]. A systematic review identified exposure duration, frequency, and gathering density as key determinants of outdoor transmission, noting that lockdown measures and mask mandates limited these factors, thereby reducing outdoor SARS-CoV-2 transmission [30]. Conversely, the vast majority (95%) of transmission clusters have occurred indoors [29,30]. Closed environments like hospitals, restaurants, offices, and public transportation have been identified as super-spreading locations [25,35–37]. Systematic and analytical reviews have examined factors contributing to this increased risk through detailed investigation of case studies [38–40]. While some factors relate to human behavior [41,42], contamination of surfaces [43], and the action of sunlight [44,45], the higher indoor transmission rates are largely attributed to inherent differences in airflow dynamics and droplet behavior between indoor and outdoor environments [38]. Thus, the complex, turbulent dispersion of exhaled droplets within confined spaces is primarily driven by intricate flow patterns arising from ventilation systems and various obstacles. Factors such as confinement, low ambient velocities, and limited air dilution contribute to elevated aerosol concentrations in closed environments, increasing the risk of both near-field and far-field exposure to pathogens [40]. The stark contrast in aerosol transmission dynamics between indoor and outdoor environments underscores the need for a comprehensive re-evaluation of the mechanisms involved in airborne disease transmission. This re-evaluation is essential for the development of effective mitigation strategies to combat future pandemics.

Similar to the several airborne diseases, SARS-CoV-2 virus propagates from an infected source to a susceptible individual through droplet (direct and indirect) and airborne transmissions, as shown in Figure 2. Airborne virus like SARS-CoV-2 primarily target and establish infection within the upper and lower respiratory tracts of the host. Expiratory activities such as talking, coughing, and sneezing [2,46] generate pathogen-laden respiratory droplets through two fundamental mechanisms [47]. The first mechanism involves the instability [48] and eventual fragmentation of the mucus lining due to airflow-induced shear stress, leading to droplet formation. The second mechanism involves the rupture of the thin liquid film lining the small airways during the cyclic opening and closing of these passages during breathing, leading to the generation of small droplets [49]. Exhaled droplets from infected individuals typically contain soluble, non-volatile matter (Na^+, K^+, Cl^-, lactate, and glycoprotein) comprising up to an approximately 0.71% mole fraction, in addition to viral particles [50]. The characteristics of exhaled droplet clouds, including velocity, size distribution, and number density, are dependent on the specific expiratory activity performed by the infected individual. Numerous studies have attempted to quantify the characteristics of exhaled droplet clouds, revealing substantial variation across different expiratory maneuvers. However, reported results exhibit significant variability due to limitations in sampling methods and instrument sensitivity, particularly for larger droplets [51–53].

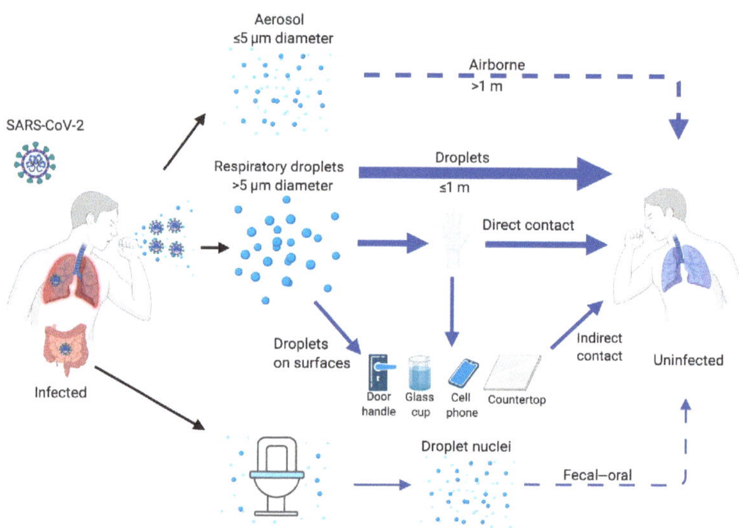

Figure 2. Transmission routes for SARS-CoV-2 virus to travel from an infected source to a susceptible individual [54].

Reported droplet sizes range from 0.1 to 1000 µm, with size distributions often described as continuous, bimodal, or trimodal [55,56]. Numerous studies have employed modern data-fitting algorithms and equations to replicate experimental data for computational, experimental, and statistical purposes [57–59]. The viscosity and surface tension of a droplet are important features that can also control droplet size distribution, particularly by coalescence and breakage phenomena. In the case of virus-laden droplets, the surface tension and density of these particles are found to be similar to those of water, and the viscosity of these droplets is higher than that of water by one or two orders of magnitude, making them less prone to coalescence and breakage [60–63]. Regarding number density and ejection velocity, a single sneeze is estimated to generate on the order of 10^4 droplets with velocities approaching 20 m/s [55]. Coughing produces 10 to 100 times fewer droplets with velocities around 10 m/s. The expiration jets associated with coughing and sneezing are turbulent, with Reynolds numbers on the order of 10^4 [64]. Just like sneezing and coughing, normal breathing and talking are acknowledged to atomize droplets. It has been discovered that half a minute of speech can release a liquid of volume equivalent to a cough [65,66] and that the ejection velocity during normal breathing and sneezing can exceed 5 m/s [67].

The subsequent behavior of exhaled droplets is governed by a complex interplay between their intrinsic properties and environmental factors such as temperature and humidity [2,68]. Depending on these factors, droplets undergo several physical phenomena, such as advection, evaporation, coalescence, and breakage, before entering into a susceptible host, as shown in Figure 3. They may either settle rapidly onto surfaces, potentially leading to contamination, or undergo rapid evaporation, remaining suspended as droplet nuclei capable of long-range transport. These droplet nuclei, comprising non-volatile matter and any entrained pathogens, represent a crucial vector for airborne disease transmission. The critical droplet size threshold differentiating settling behavior from airborne persistence has been estimated to range between 50 and 150 μm, with variations attributed to fluctuations in temperature and humidity [2,69].

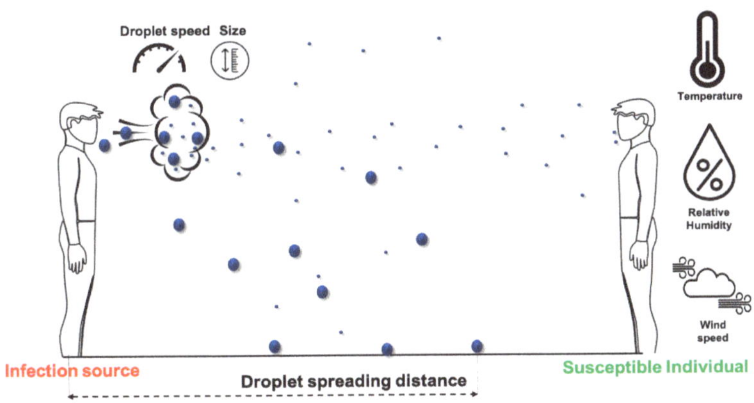

Figure 3. Factors influencing the evolution of expelled droplets during disease transmission from an infected source to a susceptible individual.

Droplets with sizes larger than the critical diameter are separated from the ejected puff of particles and end up being deposited on surfaces before becoming fully evaporated. In the context of the COVID-19 pandemic, these droplets are responsible for indirect contact transmission via contaminated surfaces and droplet transmission. A surface contaminated by SARS-CoV-2 or SARS-CoV-1 virus is a potential source of infection transmission for several hours [43]. On the other hand, the medium/small aerosols present in the puff are constantly undergoing advection and evaporation and turn into droplet nuclei with sizes smaller than 10 μm [47].

The velocity at which the ejected puff is expelled plays an important on in the advection of droplets. While the distance traveled by these large droplets is around 3–6 feet [69,70] during breathing or coughing activity, droplets during a violent respiratory act like sneezing are found to settle 20 feet from the infection source. One should note that ambient air currents in the direction of droplet expulsion can enhance the range of these settling droplets. With an air current of 4 km/h, droplets expelled from a cough in the same direction were found 20 feet from the infection source [60]. For droplet nuclei, the settling velocity is significantly lower; therefore, they remain with the puff and travel greater distances, remaining suspended in air for a longer time. When the settling velocity is less than the ambient turbulent fluctuations, the motion of particles suspended in air is governed by ambient air flow.

The rate of evaporation of a single aerosol depends on a variety of factors, such as relative humidity, air temperature, and the relative velocity between aerosols and ambient air. It plays an important role in determining the distance traveled by the droplets. Studies have shown that the droplet spreading distance is longer and the aerosolization rate is lower during winter than summer, indicating the importance of weather conditions for airborne transmission [71,72]. Experimental studies indicate that higher temperatures and lower humidities lead to higher evaporation rates, which increase the critical droplet diameter [2,69]. Computational studies support the hypothesis that the evaporation of droplets cannot be ignored while performing simulations to understand airborne transmission [73,74].

In response to these transmission dynamics, countries worldwide implemented various mitigation strategies during the COVID-19 pandemic to curb droplet and airborne transmission. While a 3–6 foot social distancing recommendation was proposed to reduce droplet transmission [75,76], its efficacy in crowded outdoor settings was limited, as wind can disperse droplets up to 6 m. Consequently, many countries resorted to complete lockdowns to suppress the spread of infection. The use of N95 and surgical masks was also recommended, as they can shield individuals from large droplets and droplet nuclei (inward protection) [77] and mitigate droplet spread during exhalation (outward protection) [78–80]. Additionally, the use of hydro-alcoholic solutions and regular hand washing were promoted to disinfect contaminated hands and reduce indirect contact transmission [76,81]. Respiratory etiquette, such as covering the mouth and nose during coughs or sneezes, was encouraged to disrupt the transmission chain, although its effectiveness in completely blocking droplet release and dispersion is limited [82]. Plexiglas walls/barriers were installed at restaurants, banks etc. to reduce droplet transmission between individuals. The success of these mitigation measures hinged on individual compliance, and they did not directly address air quality, particularly in closed environments. The WHO; the Centers for Disease Control and Prevention (CDC); and the American Society of Heating, Refrigerating, and Air-Conditioning Engineers (ASHRAE) [75,76,83] have established guidelines to control the spread of airborne infectious diseases in indoor environments such as buildings and public spaces. These guidelines focus on enhancing ventilation, improving air filtration, and maintaining hygiene practices. The WHO advocates for adequate airflow, mask wearing, and regular sanitation to minimize transmission risks. The CDC promotes a layered approach that includes proper ventilation, high-efficiency filtration systems, air-cleaning devices such as HEPA filters, physical distancing, and regular hygiene. ASHRAE Standard 241 further supports these strategies by defining specific clean airflow requirements, utilizing advanced air filtration and UV technology and implementing an infection risk management mode during high-risk periods. Together, these guidelines aim to mitigate airborne pathogen transmission and improve indoor air quality to safeguard public health. Given that aerosol dispersion is influenced by complex flow patterns in confined spaces and in line with the the WHO's declaration that "everyone has a right to breathe healthy indoor air" [84], fluid dynamics-based mitigation solutions are crucial for ensuring quality air in enclosed settings. The next section explores air quality control methods developed with the goal of providing pathogen-free breathing air.

3. Air Quality Control

In the past semi-centennial, several researchers worked on understanding and improving indoor air quality (IAQ), especially to mitigate the airborne transmission of pathogens, by conducting case studies, as well as experimental and numerical studies [85–87]. Poor IAQ leads to not only high exposure risk and infection spread but also headaches, breathing difficulties, fatigue, etc. [88,89]. The risk-reduction techniques proposed or investigated to control IAQ in any indoor setting are generalized into the following categories: ventilation (natural and mechanical) and portable air filtration.

Regular sanitation of public spaces like hospitals and subways is conducted using UV surface disinfectants or by fogging machines that release a dispersion of a fine mist of disinfectants in air, which improves air quality [90].

3.1. Ventilation

Ventilation is the process of introducing and distributing outdoor air or appropriately treated recirculated air into a building or a room [91]. Throughout the years, there have been several studies conducted on indoor ventilation and its role in the airborne transmission of diseases, as well as minimum requirements to mitigate disease transmission in different indoor settings. A first comprehensive review on ventilation and its impact on airborne transmissions until 1960 was conducted by Wells WF, Riley, and O'Grady [68,92] indicating that the infection rate among susceptible hosts is higher in poorly ventilated regions. A multi-disciplinary systematic review using the research articles published from 1960 to 2007 on the role of ventilation in airborne disease transmission was performed by Li Yuguo et al. [85]. This review highlights that indoor spaces with low air-exchange rates, specifically less than 4 Air Changes per Hour (ACH), have been linked to the transmission of tuberculosis [93], measles [94], and influenza [95] in hospitals, pediatric offices, and aircraft, respectively. An experimental study conducted on mice inside a cage with three different ventilation rates indicated that infection risk is inversely proportional to the ventilation rate [96]. Smoke puff studies conducted inside hospital wards concluded that airflow patterns are directly related to the infectious spread of diseases like chickenpox [97], tuberculosis [98], small pox [99], etc. These findings, along with those of other studies, affirm that ventilation rates and airflow patterns in enclosed spaces play a crucial role in infection transmission.

Several ventilation strategies based on such studies have come up in the past decade to improve IAQ and provide thermal comfort for different indoor settings. These ventilation strategies are primarily categorized into two categories: Mechanical Ventilation (MV) and Natural Ventilation (NV). A multi-disciplinary systematic review on different types of mechanical and natural ventilation systems and their efficiency in mitigating the exposure risk was conducted by Al-Rikabi [88].

This review highlights that, based on the inlet–outlet placement in a closed environment, there are 11 types of mechanical ventilation systems in use today. These systems can be grouped into three categories: uniform steady-state systems, such as mixing ventilation and diffuse ceiling ventilation; non-uniform steady-state systems, such as displacement ventilation and stratum ventilation; and unsteady ventilation systems, such as intermittent ventilation. Figure 4 provides an illustration of the aforementioned ventilation systems. Several numerical and experimental studies on these ventilation systems have been conducted inside various enclosed spaces, like classrooms, office space, elevators, bus cabins etc., to study the induced flow patterns and the resultant dispersion of pathogen-laden droplets. Some studies have compared their impact on the spread of airborne infection by testing different ventilation systems inside the same indoor space. To calculate the probability of infection spread or compute surface contamination due to respiratory activities such as breathing, talking, coughing, and sneezing, the Wells–Riley model [68,100] is employed.

The Wells–Riley infection prediction model is a widely-used mathematical framework for estimating the probability of airborne disease transmission in indoor environments. It calculates the infection risk (P) by considering factors such as the number of infectious individuals (I), the ventilation rate (Q), and the duration of exposure (t). A key concept in this model is the "quanta" (q), which represents the infectious dose required to cause infection in a susceptible person. The infection probability for a susceptible individual can be expressed as $P = 1 - e^{(-\frac{Iqbt}{Q})}$, where b is the breathing rate of a susceptible individual. While the Wells–Riley model provides a useful estimate for predicting infection spread, its accuracy is limited by simplifying assumptions, such as the uniform distribution of infectious particles in the room and constant quanta generation, which may not always hold true in real-world scenarios. As a result, some research studies prefer to evaluate the effectiveness of mitigation solutions based on the contaminant concentration in the room.

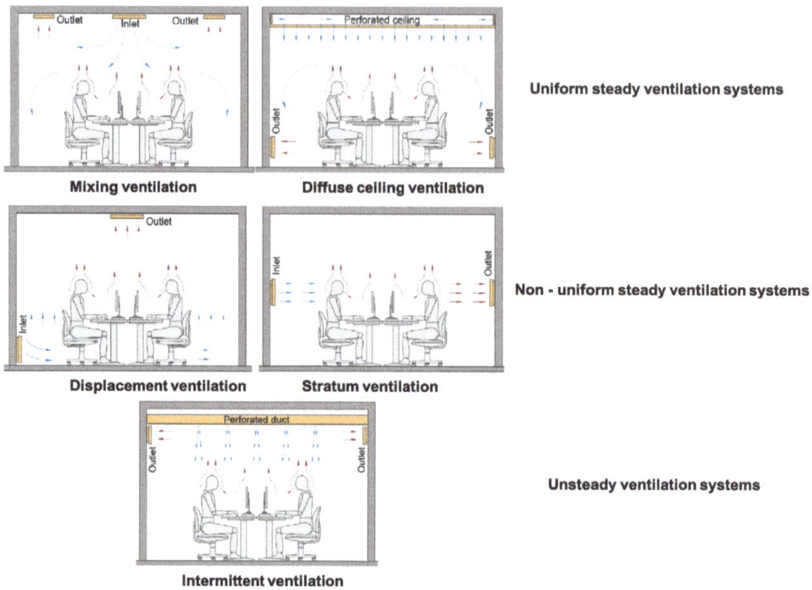

Figure 4. Illustrations of different mechanical ventilation systems [88].

Indoor environments with uniform steady ventilation systems like mixing ventilation [101] and diffuse ceiling ventilation [102] have the highest exposure risk compared to other ventilation systems [103]. It was found that high supply and exhaust pose higher infection risks with uniform steady ventilation systems [104,105]. These studies correct the ideology that a higher ventilation rate does not guarantee better mitigation of infection spread [104,106]. Non-uniform steady ventilation systems such as displacement ventilation [101] and stratum ventilation [107] show minimal contamination levels [108–111] when compared to mixing ventilation and diffuse ceiling ventilation. Unsteady ventilation systems seem to be less effective than uniform steady air systems due to periodic on/off cycles in the air supply, as the concentration of contaminants increases when the air supply is off [112].

Indoor air quality (IAQ) and thermal comfort can also be enhanced by introducing fresh air through windows, doors, and other openings. Natural ventilation, a passive strategy utilizing natural airflow, can provide outdoor air in moderate climates and mitigate airborne transmission. Common wind-driven natural ventilation strategies include single-side, cross, and wind-catcher ventilation, as illustrated in Figure 5. Many numerical and experimental studies have been carried out inside classrooms, office rooms, buses, etc., to study the efficacy of each of these strategies. Some studies have compared these strategies with other natural ventilation systems and/or with mechanical ventilation systems [88]. Cross ventilation, achieved through dual-side windows, generally outperforms single-side ventilation [113,114]. However, in specialized facilities like hospitals, not all natural ventilation systems meet stringent clinical standards [115,116]. Research suggests that simple architectural modifications, such as optimization of window placement and size, can significantly improve contaminant removal through natural ventilation [117,118]. However, this approach may be limited in areas with insufficient wind to achieve the required ventilation rate, highlighting the fact that the efficacy of natural ventilation systems cannot be guaranteed at all times.

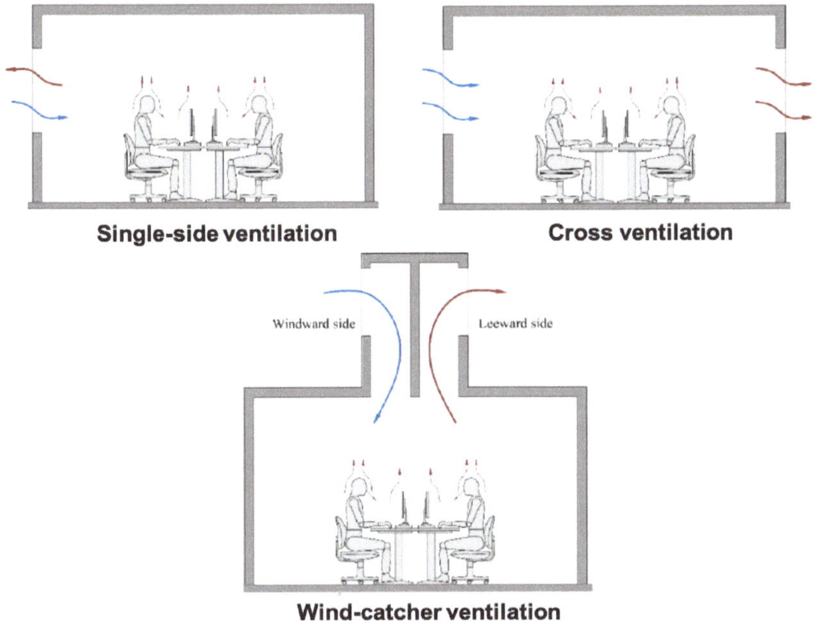

Figure 5. Illustrations of different mechanical ventilation systems [88].

Given the strengths and limitations of both mechanical and natural ventilation, a flexible strategy that integrates elements of each would be ideal for future pandemic preparedness. Studies have highlighted the importance of modifying ventilation systems in enclosed spaces to optimize airflow and reduce pathogen exposure, with recommendations to reposition exhausts and inlets as necessary [119,120]. Additionally, integrating high-efficiency particulate air (HEPA) filters or UV lamps within ventilation units can inactivate pathogens before distributing air throughout a room. Emerging technologies such as personal ventilation devices also show promise in reducing both short- and long-range transmission risks, although further research is needed to confirm their efficacy [105].

While these measures can inform the design of new indoor spaces, retrofitting existing environments remains a practical approach to infection control. However, implementing such measures may face challenges due to the resource-intensive nature of mechanical ventilation upgrades and the dependency of natural ventilation on external conditions. Models like the Wells–Riley model provide useful infection estimates but lack the ability to account for complex, dynamic airflow patterns and occupant density variations found in real-world settings. Future research could benefit from the exploration of hybrid ventilation models that combine mechanical and natural systems, creating adaptable, energy-efficient solutions for air quality and infection control. Additionally, integrating portable HEPA filters or UV air purifiers in enclosed areas offers an immediate, cost-effective way to lower infection risks, particularly in high-density or poorly ventilated environments [121].

3.2. Portable Air Cleaners (PACs)

The objective of the design of a portable air cleaner (PAC) is to enhance indoor air quality (IAQ) in enclosed spaces. In poorly ventilated environments, PACs can mitigate airborne transmission through various filtration mechanisms [92]. Commercial PACs employ diverse technologies to capture or destroy pathogen-laden particles, depending on the particle size, inactivation technology, and specific indoor environment. PACs can be classified into three categories: mechanical filtration, electrical filtration, and UV light filtration [86]. Mechanical filtration removes particles by capturing them in filter media, whereas electrical filtration utilizes electrostatic attraction to trap particles. UV air purification, on the other hand, inactivates pathogens within particles by disrupting their RNA. Among PACs with mechanical filtration devices, commonly used devices in indoor spaces are panel filters, pleated filters, and HEPA filters [86]. Among them, HEPA filters are highly recommended for capturing sub-micron-sized particles like droplet nuclei [121,122]. In the domain of electrical air filtration devices, electrostatic precipitators (ESPs) are the most commonly recommended and applied to capture micron and sub-micron particles. Among UV air purifiers, UV-C lamps are recommended to inactivate infectious airborne particles. The application of portable HEPA filters and electrostatic precipitators inside classrooms and offices have been examined by several researchers, comparing their efficiency against other filtration mechanisms. A summary of available literature on these mechanisms and the application of these devices in real-life scenarios is presented below.

3.2.1. Mechanical Air Filters: HEPA Filters

HEPA filters are typically manufactured by pleating microfiber glass or other fibrous media consisting of multiple layers of randomly arranged fibers with diameters ranging from 2 to 500 nm [123]. Particles entering the HEPA filter are entrapped by one of the following three mechanisms: (1) impaction, (2) interception, or (3) diffusion [124]. If a sieving mechanism is included in this design, they are called ultra-low particulate air (ULPA) filters [125]. This intricate design enables the capture of particles not only >1 μm but also those in the sub-micron range [124]. The particle removal rate of a HEPA filter varies depending on the size and power of the portable air cleaner (PAC). Except for sub-micron particles, HEPA-based PACs have a high removal efficiency, and they can remove all particles in a short period of time (such as 5 min) [126,127]. An efficiency study performed on a single HEPA filter-based PAC showed significant improvement in aerosol concentration when compared with a room with no filter [128]. The impact of these filters on room ventilation has been tested by varying the rate of ventilation within a closed room. Depending on the type of ventilation system, increasing the air exchange rate can either enhance or reduce the effectiveness of the filter. A higher air exchange rate may improve contaminant removal by introducing fresh air more frequently, but it can also reduce the contact time between airborne particles and the filter, thereby lowering its inactivation potential. Conversely, a low ventilation rate results in reduced airflow and longer residence times for contaminants within the room, leading to lower overall effectiveness of the HEPA filter [129]. The effectiveness of HEPA filter-based PACs also depends significantly on their placement within a closed environment. Positioning a PAC in a location where the majority of the room's air flows ensures that a larger volume of contaminated air passes through the filter. This allows the purifier to capture more airborne pathogens, thereby minimizing the risk of infection spread. Conversely, improper placement with poor airflow may reduce a purifier's performance [128]. While this filtration mechanism seems to be a viable option, one has to consider such filters become sinks of bio-hazardous pathogens [130]. Overall, experimental and numerical studies conclude that the application of HEPA filters along with mechanical or natural ventilation can effectively reduce the airborne virus concentration inside a closed environment.

3.2.2. Electronic Air Filters: ESP Filters

Electrostatic precipitators (ESPs) are the most common type of electrical filters used inside ventilation ducts and as PACs. Electrostatic precipitation is a technique to remove suspended particles in a gas using an electrostatic force [131,132]. With a high voltage supply, a strong electric field is established, which charges airborne particles and collects them near an oppositely charged plate [5,130]. These filtration devices are mainly used in industrial applications, but studies have shown their application in indoor spaces like houses, offices, and factories. ESPs are commonly classified as single-stage or two-stage filters. In single-stage filters, both charging and particle removal from a gas stream are performed by the same set of electrodes, therefore using a very high voltage of 50–70 kV [133]. If different sets of electrodes are used for charging and collecting the aerosols, it is two stage precipitator and it requires only 12–15 kV [133]. The performance of these filtration devices is commonly computed in terms of the clean air delivery rate (CADR) or particle removal rate [134]. The CADR is a measure that considers the airflow rate through the cleaner, wit the particle removal efficiency presented in m^3/h. Tests of PACs developed with ESP filtration in laboratories and in real-life scenarios show that the CADR of a PAC depends on its placement inside a closed room with respect to the infection source [135]. Experimental studies also indicate that the CADR of a PAC depends on the electric power provided. The higher the electric power, the higher the CADR [136]. But this has to be treated carefully, as it can lead to the production of hazardous byproducts in indoor and transportation environments, such as ozone from corona discharge and/or the ionization process [137]. Therefore, by regulating the amount of electric PACs with ESP filtration mechanisms, experimental and computational studies support this mechanism being used to mitigate the airborne transmission cycle.

3.2.3. Mechanical vs. Electronic Air Filters

While HEPA and ESP are the most commonly used mechanical and electrical filtration mechanisms as portable air cleaners, there are several other PACs with different filtration techniques that can serve the purpose of mitigating airborne infections. One has to identify the best and most cost-effective solution to reduce the concentration of infectious particles in any indoor environment. Therefore, the efficiency/efficacy of a PAC, as well as its effectiveness, should be tested in an enclosed space. Efficiency is the percentage of particles removed in a single pass through the filter, as explained below. Effectiveness is a dimensionless quantity that shows how effective as PAC is in reducing the concentration of contaminants in a real-life setting [130]. When HEPA filters are compared with other mechanical filters, HEPA-based PACs exhibit the highest CADR [138]. When HEPA and ESP filters are compared, some experimental studies [138–141] have revealed that HEPA-based PACs have higher CADR values, while some other studies [136] have indicated that ESP-based PACs have the highest CADR among tested filters. To compare HEPA or particle filters with electric filters, one should remember that the CADR of an electrical filter can be improved by providing more electric power. HEPA filters also require electric power to function. Therefore, to identify better PACs, the electric power used by a PAC can be normalized according to the clean air delivery rate (CADR). The specific electric power for an ESP is typically half the specific power needed for a mechanical filter [130]. But there are some deviations to this, as some studies have shown that HEPA filters achieve better performance than ESP-based PACs [138]. Therefore, its not simple to make generalizations and say which filtration mechanism is the optimal strategy. One can consider the application of an electrostatically assisted mechanical filters as a possible mitigation solution to prevent airborne transmission [142].

A comparison between HEPA and ESP filters reveals important considerations for selecting the most effective air purification approach for different indoor environments. While HEPA filters generally achieve higher clean air delivery rates (CADRs) across studies, ESP filters can offer comparable or superior CADRs when powered adequately. However, ESP systems require careful regulation due to the potential productions of ozone and other byproducts during the ionization process. This trade-off suggests that while ESP filters may provide efficiency at a lower specific energy cost, HEPA filters offer a safer, more reliable option, especially in high-occupancy settings where ozone byproducts are a concern.

Additionally, research on optimal PAC placement, airflow interaction, and power normalization highlights the complexity of finding a universally optimal solution, indicating the need for tailored approaches. For spaces where both energy efficiency and a high CADR are critical, electrostatically assisted mechanical filters may offer a promising compromise by combining HEPA and ESP technologies. This hybrid approach could provide a balanced solution, particularly in environments with stringent air quality demands, although further research is necessary to validate its effectiveness across diverse settings.

While ultraviolet (UV) disinfection PACs have been proposed as a potential solution for air purification, particularly in the context of the COVID-19 pandemic, empirical data regarding their efficacy and effectiveness in real-world settings remain limited. The pandemic has stimulated increased research into the application of UV radiation in air filters to inactivate viruses, resulting in a few studies exploring its potential. However, to comprehensively evaluate the efficacy of UV-based PACs and their effectiveness in a real-world setting, a thorough examination of existing research on UV radiation's role in microbial inactivation and various studies conducted on these portable air cleaners should be performed.

3.3. *UV Air Purification*

UV radiation in the range of 200–315 nm is considered ultraviolet germicidal irradiation (UVGI) [143] and is known to prevent the spread of infectious diseases. UV-C lamps generating wavelengths in this range are commonly used to disinfect surfaces, water, and air [144]. UVGIwas first discovered in 1845 by Downes and Blunt [145] when they observed that the growth of micro-organisms is prevented by exposure to sunlight. They then continued to study the effects of lights on different micro-organisms. Experiments on microbial inactivation by light exposure conducted by Tyndall [146] revealed that the inactivation rate varies for different durations of exposure, wavelengths used in the light spectrum, and doses of radiation provided. Initial investigations focused on the rate of inactivation of pathogens against wavelengths, testing for UV-C (100–280 nm), UV-B (280–315 nm), UV-A (315–400 nm), visible (400–700 nm), and infrared (700–10^6 nm) ranges. Geisler's experiments [147] indicated that UV radiation from sunlight and electric lamps was more effective than longer wavelengths. By increasing radiation intensity, all wavelengths have shown lethal effects with long durations of exposure. Bang [148] discovered that UV-B and UV-C wavelengths are more effective than UV-A radiation. Duclaux's work [149] revealed that bacteria, fungi, and other microbes have different sensitivities to sunlight exposure and inactivate at different rates. Later, several studies began to quantify this microbial sensitivity of different pathogens to UVGI by modifying the dose of radiation. When the photons from UV radiation are absorbed by microbes, its desoxyribose nucleic acid (DNA) is damaged and renders its ability to reproduce [150]. All these experiments were conducted with pathogens on a surface like a Petri dish or in liquid.

In 1935, Wells [151] extended the application of this method to airborne pathogen-laden aerosols and showed that UVGI can inactivate airborne infectious agents. Sharp [152] tested airborne disinfection using UVGI in an in-duct heating, ventilation, and air conditioning (HVAC) system. Since these initial investigations, more studies have come together in regards to the study of airborne pathogen disinfection through UV radiation. Jensen [153] inactivated five different viruses using UVGI and found that their sensitivities differed in air when compared to deactivated experiments performed on plates or in water. It has been established that when exposed to an adequate dose of UV-C light, airborne pathogens can be deactivated, which suggests a potential application for IAQ control. This is accomplished successfully by following ways: (1) upper-room UVGI installation or (2) UVGI-integrated HVAC systems.

3.3.1. Upper-Room UVGI

Upper-room Ultraviolet Germicidal Irradiation (UR-UVGI) technology utilizes shielded UV lamps suspended from the ceiling of populated spaces, as shown in Figure 6, to create an irradiation field above occupants, effectively inactivating airborne pathogens [154,155]. This approach takes advantage of natural convection currents to transport airborne microorganisms through the UV field, enabling the rapid disinfection of large air volumes. The effectiveness of this system in disinfecting large volumes of air depends heavily on the fluid dynamics within the room, making airflow patterns crucial in the design and operation of UR-UVGI systems. The efficacy of this technology in inactivating viruses such as chickenpox and measles in school settings was first demonstrated by Wells [156].

Subsequent experimental studies in isolation wards, commercial spaces, and residential areas have confirmed its effectiveness [157–159]. Computational fluid dynamics (CFD) studies [160–162] on naturally or mechanically ventilated hospital wards and test chambers have further demonstrated the efficiency of UR-UVGI by integrating air mixing, UV fields, and microbial degradation. These studies have also led to the development of mathematical models [163,164] to assess airborne contamination in rooms equipped with ceiling-mounted UVGI technology. Research aimed at enhancing the effectiveness of UR-UVGI has explored various strategies, including the use of ceiling fans, optimization of UV lamp placement, adjustments to room ventilation rates, modifications to ceiling height, and variations in UV lamp power [162,165–167]. The effectiveness of UR-UVGI diminishes with higher air exchange rates, as this reduces microbial exposure to the UV lamps. Rooms with higher ceilings and UR-UVGI systems have been shown to contain lower levels of contaminants compared to rooms with lower ceilings. Additionally, while 254 nm UV lamps are commonly used, 222 nm UV-C radiation has also demonstrated a high rate of pathogen inactivation. The combination of low-speed ceiling fans and UR-UVGI has proven more effective than the absence of fans. As research continues, new approaches and innovations are emerging to further enhance the efficiency of this mitigation solution [168,169].

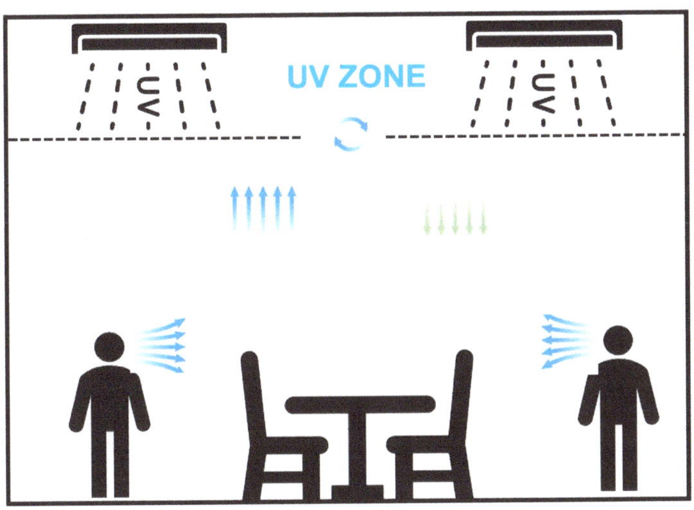

Figure 6. General representation of an upper-room UVGI (UR-UVGI) setup.

Based on the research findings, UR-UVGI technology shows significant potential as a mitigation solution to prevent airborne transmissions. However, it also comes with some inherent limitations. The effectiveness of this technology is mainly confined to the upper part of a room, typically above 7 feet, and it relies heavily on air circulation to transport pathogens to the UV-exposed zone. Poor air circulation can diminish the technology's overall impact, potentially leaving the lower, occupied zone untreated. The efficacy of UVGI depends heavily on both the intensity of the UV radiation and the duration of exposure, with certain pathogens requiring stronger or longer UV exposure to be effectively neutralized. Since this mitigation solution is designed to be used while people are present in the room, preventing UV radiation leakage from the shielded zone is critical, as exposure to UV radiation can be harmful and potentially lethal to humans. Ensuring proper shielding and safety measures is essential to maintain the effectiveness of the system without compromising occupant safety. Furthermore, installing UVGI systems, especially in existing buildings, can be complex and costly, often necessitating significant retrofitting. Despite these limitations, when properly implemented and maintained, UR-UVGI can be a valuable tool in reducing airborne pathogens.

3.3.2. UV-Integrated Ventilation

UV radiation can be integrated into HVAC systems by installing an array of UV lamps within the ventilation ducts as shown in Figure 7, where a high-intensity UV field is generated to effectively inactivate pathogen-laden particles entering the system. These HVAC systems are carefully designed to confine UV exposure to within the ducts, ensuring the safety of occupants while also disinfecting surfaces within the HVAC components. Although this approach may consume more electrical power compared to standard HVAC systems, it significantly enhances the inactivation of airborne pathogens. Additionally, proper shielding is essential to limit UV irradiation to the ducts, preventing any exposure to occupants. Experimental studies such as in ICU settings and car HVAC systems [170–172] have demonstrated improved air quality through the application of UVGI in ducts. Computational studies [173–175] have also simulated various UV lamp configurations and air velocities within ducts to evaluate their impact on inactivation rates. Both experimental and numerical findings suggest that factors such as lamp placement within the duct, duct length (from inlet to outlet), and chamber size play critical roles in determining the effectiveness of microbial inactivation.

The inactivation efficiency of pathogens within HVAC ducts is influenced by the fluid mechanics inside the ducts, which determine the exposure time of airborne pathogens to UV radiation. This makes the volumetric flow rate of the ventilation system a critical parameter. Optimal performance of UV disinfection in HVAC systems is often observed at lower airflow speeds, as lower speeds allow for longer exposure times, enhancing pathogen inactivation, as demonstrated by Yang [175]. The application of UV-C wavelengths, whether 222 nm or 254 nm, does not appear to significantly alter the inactivation rate [176]. These findings have informed the development of mathematical models and numerical approaches for predicting inactivation within HVAC ducts. To further enhance the efficiency of this mitigation strategy, the combination of UV lamps with HEPA filters and/or micro-static electricity generators is being explored [177,178].

Figure 7. General representation of an array of UV lamps installed inside a ventilation duct.

Similar to UR-UVGI, UVGI integrated into HVAC systems also presents certain limitations. The rapid movement of air through ducts limits the exposure time of pathogens to UV light, potentially reducing its effectiveness, especially against more resistant microorganisms. Shadowing from duct components like filters and bends can create areas where UV light does not reach, allowing pathogens to survive and circulate. While UVGI can inactivate most airborne pathogens entering the HVAC system, it does not address droplets suspended in the room that do not enter the HVAC system. Continuous operation is required to maintain effective UV intensity, which can lead to higher energy consumption and costs, potentially making it less economically viable for some facilities. Additionally, prolonged UV exposure can degrade certain materials within the HVAC system, leading to inefficiencies and the need for more frequent repairs. Initial installation costs can also be high, especially when retrofitting existing systems, which may be prohibitive for some facilities. There are also health risks associated with UV systems, as improper installation or leakage could expose maintenance personnel or building occupants to harmful UV radiation. Nonetheless, when properly implemented and maintained, UVGI-integrated HVAC systems offer a viable solution for reducing indoor contaminant concentrations.

3.3.3. UR-UVGI vs. UV Integrated Ventilation

Upper-room UVGI and UV-C-integrated ventilation systems both use ultraviolet light to reduce airborne pathogens, but they differ in their setup, operation, and areas of application. Upper-room UVGI systems involve the installation of UV-C lamps high on walls or ceilings, creating a disinfection zone in the upper part of a room where air circulates naturally or with the help of fans. This setup is best for smaller, localized spaces such as classrooms or offices, as it relies on air mixing within the room to move contaminated air into the UV-C-treated area. The system is easy to install, does not require modifications to HVAC systems, and minimizes UV exposure of occupants by keeping the light confined to the upper room. However, it may be less effective in larger spaces or rooms with poor air circulation, and it only disinfects the air, not surfaces.

In contrast, UV-C-integrated ventilation systems involve the installation of UV-C lamps directly in HVAC ducts or air-handling units, disinfecting air as it moves through the ventilation system before being distributed throughout the building. This approach is ideal for treating air across larger or multi-room facilities, ensuring consistent air quality throughout. While more complex and costly to install due to the need for integration with HVAC systems, it is highly effective for whole-building air disinfection and poses no risk of UV exposure to occupants, since the light is confined within the ducts. Unlike surface disinfection methods, both systems focus solely on airborne pathogens, with the choice between the two depending on the size of the space, installation feasibility, and specific air treatment needs.

In summary, both upper-room UVGI and UV-C-integrated ventilation systems provide effective solutions for reducing airborne pathogens in indoor environments, yet their optimal use varies depending on the application needs. Upper-room UVGI is particularly well suited for smaller spaces where direct integration with an HVAC system is not feasible, offering straightforward installation with minimal risk of occupant exposure. However, its efficacy is limited to areas with sufficient air circulation to bring contaminants into the UV-treated zone. On the other hand, UV-C-integrated HVAC systems excel in large, multi-room facilities, providing consistent disinfection across the entire building. Such a system is contained within the ductwork, avoiding any direct exposure to occupants but requiring careful installation and maintenance to address limitations like shadowing in the ducts and potential material degradation. Each system has unique operational benefits, and in settings where both local and central disinfection are needed, a combined approach may offer the most comprehensive solution. Future research could explore hybrid systems that leverage localized treatment by upper-room UVGI in combination with the broad reach of HVAC-integrated UV-C, providing an adaptable, layered approach to indoor air quality management.

3.3.4. UV-Based Portable Air Cleaners (UV-PACs)

Another application of UV radiation for airborne pathogen disinfection is the use of a portable air cleaner with UV air purification technology. As mentioned earlier, limited research is available on the performance of UV-based PAC technology. Therefore, an analytical review of the available UV-C radiation-based PACs in the industry, along with peer-reviewed computational and experimental studies, was conducted to assess their effectiveness and efficacy as a mitigation solution. A set of selection criteria used in several published peer-reviewed articles [85,88,179,180] involves following the four phases of a systemic review [85,181]: (1) identification, (2) screening, (3) eligibility, and (4) inclusion of studies for the review. To perform this review, Google Scholar, PubMed, and Web of Science were used as search engines to collect a list of studies conducted over time. These collected papers were then subjected to multiple steps of scrutiny to retrieve the relevant papers for this study. A flow chart was designed, as illustrated below, to explain the methodology used to perform this review (Figure 8). Using all the keywords illustrated in the flow chart, a total of 3378 papers were collected and. Then, an elimination process was conducted, comprising the elimination of papers beyond the scope of the review by reading the title and keywords. Later, by reading their abstracts, partially skimming their methodologies, etc., the set was reduced to 108 papers. Finally, after an in-depth review, nine papers were chosen for this study. The aim of this review was to compile only scientific articles on UV air purifiers, examining their potential as mitigation solutions by evaluating either experimental or computational results that qualitatively and quantitatively demonstrate their efficacy and effectiveness in real-life scenarios.

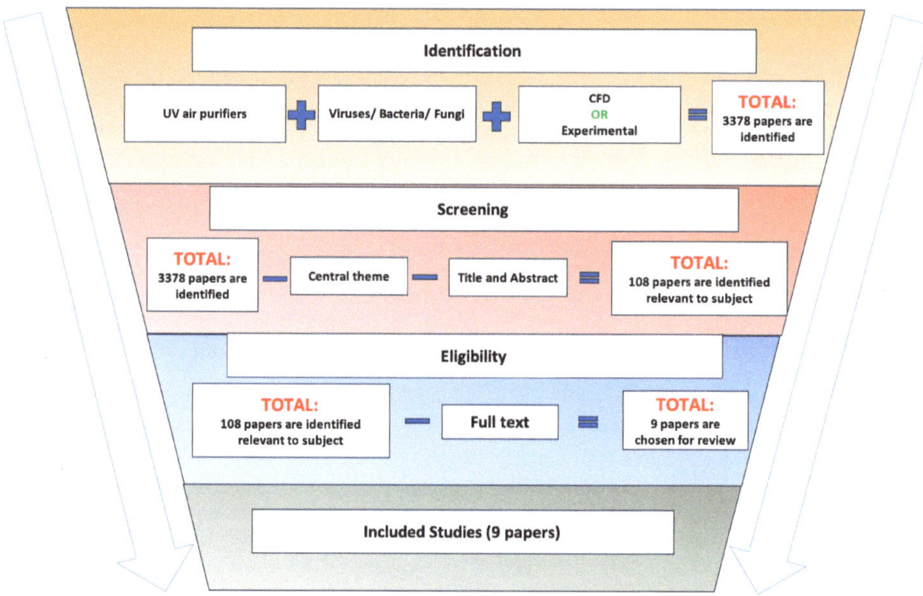

Figure 8. Flow chart explaining the review process for UV air purifiers.

Table 2 provides a summary of the reviewed studies, outlining the configurations of portable air cleaners (PACs), types of pathogens tested, methodologies applied, conclusions drawn, and limitations encountered. A general representation of these purifier designs is illustrated in Figure 9. Among the selected articles, four involve experimental studies, while the remaining focus on computational analyses. Five of these studies investigate PACs that combine particulate matter filtration with UV radiation, while the other three examine purifiers utilizing only UV radiation technology. The studied pathogens include viruses such as SARS-CoV-2 and feline coronavirus (FCoV), as well as a range of bacteria, including tuberculosis and fungi such as staphylococcus aureus.

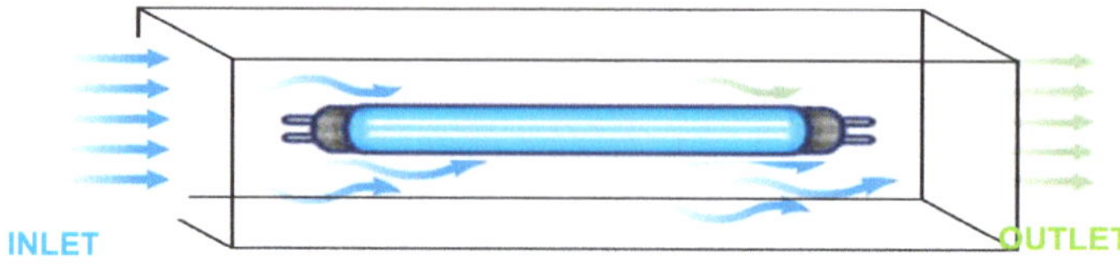

Figure 9. General representation of a UVGI portable air cleaner.

Concerning the experimental studies, three demonstrated the purifier's efficacy, and two evaluated effectiveness in real-life settings. Rausch [182] developed a protocol to rapidly assess the efficacy of UV air purifiers in laboratory conditions. The protocol involves pre-treatment of the sample with enzymes; then, the amount of surviving virus particles is determined using reverse transcription polymerase chain reaction (RT-PCR). The results were validated using UV-C air purifiers, showing their efficacy in viral inactivation. Li [178] explored the efficacy of a PAC that integrates high-efficiency filters with UV lamps, assessing the individual impact of these components on air treatment at various volumetric flow rates and identifying the necessary UV dosage for optimal performance. Samples were collected at different locations in the experimental setup, and a total plate count was performed. Their results showed that the combination of UV with MERV-8 and MERV-13 filters was the most effective configuration for the treatment of pathogen-laden environments. Garg [183] tested the efficacy of a PAC equipped with UV-C lamps both inside ducts and as a standalone unit against the SARS-CoV-2 virus. They verified the efficacy of their UV-PAC by trapping particles on a gelation filter and quantified viral RNA through an RT-PCR test. They extended these findings to study the effectiveness of the purifier in a classroom using the Wells–Riley model, demonstrating the purifier's impact when placed in various classroom locations. Similarly, Messina [184] studied the effectiveness of a PAC equipped with UV-C and HEPA filters in a hospital operating room (OR). Samples were collected using a well-calibrated particle counter. Contamination levels were measured across four different configurations (OR: ON/OFF; PAC: ON/OFF), concluding that contamination was significantly reduced when both the OR and PAC were active.

In summary, these studies collectively indicate that UV radiation technology is capable of inactivating airborne viruses, bacteria, and fungi and can be safely employed in occupied closed spaces. However, while efficacy studies are well documented, performing real-world effectiveness studies is challenging, as it is not feasible to experimentally release aerosolized pathogens in enclosed, occupied spaces.

Among the computational studies, two focus on evaluating the efficacy of a purifier using CFD simulations [185]. These simulations were performed by simulating flow fields using Reynolds-averaged Navier–Stokes (RANS) equations or filtered Navier–Stokes equations, commonly known as large eddy simulations (LESs). These simulations were coupled with the Chick-Watson disinfection model, as commonly seen in inactivation studies of airborne pathogens with UV radiation. This model provides the rate of inactivation (S) of pathogens inside a UV field (I) for an exposure time (t) as $S = 1 - \frac{N}{N_0}e^{-kIt_e}$, where k is a constant value that depicts the susceptibility of a pathogen to UV radiation and N and N_0 represent number of microbes present at times $t = 0$ and $t = t_e$, respectively. Sankurantripati [186] focused on studying UV-PAC efficacy using LES coupled with Lagrangian particle tracking to compute the turbulent dispersion of particles for different volumetric flow rates. This solver was coupled with a UV radiation model and the Chick–Watson model to determine the inactivation rate of a purifier. They provided the minimum UV dosage required for the tested purifier to inactivate SARS-CoV-2 with 95% rate of inactivation.

The remaining three studies focus on evaluating the effectiveness of purifiers using the Wells–Riley model. Kowalski [187] examined the effectiveness of an industrial purifier in a simplified residential room against a standard array of aerosolized pathogens using mathematical modeling to simulate infectivity risk for residents. The purifier's effectiveness was calculated by considering factors such as the volumetric flow rate, UV dosage, pathogen susceptibility to UV-C radiation, and the contaminant concentration in the room. Bergam [188] and Kapse [185] proposed conceptual designs for portable air cleaners that use UV radiation and high-efficiency filtration technology. Bergam focused on eliminating airborne pathogens with HEPA filters periodically cleaned by UV radiation. Their purifier's effectiveness was tested using 2D CFD simulations with the Chick–Watson model in an indoor restaurant setting. Srivastava [189] conducted RANS simulations to assess the performance of a UV purifier in an office building, simulating airflow patterns under natural and mechanical ventilation, which they validated against experimental results. By placing the UV purifier in various locations, they analyzed infectivity spread using the Wells–Riley model.

The computational methodologies used in these studies attempt to capture the physical phenomena involved in treating aerosolized pathogens with UV purifiers. However, the PAC models developed by Bergam [188], Kowalski [187], and Srivastava [189] do not adequately demonstrate purifier efficacy. Effectiveness studies do not accurately simulate the dispersion of droplets or account for factors such as temperature and humidity, which are critical to the airborne lifetime of pathogens. While Kapse's [185] efficacy study provides a good starting point, it omits accurate representations of phenomena like the heat generated by UV lamps and particle dispersion inside the PAC, which affect pathogen residence time. Current literature on UV-PACs is lacking a high-fidelity methodology that involves accurate prediction of the flow fields and small-scale turbulence that govern the dynamics of pathogen-laden particles. Such a methodology coupled with a UV radiation and disinfection model can provide accurate computation of the rate of inactivation of UV-PACs. It could also be extended to assess the effectiveness of UV-PACs inside closed environments.

The reviewed studies highlight the potential of UV-based PACs as an effective tool for airborne pathogen disinfection yet reveal significant gaps in both experimental and computational methodologies. While laboratory efficacy tests have demonstrated that UV-PACs can inactivate a range of pathogens, translating these results to real-world applications remains challenging due to practical limitations, such as the inability to safely release aerosolized pathogens in occupied spaces. Computational studies, which rely on models like the Chick–Watson disinfection framework and the Wells–Riley model, offer valuable insights but often fall short of capturing critical environmental factors such as temperature, humidity, and complex particle dispersion dynamics. Furthermore, simplifications like the use of Reynolds-averaged Navier–Stokes (RANS) equations or the omission of effects like the generation of heat by UV lamps can limit the accuracy of these models. To advance UV-PAC technology, future research should focus on developing high-fidelity simulations that incorporate small-scale turbulence and realistic environmental variables. Such advanced modeling, combined with rigorous in situ testing, would allow for a more accurate assessment of UV-PAC effectiveness in occupied indoor spaces, providing robust guidance for optimizing these systems in diverse real-world settings.

Table 2. Research on UV air purification technology (sorted by year of publication).

Author (Year)	Portable Air Cleaner (PAC)	Pathogen	Methodology	Test Location	Remarks	Limitations
Kowalski (2013) [187]	3 UV lamps with a carbon-treated particulate screen	Standard test array of residential pathogens	Mathematical model for disinfection	Simplified residential room with inlet and outlet	Tested various airborne pathogens under different UV dosages to evaluate PAC performance in a residential home	Purifier efficacy not explicitly provided, UV lamps' impact unclear. Assumptions about flow fields were made.
Bergam (2020) [188]	HEPA filters and UV-C LEDs	SARS-CoV-2	2D simulations, Chick–Watson model	Virtual indoor restaurant with 4 tables	Developed a PAC design with HEPA filters cleaned by UV-C LEDs	Efficacy not shown through computations or experiments
Messina (2020) [184]	UV-C lamp with HEPA filters	Staphylococcus aureus	Experimental study with particle counter	Tested in hospital OR settings	Highest IAQ observed with PAC active during procedures	Efficacy of purifier and UV lamps unclear. Tested only in one location.
Srivastava (2021) [189]	UV-C lamp and ventilation	SARS-CoV-2	CFD for airflow, Wells–Riley model for infection risk	Office building with computers	Validated airflow patterns and verified infection risk by altering purifier location and ventilation rates	Minimal information about the purifier. Assumed PAC efficacy is 99%.

Table 2. *Cont.*

Author (Year)	Portable Air Cleaner (PAC)	Pathogen	Methodology	Test Location	Remarks	Limitations
Garg (2022) [183]	UV-C lamps	SARS-CoV-2	Experimental study, RT-PCR test, Wiles–Riley model	Lab and classroom	Results show UV radiation doses required for residence time in purifier	Wiles–Riley model lacks full accuracy, needs high-fidelity experiments
Li (2022) [178]	16 UV-C lamps with PM filters (MERV-8, MERV-13)	Viable airborne bacteria	Total plate-count test (CFUs/ml)	Lab and poultry farm	PAC removed PM and inactivated bacteria under different UV dosages	UV lamps' effectiveness not isolated from the PAC
Rausch (2022) [182]	2 UV purifiers	Feline coronavirus	Enzyme pre-treatment and RT-PCR assays	Laboratory setup	Developed a rapid protocol to test efficacy, with a 100% inactivation rate	Focused on protocol development, not real-life scenario testing
Kapse (2023) [185]	Dust filters with UV-C LED arrays	SARS-CoV-2, tuberculosis, and influenza-A	CFD (RANS + k-ω), UV-C methods, and Chick–Watson model	Inside the purifier	Conceptual design aims to inactivate pathogen-laden particles with UV-C LEDs	Pathogen susceptibility constant from the literature for 254 nm UV-C, while LEDs were 279 nm. Study needs real-life testing
Sankurantripati (2024) [186]	2 UV-C lamps with fans	SARS-CoV-2	LES and Lagrangian tracking with a UV disinfection solver	Inside the purifier	Tested efficacy for 3 different flow rates and UV dosages, with 95% inactivation achieved	Constant particle diameter assumed and real-world validation absent

4. Conclusions

The recent COVID-19 pandemic has underscored the critical need to address airborne infections in enclosed spaces, where transmission rates are significantly higher compared to outdoor environments. Historical studies of disease transmission modes have highlighted the importance of understanding how pathogens like viruses, bacteria, and fungi spread, be it through contact, droplets, or airborne particles. This knowledge has been instrumental in shaping the mitigation strategies necessary to control the spread of infectious diseases.

While individual measures like mask wearing, social distancing, and regular sanitation help curb direct transmission, they do not address the quality of indoor air, which is a major factor in the spread of airborne pathogens. Improving indoor air quality (IAQ) through ventilation and filtration techniques is therefore crucial. Studies have shown that optimizing HVAC design and ventilation rates can significantly enhance IAQ, but retrofitting existing buildings or public transport systems is not always practical. This has led to increased interest in portable air filtration technologies.

Three primary filtration techniques—mechanical, electrical, and UV radiation—offer viable solutions for improving IAQ. Mechanical filters, such as HEPA filters, and electronic filters like electrostatic precipitators (ESPs), have been shown to effectively reduce airborne contaminants in enclosed environments. However, mechanical filters can accumulate pathogen-laden particles, eventually becoming bio-hazardous. UV radiation-based filters offer an alternative by inactivating pathogens through RNA or DNA disruption, without trapping them. Among these, UV-C technology has been successfully used in in-duct systems, upper-room UVGI setups, and portable air cleaners (UV-PACs).

Of the three UV filtration methods, UV-PACs offer the greatest flexibility, as they can be deployed in various spaces without requiring integration with existing ventilation systems. While laboratory studies have demonstrated the high efficacy of UV-PACs in neutralizing airborne pathogens, real-world effectiveness studies remain limited due to the inherent risks of testing with aerosolized pathogens in occupied spaces. In this context, computational studies are vital for assessing both the efficacy and effectiveness of UV air purifiers. High-fidelity computational models, informed by the physics of airborne transmission, are essential to understand how factors like purifier placement, airflow dynamics, and UV dosage affect performance.

This review highlights the significant potential of portable UV-PACs as a flexible, non-invasive solution for improving indoor air quality and reducing the risk of airborne pathogen transmission. UV-C-based technologies provide an alternative to traditional mechanical and electrical filtration methods, with the advantage of inactivating pathogens without trapping them, reducing bio-hazard accumulation over time. In particular, UV-PACs offer versatility, as they can be deployed across diverse environments, from hospitals and offices to public transport, without the need for extensive HVAC integration. Despite promising laboratory results demonstrating high efficacy, the real-world effectiveness of UV-PACs remains challenging to evaluate due to safety risks associated with the testing of aerosolized pathogens in occupied spaces. Current computational studies provide useful insights, but limitations in capturing complex environmental variables such as temperature, humidity, and airflow patterns underscore the need for more advanced modeling techniques. High-fidelity computational models that account for these factors are essential to accurately simulate UV-PAC performance in real-world scenarios. Moving forward, research must focus on developing and validating these computational models, which will enable more precise assessments of UV-PAC placement, UV dosage, and overall air purification efficacy in various settings. This approach is critical not only to enhance indoor air quality but also to prepare enclosed spaces for future infectious disease outbreaks, ensuring that advanced air purification technologies are both effective and broadly applicable across different indoor environments.

Author Contributions: Scientific articles were reviewed by and this paper was written by S.S. Supervision and editing were performed by F.D. All authors have read and agreed to the published version of the manuscript.

Funding: This research received no external funding.

Informed Consent Statement: Informed consent was obtained from all subjects involved in the study.

Data Availability Statement: The original contributions presented in this study are included in the article. Further inquiries can be directed to the corresponding author.

Conflicts of Interest: The authors declare no conflicts of interest.

Abbreviations

The following abbreviations are used in this manuscript:

HEPA	High-Efficiency Particulate Air
ESP	Electrostatic Precipitator
UV-C	Ultraviolet-C
PAC	Portable Air Cleaner
CADR	Clean Air Delivery Rate
RT-PCR	Reverse Transcription Polymerase Chain Reaction
SARS-CoV-2	Severe Acute Respiratory Syndrome Coronavirus 2
RNA	Ribonucleic Acid
HVAC	Heating, Ventilation, and Air Conditioning
UR-UVGI	Upper-Room Ultraviolet Germicidal Irradiation
CFD	Computational Fluid Dynamics
IAQ	Indoor Air Quality
COVID-19	Coronavirus Disease 2019
WHO	World Health Organization
HICPAC	Healthcare Infection Control Practices Advisory Committee
ASHRAE	American Society of Heating, Refrigerating and Air-Conditioning Engineers

References

1. Siegel, J.D.; Rhinehart, E.; Jackson, M.; Chiarello, L. 2007 guideline for isolation precautions: Preventing transmission of infectious agents in health care settings. *Am. J. Infect. Control* **2007**, *35*, S65–S164. [PubMed]
2. Wells, W.F. On air-borne infection: Study II. Droplets and droplet nuclei. *Am. J. Epidemiol.* **1934**, *20*, 611–618. [CrossRef]
3. Duguid, J.P. The size and the duration of air-carriage of respiratory droplets and droplet-nuclei. *Epidemiol. Infect.* **1946**, *44*, 471–479. [CrossRef]
4. Jones, R.M.; Brosseau, L.M. Aerosol transmission of infectious disease. *J. Occup. Environ. Med.* **2015**, *57*, 501–508. [CrossRef]
5. Hinds, W.C.; Zhu, Y. *Aerosol Technology: Properties, Behavior, and Measurement of Airborne Particles*; John Wiley & Sons: Hoboken, NJ, USA, 2022.
6. Furuse, Y.; Suzuki, A.; Oshitani, H. Origin of measles virus: Divergence from rinderpest virus between the 11th and 12th centuries. *Virol. J.* **2010**, *7*, 52 . [CrossRef]
7. Johnson, N.P.A.S.; Mueller, J. Updating the accounts: Global mortality of the 1918–1920 "Spanish" influenza pandemic. *Bull. Hist. Med.* **2002**, *76*, 105–115. [CrossRef]
8. Worldometer. COVID-19 Coronavirus Pandemic. *Worldometer*. Available online: https://www.worldometers.info/coronavirus/ (accessed on 28 June 2024).
9. Lewis, D. Why the WHO took two years to say COVID is airborne. *Nature* **2022**, *604*, 26–31. [CrossRef]
10. Jenner, E. *An Inquiry into the Causes and Effects of the Variolae Vaccinae, or Cow-Pox 1798*; Vaccination against Smallpox. Prometheus Books Great Minds Series; Ashley & Brewer: Springfield, MA, USA , 1996.
11. Bennett, J.H. On the parasitic vegetable structures found growing in living animals. *Earth Environ. Sci. Trans. R. Soc. Edinb.* **1844**, *15*, 277–294. [CrossRef]
12. Koch, R. *Die Aetiologie der Tuberkulose*; Robert Koch-Institut: Berlin, Germany, 2010.
13. Klebs, E. Ueber Diphtherie. *Verhandlungen Des Congr. FüR Inn. Med.* **1883**, *2*, 139–154.
14. Rouphael, N.G.; Stephens, D.S. Neisseria meningitidis: Biology, microbiology, and epidemiology. In *Neisseria Meningitidis: Advanced Methods and Protocols*; Springer: Berlin/Heidelberg, Germany, 2012; pp. 1–20.
15. Posadas, A. Un nuevo caso de micosis fungoidea con psorospermias. *An. Circ. Med. Argent.* **1892**, *15*, 585–597.
16. Lorber, B. The common cold. *J. Gen. Intern. Med.* **1996**, *11*, 229–236. [CrossRef] [PubMed]
17. Bordet, J.; Gengou, O. Le microbe de la coqueluche. *Bulletin de l'Académie Royale de Médecine de Belgique* **1906**, *20*, 731–741.
18. Darling, S.T. A protozoön general infection producing pseudotubercles in the lungs and focal necroses in the liver, spleen and lymphnodes. *J. Am. Med Assoc.* **1906**, *46*, 1283–1285. [CrossRef]
19. Morens, D.M.; Fauci, A.S. The 1918 influenza pandemic: Insights for the 21st century. *J. Infect. Dis.* **2007**, *195*, 1018–1028. [CrossRef]
20. Usman, N.; Suarez, M. Adenoviruses. In *StatPearls*; StatPearls Publishing: Treasure Island, FL, USA, 2020.
21. Weller, T.H. Observations on the behavior of certain viruses that produce intranuclear inclusion bodies in man. *Harvey Lect.* **1956**, *52*, 228–254.
22. Flint, S.J.; Racaniello, V.R.; Rall, G.F.; Hatziioannou, T.; Skalka, A.M. *Principles of Virology, Volume 2: Pathogenesis and Control*; John Wiley & Sons: Hoboken, NJ, USA, 2020.
23. World Health Organization. *WHO-Convened Global Study of Origins of SARS-CoV-2: China Part*; World Health Organization: Geneva, Switzerland, 2021.
24. Shen, Y.; Li, C.; Dong, H.; Wang, Z.; Martinez, L.; Sun, Z.; Handel, A.; Chen, Z.; Chen, E.; Ebell, M.H.; et al. Community outbreak investigation of SARS-CoV-2 transmission among bus riders in Eastern China. *Jama Intern. Med.* **2020**, *180*, 1665–1671. [CrossRef]
25. Lu, J.; Gu, J.; Li, K.; Xu, C.; Su, W.; Lai, Z.; Zhou, D.; Yu, C.; Xu, B.; Yang, Z. COVID-19 outbreak associated with air conditioning in restaurant, Guangzhou, China, 2020. *Emerg. Infect. Dis.* **2020**, *26*, 1628. [CrossRef]
26. Miron, O.; Yu, K.-H.; Wilf-Miron, R.; Davidovitch, N. COVID-19 infections following outdoor mass gatherings in low incidence areas: Retrospective cohort study. *medRxiv* **2020**. [CrossRef]
27. Zhang, R.; Li, Y.; Zhang, A.L.; Wang, Y.; Molina, M.J. Identifying airborne transmission as the dominant route for the spread of COVID-19. *Proc. Natl. Acad. Sci. USA* **2020**, *117*, 14857–14863. [CrossRef]
28. Morawska, L.; Cao, J. Airborne transmission of SARS-CoV-2: The world should face the reality. *Environ. Int.* **2020**, *139*, 105730. [CrossRef]
29. Qian, H.; Miao, T.; Liu, L.; Zheng, X.; Luo, D.; Li, Y. Indoor transmission of SARS-CoV-2. *Indoor Air* **2021**, *31*, 639–645. [CrossRef] [PubMed]
30. Bulfone, T.C.; Malekinejad, M.; Rutherford, G.W.; Razani, N. Outdoor transmission of SARS-CoV-2 and other respiratory viruses: A systematic review. *J. Infect. Dis.* **2021**, *223*, 550–561. [CrossRef] [PubMed]
31. Nishiura, H.; Oshitani, H.; Kobayashi, T.; Saito, T.; Sunagawa, T.; Matsui, T.; Wakita, T.; MHLW COVID-19 Response Team; Suzuki, M. Closed environments facilitate secondary transmission of coronavirus disease 2019 (COVID-19). *MedRxiv* **2020**. [CrossRef]
32. Lan, F.-Y.; Wei, C.-F.; Hsu, Y.-T.; Christiani, D.C.; Kales, S.N. Work-related COVID-19 transmission in six Asian countries/areas: A follow-up study. *PLoS ONE* **2020**, *15*, e0233588. [CrossRef] [PubMed]
33. Leclerc, Q.J.; Fuller, N.M.; Knight, L.E.; Funk, S.; Knight, G.M.; CMMID COVID-19 Working Group. What settings have been linked to SARS-CoV-2 transmission clusters? *Wellcome Open Res.* **2020**, *5*, 83 . [CrossRef] [PubMed]

34. Szablewski, C.M. SARS-CoV-2 transmission and infection among attendees of an overnight camp—Georgia, June 2020. *MMWR Morb. Mortal. Wkly. Rep.* **2020**, *69*, 1023–1025. [CrossRef]
35. Hamner, L. High SARS-CoV-2 attack rate following exposure at a choir practice—Skagit County, Washington, March 2020. *MMWR Morb. Mortal. Wkly. Rep.* **2020**, *69*, 606–610. [CrossRef]
36. de Man, P.; Paltansing, S.; Ong, D.S.Y.; Vaessen, N.; van Nielen, G.; Koeleman, J.G.M. Outbreak of coronavirus disease 2019 (COVID-19) in a nursing home associated with aerosol transmission as a result of inadequate ventilation. *Clin. Infect. Dis.* **2021**, *73*, 170–171. [CrossRef]
37. Park, S.Y.; Kim, Y.-M.; Yi, S.; Lee, S.; Na, B.-J.; Kim, C.B.; Kim, J.-I.; Kim, H.S.; Kim, Y.B.; Park, Y.; et al. Coronavirus disease outbreak in call center, South Korea. *Emerg. Infect. Dis.* **2020**, *26*, 1666. [CrossRef]
38. Rowe, B.R.; Canosa, A.; Drouffe, J.-M.; Mitchell, J.B. A simple quantitative assessment of the outdoor versus indoor airborne transmission of viruses and COVID-19. *Environ. Res.* **2021**, *198*, 111189. [CrossRef]
39. Ninyà, N.; Vallecillos, L.; Marcé, R.M.; Borrull, F. Evaluation of air quality in indoor and outdoor environments: Impact of anti-COVID-19 measures. *Sci. Total Environ.* **2022**, *836*, 155611. [CrossRef]
40. Beggs, C.B.; Abid, R.; Motallebi, F.; Samad, A.; Venkatesan, N.; Avital, E.J. Airborne transmission of SARS-CoV-2: The contrast between indoors and outdoors. *Fluids* **2024**, *9*, 54. [CrossRef]
41. James, A. High COVID-19 attack rate among attendees at events at a church—Arkansas, March 2020. *MMWR Morb. Mortal. Wkly. Rep.* **2020**, *69*, 632–635. [CrossRef] [PubMed]
42. Lesser, I.A.; Nienhuis, C.P. The impact of COVID-19 on physical activity behavior and well-being of Canadians. *Int. J. Environ. Res. Public Health* **2020**, *17*, 3899. [CrossRef] [PubMed]
43. Van Doremalen, N.; Bushmaker, T.; Morris, D.H.; Holbrook, M.G.; Gamble, A.; Williamson, B.N.; Tamin, A.; Harcourt, J.L.; Thornburg, N.J.; Gerber, S.I.; et al. Aerosol and surface stability of SARS-CoV-2 as compared with SARS-CoV-1. *N. Engl. J. Med.* **2020**, *382*, 1564–1567. [CrossRef]
44. Merow, C.; Urban, M.C. Seasonality and uncertainty in global COVID-19 growth rates. *Proc. Natl. Acad. Sci. USA* **2020**, *117*, 27456–27464. [CrossRef]
45. Sfîcă, L.; Bulai, M.; Amihăesei, V.-A.; Ion, C.; Ștefan, M. Weather conditions (with focus on UV radiation) associated with COVID-19 outbreak and worldwide climate-based prediction for future prevention. *Aerosol Air Qual. Res.* **2020**, *20*, 1862–1873. [CrossRef]
46. Papineni, R.S.; Rosenthal, F.S. The size distribution of droplets in the exhaled breath of healthy human subjects. *J. Aerosol Med.* **1997**, *10*, 105–116. [CrossRef]
47. Mittal, R.; Ni, R.; Seo, J.-H. The flow physics of COVID-19. *J. Fluid Mech.* **2020**, *894*, F2. [CrossRef]
48. Moriarty, J.A.; Grotberg, J.B. Flow-induced instabilities of a mucus-serous bilayer. *J. Fluid Mech.* **1999**, *397*, 1–22. [CrossRef]
49. Malashenko, A.; Tsuda, A.; Haber, S. Propagation and breakup of liquid menisci and aerosol generation in small airways. *J. Aerosol Med. Pulm. Drug Deliv.* **2009**, *22*, 341–353. [CrossRef]
50. Nicas, M.; Nazaroff, W.W.; Hubbard, A. Toward understanding the risk of secondary airborne infection: Emission of respirable pathogens. *J. Occup. Environ. Hyg.* **2005**, *2*, 143–154. [CrossRef] [PubMed]
51. Mutuku, J.K.; Hou, W.-C.; Chen, W.-H. An overview of experiments and numerical simulations on airflow and aerosols deposition in human airways and the role of bioaerosol motion in COVID-19 transmission. *Aerosol Air Qual. Res.* **2020**, *20*, 1172–1196. [CrossRef]
52. Fennelly, K.P. Particle sizes of infectious aerosols: Implications for infection control. *Lancet Respir. Med.* **2020**, *8*, 914–924. [CrossRef] [PubMed]
53. Gralton, J.; Tovey, E.; McLaws, M.-L.; Rawlinson, W.D. The role of particle size in aerosolised pathogen transmission: A review. *J. Infect.* **2011**, *62*, 1–13. [CrossRef]
54. Harrison, A.G.; Lin, T.; Wang, P. Mechanisms of SARS-CoV-2 transmission and pathogenesis. *Trends Immunol.* **2020**, *41*, 1100–1115. [CrossRef]
55. Han, Z.Y.; Weng, W.G.; Huang, Q.Y. Characterizations of particle size distribution of the droplets exhaled by sneeze. *J. R. Soc. Interface* **2013**, *10*, 20130560. [CrossRef]
56. Johnson, G.R.; Morawska, L.; Ristovski, Z.D.; Hargreaves, M.; Mengersen, K.; Chao, C.Y.H.; Wan, M.P.; Li, Y.; Xie, X.; Katoshevski, D.; et al. Modality of human expired aerosol size distributions. *J. Aerosol Sci.* **2011**, *42*, 839–851. [CrossRef]
57. Balachandar, S.; Zaleski, S.; Soldati, A.; Ahmadi, G.; Bourouiba, L. Host-to-host airborne transmission as a multiphase flow problem for science-based social distance guidelines. *Int. J. Multiph. Flow* **2020**, *132*, 103439. [CrossRef]
58. Loudon, R.G.; Roberts, R.M. Relation between the airborne diameters of respiratory droplets and the diameter of the stains left after recovery. *Nature* **1967**, *213*, 95–96. [CrossRef]
59. Duchaine, F.; Cizeron, M.; Odier, N.; Dombard, J.; Marchall, S.; Francois, N.; Poinsot, T. High-performance CFD for respiratory droplet turbulent dispersion in a ventilated city bus. *Int. J. Comput. Fluid Dyn.* **2021**, *35*, 758–777. [CrossRef]
60. Dbouk, T.; Drikakis, D. On coughing and airborne droplet transmission to humans. *Phys. Fluids* **2020**, *32*, 053310. [CrossRef] [PubMed]
61. Roccon, A.; De Paoli, M.; Zonta, F.; Soldati, A. Viscosity-modulated breakup and coalescence of large drops in bounded turbulence. *Phys. Rev. Fluids* **2017**, *2*, 083603. [CrossRef]

62. Soligo, G.; Roccon, A.; Soldati, A. Breakage, coalescence and size distribution of surfactant-laden droplets in turbulent flow. *J. Fluid Mech.* **2019**, *881*, 244–282. [CrossRef]
63. Gittings, S.; Turnbull, N.; Henry, B.; Roberts, C.J.; Gershkovich, P. Characterisation of human saliva as a platform for oral dissolution medium development. *Eur. J. Pharm. Biopharm.* **2015**, *91*, 16–24. [CrossRef]
64. Bourouiba, L.; Dehandschoewercker, E.; Bush, J.W.M. Violent expiratory events: On coughing and sneezing. *J. Fluid Mech.* **2014**, *745*, 537–563. [CrossRef]
65. Borak, J. Airborne transmission of COVID-19. *Occup. Med.* **2020**, *70*, 297–299. [CrossRef]
66. Stadnytskyi, V.; Bax, C.E.; Bax, A.; Anfinrud, P. The airborne lifetime of small speech droplets and their potential importance in SARS-CoV-2 transmission. *Proc. Natl. Acad. Sci. USA* **2020**, *117*, 11875–11877. [CrossRef]
67. Tang, J.W.; Nicolle, A.D.; Klettner, C.A.; Pantelic, J.; Wang, L.; Suhaimi, A.B.; Tan, A.Y.L.; Ong, G.W.X.; Su, R.; Sekhar, C.; et al. Airflow dynamics of human jets: Sneezing and breathing—potential sources of infectious aerosols. *PLoS ONE* **2013**, *8*, e59970. [CrossRef]
68. Wells, W.F. Airborne Contagion and Air Hygiene: An Ecological Study of Droplet Infections. *JAMA* **1955**, *159*, 90. [CrossRef]
69. Xie, X.; Li, Y.; Chwang, A.T.Y.; Ho, P.L.; Seto, W.H. How far droplets can move in indoor environments—Revisiting the wells evaporation–falling curve. *Indoor Air* **2007**, *17*, 3. [CrossRef]
70. Wei, J.; Li, Y. Enhanced spread of expiratory droplets by turbulence in a cough jet. *Build. Environ.* **2015**, *93*, 86–96. [CrossRef]
71. Ma, Y.; Zhao, Y.; Liu, J.; He, X.; Wang, B.; Fu, S.; Yan, J.; Niu, J.; Zhou, J.; Luo, B. Effects of temperature variation and humidity on the death of COVID-19 in Wuhan, China. *Sci. Total Environ.* **2020**, *724*, 138226. [CrossRef]
72. Zhao, L.; Qi, Y.; Luzzatto-Fegiz, P.; Cui, Y.; Zhu, Y. COVID-19: Effects of environmental conditions on the propagation of respiratory droplets. *Nano Lett.* **2020**, *20*, 7744–7750. [CrossRef]
73. Dbouk, T.; Drikakis, D. Weather impact on airborne coronavirus survival. *Phys. Fluids* **2020**, *32*, 093312. [CrossRef]
74. Bahramian, A.; Mohammadi, M.; Ahmadi, G. Effect of indoor temperature on the velocity fields and airborne transmission of sneeze droplets: An experimental study and transient CFD modeling. *Sci. Total Environ.* **2023**, *858*, 159444. [CrossRef]
75. Centers for Disease Control and Prevention (CDC). CDC Guidelines on Social Distancing. *CDC*. Available online: https://stacks.cdc.gov/view/cdc/90522 (accessed on 18 March 2023).
76. World Health Organization (WHO). Advice for the Public—Coronavirus Disease (COVID-19). Available online: https://www.who.int/emergencies/diseases/novel-coronavirus-2019/advice-for-public (accessed on 18 March 2023).
77. *ASTM F3502-21*; Standard Specification for Barrier Face Coverings. ASTM International: West Conshohocken, PA,USA, 2021.
78. Tang, J.W.; Liebner, T.J.; Craven, B.A.; Settles, G.S. A schlieren optical study of the human cough with and without wearing masks for aerosol infection control. *J. R. Soc. Interface* **2009**, *6*, S727–S736. [CrossRef]
79. Dbouk, T.; Drikakis, D. On respiratory droplets and face masks. *Phys. Fluids* **2020**, *32*, 063303. [CrossRef]
80. Van der Sande, M.; Teunis, P.; Sabel, R. Professional and home-made face masks reduce exposure to respiratory infections among the general population. *PLoS ONE* **2008**, *3*, e2618. [CrossRef]
81. Kohn, A.; Gitelman, J.; Inbar, M. Unsaturated free fatty acids inactivate animal enveloped viruses. *Arch. Virol.* **1980**, *66*, 301–307. [CrossRef]
82. Zayas, G.; Chiang, M.C.; Wong, E.; MacDonald, F.; Lange, C.F.; Senthilselvan, A.; King, M. Effectiveness of cough etiquette maneuvers in disrupting the chain of transmission of infectious respiratory diseases. *BMC Public Health* **2013**, *13*, 811. [CrossRef]
83. Sherman, M.; Jones, B. *ASHRAE 241—2023 Control of Infectious Aerosols*; ASHRAE: Peachtree Corners, GA, USA, 2023.
84. World Health Organization. *The Right to Healthy Indoor Air: Report on a WHO Meeting, Bilthoven, The Netherlands, 15–17 May 2000*; WHO Regional Office for Europe: Copenhagen, Denmark, 2000.
85. Li, Y.; Leung, M.; Tang, J.W.; Yang, X.; Chao, C.Y.H.; Lin, J.Z.; Lu, J.W.; Nielsen, P.V.; Niu, J.; Qian, H.; et al. Role of ventilation in airborne transmission of infectious agents in the built environment—A multidisciplinary systematic review. *Indoor Air* **2007**, *17* . [CrossRef]
86. Liu, D.T.; Phillips, K.M.; Speth, M.M.; Besser, G.; Mueller, C.A.; Sedaghat, A.R. Portable HEPA purifiers to eliminate airborne SARS-CoV-2: A systematic review. *Otolaryngol. Head Neck Surg.* **2022**, *166*, 615–622. [CrossRef] [PubMed]
87. Szczotko, M.; Orych, I.; Mąą, Ł.; Solecka, J. A review of selected types of indoor air purifiers in terms of microbial air contamination reduction. *Atmosphere* **2022**, *13*, 800. [CrossRef]
88. Al-Rikabi, I.J.; Karam, J.; Alsaad, H.; Ghali, K.; Ghaddar, N.; Voelker, C. The impact of mechanical and natural ventilation modes on the spread of indoor airborne contaminants: A review. *J. Build. Eng.* **2024**, *85* , 108715. [CrossRef]
89. Domingo, J.L.; Marquès, M.; Rovira, J. Influence of airborne transmission of SARS-CoV-2 on COVID-19 pandemic. A review. *Environ. Res.* **2020**, *188*, 109861. [CrossRef]
90. Otter, J.A.; Yezli, S.; Perl, T.M.; Barbut, F.; French, G.L. The role of 'no-touch' automated room disinfection systems in infection prevention and control. *J. Hosp. Infect.* **2013**, *83*, 1–13. [CrossRef]
91. Etheridge, D.W.; Sandberg, M. *Building Ventilation: Theory and Measurement*; John Wiley & Sons: Chichester, UK, 1996; Volume 50.
92. Riley, R.L.; O'Grady, F. *Airborne Infection: Transmission and Control*; McMillan: New York, NY, USA, 1961.
93. Menzies, D.; Fanning, A.; Yuan, L.; FitzGerald, J.M.; Canadian Collaborative Group in Nosocomial Transmission of TB*. Hospital ventilation and risk for tuberculous infection in Canadian health care workers. *Ann. Intern. Med.* **2000**, *133*, 779–789. [CrossRef]
94. Bloch, A.B.; Orenstein, W.A.; Ewing, W.M.; Spain, W.H.; Mallison, G.F.; Herrmann, K.L.; Hinman, A.R. Measles outbreak in a pediatric practice: Airborne transmission in an office setting. *Pediatrics* **1985**, *75*, 676–683. [CrossRef]

95. Moser, M.R.; Bender, T.R.; Margolis, H.S.; Noble, G.R.; Kendal, A.P.; Ritter, D.G. An outbreak of influenza aboard a commercial airliner. *Am. J. Epidemiol.* **1979**, *110*, 1–6. [CrossRef]
96. Schulman, J.L.; Kilbourne, E.D. Airborne transmission of influenza virus infection in mice. *Nature* **1962**, *195*, 1129–1130. [CrossRef]
97. Gustafson, T.L.; Lavely, G.B.; Brawner, E.R., Jr.; Hutcheson, R.H., Jr.; Wright, P.F.; Schaffner, W. An outbreak of airborne nosocomial varicella. *Pediatrics* **1982**, *70*, 550–556. [CrossRef] [PubMed]
98. Hutton, M.D.; Stead, W.W.; Cauthen, G.M.; Bloch, A.B.; Ewing, W.M. Nosocomial transmission of tuberculosis associated with a draining abscess. *J. Infect. Dis.* **1990**, *161*, 286–295. [CrossRef] [PubMed]
99. Wehrle, P.F.; Posch, J.; Richter, K.H.; Henderson, D.A. An airborne outbreak of smallpox in a German hospital and its significance with respect to other recent outbreaks in Europe. *Bull. World Health Organ.* **1970**, *43*, 669.
100. Riley, E.C.; Murphy, G.; Riley, R.L. Airborne spread of measles in a suburban elementary school. *Am. J. Epidemiol.* **1978**, *107*, 421–432. [CrossRef]
101. Bolashikov, Z.D.; Melikov, A.K. Methods for air cleaning and protection of building occupants from airborne pathogens. *Build. Environ.* **2009**, *44*, 1378–1385. [CrossRef]
102. Zhang, C.; Yu, T.; Heiselberg, P.K.; Pomianowski, M.Z.; Nielsen, P.V. *Diffuse Ceiling Ventilation: Design Guide*; Department of Civil Engineering, Aalborg University: Aalborg East, Denmark, 2016.
103. Su, W.; Yang, B.; Melikov, A.; Liang, C.; Lu, Y.; Wang, F.; Li, A.; Lin, Z.; Li, X.; Cao, G.; et al. Infection probability under different air distribution patterns. *Build. Environ.* **2022**, *207*, 108555. [CrossRef]
104. Jurelionis, A.; Gagytė, L.; Prasauskas, T.; Čiužas, D.; Krugly, E.; Šeduikytė, L.; Martuzevičius, D. The impact of the air distribution method in ventilated rooms on the aerosol particle dispersion and removal: The experimental approach. *Energy Build.* **2015**, *86*, 305–313. [CrossRef]
105. Zajas, J.; Litewnicki, M. Breathing and cross-infection risk in the microenvironment around people. *Ashrae Trans.* **2014**, *120*, T1.
106. Ye, J.; Ai, Z.; Chang, Y. Aerosol transmission in queuing and dining scenarios in canteens and the effectiveness of control measures. *J. Build. Perform. Simul.* **2023**, *16*, 660–679. [CrossRef]
107. Lin, Z.; Chow, T.T.; Tsang, C.F. Stratum ventilation—A conceptual introduction. In Proceedings of the 10th International Conference on Indoor Air Quality and Climate, Beijing, China, 4–9 September 2005.
108. Barbosa, B.P.P.; de Carvalho Lobo Brum, N. Ventilation mode performance against airborne respiratory infections in small office spaces: Limits and rational improvements for COVID-19. *J. Braz. Soc. Mech. Sci. Eng.* **2021**, *43*, 316. [CrossRef]
109. Yoo, S.-J.; Kurokawa, A.; Matsunaga, K.; Ito, K. Spatial distributions of airborne transmission risk on commuter buses: Numerical case study using computational fluid and particle dynamics with computer-simulated persons. *Exp. Comput. Multiph. Flow* **2023**, *5*, 304–318. [CrossRef] [PubMed]
110. Lu, Y.; Oladokun, M.; Lin, Z. Reducing the exposure risk in hospital wards by applying stratum ventilation system. *Build. Environ.* **2020**, *183*, 107204. [CrossRef]
111. Lu, Y.; Niu, D.; Zhang, S.; Chang, H.; Lin, Z. Ventilation indices for evaluation of airborne infection risk control performance of air distribution. *Build. Environ.* **2022**, *222*, 109440. [CrossRef] [PubMed]
112. Kabanshi, A.; Wigö, H.; Sandberg, M. Experimental evaluation of an intermittent air supply system—Part 1: Thermal comfort and ventilation efficiency measurements. *Build. Environ.* **2016**, *95*, 240–250. [CrossRef]
113. Park, S.; Choi, Y.; Song, D.; Kim, E.K. Natural ventilation strategy and related issues to prevent coronavirus disease 2019 (COVID-19) airborne transmission in a school building. *Sci. Total Environ.* **2021**, *789*, 147764. [CrossRef]
114. Mu, D.; Gao, N.; Zhu, T. Wind tunnel tests of inter-flat pollutant transmission characteristics in a rectangular multi-storey residential building, part A: Effect of wind direction. *Build. Environ.* **2016**, *108*, 159–170. [CrossRef]
115. Zhou, Q.; Qian, H.; Liu, L. Numerical investigation of airborne infection in naturally ventilated hospital wards with central-corridor type. *Indoor Built Environ.* **2018**, *27*, 59–69. [CrossRef]
116. Somsen, G.A.; van Rijn, C.; Kooij, S.; Bem, R.A.; Bonn, D. Small droplet aerosols in poorly ventilated spaces and SARS-CoV-2 transmission. *Lancet Respir. Med.* **2020**, *8*, 658–659. [CrossRef]
117. Abbas, G.M.; Gursel Dino, I. The impact of natural ventilation on airborne biocontaminants: A study on COVID-19 dispersion in an open office. *Eng. Constr. Archit. Manag.* **2022**, *29*, 1609–1641. [CrossRef]
118. Escombe, A.R.; Ticona, E.; Chávez-Pérez, V.; Espinoza, M.; Moore, D.A.J. Improving natural ventilation in hospital waiting and consulting rooms to reduce nosocomial tuberculosis transmission risk in a low resource setting. *Bmc Infect. Dis.* **2019**, *19*, 88. [CrossRef]
119. Chakroun, W.; Alotaibi, S.; Habchi, C.; Ghali, K.; Ghaddar, N. Comparison of removal effectiveness of mixed versus displacement ventilation during vacuuming session. *Build. Environ.* **2019**, *155*, 118–126. [CrossRef]
120. Zhou, Y.; Deng, Y.; Wu, P.; Cao, S.-J. The effects of ventilation and floor heating systems on the dispersion and deposition of fine particles in an enclosed environment. *Build. Environ.* **2017**, *125*, 192–205. [CrossRef]
121. Morawska, L.; Tang, J.W.; Bahnfleth, W.; Bluyssen, P.M.; Boerstra, A.; Buonanno, G.; Cao, J.; Dancer, S.; Floto, A.; Franchimon, F.; et al. How can airborne transmission of COVID-19 indoors be minimised? *Environ. Int.* **2020**, *142*, 105832. [CrossRef] [PubMed]
122. Environmental Protection Agency (EPA). Air Cleaners, HVAC Filters, and Coronavirus (COVID-19). Available online: https://www.epa.gov/indoor-air-quality-iaq/air-cleaners-hvac-filters-and-coronavirus-covid-19 (accessed on 18 March 2023).
123. Perry, J.L.; Agui, J.H.; Vijayakimar, R. *Submicron and Nanoparticulate Matter Removal by HEPA-Rated Media Filters and Packed Beds of Granular Materials*; Marshall Space Flight Center: Huntsville, AL, USA, 2016.

124. First, M.W. HEPA filters. *J. Am. Biol. Saf. Assoc.* **1998**, *3*, 33–42. [CrossRef]
125. John, D.A. Air-distribution design: HEPA or ULPA filtration. *ASHRAE J.* **2013**, *55*, 84–86.
126. Ren, Y.-F.; Huang, Q.; Marzouk, T.; Richard, R.; Pembroke, K.; Martone, P.; Venner, T.; Malmstrom, H.; Eliav, E. Effects of mechanical ventilation and portable air cleaner on aerosol removal from dental treatment rooms. *J. Dent.* **2021**, *105*, 103576. [CrossRef]
127. Qian, H.; Li, Y.; Sun, H.; Nielsen, P.V.; Huang, X.; Zheng, X. Particle removal efficiency of the portable HEPA air cleaner in a simulated hospital ward. In *Building Simulation*; Springer: Berlin/Heidelberg, Germany, 2010; pp. 215–224.
128. Bluyssen, P.M.; Ortiz, M.; Zhang, D. The effect of a mobile HEPA filter system on 'infectious' aerosols, sound and air velocity in the SenseLab. *Build. Environ.* **2021**, *188*, 107475. [CrossRef]
129. Miller-Leiden, S.; Lohascio, C.; Nazaroff, W.W.; Macher, J.M. Effectiveness of in-room air filtration and dilution ventilation for tuberculosis infection control. *J. Air Waste Manag. Assoc.* **1996**, *46*, 869–882. [CrossRef]
130. Afshari, A.; Ekberg, L.; Forejt, L.; Mo, J.; Rahimi, S.; Siegel, J.; Chen, W.; Wargocki, P.; Zurami, S.; Zhang, J. Electrostatic precipitators as an indoor air cleaner—A literature review. *Sustainability* **2020**, *12*, 8774. [CrossRef]
131. Masuda, S.; Hosokawa, S. Electrostatic precipitation. In *Handbook of Electrostatic Processes*; Marcel Dekker: New York, NY, USA, 1995; pp. 441–480.
132. Mizuno, A. Electrostatic precipitation. *IEEE Trans. Dielectr. Electr. Insul.* **2000**, *7*, 615–624. [CrossRef]
133. USEPA. *EPA Air Pollution Control Cost Manual*; United States Environmental Protection Agency: Washington, DC, USA, 2002.
134. *ANSI/AHRI*; Standard 681: Performance Rating Residential Air Filter Equipment. Air-Conditioning, Heating, and Refrigeration Institute: Arlington, VA, USA, 2009.
135. Novoselac, A.; Siegel, J.A. Impact of placement of portable air cleaning devices in multizone residential environments. *Build. Environ.* **2009**, *44*, 2348–2356. [CrossRef]
136. Kosar, D.; Novosel, D. *Reduced Energy Use Through Reduced Indoor Contamination in Residential Buildings*; NCEMBT 061101; SMACNA: Washington, DC, USA, 2006. [CrossRef]
137. Boelter, K.J.; Davidson, J.H. Ozone generation by indoor, electrostatic air cleaners. *Aerosol Sci. Technol.* **1997**, *27*, 689–708. [CrossRef]
138. Mølgaard, B.; Koivisto, A.J.; Hussein, T.; H'ameri, K. A new clean air delivery rate test applied to five portable indoor air cleaners. *Aerosol Sci. Technol.* **2014**, *48*, 409–417. [CrossRef]
139. Sultan, Z.M.; Nilsson, G.J.; Magee, R.J. Removal of ultrafine particles in indoor air: Performance of various portable air cleaner technologies. *HVAC&R Res.* **2011**, *17*, 513–525.
140. Zuraimi, M.S.; Nilsson, G.J.; Magee, R.J. Removing indoor particles using portable air cleaners: Implications for residential infection transmission. *Build. Environ.* **2011**, *46*, 2512–2519. [CrossRef]
141. Waring, M.S.; Siegel, J.A.; Corsi, R.L. Ultrafine particle removal and generation by portable air cleaners. *Atmos. Environ.* **2008**, *42*, 5003–5014. [CrossRef]
142. Tian, E.; Gao, Y.; Mo, J. Electrostatically assisted air coarse filtration for energy efficient ambient particles removal: Long-term performance in real environment and influencing factors. *Build. Environ.* **2019**, *164*, 106348. [CrossRef]
143. Kowalski, W.J.; Bahnfleth, W.P.; Witham, D.L.; Severin, B.F.; Whittam, T.S. Mathematical modeling of ultraviolet germicidal irradiation for air disinfection. *Quant. Microbiol.* **2000**, *2*, 249–270. [CrossRef]
144. Reed, N.G. The history of ultraviolet germicidal irradiation for air disinfection. *Public Health Rep.* **2010**, *125*, 15–27. [CrossRef]
145. Downes, A. Researches on the effect of light upon bacteria and other organisms. *Proc. R. Soc. Lond.* **1877**, *26*, 488–500.
146. Tyndall, J. Note on the influence exercised by light on organic infusions. *Proc. R. Soc. Lond.* **1878**, *28*, 212–213.
147. Geisler, T. Zur frage über die wirkung des licht auf bakterien. *Cent. FüR. Bakteriol. Und. Parasitenkd.* **1892**, *11*, 161–173.
148. Bang, S. Über die Wirkungen des Lichtes auf Mikroben. II. *Eine Verbesserte Untersuchungs Methode. Mitt. Finsens. Med. Lysinst.* **1903**, *3*, 97–112.
149. Duclaux, E. Influence de la lumière du soleil sur la vitalité des germes des microbes. *Compt Rendus Hebd Des Seances L'Academie Des Sci.* **1885**, *100*, 119–121.
150. Hockberger, P.E. The discovery of the damaging effect of sunlight on bacteria. *J. Photochem. Photobiol. B Biol.* **2000**, *58*, 185–191. [CrossRef]
151. Wells, W.F.; Fair, G.M. Viability of *B. coli* exposed to ultra-violet radiation in air. *Science* **1935**, *82*, 280–281. [CrossRef]
152. Sharp, D.G. A quantitative method of determining the lethal effect of ultraviolet light on bacteria suspended in air. *J. Bacteriol.* **1938**, *35*, 589–599. [CrossRef]
153. Jensen, M.M. Inactivation of airborne viruses by ultraviolet irradiation. *Appl. Microbiol.* **1964**, *12*, 418–420. [CrossRef]
154. Beggs, C.B.; Avital, E.J. Upper-room ultraviolet air disinfection might help to reduce COVID-19 transmission in buildings: A feasibility study. *PeerJ* **2020**, *8*, e10196. [CrossRef]
155. Noakes, C.J.; Beggs, C.B.; Sleigh, P.A. Effect of room mixing and ventilation strategy on the performance of upper room ultraviolet germicidal irradiation systems. In Proceedings of the ASHRAE IAQ, Tampa, FL, USA, 15–17 March 2004; pp. 1–13.
156. Wells, W.F.; Wells, M.W.; Wilder, T.S. The environmental control of epidemic contagion: An epidemiologic study of radiant disinfection of air in day schools. *Am. J. Hyg.* **1942**, *35*, 97–121.
157. Al-Rawi, M.; Ikutegbe, C.A.; Auckaili, A.; Farid, M.M. Sustainable technologies to improve Indoor air quality in a residential house—A case study in Waikato, New Zealand. *Energy Build.* **2021**, *250*, 111283. [CrossRef]

158. Bang, J.-I.; Park, J.; Choi, A.; Jeong, J.-W.; Kim, J.Y.; Sung, M. Evaluation of UR-UVGI system for sterilization effect on microorganism contamination in negative pressure isolation ward. *Sustainability* **2018**, *10*, 3192. [CrossRef]
159. Lee, L.D.; Delclos, G.; Berkheiser, M.L.; Barakat, M.T.; Jensen, P.A. Evaluation of multiple fixed in-room air cleaners with ultraviolet germicidal irradiation, in high-occupancy areas of selected commercial indoor environments. *J. Occup. Environ. Hyg.* **2022**, *19*, 67–77. [CrossRef]
160. Gilkeson, C.A.; Noakes, C.J.; Khan, M.A.I. Computational fluid dynamics modelling and optimisation of an upper-room ultraviolet germicidal irradiation system in a naturally ventilated hospital ward. *Indoor Built Environ.* **2014**, *23*, 449–466. [CrossRef]
161. Kanaan, M.; Ghaddar, N.; Ghali, K. Localized air-conditioning with upper-room UVGI to reduce airborne bacteria cross-infection. In *Building Simulation*; Springer: Berlin/Heidelberg, Germany, 2016; pp. 63–74.
162. Noakes, C.J.; Khan, M.A.I.; Gilkeson, C.A. Modeling infection risk and energy use of upper-room ultraviolet germicidal irradiation systems in multi-room environments. *Sci. Technol. Built Environ.* **2015**, *21*, 99–111. [CrossRef]
163. Yang, Y.; Lai, A.C.K.; Wu, C. Study on the disinfection efficiency of multiple upper-room ultraviolet germicidal fixtures system on airborne microorganisms. *Build. Environ.* **2016**, *103*, 99–110. [CrossRef]
164. Kanaan, M.; Ghaddar, N.; Ghali, K.; Araj, G. Upper room UVGI effectiveness with dispersed pathogens at different droplet sizes in spaces conditioned by chilled ceiling and mixed displacement ventilation system. *Build. Environ.* **2015**, *87*, 117–128. [CrossRef]
165. Rudnick, S.N.; McDevitt, J.J.; Hunt, G.M.; Stawnychy, M.T.; Vincent, R.L.; Brickner, P.W. Influence of ceiling fan's speed and direction on efficacy of upper-room, ultraviolet germicidal irradiation: Experimental. *Build. Environ.* **2015**, *92*, 756–763. [CrossRef]
166. Pichurov, G.; Srebric, J.; Zhu, S.; Vincent, R.L.; Brickner, P.W.; Rudnick, S.N. A validated numerical investigation of the ceiling fan's role in the upper-room UVGI efficacy. *Build. Environ.* **2015**, *86*, 109–119. [CrossRef]
167. Zhu, S.; Lin, T.; Wang, L.; Nardell, E.A.; Vincent, R.L.; Srebric, J. Ceiling impact on air disinfection performance of Upper-Room Germicidal Ultraviolet (UR-GUV). *Build. Environ.* **2022**, *224*, 109530. [CrossRef]
168. Nunayon, S.S.; Zhang, H.; Lai, A.C.K. Comparison of disinfection performance of UVC-LED and conventional upper-room UVGI systems. *Indoor Air* **2020**, *30*, 180–191. [CrossRef] [PubMed]
169. Singh, D.; Soorneedi, A.R.; Vaze, N.; Domitrovic, R.; Sharp, F.; Lindsey, D.; Rohr, A.; Moore, M.D.; Koutrakis, P.; Nardell, E.; et al. Assessment of SARS-CoV-2 surrogate inactivation on surfaces and in air using UV and blue light-based intervention technologies. *J. Air Waste Manag. Assoc.* **2023**, *73*, 200–211. [CrossRef] [PubMed]
170. de Souza, S.O.; Cardoso, A.A., Jr.; Sarmento, A.S.C.; d'Errico, F. Effectiveness of a UVC air disinfection system for the HVAC of an ICU. *Eur. Phys. J. Plus* **2022**, *137*, 37. [CrossRef] [PubMed]
171. Mariita, R.M.; Davis, J.H.; Lottridge, M.M.; Randive, R.V.; Witting, H.; Yu, J. Towards a Healthy Car: UVC LEDs in an Automobile's HVAC Demonstrates Effective Disinfection of Cabin Air. *Atmosphere* **2022**, *13*, 1926. [CrossRef]
172. Qiao, Y.; Yang, M.; Marabella, I.A.; McGee, D.A.J.; Aboubakr, H.; Goyal, S.; Hogan, C.J., Jr.; Olson, B.A.; Torremorell, M. Greater than 3-log reduction in viable coronavirus aerosol concentration in ducted ultraviolet-C (UV–C) systems. *Environ. Sci. Technol.* **2020**, *55*, 4174–4182. [CrossRef]
173. Brockmann, G.; Brandt, S.; Kriegel, M. Numerical approach to improve UVC radiation for air disinfection and investigation of the scalability. *E3S Web Conf.* **2023**, *396*, 03002. [CrossRef]
174. Capetillo, A.; Noakes, C.J.; Sleigh, P.A.; Khan, A. In-duct UVGI air sterilisation: Optimisation study for high performance energy efficient systems. In Proceedings of the Indoor Air, Hong Kong, 7–12 July 2014.
175. Yang, Y.; Zhang, H.; Lai, A.C.K. Lagrangian modeling of inactivation of airborne microorganisms by in-duct ultraviolet lamps. *Build. Environ.* **2021**, *188*, 107465. [CrossRef]
176. Zhang, H.; Lai, A.C.K. Evaluation of single-pass disinfection performance of far-UVC light on airborne microorganisms in duct flows. *Environ. Sci. Technol.* **2022**, *56*, 17849–17857. [CrossRef]
177. Xie, Y.; Zhu, X.; Zhang, P.; Wang, S.; Yang, J.; Li, J. Cost-effective instant air disinfection for building ventilation system by a combination of UV and micro-static electricity. *Chem. Eng. J.* **2023**, *454*, 140231. [CrossRef]
178. Li, P.; Koziel, J.A.; Macedo, N.; Zimmerman, J.J.; Wrzesinski, D.; Sobotka, E.; Balderas, M.; Walz, W.B.; Paris, R.V.; Lee, M.; et al. Evaluation of an air cleaning device equipped with filtration and UV: Comparison of removal efficiency on particulate matter and viable airborne bacteria in the inlet and treated air. *Int. J. Environ. Res. Public Health* **2022**, *19*, 16135. [CrossRef]
179. Nair, A.N.; Anand, P.; George, A.; Mondal, N. A review of strategies and their effectiveness in reducing indoor airborne transmission and improving Indoor air quality. *Environ. Res.* **2022**, *213*, 113579. [CrossRef] [PubMed]
180. Peng, Y.; Lei, Y.; Tekler, Z.D.; Antanuri, N.; Lau, S.-K.; Chong, A. Hybrid system controls of natural ventilation and HVAC in mixed-mode buildings: A comprehensive review. *Energy Build.* **2022**, *276*, 112509. [CrossRef]
181. Liberati, A.; Altman, D.G.; Tetzlaff, J.; Mulrow, C.; Gøtzsche, P.C.; Ioannidis, J.P.A.; Clarke, M.; Devereaux, P.J.; Kleijnen, J.; Moher, D. The PRISMA statement for reporting systematic reviews and meta-analyses of studies that evaluate health care interventions: Explanation and elaboration. *Ann. Intern. Med.* **2009**, *151*, W65–W94. [CrossRef]
182. Rausch, F.; Tanneberger, F.; Abd El Wahed, A.; Truyen, U. Validation of the efficacy of air purifiers using molecular techniques. *PLoS ONE* **2023**, *18*, e0280243. [CrossRef] [PubMed]
183. Garg, H.; Ringe, R.P.; Das, S.; Parkash, S.; Thakur, B.; Delipan, R.; Kumar, A.; Kulkarni, K.; Bansal, K.; Patil, P.B.; et al. UVC-based air disinfection systems for rapid inactivation of SARS-CoV-2 present in the air. *Pathogens* **2023**, *12*, 419. [CrossRef]

184. Messina, G.; Spataro, G.; Catarsi, L.; De Marco, M.F.; Grasso, A.; Cevenini, G. A mobile device reducing airborne particulate can improve air quality. *AIMS Public Health* **2020**, *7*, 469. [CrossRef]
185. Kapse, S.; Rahman, D.; Avital, E.J.; Venkatesan, N.; Smith, T.; Cantero-Garcia, L.; Motallebi, F.; Samad, A.; Beggs, C.B. Conceptual design of a UVC-LED air purifier to reduce airborne pathogen transmission—A feasibility study. *Fluids* **2023**, *8*, 111. [CrossRef]
186. Sankurantripati, S.; Duchaine, F.; Francois, N.; Marshall, S. *High Fidelity Simulations of Airborne Virus Inactivation in a UV Air Purifier: Impact of Volumetric Flow Rate and UV Radiation Intensity*; CERFACS: Toulouse, France, 2024.
187. Kowalski, W. *Dr. Residential Application of the RxAir™ UV Light Portable Air Purification Unit*; UV Flu Technology: Yarmouthport, MA, USA, 2013.
188. Bergam, N.; Chen, L.; Lende, S.; Snow, S.; Zhang, J.; DiBuono, M.J.; Calzaretto, N. Designing and simulating a smart SARS-CoV-2 air purifier. *N. J. Gov.'s School Eng. Technol.* **2020**, 1–10.
189. Srivastava, S.; Zhao, X.; Manay, A.; Chen, Q. Effective ventilation and air disinfection system for reducing coronavirus disease 2019 (COVID-19) infection risk in office buildings. *Sustain. Cities Soc.* **2021**, *75*, 103408. [CrossRef]

Disclaimer/Publisher's Note: The statements, opinions and data contained in all publications are solely those of the individual author(s) and contributor(s) and not of MDPI and/or the editor(s). MDPI and/or the editor(s) disclaim responsibility for any injury to people or property resulting from any ideas, methods, instructions or products referred to in the content.

fluids

MDPI

Article

Gradient-Based Aero-Stealth Optimization of a Simplified Aircraft

Charles Thoulon [1,2,*], Gilbert Roge [1] and Olivier Pironneau [2]

[1] Dassault-Aviation, 92210 St. Cloud, France; gilbert.roge@dassault-aviation.com
[2] Laboratoire Jacques-Louis Lions, Sorbonne Université, 75005 Paris, France; olivier.pironneau@gmail.com
* Correspondence: charles.thoulon@gmail.com

Abstract: Modern fighter aircraft increasingly need to conjugate aerodynamic performance and low observability. In this paper, we showcase a methodology for a gradient-based bidisciplinary aero-stealth optimization. The shape of the aircraft is parameterized with the help of a CAD modeler, and we optimize it with the SLSQP algorithm. The drag, computed with the help of a RANS method, is used as the aerodynamic criterion. For the stealth criterion, a function is derived from the radar cross-section in a given cone of directions and weighed with a function whose goal is to cancel the electromagnetic intensity in a given direction. Stealth is achieved passively by scattering back the electromagnetic energy away from the radar antenna, and no energy is absorbed by the aircraft, which is considered as a perfect conductor. A Pareto front is identified by varying the weights of the aerodynamic and stealth criteria. The Pareto front allows for an easy identification of the CAD model corresponding to a chosen aero-stealth trade-off.

Keywords: multidisciplinary optimization; shape optimization; gradient based; CAD based; aerodynamics; stealth; radar cross-section

1. Introduction

With the omnipresence of radar technology, from early detection to target acquisition by a hostile air defense system to active radar homing for missiles, modern fighter aircraft design needs to take stealth into consideration. The designer needs to maximize aerodynamic performances while keeping a low Radar Cross-Section (RCS).

The frequency bands commonly used by modern radars are the L, S, C, and X bands [1]. Their working frequencies range from 1 GHz to 12 GHz, corresponding to wavelengths between 0.025 m and 0.3 m, which are much smaller than typical aircraft dimensions. Therefore, the radar waves are considered high-frequency. This means that the scattering characteristics of the aircraft are heavily influenced by its geometric details. On the other hand, the aerodynamic qualities of an aircraft can be very sensitive to even small changes to its geometry. There is then a need for methodologies able to efficiently design shapes that balance aerodynamic and stealth performances.

Some studies such as [2] deal with the contradictory effects of shape on aerodynamic efficiency and stealth, but few of them use integrated aero-stealth optimization methods. Among them, Lee et al. [3] used a robust evolutionary algorithm for the wing planform and sections design. In order to mitigate the large computation costs due to the number of objective function evaluations incurred by the evolutionary algorithm, they used a potential flow and physical optics solver. Later, Wu et al. [4] used the adaptive metamodel method in order to optimize a high aspect ratio wing for the aerodynamics, stealth, and structure, while Jiang et al. [5] used the full factorial design method in combination with radial basis functions in order to optimize the aero-stealth of the rotor airfoil of a helicopter.

All those studies used the gradient-free optimization method, which has the advantage of not necessitating the costly computation of derivatives, which can be hard to derive or implement. However, they need a high number of objective function evaluations, whose computation may be prohibitive when using high-fidelity solvers. In an industrial design process, gradient-free algorithms are typically used for preliminary design with low-fidelity models, while gradient-based methods are more adapted to large-scale aircraft shape design with mid- to high-fidelity models. Furthermore, surrogate-based algorithms are known to struggle with constraints handling, satisfying them with a poor accuracy and struggling to meet the stop criteria [6].

1.1. Gradient-Based Method

It is well known that gradient-based methods usually necessitate significantly less function evaluations than their gradient-free counterpart. For example, Lyu et al. [7] benchmarked several optimization algorithms, both gradient-free and gradient-based, on three different problems with increasing complexity. They showed that the number of evaluations required by gradient-free methods tend to grow as the square or the cube of the number of optimization parameters, while the gradient-based methods keep a linear behavior. It is clear form their work that as soon as the design space has a higher dimension than ten, gradient-free algorithms require tens or hundreds of thousands of function evaluation while gradient-based algorithms using an analytic method such as the adjoint formulation to compute the gradient require less than a hundred.

The adjoint formulation of the problem allows for an efficient computation of the gradient of the objective and constraints with respect to the optimization parameters. For the applications of this study, the cost of the gradient derivation is roughly linear with respect to the number of design variables and the size of the mesh. Various publications have already covered the application of the adjoint method to gradient-based optimization. For instance, Martin applied the adjoint method to robust airfoil optimization [8]. Martin used the Automatic Differentiation (AD) tool TAPENADE [9] to develop his discrete adjoint solver which drastically reduced the implementation time. Lyu et al. applied their own adjoint solver to the optimization of a blended-wing-body aircraft [10] and to the Common Research Wing [11].

However, it is well known that gradient-based algorithms have a lower probability of converging towards a global minimum as opposed to a local minimum than global, gradient-free algorithms and are very sensitive to the starting point in the case where the response surface in the design space is multimodal. Furthermore, there is increasing evidence that the aerodynamic shape optimization is multimodal [12], thus making gradient-based algorithms ineffective at exploring unusual yet potentially revolutionary designs. In an industrial framework, this difficulty is solved by a back-and-force between the early design phase where global algorithms are used in conjunction with lower-fidelity methods in order to select a limited number of interesting designs, which are then investigated further during the preliminary design phase with a higher fidelity method and local, gradient-based optimization algorithms. In this paper, we study a use-case representative of the preliminary design phase.

1.2. Aero-Stealth Optimization

As for aero-stealth optimization, gradient-free algorithm have proven very popular in the past few years. For instance, Garnier et al. [13] used an algorithm based on the coupling of the NSGA-II genetic algorithm and the multi-objective efficient global optimization approach [14] to optimize an UCAV planform for aero-stealth. Liu et al. [15] used hierarchical kriging models and other surrogate models for the aero-stealth optimization of a flying-wing aircraft. Zhou and Huang [16] used a two-particle search algorithm to optimize the aero-stealth performances of a vertical tail. Finally, Taj et al. [17] used a machine learning algorithm called a Gaussian process regression to perform the design exploration of aerodynamics and RCS for a fighter aircraft and a genetic algorithm to minimize drag and RCS. Because they used the gradient-free algorithm, these works are restricted to using surrogate models or low-fidelity methods that are not suited for the level of precision required at the stage of preliminary design in an industrial process.

Few studies, however, have dealt with the gradient-based aero-stealth optimization. Pan et al. [18] performed a CAD-based, gradient-based aero-stealth optimization of a flying-wing aircraft, solving the Prandtl–Glauert equation to compute the pressure coefficients of the aircraft and a Physical Optics (PO) method to predict RCS. Li et al. [19] used a RANS and a PO solver as well as a free-form deformation algorithm in order to perform a CAD-free, gradient-based aero-stealth optimization of a flying-wing aircraft. They performed various optimizations using the aerodynamic and stealth criteria alternatively as the objective and a constraint and showed that the better-performing optimization result is obtained when using an objective function built as a trade-off of the aerodynamic and stealth criteria.

1.3. Contribution

The purpose of this study is to showcase a gradient-based optimization framework in an industrial context to efficiently explore aero-stealth design trade-offs using our in-house CAD modeler GANIMEDE [20], a RANS solver [21], a physical optics solver, and the Sequential Least Square Quadratic Programming (SLSQP) algorithm [22]. We used a realistic criterion as a PO objective that is representative of modern stealth shape design consideration. We demonstrate our methodology on a simplified aircraft shape with 14 design variables. Because we use a RANS solver, which is a medium-fidelity solver whose computation time is about 14 min in the case we consider, a design space with more than ten design variable is too expensive for a gradient-free algorithm to explore. We identify a Pareto front to showcase various optimal shapes corresponding to various aero-stealth trade-offs.

The use of CAD-based optimization offers a lot of advantages in an industrial framework. First, it greatly reduces the number of design variables and constraints needed to parameterize and manipulate a realistic aircraft shape and allows us to use variables that are independent of one another. It also allows us to easily specify and compute geometric constraints such as the wing surface reference (see Section 4.4) or the wing absolute thickness. Furthermore, it is possible to define design variable that make sense from an aeronautic designer's point of view, such as the leading edge sweep angle or the trailing edge camber. Lastly, the CAD geometry is an important piece of the aircraft design process, being a tool of communication between the different disciplines at each step of the design process, from the early design step to the manufacturing step.

This paper is organized as follows. In Section 2, we present the gradient-based aero-stealth coupled optimization process. Figure 1 shows the process as a Design Structure Matrix (DSM). We also describe some numerical methods used throughout this study. Next, the results of the code validation are shown in Section 3 and, in Sections 4 and 5, we introduce the problem formulation, baseline geometry of the flying wing, meshes, and shape design variables. In Section 6, the accuracy of the sensitivities is tested. After that, in Section 7, a Pareto front is showcased and particularly interesting geometries are highlighted. Finally, in Section 8, we discuss the conclusions based on the results.

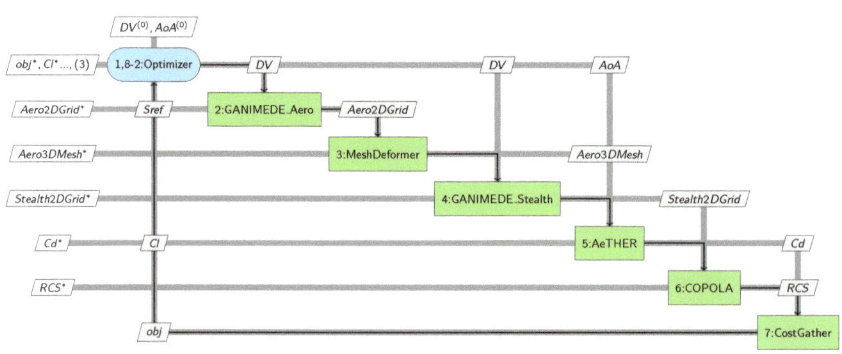

Figure 1. DSM representation of the optimization process.

2. Methodology

2.1. The Physical Optics Approximation

In an aircraft, stealthiness can mean a number of different things. Here, we understand it as the capacity of the aircraft to scatter a limited amount of energy back into the direction of the emitter. We consider a monostatic radar, which represents the large majority of the systems in use. This means that the emitting antenna and the receiving antenna are collocated. A popular measure of the scattered electromagnetic energy is the radar cross-section (RCS) σ defined as follows:

$$\sigma(\theta, \phi) = \lim_{R \to \infty} R^2 \frac{|\mathbf{E}_P|^2}{|\mathbf{E}_i|^2}, \tag{1}$$

where $\mathbf{E}_P$ is the scattered electric fields at a point $P(R, \theta, \phi)$ in polar coordinates (defined in Figure 2) and $\mathbf{E}_i$ is the incident electric field at the target, whose norm is considered to be 1 V/m in practice. Usually, the RCS is expressed in decibels:

$$\sigma_{\text{dB}} = 10 \log 10(\sigma).$$

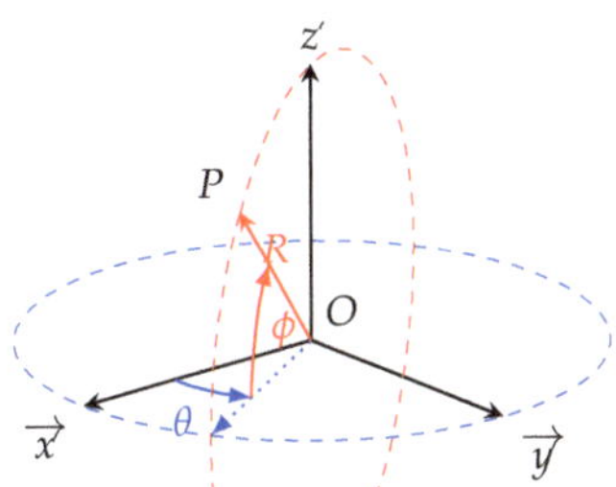

Figure 2. Definition of the bearing (θ) and elevation (ϕ) angles.

We restrict this study to radar bands L (1–5 GHz, used for moderate-distance surveillance) through X (around 10 GHz, used for missile guidance systems). These frequency bands produce on a target the size of an aircraft of predominantly high-frequency scattering [1]. Thus, an asymptotically exact method such as a Boundary Element Method (BEM) or Fast Multipole Method (FMM) would be too expensive for an optimization process. PO is one of the most common high-frequency approximate methods for RCS prediction. It is much cheaper than BEM, and it gives a better prediction than a purely optical approximation [23].

For such frequencies, the size of the aircraft is large with respect to the wavelength. Another reasonable assumption is that the distance between the aircraft and the radar antenna is also large with respect to the wavelength, so that the incident electric field can be considered a plane wave, and with respect to the size of the aircraft, so the far-field Green's function approximation to the Stratton–Chu [24] integral equation is valid. Together with a tangent plane approximation, these can be simplified, and the scattered electric field $\mathbf{E}_P$ at a point P can be expressed as follows:

$$\mathbf{E}_P = -j \frac{k Z_0}{4\pi} \frac{e^{-jkR}}{R} \int_S \mathbf{u}_s \times (\mathbf{u}_s \times \mathbf{J}) \exp(jk\mathbf{R} \cdot (\mathbf{u}_i - \mathbf{u}_s)), \tag{2}$$

with $j = \sqrt{-1}$, k the wave number, Z_0 the impedance of free space, S the surface of the target, $\mathbf{u}_i$ and $\mathbf{u}_s$ the unit direction vectors for the incident and scattered fields, respectively, $\mathbf{R} = \overrightarrow{OP}$, and $R = |\mathbf{R}|$, and where J is the surface current density vector. In the PO approximation,

$$\mathbf{J} = \begin{cases} \frac{2}{Z_0}\mathbf{n} \times \mathbf{u}_i \times \mathbf{E}_i & \text{in the illuminated zone,} \\ \mathbf{0} & \text{in the shadow zone,} \end{cases} \tag{3}$$

where $\mathbf{n}$ is the local normal unit vector to the surface.

Note that by (2),

$$\mathbf{E}_P = \mathcal{O}\left(\frac{1}{R}\right) \qquad \text{as } R \to \infty.$$

Therefore, the RCS does not depend on the range.

Although this approximate method does not account for diffraction, and in particular wedge diffraction, extensions exist to correct the induced current along wedges. One such extension is called Partial Theory of Diffraction (PTD) [25] and has been added to PO to give a better prediction of the RCS of an angular shape, such as the trailing edge of a wing.

The solver including PO and PTD along with their differentiation, described on the next section, is dubbed COPOLA.

2.2. Automatic Differentiation of Physical Optics

If the characteristics of the radar wave (wavelength, polarization, and direction) are given, the RCS depends only on $\mathbf{x}_e$, the coordinates of the vertices of the surface grid used for the electromagnetism problem.

If

$$F_e(\nu) = \hat{F}_e(\sigma(\mathbf{x}_e(\nu)))$$

is a function depending explicitly only on the RCS σ, then the directional derivative of F_e with respect to a set of geometric variables ν (e.g., sweep angle, chord length) is

$$\mathrm{d}F_e = \hat{F}_e'(\sigma) \cdot \sigma'(\mathbf{x}_e) \cdot \mathbf{x}_e'(\nu) \cdot \mathrm{d}\nu. \tag{4}$$

We used the Automatic Differentiation (AD) tool TAPENADE version 3.16 [9] in order to generate a differentiated program from the original code, which is able to compute the terms $\mathrm{d}F_e/\mathrm{d}\sigma$ and $\mathrm{d}\sigma/\mathrm{d}\mathbf{x}_e$. With this tool, it is possible to save a lot of development and debugging time. TAPENADE offers two modes: tangent (or direct) and reverse. The tangent mode computes the directional derivative of F_e in the direction $\mathrm{d}\nu$ using the chain rule, from left to right. This mode is immediate to use as it only needs to create a copy of the input code, interleaved with the derivative instructions. The reverse mode is an efficient way of computing the gradient of $\nabla F_e \cdot \mathrm{d}\nu = \mathrm{d}\nu^* \cdot \mathbf{x}_e'(\nu)^* \cdot \sigma'(\mathbf{x}_e)^* \cdot \hat{F}e'(\sigma)^*$. When F_e is scalar-valued and the dimension of ν is large, and assuming that the storage capacity is available, it is theoretically more efficient to compute the Jacobian matrix ∇F_e of F_e row by row and then compute $\mathrm{d}F_e = \nabla F_e \mathrm{d}\nu$ than to compute it by the chain rule with the tangent mode.

Here, however, all the steps of the chain rule computation are cheap to compute since they do not require us to solve a system. Therefore, the actual gain in efficiency from the reverse mode is expected to be small, and since it is much more difficult to implement than the tangent mode, we did not deem it worth the trouble.

The term $\mathrm{d}\mathbf{x}_e/\mathrm{d}\nu$, which is the derivative of the vertex coordinates of the surface grid with respect to a set of geometric variables, is computed by the CAD modeler (see Section 4.1) and is given as an input to the differentiated PO program.

This allows us to compute the cost function F_e as well as its derivative $\mathrm{d}F_e/\mathrm{d}\nu$ in a single call of the PO code.

2.3. The Aerodynamic Solver

Our in-house Navier–Stokes AETHER code [21,26] solves the 2D, axisymmetric, and 3D compressible Navier–Stokes equations. It uses a finite element approach based on a symmetric form of the equations written in terms of entropy variables. The advantages of this change of variables are numerous: in addition to the strong mathematical and numerical coherence they provide such as dimensionally correct dot product, symmetric operators with positive properties and efficient preconditioning, entropy variables yield further improvements over the usual conservation variables, particularly in the context of chemically reacting flows.

The code can handle the unstructured mixture of numerous types of elements (triangles and quadrilaterals in 2D; tetrahedral, bricks, and prisms in 3D).

Several one- and two-equation Reynolds-averaged turbulence models are available: Spalart–Allmaras, (k, ε), (k, ω), (k, kl), DRSM… Additional equations devoted to the evaluation of the laminar–turbulence transition can be activated. These models are integrated down to the wall. Extensions to DES and LES are also available.

The linearization of Navier–Stokes equations and turbulence equations requires the differentiation of all the routines involved in the aerodynamic computation. These differentiated routines were obtained using TAPENADE, an Automatic Differentiation tool. This tool provides the differentiated Fortran routines with respect to input variables prescribed by the user.

We used a compressible Reynolds-Averaged Navier–Stokes [27] method with the Spalart–Allmaras [28] turbulence model in order to compute the aerodynamic criterion and constraints.

2.4. Adjoint Approach for CFD

Let W be a vector state solution of the non-linear RANS equation system which we note as

$$\mathcal{R} = 0. \tag{5}$$

Finally, let F_a be an observation function of the field W. For the process of optimization, we need to compute the derivatives $\frac{\mathrm{d}F_a}{\mathrm{d}\xi}$ and $\frac{\mathrm{d}F_a}{\mathrm{d}\nu}$ with respect to a set of aerodynamic variables ξ (e.g., the angle of attack) and geometric variables ν (e.g., the wing sweep angle), respectively. The variables ν and ξ are independent of one another.

In order to mitigate the cost of computing the gradient of the aerodynamic criterion with respect to many variables, we use a double adjoint formulation [29].

When a surface grid of coordinates $\mathbf{x}_a$ is subjected to a deformation owing to the CAD modeler, the deformation is propagated to the 3D grid of coordinates $\mathbf{X}_a$ with the help of an elliptic operator described by the system:

$$\mathcal{L}(\mathbf{X}_a, \mathbf{x}_a) = 0. \tag{6}$$

Let us write F_a and $\mathcal{R}$ as functions of ξ, $\mathbf{X}_a$, and W. Note that ξ is an independent variable but $\mathbf{X}_a$ depends on ν and W depends on both ξ and ν.

2.4.1. Gradient with Respect to Aerodynamic Variables

First, the derivative of F_a and $\mathcal{R}$ with respect to ξ can be written as

$$\mathrm{d}F_a = \left(\left.\frac{\partial F_a}{\partial \xi}\right|_W - \Psi^{\mathrm{T}} \left.\frac{\partial \mathcal{R}}{\partial \xi}\right|_W \right) \mathrm{d}\xi. \tag{7}$$

where Ψ is the solution to the aerodynamic adjoint system:

$$\left(\left.\frac{\partial \mathcal{R}}{\partial W}\right|_\xi \right)^{\mathrm{T}} \Psi = \left(\left.\frac{\partial F_a}{\partial W}\right|_\xi \right)^{\mathrm{T}}. \tag{8}$$

2.4.2. Gradient with Respect to Geometric Variables

The derivative of $\mathcal{R}$ and F_a with respect to ν is

$$\mathrm{d}F_a = \left(\left.\frac{\partial F_a}{\partial \mathbf{X}_a}\right|_W - \Psi^{\mathrm{T}} \left.\frac{\partial \mathcal{R}}{\partial \mathbf{X}_a}\right|_W \right) \frac{\partial \mathbf{X}_a}{\partial \nu} \mathrm{d}\nu, \tag{9}$$

where Ψ is the same as defined in Equation (8).

In this case, it is not enough, though, as $\frac{\partial \mathbf{X}_a}{\partial \nu}$ is expensive to compute. In order to find a more convenient expression, we must write an adjoint problem of the system, which finally yields

$$dF_a = -\Phi^{\mathrm{T}} \frac{\partial \mathcal{L}}{\partial \mathbf{x}_a}\bigg|_{\mathbf{X}_a} \frac{\partial \mathbf{x}_a}{\partial \nu} d\nu, \tag{10}$$

where Φ is the solution of the mesh adjoint equation system:

$$\left(\frac{\partial \mathcal{L}}{\partial \mathbf{X}_a}\bigg|_{\mathbf{x}_a} \right)^{\mathrm{T}} \Phi = \left(\frac{\partial F_a}{\partial \mathbf{X}_a}\bigg|_{\mathbf{x}_a} \right)^{T} - \left(\frac{\partial \mathcal{R}}{\partial \mathbf{X}_a}\bigg|_{\mathbf{x}_a} \right)^{T} \Psi. \tag{11}$$

3. Validation

3.1. Validation of the COPOLA Code

We used a sphere geometry as a test case for the validation of COPOLA. According to Mie's theory, if a sphere of radius R is illuminated by a wave of wavelength λ, then the RCS σ converges to the section area of the sphere as the wavelength decreases:

$$\lim_{\lambda \to 0} \sigma = \pi R^2.$$

Moreover, due to the symmetric properties of the sphere, its RCS does not depend on the bearing angle.

Since the PO method is a high-frequency approximation, it should predict an RCS close to the analytical value. Figure 3 shows the PO prediction of the RCS of a sphere of radius 1 m with respect to the bearing angle (Figure 3a) and frequency (Figure 3b). The plot in Figure 3b is computed with 2000 frequencies between 10 MHz and 20 GHz.

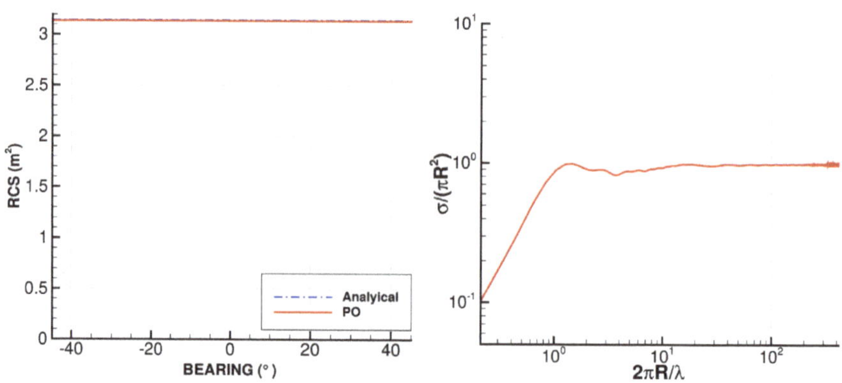

(a) (b)

Figure 3. The RCS of the sphere of radius $R = 1$ m. (**a**) The RCS function of the bearing angle. $f = 5$ GHz. (**b**) Dimensionless RCS function of the radius-to-wavelength ratio.

In Figure 3a, the RCS is independent of the bearing angle, as it should be, and the error of the analytical value is around 0.15%. Figure 3b shows that when the frequency is too low ($\frac{2\pi f R}{c} < 50$, where c is the speed of light), the predicted RCS is very far from the analytical value, and the PO approximation is not trustworthy. When the frequency is too high ($\frac{2\pi f R}{c} > 100$), the prediction starts to deteriorate again due to the mesh being too coarse for the higher frequencies.

Figure 4 shows the comparison between an RCS simulation and a BEM computation on the GFA geometry (described in Section 4). The BEM computation was performed owing to the SPECTRE solver [30,31].

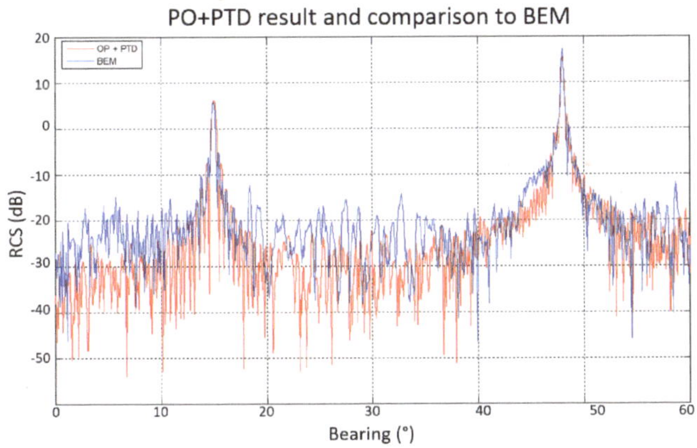

Figure 4. Comparison of the RCS function of the bearing angle predicted by the COPOLA and BEM codes on a GFA geometry (see Figure 5).

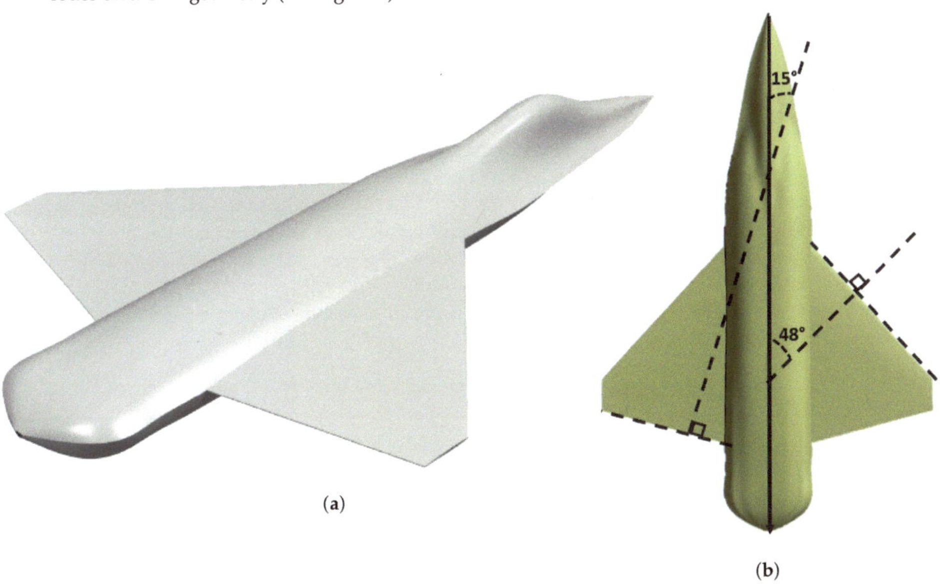

Figure 5. The Generic Fighter Aircraft. (**a**) 3D-view. (**b**) Bottom view with the leading and trailing edge sweep angles.

PO is able to predict accurately both the height and the width of the two main peaks. Away from of these peaks, the RCS is very low (around −20 dB). The precision of PO is not as good in these zones because low-frequency effects such as creeping waves are predominant, but since the RCS is so much lower than that of the peaks, it is negligible for the purpose of this study.

We present the validation of the sensitivities in Section 6.

3.2. *Validation of the Aerodynamic Tool*

AETHER is the solver used by Dassault-Aviation in an industrial context. It has been validated extensively both with numerical benchmark cases and against experimental data. For example, see [32] or [21].

4. Baseline Description

The geometry we use is a simplified shape of a Generic Fighter Aircraft (GFA, see Figure 5). It features the fuselage without air intake and a pair of wings. The wings have a trapezoidal shape without leading edge extension. It does not have a tail fin, horizontal stabilizer, or canard. The front of the fuselage has a ridge on both sides. We did not use a realistic geometry for the rear of the fuselage, so the physics of the wake is not accurately represented. Since we only performed supersonic simulations, this does not disturb the flow upstream. This also allows us to be lenient with meshing the downstream region, thus saving computation time.

4.1. The CAD Modeler

In order to model and modify the geometry, we used the CAD tool GANIMEDE [20]. In this study, we used some functionalities of GANIMEDE that have not yet been differentiated, which prevented us from using the exact differentiation presented in [20]. The surface is represented by a set of NURBS grids and parameterized with control points. GANIMEDE uses both global and local variables as input. Local variables are the osculating parameters (i.e., position, tangent, and curvature) at a control point. Global variables allow us to modify usual aeronautical design variables such as sweep and twist angles, or the chord of a wing. In this study, we used only global variables.

If a surface mesh is associated with the baseline geometry, GANIMEDE projects the mesh onto the deformed geometry and computes the derivatives of the mesh node coordinates with respect to the design variables $\frac{d\mathbf{x}}{d\nu}$ using centered finite difference. Because of the smoothing and other refinement operations performed by the CAD modeler on the NURBS geometry, too small a step size can lead to a bad gradient, so we have to find an optimal step size for each design variable. The optimal step sizes are determined on an individual basis by analyzing the field of the second derivative of the displacement of the geometry with respect to each design variable. This field should preserve the spatial symmetries or anti-symmetries of the original design variable and should not be too distorted. The final step sizes that we determined for each design variable are presented in Table 1, which summarizes the chosen design variables.

Table 1. Summary of the values of the design variables for a few chosen optimal shapes.

Design Variable	Unit	Section	Index	Step Size	Min	Max	Baseline	Stealth-Optimal Shape	Aero-Optimal Shape
Sweep angle	°	2	1	0.05	28	58	48	30	58
		4	2	0.05	−10	10	0	0	10
Chord length	mm	2	3	10^{-3}	5991	7863	7489	7059	7863
		4	4	10^{-3}	2643	4845	4405	1211	4845
		6	5	10^{-3}	943	1238	1179	1123	1238
Span	mm	2–6	6	10^{-3}	4864	6675	6230	6674	5737
Twist angle	°	2	7	0.05	0	2	0	−0.5	−0.3
		3	8	0.05	−2	2	0	−0.8	−0.6
		5	9	0.05	−2	2	0	−0.6	−1.8
		6	10	0.05	−2	2	0	−0.8	−1.9
Leading edge camber	% chord length	3	11	10^{-3}	−1	1	0	−1	−0.69
		5	12	10^{-3}	−1	1	0	0	0.64
		6	13	10^{-3}	−1	1	0	1	0.63
AoA	°	–	14	0.01	1	3	2	2.3	2.7

4.2. Aerodynamic Analysis of the Baseline

We performed a CFD simulation of the GFA geometry. The upstream Mach number is 1.8, with an angle of attack of 2°. The reference surface used for the computation of the aerodynamic coefficients is 68 m^2. The overall drag coefficient is a 387.5 drag count and the lift coefficient is 0.113. These values are presented in Table 2.

Table 2. Summary of the cost and constraint values for various shapes in the Pareto front.

w	C_D	C_L	F_{RCS}	$\hat{\sigma}$	S_{ref}	N Cycle
1.0	346.97	$112.5 \cdot 10^{-3}$	$186.57 \cdot 10^{-3}$	$8.4 \cdot 10^{-3}$	68.6	41
0.995	352.16	$112.5 \cdot 10^{-3}$	$144.41 \cdot 10^{-3}$	$6.9 \cdot 10^{-3}$	68.6	63
0.99	354.34	$112.5 \cdot 10^{-3}$	$104.80 \cdot 10^{-3}$	$6.0 \cdot 10^{-3}$	68.6	62
0.95	365.58	$112.5 \cdot 10^{-3}$	$94.15 \cdot 10^{-3}$	$6.3 \cdot 10^{-3}$	68.6	40
0.9	367.49	$112.5 \cdot 10^{-3}$	$84.99 \cdot 10^{-3}$	$7.8 \cdot 10^{-3}$	68.6	59
0.8	372.88	$112.5 \cdot 10^{-3}$	$78.02 \cdot 10^{-3}$	$25.3 \cdot 10^{-3}$	68.6	47
0.75	380.85	$112.5 \cdot 10^{-3}$	$40.75 \cdot 10^{-3}$	$81.6 \cdot 10^{-3}$	68.6	69
0.52	381.72	$112.5 \cdot 10^{-3}$	$40.31 \cdot 10^{-3}$	$84.4 \cdot 10^{-3}$	68.6	65
0.51	426.99	$112.5 \cdot 10^{-3}$	$2.24 \cdot 10^{-3}$	$20.9 \cdot 10^{-3}$	68.6	91
0.5	427.22	$112.5 \cdot 10^{-3}$	$2.23 \cdot 10^{-3}$	$21.0 \cdot 10^{-3}$	68.6	98
0.0	432.54	$112.5 \cdot 10^{-3}$	$1.74 \cdot 10^{-3}$	$23.3 \cdot 10^{-3}$	68.6	96
baseline	387.46	$112.6 \cdot 10^{-3}$	$82.27 \cdot 10^{-3}$	$7.4 \cdot 10^{-3}$	68.6	–

The field of the local Mach number in the symmetry plane and in the planes parallel to the symmetry plane at $Y/b = 0.45$ and $Y/b = 0.8$ are displayed in Figure 6, Figure 7, and Figure 8, respectively.

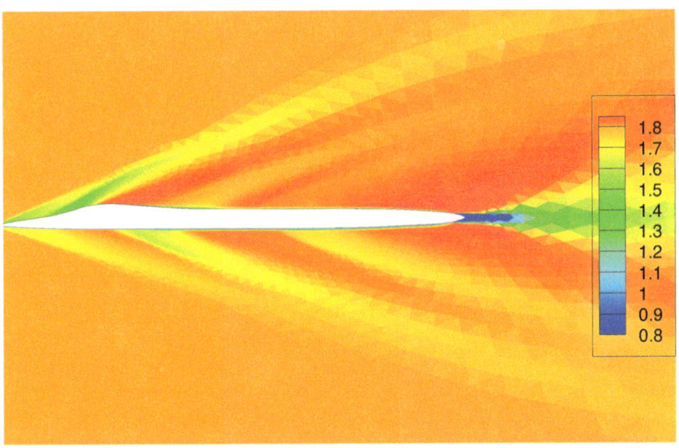

Figure 6. Distribution of local Mach number around the fuselage of the GFA in the symmetry plane.

In the symmetry plane (Figure 6), we can clearly see the weak, attached shock on the nose of the GFA. A large detached area is present at the rear of the aircraft, which is expected since no attention was directed toward this area.

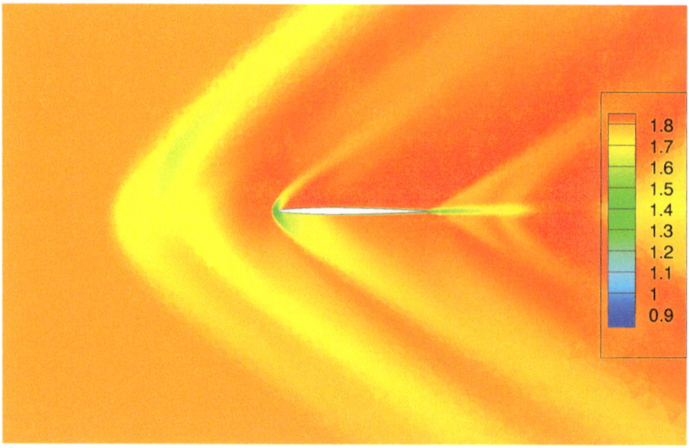

Figure 7. Distribution of local Mach number around the cross-section of the wing of the GFA at $Y/b = 0.45$.

In Figures 7 and 8, directly upstream of the wing leading-edge, we can see a weak bow shock.

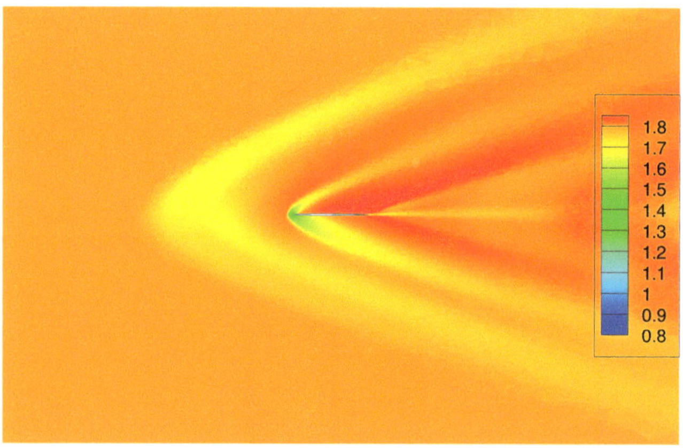

Figure 8. Distribution of local Mach number around the cross-section of the wing of the GFA at $Y/b = 0.8$.

4.3. Stealth Analysis of the Baseline

The span of the baseline shape being 6 m, the frequency should be higher than 25 MHz in order to meet the criterion of Section 3:

$$50 < \frac{2\pi f L_{ref}}{c} < 100$$

The COPOLA simulation was performed on a range of bearing angles from 0 to 70° with a step of 0.01° and on a range of elevation angles from −20 to −10° with the same step in order to simulate an aircraft reflecting hostile waves from a radar on the ground far in front of the plane. For each line of sight, the simulation is run for 11 frequencies around a main frequency at 1 GHz and spanning 100 MHz. The step for the bearing and elevation angles has been chosen according to the Shannon criterion.

In Figure 9a,b, a peak is visible around 48° of bearing. This corresponds to the direction the leading edge is facing. This peak is the specular reflection of the incident wave onto the leading edge. A second peak appears at around −15°. This peak has a much lower intensity compared to the specular reflection peak due to the non-continuity of the normal at the trailing edge wedges. Physically, this means that the trailing edge wedges diffract the incident wave. However, the peak appears to have the same intensity no matter the polarization, which, given the orientation of the trailing edge (see Figure 5) should not be the case: the trailing edges being horizontal, the diffraction should be negligible when observed with a horizontal polarization [33]. We conclude that PO alone is not enough to compute the trailing edge diffraction, and we need to add the PTD extension to our simulation to adequately predict the RCS. When taking diffraction into account, the second peak appears much brighter with the vertical polarization (Figure 9c), only a few dB less intense than the specular reflection peak, and it is almost absent with the horizontal polarization (Figure 9d).

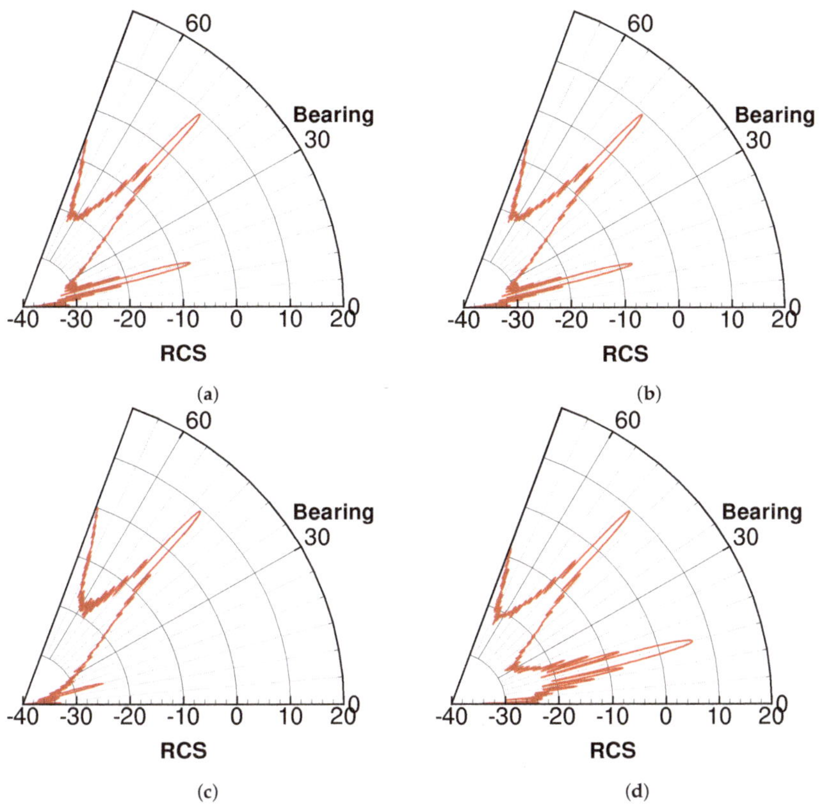

Figure 9. Profile of RCS with respect to the bearing angle on the GFA (elevation : $\phi = -20°$). (**a**) Purely PO simulation with horizontal polarization. (**b**) Purely PO simulation with vertical polarization. (**c**) PO simulation with PTD extension with horizontal polarization. (**d**) PO simulation with PTD extension with vertical polarization.

Over the rest of the observation window, the RCS is very low, around −30 to −20 dB, which is practically negligible but increases as the bearing angle comes close to 70°. This is likely due to the influence of the fuselage, but as it is far away from the axis of the plane, it is not relevant to this study. One might note, however, that the value around 70° is not the same with or without the PTD extension. This is likely due to a diffraction by the wing tip of wedges on the fuselage that are overestimated by PO alone.

Since the vertical polarization is strictly richer than the horizontal polarization in our case, it is the one we used for the RCS simulation in the rest of this study.

4.4. Optimization Parameters

The wing of the GFA is parameterized with the help of 6 control sections, with 15 control points along each of them (see Figure 10). All the sections are and stay parallel to the symmetry plane. Section number 1 is at the symmetry plane and is not modified. Section 2 is a few centimeters away from the root of the wing. At Section 4, the continuity of the tangent is not ensured, allowing for a sharp angle on the wing. The fuselage is not modified.

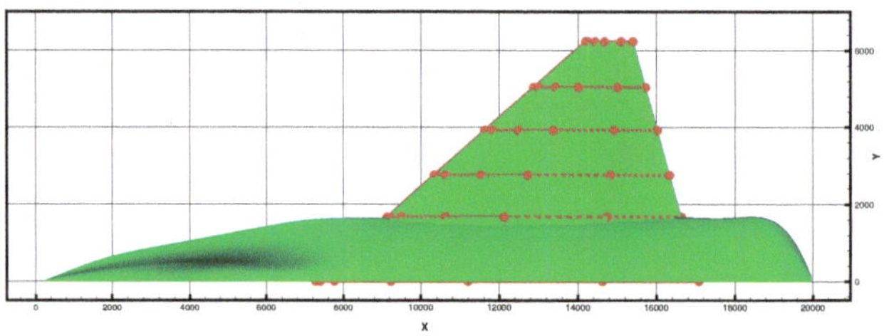

Figure 10. The wing parameterized with the control sections (red lines) and control points (red dots).

The geometry is parameterized with the help of 13 global variables:

- The leading edge sweep angle in Sections 2 and 4;
- The chord of Sections 2, 4 and 6;
- The distance between Sections 2 and 6;
- The twist angle of Sections 2, 3, 5 and 6;
- The leading edge camber of Sections 3, 5 and 6.

The chords of Sections 3 and 5 are interpolated between those of Section 2, Section 4, and Section 6, respectively. The twist angle for a given section is a rotation along an axis perpendicular to the section going through a point at 25% chord and 50% thickness. Both the leading edge camber and the twist angle in Section 4 are interpolated between Sections 3 and 5.

Finally, the angle of attack is the 14th optimization parameter. It is private to the aerodynamic process.

In Table 1, we summarize the design variable as well as various pieces of information such as their index, their bounds, the step size used for the computation of the gradient of the displacement, their baseline values, and their value for the monodiscipline optimized shapes.

5. Cost and Constraints

In order to explore multiple trade-offs between aerodynamic and stealth optima, we define the cost function as a weighed sum of an aerodynamic criterion and a stealth criterion:

$$F = wF_a + (1-w)F_e, \quad (12)$$

with $0 \leq w \leq 1$.

5.1. Aerodynamic Criterion

For the aerodynamic optimization, we minimize the drag coefficient (C_D):

$$F_a = \frac{C_D}{{C_D}^{ref}}. \quad (13)$$

This is somewhat simplistic and will lead to unrealistic shapes. Practically, a criterion based on the pitching moment coefficient should be added to the drag criteria, as well as a criterion based on the rolling stability or maneuverability.

However, since the goal of this study is to showcase the methodology rather than find a realistic shape, we settled for this simplified criterion.

5.2. Stealth Criterion

Typically, a stealth aircraft shape is designed to scatter radar wave away from any hostile radar antenna, which is usually located roughly in front of the aircraft and slightly below. For example, the upward facing planes of the F117 scatter the radar wave into the sky, while more modern aircraft, like the B2, are built to concentrate the reflected radar wave only as a very narrow signal in a few specific directions ("peak directions"). The stealth criterion is built to reflect and favor this behavior: we want to minimize the RCS in every direction in a given sector $(\theta_0, \theta_1) \times (\phi_0, \phi_1)$, except for a limited interval around a peak direction where the RCS can be high. We built an electromagnetic criterion in order to reflect this design methodology as follows.

In order to manage dimensionless quantities, the electromagnetic criterion is defined as follows:

$$F_e = \frac{\tilde{\sigma}}{\tilde{\sigma}^{ref}}, \tag{14}$$

where $\tilde{\sigma}$ is a weighed average of a function of σ_{dB}:

$$\tilde{\sigma} = \int_{\theta_0}^{\theta_1} \int_{\phi_0}^{\phi_1} \Gamma_d(\theta)\delta_s(\sigma_{\mathrm{dB}}) \sin(\phi) \mathrm{d}\theta \mathrm{d}\phi. \tag{15}$$

The function δ_s is a differentiable threshold function built to cut the low-intensity variations that we note in Figure 9, which have a negative impact on the computation of the gradient of the cost function:

$$\delta_s(x) = \frac{x - s}{1 + e^{-\tau(x-s)}}, \tag{16}$$

where τ is a parameter that allows us to control the sharpness of the peak. The threshold $s = -5$ dB was chosen.

The function Γ_d is a weighting function that we chose in order to favor the peak directions. It is close to 1 everywhere, except in the neighborhood of the direction d where it decreases quickly:

$$\Gamma_d(\theta) = 1 - \frac{1}{2} \sum_{i=0}^{4} \frac{6 - i}{10} \exp\left(- \left(\frac{\theta - d}{10^i \tau} \right)^2 \right). \tag{17}$$

Graphs of the function Γ_d are shown in Figure 11 for various values of τ. For this study, we chose, somewhat arbitrarily, a peak direction at a bearing of $d = 30°$, and $\tau = 0.1$.

5.3. Constraints

In order to ensure that the reduced drag coefficient does not come at the cost of a lower lift, we constrain the optimization to work at a constant lift coefficient (C_L).

Furthermore, we want to preserve the reference surface (S_{ref}) of the wing. S_{ref} is defined in Figure 12.

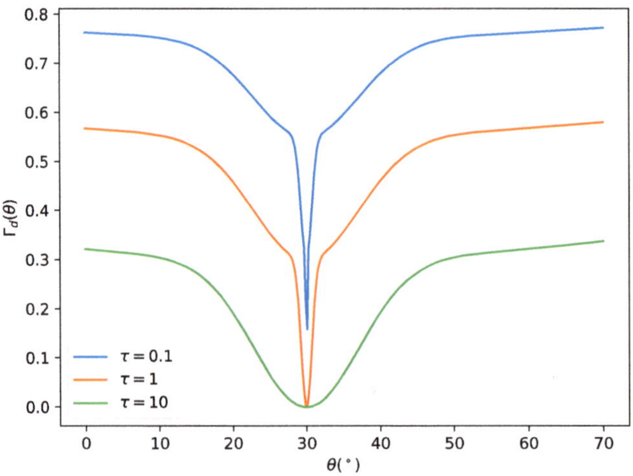

Figure 11. The function $\Gamma_d(\theta)$ with $d = 30^\circ$ for various values of τ.

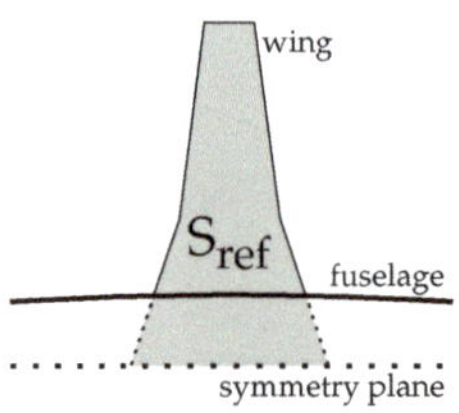

Figure 12. Definition of S_{ref}.

6. Sensitivity Validation

6.1. RCS with Respect to Geometric Variables

We validated the RCS gradient with respect to the 13 shape design variables by comparing its finite differences (FDs) for various step sizes ranging from 1 to 10^{-12} in the unit of each individual variable. For the sake of simplicity, each FD gradient is computed with the same step size for each design variable. In practice, of course, we should fine tune a different step size for each individual variable in order to get a more precise gradient.

If the gradient is properly computed, we expect that, if the step size is small enough, the error decreases with the step size, up until a point where the numerical error on the function is of the same order of magnitude as the difference between the values of the RCS function at ν and $\nu + \epsilon\hat{d}$ (where $\hat{d}$ is an arbitrary direction unit vector). After this critical point, the error should increase roughly linearly as the step size continues decreasing.

Let f denote a real-valued function that is at least three times differentiable, f' its derivative, and $\Delta_\epsilon f$ and $\delta_\epsilon f$ its derivative approximated with forward and centered FDs, respectively, with a step size ϵ. It is well known that

$$\Delta_\epsilon f - f' = o(\epsilon) \quad \text{as } \epsilon \to 0, \tag{18}$$

$$\delta_\epsilon f - f' = o\left(\epsilon^2\right) \quad \text{as } \epsilon \to 0. \tag{19}$$

Hence, the logarithm of the error between the exact gradient and the FD should be linear with respect to the step size if the step size is small enough but bigger than the critical step size, and its slope should be 1 for the forward FD and 2 for the centered FD. We consider that the minimal error should be at most 10^{-3} for the forward FD and 10^{-6} for the centered FD in order to deem the gradient acceptable.

Figure 13 shows the norm of the difference (the solid line) between the gradient computed by COPOLA and the FD gradients function of the step size for both forward (in red) and centered (in green) differences, as well as the slope of the logarithm of the error (the dashed line).

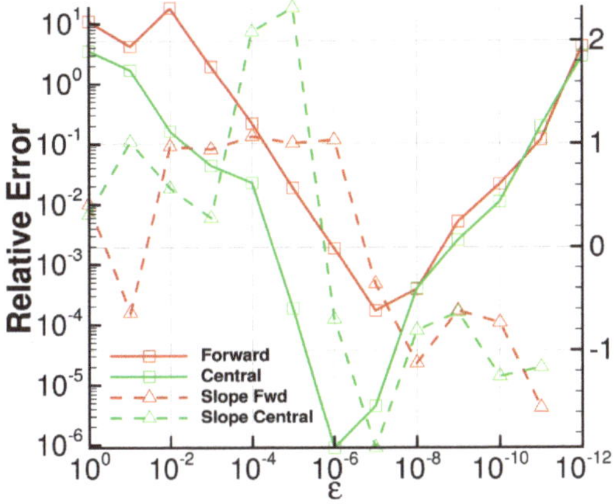

Figure 13. Error between the gradient of the stealth cost function and the same computed by the FD (both forward and central) function of the FD step size (ϵ).

The forward FD behaves very well: it is linear with a slope of 1 for a wide range of step sizes, and the minimal error is a few 10^{-4}. The centered difference is also acceptable with a minimal error of around 10^{-6} and a slope of 2 right before the critical step size.

6.2. C_D with Respect to Geometric and Aerodynamic Variables

Figure 14 shows the error between the adjoint gradient and the finite differences, both centered and forward, for the optimal step size determined for the CAD gradient. When the value of the adjoint gradient is high enough (more than 10^{-3}), the relative error is used:

$$\frac{|\delta F_a/\delta v - \partial F_a/\partial v|}{\partial F_a/\partial v}.$$

When the gradient is too small, the relative error may be misleading, so the absolute error is used instead:

$$|\delta F_a/\delta v - \partial F_a/\partial v|.$$

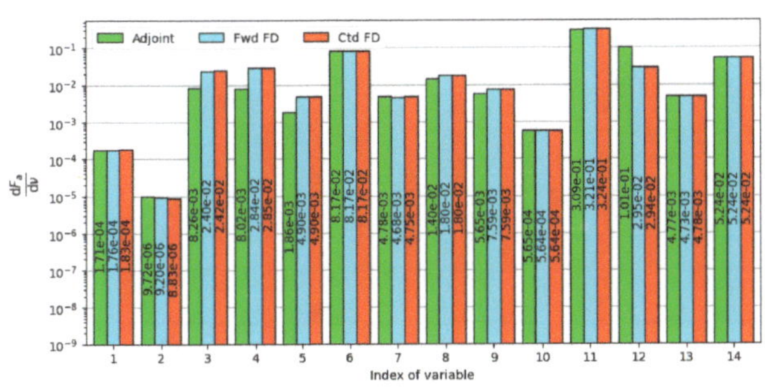

Figure 14. Comparison between the value of the gradient of the aerodynamic drag with respect to the geometric variables v as computed by the adjoint formulation and using the forward (Fwd) and centered (Ctd) FDs for the optimal step size determined for the CAD gradient. The variable indices and step sizes are summarized in Table 1.

Most of the points show an accurate gradient, with an error of at most 10^{-3}. The only exceptions are the gradients with respect to the leading edge camber of the 3rd and 5th sections, but the objective function is less sensitive to these variables. In these cases, the forward and central finite differences are not in good agreement with each other, which indicates that the step size imposed by the CAD gradient is not small enough to compute an accurate aerodynamic gradient. In future developments, the new features of the CAD modeler will be differentiated analytically, and the inaccuracy due to the CAD finite differences will not plague us anymore.

A more accurate validation of the aerodynamic gradient would be achievable with a methodology similar to that used in Section 6.1 in order to free ourselves from the step size determined by the CAD gradient, but this was not feasible for the CFD code.

7. Optimization Results

In this section, we present the results of the various optimizations we performed. We analyze the qualities of both the optimized aerodynamic and stealth shapes with respect to the baseline and each other. We then draw a Pareto front to illustrate the various trade-offs between stealth and aerodynamics. Furthermore, we present two shapes corresponding to two trade-offs in particular.

Figure 15 shows the optimization history of those four optimizations, with the evolution of the objective function and the normalized constraints with the iterations. The aerodynamic optimization (Figure 15a) needs 18 iterations to converge, meaning 18 gradient evaluations and 22 objective function evaluations, which is standard for this type of optimization. Because the objective varies a lot with a small shape variation, the stealth optimization (Figure 15b) needs more than double that number, with 40 iterations or 40 gradient evaluations and 56 objective function evaluations. This is still an acceptable performance for a gradient-based process. The trade-offs shape fall in between, as could be expected. The $w = 0.8$ shape (Figure 15c) took 16 gradient evaluations and 31 objective function evaluations (16 iterations) to converge while the $w = 0.75$ shape (Figure 15d) needed 21 gradient evaluations and 48 objective function evaluations (21 iterations).

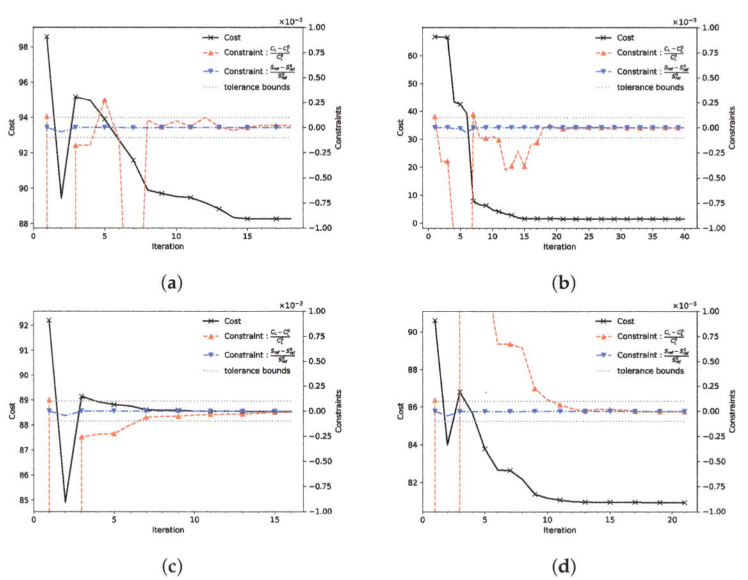

Figure 15. Optimization history. The cost is read on the left axis. The normalized constraints are read on the right axis. The tolerance for the constraints is 10^{-4}. (**a**) Aerodynamic optimization history. (**b**) Stealth optimization history. (**c**) $w = 0.8$ shape optimization history. (**d**) $w = 0.75$ shape optimization history.

In each case, the constraints are satisfied to a precision that is an order of magnitude better than the constraint that we had set (10^{-4}).

7.1. Aerodynamic Optimization

First, we performed an optimization with a purely aerodynamic objective function ($w = 1$). The planform of the optimal shape compared to that of the baseline is presented in Figure 16. The corresponding values of the design variables can be found in Table 1. Most notably, the leading edge sweep angles reach their respective upper-bounds, leading to an unrealistic shape. As discussed in Section 5.1, this was expected because of a simplistic criterion, relying only on the drag coefficient rather than dealing with any lateral stability criterion. Since the simulation is supersonic, the main contributor to the drag is the wave drag due to the shock waves developing on the nose of the aircraft and the wing leading edge. Increasing the leading edge sweep angle decreases the shock strength and the wave drag.

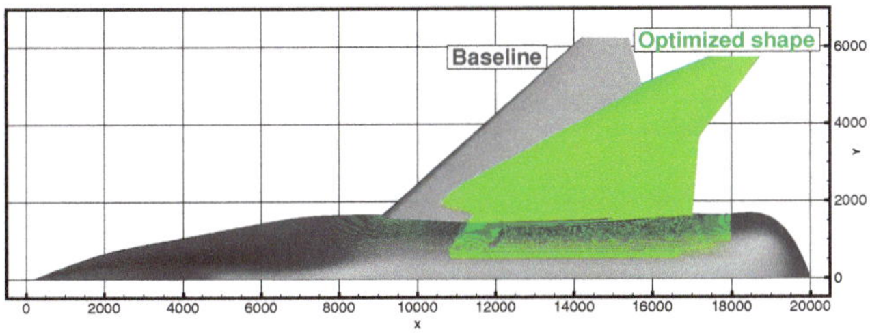

Figure 16. Optimal shape from a purely aerodynamic standpoint.

Figures 17 and 18 show a comparison of the fields of local Mach number around the cross-section of the wing of the baseline (top) and the aerodynamic optimal shape (bottom) at $Y/b = 0.45$ and 0.8, respectively. The decreasing strength of the bow shockwave can be clearly seen at both positions.

Figure 17. Comparison of the distributions of local Mach number around the cross-section of the wing profile at $Y/b = 0.45$ between the baseline (**top**) and the optimized aerodynamic shape (**bottom**).

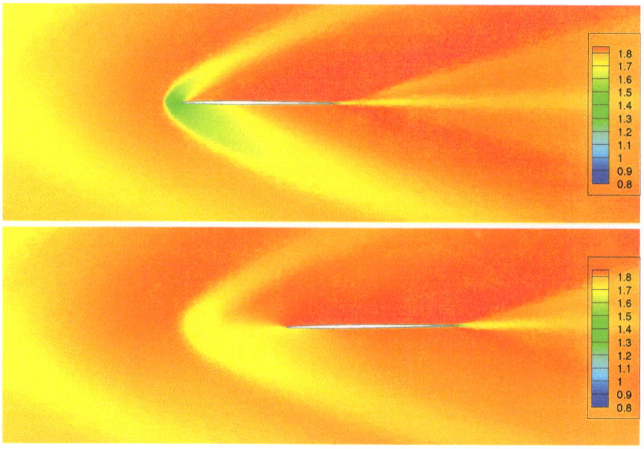

Figure 18. Comparison of the distributions of local Mach number around the cross-section of the wing profile at $Y/b = 0.8$ between the baseline (**top**) and the optimized aerodynamic shape (**bottom**).

The reduction in wave drag owing to the weakening of the bow shock leads to a reduction of 40 drag counts, a little more than 10% of the baseline. The lift coefficient stays the same, at 0.0113, as per the constraint.

Figure 19a shows the integration of the total drag coefficients along the X axis for different shapes. The fuselage is unmodified by the optimization process, so no difference can be seen for $X < X_{LE}^{root}$, but the drag of the optimized aerodynamic shape (red line) starts increasing later and more gradually than for the other shapes, which translates to the gain in wave drag because of the higher leading sweep angles. Figure 19b shows the distribution of the wing lift coefficient with respect to the normalized Y coordinate for different shapes. The lift is better distributed along the wing.

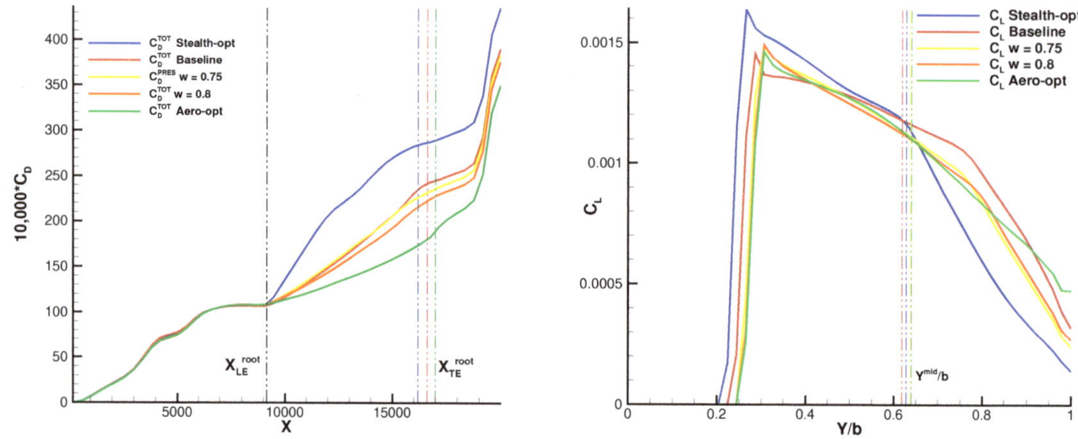

(**a**) Integration of C_D along the X axis. (**b**) Distribution of the wing C_L.

Figure 19. Detailed analysis of the distribution of aerodynamic coefficients along their respective axes.

On the other hand, the multiple values of leading and trailing edge sweep angles have an adverse effect on the stealth criterion. This is because this leads to multiple RCS peaks as well as a wider spread of the RCS due to wedge diffraction, where the stealth criterion privileges a few, very sharp peaks.

In Figure 20, three peaks are visible. The very wide peak around 5° corresponds to the diffraction on the first section of the trailing edge wedge, and the peak around 36° corresponds to the second section of the trailing edge. The more intense peak around 58° corresponds to the specular reflection on the first section of the leading edge. The peak corresponding to the specular reflection on the second section of the leading would be in the 63° direction and is not visible in this figure. This yields a value of the stealth criterion more than twice that of the baseline.

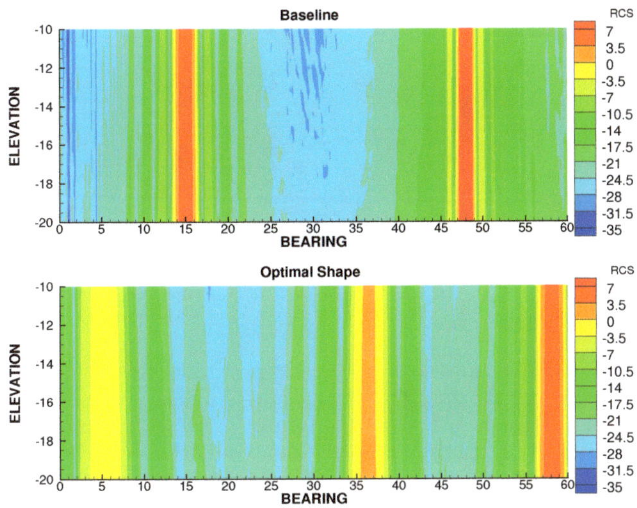

Figure 20. RCS of the optimized aerodynamic shape compared to that of the baseline.

Both the drag coefficient and the stealth criterion value can be read in Table 2 along with other values for various shapes.

7.2. Stealth Optimization

We then performed an optimization with a purely stealth cost function ($w = 0$). The planform of the optimal shape compared to that of the baseline is presented in Figure 21. The corresponding values of the design variables can be found in Table 1. The leading and trailing edges are aligned with a sweep angle of $\pm 30°$, which is the result expected from the choice of the stealth criterion. This allows for the RCS to be concentrated in a sharp peak around a bearing angle of 30°, as seen in Figure 22. As a result, the stealth criterion value for the optimized shape is almost 50 times lower than that of the baseline (see Table 2).

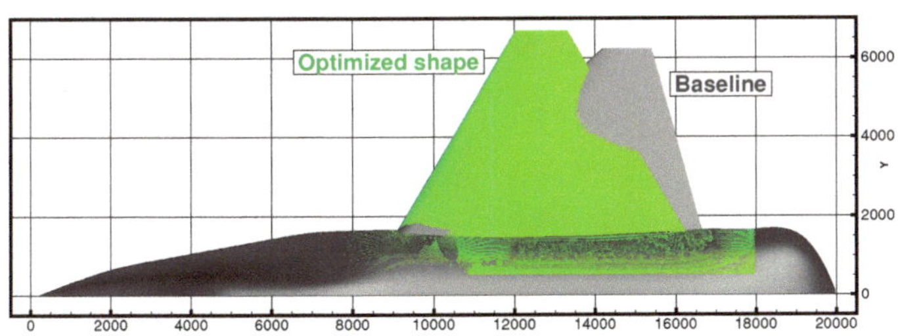

Figure 21. Optimal shape from a purely low-observability standpoint.

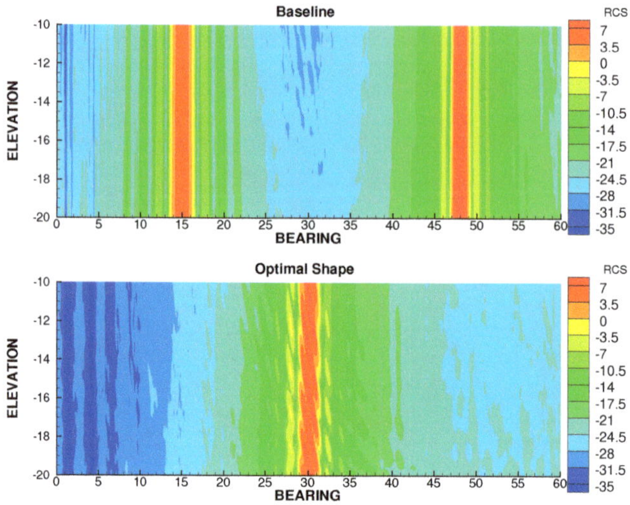

Figure 22. RCS of the stealth-optimal shape compared to that of the baseline.

However, the decrease in leading edge sweep angle is very detrimental to the aerodynamic drag and particularly to the wave drag: the drag count is 55 points greater than that of the baseline (see Table 2). This can be seen clearly in Figures 23 and 24, where a strong shock appears just upstream of the leading edge, which was absent from the other shapes. This is confirmed in Figure 19a, which shows a much steeper and earlier increase in C_D than for the other shapes just downstream of the leading edge position. The pressure drag of the stealth-optimal shape alone is greater than the total drag on most of the baseline wing and on the totality of the aerodynamic optimal shape. The contribution of the pressure drag to the total drag is also much more important for the stealth-optimal shape than the other shapes.

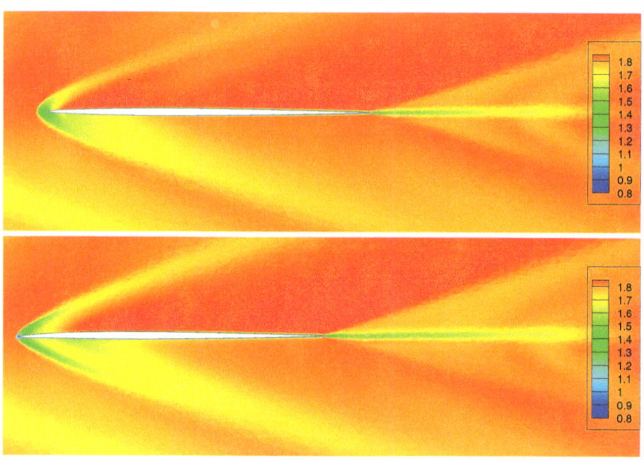

Figure 23. Comparison of the distributions of the local Mach number around the cross-section of the wing profile at $Y/b = 0.45$ between the baseline (**top**) and the stealth-optimal shape (**bottom**).

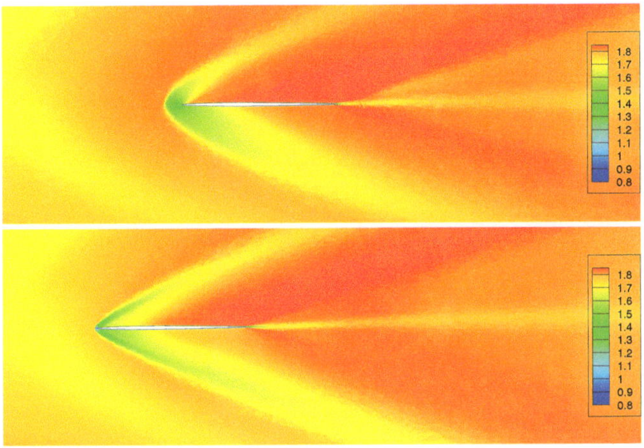

Figure 24. Comparison of the distributions of the local Mach number around the cross-section of the wing profile at $Y/b = 0.8$ between the baseline (**top**) and the stealth-optimal shape (**bottom**).

In Figure 19b, we can see that the lift is more concentrated next to the root for the stealth-optimal shape than for the other shapes. On a realistic aircraft, such a behavior would be detrimental to stalling resistance at great AoA maneuvers, for example.

7.3. Pareto Optimality

In the two previous sections, we show that the requirements for a good aerodynamic shape and a good stealth shape are mainly conflicting. Therefore, the need for compromise arises. We explore different combinations and draw a Pareto front in Figure 25. We present a few selected shapes. The values of the criteria and constraints for these shapes are summarized in Table 2.

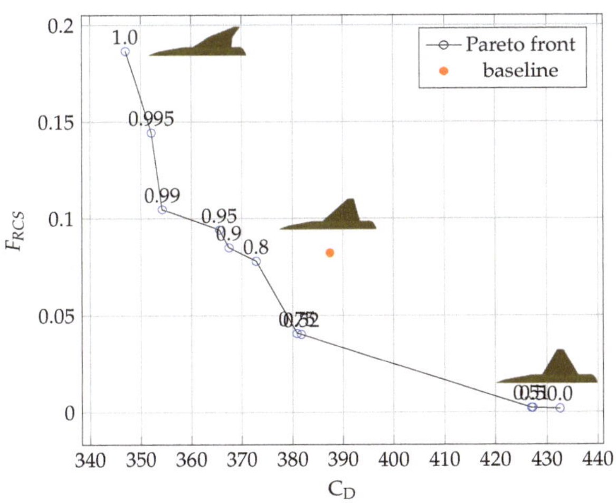

Figure 25. Pareto front.

Note that the Pareto front features three main attraction points, around $w = 0.9$, $w = 0.75$, and $w \in [0, 0.5]$. These attractors are probably due to the steepness of the weighting used in the RCS cost function (see Figure 11). The shapes around $w = 0.5$ and below are essentially the same as the purely optimized stealth shape (Figure 21).

The shape corresponding to $w = 0.75$ and $w = 0.8$ is showcased in Figure 26 and Figure 27, respectively.

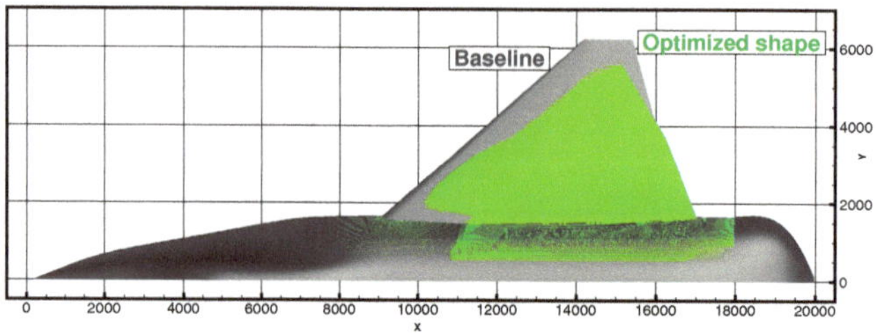

Figure 26. Shape corresponding to the $w = 0.8$ point.

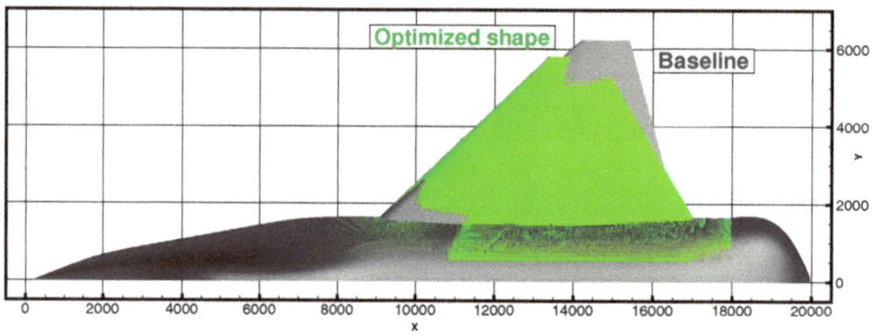

Figure 27. Shape corresponding to the $w = 0.75$ point.

The drag difference between these shapes is mild (8 drag count) and is largely due to the sweep angle as can be seen in the difference in pressure drag in Figure 19a. This can be explained by the variation in intensity of the bow shock at the leading of the wings, because of the different sweep angles (see Figures 28, 29, 30 and 31). The difference in RCS cost function is more notable. Although the RCS cost function is lower for the $w = 0.75$ shape than the $w = 0.8$, its average RCS is higher. This is because, on the $w = 0.8$, the trailing edge exhibits two different sweep angles, thus creating a secondary diffraction peak at 20° (see Figure 32), while there is only one trailing edge sweep angle on the $w = 0.75$ shape and thus only one diffraction peak at 30° (see Figure 33). These two shapes are also better than the baseline both in terms of aerodynamic drag (7 and 15 drag counts lower than the baseline respectively) and RCS function, though the $w = 0.8$ shape is barely better than the baseline (see Table 2).

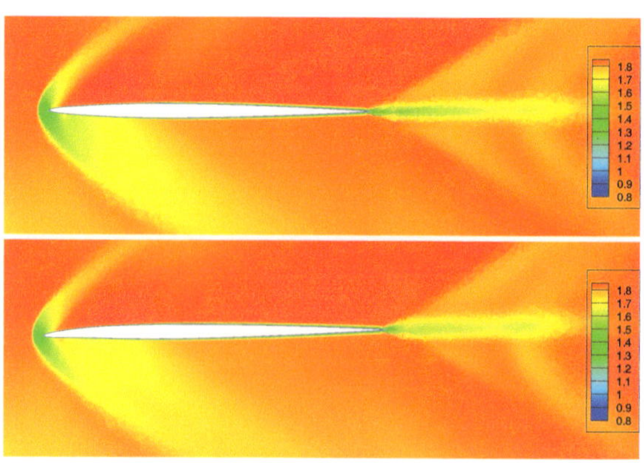

Figure 28. Comparison of the distributions of local Mach number around the cross-section of the wing profile at $Y/b = 0.45$ between the baseline (**top**) and the $w = 0.8$ shape (**bottom**).

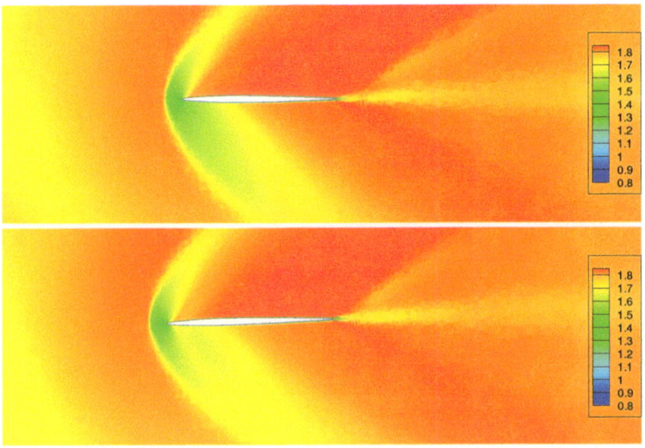

Figure 29. Comparison of the distributions of local Mach number around the cross-section of the wing profile at $Y/b = 0.8$ between the baseline (**top**) and the $w = 0.8$ shape (**bottom**).

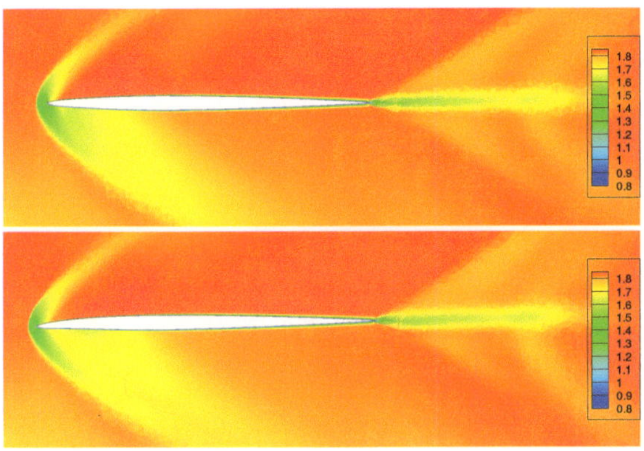

Figure 30. Comparison of the distributions of the local Mach number around the cross-section of the wing profile at $Y/b = 0.45$ between the baseline (**top**) and the $w = 0.75$ shape (**bottom**).

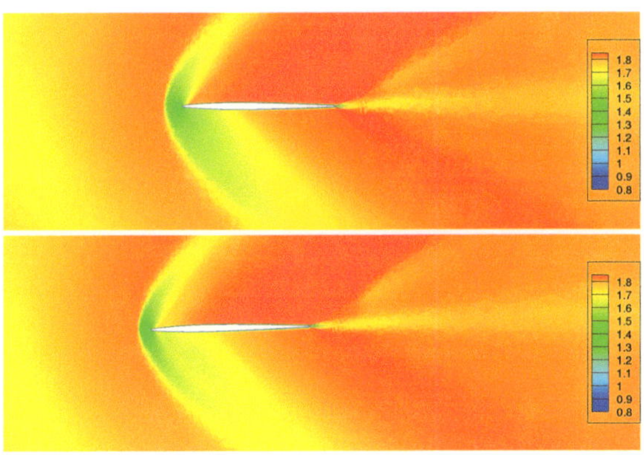

Figure 31. Comparison of the distributions of local Mach number around the cross-section of the wing profile at $Y/b = 0.8$ between the baseline (**top**) and the $w = 0.75$ shape (**bottom**).

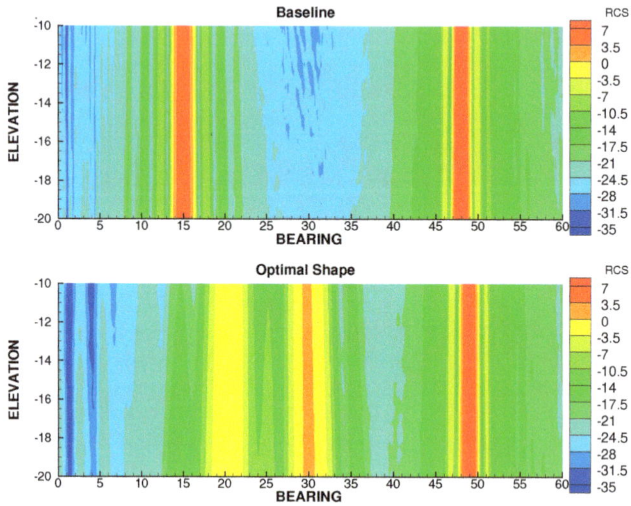

Figure 32. RCS of the $w = 0.8$ shape compared to that of the baseline.

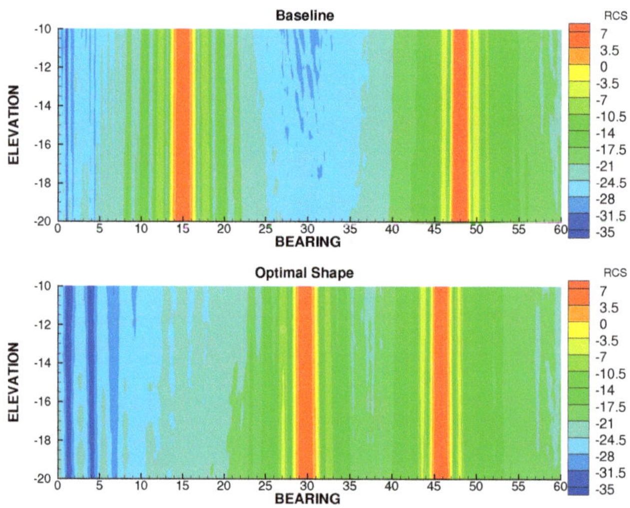

Figure 33. RCS of the $w = 0.75$ shape compared to that of the baseline.

8. Conclusions

In this paper, we present a constrained, gradient-based aero-stealth optimization framework in order to determine an optimal shape for an aircraft. For the sake of keeping computation time low, we showcased our method on a simplified shape with fourteen parameters, but it is able to deal with a more complex shape parameterized with hundreds of design variables.

We computed the aerodynamic gradient through an adjoint formulation and the RCS gradient with the help of the AD tool TAPENADE. We validated these gradients by comparing them to finite differences. The results showed an acceptable difference.

After developing and validating the PO solver, we optimized a simplified aircraft geometry, with a weighted sum of the aerodynamic drag and a criterion based on the RCS distribution as the cost function. We varied the weights in order to identify a Pareto front. We extracted four of those optimal shapes and analyzed them.

The algorithm optimized the RCS mainly by tweaking the leading edge sweep angle and the chord length, which had an impact on the trailing edge sweep angle. This is because the stealth criterion depended heavily on the angular distribution of the RCS rather than merely its intensity.

When the weight is 1, the cost function is purely the aerodynamic drag. The optimal shape features a 40 drag count decrease but at the cost of spreading the RCS across the observation window, which is detrimental to the stealth criterion. On the other hand, when the weight is 0, the cost function is purely the stealth criterion. The algorithm is able to improve the RCS of the shape across most of the observation window by concentrating the scattered wave in a single direction, but this has a dramatic impact on the wave drag of the aircraft.

With weights between 0.5 and 0.9, the optimizer was able to find trade-offs that were strictly better than the baseline shape.

The framework allowed for efficient optimization, each one converging in less than 100 objective function and gradient evaluations and taking less than 10,000 CPU h to complete, which is consistent with the standards of industrial aerodynamic design. Furthermore, the use of a CAD facilitates interdisciplinary communication and is easy to set up by aerodynamic engineers, even when they are not shape parameterization experts. This makes it very useful in an industrial context.

Going forward, a more advanced method to identify the Pareto front, such as the one proposed by Desideri [34], would be beneficial.

Author Contributions: Conceptualization, C.T., G.R. and O.P.; Methodology, C.T. and G.R.; Software, C.T.; Validation, C.T.; Writing—original draft preparation, C.T.; Writing—review, G.R. and O.P.; Supervision, O.P. All authors have read and agreed to the published version of the manuscript.

Funding: This research received no external funding.

Data Availability Statement: The original contributions presented in the study are included in the article, further inquiries can be directed to the corresponding author.

Acknowledgments: Our thanks to Quentin Carayol, Hervé Steve, and Nicolas Réau for their help in modifying COPOLA; to Steven Kleinveld for his help with GANIMEDE; and to Laurent Hascoët from INRIA for his help with TAPENADE. Finally, we are grateful to Sarah Julisson for maintaining and adapting the optimization system of Dassault-Aviation to our needs.

Conflicts of Interest: Authors Charles Thoulon and Gilbert Roge were employed by the company Dassault-Aviation. The remaining author declares that the research was conducted in the absence of any commercial or financial relationship that could be construed as a potential conflict of interest.

Abbreviations

The following abbreviations are used in this manuscript

AD	Automatic Differentiation
AETHER	In-house CFD solver
CAD	Computer-Assisted Design
CFD	Computational Fluid Dynamics
COPOLA	In-house differentiated PO and PTD solver
GANIMEDE	In-house CAD modeler
GFA	Generic Fighter Aircraft (baseline geometry)
PO	Physical Optics
PTD	Physical Theory of Diffraction
RCS	Radar Cross-Section
TAPENADE	Automatic Differentiation tool developed by INRIA
DSM	Design Structure Matrix

References

1. Knott, E.F.; Schaeffer, J.F.; Tulley, M.T. *Radar Cross Section*; SciTech Publishing: Raleigh, NC, USA, 2004.
2. Aliaga, C.; Kopp, M.; Salui, K.B.; Mahapatra, D.; Sardesai, A. Aerodynamic and stealth studies of canard-wing Configurations at Transonic Speeds using Ansys Fluent & Ansys HFSS SBR+. In Proceedings of the AIAA AVIATION 2022 Forum, Chicago, IL, USA, 27 June–1 July 2022; p. 3322.
3. Lee, D.; Gonzalez, L.F.; Srinivas, K.; Auld, D.; Periaux, J. Multi-objective/multidisciplinary design optimisation of blended wing body UAV via advanced evolutionary algorithms. In Proceedings of the 45th AIAA Aerospace Sciences Meeting and Exhibit, Reno, NV, USA, 8 January 2007–11 January 2007; p. 36.
4. Wu, D.; Long, T.; Li, Y.; Jiang, M.; Huang, B. Aero-structure-stealth coupled optimization for high aspect ratio wing using adaptive metamodeling method. In Proceedings of the 15th AIAA/ISSMO Multidisciplinary Analysis and Optimization Conference, Atlanta, GA, USA, 16–20 June 2014; p. 2304.
5. Jiang, X.; Zhao, Q.; Zhao, G.; Meng, C. Numerical Optimization on Aerodynamic/Stealth Characteristics of Airfoil Based on CFD/CEM Coupling Method. *Trans. Nanjing Univ. Aeronaut. Astronaut.* **2016**, *33*, 274–284.
6. Parr, J.; Holden, C.M.; Forrester, A.I.; Keane, A.J. Review of efficient surrogate infill sampling criteria with constraint handling. In Proceedings of the 2nd International Conference on Engineering Optimization, Lisbon, Portugal, 6–10 September 2011.
7. Lyu, Z.; Xu, Z.; Martins, J. Benchmarking optimization algorithms for wing aerodynamic design optimization. In Proceedings of the 8th International Conference on Computational Fluid Dynamics, Chengdu, China, 11–13 November 2014; Citeseer: Princeton, NJ, USA, 2014; Volume 11, p. 585.
8. Martin, L. Conception Aérodynamique Robuste. Ph.D. Thesis, Université de Toulouse, Université Toulouse III-Paul Sabatier, Toulouse, France, 2010. Available online: https://core.ac.uk/download/pdf/12095692.pdf (accessed on 29 July 2024).
9. Hascoët, L.; Pascual, V. The Tapenade Automatic Differentiation tool: Principles, Model, and Specification. *ACM Trans. Math. Softw.* **2013**, *39*, 1–43. Available online: https://gitlab.inria.fr/tapenade/tapenade (accessed on 29 July 2024). [CrossRef]
10. Lyu, Z.; Martins, J.R. Aerodynamic design optimization studies of a blended-wing-body aircraft. *J. Aircr.* **2014**, *51*, 1604–1617. [CrossRef]
11. Lyu, Z.; Kenway, G.K.; Martins, J.R. Aerodynamic shape optimization investigations of the common research model wing benchmark. *AIAA J.* **2015**, *53*, 968–985. [CrossRef]
12. Bons, N.P.; He, X.; Mader, C.A.; Martins, J.R. Multimodality in aerodynamic wing design optimization. *AIAA J.* **2019**, *57*, 1004–1018. [CrossRef]
13. Garnier, E.; Langlet, S.; Klotz, P.; Cadillon, J.; Simon, J.; Castelli, J.C.; Levadoux, D. Surrogate-based conception for aerodynamic/stealth compromise. In Proceedings of the AVT-324 Specialists' Meeting on Multidisciplinary Design Approaches and Performance Assessment of Future Combat Aircraft, virtuel, Italy, 28–30 September 2020.
14. Guerra, J. Optimisation Multi-Objectif Sous Incertitudes de Phénomènes de Thermique Transitoire. Ph.D. Thesis, Institut Superieur de L'aeronautique et de L'espace (ISAE), Toulouse, France, 2016.
15. Liu, Z.; Song, W.; Han, Z.; Wang, Y. Multifidelity Aerodynamic/Stealth Design Optimization Method for Flying Wing Aircraft. In Proceedings of the 32th Congress of the International Council of the Aeronautical Sciences, ICAS, Shanghai, China, 6–10 September 2021.
16. Zhou, Z.; Huang, J. Quantitative weight and two-particle search algorithm to optimize aero-stealth performance of a backward inclined vertical tail. *Aerospace* **2023**, *10*, 345. [CrossRef]
17. Taj, Z.U.D.; Bilal, A.; Awais, M.; Salamat, S.; Abbas, M.; Maqsood, A. Design exploration and optimization of aerodynamics and radar cross section for a fighter aircraft. *Aerosp. Sci. Technol.* **2023**, *133*, 108114. [CrossRef]
18. Pan, Y.; Huang, J.; Li, F.; Yan, C. Integrated design optimization of aerodynamic and stealthy performance for flying wing aircraft. In Proceedings of the International MultiConference of Engineers and Computer Scientists, Hong Kong, 15–17 March 2017; Volume 2, pp. 15–17.
19. Li, M.; Bai, J.; Li, L.; Meng, X.; Liu, Q.; Chen, B. A gradient-based aero-stealth optimization design method for flying wing aircraft. *Aerosp. Sci. Technol.* **2019**, *92*, 156–169. [CrossRef]
20. Kleinveld, S.; Rogé, G.; Daumas, L.; Dinh, Q. Differentiated parametric CAD used within the context of automatic aerodynamic design optimization. In Proceedings of the 12th AIAA/ISSMO Multidisciplinary Analysis and Optimization Conference, Victoria, BC, Canada, 10–12 September 2008; p. 5952.
21. Chalot, F. Industrial Aerodynamics. In *Encyclopedia of Computational Mechanics*; Stein, E., de Borst, R., Hughes, T.J.R., Eds.; John Wiley & Sons, Ltd.: Hoboken, NJ, USA, 2004; Volume 3: Fluids, Chapter 12.
22. Barclay, A. *SQP Methods for Large-Scale Optimization*; University of California: San Diego, CA, USA, 1999.
23. Laval, D. Application de Méthodes Asymptotiques à la Simulation de la Diffraction électromagnétique par un Corps Régulier. Ph.D. Thesis, Université Sciences et Technologies-Bordeaux I, Bordeaux, France, 2006. Available online: https://theses.hal.science/tel-00197165/document (accessed on 29 July 2024).
24. Stratton, J.A.; Chu, L. Diffraction theory of electromagnetic waves. *Phys. Rev.* **1939**, *56*, 99. [CrossRef]
25. Michaeli, A. Elimination of infinities in equivalent edge currents, Part I: Fringe current components. *IEEE Trans. Antennas Propag.* **1986**, *34*, 912–918. [CrossRef]

26. Chalot, F.; Johan, Z.; Mallet, M.; Billard, F.; Martin, L.; Barré, S. Extension of methods based on the stabilized finite element formulation for the solution of the Navier–Stokes equations and application to aerodynamic design. *Comput. Methods Appl. Mech. Eng.* **2023**, *417*, 116425. [CrossRef]
27. Alfonsi, G. Reynolds-averaged Navier–Stokes equations for turbulence modeling. *Appl. Mech. Rev.* **2009**, *62*, 040802. [CrossRef]
28. Spalart, P.R.; Allmaras, S.R. One-equation turbulence model for aerodynamic flows. *La Rech. Aerosp.* **1994**, *1*, 5–21.
29. Martin, L.; Roge, G. Calcul de la sensibilité d'ordre deux d'une observation aérodynamique. In *Proceedings of the ESAIM: Proceedings*; EDP Sciences: Les Ulis, France, 2009; Volume 27, pp. 138–155.
30. Poggio, A.J.; Miller, E.K. *Integral Equation Solutions of Three-Dimensional Scattering Problems*; MB Assoc.: Winnipeg, MB, Canada, 1970; Chapter 4, pp. 195–264.
31. Carayol, Q. Développement et Analyse D'une Méthode Multipôle Multiniveau Pour L'électromagnétisme. Ph.D. Thesis, Université Paris 6, Paris, France, 2002. Available online: https://theses.fr/2002PA066485 (accessed on 19 May 2024).
32. Chalot, F.; Mallet, M.; Ravachol, M. A comprehensive finite element Navier-Stokes solver for low-and high-speed aircraft design. In Proceedings of the 32nd Aerospace Sciences Meeting and Exhibit, Reno, NV, USA, 10–13 January 1994; p. 814.
33. Jenn, D. *Radar and Laser cross Section Engineering*; American Institute of Aeronautics and Astronautics, Inc.: Reno, NV, USA, 2005.
34. Désidéri, J.A. *MGDA Variants for Multi-Objective Optimization*; Institut National de Recherche en Informatique et en Automatique: Sophia-Antipolis, France, 2012.

Disclaimer/Publisher's Note: The statements, opinions and data contained in all publications are solely those of the individual author(s) and contributor(s) and not of MDPI and/or the editor(s). MDPI and/or the editor(s) disclaim responsibility for any injury to people or property resulting from any ideas, methods, instructions or products referred to in the content.

MDPI AG
Grosspeteranlage 5
4052 Basel
Switzerland
Tel.: +41 61 683 77 34

Fluids Editorial Office
E-mail: fluids@mdpi.com
www.mdpi.com/journal/fluids

Disclaimer/Publisher's Note: The title and front matter of this reprint are at the discretion of the Guest Editors. The publisher is not responsible for their content or any associated concerns. The statements, opinions and data contained in all individual articles are solely those of the individual Editors and contributors and not of MDPI. MDPI disclaims responsibility for any injury to people or property resulting from any ideas, methods, instructions or products referred to in the content.

www.ingramcontent.com/pod-product-compliance
Lightning Source LLC
Chambersburg PA
CBHW042013260926
51850CB00020B/3376